Hormones in Health and Disease

Series Editor
V.K. Moudgil

Hormones and Cancer

Wayne V. Vedeckis
Editor

Birkhäuser
Boston • Basel • Berlin

Wayne V. Vedeckis
Department of Biochemistry and Molecular Biology
Louisiana State University Medical Center and
Stanley S. Scott Cancer Center
1901 Perdido Street
New Orleans, LA 70112-1393
USA

Library of Congress Cataloging-In-Publication Data

Hormones and cancer / Wayne V. Vedeckis, editor.
p. cm. -- (Hormones in health and disease)
Includes bibliographical references and index.
ISBN 0-8176-3797-4 (alk. paper). – ISBN 3-7643-3797-4 (alk. paper)
1. Cancer–Endocrine aspects. 2. Cancer–Hormone therapy.
I. Vedeckis, Wayne V., 1947– . II. Series.
[DNLM: 1. Neoplasms, Hormone-Dependent–physiopathology.
2. Neoplasms, Hormone-Dependent–therapy. 3. Hormones–physiology.
4. Hormones–therapeutic use. 5. Breast Neoplasms. 6. Prostatic Neoplasms.
7. Leukemia. QZ 200 H8121 1996]
RC268.2.H672 1996
616.99'4071–dc20
DNLM/DLC
for Library of Congress 96-4653
CIP

Printed on acid-free paper

ISBN 0-8176-3797-4
ISBN 3-7643-3797-4

Cover design by David Gardner, Dorchester, MA.
Typeset in Great Britain by Alden, Oxford, Didcot and Northampton.
Printed and bound by Maple-Vail, York, PA.
Printed in the U.S.A.
9 8 7 6 5 4 3 2 1

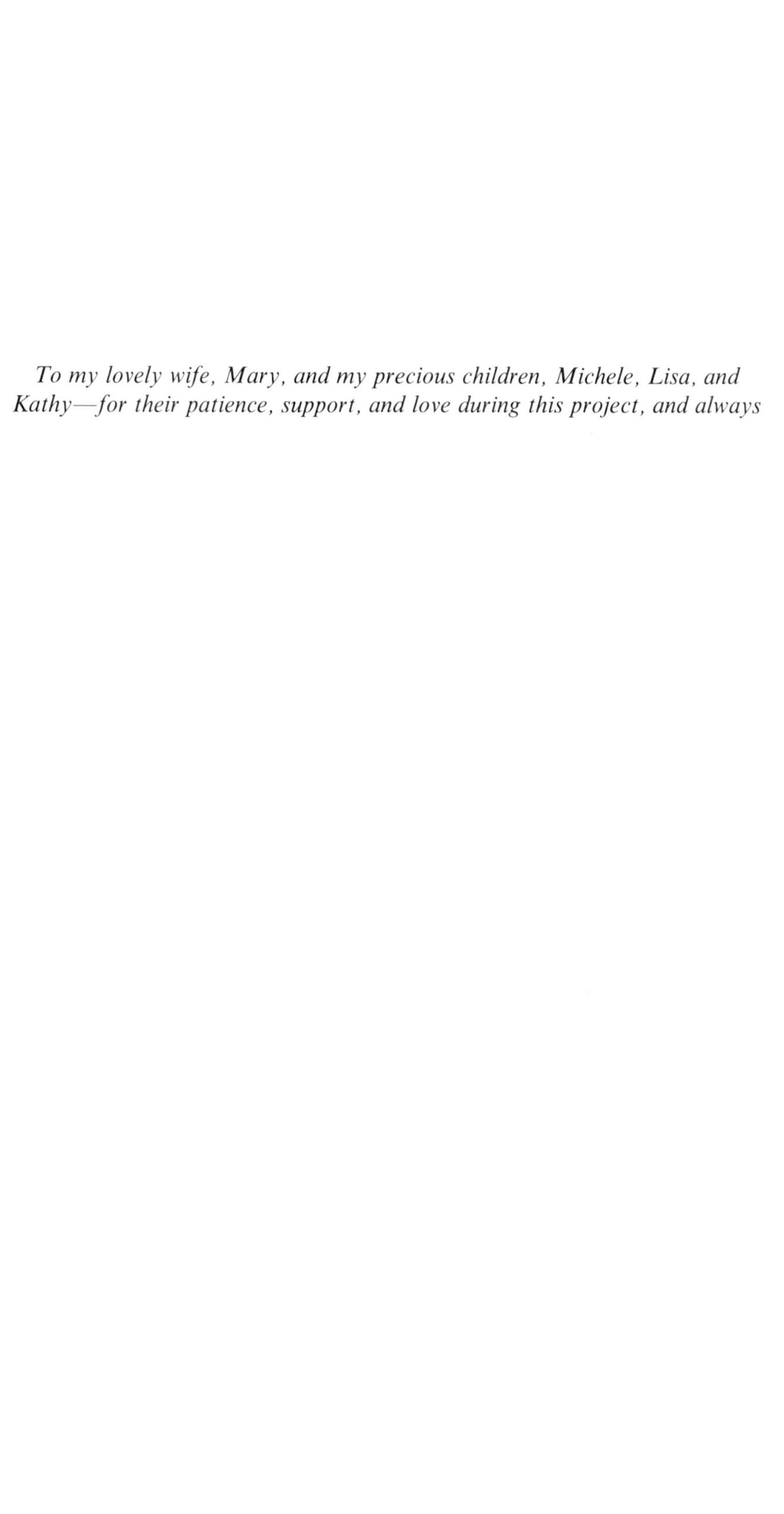

To my lovely wife, Mary, and my precious children, Michele, Lisa, and Kathy—for their patience, support, and love during this project, and always

Contents

PART II: *BREAST CANCER*

PART III: *PROSTATE CANCER*

PART IV: *HEMATOLOGICAL MALIGNANCIES*

Foreword

The series, *Hormones in Health and Disease*, was launched in 1993 to provide a scientific platform for investigators engaged in research on the biological actions of hormones and to anticipate relevance for their findings in clinical applications. The first volume of the series was dedicated to the discussion and understanding of molecular mechanisms by which steroid hormones influence target cells in normal and pathological conditions. With the diversity of information and the vast amount of literature on steroid hormone physiology, a more thorough treatment of *Hormones and Cancer* was identified as a timely topic.

In this second volume in the series, Dr. Wayne V. Vedeckis has successfully undertaken the monumental task of editing the findings of the leading investigators in hormone and cancer research. Dr. Vedeckis brings to this project two decades of research experience in hormone action; he is actively engaged in elucidating hormone and cancer interrelations. It is a pleasure to welcome him to the series as an editor and congratulate him and all contributors in presenting this comprehensive treatise.

The 20 chapters include discussions on contemporary topics relating control of cell division and signal transduction to the basic mechanisms of carcinogenesis by cloning patient genes, and recognizing the importance of steroid receptors in treatment protocols of various endocrine abnormalities. The organization and presentation of different chapters, Dr. Vedeckis' experience and expertise, and the quality of contributions make *Hormones and Cancer* a most welcome addition to the series. It is hoped that readers will find this volume to be a valuable guide and a useful resource for fundamental processes as well as the current thinking of active researchers who contributed to this endeavor.

For their continued support and recognition of the importance of this undertaking, Birkhäuser Boston has my gratitude and appreciation. Future volumes are planned to address emerging needs and evolving biomedical issues that deal with *Hormones in Health and Disease*.

V.K. Moudgil
Series Editor

Preface

Beginning early in the next century, cancer will overtake cardiovascular disease as the number one cause of death from disease in America. With the increase in the overall life span and age of citizens in developed countries, the prevalence of cancer and its impact will continue to grow. Doubtless to say, all readers of this volume have had, or will have, their lives touched either directly or indirectly by the collection of diseases we call cancer.

So much for the bad news. The good news is that we are making unprecedented progress in understanding the biochemistry, molecular biology, and cell biology of the cancer. The identification of oncogenes, tumor suppressor genes, signal transduction pathways controlling the cell cycle, and the components of the cell cycle machinery itself has resulted in a much clearer, albeit complex, vision of normal and abnormal cell division. This information has allowed for the improved diagnosis of cancer, as well as elucidated specific potential targets for directed therapy.

One bright light in our battle against certain types of cancers is the use of hormonal manipulations for intervention. While not a panacea, this approach has extended the arsenal available for a rational approach to cancer therapy. Most notable is the acceptance of anti-estrogen therapy, primarily the use of tamoxifen, in the treatment of hormone-responsive breast tumors. The incorporation of aromatase inhibitors for treating metastatic breast disease has also been notable. On the other hand, anti-androgen treatment of prostate cancer has so far been very disappointing; however, on a more positive note, recent results that include an analysis of androgen receptor mutations and the recognition of important stromal-epithelial interactions raises hope that hormonal intervention will find more success in the treatment of prostate cancer. Similarly, the use of corticosteroids in the treatment of acute lymphoblastic leukemia continues to be of significant patient benefit, and the recent clinical complete remissions obtained in acute promyelocytic leukemia patients treated with all-*trans* retinoic acid have been truly impressive.

It is with these facts in mind that I accepted Dr. Virinder Moudgil's gracious invitation to edit a volume for his series, *Hormones in Health and Disease*. My primary consideration in choosing the subjects for *Hormones and Cancer* was the need for clear clinical relevance to human disease. Not that animal or tissue culture results are unimportant—they are indispensable. But I wanted to make this volume valuable to both clinicians and basic scientists alike. Thus, I asked the basic scientists to correlate the results in the laboratory to those in the clinic whenever possible. I am extremely pleased with the results. Only when the basic science laboratory findings are translated to the clinic do we completely experience the fruits of our labors as scientists and clinicians.

The first section of this book is composed of five chapters that give the basic science foundation necessary to interpret both laboratory and clinical results. They discuss the basic cell signaling pathways that result in changes in cellular physiology and proliferation after an environmental stimulus. The authors have done an admirable job in presenting these complicated systems in an understandable manner.

The next three sections of the book address the major neoplasms that are regulated by hormonal stimuli—breast cancer, prostate cancer, and leukemia. Obviously, other cancers are affected by hormones, but some limit in scope was necessary. These three cancers were chosen because of their prevalence, the clear role of hormones in their development, and the utility of hormonal manipulation in their treatment. Each section contains a description of the disease from a clinical viewpoint. I felt it was crucial that the basic scientists who read this volume be informed of the biological and clinical aspects of the diseases that they are studying. It is easy for the basic scientist (myself included) to become insulated in the laboratory, to the point that the health relevance of their research can sometimes be forgotten. The clinical chapters focus the reader upon the disease process itself and how it is diagnosed and treated.

A second major focus in each section is the molecular genetics of the disease. The recent spectacular advances in dissecting the molecular bases for disease using this approach clearly warrants these chapters. Although steroid hormones and retinoids play major roles in the cancers discussed, it is becoming increasingly obvious that the interplay between the pathways regulated by these molecules and those controlled by growth factors must be addressed for a more complete understanding of the disease. Thus, the role of growth factors and their signaling pathways is included. Finally, the well known ways in which steroid hormones and retinoids (and their receptors) affect disease progression and treatment are described in detail.

I am deeply grateful to the authors who have given so unselfishly in this effort. The fact that acknowledged experts in these fields have devoted so much time and effort to produce the outstanding chapters in this book, with only a modest tangible reward, has reminded me of the idealism that drew me to science in the first place. I also wish to thank the staff at Birkhäuser for their tireless efforts in bringing this volume to fruition. They were all a pleasure to work with, and their enduring patience is appreciated.

Finally, I would like to offer this volume in memory of all victims of cancer. In particular I would like to remember: my mother, Helen Vedeckis, who died of a metastatic, undifferentiated carcinoma from an unknown primary; my mother-in-law, Mary Lavieri, who perished from bone metastases secondary to renal cell carcinoma; and my dear friend, Charmaine Trainor, who succumbed to breast cancer. I also would like to call to memory the family members and friends of the contributing authors and the readers of this book who have died of cancer. The suffering and courage of all cancer patients are the main reasons that I hope that this volume will hasten progress to improve the prevention and treatment of this disease.

Wayne V. Vedeckis
Editor

Contributors

Evelyn R. Barrack, Ph.D., Department of Urology, The Johns Hopkins University School of Medicine, and The Brady Urological Institute, The Johns Hopkins Hospital, 600 N. Wolfe Street/Marburg 115, Baltimore, MD 21287-2101, USA

R. Daniel Beauchamp, M.D., Department of Surgery, Vanderbilt University Medical Center, Nashville, TN 37232, USA

Angela M.H. Brodie, Ph.D., Department of Pharmacology and Experimental Therapeutics, University of Maryland School of Medicine, 655 West Baltimore Street, Baltimore, MD 21202, USA

Carl G. Castles, The University of Texas Health Science Center at San Antonio, Division of Medicine/Medical Oncology, 7703 Floyd Curl Drive, San Antonio, TX 78284-7884, USA

William H. Catherino, Department of Human Oncology, University of Wisconsin Comprehensive Cancer Center, 600 Highland Avenue, Laboratory K4/653, Madison, WI 53792, USA

John A. Cidlowski, Ph.D., Laboratory of Integrative Biology, National Institute of Environmental Health Science, P.O. Box 12233, 111 T.W. Alexander Drive, Research Triangle Park, NC 27709, USA

Claudia S. Cohn, Ph.D., Department of Anatomy, Tulane University Medical Center, 1430 Tulane Avenue, New Orleans, LA 70112, USA

Sami G. Diab, The University of Texas Health Science Center at San Antonio, Division of Medicine/Medical Oncology, 7703 Floyd Curl Drive, San Antonio, TX 78284-7884, USA

Clark W. Distelhorst, M.D., Division of Hematology/Oncology, Case Western Reserve University, 10900 Euclid Avenue, Cleveland, OH 44106, USA

Suzanne A.W. Fuqua, Ph.D., The University of Texas Health Science Center at San Antonio, Division of Medicine/Medical Oncology, 7703 Floyd Curl Drive, San Antonio, TX 78284-7884, USA

Julia M.W. Gee, Ph.D., Breast Cancer Research Laboratory, Tenovus Cancer Research Centre, Tenovus Building, University of Wales College of Medicine, The Heath, Cardiff, CF4 4XX, United Kingdom

Antonio Giordano, M.D., Ph.D., Departments of Microbiology/Immunology and Pathology, Jefferson Cancer Institute, Thomas Jefferson University, 233 S. 10th Street, Rm 531 , Philadelphia, PA 19107, USA

Paul Goodfellow, Ph.D., Department of Surgery, Washington University School of Medicine, St. Louis, MO 63110, USA

William J. Gradishar, M.D., Division of Hematology/Medical Oncology, Department of Medicine, Robert Lurie Cancer Center, Northwestern University Medical School, Chicago, IL 60611, USA

Francesco Grignani, Laboratorio di Biologia Molecolare, Istituto di Medicina Interna e Scienze Oncologiche, Perugia University, Policlinico Monteluce, Via Brunamonti, 51, 06100 Perugia, Italy

Janette M. Hakimi, Ph.D., Department of Urology, The Johns Hopkins University School of Medicine, Baltimore, MD 21287-2101, USA

Steven Hill, Ph.D., Department of Anatomy, Tulane University Medical Center, 1430 Tulane Avenue, New Orleans, LA 70112, USA

Kathryn B. Horwitz, Ph.D., University of Colorado Health Sciences Center, Departments of Medicine and Pathology, and the Molecular Biology Program Division of Endocrinology, 4200 E. 9th Avenue, Campus Box B151, Denver, CO 80262, USA

Candace M. Howard, Departments of Microbiology/Immunology and Pathology, Jefferson Cancer Institute, Thomas Jefferson University, 233 S. 10th Street, Rm 531 , Philadelphia, PA 19107, USA

V. Craig Jordan, Ph.D., D.Sc., Robert H. Lurie Cancer Center, Northwestern University School of Medicine, Olson 8258, 303 E. Chicago Avenue, Chicago, IL 60611, USA

Diane M. Klotz, Ph.D., Department of Anatomy, Tulane University Medical Center, 1430 Tulane Avenue, New Orleans, LA 70112, USA

Tien C. Ko, M.D., Department of Human Biological Chemistry and Genetics, and Department of Surgery, University of Texas Medical Branch, Galveston, TX 77550-0645, USA

Xiaohua Leng, Ph.D., Department of Cell Biology, Baylor College of Medicine, One Baylor Plaza, Houston, TX 77030, USA

Jill Macoska, Ph.D., The Michigan Prostate Center and Department of Surgery, Section of Urology, 5510A MSRB I, 1150 West University Drive, Ann Arbor, MI 48109-0680, USA

Jeffrey F. Moley, M.D., St Louis Veteran's Administration Medical Center, St. Louis, MO, and Department of Surgery, Washington University School of Medicine, 1 Barnes Hospital Plaza, 5108 Queeny Tower, St. Louis, MO 63110, USA

Jennifer W. Montague, Laboratory of Integrative Biology, National Institute of Environmental Health Science, P.O. Box 12233, 111 T.W. Alexander Drive, Research Triangle Park, NC 27709, USA

Monica Morrow, M.D., Department of Surgery, Robert Lurie Cancer Center, Northwestern University Medical School, 250 East Superior Street, Wesley #201, Chicago, IL 60611, USA

Robert I. Nicholson, Ph.D., Breast Cancer Research Laboratory, Tenovus Cancer Research Centre, Tenovus Building, University of Wales College of Medicine, The Heath, Cardiff, CF4 4XX, United Kingdom

Donna M. Peehl, Ph.D., Department of Urology, Stanford University School of Medicine, Stanford, CA 94305-5118, USA

Pier Giuseppe Pelicci, Laboratorio di Biologia Molecolare, Istituto di Medicina Interna e Scienze Oncologiche, Perugia University, Policlinico Monteluce, Via Brunamonti, 51, 06100 Perugia, Italy; and Istituto Europeo di Oncologia, Via Ripamonti, 435, 20141 Milano, Italy

Magnus Pfahl, Ph.D., Sidney Kimmel Cancer Center, 11099 North Torrey Pines Road, Suite 250, La Jolla, CA 92037, USA

Rachel H. Rondinelli, PhD., Department of Urology, The Johns Hopkins University School of Medicine, Baltimore, MD 21287-2101, USA

Mark P. Schoenberg, M.D., Department of Urology, The Johns Hopkins University School of Medicine, and The Brady Urological Institute, The Johns Hopkins Hospital, Baltimore, MD 21287-2101, USA

Joseph A. Smith, Jr., M.D., Department of Urologic Surgery, D-4314, A-1302 Medical Center North, Vanderbilt University School of Medicine, Nashville, TN 37232-2765, USA

Glenn S. Takimoto, Ph.D., University of Colorado Health Sciences Center, 4200 E. 9th Avenue, Denver, CO 80262, USA

Martin S. Tallman, M.D., Division of Hematology/Oncology, Northwestern University Medical School, Robert H. Lurie Cancer Center of Northwestern University, 233 East Erie, Suite 700, Chicago, IL 66011, USA

E. Aubrey Thompson, Jr., Ph.D., Department of Human Biological Chemistry and Genetics, University of Texas Medical Branch, F45 HBC&G, Galveston, TX 77550-0645, USA

Ming-Jer Tsai, Ph.D., Department of Cell Biology, Baylor College of Medicine, One Baylor Plaza, Houston, TX 77030, USA

Sophia Y. Tsai, Ph.D., Department of Cell Biology, Baylor College of Medicine, One Baylor Plaza, Houston, TX 77030, USA

Lin Tung, Ph.D., University of Colorado Health Sciences Center, 4200 E. 9th Avenue, Denver, CO 80262, USA

Wayne V. Vedeckis, Ph.D., Department of Biochemistry and Molecular Biology, Louisiana State University Medical Center, and Stanley S. Scott Cancer Center, 1901 Perdido Street, New Orleans, LA 70112-1393, USA

Part I

Underlying Mechanisms and Principles

Perhaps no areas of biology have experienced greater fundamental advances in the last 15 years than those summarized in this section. Chapter 1 recounts the phenomenal progress that has been made in understanding the cellular components that control cell division. The elucidation of the roles for the cyclins, cyclin-dependent kinases (Cdks), Cdk-activating kinase (CAK), and cyclin kinase inhibitors in controlling progression through the cell cycle has been truly remarkable. The interaction of a cyclin-Cdk kinase with a tumor suppressor gene product, the retinoblastoma protein, shows directly how a tumor suppressor controls transcription of genes that promote cell cycle phase transition. This chapter also introduces p53: it can act as a tumor suppressor or oncogenic protein and it plays a crucial role in apoptosis and as a checkpoint regulator in preventing cells with highly damaged DNA (which could contribute to oncogenesis via aberrant gene expression) from continuing through the cell cycle. Finally, a potent inhibitor of epithelial cell division, TGF-β, and its interaction with the cell cycle machinery, is introduced.

Chapter 2 expands on the theme of how extracellular signals, such as growth factors, work through intracellular signaling pathways to control cell division, and the complex role of TGF-β is discussed. This chapter epitomizes why it is not easy to predict what types of studies will be most fruitful in increasing our knowledge of the process of neoplastic transformation. Much of the research that has guided studies on growth factor regulation of cell division is based on the mechanism by which a mating factor inhibits yeast cell division! These results are correlated with those on the role of TGF-β in controlling epithelial cell division.

Another important cell signaling pathway involves the tyrosine protein kinases, many of which are transmembrane proteins that bind extracellular growth factors. The Ret protein is such a receptor, and it is discussed in

Chapter 3. This is an example of a field in which a clinical disease, multiple endocrine neoplasia 2A (MEN 2A), led to the discovery of the biochemistry involved; that is, translational research from the clinical setting to the basic science laboratory. This chapter also introduces the most exciting research approach in current use in medicine—positional cloning of patient genes to identify the proteins involved in disease. This theme recurs throughout this volume, and it points to molecular genetics as the premier frontier for future medical research. How the endocrine imbalances that result from MEN 2A are dealt with clinically is also discussed in this chapter.

Truly remarkable advances have been made in the field described in Chapters 4 and 5. The structure, function, and molecular mechanism of action of the nuclear receptors (for steroids, thyroid hormone, the retinoids, and vitamin D_3) have now been well defined. Mutations in these ligand-activated transcription factors can directly contribute to disease, including cancer. (The importance of the receptors for estrogens, progestins, androgens, corticosteroids, and retinoids in cancer is a major focus of the remaining chapters in this volume.) Chapter 4 gives an overview of the structure and function of the nuclear receptors and describes recent studies on how these receptors interact with the basal transcription machinery in the cell. Chapter 5 expands on the role of retinoid receptors in promoting cellular differentiation and inhibiting neoplastic transformation. It also points to an exciting area for future research—the development of isoform- and receptor-specific ligands for the targeted treatment of disease. This shows the importance of basic science in directing the rationale design of therapeutic drugs based on our knowledge of protein structure and function.

Thus, this first section lays the necessary, fundamental groundwork for the study of cancer and cell signaling pathways. An understanding of these principles and mechanisms is important for fully appreciating the clinical and basic science of hormones and cancer.

1

Neoplastic Transformation: Oncogenes, Tumor Suppressors, Cyclins, and Cyclin-Dependent Kinases

CANDACE M. HOWARD AND ANTONIO GIORDANO

INTRODUCTION

Cancer is a genetic disease that results from multiple genomic changes. These ultimately lead to the deregulation of the cell cycle machinery and to autonomous cell proliferation. Neoplastic transformation involves four sets of genes: 1) oncogenes, 2) tumor-suppressor genes, 3) mutator genes, and 4) apoptotic genes. In the hematopoietic system, the first step in oncogenesis is the activation of an oncogene that may then be followed by the activation of an additional oncogene and/or the loss of function of a tumor-suppressor gene. The activation of oncogenes may play a predominant role in the formation of sarcomas as well. Tumors of both the hematopoietic system and soft tissues exhibit a karyotype close to normal. On the other hand, carcinomas, the most prevalent forms of cancer, are predominantly due to the loss of function of tumor-suppressor genes with multiple sites of loss of heterozygosity (LOH), and they have dramatic alterations in the karyotype (Rabbitts, 1994).

Oncogene activation may occur by gene translocation, amplification, or mutation. Translocations of proto-oncogenes are found in various forms of leukemia, lymphomas, and sarcomas. The most common form of Burkitt's lymphoma exemplifies an enhancement of the *myc* gene expression due to its translocation from chromosome 8 to chromosome 14, t(8;14). This places the *myc* gene under the control of the immunoglobulin heavy chain promoter, thus producing constitutively high levels of expression of this growth promoting gene which is no longer subject to down-regulation or

Hormones and Cancer
Wayne V. Vedeckis, Editor

suppression. Less commonly, Burkitt's lymphoma results from the juxtaposition of the c-*myc* gene with a portion of the gene encoding the kappa or lambda immunoglobulin molecules (kappa -2; lambda -22) by the translocation of t(2;8) or t(8;22). A similar situation occurs in follicular tumors. These low-grade malignancies involve a translocation, t(14;18), and subsequent transcriptional up-regulation of the *bcl*-2 gene, the master gene in control of apoptosis (programmed cell death). Overexpression of *bcl*-2 inhibits apoptosis and therefore creates a tumor that is more resistant to chemotherapy. The first oncogene was discovered by studying the Philadelphia chromosome in chronic myelogenous leukemia, where a translocation between chromosomes 9 and 22, t(9;22), results in the *bcr-abl* chimerization, a gene fusion. This causes an increase in the activation of a tyrosine-specific protein kinase (*abl*), which promotes cellular growth (Rabbitts, 1994). Gene amplification is also found in solid tumors. The *myc* and *erbB*-2 oncogenes can each be found to be amplified in 10–20% of breast cancers, and this leads to a poor prognosis (Gaffey et al., 1993).

Current evidence points to the multistep theory of carcinogenesis, which states that the final stages of neoplasia and malignancy arise only after the "independent" mutations of several disparate genes within the progenitor cell (Renan, 1993). The fidelity of the human replicative machinery is very high, with the occurrence of a single mutation being a rare event—on the order of 10^{-9} per cell per generation. The evolution of colon tumors involves the occurrence of at least 6 independent mutations in the genome (Fearon and Vogelstein, 1990). The normal low frequency of spontaneous mutations cannot account for these phenomena. However, if something were to disrupt the fidelity of replication resulting in genomic instability, this would then allow for the possibility of multiple mutations to become a more feasible event. The discovery of mutator genes solved this scientific and statistical dilemma. The human mutator gene homolog, hMSH2, was recently cloned and found to be associated with hereditary nonpolyposis colon cancer (HNPCC). By virtue of the proximity of the hMSH2 locus to that of the HNPCC gene and the finding of an instability of dinucleotide repeats in both HNPCC and in the *Saccharomyces cerevisiae* msh2 mutations, the hMSH2 and HNPCC genes are thought to be the same. The hMSH2 gene is speculated to have a role in oncogenesis by its mutation; this produces a dominantly acting defect in the cells mismatch repair system, which results in an enhanced rate of spontaneous mutation and genomic instability (Fishel et al., 1993).

For the various genomic alterations mentioned to elicit their effects in cell immortalization and uncontrolled proliferation, they must in some way disturb the normal processes controlling the cell cycle machinery. The cell then becomes an autonomous replicating unit, no longer relying on the external environment of hormones and paracrines to provide the cell with the instructions of when to divide, to arrest, to differentiate, or to apoptose. At the heart of all neoplasia is an ultimate disturbance in the positive regulators (proto-oncogenes), negative regulators (tumor-suppressor genes), and/or apoptotic regulators of cellular growth and proliferation. The involvement of specific molecules, in particular pathways and in unique combinations, accounts for the phenotypic variability and individuality of each type of malignancy, the various degrees of adaptation by certain tumors in developing resistance to the host immune system, and the different therapeutic modalities useful against each cancer. To develop more effective strategies for the management of cancer, we must first obtain a better comprehension of the natural mechanisms of the cell cycle machinery and how these processes interlink and how they can become disturbed and deregulated. To manipulate a system better, one must first understand how the system works and what are its limitations.

The discovery of tumor-suppressor genes revolutionalized the field of cancer research and cell cycle molecular biology. According to Knudson's "two-hit" hypothesis (Knudson, 1971), the development of several human cancers is thought to involve loss of heterozygosity (LOH) of putative tumor-suppressor genes, several of which are not yet identified. Many forms of malignancies have been linked to mutations in the retinoblastoma susceptibility gene (RB), the first tumor-suppressor gene identified. RB provides a link between the cell cycle machinery, the G1/S cyclins and cyclin-dependent kinases (Cdks), and the transcription of genes necessary for entry into S phase. Another tumor suppressor, p53, is found mutated in more than 50% of human tumors (Bienz-Tadmor et al., 1985). Both tumor-suppressor genes, RB and p53, are negative regulators of the cell cycle and are involved in growth arrest. p53 is also implicated in the recognition and repair of DNA damage, as well as in the induction of programmed cell death, apoptosis. Recently, it has become increasingly clear that these two proteins interact with each other through various pathways in order to exert their effects on cellular events.

CELL CYCLE CLOCK

The mammalian cell cycle machinery involves the formation, activation, inactivation, and degradation of sequential complexes of cyclins with cyclin-dependent kinases (Cdks). This process is governed by positive and negative feedback regulations at various critical transition points to ensure that the necessary molecular events occur in each phase of the cell cycle before progression to the next stage. Uncoupling of this intricately controlled regulatory process often results in cellular transformation and oncogenesis. Figure 1 demonstrates a generalized model of the currently known major proteins of the cell cycle machinery.

Disruptions in the cell cycle machinery have been implicated in the development of certain cancers. Cyclins are cell cycle regulators first identified in marine embryos and named for their cyclical accumulation and degradation

The Cell Cycle Clock

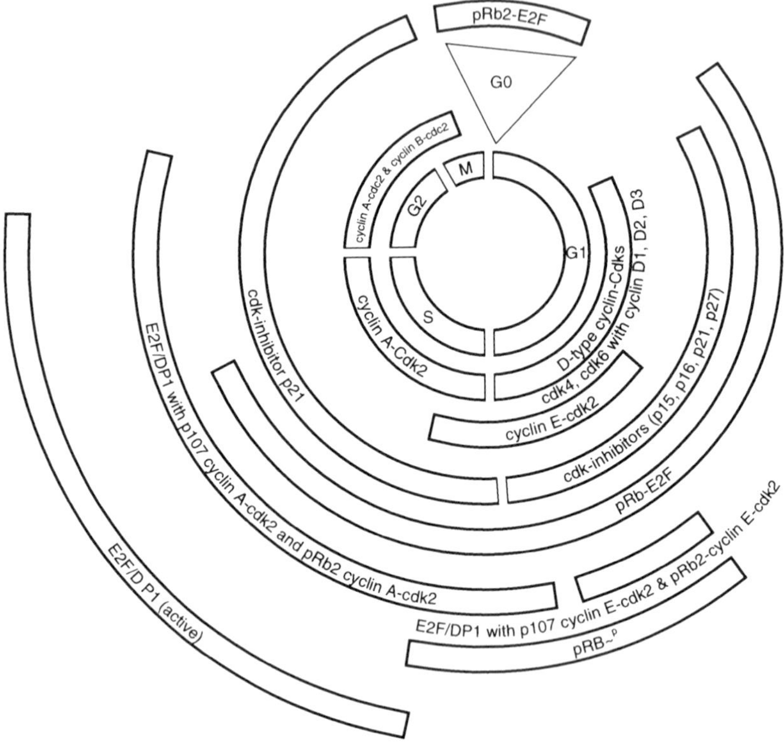

Figure 1. Generalized model of the major proteins of the cell cycle machinery and the phases in which they are found.

at each state of the cell cycle. In humans, the overexpression of certain G1 cyclins has been associated with oncogenesis. Hepatitis B viral integration into the cyclin A gene has been found in a case of hepatocellular carcinoma (Wang et al., 1990). Elevated levels of cyclin A have also been correlated with hematologic malignancies (Paterlini et al., 1993). Gene derangement and aberrant expression of cyclins have been reported in breast, esophageal, and skin malignancies (Jiang et al., 1992; Bianchi et al., 1993; Keyomarsi and Pardee, 1993). The candidate oncogene, PRAD1, found to be over-expressed and rearranged in parathyroid tumors, actually encodes for the cyclin D1 (Motokura, 1991). Certain colorectal cancer cell lines have been reported to possess an amplification of cyclin D2 and cyclin E (Leach et al., 1993). The recent chromosomal mapping of Cdk3 and two putative Cdks, PISSLRE and PITALRE, have localized them to regions previously shown to exhibit loss of heterozygosity (LOH) in breast, bladder, and other malignant tumors (Grana et al., 1994a,b; Bullrich et al., 1995; Li et al., 1995). The transforming protein E1A, from a DNA oncovirus, has been found to associate with cyclin A in adenovirus-transformed cells (Giordano et al., 1989, Pines and Hunter, 1990). This was the first evidence for the linkage between the cell cycle and the process of neoplastic transformation. The binding of E1A may promote cellular proliferation by disrupting the normal interaction between cyclin A and the E2F/DRTF1 transcription factor (Chellappan, 1994). DNA oncoviruses, via their oncoproteins targeting key regulatory and proliferative proteins of the cell cycle, serve as valuable tools in the attempt to elucidate the complex mechanisms governing the cell cycle machinery.

It is well established that $p34^{cdc2}$protein kinase plays an important role in the regulation of the cell division cycle. In fission yeast, cdc2 function is required at two points of the cell division cycle: before the initiation of DNA replication and at the time of initiation of mitosis. In *Xenopus*, the cdc2 gene product has been identified as a component of MPF (M-phase promoting factor), a factor that causes an interphase cell to enter mitosis; other components include the product of the Suc1 gene and p62, homologous to the human cyclin B gene. The MPF complex also acts as a histone H1 kinase that is maximally active at mitotic metaphase. In mammalian cells the cdc2 kinase is required for mitosis and for the transition from G1 to S phase (DNA replication) (King et al., 1994; MacLachlan et al., 1995). At least two complexes of mammalian $p34^{cdc2}$ have been identified, the p62 (cyclin B)-$p34^{cdc2}$complex, and the p60 (cyclin A)-$p34^{cdc2}$ complex (Draetta and Beach, 1988; Giordano et al., 1989). The p60 (cyclin A)-$p34^{cdc2}$ complex also displays a cell cycle–dependent histone H1 kinase activity, but it shifts the timing of maximal activity to interphase (Giordano et al., 1989). Recently, another complex has been identified between p50 (cyclin E) and Cdk2, a member of the extended cdc2-related protein family (Koff et al., 1992). These three complexes (cyclin B-Cdc2, cyclin A-Cdc2, cyclin E-Cdk2) display a cell cycle–dependent histone H1 kinase activity that has different timings of activation during the cell division cycle (Giordano et al., 1989; Draetta and Beach, 1988; Koff et al., 1992).

G1 Phase

Throughout the G1 phase of the cell cycle, growth factors exhibit their effects by binding to specific surface receptors. This leads to the activation of signaling cascades, which control the transcription of immediate and delayed early response genes. Even within cells that ultimately undergo divergent long-term responses, many of the same intracellular signaling molecules exhibit uniform activation. Withdrawal of growth factors in mammalian cells leads to arrest of the cells in early G1 into what is known as the G0 phase. For cells to reenter the cell cycle from quiescence, G0, the Cdk complexes in the cells must be activated (Fig. 1). Activation of the Cdk complexes requires the aggregation of cyclins and Cdk subunits, the derangement of inhibitory molecules such as p15, p16, p21 and p27 (Serrano et al., 1993; Kato et al., 1994a; Polyak et al., 1994a,b; Toyoshima and Hunter, 1994) from the Cdk complexes and phosphorylation by the Cdk-activating kinase (CAK) (Fisher and Morgan, 1994; Kato et al., 1994a; Matsuoka et al., 1994). The cyclins involved in G1 of mammalian cells include cyclins C, D, and E. In yeast, these cyclins are encoded by CLN genes whose protein products regulate cdc2 activity at START, the point at which a cell commits to DNA replication (Hunter and Pines, 1991). Neither cyclin A nor B is available before S phase. Throughout the cell cycle, cyclin C levels do not oscillate very much, exhibiting only a slight increase in early G1. However, this is not the case for cyclin E, which has periodic expression and peaks at the G1/S transition, suggesting a possible role for it in the regulation of S-phase entry (Koff et al., 1991, 1992).

There are three D-type cyclins that vary in their relative levels among cell types (Fig. 1; Draetta, 1994). Following serum addition to quiescent cells, D-type cyclins are induced. Subsequent induction and maintenance of D-type cyclin mRNAs depend on the presence of growth factors (Matsushime et al., 1991a,b). The D-type cyclins pose a possible link between cell cycle, signal transduction, and oncogenesis because the overexpression of cyclin D1 has been found in some tumors. The D-type cyclins have been shown to associate with four members of the Cdk protein family: Cdk2, 4, 5, and 6; however, the major catalytic partners of the D-type cyclins are Cdk4 and Cdk6 (Bates et al., 1994; Matsushime et al., 1994). Like all nascent cyclin-Cdk complexes, those containing cyclin D become activated only when phosphorylated by CAK. In mammalian cells CAK has recently been shown to complex with cyclin H (thus termed Cdk7) and then to phosphorylate and activate Cdk2, Cdc2 and Cdk4 complexes (Fisher et al., 1994; Matsuoka et al., 1994). Depletion of CAK from the cell lysates prevents the activation of cyclin D–Cdk4 complexes (Matsuoka et al., 1994). This demonstrates that the CAK protein is a necessary component of the Cdk4-activation kinase.

The three D-type cyclins may not be functionally equivalent because cyclin D1 forms a much less stable complex with pRb than either cyclins D2 or D3 (Ewen et al., 1993a; Kato et al., 1993a,b). Unlike cyclin E or A, the D-type cyclins share the sequence Leu-X-Cys-X-Glu (LXCXE) near their amino termini with the DNA viral oncoproteins, which bind pRb and pRb-related p107. Point mutations within this region disrupt the ability of D-type cyclins to bind to pRb and to p107 *in vitro;* thus, D-type cyclins can be more effectively competed by oncoprotein-derived peptides containing the LXCXE residues (Dowdy et al., 1993; Ewen et al., 1993a). Additional evidence of the possible functional divergences between the D-type cyclins comes from the observation that the overexpression of cyclins D2 and D3 in hematopoietic cell lines inhibits differentiation of granulocytes; however, overexpression of cyclin D1, which is not a normal constituent of these cells, does not affect their differentiation (Kato and Sherr, 1993). The effects of the D-type cyclins are cell-type specific, because in some cell types their overexpression results in an enhanced capacity of the cells to enter S phase and lowers the cells' requirements for growth factors (Quelle et al., 1993; Musgrove et al., 1994).

Microinjection of antisense plasmids or antibodies of cyclin D1 during mid-G1 into quiescent fibroblasts is inhibitory to cell cycle progression upon serum stimulation, but it is ineffectual when administered near the G1/S boundary (Baldin et al., 1993). This indicates a functional role for the D-type cyclins occurring after the G0/G1 transition. Recent studies have pointed to the possible necessity of down-regulation of cyclin D1 levels and/or its removal from the nucleus at the end of G1, which more precisely delineates the timing of the functional activity of the D-type cyclins. This conjecture originates from the findings that as cells enter S phase, cyclin D1 vanishes from the nucleus and the total concentration of the protein markedly decreases (Baldin et al., 1993). This notion is also supported by the observation that the ectopic expression of cyclin D1 arrests fibroblasts before entering S phase and prevents their repair of damaged DNA (Pagano et al., 1993). When pRb becomes hyperphosphorylated by Cdk4, complexes between pRb and cyclin D2 and D3 become destabilized. However, this disruption may be overridden by the expression of a kinase-defective Cdk4 mutant instead of the wild type, and stable ternary complexes are formed once again (Kato et al., 1993a). This suggests that the D-type cyclins may serve as multifunctional regulators by also targeting Cdk4 and other partners to specific substrates.

The D-type cyclins as well as their Cdk partners associate with proliferating cell nuclear antigen (PCNA), the auxiliary subunit of DNA polymerase delta, and with a 21-kDa protein, p21 (Xiong et al., 1992a). The D-type cyclins along with cyclins E and A, are implicated in the regulation of DNA replication. PCNA is not a phosphoprotein; however, other proteins present at replication origins are Cdk substrates (Hechman and Roberts 1994). The polypeptide p16, a Cdk inhibitor, binds Cdk4, resulting in destabilization of cyclin D complex formation and thus halts Cdk4 activity. The presence of p16 prevents the phosphorylation of pRb by the Cdk4–cyclin D complex, which prevents the stimulation of releasing factors that would otherwise turn on DNA transcription and/or replication. This autoregulatory loop then comes full circle as the inactivation of pRb leads to the enhanced expression of p16 as cells approach S phase (Serrano et al., 1993; Tam et al., 1994).

Current evidence indicates that the D-type cyclins serve primarily as stimulants for the progression through the G1 stage of the cell cycle instead of promoters of the G1/S transition. This is supported by the timing of D-type cyclin's accumulation in relation to cyclin E. In cells of non-lymphoid origin, before the arrival of cyclin E, both cyclin D1 and Cdk4 amass in mid-G1. Cyclin D2 and its associated kinases, Cdk4 and Cdk6, accumulate early in G1 in stimulated human T lymphocytes (Matsushime et al., 1991b;

Surmacz et al., 1992; Ajchenbaum et al., 1993; Sewing et al., 1993; Meyerson and Harlow, 1994). Cyclin D3 follows D1 and D2 in its appearance (Musgrove et al., 1993; Meyerson and Harlow, 1994; Tam et al., 1994). Most importantly, removal of growth factors from macrophages in S phase results in a rapid decrease in cyclin D1–Cdk4 complexes, yet the cells are still capable of the completion of S phase and division (Matsushime et al., 1991a). The importance of the timing of the microinjection of antisense constructs or antibodies to cyclin D1 early in G1 to elicit an inhibitory effect, as mentioned earlier, also lends further credence to the notion that the D-type cyclins are stimulants for the progression through the G1 phase, rather than promoters of the G1/S transition.

The G1/S Transition

The rate-limiting factor involved in the G1/S transition is hypothesized to be cyclin E (Fig. 1). Cyclin E associates primarily with Cdk2 (Dulic et al. 1992; Koff et al., 1992). Currently, it is unclear as to the precise mechanisms involved in the inactivation of cyclin E. The short half-life of the protein is thought to be mediated by its PEST sequences (Koff et al., 1991; Lew et al., 1991), because the removal of such sequences from yeast G1 cyclins increases their half-life (Nash et al., 1988). In human foreskin fibroblasts and in Rat-1 cells, the stable, ectopic overexpression of human cyclin E has been found to shorten their G1 interval, decrease cell size, and reduce their serum dependence for accomplishing the transition from G1to S phase without altering the generation time. Thus, the shortened G1 interval was accompanied by lengthened S and G2phases. The reduction in cell size suggests that the size of the cell may be more dependent on G1 cyclin levels than on the length of the cell cycle. Stepwise withdrawal of serum from the cells slowed the rate of decline in their S phase fraction and resulted in a less profound lengthening of their G1 interval. The generation times of the cyclin E overexpressors and the control group increased in parallel, and both were growth arrested in 0.1% serum. This meant that cyclin E did not induce cell transformation and, further, that overexpression of cyclin E retains the requirement for serum-dependent signals to continue cellular proliferation. Thus, this study demonstrated that cyclin synthesis in mammalian cells is rate limiting for the G1/S transition. Possible oncogenic properties for cyclin E have been uncovered by the discovery of cyclin E overexpression in breast cancer resulting from a gene rearrangement (Keyomarsi et al., 1994).

Entry into S phase does not depend on the presence of cyclins A or B (Fang and Newport, 1991). The initiation of DNA replication, conversely, depends on the presence of functional cyclin A (Girard, 1991), and a disruption of cyclin A function may upset the normal timely sequential relationship between the S and M phases (Walker and Maller, 1991). Both cyclins A and B are first synthesized after the start of S phase, and the synthesis and destruction of cyclin A occurs before that of cyclin B (Minshull et al., 1990; Pines and Hunter, 1990). In humans, cyclin A localizes to the nucleus in interphase and is first detected near the G1/S transition, whereas cyclin B accumulates in the cytoplasm and does not enter the nucleus until just before the breakdown of the nuclear membrane (Pines and Hunter, 1990). Cyclin A–Cdk2 as well as cyclin E–Cdk2 associated subunits are able to form higher order complexes with pRb, p107, pRb2/p130 (Rb-related), and the transcription factor E2F. Such dynamic associations infer that Cdk2 may be functioning on a secondary level as a temporal regulator of gene transcription during the G1 and S phases of the cell cycle (Nevins, 1992), depending on which complexes it forms.

Mitosis

Cyclin B bound to $p34^{cdc2}$ is linked to entry of the cell into the mitotic phase; however, the complex is present throughout the S and G2 phases (Fig. 1). Hyperphosphorylation of the cyclin B–Cdc2 complex at Thr-14 and Tyr-15 within the $p34^{cdc2}$ adenosine triphosphate (ATP)-binding site inactivates the kinase to allow normal progression of the cell through the S and G2 phases. Dephosphorylation of the previously mentioned residues by the appropriate phosphatases and phosphorylation of the $p34^{cdc2}$ at Thr-161 (thought to stabilize cyclin B binding) lead to activation of the $p34^{cdc2}$ kinase and stimulate entry into mitosis (Hartwell and Weinert, 1989). Recent evidence, however, shows that the Thr-161 phosphorylation of $p34^{cdc2}$ kinase has no effect on cyclin B binding ability, leaving the biological significance of this event undefined (Desai et al., 1995). This process does exhibit negative regulation in some systems. Unreplicated DNA in *Xenopus* egg extracts has been shown to block the activation of the cyclin B–cdc2 complex by preventing the removal of the phosphates from the inhibitory sites in cdc2 (Tyr-15 and Thr 14). The inhibition of the cdc2 kinase occurs by the activation of the Wee1-Mik-1–related protein kinases that phosphorylate these residues (Smythe and Newport, 1992). Following successful DNA replication and progression past the G2/M transition, the cell divides and is able to exit mitosis by the abrupt degradation of cyclin B through ubiquitin mediation in anaphase. This releases the $p34^{cdc2}$ as a monomer, which is subsequently inactivated (Hartwell and Weinert, 1989).

Cdk Inhibitors

One of the most exciting findings in the area of the cell cycle and its connection to tumorigenesis has been the isolation of a family of small cyclin–cyclin dependent kinase (Cdk) inhibitor proteins. This discovery has uncovered an additional layer of regulation to the cell cycle machinery.

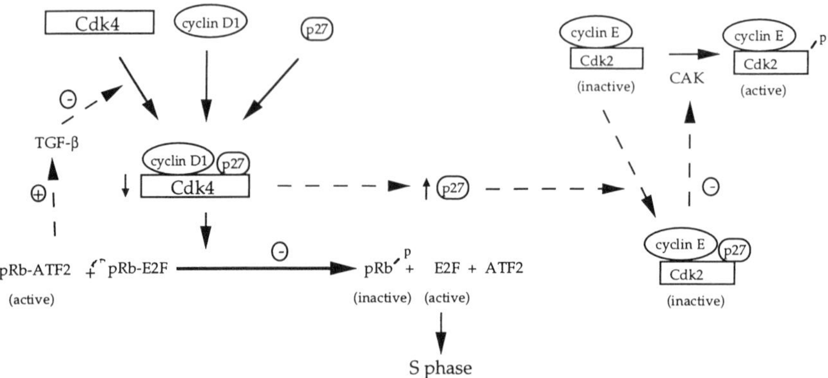

Figure 2. A model of the actions of the growth inhibitor transforming factor beta (TGF-β). TGF-β inhibits the expression of Cdk4 in the G1 phase of the cell cycle, leading to a decrease in the number of cyclin D–Cdk4 complexes. Since most of the CDK-inhibitor p27 in cycling cells is found juxtaposed in the cyclin D–Cdk4 complexes, this results in an indirect induction of p27. The association of p27 with newly formed cyclin E–Cdk2 complexes inhibits their activation by CAK. The suppression and inhibition of cyclin D–Cdk4 and cyclin E–Cdk2 prevents the phosphorylation of pRb and the release of the E2F transcription factor required for the progression into S phase. pRb is also able to bind and enhance the activity of the transcription factor ATF-2. This complex may stimulate the expression of TGF-β, thus serving as a positive feedback loop for G1 growth arrest.

p27^{KIP1}. Cells are arrested in late G1 by the inhibitor, transforming growth factor beta (TGF-β) before the phosphorylation of the retinoblastoma protein (Laiho et al., 1990). TGF-β inhibits the expression of Cdk4 in G1 but has no such effect on cyclin D; however, this blocks the activation of the cyclin D–Cdk4 complex (Fig. 2; Ewen et al., 1993b). Since overexpression of Cdk4 in TGF-β–responsive cells negates the TGF-β block of the cell cycle, this suggests that the inhibition of Cdk4 expression is an important event in TGF-β–mediated cell cycle arrest (see Chapter 2, this volume). TGF-β has no effect on the expression levels of cyclin E or Cdk2; however, active cyclin E–Cdk2 complexes fail to form in cells treated with TGF-β (Koff et al., 1993).

The TGF-β–mediated cell cycle arrest involves the *indirect* induction of an inhibitor of cyclin-Cdk complexes, p27$^{\mathrm{KIP1}}$, which has recently been cloned (Polyak et al., 1994a; Toyoshima and Hunter, 1994). Most p27 found in cycling cells is juxtaposed in the cyclin D–Cdk4 complexes (Fig. 2; Ewen et al., 1993b). TGF-β has not been shown to induce the transcription of p27 (Polyak et al., 1994a). Rather, by repressing the synthesis of Cdk4, TGF-β reduces the number of cyclin D–Cdk4 complexes, which then possibly frees p27 to inhibit other cyclin-dependent kinases. The inactive form of cyclin E–Cdk2 in TGF-β–treated cells is found to complex with p27 (Koff et al., 1993; Polyak et al., 1994b). Newly formed complexes of cyclin E–Cdk2 are activated by phosphorylation of Cdk2 on the residue Thr-160 by the activation enzyme CAK; however, the association of p27 to these new cyclin E–Cdk2 complexes effectively inhibits CAK and thus blocks the activation of the new complexes (Polyak et al., 1994a). Both events—the decrease in number of the cyclin D–Cdk4 complexes and the inactivation of the cyclin E–Cdk2 complexes—block the phosphorylation of not only pRb but also of other regulatory factors by cyclin-Cdks, which are necessary events for the transition of cells into S phase.

$p15^{INK4B/MTS2}$. The TGF-β pathway may also involve the induction of a Cdk inhibitor, p15, a p16 homolog. TGF-β transactivates p15 in keratinocytes, which inhibits the D-type Cdk complexes (Cdk4/Cdk6) (Sherr, 1994). This releases additional p27 to inhibit other Cdks and thus promotes cell cycle arrest.

p21. Other mammalian Cdk inhibitors in addition to p27 have also been found, including p21 and p16. In fact, the first Cdk inhibitor to be isolated was p21. p21 was simultaneously discovered by several laboratories using various approaches and has been termed Cip1, p21, $p20^{CAP1}$, Waf1, and Sdi1 (El-Deiry et al., 1993; Gu et al., 1993; Harper et al., 1993; Xiong et al., 1993; Noda et al., 1994). The importance of p21 currently seems to lie in three areas: 1) the p21 gene is transcriptionally regulated by the tumor-suppressor protein p53 (El-Deiry et al., 1993), 2) p21 has been identified as a Cdk inhibitor (Gu et al., 1993; Harper et al., 1993), and 3) p21 may block the entry of senescent cells into the cell cycle (Noda et al., 1994). Interestingly, p21 induction during initiation of terminal differentiation occurs in a p53-independent manner (Jiang et al., 1994). This hints to a more universal function for p21 in differentiation as well as growth control. The Cdk inhibitors p21 and p27 share 44% identity in their amino terminal regions and, *in vitro*, p27 is able to inhibit the same gamut of Cdks as p21 (Polyak et al., 1994a). Normal fibroblasts exhibit a wide range of cyclin-Cdk complexes associated with p21. Strong associations occur with cyclin D–Cdk4 and cylin E–Cdk2 complexes in the G1/S transition, as well as to the cyclin A–Cdk2 complex during mitosis (Xiong et al., 1992a,b; Harper et al., 1993; Xiong et al., 1993). However, cyclin B–Cdc2 forms only a weak partnership with p21 in M phase (Xiong et al., 1992a). The strength of association of the cyclin-Cdk complexes with p21 is directly proportional to the ability of p21 to inhibit their functions *in vitro*.

In response to DNA damage, the expression levels of p53 dramatically increase, and this leads to the transcriptional induction of DNA damage–inducible genes. Within the promoter of the p21 gene are two consensus p53-binding sites (El-Deiry et al., 1993). Exposure of cells to DNA damaging agents enhances transcription of p21, leading to the agglomeration of p21 in cyclin E–Cdk2 immune complexes. This culminates in decreased kinase activity that is thought to mediate the G1 arrest of the cell cycle (Dulic et al., 1994; El-Deiry et al., 1994). This scenario occurs only if the cells have wild-type p53 (and fails with mutant p53), thus supporting the hypothesis that p21 is the primary mediator of p53-dependent G1 arrest. Ectopic over-expression of p21 alone in normal fibroblasts is also able to elicit growth arrest (Harper et al 1993; Noda et al., 1994). Basal levels of p21 may also depend on the functional activity of p53, since p21 is absent from the Cdk complexes of various transformed cell lines lacking an active p53 (Xiong et al., 1992a; El-Deiry et al., 1993, 1994). To date, p21 mutants have not been identified in human cancers, thus leading medical researchers to look for additional p53-inducible genes involved in growth regulation.

In addition to serving in DNA-damage checkpoint regulation by causing G1 arrest through the inhibition of Cdks, the p21 protein has also been implemented in the cells proofreading system during cellular replication to improve fidelity. PCNA enhances processivity of DNA polymerase delta (pol D) by forming a homotrimer that holds the complex of pol D with its replication factor C (RFC) to the DNA template to allow synthesis of both the leading and lagging strands of DNA. *In vitro* studies show that p21 can effectively associate with PCNA and block pol D–dependent DNA replication, which may be overcome only by the addition of PCNA (Flores-Rozas et al., 1994; Waga et al., 1994). p21 may serve important functions in this manner. p21 may act to reduce the processivity of the lagging strand, which is desirable because the rate-limiting step here is the dissociation of the DNA polymerase at the end of the Okazaki fragment. By inhibiting the elongation step of polymerization (Flores-Rozas et al., 1994), p21 is effectively slowing the replicative process to allow for the proofreading mechanisms of the replicative machinery to correct any errors in replication. The *in vivo* verification of these results is still pending. Whether or not p21 is able to affect other functions of PCNA is not currently known. This is of interest because PCNA is also involved in excision repair. Here one may again hypothesize a role for p21 in decreasing the processivity of the DNA polymerase repair complex as this would be beneficial in filling in the excision repair gaps. If this is the case, this would give eukaryotes another similarity with bacteria, because bacteria alter the processivity of polymerase III by interactions with proteins where eukaryotes were thought before to augment processivity only by the implementation of three different DNA polymerases. This new model calls for a higher order of regulation of processivity of DNA polymerase than once thought.

$p16^{INK4}$. The Cdk inhibitor p16, also known as multiple tumor suppressor 1 (MTS1), has recently been cloned and implicated as a tumor-suppressor gene directly linked to the cell cycle (Kamb et al., 1994). In certain transformed cells Cdk4, lacking the normal components of the Cdk4 complexes, such as the D-type cyclins, PCNA, and sometimes p21 (Xiong et al., 1992 a,b) is found in association with p16. Complex formation of p16 with Cdk4 is thought to block or upset the D-type cyclin association (Serrano et al.,

1993). This dissociation is in contrast to the inhibitory mechanisms employed by the Cdk inhibitors p21 and p27. p16 is the first Cdk inhibitor in which structural homologs such as MTS2 (p15) have been indicated (Kamb et al., 1994). Also, p16 seems to have a higher selectivity than the other two Cdk inhibitors, since p16 appears specifically to inhibit the D-type cyclin-dependent kinases, Cdk4 and Cdk6. These unique characteristics of p16 suggest that this protein may play a specialized and important role in growth regulation. In addition, the human p16 gene has been mapped to chromosome 9p21 (MTS1 locus), a region that is frequently deleted in gliomas and melanomas (Kamb et al., 1994; Nobori et al., 1994). MTS1 has been found homozygously deleted in cell lines derived from tumors of lung, breast, brain, bone, skin, bladder, kidney, ovary, and lymphocyte. Nonsense, missense, or frameshift mutations were also found in at least one copy of the gene in melanoma cell lines (Kamb et al., 1994). However, only 10–20% of primary tumors demonstrate an alteration in the locus (Kamb et al., 1994; Spruck et al 1994).

Recently, direct evidence of the inhibitory ability of p16 on cell growth has been provided. Ectopic expression of p16 effectively blocked entry into S phase of cells induced to proliferate by the H-*ras* oncogene and suppressed cellular transformation by oncogenic H-Ras and Myc (Serrano et al.1995).

TUMOR SUPPRESSORS

The Retinoblastoma Protein (pRb)

The genetic characteristics and the epidemiologic studies of Knudson concerning familial and nonfamilial retinoblastoma lead to the "Knudson two-hit hypothesis." This laid the groundwork for the discovery of tumor-suppressor genes and for a revolutionary new way of thinking about oncogenesis. Knudson concluded that the familial form of retinoblastoma results when an individual inherits a mutant (nonfunctional) allele from the affected parent and the childhood tumor manifests itself when the wild-type (normal) allele of the same gene is functionally "knocked out" by a somatic event, resulting in a LOH. The nonfamilial form of retinoblastoma requires two separate somatic mutations of both alleles of the gene and thus accounts for the sporadic nature of this form of retinoblastoma and its appearance later in life (Knudson, 1971). Subsequent studies confirmed that all retinoblastomas carry mutations in both alleles of the same gene. Before this, oncogenesis was thought to evolve by dominantly acting oncogenes that could only occur by somatic mutations because the dominant oncogenic phenotype would result in the gestational death of the embryo if it were acquired from the germline. So, for the first time, a recessive mutation conferring an inherited increased risk of neoplasia could be envisioned. The finding of deletions on chromosome 13q14 in both familial and sporadic retinoblastomas (Knudson, 1984; Cavenee et al., 1985) supported the Knudson hypothesis and led to the eventual cloning of the first tumor-suppressor gene, the retinoblastoma gene, RB (Friend et al., 1986; Fung et al., 1987; Lee et al., 1987).

The retinoblastoma gene encodes a 105-kDa protein product, pRb. The nuclear phosphoprotein, pRb, serves as a negative regulator of cell cycle progression, at least in some specific cell types. RB transcripts are present in all normal tissues examined (Lee et al., 1987). In cell lines that lack a functional pRb protein, the restoration of pRb function, either by microinjection of the protein in early to mid-G1 or by transfection of the cDNA, arrests cell growth in the G1 phase of the cell cycle (Goodrich et al., 1991; Hinds et al., 1992; Zhu et al., 1993). This inhibition of cell growth by pRb depends on the sequences necessary for the interaction of pRb with the transcription factor E2F, as well as a number of oncoproteins from human DNA viruses such as E1A, T antigen, and E7 [from adenovirus, simian virus 40 (SV40), and human papillomavirus, respectively] (Chellapan et al., 1991; Hiebert et al., 1992; Qian et al., 1992; Qin et al., 1992). Binding of pRb to this set of associated proteins occurs at a specific "pocket region" in pRb.

The "pocket region" of pRb is also a common site of mutation in many human tumors, and it has been found altered or deleted in all pRb mutants from the tumors thus far examined (Knudson, 1984; Weinberg, 1991). Inactivation of RB function involves various genetic lesions including large-scale deletions and splicing mutations resulting in the deletion of an exon (Horowitz et al., 1989; Kaye et al., 1990). Smaller deletions in the promoter region of RB and point mutations have also been found (Bookstein et al., 1990a,b). Genetic mutations in RB have been demonstrated in several human tumors, such as osteosarcomas, bladder carcinomas, prostate carcinomas, breast carcinomas, small cell lung carcinomas, cervical carcinomas, and leukemias (Horowitz et al., 1990).

pRb and the Cell Cycle

Post-transcriptional regulation of pRb involves changes in the phosphorylation state of pRb throughout the cell cycle (Chen et al., 1989; De Caprio et al., 1992). During the G0 and G1 phases of the cell cycle, pRb is found in a hypophosphorylated form that becomes increasingly phosphorylated on serine and threonine residues as the cell cycle progresses. The cell cycle–dependent phosphorylation of pRb may occur in several steps. In primary human T lymphocytes, pRb probably undergoes phosphorylation at discrete residues during 3 phases of the cell cycle: late G1, S phase, and again in G2/M (DeCaprio et al., 1992). All phosphorylations of pRb are on serine/threonine residues; to date, no evidence exists for phosphorylation on tyrosine residues in pRb (Chen et al., 1989; Ludlow et al., 1989). As the cells emerge from mitosis and enter into the G1 phase again, or become senescent in G0, pRb is dephosphorylated back to its hypophosphorylated form possibly by pp1 phosphatase (Durfee et al., 1993). The viral oncoproteins and the transcription factor E2F bind specifically to hypophosphorylated pRb (Chellappan et al., 1991; Hiebert et al., 1992; Qian et al., 1992; Qin et al., 1992). The hypophosphorylated pRb is thought to be the physiologically active form; it exhibits the growth-suppressive function that can be inactivated in G1 by phosphorylation or by dimerization with viral oncoproteins to allow DNA replication.

The specific kinase(s) responsible for the *in vivo* phosphorylation of pRb is/are currently not known. Several Cdks have been found in association with pRb *in vivo* in a region similar to "the pocket". The A, B, D, and E cyclin–dependent kinases as well as the recently discovered Cdc2-related kinase PITALRE (Grana et al., 1994a) phosphorylate pRb *in vitro*. Considering the timing of phosphorylation of pRb as well as the peak activity intervals of the Cdks and their interactive abilities, it seems plausible that pRb is phosphorylated by the D-type–dependent kinases in early G1, cyclin E–Cdk2 at the G1/S transition, and by the cyclin A–Cdc2 complex at the G2/M transition. Since interleukin 2 is stimulatory and TGF-β is inhibitory to cyclin E–Cdk2 activity, this offers one possible route by which exogenous growth-regulatory signals influence pRb and the cell cycle machinery (Koff et al., 1993; Polyak et al., 1994b).

Phosphorylation of pRb by the D-type–dependent kinases may play a key role in overriding the G1 cell cycle arrest of pRb. Microinjection of cyclin D antibodies into RB +/+ cells but not into the RB −/− tumor cell line, SAOS-2 (human osteosarcoma), blocks entry into S phase (Baldin et al., 1993). This suggests that pRb may indeed be the key substrate for the D cyclins. Several cell lines harboring mutated, or pRb inactivated by the expression of the oncoprotein SV40 large T antigen, exhibit an enhanced expression of the specific cyclin D–Cdk4 kinase inhibitor, p16 (Serrano et al., 1993). Even though it seems counterintuitive to consider Cdk inhibitors oncogenic, if pRb is the sole substrate for cyclin D–Cdk4, then overexpression of p16 may be permissive for continued proliferation, possibly in a cell type–specific manner (Serrano et al., 1993).

pRb Cellular Targets

Functioning as a transcriptional regulator through the modulation of the activity of several transcription factors, pRb exerts its growth-suppressive effects. pRB binds directly to the transcription factor ATF-2 and enhances its activity (Kim et al., 1992). ATF-2 is thought to be involved in the expression of TGF-β. In this example, pRb may actually up-regulate the expression of TGF-β and thereby lead to growth suppression through the expression of this growth inhibitor. As discussed previously, TGF-β would in turn disrupt and inactivate the cyclin D–Cdk4 complexes, leading to the release of the Cdk inhibitor p27 and the effective inhibition of cyclin E–Cdk2 (Fig. 2).

This would prevent the phosphorylation of pRb and thus serve as a positive feedback loop to block progression into S phase.

The transcription factor E2F, first identified as a DNA-binding protein required for E1A-mediated induction of the adenovirus E2 promoter, binds to hypophosphorylated pRb in the "pocket region" during the G0 and G1 phases of the cell cycle (Bandara et al., 1991b; Chellappan et al., 1991; Raychaudhuri et al., 1991; Nevins, 1992; Chellapan et al., 1994). E2F binding sites are located in the promoters of multiple growth-promoting and growth-responsive genes such as c-*myc*, c-*myb*, dihydrofolate reductase (DHFR), thymidine kinase (TK), thymidine synthetase, DNA polymerase alpha, RB, cyclin A, cyclin D1, cdc2, and E2F itself (Chellapan et al., 1994; Sala et al., 1994). The heterodimer of pRb2-E2F is thought to sequester the active form of E2F and in effect block the transactivation of genes whose products are necessary for DNA synthesis (Nevins, 1992; Sala et al., 1994). Association with the viral oncoproteins or the phosphorylation of pRb2 releases E2F in its functional form, which is then able to promote the transcription of genes required for S phase entry and completion.

Recently, four E2F-related proteins (E2F2, E2F3, E2F4, and E2F5) have been isolated and shown to share homologous domains with E2F (Ivey-Hoyle et al., 1993; Beijersbergen et. al 1994; Ginberg et al., 1994). This confirmed the speculation that an E2F family may exist. E2F binds as a heterodimer to DNA with dimerization partner-1 (DP1) (Fig. 1; Bandara et al., 1993; Huber et al., 1993). Besides the sequestering of E2F by pRb, the activity of the E2F/DP1 heterodimer is further regulated during the cell cycle by phosphorylation of DP1. As the phosphorylation level of DP1 increases the DNA binding activity of E2F/DP1 decreases proportionally (Bandara et al., 1994). Since a cyclin A binding domain is located in the amino terminus of E2F1, cyclin A–Cdk2 is the most likely candidate for the late S phase phosphorylation of DP1 (Krek et al., 1994). The regulation of transcriptional activity is made even more complex when one considers the involvement of the Rb family proteins, p107 and p130/pRb2. The Rb family is capable of association with the E2F family. This interaction seems to be temporally modulated and varies between family members. Several sources of data suggest that pRb and p107 associate with distinct E2F species. The *in vivo* results demonstrate that E2F1, E2F2, and E2F3 all form complexes with pRb but do not interact with p107 (Chittenden et al., 1993). E2F4 and E2F5 were both cloned by their ability to interact with pRb2/p130. E2F4 undergoes complex formation *in vivo* with pRb2/p130 during G0 and G1, then later associates with p107 in late G1 and S phases (Beijersbergen 1994; Ginsberg et al., 1994).

pRb and Cellular Differentiation

In addition to the role of pRb as a checkpoint regulator of the cell cycle, the protein is also implicated in cellular differentiation through complex formation and subsequent modulation of various transcription factors. Cell lines that lack a functional pRb are unable to undergo myogenic conversion. This points to a possible role of pRb in the myogenic activation of Myo D (Gu et al., 1993). Evidence directly links pRb with several other transcription factors that are involved in the withdrawal from the cell cycle and the induction of differentiation (Gu et al., 1993; Dunaief et al., 1994), such as the brahma-related gene 1 (BRG1). Perhaps the most convincing evidence for the importance of pRb in cellular differentiation and specialization stems from the studies of RB knockout mice. Homogeneous germline disruptions of the RB gene (RB −/−) proved to be a lethal mutation, with the mice dying *in utero* by the 14th day of gestation and manifesting gross defects in the development of the hematopoietic and central nervous systems. The heterozygotic mice (RB +/−) matured normally until 9 months of age, when they developed thyroid and intermediate lobe pituitary tumors but, surprisingly, showed no indication of retinoblastoma. The tumors in the heterozygotes had undergone a loss heterozygosity and were RB −/− (Clarke et al., 1992; Lee et al., 1992). The fact that the vast majority of cell division cycles in the RB −/− embryos occurred without the presence of pRb suggests that other proteins, possibly the Rb family members, can complement pRb function, or perhaps the cell can elicit a different signal transduction pathway to regulate the activities of proliferation. These results also point to pRb involvement in neurogenesis and hematopoiesis. The link between pRb and hematopoiesis is supported by recent transcriptional studies indicating that the RB gene plays an erythroid- and stage-specific functional role in normal human adult hematopoiesis, particularly at the level of late erythroid hematopoietic progenitor cells (Condorelli et al., 1995).

RB Family

The "pocket region" of pRb is shared by two additional E1A associated proteins, and this has led to the identification of two members of the Rb family, p107 and p130/pRb2 (Ewen et al 1991; Mayol et al., 1993; Zhu et al, 1993). Recent functional studies of p107 and p130/pRb2 have indicated that although the Rb family members may be able to complement each other, the proteins are not fully functionally redundant. Both p107 and p130/pRb2 form stable complexes with cyclin A–Cdk2 and cyclin E–Cdk2. As with pRb, p107 phosphorylation is cell cycle regulated, with a high level of hyperphosphorylated p107 during S phase and at the G2/M transition (Shirodkar et al., 1992). Various phosphorylated forms of p130/pRb2 are found *in vivo*. A hyperphosphorylated p130/pRb2 occurs at the G1/S transition (Baldi et al., 1995). Like pRb, ectopic expression of p107 and p130/pRb2 is able to suppress the growth of the osteosarcoma cell line SAOS-2 (Hinds et al., 1992; Quin et al., 1992; Zhu et al., 1993; Claudio et al., 1994). Additionally, pRb, p107, and p130/pRb2 all form complexes with E2F (Chellappan et al., 1991; Cao et al., 1992; Shirodkar et al., 1992; Chitteden et al., 1993). However, the temporal order of complex formation appears to vary. For example, the binding of p107 to E2F is first detected at the G1–S boundary and remains stable throughout S phase (Cao et al., 1992, Shirodkar et al., 1992). On the other hand, pRb-E2F complexes are found in the G1 phase and then dissociate in late G1 (Chellappan et al., 1991, Hiebert et al., 1992). The main form of E2F detected in the G0–G1 phases in primary mouse fibroblasts is E2F bound to p130/pRb2, which is then replaced by p107-E2F complexes in late G1 (Cobrinik et al., 1993).

In addition to these differences, there is no evidence to support the notion that p107 is normally a tumor suppressor. To date, there are no examples of naturally occurring mutations of p107, despite the testing of several hundred human tumor samples and cell lines (M. Ewen, personal communication). In addition, p107 has been mapped to a chromosome region not commonly found to be cytogenetically altered in human neoplasms (Ewen et al., 1991). Furthermore, the expression of p107 in cell lines derived from retinoblastoma tumors implies that p107 is unable to complement the lack of pRb and to suppress tumor formation (Ewen et al., 1989). However, p130/pRb2 has been mapped to human chromosome 16q12.2 (Yeung et al, 1993), an area in which deletions have been found in several human neoplasms including breast, ovarian, hepatic, and prostatic cancers (Yeung et al, 1993). This suggests a role for the p130/RB2 gene as a tumor-suppressor gene in human cancer.

Amino acid comparisons of the three protein sequences of the Rb family suggest a closer relationship of p130/pRb2 to p107 (Mayol et al., 1993). The E1A mutant pm928, which binds p107 but fails to bind pRb and p130/pRb2, is able to initiate DNA synthesis, but the transformation property of the oncoprotein is abrogated (Giordano et al., 1991a,b; Moran, 1993). This implies that p130/pRb2 and pRb may elicit repression of cell cycle progression at similar checkpoints. Recently, it was discovered that p130/pRb2 binds *in vitro* and *in vivo* to the oncoprotein E7 from the high risk human papillomavirus type 16 (HPV16) (M. Tommasino et al., personal communication). This imparts a clinical significance to the presence of a functional p130/pRb2, and points to a possible future role for it as a prognostic indicator.

The introduction of p130/pRb2 in HONE-1 cells, a cell line of human nasopharyngeal carcinoma (NPC) that expresses p130/pRb2 at a low level, causes a significant reduction in cell proliferation and a change in morphology (Claudio et al., 1994). The HONE-1 cell line expresses the RB gene product with normal size and abundance, and no point mutation has been detected in the common sites for RB mutations (Sun et al., 1993). Previous evidence has led to the conclusion that nasopharyngeal carcinogenesis shows no detectable retinoblastoma susceptibility gene alterations (Sun et al., 1993). These recent findings hint to a possible involvement of p130/pRb2 in nasopharyngeal carcinogenesis in the face of a functionally intact

RB gene. NPC is a rare disease in most parts of the world; however, the disease has a racial and geographical distribution. The people of Southern China are among those that deviate from the low-risk profile so much so that NPC is the most common cancer in the city of Guangzhow (Canton) and constitutes 32% of all cancer (Yan et al., 1988; Mascolo et al., 1992).

Most interestingly, the T98G human glioblastoma cell line, which is resistant to the suppression effect of both pRb and p107 (Zhu et al., 1993), demonstrated a drastic reduction in cellular proliferation upon overexpression of p130/pRb2 (Claudio et al., 1994). This suggests that p130/pRb2 may function in a completely different pathway than that of p107 or pRb, and that a certain mutation(s) within the T98G cells predisposes them to be sensitive to the effects of p130/pRb2 but not to those of p107 or pRb. An alternate explanation may be that the three proteins may share functional properties, whereas p130/pRb2 has an additional property that p107 and pRb are unable to complement in this particular cell line.

THE TUMOR SUPPRESSOR p53

The p53 gene (and its protein product p53) is one of the most studied tumor suppressors. p53 has been implicated in growth arrest, differentiation, DNA repair and genomic stability, and in apoptosis. Figure 3 serves as a simplified model of the p53-regulated pathways. Holding true to the definition of a tumor-suppressor gene, the overexpression of wild-type p53 inhibits the growth of both normal and transformed cells (Finlay et al., 1989; Mercer et al., 1990). Like pRb, p53 elicits many of its effects by transcriptional regulation. However, p53 itself is a sequence-specific DNA binding protein that can transactivate or repress specific genes. pRb is surmised to express its inhibitory effects on transcription by binding and sequestering specific positive transcription factors. p53, on the other hand, functioning as a negative transcriptional regulator, has been reported to associate with the basic transcription machinery, the TATA box binding proteins (TBP), by two terminal regions (Seto et al., 1992; Mack et al., 1993) and to repress, nonspecifically, gene expression. Additionally, unlike the RB gene mutations that are recessive and only manifest disease on LOH, p53 mutations are dominant. Since oligomerization of p53 is required for its normal function as a transcription factor, complex formation between wild-type and mutant p53 may not be functional. This offers a possible explanation of the dominant negative effects of mutant p53.

The importance of the p53 gene in oncogenesis is exemplified by the fact that more than 50% of human tumors lack a copy of the wild-type functional p53 gene. Point mutations are the most common mutations found in the p53 gene; however, others include deletion, insertion, truncation, viral genome

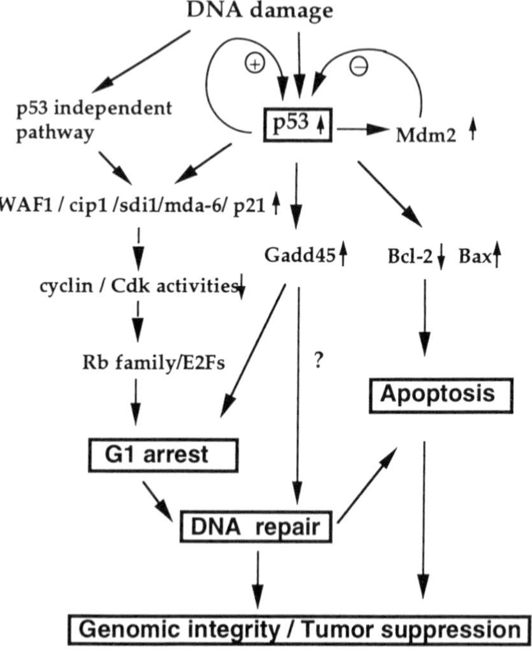

Figure 3. Simplified model of the p53 pathways (Reprinted with permission of CRC Press, Inc., from Sang et al., 1995).

insertion and disruption, rearrangements, and translocation (Prokocimer and Rotter, 1994). One of the most frequently mutated regions in tumors is the p53 DNA binding domain. The viral oncoprotein, SV40 large T antigen, which mediates transformation via physical interaction with and inactivation of p53, binds to p53 in a region overlapping the p53 DNA-binding domain (Jenkins et al., 1988). This points to the significance of this particular region in tumor suppression function. However, the integrity of the highly charged basic region of the carboxy terminus is also necessary for sequence-specific DNA-binding activity by mediating oligomerization and nuclear localization of p53 (Sturzbecher et al., 1992; Clore et al., 1994). The oncoproteins from HPV16 and HPV18 (Werness et al., 1990) target the carboxy termini and effectively disrupt p53 DNA-binding regulation. Adenovirus oncoprotein E1B's binding domain maps to the highly charged acidic transactivation domain at the p53 amino terminus (Sarnow et al., 1982; Fields and Jang, 1990; Raycrof et al., 1990) suggesting that E1B aids in cellular transformation by inhibiting the transactivation function of p53. DNA tumor viruses that harbor mutant oncoproteins unable to bind p53 are defective in their transforming activity (Jenkins et al., 1988), testifying to the critical role that the blocking of p53 transactivation function plays in neoplastic transformation, and indicating the importance of the p53 target genes in the control of cellular proliferation.

Regulation of p53

Regulation of p53 activity occurs at the transcriptional and the post-translational levels. Several elements are contained within the p53 promoter, such as the SP1, PF-1, NF-1, and NF-kappa-like elements (Ginsberg et al., 1990; Deffie et al., 1993). The promoter also contains a basic helix-loop-helix (bHLH) binding domain (Ronen et al., 1991). Various signals could regulate a promoter arrangement such as this. The differentiation-related induction of p53 is transcriptionally regulated (Reisman and Rotter, 1989). As normal hematologic cell lines mature and specialize, the expression level of p53 is markedly increased (Prokocimer and Rotter, 1994). In agreement with this model of p53 involvement in the development of the hematopoietic system, constitutive expression of p53 in some hematologic cell lines progresses the cells along the lines of cellular differentiation (Kastan et al., 1991a,b). The ectopic overexpression of wild-type p53 in the established cell lines L12, HL-60, and K462 (which lack a functional p53) promotes cellular differentiation (Pinhasi-Kimhi et al., 1986; Shaulsky et al., 1991; Feinstein et al., 1992). However, post-transcriptional alterations of p53 seem to be the most important means of regulation in the p53 DNA damage response pathway.

p53 undergoes extensive post-translational modification, including multiple site phosphorylation by several protein kinases within the alpha (transactivation domain) and carboxy (oligomerization and nuclear localization domains) termini (Meek, 1994). The double-stranded DNA-activated kinase (DNA-PK) is activated by strand breaks or single-stranded gaps in the DNA. DNA-PK phosphorylates p53 serine residues at positions 15 and 37. Mutational analysis of p53 has shown that substitution of serine 15 by alanine partially abrogates the ability of the mutant p53 to induce G1 arrest (Lees-Miller et al., 1992). This suggests a possible signal cascade elicited to mediate the p53 G1 arrest in response to DNA damage. Exposure of cells to DNA-damaging agents activates the mitogen-activated protein (MAP) kinase (Devary et al., 1992; Radler-Pohl et al., 1993). Serum stimulation and activation of quiescent cells activates two major p53 kinase activities exhibiting identical biochemical parameters as the MAP kinase (Milne et al., 1994); therefore, the MAP kinase may be important in modulating the response of p53 to specific forms of DNA damage. Studies suggest that p53 may also be regulated through phosphorylation by the cell cycle–dependent kinases because both cyclin A– and cyclin B–Cdc2 kinases phosphorylate p53 *in vitro* (Bischoff et al., 1990; Sturzbecher et al., 1990). Upon transition into S phase, hyperphosphorylated p53 as well as p53 in complex with a Cdk-like kinase have been demonstrated by immunoprecipitation (Milner et al., 1990; Sturzbecher et al., 1990). The serine/threonine kinase casein kinase II (CKII) is thought to phosphorylate p53 at its carboxy terminus (at serine 392) thereby regulating the stability of the p53 protein (Herrmann et al., 1991; Hupp et al., 1992).

The activity of p53 is also modulated by a negative feedback regulatory loop. p53 directly transactivates the mdm2 gene. The mdm2 gene product,

Mdm2, then targets and inactivates p53 at its amino terminus, the transactivation domain (Barak and Oren, 1992; Momand et al., 1992). Overexpression of the proto-oncogene mdm2 could result in the inhibition of p53, leading to the first steps of oncogenesis.

Response to DNA Damage

It has been proposed that p53 functions as a guardian of the genome (Lane, 1992). The G1 growth arrest response of p53 mediated by the Cdk inhibitor p21 has been discussed previously (Fig. 3). DNA strand breaks also induce p53 to transactivate the growth arrest DNA damage (gadd) gene (Fig. 3; Fornace et al., 1989). Ectopic overexpression of gadd45 also demonstrates growth-suppressive properties (Zhan et al., 1994). Upon detection of low to moderate levels of DNA damage, it is thought that p53 transactivates the p21 and gadd45 genes to arrest the cells before the replication of damaged DNA. The cells are then given time to repair their mutated genome. As discussed earlier, the p21 protein may also assist in this function by decreasing the processivity of DNA polymerase and allowing the proofreading mechanisms time to correct any mistakes. Additionally, gadd45 may be instrumental in this process by stimulating the DNA excision repair pathway and interacting with PCNA (Smith et al., 1994).

Apoptosis

In a multicellular organism it is better to sacrifice a cell with irreparable genomic damage, as opposed to a unicellular organism that elicits a cascade to mutate randomly the genome in an attempt for survival at any cost. Genomic instability leads to an enhanced risk of neoplasia that is devastating to a multicellular organism while the loss of a single cell is insignificant. p53 scans the genome, and if high levels of DNA damage are detected, p53 will transactivate the *bax* gene and down-regulate the *bcl*-2 gene and effectively induce apoptosis (Fig. 3; Selvakumaran et al., 1994). The p53-dependent apoptosis pathway can be induced not only by DNA damage but also by viral infection, oncoproteins, and chemotherapeutic agents directed against cancer (Karp and Broder, 1994). Therefore, a successful oncovirus must find a way to circumvent this pathway to transform the cell. Bcl-2 and the adenovirus oncoprotein E1B are able to block the p53-dependent and independent pathways of apoptosis (Boyd et al., 1994; Chiou et al., 1994). The down-regulation of Bcl-2 by p53 is thought to occur by an indirect mechanism. Three novel proteins thought to be involved in the regulation of programmed cell death have recently been cloned: Nip1, Nip2, and Nip3. E1B must be capable of interacting with the Nip proteins to suppress p53-mediated apoptosis (Boyd et al., 1994).

The Interconnection between p53 and pRb

The fact that DNA tumor viruses simultaneously evolved the ability to repress both p53 and pRb function to accomplish cellular transformation suggests a cooperation between the two proteins in their strategies to regulate proliferation. Just how this is translated to normal cellular processes was not known until the transactivation products of p53 were identified and characterized. The Cdk inhibitor, p21, links the DNA damage response of p53-mediated G1 arrest to the inhibition of phosphorylation of pRb, a necessary event to enter S phase. This coordination of activities between the two proteins is thought to extend to the induction of apoptosis as well. The E2F1 transcription factor, regulated by complex formation with pRb, cooperates with p53 to mediate apoptosis (Wu and Levine, 1994).

Convincing support of the cooperation between p53 and pRb comes from the study of gene knockout mice. As expected, germline mutations in RB or p53 predispose the mice to malignancy (Donehower, 1992; Jacks et al., 1992; Lee et al., 1992). As mentioned previously, mice with RB −/− genotype die *in utero* because of gross defects in hematopoiesis and rampant neuronal cell death. In the early embryos of the RB −/− mice, disorganized uncontrolled lens proliferation and a failure to express the proper pattern of late differentiation products led to a large amount of apoptotic cells in the lens fibers. However, these apoptotic cells are absent in the abnormally developed lenses of mice null for RB and p53 (RB −/− p53 −/−) (Morgenbesser et al., 1994; Williams et al., 1994). This suggests coordination between the functions of pRb and p53. The lack of functional pRb results in a failure of terminal differentiation, possibly leading to a hyperproliferative state. This may be held in balance by p53 induction of apoptosis to orderly dispose the abnormally differentiated cells. This supports the model first conjectured by the oncovirus studies: that the ultimate manifestation of neoplasia requires the functional loss of both pRb and p53. That is, cells fail to differentiate properly (pRb loss of function) and do not apoptose (p53 loss of function). The analysis of the mice null for both RB and p53 genes revealed the typical tumors of each individual gene knock-out as well as novel tumors when compared with the viable RB +/− mice and the p53 −/− mice (Li et al., 1994). This suggests that in some tissues pRb and p53 can complement the tumor suppression function of the other and that some tissues require the inactivation of both proteins for neoplastic transformation.

FUTURE PROSPECTS

To combat cancer more effectively, pharmaceutical design should take advantage of the key molecules governing cellular processes. Many of the chemotherapeutic agents used today work by inducing apoptosis. Knowledge of the apoptotic pathways and their regulation should lead to the design of more effective drugs with less harmful side effects. Indeed, the drug-resistance patterns of the neoplasms may be anticipated and effectively circumvented. For example, the directed inhibition of Bcl-2 by disruption of its interaction with the Nip proteins in conjunction with agents that induce apoptosis may prevent or delay the development of chemoresistant neoplasms. Since the molecules of the cell cycle machinery interact with each other through different signaling cascades, combination therapy targeting parallel functional pathways may decrease the tactical options of the cancerous cell for survival. The neoplastic cells may then be unable to elicit compensatory pathways before cell death.

Much can be learned by the strategies employed by viruses in their attempt to take over the cell cycle machinery. Mutated viral particles serving as vectors may, in the future, be used to replace mutated genes, especially the tumor-suppressor genes. Owing to the frequency of mutation in human tumors and the interactive nature of the products of the RB and p53 genes, simultaneously targeting RB and p53 genes for therapy may prove to be more efficient, and have a wider range of applications than restoration of only one or the other gene. Of course, the multistep nature of oncogenesis decreases the likelihood that gene therapy will be the panacea for cancer as was once thought. However, upon the eventual refinement of delivery and targeting systems, gene therapy should offer an effective alternative therapeutic regime that may prove to be much less toxic and devastating than the current options of surgery, radiation, and chemotherapy.

Site-specific recombination and DNA repair are being investigated to operate as a sort of "gene surgery". "Gene surgery" would be best suited for correcting point mutations or for mutations involving short regions of insertion or deletion. Such a system would be ideal for correcting the common point mutations found in retinoblastoma and the p53 gene in various human tumors. The same system could also be used to "knock out" oncogenic genes by inducing frameshift or nonsense mutations, which produce a nonfunctional transcript. Unfortunately, this approach is still highly theoretical.

The benefits of cell cycle research will be realized fastest in the realms of molecular diagnosis and prognostic indication. Already markers for various cell cycle regulatory proteins and PCR protocols are being developed and implemented for early diagnosis and guidance in cancer management. Molecular biological techniques may maximize the parameters of sensitivity and specificity in cancer screening and prevent the extensive work-ups sometimes required for the diagnosis of a neoplasm. Additionally, the potential for decreases in morbidity and mortality through early recognition and intervention would prove to be cost effective in the long term.

Early detection may also allow for more effective immunotherapy for cancer. Immunotherapy requires an immunocompetent host, low tumor burden, and an immunogenic tumor. Mouse therapy experiments have shown that a tumor burden of 10^6 to 10^7 tumor cells can be effectively managed by immunotherapy. In humans 10^6 to 10^9 cells constitutes a small tumor burden, since a small 1-cc tumor mass is composed of 10^9 cells. A terminally ill patient has around 10^{11} tumor cells. Molecular diagnostics increases the chances of being able to detect the cancer before the host becomes severely immunocompromised and while the tumor burden is within the effective range of immunotherapy.

Neoplasms are classified morphologically by tissue of origin and cytopathology. Reclassification by molecular alterations in the oncogenes (cyclins and cyclin-dependant kinases), tumor-suppressor genes, apoptotic genes, and/or mutator genes would be beneficial in guiding therapy and in evaluating tumor epidemiology. The current grading system could be complemented by molecular biological techniques to determine genomic stability. This system could eventually be automated and result in diagnoses that are less subject to interpreter bias.

ACKNOWLEDGMENT

We wish to acknowledge the support of the Sbarro Institute for Cancer Research and Molecular Medicine, The Council for Tobacco Research, and NIH Grant CA 60999.

REFERENCES

Aaronson SA (1991): Growth factors and cancer. *Science* 254:1146–1153

Ajchhenbaum F, Ando K, De Caprio JA, Griffin JD (1993): Independent regulation of human D-type cyclin gene expression during G1 phase in primary human T lymphocytes. *J Biol Chem* 268:4113–4119

Anderson C.W. (1993): DNA damage and the DNA-activated protein kinase. *TIBS* 18:433–437

Arnold A, Kim HG, Gaz RD, Eddy RL, Fukushima Y, Byers MG, Shows TB, Kronenberg HM (1993): Molecular cloning and chromosomal mapping of DNA rearranged with the parathyroid hormone gene in a parathyroid adenoma. *J Clin Invest* 83:2034–2040

Bagchi S, Weinmann R, Raychaudhuri P (1991): The retinoblastoma protein copurifies with E2F-1, and E1A-regulated inhibitor of the transcription factor E2F. *Cell* 65:1063–1072

Baldi A, De Luca A, Claudio PP, Baldi F, Giordano GG, Tommasino M, Paggi MG, Giordano A (1995): The Rb2/p130 gene product is a nuclear protein whose phosphorylation is cell cycle regulated. *J Cell Biochem* 59:402–408

Baldin V, Lukas J, Marcote MJ, Pagano M, Draetta G (1993): Cyclin D1 is a nuclear protein required for cell cycle progression in G1. *Genes Dev* 7:812–821

Bandara LR, La Thangue NB (1991a): Adenovirus E1a prevents the retinoblastoma gene product from complexing with a cellular transcription factor. *Nature* 351:494–497

Bandara LR, Adamczewski J, Hunt T, La Thangue NB (1991b): Cyclin A and the retinoblastoma gene product complex with a common transcription factor. *Nature* 352:249–251

Bandara LR, Buck VM, Zamanian M, Johnston LH, La Thangue, NB (1993): Functional synergy between DP-1 and E2F-1 in the cell cycle-regulating transcription factor DRTF1/E2F. *EMBO J* 13:4317–4324

Bandara LR, Lam EW-F, Sorensen TS, Zamanian M, Girling R, La Thangue NB (1994): DP-1: a cell cycle-regulated and phosphorylated component of transcription factor DRTF1/E2F which is functionally important for recognition by pRb and the adenovirus E4 orf 6/7 protein. *EMBO J* 13:3104–3114

Barak Y, Oren M. (1992): Enhanced binding of a 95 kDa protein to p53 in cells undergoing p53 mediated growth arrest. *EMBO J* 11:2115–2121

Barak Y, Juven T., Haffner R., Oren, M (1993): mdm2 expression is induced by wild type p53 activity. *EMBO J* 12:461–468

Bates S, Bonetta L, MacAllan D, Parry D, Holder A, Dickson C, Peters G (1994): Cdk6 (PLSTIRE) and Cdk4 (PSK-J3) are a distinct subset of the cyclin-dependent kinases that associate with Cyclin D1. *Oncogene* 9:71–79

Beijersbergen RL, Kerkhoven RM, Zhu L, Voorhoeve FM, Bernards R (1994): E2F-4, a new member of the E2F gene family, has oncogenic activity and associates with p107 in vivo. *Genes Dev* 8:2680–2690

Bianchi AB, Fischer SM, Robles AI, Rinchik EM, Conti CJ (1993): Overexpression of cyclin D1 in mouse skin carcinogenesis. *Oncogene* 8:1127–1133

Bienz-Tadmor B, Zakut-Houri R, Libresco S, Givol D, Oren M (1985): The 5′ region of the p53 gene: evolutionary conservation and evidence for a negative regulatory element. *EMBO J* 4:3209–3213

Bischoff JR, Friedman PN, Marshak DR, Prives C, Beach D (1990): Human p53 is phosphorylated by p60-cdc2 and cyclin B-cdc2. *Proc Natl Acad Sci USA* 87:4766–4770

Bookstein R, Shew JY, Chen PL, Scully P, Lee WH (1990 a):Suppression of tumorigenicity of human prostate carcinoma cells by replacing a mutated RB gene. *Science* 247:712–715

Bookstein R, Rio P, Madreperla SA, Hong F, Allred C, Grizzle WE, Lee, WH (1990b): Promoter deletion and loss of retinoblastoma gene in human prostate carcinoma. *Proc Natl Acad Sci USA* 87:7762–7766

Boyd JM, Malstrom S, Subramanian T, Venkatesh LK, Schaeper U, Elangovan B, D'Sa-Eipper C, Chinnadurai G (1994): Adenovirus E1B 19 KDa and Bcl-2 proteins interact with a common set of cellular proteins. *Cell* 79:341–351

Bullrich F, MacLachlan TK, Sang N, Druck T, Veronese ML, Chiorazzi N, Koff A, Heubner K, Croce CM, Giordano A (1995): Chromosomal mapping of members of the cdc2 family of protein kinases, cdk3, cdk6, PISSLRE and PITALRE and a cdk inhibitor, p27Kip1, to regions involved in human cancer. *Cancer Res* 55:1199–1205

Cao L, Faha B, Dembski M, Tsai LH, Harlow E, Dyson N (1992): Independent binding of the retinoblastoma protein and p107 to the transcription factor E2F. *Nature* 355:176–179

Cavenee WK, Hansen MF, Nordenskjold M, Kock E, Maumenee I, Squire, J, Philips RA, Gallie BL (1985): Genetic origin of mutations predisposing to retinoblastoma. *Science* 228:501–503

Chellappan S, Hiebert S, Mudryj M, Horowitz J, Nevins J (1991): The E2F transcription factor is a cellular target for the RB protein. *Cell* 65:1053–106

Chellappan SP (1994): The E2F transcription factor: role in cell cycle regulation and differentiation. *Mol Cell Diff* 2:201–220

Chen P, Scully P, Shew J, Wang JYJ, Lee W (1989): Phosphorylation of the retinoblastoma gene product is modulated during the cell cycle and cellular differentiation. *Cell* 58:1193–1198

Chiou S-K, Rao L, White E (1994): Bcl-2 blocks p53-dependent apoptosis. *Mol Cell Biol* 14:2556–2563

Chittenden T, Livingston DM, Kaelin WG (1991): The T/E1A-binding domain of the retinoblastoma product can interact selectively with a sequence-specific DNA-binding protein. *Cell* 65:1071–1082

Chittenden T, Livingston DM, DeCaprio JA (1993): Cell cycle analysis of E2F in primary human T cells reveals novel E2F complexes and biochemically distinct forms of free E2F. *Mol Cell Biol* 13:3975–3983

Clarke AR, Mandag AR, Roon MV, Lugt NM, Valk MV, Hooper ML, Berns A, Riele HT (1992): Requirement of a functional Rb-1 gene in murine development. *Nature* 359:328–330

Claudio PP, Howard CM, Baldi A, De Luca A, Fu Y, Condorelli G, Sun Y, Colburn N, Calabretta B, Giordano A (1994): p130/pRb2 has growth suppressive properties similar to yet distinctive from those of retinoblastoma family members pRb and p107. *Cancer Res* 54:5556–5560

Clore G, Omichinski JG, Sakaguchi K, Zambrano N, Sakamoto H, Appella E, Groneborn AM (1994): High-resolution structure of the oligomerization domain of p53 by multi-dimensional NMR. *Science* 265:386–391

Cobrinik D, Whyte P, Peeper DS, Jacks T, Weinberg RA (1993): Cell cycle-specific association of E2F with the p130 E1A-binding protein. *Genes Dev* 7:2392–2404

Condorelli G, Testa U, Valtieri M, De Luca A, Barberi T, Vitelli L, Montesoro E, Campisi S, Giordano A, Peschle C. (1995): Developmental regulation of Rb in normal adult hematopoiesis: peak expression and functional role in advanced erythroid differentiation. *Proc Natl Acad Sci USA* (in press).

Dang CV, and Lee WMF, (1989): Nuclear and nucleolar targeting sequences of c-erb, c-myc, N-myc, p53, HSP70 and HIV tat proteins. *J Biol Chem* 264:18019–18023

Debbas M, and White E (1993): Wild-type p53 mediates apoptosis by E1A which is inhibited by E1B. *Genes Dev* 7:546–554

De Caprio JA, Ludlow JW, Figge J, Shew JY, Huang CM, Lee WH, Marsilio E, Paucha E, Livingston DM, (1988): SV40 Large Tumor Antigen forms a specific complex with the product of the retinoblastoma susceptibility gene. *Cell* 54:275–283

De Caprio JA, Furukawa Y, Ajchenbaum F, Griffin JD, Livingston DM, (1992): The retinoblastoma-susceptibility gene product becomes phosphorylated in multiple stages during cell cycle entry and progression. *Proc Natl Acad Sci USA* 89:1795–1798

Deffie A, Wu H, Reinke V, Lozano G, (1993): The tumour suppressor p53 regulates its own transcription. *Mol Cell Biol* 13:3415–3423

Desai D, Wessling HC, Fisher RP, Morgan DO (1995): Effects of phosphorylation by CAK on cyclin binding by CDC2 and CDK2. *Mol Cell Biol* 15:345–350

Devary Y, Gottlieb RA, Smeal T, Karin M, (1992): The mammalian ultraviolet response is triggered by activation of Src tyrosine kinases. *Cell* 71:1081–1091

Donehower LA, Harvey M, Slagle BL, McArthur MJ, Montgomery CAJ, Butel JS, Bradley A, (1992): Mice deficient for p53 are developmentally normal but susceptible to spontaneous tumours. *Nature* 356:215–221

Dowdy SF, Hinds PW, Lojuie K, Reed SI, Arnold A, Weinberg RA (1993): Physical interaction of the retinoblastoma protein with human D cyclins. *Cell* 73:499–511

Downes CS, and Wilkins AS, (1994): Cell cycle checkpoints, DNA repair and DNA replication strategies. *Bio Essays* 16:75–79

Draetta G, Beach D (1988): Activation of cdc2 protein kinase during mitosis in human cells: Cell cycle-dependent phosphorylation and subunit rearrangement. *Cell* 54: 17–26

Draetta G (1994): Mammalian G1 cyclins. *Curr Opinion Cell Biol* 6:842–846

Dulic V, Lees E, Reed SI (1992): Association of human cyclin E with a periodic G1-S phase protein kinase. *Science* 257:1958–1961

Dulic V, Kaufman WK, Wilson S, Tisty TD, Lees E, Harper JW, Elledge SJ, Reed SI (1994): p53-dependent inhibition of cyclin dependent kinase activities in human fibroblasts during radiation-induced G1 arrest. *Cell* 76:1013–1023

Dunaief JL, Strober BE, Guha S, Khavari PA, Alin K, Luban J., Begemann M, Crabtree GR, Goff SP (1994) The retinoblastoma protein and BRG1 form a complex and cooperate to induce cell cycle arrest. *Cell* 79:119–130

Durfee T, Becherer K, Chen P-L, Yeh S-H, Yang Y, Kilburn AE, Lee W-H, Elledge DSJ (1993): The retinoblastoma protein associates with the protein phosphatase type-1 catalytic subunit. *Genes Dev* 7:555–569

Dyson N, Howley PM, Muenger K, Harlow E (1989): The human papilloma virus-16 E7 oncoprotein is able to bind to the retinoblastoma gene product. *Science* 243: 934–937

El-Deiry WS, Tokino T, Velculescu VE, Levy DB, Parsons R, Trent JM, Lin D, Mercer WE, Kinzler KW, Vogelstein B (1993): WAF1, a potential mediator of p53 tumor suppression. *Cell* 75:817–825

El-Deiry WS, Harper JW, O'Connor PM, Velculescu V, Canman CE, Jackman J, Pietenpol J, Burrel M, Hill DE, Wang Y, Wilman KG. (1994): WAF1/CIP1 is induced in p53 mediated G1 arrest and apoptosis. *Cancer Res* 4:1169–1174

Eliyahu D, Raz A, Gruss P, Givol D, Oren M (1984): Participation of p53 cellular tumor antigen in transformation of normal embryonic cells. *Nature* 312:646–649

Elledge SJ, and Harper W (1994): Cdk inhibitors: on the threshold of checkpoints and development. *Curr Opinion Cell Biol.* 6:847–852

Ewen ME, Ludlow JW, Marsilio E, De Caprio JA, Millikan RC, Cheng SH, Paucha E, Livingston DM (1989): An N-terminal transformation-governing sequence of SV40 large T antigen contributes to the binding of both p110 Rb and a second cellular protein p120. *Cell* 58:257–267

Ewen ME, Xing Y, Lawrence JB, Livingston DM (1991): Molecular cloning, chromosomal mapping, and expression of the cDNA for p107, a retinoblastoma gene product-related protein. *Cell* 66:1155–1164

Ewen ME, Sluss HK, Sherr CJ, Matsushime H, Kato J, Livingston DM. (1993a): Functional interactions of the retinoblastoma protein with mammalian D-type cyclins. *Cell* 73:487–497

Ewen ME, Sluss L, Whitehouse L, Livingston D (1993b): Cdk4 modulation by TGF-β leads to cell cycle arrest. *Cell* 74:1009–1020

Ewen ME (1994): The cell cycle and the retinoblastoma protein family. *Cancer Metastasis Rev.* 13:45–66

Fang F, Newport J (1991): Evidence that the G1-S and G2-M transitions are controlled by different Cdc2 proteins in higher eukaryotes. *Cell* 66:731–742

Fearon ER, Vogelstein B (1990): A genetic model for colorectal tumorigenesis. *Cell* 61:759–767

Feinstein E, Gale RP, Reed J, Canaani E (1992): Expression of the normal p53 gene induces differentiation of K562 cells. *Oncogene* 7:1853–1857
Fields S, Jang SK (1990): Presence of a potent transcription activating sequence in the p53 protein. *Science* 249:1046–1049
Finlay CA, Hinds PW, Tan TH, Eliyahu D, Oren M, Levine AJ (1988): Activating mutations for transformation by p53 produce a gene product that forms an hsp-70-p53 complex with an altered half life. *Mol Cell Biol* 8:531- 539
Finlay CA, Hinds PW, Levine AJ (1989): The p53 proto-oncogene can act as a suppressor of transformation. *Cell* 57:1083–1093
Firpo EJ, Koff A, Solomon MJ, Roberts JM (1994): Inactivation of a Cdk2 inhibitor during interleukin 2-induced proliferation of human T lymphocytes. *Mol Cell Biol* 14:4889–4901
Fishel R, Lescoe MK, Rao MRS, Copeland NG, Jenkins NA, Garber J, Kane M, Kolodner R (1993): The human mutator gene homolog MSH2 and its association with hereditary nonpolyposis colon cancer. *Cell* 75:1027–1038
Fisher RP, Morgan DO (1994): A novel cyclin associates with MO15/CDK7 to form the CDK-activating kinase. *Cell* 78:713–724
Flores-Rosaz H, Kelman Z, Dean F, Pan Z-Q, Harper JW, Elledge SJ, O'Donnell M, Hurwitz J (1994): Cdk-interacting protein-1 (CIP1, Waf1) directly binds with proliferating cell nuclear antigen and inhibits DNA replication catalyzed by the DNA polymerase d holoenzyme. *Proc Natl Acad Sci USA* 91:8655–8659
Fornace Jr AL, Nebet DW, Holander MC, Lenthy JD, Papathanasiou M, Fargnoli J, Holbrook NJ (1989): Mammalian genes coordinately regulated by growth arrest signals and DNA-damaging agents. *Mol Cell Biol* 9:4196–4203
Friend SH, Bernards R, Rogelj S, Weinberg RA, Rapaport JM, Albert DM, Dryja TP (1986): A human DNA segment with properties of the gene that predisposes to retinoblastoma and osteosarcoma. *Nature* 323:643–646
Fung YK, Murphree AL, T'Ang A, Qian J, Hinrichs SH, Benedict WF (1987): Structural evidence for the authenticity of the human retinoblastoma gene. *Science* 236:1657–1661
Gaffey MJ, Frierson HF Jr, William ME (1993): Chromosome 11q13, c-erb-2, and c-myc amplification in invasive breast carcinoma; clinicopathologic correlations. *Mod Pathol* 6:654–659
Ginsberg D, Oren M, Yaniv M, Piette J (1990): Protein-binding elements in the promoter regions of the mouse p53 gene. *Oncogene* 5:1285–1290
Ginberg D, Vairo G, Chittenden T, Xiao Z-X, Xu G, Wydner KL, DeCaprio JA, Lawrence JB, Livingston DM (1994): E2F4, a new member of the E2F transcription factor family, interacts with p107. *Genes Dev* 8:2665–2679
Giordano A, Whyte P, Harlow E, Franza BR Jr, Beach D, Draetta G (1989): A 60 kd cdc2-associated polypeptide complexes with the E1A proteins in adenovirus infected cells. *Cell* 58:981–990
Giordano A, McCall C, Whyte P, Franza BR (1991a): Human cyclin A and the retinoblastoma protein interact with similar but distinguishable sequences in the adenovirus E1A gene product. *Oncogene* 6:481–485
Giordano A, Lee J, Scheppler JA, Herrmann C, Harlow E, Deuschle U, Beach D, Franza BR (1991b): Cell-cycle regulation of histone H1 kinase activity associated with the adenoviral protein E1A. *Science* 253:1271–1275
Girard F, Strausfeld U, Fernandez A, Lamb N (1991): Cyclin A is required for the onset of DNA replication in mammalian fibroblasts. *Cell* 67:1169–1179.

Girling R, Partridge JF, Bandara LR, Burden R, Totty NF, Hsuan JJ, La Thangue NB (1993): A new component of the transcription factor DRTF1/E2F. *Nature* 362:83–87

Goodrich DW, Wang NP, Qian YW, Lee EYH, Lee W-H (1991): The retinoblastoma gene product regulates progression through the G1 phase of the cell cycle. *Cell* 67:297–302

Grana X, De Luca A, Sang N, Fu Y, Claudio PP, Rosenblatt J, Morgan DO, Giordano A (1994a):PITALRE, a nuclear CDC2-related protein kinase that phosphorylates the retinoblastoma protein in vitro. *Proc Natl Acad Sci USA* 91:3834–3838

Grana X, Claudio PP, De Luca A, Sang N, Giordano A (1994b): PISSLRE, a human novel CDC-2 related protein kinase. *Oncogene* 9:2097–2103

Gu W, Schneider JW, Condorelli G, Kaushal S, Mahdavi V, Nadal-Ginard B (1993): Interaction of mitogenic factors and the retinoblastoma protein mediates muscle cell commitment and differentiation. *Cell* 72:309–324

Gu Y, Turck CW, Morgan DO (1993): Inhibition of CDK2 activity in vivo by an associated 20K regulatory subunit. *Nature* 366:707–710

Guan K-L, Jenkins CW, Li Y, Nichols MA, Wu X, O'Keefe CL, Matera AG, Xiong Y (1994): Growth suppression by p18, a p16INK4/MTS1-and p14INK4B/MTS2-related CDK6 inhibitor, correlates with wild-type pRb function. *Genes Dev* 8: 2939–2952

Harlow E, Williamson NJ, Ralston R, Helfman DM, Adams TE (1985): Molecular cloning and in vitro expression of a cDNA clone for human cellular tumor antigen p53. *Mol Cell Biol* 5:1601–1610

Harper JW, Adami G, Wei N, Keyomarsi K, Elledge SJ (1993): The 21 kd Cdk interacting protein Cip 1 is a potent inhibitor of G1 cyclin-dependent kinases. *Cell* 75:805–816

Hartwell LH, Weinert TA (1989): Checkpoints: controls that ensure the order of cell cycle events. *Science* 246:629–634

Hartwell LH (1992): Defects in a cell cycle checkpoint may be responsible for the genomic instability of cancer cells. *Cell* 71:543–546

Hechman KA, Roberts J M (1994): Rules to replicate by. *Cell* 79:557–562

Helin K, Wu CL, Fattaey A, Lees JA, Dynlacht BD, Ngwu C, Harlow E (1993): Heterodimerization of the transcription factors E2F-1 and DP-1 leads to cooperative trans-activation. *Genes Dev* 7:1850–1861

Herrmann CPE, Kraiss S, Montenarh M (1991): Association of casein kinase II with immunopurified p53. *Oncogene* 6:877–884

Hiebert SW, Chellappan SP, Horowitz JM, Nevins J (1992): The interaction of RB with E2F coincides with an inhibition of the transcriptional activity of E2F. *Genes Dev* 6:177–185

Hinds PW, Mittnacht S, Dulic V, Arnold A, Reed SI, Weinberg RA (1992): Regulation of retinoblastoma protein function by ectopic expression of human cyclins. *Cell* 70:993–1006

Hoffman B., Liebermann DA (1994): Molecular controls of apoptosis: differentiation/ growth arrest primary response genes, proto-oncogenes, and tumor suppressor genes as positive and negative modulators. *Oncogene* 9:1807–1812

Hollingsworth RE Jr, Hensey CE, Lee W-H (1993): Retinoblastoma protein and the cell cycle. *Curr Opinion Genet Dev* 3:55–62.

Horowitz JM, Yandell DW, Park SH, Canning S, Whyte P, Buchkovich K, Harlow E, Weinberg RA, Dryja TP (1989): Point mutational inactivation of the retinoblastoma antioncogene. *Science* 243:937–940

Horowitz JM, Park SH, Bogenmann E, Cheng JC, Yandell DW, Kaye FJ, Minna JD, Dryja TP, Weinberg RA (1990): Frequent inactivation of the retinoblastoma anti-oncogene is restricted to a subset of human tumor cells. *Proc Natl Acad Sci USA* 87:2775–2790

Howes KA, Ransom N, Papermaster DS, Lasudry JGH, Albert DM, Windle JJ (1989): Apoptosis or retinoblastoma: alternative fates of photoreceptors expressing the HPV-16 E7 gene in the presence or absence of p53. *Genes Dev* 8:1300–1310

Huang HJ, Yee JK, Shew JY, Chen PL (1988): Suppression of the neoplastic phenotype by replacement of the RB gene in human cancer cells. *Science* 242:1563–1566

Huber HE, Edwards G, Goodhart PJ, Patrick DR, Huang PS, Iveyhoyle M, Barnett SF, Oliff A, Heimbrook DC (1993): Transcription factor E2F binds DNA has heterodimer. *Proc Natl Acad Sci USA* 90:3525–3529

Hunt T. (1989): Maturation promoting factor, cyclin and the control of M-phase. *Curr Opinion Cell Biol* 1:268–274

Hunter T, Pines J (1991): Cyclins and cancer. *Cell* 66:1071–1074

Hunter T, Pines J (1994): Cyclins and Cancer II:cyclin D and CDK inhibitors come of age. *Cell* 79:573–582

Hupp TR, Meek DW, Midgely CA, Lane DP (1992): Regulation of the specific DNA binding function of p53. *Cell* 71:875–886

Ivey-Hoyle M, Conroy R, Huber HE, Goodhart PJ, Oliff A, Heimbrook DC, (1993): Cloning and characterization of E2F-2, a novel protein with the biochemical properties of transcription factor E2F. *Mol Cell Biol* 13:7802–7812

Jacks T, Fazeli A, Schmitt EM, Bronson RT, Goodell MA, Weinberg RA (1992): Effects of an Rb mutation in the mouse. *Nature* 395:295–300

Jenkins JR, Rudge K, Currie GA (1984): Cellular immortalization by a cDNA clone encoding the transformation-associated phosphoprotein p53. *Nature* 312:651–654

Jenkins JR, Chumakov P, Addison C, Sturzbecher HW, Wade-Evans A (1988): Two distinct regions of the murine p53 primary amino acid sequence are implicated in stable complex formation with simian virus 40 T antigen. *J Virol* 62:3903–3906

Jiang H, Lin J, Su ZZ, Collart FR, Huberman E, Fisher PB (1994): Induction of differentiation in human promyelocytic HL-60 leukemia cells activates p21, WAF1/CIP1, expression in the absence of p53. *Oncogene* 9:3397–3406

Jiang W, Kahn SM, Tomita N, Zhang YJ, Lu SH, Weinstein IB (1992): Amplification and expression of the human cyclin D gene in esophageal cancer. *Cancer Res* 52:2980–2983

Johnson GL, Vaillancourt RR (1994): Sequential protein kinase reactions controlling cell growth and differentiation. *Curr Opinion Cell Biol* 6:230–238

Kamb A, Gruis NA, Weaver-Feldhaus J, Liu Q, Harshman K, Tavtigan SV, Stockert E, Day RSI, Johnson BE, Skolnick MH (1994): A cell cycle regulator potentially involved in the genesis of many tumor types. *Science* 264:436–440

Karp JE, and Broder S (1994): New directions in molecular medicine. *Cancer Res* 54:653–665

Kastan MB, Radin AI, Kuerbitz J, Onyekwere O, Wolkow CA, Civin CI, Stone KD, Woo T, Ravindranath Y, Craig RW (1991a): Levels of p53 increase with maturation in human hematopoietic cells. *Cancer Res* 51:4279–4286.

Kastan MB, Zhan Q, El-Deiry WS, Carrier F, Jacks T, Walsh WV, Plunkett BS, Vogelstein B, Fornace AJ Jr (1991b): A mammalian cell cycle checkpoint pathway utilizing p53 and GADD45 is defective in ataxia-telangiectasia. *Cell* 71:587–589

Kato J, Matsushime H, Hiebert SW, Ewen ME, Sherr CJ (1993a): Direct binding of cyclin D to the retinoblastoma product (pRb) and pRb phosphorylation by the cyclin D-dependent kinase CDK4. *Genes Dev* 7:331–342

Kato J, Sherr CJ (1993b): Inhibition of granulocyte differentiation by G1 cyclins, D2 and D3, but not D1. *Proc Natl Acad Sci USA* 90:11513–11517

Kato J, Matsuoka M, Polyak K, Massague J, Sherr CJ (1994a): Cyclic AMP-induced G1 phase arrest mediated by an inhibitor (p27 kip1) of cyclin-dependent kinase-4 activation. *Cell* 79:487–496

Kato J, Matsuoka M, Strom D, Sherr CJ (1994b): Regulation of cyclin D-dependent kinase (Cdk4) by Cdk4 activating kinase (Cak). *Mol Cell Biol* 14:2713–2721

Kaye FJ, Kratzke RA, Gerster JL, Horowitz JM (1990): A single amino acid substitution results in a retinoblastoma protein defective in phosphorylation and oncoprotein binding. *Proc Natl Acad Sci USA* 87:6922–6926

Keyomarsi K, Pardee AB (1993): Redundant cyclin overexpression and gene amplification in breast cancer cells. *Proc Natl Acad Sci USA* 90:1112–1116

Keyomarsi K, O'Leary N, Molnar G, Lees E, Fingert HJ, Pardee AB (1994): Cyclin E, a potential prognostic marker for breast cancer. *Cancer Res* 54:380–385

Kim S, Wagner S, Liu F, O'Reilly MA, Robbins PD, Green MR (1992): Retinoblastoma gene product activates expression of the human TGF-β gene through transcription factor ATF-2. *Nature* 358:331–334

King RW, Jackson PK, Kirschner MW (1994): Mitosis in Transition. *Cell* 79:563–571)

Knudson AG Jr (1971): Mutation and cancer: statistical study of retinoblastoma. *Proc Natl Acad Sci USA* 68:820–823

Knudson AG (1984): Retinoblastoma: clues to human oncogenesis. *Science* 228: 1028–1033

Koff A, Cross F, Fisher A, Schumacher J, Leguellec K, Philippe M, Roberts JM (1991): Human cyclin E, a new cyclin that interacts with two members of the CDC2 gene family. *Cell* 66:1197–1228

Koff A, Giordano A, Desai D, Yamashita K, Harper W, Elledge S, Nishimoto T, Morgan D, Franza R, Roberts J (1992): Formation and activation of a cylin E-Cdk2 complex during the G1 phase of the human cell-cycle. *Science* 257: 1689–1693

Koff A, Ohtsuki M, Polyak K, Roberts JM, Massague J (1993): Negative regulation of G1 in mammalian cells: inhibition of cyclin E-dependent kinase by TGF-β. *Science* 260:536–539

Kraiss S, Barnekow A, Montenarh M (1991): Protein kinase activity associated with immunopurified p53 protein. *Oncogene* 5:845–855

Krek W, Ewen ME, Shirodkar S, Arany Z, Kaelin WG, Livingston DM (1994): Negative regulation of the growth-promoting transcription factor E2F-1 by a stably bound cyclin A-dependent protein kinase. *Cell* 78:161–172

Laiho M, DeCaprio JA, Ludlow JW, Livingston DM, Massague J (1990): Growth inhibition by TGF-ß linked to suppression of retinoblastoma protein phosphorylation. *Cell* 62:175–185

Lane DP, Crawford LV (1979): T antigen is bound to host protein in SV40- transformed cells. *Nature* 278:261–263

Lane DP (1992): p53:Guardian of the genome. *Nature* 358:15–16

Leach FS, Elledge SJ, Sherr CJ, Willson JK, Markowitz S, Kinzler KW, Vogelstein B (1993): Amplification of cyclin genes in colorectal carcinomas. *Cancer Res* 53: 1986–1989

Lee EY, Chang CY, Hu N, Wang YC, Lai CC, Herrup K, Lee WH, Bradley A (1992): Mice deficient for RB are nonviable and show defects in neurogenesis and hematopoiesis. *Nature* 359:288–294

Lee WH, Bookstein R, Hong F, Young LJ, Shew JY, Lee EY (1987): Human retinoblastoma susceptibility gene: cloning, identification, and sequence. *Science* 235: 1394–1399

Lees JA, Saito M, Vidal M, Valentine M, Look T, Harlow E, Dyson N, Helin K (1993): The retinoblastoma protein binds to a family of E2F transcription factors. *Mol Cell Biol* 13:7813–7825

Lees-Miller SP, Sakaguchi K, Ullrich SJ, Appella E, Anderson CW (1992): Human DNA-activated protein kinase phosphorylates serines 15 and 37 in the amino-terminal transactivation domain of human p53. *Mol Cell Biol* 12:5041–5049

Lew DJ, Dulic V, Reed SI (1991): Isolation of three novel human cyclins by rescue of G1 cyclin (cln) function in yeast. *Cell* 66:1197–1206

Lew J, Huang Q-Q, Qi A, Winkfein RJ, Aebersold R, Hunt T, Wang JH (1994): A brain-specific activator of cyclin-dependent kinase 5. *Nature* 371:422–426

Li S, MacLachlan TK, De Luca A, Claudio PP, Condorelli G, Giordano A (1995): The cdc2-related PISSLRE is essential for cell growth and acts in G2 phase of the cell cycle. *Cancer Research* 55:3992–3996

Li Y, Nichols MA, Shay JW, Xiong Y (1994): Transcriptional repression of the D type cyclin-dependent kinase inhibitor p16 by the retinoblastoma susceptibility gene product pRb. *Cancer Res* 54:6078–6082

Lin W-C, Desidero S (1993): Regulation of V(J)D recombination activator protein RAG-2 by phosphorylation. *Science* 260:953–959

Ludlow JW, DeCaprio JA, Huang C-M, Lee W-H, Paucha E, Livingston DM (1989): SV40 Large T antigen binds preferentially to an underphosphorylated member of the retinoblastoma susceptibility gene product family. *Cell* 56:57–65

Mack DH, Vartikar J, Pipas JM, Laimins LA (1993) Specific repression of TATA-mediated but not initiator-mediated transcription by wild-type p53. *Nature* 363:281–283

MacLachlan TK, Sang N, Giordano A (1995): Cyclins and cyclin-dependent kinases and CdK inhibitors: implications in cell cycle control and cancer. *Critical Reviews in Eukaryotic Gene Expression* 5:127–156

Mascolo A, Levin S, Giordano A. (1992): A molecular biological approach to the study of nasopharyngeal cancer in Chinese Garment workers. *Ramazzini Newsletter* 2:54–58

Matlashewski G, Lamb P, Pim D, Peacock J, Crawford G, Benchimol S (1984): Isolation and characterization of a human p53 cDNA clone: expression of the human p53 gene. *EMBO J* 3:3257–3262

Matsuoka M, Kato J, Fisher RP, Morgan DO, Sherr CJ (1994): Activation of cyclin-dependent kinase-4 (Cdk4) by mouse MO15-associated kinase. *Mol Cell Biol* 14:7265–7275

Matsushime H, Roussel M, Asmun R, Sherr CJ (1991a): Colony-stimulating factor 1 regulates novel cyclins during the G1 phase of the cell cycle. *Cell* 65:701–713

Matsushime H, Roussel MF, Sherr CJ (1991b): Novel mammalian cyclins (CYL genes) expressed during G1. *Cold Spring Harbor Symposia* 56:69–74

Matsushime H, Quelle DE, Shurtleff SA, Shibuya M, Sherr CJ, Kato J (1994): D-type cyclin-dependent kinase activity in mammalian cells. *Mol Cell Biol* 14:2066–2076

Mayol X, Graña X, Baldi A, Sang N, Hu Q, Giordano A (1993): Cloning of a new member of the retinoblastoma gene family (pRb2), which binds to the E1A transforming domain. *Oncogene* 8:2561–2566

Meek DW (1994): Post -translational modification of p53. *Semin Cancer Biol* 5:203–210

Mercer WE, Avignolo C, Baserga R (1984): Role of p53 protein in cell proliferation as studied by microinjection of monoclonal antibodies. *Mol Cell Biol* 4:276–281

Mercer WE, Shields MT, Amin M, Sauve GJ, Appella T, Romano JW, Ullrich SJ (1990): Negative growth regulation in a glioblastoma tumor cell line that conditionally expresses human wild-type p53. *Proc Natl Acad Sci USA* 87:6166–6170

Meyerson M, Enders GH, Wu C-L, Su L-K, Gorka C, Nelson C, Harlow E, Tsai L-H (1992): A family of human cdc2-related protein kinases. *EMBO J* 11:2909–2917

Meyerson M, Harlow E (1994): Identification of G1 kinase activity for Cdk6, a novel cyclin D partner. *Mol Cell Biol* 14:2077–2086

Milne DM, Palmer RH, Campbell DG, Meek DW (1992): Phosphorylation of the p53 tumour suppressor protein at three N-terminal residues by a novel casein kinase I-like enzyme. *Oncogene* 7:1361–1369

Milne DM, Campbell DG, Caudwell FB, Meek DW (1994): Phosphorylation of the tumour suppressor protein p53 by mitogen activated protein (MAP) kinases. *J Biol Chem* 269:9253–9260

Milner J, Cook A, Mason J (1990): p53 is associated with p34CDC-2 in transformed cells. *EMBO J* 9:2885–2889

Minshull J, Golsteyn R, Hill CS, Hunt T (1990): The A- and B-type cyclin associated cdc2 kinases in Xenopus turn on and off at different times in the cell cycle. *EMBO J* 9:2865–2875

Momand J, Zambetti GP, Olson D, George D, Levine AJ (1992): The mdm-2 oncogene product forms a complex with the p53 protein and inhibits p53-mediated transactivation. *Cell* 69:1237–1245

Moran E (1993): DNA tumor virus transforming proteins and the cell cycle. *Curr Opinion Genet Dev* 3:63–70

Morgenbesser SD, Williams BO, Jacks T, Depino RA (1994): p53-dependent apoptosis produced by Rb-deficiency in the eye. *Nature* 371:72–74

Motokura T, Bloom T, Kim YG, Jueppner H, Ruderman J, Kronenberh H, Arnold A (1991): A novel cyclin encoded by a bcl-1–linked candidate oncogene. *Nature* 350:512–515

Murray AW, and Kirschner MW (1989): Dominoes and clocks: the union of two views of the cell cycle. *Science* 246:614–621

Murray AW (1992): Creative blocks: cell cycle checkpoints and feedback control. *Nature* 359:599–604

Murray AW (1994): Cell cycle checkpoints. *Curr Opinion Cell Biol* 6:872–876

Musgrove EA, Hamilton JA, Lee CS, Sweeney KJE, Watts CK, Sutherland RL (1993): Growth factor, steroid, and steroid antagonist regulation of cyclin gene expression associated with changes in T-47D human breast cancer cell cycle progression. *Mol Cell Biol* 13:3577–3587

Musgrove EA, Lee CS, Buckley MF, Sutherland RL (1994): Cyclin D1 induction in breast cancer cells shortens G(1) and is sufficient for cells arrested in G(1) to complete the cell cycle. *Proc Natl Acad Sci USA* 91:8022–8026

Nash R, Tokiwa G, Anand S, Erickson K, Futcher AB (1988): The WHI1+ gene of Saccharomyces cerevisiae tethers cell division to cell size and is a cyclin homolog. *EMBO J* 7:4335–4336

Nasmyth K (1993): Control of the yeast cell cycle by the Cdc28 protein kinase. *Curr Opinion Cell Biol* 5:166–179

Nelson WG, Kastan MB (1994): DNA strand breaks: the DNA template alterations that trigger p53-dependent DNA damage response pathways. *Mol Cell Biol* 14: 1815–1823

Nevins JR (1992): E2F:A link between the Rb tumor suppressor protein and viral oncoprotein. *Science* 258:424–429

Nobori T, Miura K, Wu DJ, Lois A, Takabayashi K, Carson DA (1994): Deletion of the cyclin-dependent kinase-4 inhibitor gene in multiple human cancers. *Nature* 368:753–756

Noda A, Ning Y, Venable SF, Pereira-Smith OM, Smith JR (1994): Cloning of senescent cell-derived inhibitors of DNA synthesis using an expression screen. *Exp Cell Res* 211:90–98

O'Connell MJ, and Nurse P (1994): How cells know they are in G1 or G2. *Curr. Opinion Cell Biol* 6:867–871

Oliner JD, Inzler KW, Meltzer PS, George DL, Vogelstein B (1992): Amplification of a gene encoding a p53-associated protein in human sarcomas. *Nature* 358:80–83

Oliner JD, Pietenpol JA, Thiagalingam S, Gyuris J, Kinzler KW Vogelstein B (1993): Oncoprotein MDM2 conceals the activation domain of tumor suppressor p53. *Nature* 362:857–860

Oltvai ZN, Milliman CL, Korsmeyer SJ (1993): Bcl-2 heterodimerizes in vivo with a conserved homolog, Bax, that accelerates programmed cell death. *Cell* 74:609–619

Pagano M, Theodoras AM, Tam SW, Draetta G (1993): Cyclin D1-mediated inhibition of repair and replicative DNA synthesis in human fibroblasts. *Genes Dev* 8: 1627–1639

Pan H, Griep AE (1994): Altered cell cycle regulation in the lens of HPV-16 E6 or E7 transgenic mice: implications for tumor suppressor gene function in development. *Genes Dev* 8:1285–1299

Parada LF, Land H, Weinberg RA, Wolf D, Rotter V (1984): Cooperation between the gene encoding p53 tumor antigen and ras in cellular transformation. *Nature* 312:649–651

Pardee AB (1989): G1 events and regulation of cell proliferation. *Science* 246:603–608

Paterlini P, Suberville AM, Zindi F, Melle J, Sonnier M, Marie JP, Dreyfus F, Brechot C (1993): Cyclin A expression in human hematological malignancies: a new marker of cell proliferation. *Cancer Res* 53:235–238

Peter M, Herskowitz I (1994): Joining the complex: cyclin-dependent kinase inhibitory proteins and the cell cycle. *Cell* 79:181–184

Pines J, Hunter T (1990): Human cyclin A is adenovirus ElA-associated protein p60 and behaves differently from cyclin B. *Nature* 346:760–763

Pines J (1994): Arresting development in cell-cycle control. *TIBS* 19:143–145

Pinhasi-Kimhi O, Michalovitz D, Ben-Ze'ev A, Oren M (1986): Specific interaction between the p53 cellular tumor antigen and major heat shock proteins. *Nature* 320:182–184

Polyak K, Lee M, Erdjement-Bromage H, Koff A, Roberts J, Tempst P, Massague J (1994a): Cloning of p27 kip1, a cyclin-dependent kinase inhibitor and a potential mediator of extracellular antimitogenic signals. *Cell* 78:59–66

Polyak K, Kato M, Solomon MJ, Sherr CJ, Massague J, Roberts JM, Koff A (1994b): p27 kip1 and Cyclin D-Cdk4 are interacting regulators of cdk2, and link TGF-β and contact inhibition to cell cycle arrest. *Genes Dev* 8:9–22

Prokocimer M, Rotter V (1994): Structure and function of p53 in normal cells and their aberrations in cancer cells: projection on the hematologic cell lineages. *Blood* 84:2391–2411

Qian Y, Luckey C, Horton L, Esser M, Templeton DJ (1992): Biological function of the retinoblastoma protein requires distinct domains for hyperphosphorylation and transcription factor binding. *Mol Cell Biol* 12:5363–5372

Quelle D, Ashmun R, Shurtleff S, Kato J, Bar-Sagi D, Roussel M, Sherr C (1993): Overexpression of mouse D-type cyclins accelerates G1 phase in rodent fibroblasts. *Genes Dev* 7:1559–1571

Quin XQ, Chittenden T, Livingston DM, Kaelin WG. (1992): Identification of a growth suppression domain within the retinoblastoma gene product. *Genes Dev* 6:953–964

Rabbitts TH (1994): Chromosomal translocations in human cancer. *Nature* 372:143–149

Radler-Pohl A, Sachsenmeier C, Gebel S, Auer H-P, Bruder JT, Rapp U, Angel P, Rahmsdorf HJ, Herrlich P (1993): UV-induced activation of AP-1 involves obligator extranuclear steps including Raf-1 kinase. *EMBO J* 12:1005–1012

Raychaudhuri P, Bagchi S, Devoto SH, Kraus VB, Moran E, Nevins JR (1991): Domains of the adenovirus E1A protein that are required for oncogenic activity are also required for dissociation of E2F transcription factor complex. *Genes Dev* 5:1200–1211

Raycrof L, Wu H, Lozano G (1990): Transcriptional activation by wild type but not transforming mutants of the p53 anti-oncogene. *Science* 249:1049–1051

Reed DSI, Wittenberg C, Lew DJ, Dulic V, Henze M (1991): G1 control in yeast and animal cells. *Cold Spring Harbor Symposia* 56:61–67

Reisman D, Rotter V (1989): Two promoters that map to 5'-sequences of the human p53 gene are differentially regulated during terminal differentiation of human leukemic cells. *Oncogene* 4:945–953

Renan MJ (1993): How many mutations are required for tumorigenesis?: Implications from human cancer data. *Mol Carcinogen* 7:139–146

Ronen D, Rotter V, Reisman D (1991): Expression from murine p53 promoter is mediated by factor-binding to a down stream helix-loop-helix recognition motif. *Proc Natl Acad Sci USA* 88:4128–4132

Sala A, Nicolaides NC, Engelhard A, Bellon T, Lawe DC, Arnold A Grana X, Giordano A, Calabretta B (1994): Correlation between E2F-1 requirement in the S phase and E2F-1 transactivation of cell cycle-related genes in human cells. *Cancer Res* 54:1402–1406

Sang N, Baldi A, Giordano A (1995): The roles of tumor suppressors pRb and p53 in cell proliferation and cancer. *Mol Cell Differ* 1:1–29

Sarnow P, Ho YSH, Williams J, Levine AJ (1982): Adenovirus ElB-58 kd tumor antigen and SV40 large tumor antigen are physically associated with the same 54 kd cellular protein in transformed cells. *Cell* 28:387–396

Scheffner M, Werness BA, Hulbregtse JM, Levine AJ, Howley PM (1990): The E6 oncoprotein encoded by human papillomavirus types 16 and 18 promotes the degradation of p53. *Cell* 63:1129–1136

Schwarz JK, Devoto SH, Smith EJ, Chellappan SP, Jakoi L, Nevins JR (1993): Interactions of the p107 and Rb proteins with E2F during the cell proliferation response. *EMBO J* 12:1013–1020

Selvakumaran M, Lin HK, Miyashita T, Wang HG, Krajewski S, Reed JC, Hoffman B, Liebermann DA (1994): Immediate early up-regulation of bax expression by p53 but not TGF beta 1:a paradigm for distinct apoptotic pathways. *Oncogene* 9:1791–1798

Serrano M, Hannon GJ, Beach D (1993): A new regulatory motif in cell-cycle control causing specific inhibition of cyclin D/Cdk4. *Nature* 366:704–707

Serrano M, Gomez-Lahoz E, DePinho RA, Beach D, Bar-Sagi D (1995): Inhibition of Ras-induced proliferation and cellular transformation by p16INK4. *Science* 267: 249–252

Seto E, Usheva A, Zambetti GP, Momand J, Horikoshi N, Weinman R, Levine AJ, Shenk T (1992): Wild-type p53 binds to the TATA-binding protein and represses transcription. *Proc Natl Acad Sci USA* 89:12028–12032

Sewing A, Burger C, Brusselbach S, Schalk C, Lucibiello F, Muller R (1993): Human cyclin D1 encodes a labile nuclear protein whose synthesis is directly induced by growth factors and suppressed by cyclic AMP. *J Cell Sci* 104:545–555

Shaulsky G, Goldfinger N, Peled A, Rotter V (1991): Involvement of wild type p53 in pre B-cell differentiation in vitro. *Proc Natl Acad Sci USA* 88:8982–8986

Shaulsky G, Goldfinger N, Tosky MS, Levine AJ, Rotter V (1991): Nuclear localization is essential for the activity of p53 protein. *Oncogene* 6:2055–2065

Sherr CJ (1994): G1 phase progression: cycling on cue. *Cell* 79:551–555

Shirodkar S, Ewen M, DeCaprio JA, Morgan J, Livingston DM, Chitteden T, (1992): The transcription factor E2F interacts with the retinoblastoma product and a p107-cyclin A complex in a cell cycle-regulated manner. *Cell* 66:157–166

Slebos RJ, Lee MH, Plunkett BS, Kessis TD, Williams BO, Jacks T, Hedrick L, Kastan MB, Cho KR (1994): p53-dependent G1 arrest involves pRb-related proteins and is disrupted by the human papillomavirus 16 E7 oncoprotein. *Proc Natl Acad Sci USA* 91:5320–5324

Smith ML, Chen IT, Ahan Q, Bae I, Chen C-Y, Gilmer TM, Kastan MB, O'Connor PM, Fornace AJ Jr (1994): Interaction of the p53-regulated protein Gadd45 with proliferating cell nuclear antigen. *Science* 266:1376–1380

Smythe C, Newport J (1992): Coupling of mitosis to the completion of S phase in Xenopus occurs via modulation of the tyrosine kinase that phosphorylates p34CDC2. *Cell* 68:787–797

Solomon E, Borrow J, Goddard AD (1991): Chromosome aberrations and cancer. *Science* 254:1153–1160

Solomon MJ, Lee T, Kirschner MW (1992): Role of phosphorylation p34cd2 activation: identification of an activating kinase. *Mol Biol Cell* 3:13–27

Spruck CH, Gonzalez-Zurueta M, Shibata A, Simoneau AR, Lin MF, Gonzales F, Tsai YC, Jones P (1994): p16 gene in uncultured tumors. *Nature* 370:183–184

Sturzbecher HW, Maimets T, Chumakov P, Brain R, Addison C, Simanis V, Rudge K, Philp R, Grimaldi M, Court W, Jenkins JR (1990): p53 interacts with p34cdc-2 in mammalian cells: implications for cell cycle control and oncogenesis. *Oncogene* 5:795–801

Sturzbecher H-W, Brain R, Addison C, Rudge K, Remm M, Grimaldi M, Keenan E, Jenkins JR (1992): A C-terminal α-helix plus basic region motif is the major structural determinant of p53 tetramerization. *Oncogene* 7:1513–1523

Sun Y, Hegamyer G, Colburn N (1993): Nasopharyngeal carcinoma shows no detectable retinoblastoma susceptibility gene alterations. *Oncogene* 8:791–795

Surmacz E, Reiss K, Sell C, Baserga R (1992): Cyclin D1 messenger RNA is inducible by platelet-derived growth factor in cultured fibroblast. *Cancer Res* 52:4522–4525

Symonds H, Krail L, Remington L, Saenz-Robles M, Lowe S, Jacks T, Van Dyke T (1994): p53-dependent apoptosis suppresses tumor growth and progression *in vivo*. *Cell* 78:703–712

Tam SW, Shay JW, Pagano M (1994): Differential expression and cell cycle regulation of the cyclin dependent kinase 4 inhibitor p16 INK4. *Cancer Res* 54:5816–5820

Toyoshima H, Hunter T (1994): p27, a novel inhibitor of G1-cyclin Cdk protein kinase activity, is related to p21 *Nature* 78:67–74

Tsai L-H, Delalle I, Cavins VS Jr, Chae T, Harlow E (1994): p35 is neural-specific regulatory subunit of cyclin-dependent kinase 5. *Nature* 371:419–422

Waga S, Hannon GJ, Beach D, Stillman B (1994): The p21 inhibitor of cyclin-dependent kinases controls DNA replication by interaction with PCNA. *Nature* 369:574–578

Walker DH, Maller JL (1991): Role for cyclin A in the dependence of mitosis on completion of DNA replication. *Nature* 354:314–317

Wang J, Chenivesse X, Henglein B, Brechot C (1990): Hepatitis B virus integration in a cyclin A gene in a hepatocelluar carcinoma. *Nature* 343:555–557

Weinberg RA (1991): Tumor suppressor genes. *Science* 254:1138–1145

Werness BA, Levine AJ, Howley PM (1990): Association of human papillomavirus type 16 and 18 E6 proteins with p53. *Science* 248:76–79

Whyte P, Buckovich KJ, Horowitz JM, Friend SH, Raybuck M, Weinberg RA, Harlow E (1988): Association between an oncogene and an anti-oncogene: the adenovirus E1A proteins bind to the retinoblastoma gene product. *Nature* 334: 124–127

Williams BO, Remington L, Albert DM, Mukai S, Bronson RT, Jacks T (1994): Cooperative tumorigenic effects of germline mutation in RB and p53. *Nature Genet* 7:480–484

Wu X, Levine AJ (1994): p53 and E2F-1 cooperate to mediate apoptosis. *Proc Natl Acad Sci USA* 91:3602–3606

Xiong Y, Zhang H, Beach D (1992a): Subunit rearrangment of the cyclin dependent kinases is associated with cellular transformation. *Genes Dev* 7:1572–1583

Xiong Y, Zhang H, Beach D (1992b): D-type cyclins associate with multiple protein kinases and the DNA replication and repair factor PCNA. *Cell* 71:505–514

Xiong Y, Hannon GJ, Zhang GJ, Casso D, Kobayashi R, Beach D (1993): p21 is a universal inhibitor of cyclin kinases. *Nature* 366:701–704

Yan L, Xi Z, Drettener B (1988): Epidemiological studies of nasopharyngeal cancer in the Guangzhou area, China. Preliminary report. *Acta Otolaryngol* 107:424–427

Yeung RS, Bell DW, Testa JR, Mayol X, Baldi A, Grana X, Klinga-levan, K, Knudson AG, Giordano A (1993): The retinoblastoma-related gene, RB2, maps to human chromosome 16q12 and rat chromosome *Oncogene* 8:3465–3468

Zamanian M, La Thangue NB (1992): Adenovirus E1a prevents the retinoblastoma gene product from repressing the activity of a cellular transcription factor. *EMBO J* 11:2603–2610

Zhan Q, Lord KA, Alamo I Jr, Hollander MC, Carrier F, Ron D, Kohn KW, Hoffman B, Liebermann DA, Fornace AJ Jr (1994): The gadd and MyD genes define a novel set of mammalian genes encoding acidic proteins that synergistically suppress cell growth. *Mol Cell Biol* 14:2361–2371

Zhu L, Van Den Heuvel S, Helin K, Fattaey A, Ewen M, Livingston D, Dyson N, and Harlow E (1993): Inhibition of cell proliferation by 107, a relative of the retinoblastoma protein. *Genes Dev* 7:1111–1125

Zlegler A, Jonason AS, Leffell DJ, Simon JA, Sharma HW, Kimmelman J, Remington L, Jacks T, Brash DE (1994): Sunburn and p53 in the onset of skin cancer. *Nature* 372:773–776

2

TGF-β Inhibition of Intestinal Epithelial Cell Proliferation: G1 Cyclins, Cyclin-Dependent Kinases, and Cyclin Kinase Inhibitors

E. AUBREY THOMPSON, TIEN C. KO AND R. DANIEL BEAUCHAMP

Colon carcinogenesis is a multistep process involving alterations in the normal replication of intestinal epithelial cells (Vogelstein et al., 1988; Cameron et al., 1990; Vogelstein, 1990). In recent years, there has been a significant increase in our understanding of the basic molecular mechanisms that underlie regulation of cell proliferation. This new insight has provided a clearer picture of gut epithelial cell replication and of the changes that lead to breakdown of these regulatory mechanisms during tumor initiation or progression. Analysis of gut epithelial cell proliferation also promises to provide fundamental new insight into hormonal inhibition of cell proliferation and into the relationship between hormone insensitivity and transformation.

TGF-β AND PROLIFERATION OF THE GUT EPITHELIUM

The intestinal epithelium is one of the most rapidly proliferating tissues in the body, capable of complete cellular renewal every 3 to 8 days (Cairnie et al., 1965; Cheng and LeBlond, 1974; Babyatsky and Podolsky, 1991). There is normally a dynamic equilibrium between cell proliferation at the base of the crypt and cell extrusion at the tip of the villus. Cells in the crypt proliferate rapidly and are less differentiated, whereas cells nearer the villus tip are terminally differentiated and do not divide. This continual process of mucosal renewal is tightly regulated and is altered during malignant transformation. In experimental models of colon carcinogenesis,

Hormones and Cancer
Wayne V. Vedeckis, Editor

there is an increase in the proliferative compartment that is thought to be due to loss of normal inhibitory controls (Sporn and Roberts, 1985). This aberrant regulation leads to intestinal hyperplasia and subsequent formation of adenoma, carcinoma *in situ*, and finally invasive carcinoma (Cameron et al., 1990).

Intestinal crypt cell proliferation is thought to be regulated by autocrine and paracrine peptide growth factors (Barnard et al., 1989; Ko et al., 1993) that stimulate or inhibit proliferation. Among these factors, the transforming growth factor beta family of peptides (TGF-β) plays a major role in differentiation and growth inhibition of the intestinal epithelium. TGF-βs are 25 kDa homodimeric polypeptides belonging to a superfamily of growth-regulatory molecules that includes müllerian inhibiting substance, activins, inhibins, and bone morphogenetic proteins, among others. Three isoforms of TGF-β (TGF-β1, TGF-β2, and TGF-β3) have been identified. These share 70–75% amino acid sequence identity, bind to and activate the same receptors, and have identical biological activities, with few exceptions. In this chapter the term "TGF-β" refers to any or all three of the isoforms. Transforming growth factors originally were named for the ability to induce the transformed phenotype in fibroblasts grown in monolayer culture and to stimulate colony formation in soft agar. TGF-β is now known to regulate a diversity of biological processes that include cellular proliferation, cell migration, differentiation, and extracellular matrix deposition. The growth-regulatory effects of TGF-β appear to be cell-type specific: TGF-β promotes proliferation of many types of mesenchymal cells, whereas it is a potent inhibitor of proliferation of most epithelial, endothelial, and hematopoietic cell types.

TGF-β is a potent inhibitor of cultured rat intestinal crypt cell proliferation (Kurokawa et al., 1987; Barnard et al., 1989; Filmus et al., 1992; Ko et al., 1993). There is increased expression of TGF-β1 mRNA in rat intestinal cells as they migrate from crypt to the villus tip (Barnard et al., 1989). This inverse relationship between TGF-β expression and mitotic activity has been confirmed by immunohistochemical studies that demonstrate abundant TGF-β in the nonproliferating enterocytes near the villus tip, with little or no TGF-β in the rapidly dividing crypt cells of both small bowel and colon (Barnard et al., 1993). TGF-β1 mRNA is expressed and TGF-β bioactivity is secreted by a normal rat intestinal crypt cell line, IEC-6 (Barnard et al., 1989; Ko et al., 1992). TGF-β1 inhibits proliferation of IEC-6 cells and induces differentiation, as evidenced by stimulation of sucrase activity (Kurokawa et al., 1987), IgA secretory component production, and major histocompatibility complex class I antigen expression (McGee et al., 1991). These findings support the current view that TGF-β is the major autocrine or paracrine regulator of intestinal cell proliferation and differentiation, and that this hormone accounts for the phenotypic changes that occur as intestinal epithelial cells migrate from the crypt toward the villus.

TGF-β RESPONSIVENESS AND MALIGNANT TRANSFORMATION OF GUT EPITHELIAL CELLS

TGF-β regulation of intestinal cell proliferation is altered during malignant transformation. TGF-β is a potent growth suppressor of nontumorigenic human colonic adenoma cells. However, conversion of an adenoma to a tumorigenic adenocarcinoma is associated with a decreased response to the inhibitory actions of TGF-β (Hoosein et al., 1989; Manning et al., 1989). Transformation of a normal rat intestinal epithelial cell line (IEC-18) to a tumorigenic phenotype by overexpression of activated *H-ras* genes (e.g., [Val-12]Ras) results in a loss of TGF-β–mediated growth regulation (Filmus et al., 1992). Studies in human colon carcinoma cell lines have demonstrated a strong correlation between the degree of differentiation of the tumor and sensitivity to the antiproliferative and differentiation-promoting effects of TGF-β (Hoosein et al., 1989). Thus, loss of responsiveness to the inhibitory effects of TGF-β appears to be an important event in the loss of normal growth control that attends malignant transformation of the gut epithelium.

Considering the importance of TGF-β1 as a regulator of both normal and neoplastic intestinal cell proliferation, it may seem surprising that the mechanism is not better understood. TGF-β arrests growth of keratinocytes (Howe et al., 1991), lung epithelial cells (Laiho et al., 1990), and intestinal epithelial cells (Ko et al., 1994) by blocking cell cycle progression in late G1 (Pietenpol et al., 1990; Laiho et al., 1990; Howe et al., 1991; Ko et al., 1994). G1 arrest is associated with a decrease in the activity of cyclin E–associated histone H1 kinase activity (Koff et al., 1993) and inhibition of retinoblastoma protein (pRb) phosphorylation (Laiho et al., 1990). Since TGF-β appears to function primarily to arrest epithelial cells in the G0/G1 phase of the cell cycle, it is reasonable to assume that the hormone regulates some key step in progression through G1 and entry into S phase. The nature of those principles that regulate cell cycle progression is beginning to emerge from recent genetic and biochemical studies of yeast.

G1 PROGRESSION AND CELL DIVISION CONTROL IN YEAST

Initiation of DNA replication (entry into S phase) as well as mitotic division (entry into M phase) are controlled by sequential activation and inactivation of one or more serine/threonine protein kinases. Among these, the prototype is encoded by the *cdc2* gene of the fission yeast *Schizosaccharomyces pombe* and by the homologous *CDC28* gene of the budding yeast *Saccharomyces cerevisiae* (reviewed in Meyerson et al., 1991; Kidd, 1992; Norbury and Nurse, 1992). There are striking similarities between cell cycle control genes and cell cycle regulatory mechanisms in all eukaryotic cells; therefore, it is appropriate to consider briefly how cell cycle progression is regulated in budding yeast.

There are several kinds of regulatory subunits that associate with CDC28. One major class of regulatory subunits includes the cyclins, which are essential for activation of the catalytic subunit. At least nine potential cyclins have been identified, of which three, CLN1, CLN2, and CLN3, are essential for progression through G1 phase (Richardson et al., 1989). The CLN/CDC28 complex catalyzes the rate-limiting step in G1 progression in yeast. As discussed below, physiological regulation of CLNs provides for a response that is analogous to hormonal regulation of cell proliferation in mammalian cells.

Most, if not all, of the genes that are required for initiation of replicative DNA synthesis in budding yeast are regulated by one of two transcription factor complexes called SBF and MBF (Nasmyth and Shore, 1987; Herskowitz, 1989; Gordon and Campbell, 1991; McIntosh et al., 1991). As illustrated in Figure 1, these factors associate with cell cycle specific promoters, interacting with specific DNA sequences called SCB and MCB. (For a more detailed review of these factors and their regulation, see Moll et al., 1993.) SBF and MBF are activated by CLN/CDC28 complexes in late G1 (Breeden and Nasmyth, 1987), thereby inducing the synthesis of the machinery that is required for DNA replication. The *CLN1* and *CLN2* genes are themselves regulated by SBF (Nasmyth and Dirick, 1991; Ogas et al., 1991), so as to provide a positive feedback loop in which activation of the CLN/CDC28 kinase induces *CLN* gene transcription (Cross and Tinkelenberg, 1991). *CLN3*, on the other hand, is not regulated by SBF or MBF, although CLN3/CDC28 can activate SBF, as shown in Figure 1. CLN3 expression appears to be determined by the size of the cell, such that CLN3 abundance is low in newly budded daughter cells and increases as such cells traverse G1 phase (Nash et al., 1988; Schwob et al., 1994), as shown in Figure 2. It has been proposed that CLN3 may trigger the initial activation of CDC28 when yeasts reach a size that is appropriate for budding (Cross, 1988; Nash et al., 1988). According to this hypothesis, CLN3

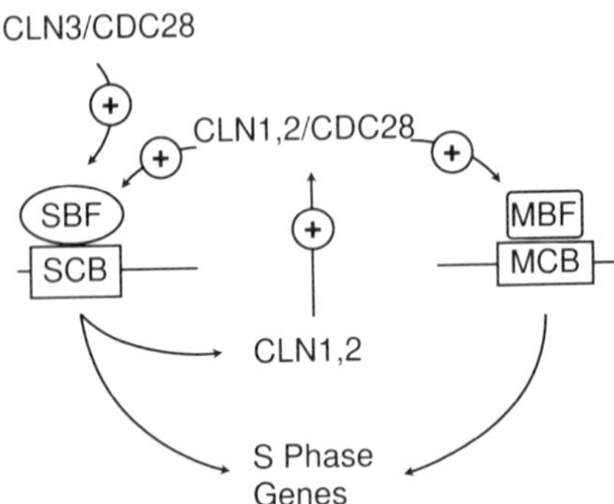

Figure 1. Yeast G1 cyclins (CLNs) activate transcription of S phase genes, including *CLN1* and *CLN2* genes.

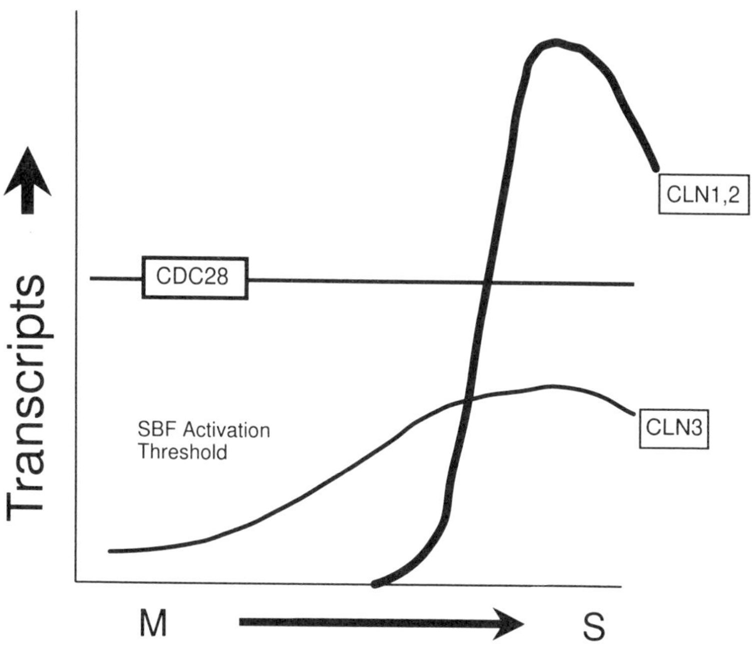

Figure 2. CLN3 mRNA increases as a function of yeast cell size and activates transcription of *CLN1* and *CLN2* once a threshold of CLN3 abundance is exceeded.

activates CDC28 once a certain threshold of CLN3 concentration is exceeded. The CLN3/CDC28 kinase complex probably triggers the initial activation of SBF, which then stimulates transcription of *CLN1* and *CLN2*. As a result of induction of CLN1 and CLN2, the activity of the CLN/CDC28 kinase increases. The activity of SBF increases in consequence, and so on. This autostimulatory loop accounts for the rapid increase in expression of CLN1 and CLN2 expression that occurs in late G1 (Fig. 2), and activation of CLN/CDC28 kinase activity irreversibly commits the cell to initiate S phase.

Two important concepts emerge from this consideration of G1 progression in yeast. The first of these hinges on the observation that CDC28 kinase is not activated as a direct function of the intracellular abundance of the CLN3 regulatory subunit. Activation of SBF occurs only after the abundance of CLN3 exceeds a certain threshold. This threshold phenomenon implies that other regulatory principles must constrain the activity of CLN/CDC28 complexes until a minimum abundance of the complex obtains. The second important concept illustrated by this consideration of yeast G1 progression is that of irreversibility. Once CLN/CDC28 is activated, initiation of DNA replication is entrained. This replicative "point of no return" has been called "START" in budding yeast (Hartwell, 1974)), and is conceptually homologous to the "restriction point" described for serum-stimulated mammalian cells (Pardee, 1989). Progression through the mammalian restriction point is regulated by a wide variety of hormones and hormone-like substances. In yeast, progression through START is regulated by hormones, or something very much akin to hormones.

HORMONAL REGULATION OF YEAST PROLIFERATION

Expression of yeast *CLN* genes is regulated by yeast-mating factors, which may be considered as hormones that inhibit cell proliferation. Yeast-mating factor α arrests cell proliferation at START (reviewed in Marsh et al., 1991; Reed, 1991). The signal transduction pathway leading from α-mating factor to mitotic arrest is strikingly familiar to endocrinologists, as may be perceived from the outline given in Figure 3. The α-mating factor receptor (MFR in Fig. 3) is a transmembrane protein linked to a heterotrimeric guanosine triphosphate (GTP)-binding protein (reviewed in Sprague, 1992). Binding of the polypeptide ligand induces dissociation of the receptor-bound G protein. The $G\beta\gamma$ complex activates a protein kinase cascade, which leads through several steps to activation of a mitogen-activated protein (MAP) kinase homolog, called FUS3 (reviewed in Pelech and Sanghera, 1992). FUS3 phosphorylates and activates a transcription factor (STE12), which induces transcription of a large number of genes (Dolan et al., 1989; Errede and Ammerer, 1989). Among these STE12-induced genes are *FUS3* itself and another gene called *FAR1*, which encodes a prototypic cyclin kinase inhibitor (Peter et al., 1993).

Phosphorylated FAR1 binds stoichiometrically to the CLN/CDC28 complex and inhibits kinase activity, as illustrated in Figure 3. Inhibition of CLN/CDC28 activity results in a decrease in the activity of the transcription factors SBF and MBF. Transcription of *CLN1* and *CLN2* ceases, as does transcription of all genes that are required for DNA replication. Cell cycle arrest ensues.

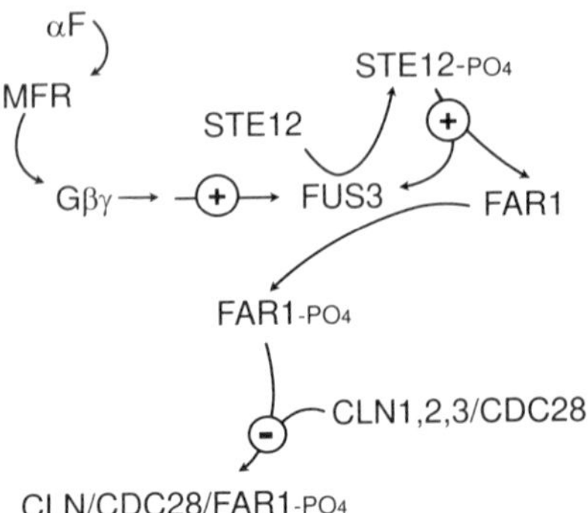

Figure 3. Yeast-mating factor α triggers a MAP kinase cascade that activates an inhibitor if CLN/CDC28 kinase activity.

Studies of yeast cell proliferation have much to say to those who are interested in regulation of mammalian cell proliferation. If one is willing to stretch a point, yeast cells may yield important clues as to how the mitotic activity of metazoan cells is regulated by hormones. It is obvious that the overall structure of the signal transduction pathways is highly conserved in yeast and mammals. We know that many of the regulatory factors are interchangeable (e.g., mammalian G1 cyclins will substitute for CLNs). What then are the salient features that may be discerned in yeasts, and what hypotheses may be built on those features?

It is clear that regulation of transcription of G1 cyclins is central to G1 progression in yeast. It is also clear that G1 cyclins regulate their own expression, so that kinase activity and cyclin gene expression are directly related. Finally, cell division control (CDC) kinase activity is regulated by inhibitors, including but not limited to FAR1. When one considers the extraordinary degree of conservation of structure and function among the cell division control components of yeasts, invertebrates, and mammals, it is altogether plausible that vertebrate hormones may regulate cell proliferation by mechanisms that resemble those that impinge on G1 → S phase progression in yeasts.

CYCLIN D/Cdk4 AND CYCLIN E/Cdk2 COMPRISE THE MAMMALIAN G1 CELL DIVISION CONTROL KINASES

The vertebrate homologs of *CLN1,2,3* are three D-type cyclins encoded by the *CcnD1, CcnD2,* and *CcnD3* loci (Lew et al., 1991; Matsushimi et al., 1991; Xiong et al., 1991). The nomenclature of the G1 cyclins that will be used in this chapter is given in Figure 4, which also outlines the major interactions between the G1 cyclins and their catalytic subunits, the cyclin-dependent kinases (Cdks). Like the CLNs, D-type cyclins are regulated by growth factors (Matsushimi et al., 1991; Surmacz et al., 1992; Won et al., 1992; Ajchenbaum et al., 1993; Winston and Pledger, 1993). Cyclin D1 (encoded by *CcnD1*) is not

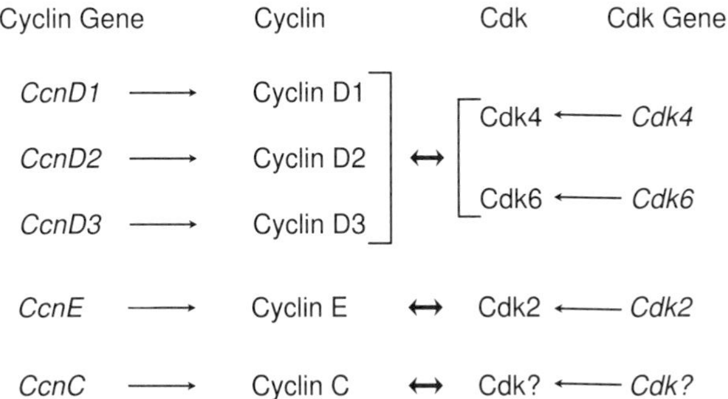

Figure 4. The relationship between G1 cyclin genes (*Ccns*), cyclin-dependent kinase genes (*Cdks*), their products, and cyclin/Cdk complexes.

expressed in quiescent macrophages, but is induced rapidly upon mitotic stimulation with CSF-1 (Matsushimi et al., 1991). Withdrawal of CSF-1 causes cyclin D1 to disappear and macrophages to re-enter the quiescent state. Cyclin D1 is also induced upon serum stimulation of quiescent fibroblasts (Surmacz et al., 1992; Won et al., 1992), and microinjection of CcnD1 antisense expression vectors into serum-stimulated fibroblasts blocks the mitogenic response (Baldin et al., 1993). Mitotic stimulation of quiescent lymphocytes results in a rapid increase in CcnD2 expression, which is followed by induction of CcnD3 (Ajchenbaum et al., 1993).

The D cyclins associate with and activate the product of the *Cdk4* locus, although Cdk2 may also be activated to an extent (Matsushimi et al., 1992; Xiong et al., 1992). In lymphoid cells, Cdk6 may be the major catalytic partner of the D cyclins (Meyerson and Harlow, 1994). There are at least two additional cyclins that are involved in initiation of DNA replication in metazoan cells, as shown in Figure 4. Cyclin E associates with Cdk2 and serves essential but unknown functions in very late G1 or early S phase (Dulic et al., 1992; Koff et al., 1992; Pagano et al., 1992; Rosenblatt et al., 1992). Cyclin C (CcnC) mRNA appears early in G1, but the catalytic partner of cyclin C and the role of cyclin C in G1 progression are unknown (Lew et al., 1991).

Cdk2-containing cell division control kinases phosphorylate histone H1 *in vitro*, although the physiological substrates are unknown (Draetta, 1990). The only known substrate for cyclin D–Cdk4 kinases is pRb, the product of the retinoblastoma susceptibility gene *Rb-1* (Ewen et al., 1993a; Kato et al., 1993). Hyperphosphorylation of pRb occurs in late G1 and appears to be required for entry into S phase; accumulation of underphosphorylated pRb prevents DNA synthesis (Buchkovich et al., 1989; DeCaprio et al.; Ludlow et al., 1989, 1990, 1991). Cyclin D–dependent kinases may also regulate the activity of the Rb-related, cell cycle regulatory proteins p107 and p130 (Lees et al., 1992; Mayol et al., 1993). Such observations suggest that one function of D-cyclin/Cdks might be to inactivate tumor suppressor gene products so as to facilitate entry into S phase. It is tempting to speculate that phosphorylation of proteins such as pRb may result in activation of transcription factors that control genes whose products are required for DNA replication. In this context, the mammalian transcription factor E2F and its relatives would be homologous to SBF and MBF of yeast (as reviewed in Draetta, 1994). See Chapter 1 for a more detailed review of the role of pRb in cell cycle regulation.

CYCLIN-ACTIVATING KINASE ACTIVATES G1 CYCLIN/Cdk COMPLEXES

The activity of the mammalian G1 CDC kinases is determined, in part, by phosphorylation of critical threonine residues of the Cdk subunit. Cdk4 is

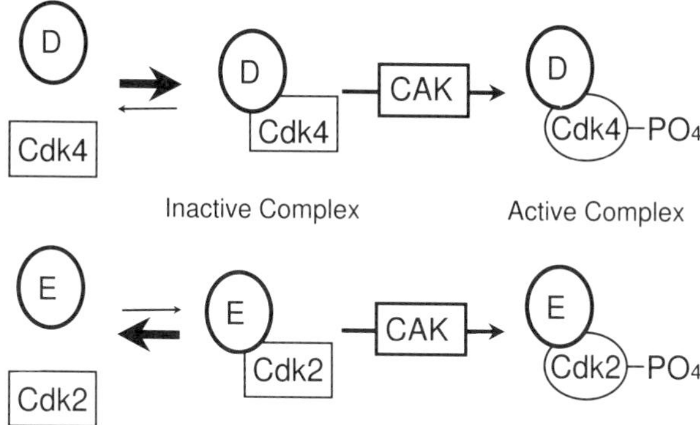

Figure 5. Activation of cyclin D/Cdk4 and cyclin E/Cdk2 complexes requires threonine phosphorykation by Cdk-activating kinase (CAK).

inactive unless it is phosphorylated on Thr-172 (Kato et al., 1994a; Matsuoka et al., 1994). Likewise, Cdk2 activity requires phosphorylation on Thr-160 (Gu et al., 1992; Solomon et al., 1992; Kato et al., 1994a). As shown in Figure 5, both Cdks are phosphorylated by an enzyme called cyclin activating kinase (CAK), which is, itself, a cyclin-dependent kinase (Fisher and Morgan, 1994). CAK will phosphorylate only the cyclin/Cdk complex. Phosphorylation stabilizes the complex, particularly the cyclin E/Cdk2 complex, which appears to be relatively unstable unless Cdk2 is phosphorylated (Polyak et al., 1994a; Singerland et al., 1994). The activity of CAK does not appear to be regulated during the cell cycle or in response to changes in mitotic activity (Matsuoka et al., 1994). Regulation of Cdk phosphorylation by CAK depends on changes in cyclin/Cdk complex formation as well as the accessibility of the complex to phosphorylation by CAK. Both these parameters are determined, at least in part, by proteins that were first identified as inhibitors of cyclin dependent kinase activity.

CYCLIN KINASE INHIBITORS IN MAMMALIAN CELLS

In yeast, cell cycle progression is regulated, at least in part, by changes in the expression of cyclin kinase inhibitors. FAR1 regulates CLN/CDC28, as described above, and SIC1 regulates kinase activity of CLB5,6/CDC28, which is necessary for DNA replication in *S. cerevisiae* (Schwob et al., 1994). The product of the *rum1* gene is the most thoroughly studied cyclin kinase inhibitor from *S. pombe* (Moreno and Nurse, 1994). Mammalian cells contain several cyclin kinase inhibitors. (For a more detailed review of cyclin kinase inhibitors, see Hunter, 1993; Elledge and Harper, 1994; Peter and Herskowitz, 1994; Sherr, 1994). Four cyclin kinase inhibitors have been studied in some detail, but experimental observations suggest that there may be several additional cyclin kinase inhibitors. The four relatively

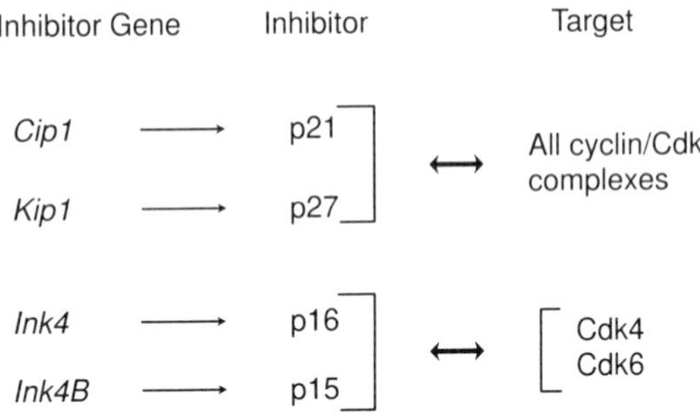

Figure 6. The mammalian cyclin kinase inhibitors fall into two distinct classes, based on their effects on different Cdks.

well-characterized cyclin kinase inhibitors fall into two distinct groups, as illustrated in Figure 6. One group includes the product of the *Cip1* gene which is also known as *Waf1*, *Sdi1*, and *Cap20* (El-Deiry et al., 1993; Gu et al., 1993; Harper et al., 1993; Noda et al., 1993; Xiong et al., 1993a).

Cip1 encodes a protein of 164 amino acids, which exhibits an apparent molecular weight of 21 kDa and will be called p21 in this chapter (Harper et al., 1993). *Kip1* encodes a related protein of 197 amino acids (Polyak et al., 1994b; Toyoshima and Hunter, 1994). Kip1 has an apparent molecular weight of 27 to 28 kDa and will be called p27. The p21 and p27 proteins exhibit some structural similarities. The amino terminal 60 amino acids are 44% identical, but there is little sequence conservation beyond the N terminus. Nevertheless, the properties of these two proteins are similar, if not identical.

Figure 7 attempts to summarize our current understanding of the mechanism of action of p27. Insofar as can be determined by available data, p21 acts in exactly the same way. For the purposes of this discussion, we will assume that p21 and p27 are functionally equivalent, although this will probably prove to be incorrect. Both p21 and p27 bind to and inhibit cyclin/Cdk complexes. The mechanism appears to be complex. Both p21 and p27 inhibit CAK phosphorylation when bound to either cyclin D/Cdk4 (Fig. 7A) or cyclin E/Cdk2 complexes (Fig. 7B). Inhibition of phosphorylation by CAK appears to occur at relatively low stoichiometry of inhibitor to cyclin/Cdk complex (Slingerland et al., 1994), although binding of one p21 or one p27 does not appear to inhibit active (Thr-172–phosphorylated) cyclin D/Cdk4 or active (Thr-160–phosphorylated) cyclin E/Cdk2 kinases (Zhang et al, 1994). At higher concentrations, both p21 and p27 will bind to and inhibit the active (i.e., phosphorylated) cyclin/Cdk complexes (Xiong et al., 1993b). Inhibition of phosphorylated Cdks appears to require binding of multiple inhibitors.

The p21/p27 family of cyclin kinase inhibitors appears to prevent activation of cyclin/Cdk complexes at low concentration and to inhibit activated cyclin/Cdk complexes at high concentrations. Although serum-starved

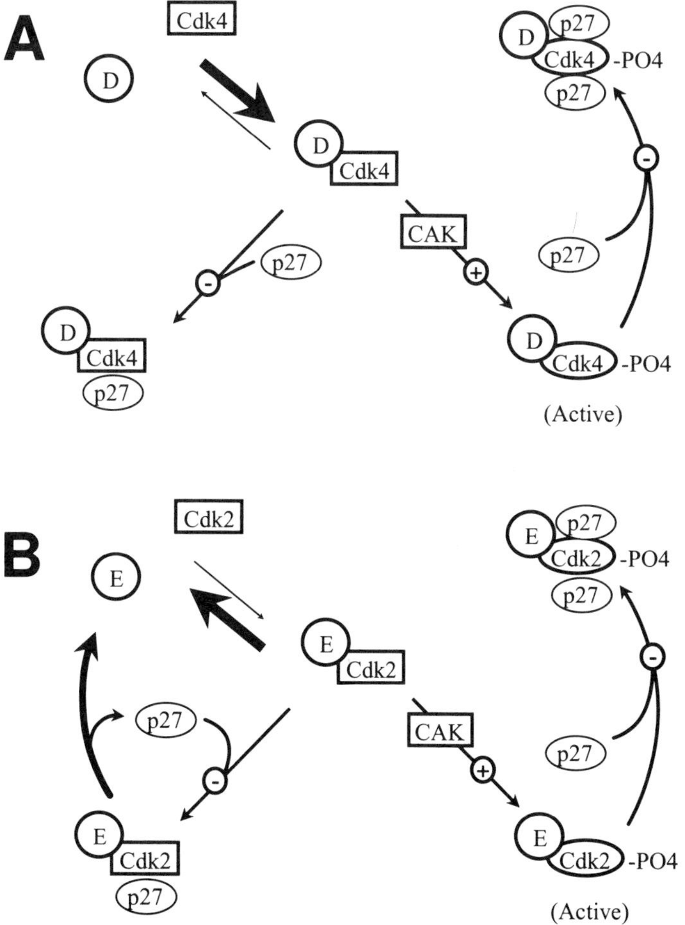

Figure 7. Inhibition of cyclin D/Cdk4 (A) and cyclin E/Cdk2 (B) complexes by p27 involves inhibition of CAK as well as inhibition of active cyclin/Cdk complexes.

fibroblasts contain more p21 mRNA than do dividing cells (Noda et al., 1993; Li et al., 1994), the abundance of p21 and p27 does not appear to be regulated during cell cycle progression in dividing cells (Li et al., 1994; Toyoshima and Hunter, 1994). Constitutive expression of p21/p27 provides a reservoir of these inhibitors that must be neutralized before newly synthesized G1 cyclin/Cdk complexes can be activated (Hunter, 1993; Elledge and Harper, 1994). Sequestration of these inhibitors appears to result from association with cyclin/Cdk complexes or perhaps other, as yet unidentified, proteins (Singerland et al., 1994). Most of the p21/p27 in actively dividing cells is probably sequestered by the more stable cyclin D/Cdk4 complex (Koff et al., 1993; Polyak et al., 1994a; Singerland et al., 1994). This may be due to the inherent instability of the unphosphorylated cyclin E/Cdk2 complex. Indeed, by preventing Thr-160 phosphorylation by CAK, p27 may contribute to the instability of the cyclin E/Cdk2 complex (as illustrated in Fig. 7B).

The p21/p27 family of cyclin kinase inhibitors defines a threshold for cyclin/Cdk activation during cell cycle progression. No activation of Cdk2 or Cdk4 can occur until the abundance of the cyclin/Cdk complex exceeds that of p21 plus p27. For this reason, the rather modest (two- to threefold) increases in cyclin D expression that attend G1 progression in cycling cells may be significant in terms of activation of cyclin D and cyclin E–associated kinases. By the same analogy, an inhibitor that causes even a modest decrease in the abundance of the cyclin D/Cdk4 complex would result in a proportionate increase in the intracellular "activity" of p21 and p27. A slight increase in inhibitor "activity" or abundance can inhibit phosphorylation of Cdks and thereby prevent recruitment of newly activated cyclin/Cdk complexes. Prolongation of G1, if not outright G1 arrest, could therefore result from a relatively minor perturbation in the cyclin D/Cdk4::cyclin kinase inhibitor stoichiometry. This is consistent with the observation that modest overexpression of cyclin D1 or cyclin E shortens the G1 interval (Ohtsubo and Roberts, 1993; Quelle et al., 1993). A more robust "activation" of these inhibitors, as might attend induction of either inhibitor (Dulic et al., 1994; El-Deiry et al., 1993, 1994; Kato et al., 1994b; Leonard et al., 1994) or profound inhibition of cyclin D or Cdk4 expression, would release p21 and/or p27 in quantities sufficient for quantitative inhibition of active cyclin/Cdk complexes. Consequently, one might expect G1 arrest to come about as a result of dramatic changes in the expression of the D cyclins, Cdk4, p21, or p27.

Stated more concisely, the cyclin D/Ckd4::p21/p27 relationship defines a threshold for cell cycle progression. If the stoichiometry falls slightly below this threshold, neither cyclin D/Cdk4 nor cyclin E/Cdk2 can be activated. If the stoichiometry falls far below this threshold, activated cyclin E/Cdk2 and cyclin D/Cdk4 are inhibited.

There is a second family of cyclin kinase inhibitors, the products of the *Ink4* (Serrano et al., 1993) and *Ink4B* (Hannon and Beach, 1994) genes, as illustrated in Figure 6. The products of these genes have apparent molecular weights of 16 kDa ($p16^{Ink4}$) and 15 kDa ($p15^{Ink4B}$). The *INK4* and *INK4B* genes are linked in tandem on the genome (Hannon and Beach, 1994) and the two inhibitors share extensive structural similarity. The first 50 amino acids are 44% identical, the next 81 amino acids are 97% identical, and the primary sequences diverge thereafter. Both proteins can be subdivided into four domains that resemble ankyrin repeats (Serrano et al., 1993; Hannon and Beach, 1994). The properties of p15 and p16 appear to be similar, if not identical. Unlike p21/p27, p15/p16 do not inhibit Cdk2-containing kinase complexes; both p15 and p16 bind to Cdk4 and Cdk6, causing dissociation of the cyclin D/Cdk4 complex (Serrano et al., 1993; Hannon and Beach, 1994), as illustrated in Figure 8. The immediate result is to block phosphorylation of Cdk4 by CAK and to inhibit activated cyclin D/Cdk4 complexes.

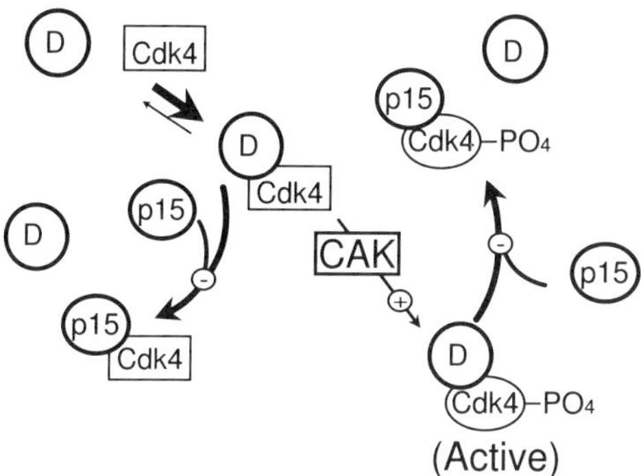

Figure 8. Inhibition of cyclin D/Cdk4 by p15 involves dissociation of the cyclin/Cdk complex.

A secondary effect is to reduce the amount of cyclin D/Cdk4 complex that is available to bind p21 and/or p27. The attendant increase in the "activity" of p21/p27 would cause inhibition of cyclin E/Cdk2 complexes. According to this hypothesis, expression of p15 and/or p16 would effectively increase the amount of cyclin D/Cdk4 that is required to exceed the threshold for activation of G1 cyclin/Cdk complexes. By the same analogy, induction of p15 or p16 would be equivalent to inhibition of expression of cyclin D or Cdk4.

TGF-β PERTURBS THE CYCLIN D/Cdk4::CYCLIN KINASE INHIBITOR STOICHIOMETRY IN EPITHELIAL CELLS

The molecular mechanism of action of TGF-β has probably been studied in greatest detail in lung epithelial cells and in keratinocytes. We have recently extended these studies to intestinal epithelial cells. All three cell types are epithelial in origin, and all undergo cell cycle arrest in late G1 when exposed to TGF-β. In a broad sense, all three cells exhibit the same response: TGF-β causes a decrease in the abundance of the cyclin D/Cdk4 complex. This prevents phosphorylation of pRb (Laiho et al., 1990) and other relevant substrates during G1 progression, and p27/p21 are released to inhibit "downstream" CDC kinases such as cyclin E/Cdk2 (Polyak et al., 1994a, 1994b; Singerland et al., 1994; Toyoshima and Hunter, 1994). Although one can, with a satisfying degree of confidence, conclude that the cyclin D/Cdk4 complex is the principle target of TGF-β, it is impossible to escape the rather unsettling conclusion that entirely different mechanisms appear to converge on the cyclin D/Cdk4 complex in lung, gut, and keratinocyte epithelial cells.

In keratinocytes, TGF-β causes a rapid increase in the abundance of p15 mRNA and protein (Hannon and Beach, 1994). Although it has not been shown that TGF-β activates transcription of *Ink4B*, this is a plausible hypothesis. As p15 accumulates, the cyclin D/Cdk4 complex dissociates (Hannon and Beach, 1994). This has two obvious effects: cyclin D/Cdk4 kinase activity is inhibited, and, to the extent that p21 and p27 are sequestered by cyclin D/Cdk4 complexes, these inhibitors are released to act on cyclin E/Cdk2. Induction of p15 therefore increases the "activity" of p21 and p27 (Polyak et al., 1994a; Singerland et al., 1994). The induction of p15 by TGF-β in some respects resembles the induction of FAR1 by yeast mating factors, although transcription of D cyclin genes in keratinocytes is not inhibited by TGF-β, in contrast to the response of *CLN* genes to mating factors.

It has been determined that p15 expression is elevated in serum-starved keratinocytes, and p15 abundance decreases rapidly when such cells are mitotically stimulated (Hannon and Beach, 1994). Figure 9A summarizes the responses that are observed when G0 keratinocytes are stimulated to re-enter the cell cycle. In addition to a decrease in the abundance of p15, a modest increase (two- to threefold) in expression of D cyclin mRNAs is observed, with a more robust induction of Cdk4 (Geng and Weinberg, 1993). The net result is a redistribution of cyclin D and Cdk4 proteins. In the G0 state, cyclin D/Cdk4 complex formation is inhibited by p15. Cyclin D may be associated with p27 under these conditions (Toyoshima and Hunter, 1994). As p15 decreases and cyclin D and Cdk4 increase, formation of the cyclin D/Cdk4 complex is facilitated. This provides a "sink" for p27 (and/or p21). Once the size of this "sink" exceeds the abundance of p27, active cyclin D/Cdk4 complex can be generated by action of CAK.

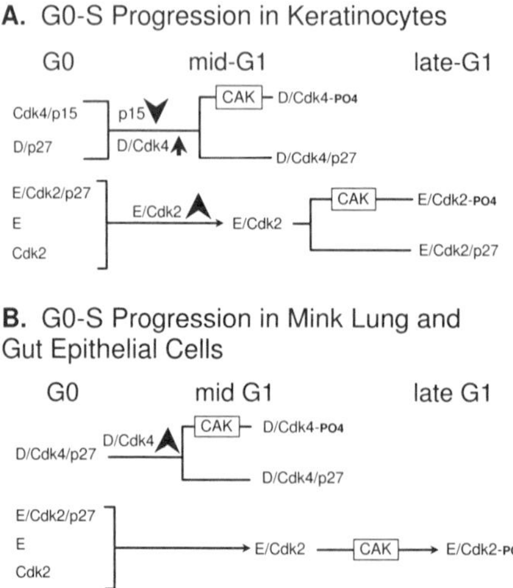

Figure 9. Progression from G0 to late G1 in serum-stimulated keratinocytes (A) or in serum-stimulated mink lung or rat intestinal epithelial cells (B) appears to involve significantly different mechanisms.

Keratinocytes also exhibit dramatic induction of cyclin E and Cdk2 upon serum stimulation (Geng and Weinberg, 1993). This induction may provide an additional means to sequester p27. The net result of these processes is generation of active cyclin D/Cdk4 and cyclin E/Cdk2 complexes as serum-stimulated keratinocytes approach late G1 (Fig. 9A).

Mink lung and rat intestinal epithelial cells exhibit a different response to TGF-β than do keratinocytes. There is little evidence that TGF-β regulates expression of any known cyclin kinase inhibitor in such cells. All available data indicate that the hormone inhibits synthesis of one or the other component of the cyclin D/Cdk4 complex. These proteins are unstable, and their abundance decreases rapidly on addition of TGF-β. This results in liberation of p27 (and/or p21), and inhibition of downstream kinases ensues. Obviously, a decrease in the cyclin D/Cdk4 complex also inhibits phosphorylation of pRb and other relevant G1 substrates.

Although both mink lung and rat intestinal cells exhibit the same response to TGF-β, different mechanisms provide for a decrease in the abundance of the cyclin D/Cdk4 complex. In gut epithelial cells, one observes a decrease in the abundance of the principle D cyclin mRNA, that encoding cyclin D1 (Ko et al., 1995). This appears to be due to specific inhibition of transcription of *CcnD1*. The abundance of Cdk4 protein decreases only slightly in TGF-β–treated intestinal epithelial cells. In mink lung epithelial cells, the primary response is a decrease in Cdk4 protein (Ewen et al., 1993b). This is not attended by a corresponding change in Cdk4 mRNA abundance, suggesting that the hormone must regulate either the translational efficiency of Cdk4 mRNA or the stability of Cdk4 protein. The synthesis of cyclin D1 also decreases by about 50% (Ewen et al., 1993b), and this may be due to a corresponding change in cyclin D1 mRNA abundance.

The data indicate that TGF-β regulates both cyclin D1 and Cdk4 expression in mink lung and rat intestinal epithelial cells. The effect of TGF-β on Cdk4 is greater in mink cells, and the effect on cyclin D1 is greater in the rat. The effect on Cdk4 is clearly post-transcriptional, whereas regulation of cyclin D1 appears to involve changes in mRNA abundance. These may reflect rather subtle differences in the manner in which these cells respond to the hormone.

There is no evidence that p15 plays a role in TGF-β responsiveness in gut epithelial cells. We have found it difficult to detect p15 mRNA expression in rat intestinal epithelial cells using human cDNA probes. Antibodies against human p15 do not detect a protein of the appropriate size in rat intestinal epithelial cells in mid-log phase growth or in cells that have been arrested by TGF-β or contact inhibition (Beauchamp and Thompson, unpublished data). Expression of p15 has not (to our knowledge) been reported in mink lung epithelial cells, but it is known that TGF-β–treated mink lung epithelial cells maintain high levels of cyclin D/Cdk4/p27 complexes (Polyak et al., 1994a). Since p15 would dissociate cyclin D/Cdk4 complexes, it seems unlikely that TGF-β induces p15 in mink lung cells.

It remains to be shown that p15 plays a role in serum arrest of mink lung or rat intestinal epithelial cells. As illustrated in Figure 9B, these cells maintain high levels of the cyclin D/Cdk4/p27 complex in G0 (Polyak et al., 1994a). Serum stimulation increases expression of both cyclin D1 and Cdk4. The primary response in mink lung cells is stimulation of Cdk4, whereas serum stimulation of quiescent rat intestinal cells results in a more dramatic induction of cyclin D1. The p27 inhibitor is titrated as the cyclin D/Cdk4 complexes accumulate. Cdk4 phosphorylation and activation of the cyclin D/Cdk4 complex accompanies escape from the p27 threshold. Once this threshold is exceeded, formation and phosphorylation of cyclin E/Cdk2 complexes can also occur. It may be notable that there is no induction of cyclin E or Cdk2 protein in serum-stimulated mink lung (Polyak et al., 1994a; Slingerland et al., 1994) or rat epithelial cells (unpublished observations from our laboratory), in contrast to the marked induction of these proteins during mitotic stimulation of keratinocytes (Geng and Weinberg, 1993).

At this time, it is not possible to discern unifying principles that underlie TGF-β inhibition of epithelial cell proliferation. Broadly, the target is the cyclin D/Cdk4 complex. Beyond that lies confusion concerning the mechanistic details. Subtle differences, such as those observed when comparing mink lung and rat epithelial cells, may be discounted as cell-specific responses, which may be interesting but perhaps not ultimately pregnant with fundamental new information on cell cycle regulation. The apparent lack of a role for p15 in mink lung and rat intestinal epithelial cells is much more paradoxical.* The difference may be trivial. No single laboratory has yet undertaken a rigorous parallel comparison of all three cell lines. The differences may be more apparent than real when differences in cellular manipulation are taken into account. The lack of homologous probes for the cyclin kinase inhibitors may also account for our inability to detect p15 mRNA in rat cells. It is also possible that different cell types may induce different inhibitors in response to TGF-β, so that a more detailed analysis of mink lung or intestinal epithelial cells may reveal heretofore uncharacterized cyclin kinase inhibitors. There is also the possibility that many of these cell-specific differences in response to TGF-β may reflect changes in regulatory pathways that occur during the process of immortalization of the cell lines.

There can be little doubt that cells achieve immortality by virtue of mutation. Some of these mutations alter cell cycle regulatory properties. For example, it has been determined that even malignant cells (which one presumes to be immortal to begin with) undergo *Ink4* (p16) deletion when placed in culture (Kamb et al., 1995). There may be several different routes to immortality, such that different cell lines may have adopted different regulatory strategies to escape proliferative senescence. This sort of reasoning leads to the notion that immortal (albeit nonmalignant) cells such as those used most frequently

* Since this chapter went to galley proofs, a paper has been published showing that p15^{Ink4} is induced upon addition of TGF-β to mink lung epithelial cells (Reynisdottir I, Polyak K, Iavarone A, Massague, J (1995): KIP/CIP and INK4 CDK inhibitors cooperate to induce cell cycle arrest in response to TGF-β. *Genes & Development* 9:1831–1845).

to study TGF-β may manifest only parts of the regulatory response that prevails in mortal cells. This is not to say that we cannot learn a great deal by studying established cell lines. Indeed, the ease with which such cells can be manipulated makes them ideal for developing hypotheses concerning cell cycle regulation in mammalian cells. The point should be made, however, that these hypotheses must ultimately be confirmed in primary cells in which all the cell cycle regulatory machinery is (presumably) intact.

The cell cycle regulatory machinery is intact in yeast, and it is partly for this reason that we are attracted by the idea that yeast may provide the models on which we can begin to understand hormonal inhibition of mammalian cell proliferation. In a general sense, concepts such as inhibitor thresholds and regulation of inhibitor amount and/or activity apply equally to epithelial and yeast cells. There appear to be important mechanistic differences. The most notable of these seems to be the apparent lack of a more cogent link between cyclin/Cdk kinase activity and cyclin gene transcription in mammalian cells. This link is strictly maintained with *CLN* genes, whereas there are many cases in which changes in cyclin/Cdk activity in mammalian cells are not associated with changes in the abundance of the corresponding cyclin mRNAs. This requires additional study, as do those mechanisms that underlie maintenance and/or induction of cyclin kinase inhibitors. Finally, there are probably uncharacterized inhibitors that may prove to play more proximal roles in TGF-β regulation. If this is so, then these apparent differences in the response to TGF-β may be downstream of the primary response mechanism. Intellectually, one is entitled to hope that this is the case, and that unifying principles will emerge as a result of the intense effort that is currently being expended to understand how TGF-β inhibits mammalian cell proliferation.

ACTIVATED *ras* ONCOGENES, CYCLIN D1, AND EPITHELIAL HYPERPLASIA

The relationship between D-type cyclin expression and growth factor stimulation, as well as the role of D cyclins in G0 arrest, suggests that D cyclins may play important roles in development, differentiation, and malignant transformation. This proposition is supported by observations that both cyclins D1 and D2 are proto-oncogenes (Rosenberg et al., 1991a; Szepetowski, 1991). The oncogenic counterpart of cyclin D1, *PRAD1/Bcl1*, has been implicated in the etiology of a variety of epithelial and B-lymphocytic malignancies (Arnold et al., 1989; Rosenberg et al., 1991a,b; Withers et al., 1991). Chromosomal rearrangement of the 11q13 locus results in overexpression of cyclin D1 in some forms of B-lymphocytic malignancies and in parathyroid adenomas. Approximately 20% of primary breast cancers and a higher percentage of squamous cell carcinomas of the head and neck exhibit amplification of *CcnD1* and overexpression of cyclin D1. Overexpression of cyclin D1 mRNA has been correlated with progression of experimentally induced skin tumors. Cyclin D2 has been identified as the proto-oncogene *vin1* (Hanna et al., 1993). The oncogenic potential of *CcnD3* remains to be demonstrated, but it will be surprising if this evidence is not shortly forthcoming.

Changes in growth factor requirements constitute one of the hallmarks of malignant transformation, and D cyclins are the primary focus for growth factor regulation of cell proliferation (see Motokura and Arnold 1993; Sherr, 1993, 1994; Hartwell and Kastan, 1994; Hunter and Pines, 1994, and references therein). Loss of TGF-β responsiveness in gut epithelial cells accompanies progression from adenoma to adenocarcinoma (Hoosein et al., 1989; Manning et al., 1991), and we have shown that inhibition of cyclin D1 expression accounts for TGF-β regulation of intestinal epithelial cell proliferation in culture (Ko et al., 1995). These data suggest that there may be a causal relationship between malignant transformation and resistance to TGF-β–mediated inhibition of cyclin D1 expression. This hypothesis is reinforced by several recent observations concerning the effect of activated *ras* oncogenes in gut epithelial cells.

Mutations in *ras* (reviewed in Townsend and Beauchamp, 1995) and TGF-β resistance (Hoosein et al, 1989; Manning et al., 1991) are characteristic of malignant gut epithelial cells. TGF-β activates *ras* in nontransformed rat intestinal and mink lung epithelial cells (Mulder and Morris, 1992). Overexpression of cyclin D1 blocks the inhibitory effects of TGF-β on esophageal epithelial cells, and such cells exhibit aspects of the transformed phenotype (Okamoto et al., 1994). Overexpression of activated *ras* genes transforms rat intestinal epithelial cells (Filmus et al., 1992), and activated *ras* can induce expression of cyclin D1 in intestinal cells (Filmus et al, 1994). These observations indicate that there is a signal transduction pathway that runs from TGF-β through *ras* to cyclin D1. It is consistent with one of the themes of this chapter to point out that *ras* has been recently shown to regulate transcription of *CLN* genes in yeast (Baroni et al., 1994; Tokiwa et al., 1994). Disruption of the TGF-β/*ras*/cyclin D1 regulatory axis may block an early step in differentiation of intestinal epithelial cells and may contribute to, or even cause, the increase in proliferative state that is characteristic of progression from adenomatous hyperplasia to carcinoma.

TGF-β RII RECEPTOR GENE, CELL PROLIFERATION, AND ADENOCARCINOMA

There are three classes of TGF-β receptors, RI, RII and RIII. The RI and RII receptors function as a heteromeric complex that is responsible for TGF-β signal transduction. RIII lacks a cytoplasmic signaling domain, and it has been proposed that this receptor may function to present TGF-β to the RI/RII complex (Lopez-Casillas et al., 1993). Loss of RII abrogates the growth-inhibitory response to TGF-β in several cell types (Boyd and Massague, 1989; Laiho et al., 1991; Wrana et al., 1992; Geiser et al., 1994).

There are several observations that suggest a relationship between the RII receptor, cyclin D/Cdk4 kinase activity, and malignant transformation of gut epithelial cells. The RII receptor is involved in TGF-β–mediated inhibition of pRb phosphorylation (Chen et al., 1993). This observation is consistent with the conclusion that TGF-β regulation of cyclin D/Cdk4 kinase activity is mediated through RII. Overexpression of Ha-*ras* induces cyclin D1 in intestinal epithelial cells (Filmus et al., 1994), and activated Ha-*ras* alleles down-regulate RII (Filmus et al., 1992). Moreover, overexpression of cyclin D1 inhibits RII expression in esophageal epithelial cells (Okamato et al., 1994). These observations hint at a complex regulatory relationship between expression of RII and the activity of cyclin D/Cdk4 kinase. Because pRb is, itself, involved in regulation of TGF-β1 gene expression (Kim et al., 1991), it may contribute to this regulatory loop in gut epithelial cells, which both produce and respond to TGF-β.

TGF-β regulation of intestinal cell proliferation is altered during malignant transformation. Conversion of a gut epithelial adenoma to a tumorigenic adenocarcinoma is associated with a decreased response to TGF-β (Manning et al., 1991; Markowitz et al., 1994), and there is a strong correlation between the degree of differentiation of human colorectal carcinoma cell lines and their sensitivity to the antiproliferative and differentiation-promoting effects of TGF-β (Hoosein et al., 1989). This relationship prevails with many different epithelial cancer cells. Many breast, hepatocellular, and gastric tumor cell lines lack RII receptors and are insensitive to growth inhibition by TGF-β (Inagaki et al., 1993; Park et al., 1994; Sun et al., 1994). Restoration of the wild-type RII restores TGF-β responsiveness and suppresses the tumorigenicity of some RII-negative cancer cell lines *in vivo* (Inagaki et al., 1993; Sun et al., 1994). It is of particular interest that Markowitz et al. (1995) recently demonstrated that RII receptors were frequently inactivated in human colon cancer cell lines that exhibited microsatellite instability (9 of 11 cell lines), but not in colon cancer cells that did not display microsatellite instability (only 3 of 27 cell lines). Microsatellite instability is a marker for a defect in DNA mismatch repair that leads to hereditary nonpolyposis colorectal cancer (HNPCC) and some nonhereditary cancers. The observation that cells that lack RII are defective in DNA repair suggests that RII, like p53, may be involved in maintaining genomic stability as well as controlling cell proliferation. In this context, it is intriguing to note that both RII and p53 share the ability to regulate the amount or activity of cyclin D/Cdk4 kinase, although it remains to be determined if regulation of cyclin-dependent kinase activity is directly related to maintenance of the genome in an intact state.

SUMMARY

In this chapter, we have summarized recent findings concerning hormonal inhibition of cell proliferation. Two model systems have been considered.

We have presented a brief overview of cell cycle progression in yeast and how G1 arrest is provoked by mating factors. This material may seem misplaced in a volume dealing with hormones and cancer, but there are several compelling reasons to consider yeast as a model system for hormonal inhibition of cell proliferation. Not least among the virtues of yeast is the means to manipulate the genetics of the organism. We know so much more about yeast because of the power of yeast genetics. Beyond that, yeast represent a true "wild-type" cell population that retains intact all the normal physiological regulatory circuitry that governs cell proliferation. This is probably not the case with any immortalized mammalian cell line, and loss of cell cycle regulatory constraints during selection for immortality in culture almost certainly introduces variables that we do not yet completely recognize. (A whimsical but entirely appropriate example of this phenomenon is described in Sheaff and Roberts, 1995.) It may be that unperceived variables of this sort may account for the fact that we cannot discern a unified molecular mechanism of action of TGF-β. Finally, yeast utilize signal transduction pathways that are virtually identical to those observed in mammalian cells. This high degree of conservation of the overall structure of the pathways, as well as the interchangeable nature of the many individual components of the pathways, strongly suggest that we may use our knowledge of yeast cell cycle control mechanisms to build hypotheses concerning those more complex regulatory phenomena that are likely to prevail in metazoan cells. For the purposes of this chapter, we have attempted to apply this concept to the second model system that we have described, TGF-β inhibition of epithelial cell proliferation.

Like yeast-mating factors, TGF-β is an inhibitor of G1 cyclin/Cdk kinase activity. Specifically, the target of TGF-β is one or more members of the family of D cyclin/Cdk4 kinase complexes, the metazoan homologs of the CLN/CDC28 kinases of budding yeast. Cyclin kinase inhibitors play critical roles in the molecular mechanisms of action of both TGF-β and yeast-mating factors, although the functional homolog of yeast FAR1 may yet be unidentified. It remains to be shown that TGF-β induces an inhibitor such as p15 in all TGF-β sensitive cells. Nevertheless, it is likely that many of the proliferative effects of the hormone-dependent changes in the cyclin D/Cdk4::p27 threshold result from redistribution of inhibitors such as p27 and p21. Although we have, in a broad sense, begun to understand how TGF-β inhibits epithelial cell proliferation, we are forced to conclude that the first principles of TGF-β action are not yet apparent. It may be that we have not yet defined the primary targets. If this is so, then cyclin D1, Cdk4, and p15 are all downstream of the primary event, and the identification of primary targets must define the future emphasis for analysis of TGF-β action.

Apropos of this volume, we have reviewed the data that underlie the hypothesis that TGF-β plays a major role in regulating proliferation and/or differentiation of normal intestinal epithelial cells. The ability of TGF-β

to control gut epithelial cell proliferation involves a regulatory axis that links the TGF-β receptor, through *ras*, to cyclin D1. Mutations of *ras* disrupt this pathway. Such mutations are prevalent in colon cancer, and this may account for the loss of TGF-β responsiveness that attends progression from adenoma to adenocarcinoma in the colon. Alternatively, loss of TGF-β responsiveness and unrestrained expression of cyclin D1 may inhibit differentiation of intestinal epithelial cells. Loss of the ability to differentiate upon exit from the intestinal crypts might be an early, premalignant event that would increase the probability of malignant transformation. A third alternative is implied from the association between loss of the TGF-β RII receptor and defects in DNA repair (Markowitz et al., 1995). There are a number of testable predictions that emanate from these hypotheses concerning TGF-β regulation of cyclin D1 expression in normal gut differentiation and in neoplastic transformation. The immediate challenge would seem to be the elucidation of a unifying hypothesis to account for TGF-β's effects on all epithelial cells. Considering the amount of effort that is currently committed to studying TGF-β inhibition of cell proliferation, such a unifying hypothesis should emerge rather soon. It may be more challenging to design experiments to determine exactly how loss of TGF-β responsiveness relates to expression of cyclin D1 and tumor progression *in vivo*.

ACKNOWLEDGMENTS

The preparation of this chapter and many of the experiments described above were supported in part by grants CA24347, CA64701, and CA64750 from the National Cancer Institute of the National Institutes of Health and grant no. 15860 from the Shriners' Hospital for Crippled Children, Schriners' Burns Institute of the University of Texas Medical Branch. The James E. Thompson Memorial Fund also contributed support. The authors would like to express their appreciation to J. Wade Harper and Stephen J. Elledge for many helpful discussions and for their generosity in providing a number of nucleotide sequences and reagents before publication.

REFERENCES

Ajchenbaum F, Ando K, DeCaprio JA, Griffin JD (1993): Independent regulation of human D-type cylcin gene expression during G1 phase in primary human T lymphocytes. *J Biol Chem* 268:4113–4119

Arnold A, Kim HG, Gaz RD, Eddy RL, Fukushima Y, Byers MG, Shows TB, Kronenberg HM (1989): Molecular cloning and chromosomal mapping of DNA rearranged with the parathyroid hormone gene in a parathyroid adenoma. *J Clin Invest* 83:2034–2040

Babyatsky MW, Podolsky DK (1991): Growth and development in the gastrointestinal tract. In: *Textbook of Gastroenterology*, Yamada T, Alpers DH, Owyang C, Powell DW, Silverstein FE, eds. Philadelphia: JB Lippincott

Baldin V, Lukas J, Marcote MJ, Pagano M, Draetta G (1993): Cyclin D1 is a nuclear protein required for cell cycle progression in G1. *Genes Dev* 7:812–821

Barnard JA, Beauchamp RD, Coffey RJ, Moses HL (1989): Regulation of intestinal epithelial cell growth by transforming growth factor type β. *Proc Natl Acad Sci USA* 86:1578–1582

Barnard JA, Warwick GJ, Gold LI (1993): Localization of transforming growth factor β isoforms in normal murine small intestine and colon. *Gastroenterology* 105:67–73

Baroni MD, Monti P, Alberghina L (1994): Repression of growth-regulated G1 cyclin expression by cAMP in budding yeast. *Nature* 371:339–342

Boyd TF, Massague J (1989): Transforming growth factor-β inhibition of epithelial cell proliferation linked to the expression of a 53-kDa membrane receptor. *J Biol Chem* 264:2272–2278

Breeden L, Nasmyth K (1987): Cell cycle control of the yeast HO gene: *cis*- and *trans*-acting regulators. *Cell* 48:389–397

Buchkovich K, Duffy LA, Harlow E (1989): The retinoblastoma protein is phosphorylated during specific phases of the cell cycle. *Cell* 58:1097–1105

Cairnie AB, Lamerton LF, Steel GG (1965): Cell proliferation studies in the intestinal epithelium of the rat. I. Determination of the kinetic parameters. *Exp Cell Res* 39:528–538

Cameron IL, Ord VA, Hunter KE, Heitman DW (1990): Colon carcinogenesis: modulation of progression. In: *Colon Cancer Cells*, Moyer MP, Poste GH, eds. San Diego, CA: Academic Press

Chen R-H, Ebner R, Derynck R (1993): Inactivation of the type II receptor reveals two receptor pathways for the diverse TGF-b activities. *Science* 260:1335–1338

Cheng H, LeBlond CP (1974): Origin, differentiation and renewal of the four main epithelial cell types in the mouse small intestine. I. columnar cell. *Am J Anat* 141:461–479

Cross FR (1988): *DAF1*, a mutant gene affecting size control, pheromone arrest and cell cycle kinetics of Saccharomyces cerevisiae. *Mol Cell Biol* 8:4675–4684

Cross FR, Tinkelenberg AH (1991): A potential positive feedback loop control CLN1 and CLN2 gene expression at start of the yeast cell cycle. *Cell* 65:875–883

DeCaprio JA, Ludlow JW, Lynch D, Furukawa Y, Griffin J, Piwnica-Worms H, Huang C-M, Livingston DM (1989): The product of the retinoblastoma susceptibility gene has properties of a cell cycle regulatory element. *Cell* 58:1085–1095

Dolan JW, Kirkman C, Fields S (1989): The yeast STE12 protein binds to the DNA sequence mediating pheromone induction. *Proc Natl Acad Sci USA* 86:5703–5707

Draetta G (1990): Cell cycle control in eukaryotes: molecular mechanisms of *cdc2* activation. *Trends Biol Sci* 15:378–383

Draetta G (1994): Mammalian G1 cyclins. *Curr Opin Cell Biol* 6:842–846

Dulic V, Lees E, Reed SI (1991): Association of human cyclin E with a periodic G1-S phase protein kinase. *Science* 257:1958–1961

Dulic V, Kaufmann WK, Wilson SJ, Tisty TD, Lees E, Harper JW, Elledge SJ, Reed SI (1994): p53-dependent inhibition of cyclin-dependent kinase activities in human fibroblasts during radiation-induced G1 arrest. *Cell* 76:1013–1023

El-Deiry WS, Tokino T, Velculescu VE, Levy DB, Parsons R, Trent JM, Lin D, Mercer D, Kinzler KW, Vogelstein B (1993): WAFl, a potent mediator of p53 tumor suppression. *Cell* 75:817–825

El-Deiry WS, Harper JW, O'Connor PM, Velculescu VE, Ganman CE, Jackman J, Pietenpol JA, Burrell M, Hill DE, Wang Y, Winman KG, Mercer WE, Kastan MB, Konh KW, Elledge SJ, Kinzler, KW, Vogelstein B (1994): WAF1k/CIP1 is induced in p53-mediated G1 arrest and apoptosis. *Cancer Res* 54:1169–1174

Elledge SJ, Harper JW (1994): Cdk inhibitors: on the threshold of checkpoints and development. *Curr Opinion Cell Biol* 6:847–852

Errede B, Ammerer G (1989): STE12, a protein involved in cell-type-specific transcription and signal transduction in yeast, is part of protein-DNA complexes. *Genes Dev* 3:1349–1361

Ewen ME, Sluss HK, Sherr CJ, Matsushime H, Kato J-Y, Livingston D (1993a): Functional interactions of the retinoblastoma protein with mammalian D type cyclins. *Cell* 73:487–497

Ewen ME, Sluss HK, Whitehouse LL, Livingston DM (1993b): TGFβ inhibition of Cdk4 synthesis is linked to cell cycle arrest. *Cell* 74:1009–1020

Filmus J, Zhao J, Buick RN (1992): Overexpression of H-*ras* oncogene induces resistance to the growth-inhibitory action of transforming growth factor beta-1 (TGF-β1) and alters the number and type of TGF-β1 receptors in rat intestinal epithelial cell clones. *Oncogene* 7:521–526

Filmus J, Robles AI, Shi W, Wong MJ, Colombo LL, Conti CJ (1994): Induction of cyclin D1 overexpression by activated *ras*. *Oncogene* 9:3627–3633

Fisher RP, Morgan DO (1994): A novel cyclin associates with MO15/CDK7 to form the CDK-activating kinase. *Cell* 78:713–724

Geiser AG, Burmester JK, Webbnink R, Roberts AB, Sporn MB (1994): Inhibition of growth by transforming growth factor-beta following fusion of two nonresponsive human carcinoma cell lines. Implication of the type II receptor in growth inhibitory responses. *J Biol Chem* 267:2588–2593

Geng Y, Weinberg RA (1993): Transforming growth factor β effects on expression of G1 cyclins and cyclin-dependent kinases. *Proc Natl Acad Sci USA* 90:10315–10319

Goodrich DW, Wang NP, Qian YW, Lee EYHP, Lee WH (1991): The retinoblastoma gene product regulates progression through the G1 phase of the cell cycle. *Cell* 67:293–302

Gordon CB, Campbell JL (1991): A cell cycle-responsive transcriptional control element and a negative control element in the gene encoding DNA polymerase alpha in *S. cerevisiae*. *Proc Natl Acad Sci USA* 88:6058–6062

Gu Y, Rosenblatt J, Morgan DO (1992): Cell cycle regulation of CDK2 by phosphorylation of THR160 and Tyr15. *EMBO J* 11:3995–4005

Gu Y, Turck CW, Morgan DO (1993): Inhibition of CDK 2 *in vivo* by an associated 20K regulatory subunit. *Nature* 366:707–710

Hanna Z, Jankowski M, Tremblay P, Jiang X, Milatovich A, Francke U, Jolicoeur, P (1993): The Vin-1 gene, identified by provirus insertional mutagenesis, is the cyclin D2 gene. *Oncogene* 8:1661–1666

Hannon GJ, Beach D (1994): $p15^{INK4B}$ is a potential effector of TGF-β-induced cell cycle arrest. *Nature* 371:257–261

Harper JW, Adami GR, Wei N, Keyomarsi K, Elledge SJ (1993): The p21 Cdk-interacting protein Cipl is a potent inhibitor of Gl cyclin-dependent kinases. *Cell* 75:805–816

Hartwell L (1974): *Saccharomyces cerevisiae* cell cycle. *Bacteriol Rev* 38:164–198

Hartwell LH, Kastan MB (1994): Cell cycle control and cancer. *Science* 266:1821–1827

Herskowitz I (1989): A regulatory hierarchy for cell specialization in yeast. *Nature* 342:749–757

Hoosein NM, McKnight MK, Levine AE, Mulder KM, Childress KE, Brattain DE, Brattain MG (1989): Differential sensitivity of subclasses of human colon carcinoma cell lines to the growth inhibitory effects of transforming growth factor-β1. *Exp Cell Res* 181:442–453

Howe PH, Draetta G, Leof EB (1991): Transforming growth factor β1 inhibition of $p34^{cdc2}$ phosphorylation and histone H1 kinase activity is associated with G1/S-phase growth arrest. *Mol Cell Biol* 11:1185–1194

Hunter T (1993): Braking the cycle. *Cell* 75:839–841

Hunter T, Pines J (1994): Cyclins and cancer II: cyclin D and CDK inhibitors come of age. *Cell* 79:573–582

Inagaki M, Moustakas A, Lin HY, Lodish HF, Carr BI (1993): Growth inhibition by transforming growth factor β (TGF-β) type I is restored in TGF-β-resistant hepatoma cells after expression of TGF-β receptor type II cDNA. *Proc Natl Acad Sci USA* 90:5359–53363

Kamb A, Gruis NA, Weaver-Feldhaus J, Liu Q, Harshman K, Tavtigan SV, Stockert E, Day RS, Johnson BE, Skolnik MH (1994): A cell cycle regulator potentially involved in genesis of many tumor types. *Science* 264:436–440

Kato JY, Matsushime H, Hiebert SW, Ewen ME, Sherr CJ (1993): Direct binding of cyclin D to the retinoblastoma gene product (pRb) and pRb phosphorylation by the cyclin D-dependent kinase CDK4. *Genes Dev* 7:331–342

Kato JY, Matsuoka M, Strom DK, Sherr CJ (1994a): Regulation of cyclin D-dependent kinase 4 (cdk4) by cdk4-activating kinase. *Mol Cell Biol* 14:2713–2721

Kato JY, Matsuoka M, Polyak K, Massague J, Sherr CJ (1994b): Cyclic AMP-induced G1 phase arrest mediated by an inhibitor ($p27^{Kip1}$) of cyclin-dependent kinase 4 activation. *Cell* 79:487–496

Kidd VJ (1992): Cell division control-related protein kinases: Putative origins and functions. *Molec Carcinogen* 5:95–101

Kim S-J, Lee H-D, Robbins PD, Busam K, Sporn MB, Roberts AB (1991): Regulation of transforming growth factor β1 gene expression by the product of the retinoblastoma-susceptibility gene. *Proc Natl Acad Sci USA* 88:3052–3056

Ko TC, Beauchamp RD, Ishizuka J, Townsend CM Jr, Thompson JC (1992): Mechanism of action of transforming growth factor-β on rat jejunal cells. *Surg Forum* 43:135–137

Ko TC, Beauchamp RD, Townsend CM Jr, Thompson JC (1993): Glutamine is essential for EGF-stimulated intestinal cell proliferation. *Surgery* 114:147–154

Ko TC, Beauchamp DR, Townsend CM, Thompson EA, Thompson JC (1994): Transforming growth factor-β inhibits rat intestinal cell growth by regulating cell cycle specific gene expression. *American J Surg* 167:14–20

Ko TC, Sheng HM, Reisman D, Thompson EA, Beauchamp RD (1995): Transforming growth factor-β1 inhibits cyclin D1 expression in intestinal epithelial cells. *Oncogene* 10:177–184

Koff A, Giordano A, Desai, D, Yamashita K, Harper JW, Elledge S, Nishimoto T, Morgan DO, Franza BR, Roberts JM (1992): Formation and activation of a cyclin E-cdk2 complex during the G1 phase of the human cell cycle. *Science* 257:1689–1694

Koff A, Ohtsuki M, Polyak K, Roberts JM, Massague J (1993): Negative regulation of G1 in mammalian cells: inhibition of cyclin E-dependent kinase by TGF-β. *Science* 260:536–539

Kurokowa M, Lynch K, Podolsky DK (1987): Effects of growth factors on an intestinal epithelial cell line: transforming growth factor β inhibits proliferation and stimulates differentiation. *Biochem Biophys Res Commun* 142:775–782

Laiho M, DeCaprio J, Ludlow J, Livingston D, Massague J (1990): Growth inhibition by TGF-β linked to suppression of retinoblastoma protein phosphorylation. *Cell* 62:175–185

Laiho M, Weis FM, Boyd FT, Ignotz RA, Massague J (1991): Responsiveness to transforming growth factor-β (TGF-β) restored by genetic complementation between cells defective in TGF-β receptors I and II. *J Biol Chem* 266:9108–9112

Lees E, Faha B, Dulic V, Reed SI, Harlow E (1992): Cyclin E/cdk2 and cyclin A/cdk2 kinases associate with p107 and E2F in a temporally distinct manner. *Genes Dev* 6:1874–1885

Leonard AD, Linke SP, Clarkin K, Wahl GM (1994): DNA damage triggers a prolonged p53-dependent arrest and long-term induction of Cip1 in normal human fibroblasts. *Genes Dev* 8:2540–2551

Lew DJ, Dulic V, Reed SI (1991): Isolation of three novel human cyclins by rescue of G1 cyclin (Cln) function in yeast. *Cell* 66:1197–1206

Li Y, Jenkins CW, Nichols MA, Xion Y (1994): Cell cycle expression and p53 regulation of the cyclin-dependent kinase inhibitor p21. *Oncogene* 9:2261–2268

Lopez-Casillas F, Wrana JL (1993): Betaglycan presents ligand to the TGF beta signaling receptor. *Cell* 773:1435–1444

Ludlow JW, DeCaprio JA, Huang CM, Lee WH, Paucha E, Livingston DM (1989): SV40 large T antigen binds preferentially to an underphosphorylated member of the retinoblastoma susceptibility gene product family. *Cell* 56:57–65

Ludlow JW, Shon J, Pipas JM, Livingston DM, DeCaprio JA (1990): The retinoblastoma susceptibility gene product undergoes cell cycle-dependent dephosphorylation and binding to and release from SV40 large T. *Cell* 60:387–396

Manning AM, Williams AC, Game SM, Paraskeva C (1991): Differential sensitivity of human colonic adenoma and carcinoma cells to transforming growth factor β (TGF-β): conversion of an adenoma cell line to a tumorigenic phenotype is accompanied by a reduced response to the inhibitory effects of TGF-β. *Oncogene* 6:1471–1476

Markowitz SD, Myeroff L, Cooper MJ, Traicoff J, Kochera M, Lutterbaugh J, Swiriduk M, Willson JK (1994): A benign cultured colon adenoma bears three genetically altered colon cancer oncogenes, but progresses to tumoriegnicity and transforming growth factor-beta independence without inactivating the p53 tumor suppressor gene. *J Clin Invest* 93:1005–1013

Markowitz S, Wang J, Myeroff L, Parson R, Sun L-z, Lutterbaugh J, Fan RS, Zborowska E, Kinzler KW, Vogelstein B, Brattain M, Willson JK (1995): Inactivation of the type ll TGF-β receptor in colon cancer cells with microsatellite instability. *Science* 268:1336–1338

Marsh L, Neiman AM, Herskowitz I (1991): Signal transduction during pheromone response in yeast. *Annu Rev Cell Biol* 7:699–728

Matsuoka M, Kato J, Fisher RP, Morgan DO, Sherr CJ (1994): Activation of cyclin-dependent kinase-4 (CDK4) by mouse MO15-associated kinase. *Mol Cell Biol* 14:7265–7275

Matsushime H, Roussel MF, Ashmun RA, Sherr CJ (1991): Colony-stimulating factor 1 regulates novel cyclins during the G1 phase of the cell cycle. *Cell* 65:701–713

Matsushime H, Ewen ME, Strom DK, Kato J, Hanks SK, Roussel MF, Sherr CJ (1992): Identification and properties of an atypical catalytic subunit ($p34^{PSK-J3}$/cdk4) for mammalian D type G1 cyclins. *Cell* 71:323–334

Mayol X, Grana X, Baldi A, Sang N, Hu Q, Giordano A (1993): Cloning of a new member of the retinoblastoma gene family (pRb2) which binds to the E1A transforming domain. *Oncogene* 8:2561–2566

McGee DW, Aicher WK, Eldridge JH, Peppard JV, Mestecky J, McGhee JR (1991): Transforming growth factor-β enhances secretory component and major histocompatibility complex class I antigen expression on rat IEC-6 intestinal epithelial cells. *Cytokine* 3:543–550

McIntosh E, Atkinson T, Storms R, Smith M (1991): Characterization of a short, cis-acting DNA sequence which conveys cell cycle stage-dependent transcription in *Saccharomyces cerevisiae*. *Mol Cell Biol* 11:329–337

Meyerson M, Faha B, Su L-K, Harlow E, Tsai L-H (1991): The cyclin dependent kinase family. *Cold Spring Harbor Symp Quant Biol* LVI:177–186

Meyerson, M, Harlow, E (1994): Identification of Gl kinase activity for cdk6, a novel cyclin D partner. *Mol Cell Biol* 14:2077–208

Moll T, Schwob E, Koch C, Moore A, Auer H, Nasmyth K (1993): Transcription factors important for starting the cell cycle in yeast. *Philos Trans R Soc Lond Biol* 340:351–360

Moreno S, Nurse P (1994): Regulation of progression through the G1 phase of the cell cycle by the *rum*1[1] gene. *Nature* 367:236–242

Motokura, T, Arnold, A (1993): Cyclins and oncogenesis. *Biochim Biophys Acta* 1155:63–78

Mulder KM, Morris SL (1992): Activation of $p21^{ras}$ by transforming growth factor β in epithelial cells. *J Biol Chem* 267:5029–5031

Nash R, Tokiwa G, Anand, S Erickson K, Futcher AB (1988): The *WHI*1$^+$ gene of *Saccharomyces cerevisiae* tethers cell division to cell size and is a cyclin homologue. *EMBO J* 7:4335–4346

Nasmyth K, Dirick L (1991): The role of *SWI4* and *SWI6* in the activity of G1 cyclins in yeast. *Cell* 66:995–1013

Nasmyth K, Shore D (1987): Transcriptional regulation in the yeast cell cycle. *Science* 237:1162–1170

Noda A, Nig Y, Venable SF, Pereira-Smith OM, Smith JR (1993): Cloning of senescent cell-derived inhibitors of DNA synthesis using an expression screen. *Exp Cell Res* 211:90–98

Norbury C, Nurse P (1992): Animal cell cycles and their control. *Annu Rev Biochem* 61:441–470

Ogas J, Andrews BJ, Herskowitz I (1991): Transcriptional activation of CLN1, CLN2, and a putative new G1 cyclin (HSC26) by SWI4, a positive regulator of G1-specific transcription. *Cell* 66:1015–1026

Ohtsubo M, Roberts JM (1993): Cyclin-dependent regulation of G1 in mammalian fibroblasts. *Science* 259:1908–1912

Okamoto A, Jiang W, Kim S-J, Spillare EA, Stoner GD, Weinstein IB, Harris CC (1994): Overexpression of human cyclin D1 reduces the transforming growth factor β (TGF-β) type II receptor and growth inhibition by TGF-β1 in an immortalized human esophageal epithelial cell line. *Proc Natl Acad Sci USA* 91:11576–11580

Pagano M, Draetta G, Jansen-Durr P (1992): Association of cdk2 kinase with the transcription factor E2F during S phase. *Science* 255:1144–1147

Pardee AB (1989): G1 events and regulation of cell proliferation. *Science* 246:603–608

Park K, Kim SJ, Bang YJ, Park JG, Kim NK, Roberts AB, Sporn MB (1994): Genetic changes in the transforming growth factor beta (TGF-beta) type II receptor gene in human gastric cancer cells: correlation with sensitivity to growth inhibition by TGF-beta. *Proc Natl Acad Sci USA* 91:8872–8776

Pelech SL, Sanghera JS (1992): Mitogen-activated protein kinases: versatile transducers for cell signalling. *Trends Biochem Sci* 17:233–239

Peter, M, Gartner A, Horecka J, Ammerer G, Herskowitz I (1993): FAR1 links the signal transduction pathway to the cell cycle machinery in yeast. *Cell* 73: 747–760

Peter M, Herskowitz I (1994): Joining the complex: cyclin-dependent kinase inhibitory proteins and the cell cycle. *Cell* 79:181–184

Pietenpol JA, Stein RW, Moran E, Yacuik P, Schlegel R, Lyons RM, Pittelkow MR, Munger K, Howley PM, Moses HL (1990): TGF-β1 inhibition of c-myc transcription and growth in keratinocytes is abrogated by viral transforming proteins with pRb binding domains. *Cell* 61:777–785

Polyak K, Kato J-Y, Solomon MJ, Sherr CJ, Massague J, Roberts JM, Koff A (1994a): $p27^{Kip1}$, a cyclin-Cdk inhibitor, links transforming growth factor-β and contact inhibition to cell cycle arrest. *Genes Dev* 8:9–22

Polyak K, Lee M-H, Erdjument-Bromage H, Koff A, Roberts JM, Tempst P, Massague J (1994b): Cloning of $p27^{Kip1}$, a cyclin-dependent kinase inhibitor and a potential mediator of extracellular antimitogenic signals. *Cell* 78: 59–66

Quelle DE, Ashmun RA, Shurtleff SA, Kato J-Y, Bar-Sagi D, Roussel MF, Sherr CJ (1993): Overexpression of mouse D-type cyclins accelerates G1 phase in rodent fibroblasts. *Genes Dev* 7:1559–1571

Reed SI (1991): Pheromone signaling pathways in yeast. *Curr Opin Genet Dev* 1:391–396

Richardson HE, Wittenberg C, Cross R, Reed SI (1989): An essential G1 function for cyclin like proteins in yeast. *Cell* 59:1127–1133

Rosenberg CL, Kim HG, Shows TB, Kronenberg HM, Arnold A (1991a): Rearrangement and overexpression of D11S287E, a candidate oncogene on chromosome 11q13 in benign parathyroid tumors. *Oncogene* 6:449–453

Rosenberg CL, Wong E, Petty EM, Bale AE, Tsujimoto Y, Harris NL, Arnold A (1991b): PRAD1, a candidate BCL1 oncogene: mapping and expression in centrocytic lymphoma. *Proc Natl Acad Sci USA* 88:9638–9642

Rosenblatt J, Gu Y, Morgan DO (1992): Human cyclin-dependent kinase 2 (CDK2) is activated during the S and G2 phases of the cell cycle and associates with cyclin A. *Proc Natl Acad Sci USA* 89:2824

Schwob E, Bohm T, Mendenhall, MD, Nasmyth K (1994): The B-type cyclin kinase inhibitor $p40^{SIC1}$ controls the G1 to S transition in *S. cerevisiae*. *Cell* 79:233–244

Serrano M, Hannon GJ, Beach D (1993): A new regulatory motif in cell-cycle control causing specific inhibition of cyclin D/CDK4. *Nature* 366:704–707

Sheaff RJ, Roberts JM (1995): Lessons from phylum Falconium. *Current Biol* 5:28–31

Sherr CJ (1993): Mammalian G1 cyclins. *Cell* 73:1059–1065

Sherr CJ (1994): G1 phase progression: cycling on cue. *Cell* 79:551–555

Singerland JM, Hengst L, Pan C-H, Alexander D, Stampfer MR, Reed SI (1994): A novel inhibitor of cyclin Cdk activity detected in transforming growth factor β–arrested epithelial cells. *Mol Cell Biol* 14:3683–3694

Solomon MJ, Lee T, Kirschner MW (1992): Role of phosphorylation in $p34^{cdc2}$ activation: identification of an activating kinase. *Mol Biol Cell* 3:13–27

Sporn MB, Roberts AB (1985): Autocrine growth factors and cancer. *Nature* 313:745–747

Sprague GF, Jr (1992): Kinase cascade conserved. *Curr Biol* 2:587–589

Surmacz E, Reiss K, Sell C, Baserga R (1992): Cyclin D1 messenger RNA is inducible by platelet-derived growth factor in cultured fibroblasts. *Cancer Res* 52:4522–4525

Sun L-z, Wu,G, Willson,JK, Zborowska E, Yang J, Rajkarunanayake I, Wange J, Gentry LE, Wang X-f, Brattain MG (1994): Expression of transforming growth factor β type II receptor leads to reduced malignancy in human breast cancer MCF-7 cells. *J Biol Chem* 269:26449–26455

Szepetowski P, Nguyen C, Perucca-Lostanlen D, Carle GF, Tsujimoto Y, Birnbaum D, Theillet C, Gaudray P (1991): D11S146 and BCL1 are physically linked but can be discriminated by their amplification status in human breast cancer. *Genomics* 10:410–416

Tokiwa G, Tyers M, Volpe T, Futcher B (1994): Inhibition of G1 cyclin activity by the Ras/cAMP pathway in yeast. *Nature* 371:342–345

Townsend CM Jr, Beauchamp RD (1995): New developments in colorectal cancer. *Curr Opin Gastroenterol* 11:36–42

Toyoshima H, Hunter T (1994): p27, a novel inhibitor of G1 cyclin-Cdk protein kinase activity, is related to p21. *Cell* 78:64–67

Vogelstein B, Fearon ER, Hamilton SR, Kern SE, Preisinger AC, Leppert M, Nakamura Y, White R, Smits AM, Boss JL (1988): Genetic alterations during colorectal-tumor development. *N Engl J Med* 319:525–532

Vogelstein B (1990): Cancer. A deadly inheritance. *Nature* 348:681–682

Winston, JT, Pledger, WJ (1993): Growth factor regulation of cyclin Dl mRNA expression through protein synthesis-dependent and-independent mechanisms. *Mol Biol Cell* 4:1133–1144

Withers DA, Harvey RC, Faust JB, Melnyk O, Carey K, Meeker TC (1991): Characterization of a candidate *bcl*-1 gene. *Mol Cell Biol* 11:4846–4853

Won KA, Xiong Y, Beach D, Gilman MZ (1992): Growth-regulated expression of D-type cyclin genes in human diploid fibroblasts. *Proc Natl Acad Sci USA* 89:9910–9914

Wrana JL, AttisanoL, Carcamo J, Zentella A, Doddy J, Laiho M, Wang, XF, Massague J (1992): TGF-beta signals through a heteromeric protein kinase receptor complex. *Cell* 71:1003–1014

Xiong Y, Connolly T, Futcher B, Beach, D (1991): Human D-type cyclins. Cell 65:691–699

Xiong Y, Zhang H, Beach D (1992): D type cyclins associate with multiple protein kinases and the DNA replication and repair factor PCNA. *Cell* 71: 505–514

Xiong Y, Hannon GJ, Zhang H, Casso P, Kobayashi R, Beach D (1993a): P21 is a universal inhibitor of cyclin kinases. *Nature* 366:701–704

Xiong Y, Hannon GJ, Zhang H, Casso D, Kobayashi R, Beach D (1993b): p21 is a universal inhibitor of cyclin kinases. *Nature* 366:701–704

Zhang H, Hannon G, Beach D (1994): p21-containing cyclin kinases exist in both active and inactive states. *Genes Dev* 8:1750–1758

3

RET Proto-Oncogene and Its Role in Multiple Endocrine Neoplasia and Medullary Thyroid Cancer

PAUL GOODFELLOW AND JEFFREY F. MOLEY

INTRODUCTION

The RET proto-oncogene encodes a transmembrane protein receptor tyrosine kinase, the function of which is unknown at present. The inclusion of a discussion of this gene in a book on hormones and cancer is appropriate because of the recently described association of mutations in RET and the hereditary cancer syndromes multiple endocrine neoplasia (MEN) type 2A, MEN 2B, and familial non-MEN medullary thyroid cancer (FMTC). In these disorders, which are inherited in an autosomal dominant fashion, individuals inherit a predisposition to medullary thyroid cancer (MTC), a malignancy of neuroendocrine cells (C-cells) that reside in the parafollicular areas of the thyroid gland. This trait is expressed with virtually 100% penetrance. MEN 2A, MEN 2B, and FMTC are clinically distinct diseases. (Schimke, 1984, Farndon et al., 1986). MEN 2A and MEN 2B are syndromes characterized by neoplasms involving more than one endocrine tissue or cell type. FMTC, on the other hand, is a nonsyndromic Mendelian trait characterized by the development of MTC in the absence of any other recognizable abnormalities (Table 1). Through DNA marker linkage studies, the gene(s) responsible for MEN 2A, MEN 2B, and FMTC were mapped to the same region of chromosome 10 (Mathew et al., 1987a; Simpson et al., 1987; Norum et al., 1990a; Lairmore et al., 1992). The coincidental localization of the predisposition loci to the same region of the genome, led to speculation that MEN 2A, MEN 2B, and FMTC might represent components of a contiguous gene

Hormones and Cancer
Wayne V. Vedeckis, Editor

Table 1. Clinical presentations of medullary thyroid cancer

	Characteristics of tumor	Associated abnormalities	Underlying genetic defect
Sporadic	Unilateral	None	Tumors may have mutations of RET in the tyrosine kinase domain
MEN 2A	Multifocal, bilateral	Hyperparathyroidism, pheochromocytomas	Missense mutation in extracellular domain of RET
MEN 2B	Multifocal, bilateral	Pheochromocytomas, mucosal neuromas, characteristic appearance, megacolon, skeletal abnormalities	Missense mutation in RET tyrosine kinase domain
FMTC	Multifocal, bilateral	None	Missense mutation in extracellular domain of RET

MEN, multiple endocrine neoplasia; FMTC, familial non-MEN medullary thyroid cancer.

syndrome or, alternatively, that they represent distinct disease alleles at a single locus.

New technologies for gene mapping helped to define more precisely the region of chromosome 10 that includes the mutations that result in MEN 2A, MEN 2B, and FMTC (Lairmore et al., 1993; Gardner et al., 1993). Clone contigs and physical maps for the candidate region were devised Lairmore et al., 1993; Brooks-Wilson et al., 1993). The RET proto-oncogene was shown to map within the minimum critical region as defined by both physical and genetic mapping experiments. Detailed analysis of the RET locus proved that all three disorders were associated with constitutional (germ-line) mutations in the gene (Mulligan et al., 1993a; Donis-Keller et al., 1993; Hofstra et al., 1994b; Carlson et al., 1994a). The discovery of the gene defects that underlie MEN 2A, MEN 2B, and FMTC has already had a major impact on the management of patients with inherited forms of MTC (Wells et al., 1994; Lips et al., 1994a).

MAPPING MEN 2A, MEN 2B, AND FMTC

DNA Marker Linkage Studies in Families with MEN 2A, MEN 2B, and FMTC

The search for the genes responsible for inherited forms of MTC began with DNA marker linkage studies in MEN 2A families. In 1987 the gene responsible for MEN 2A was assigned to the pericentromeric region of chromosome 10 (Mathew et al., 1987b; Simpson et al., 1987). Next MEN 2B was mapped to chromosome 10 (Norum et al., 1990b) and soon thereafter the FMTC locus was also assigned to the same region (Lairmore et al., 1991). Over a period of 6 years the location of the MEN 2 gene was more precisely defined with the discovery of the markers D10S94, D10S102, D10S97 and the RET gene, all tightly linked to MEN 2A (Goodfellow et al., 1990, Matlew et al., 1991b, Lichter et al., 1991, Mulligan et al., 1993b). All the markers that failed to recombine with MEN 2A mapped to 10q11.2. The proximal portion of the long arm of the chromosome seemed the most likely localization of the gene(s) responsible for MEN 2A, MEN 2B, and FMTC.

Tumor Studies

Most of inherited forms of cancers that have been studied are associated with mutations in tumor-suppressor genes. An inherited mutation in a tumor-suppressor gene predisposes to tumor development. The tumors in inherited cancer syndromes are frequently characterized by loss of heterozygosity (LOH) for sequences that include the suppressor gene wild-type allele. It was long assumed that MEN 2A, MEN 2B, and FMTC phenotypes were associated with mutations in a tumor-suppressor gene or genes and that deletion mapping in tumors might help point to the location of the disease gene. LOH studies and linkage (family) studies are complementary approaches in attempting to map the location of a tumor-suppressor gene.

LOH of chromosome 10 sequences was not seen in MTC or pheochromocytomas from MEN 2 patients, nor was it observed in sporadic tumors (Landsvater et al., 1989; Mathew et al., 1987c; Nelkin et al., 1989; Mulligan et al., 1993b). Tumor deletion mapping studies were, therefore, of no value in attempting to refine the localization of the gene(s) responsible for MEN 2A, MEN 2B, and FMTC.

Fine-Mapping in the Disease Gene Region

DNA markers tightly linked to MEN 2A (Goodfellow et al., 1990; Mathew et al., 1991a; Mulligan et al., 1991; Lichter et al., 1991) were used as probes for physical mapping studies and as entry points for the development of clone contigs in the disease gene region. Large insert clone contigs were developed by several groups through the use of yeast artificial chromosomes (YACs) (Lairmore et al., 1993; Brooks-Wilson et al., 1993; Mole et al., 1993). A number of the markers demonstrated to be nonrecombinant with MEN 2A were physically linked (RET-D10S94-D10S102) within 1.5 megabases, and the order of markers was determined to be centromere-RET-D10S94-D10S102-telomere based on a combination of genetic and physical mapping data (Brooks-Wilson et al., 1993; Lairmore et al., 1993; Mole et al., 1993; Lichter et al., 1992).

Highly informative simple sequence repeat markers helped to narrow the region including the MEN genes. The D10S141 marker was demonstrated to recombine with both MEN 2A and MEN 2B. A recombination event involving D10S94 and MEN 2A was also identified. The MEN 2A gene was this localized to a small interval between D10S141 and D10S94 (Gardner et al., 1993). Physical mapping studies showed that this region comprised 480kb and included the RET proto-oncogene (Mole et al., 1993).

Discovery of RET Mutations Associated with MEN 2A and FMTC

The RET gene was first identified with an NIH 3T3 cell transformation assay (Takahashi and Cooper, 1987). The transforming sequences recovered resulted from a rearrangement between RET and another gene. Sequence analysis of RET revealed that it is a member of the receptor tyrosine kinase gene family (Takahashi et al., 1988; Fig. 1). RET was mapped to the proximal region of the long arm of chromosome 10 in 1989 (Ishizaka et al., 1990) and was shown to be expressed at high levels in both MTCs and pheochromocytomas (Santoro et al., 1990). Based on genetic and physical mapping evidence (Gardner et al., 1993; Mole et al., 1993) and its demonstrated oncogenic

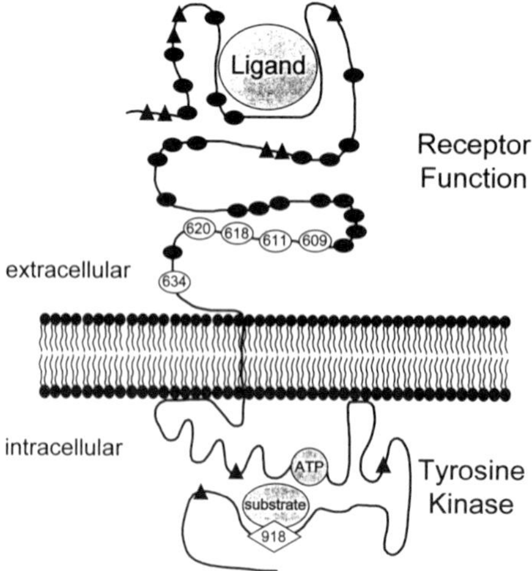

Figure 1. Diagrammatic representation of RET proto-oncogene. Codons affected by mutations in MEN 2A (609, 611, 618, 620, and 634), and MEN 2B (918) are noted. The filled ellipses represent cysteines in the extracellular cysteine-rich domain and the triangles are regions that are mutated in Hirschsprung's disease.

potential, RET was an ideal candidate for MEN 2A, MEN 2B, and FMTC. No gross structural rearrangements at the RET locus were observed (Mulligan et al., 1993b) and detailed analysis of RET in the constitutional DNA and tumors from MEN 2A, MEN 2B, and FMTC patients was undertaken in several laboratories.

RET mutations were identified in the constitutional DNA of MEN 2A and FMTC patients by two groups working independently (Donis-Keller et al., 1993; Mulligan et al., 1993a). Mulligan and co-workers (1993a) identified RET mutations in association with MEN using an expression-based mutational analysis system. RNA from MEN 2 tumors was reverse transcribed, the resulting first-strand cDNA synthesis products used as template for polymerase chain reaction (PCR) amplification of RET sequences, and the PCR products analyzed for sequence variation using the chemical cleavage mismatch procedure (CCM). The novel cleavage products that were identified by CCM were sequenced. Seven tumors from 6 MEN 2A patients were analyzed along with one sporadic MTC and one sporadic pheochromocytoma. All the MEN 2A tumors were characterized by CCM and sequence variants. Analysis of genomic DNA from MEN 2A family members proved that the variants identified in the cDNAs were also in the constitutional DNA of affected family members and not in individuals unaffected by MEN 2A. All mutations resulted in substitution of cysteine residues clustered near the transmembrane domain of RET (Table 2; Fig. 1). In all but one MEN 2A, tumor heterozygosity for the mutant and wild-type RET allele was retained. Mulligan and colleagues (1993a) suggested a dominant or dominant/negative mechanism for RET mutation in the development of MTC and pheochromocytomas in MEN 2A.

Our group used a PCR amplification, single-strand conformation variant (SSCV) and sequence analysis approach to identify mutations in RET in the constitutional DNA of patients with MEN 2A and FMTC. Both MEN 2A and FMTC were shown to be associated with mutations that result in substitution of cysteine residues in the extracellular portion of RET, immediately adjacent to the transmembrane domain. Unexpectedly, the same mutations were found to characterize MEN 2A and FMTC kindreds. No mutations were found in the cysteine-rich region of RET in DNA from MEN 2B patients. The evidence in favor of RET as the gene responsible for MEN 2A and FMTC was overwhelming, yet the identity of the MEN 2B remained unknown. The genomic structure of RET was determined by both the British and American groups. A complete genomic organization was reported by Kwok et al. (1993).

In a study of 118 families with inherited MTC (MEN 2A, MEN 2B, FMTC, and families with less certain diagnoses), Mulligan and co-workers (1993a) identified additional mutant RET alleles in MEN 2A and FMTC. Sixty-eight of families with a diagnosis of MEN 2A and six of seven families with FMTC had mutation at one of five cysteine codons in exons 10 or 11 or RET. For a group of 19 families in which the diagnosis of MEN 2A or

Table 2. MEN 2A, MEN 2B, and FMTC Mutations

Codon	Amino acid substitutions	Disease
609	Cys to Tyr	MEN 2A
611	Cys to Trp	FMTC
	Cys to Tyr	MEN 2A
618	Cys to Arg	MEN 2A
	Cys to Ser	MEN 2A
		FMTC
	Cys to Tyr	FMTC
	Cys to Gly	MEN 2A
620	Cys to Arg	MEN 2A
		FMTC
	Cys to Tyr	MEN 2A
634	Cys to Arg	MEN 2A
	Cys to Ser	MEN 2A
	Cys to Gly	MEN 2A
	Cys to Tyr	MEN 2A
		FMTC
	Cys to Phe	MEN 2A
		FMTC
	Cys to Trp	MEN 2A
918	Met to Thr	MEN 2B

Data on mutations taken from: Donis-Keller et al., 1993; Mulligan et al., 1993a; Carlson et al., 1994a; Hofstra et al., 1994a; Mulligan et al., 1994. Additional amino acid substitutions may be included in the study of Mulligan et al., 1994, but these were not defined specifically.
MEN, multiple endocrine neoplasia; FMTC, familial non-MEN medullary thyroid cancer.

FMTC was less certain, 13 were found to have RET mutations (Mulligan et al., 1994). All possible base changes at codon 634 that result in amino acid substitution have been observed in association with disease. These data suggest that there is zero tolerance for substitution of a single cysteine residue in the RET gene. All the mutational analyses strongly support the notion that RET mutations are causally associated with disease.

RET Genotype is Correlated with Disease Phenotype

Mulligan and co-workers (1994) have demonstrated that there is a correlation between particular classes of RET mutation and the disease phenotype in patients with MEN 2A and FMTC. It was shown that patients with the codon 634 TGC → CGC mutation (cys to arg) have a greater risk of developing parathyroid disease than do individuals with any of the other

RET mutations identified. The finding that some FMTC kindreds have RET mutations that are associated with MEN 2A in other families is not easily explained given the data currently available. The local chromosomal environment may exert some effect on specific RET mutations.

Discovery of a Single Missense Mutation Associated with MEN 2B

The RET mutations associated with MEN 2A and FMTC were not observed in constitutional DNA from MEN 2B patients (Donis-Keller et al., 1993; Mulligan et al., 1994). Clinical similarities between MEN 2A and MEN 2B and the fact that the inherited defects mapped to the same region of chromosome 10 suggested that MEN 2A and MEN 2B might result from different mutations in the same gene. A number of laboratories intensified the search for mutations in RET in DNA from MEN 2B patients.

Three groups reported identification of the same RET mutation associated with MEN 2B (Hofstra et al., 1994a; Carlson et al., 1994a; Eng et al., 1994). The MEN 2B mutation identified is a T→C transition, which results in substitution of a threonine residue for a methionine in RET exon 16 (Met-918; Fig. 1 and Table 2). The mutation was shown to be absent from the DNA in normal parents of patients with *de novo* MEN 2B and a sizable number of normal controls. The large number of unrelated MEN 2B patients in which the identical point mutation was observed, along with the fact that the mutation was demonstrated to have arisen *de novo* in association with disease, was taken as proof that the methionine→ threonine amino acid substitution at position 918 results in the MEN 2B phenotype. MEN 2B is unique among inherited cancer syndromes in that a single genetic defect is associated with the disease state. It is also unusual in that more than half of all patients with MEN 2B have *de novo* disease and that, in all cases reported to date, the RET mutation arose on the paternally derived chromosome (Carlson et al., 1994b). De novo MEN 2B mutation is associated with increased paternal age (Carlson et al., 1994b).

Sporadic Medullary Thyroid Cancers and Pheochromocytomas

Some 20–25% of MTC appears as inherited disease (Saad et al., 1984). "Inherited" MTC includes those tumors seen in the context of the MEN 2 syndromes and FMTC. The constitutional DNA of affected members of most MEN 2A, MEN 2B, and FMTC families is characterized by RET mutation. It is therefore not surprising that somatic mutations in RET have been observed in sporadic MTC. Donis-Keller et al. (1993) described a 6-bp deletion involving a cysteine codon in the same region of the gene in which MEN 2A and FMTC mutations are clustered. That deletion was not present in the patient's normal (constitutional) DNA (Table 2). Hofstra et al. (1994a) and Eng et al. (1994) found the RET codon 918 methionine → threonine mutation associated with MEN 2B in a number of sporadic MTCs.

To date there have been no reports of RET mutation in association with sporadic pheochromocytoma, the other tumor type common to the MEN 2 syndromes. Hofstra and colleagues (1994a) demonstrated that codon 918 was unaltered in the DNA from 5 sporadic pheochromocytomas. Pheochromocytomas are seen in a number of genetic diseases (Von Hippel-Lindau disease, neurofibromatosis type 1) and as such it seems likely that they will show a greater etiologic and genetic heterogeneity than might be the case for MTC. To date, characterization of the RET locus in sporadic tumors has been for the most part restricted to evaluation of the regions of the gene in which germline mutations have been identified. There are potentially very different biological constraints determining the spectrum of mutations in MEN 2A, MEN 2B, and FMTC and those associated with "somatic" or sporadic disease. Germ-line mutations must be compatible with a near normal pattern of development and viability of the organism as a whole. Somatic mutations need only be transforming and compatible with survival of a single cell type that will give rise to a tumor. Until a comprehensive evaluation of RET sequences in sporadic MTCs and pheochromocytomas is undertaken, it is difficult to speculate as to the frequency with which RET mutation is associated with the development of these tumors.

FUNCTIONAL STUDIES OF RET

The MEN 2B mutation and three MEN 2A mutations that have been studied in cell culture assays have dominant transforming activity (Santoro et al., 1995). Mutant RET was shown to efficiently transform NIH 3T3 cells and to be highly clonogenic in soft agar assays, whereas the wild-type RET was not. Wild-type RET, as well as the MEN 2A and MEN 2B mutant forms of the protein expressed in NIH 3T3 cells, were all shown to have apparent molecular weights of 145 and 160 kDa. MEN 2A and MEN 2B RET showed extensive tyrosine phosphorylation. The wild-type or proto-RET protein, on the other hand, had no detectable tyrosine phosphorylation in the cell culture assay (Santoro et al., 1995). Receptor autophosphorylation and activation would be expected to result in the high phosphotyrosine content of the MEN 2A and MEN 2B forms of RET. Investigation of the kinase activity of the three forms of RET expressed in transfected cells proved that wild-type RET had little or no kinase activity, whereas MEN 2A and MEN 2B forms of RET were autophosphorylated. RET–MEN 2B showed much less autophosphorylation than was the case for RET–MEN 2A (Santoro et al., 1995). All the changes observed for the mutant forms of RET would be expected to be dominant at the level of the cell.

To confirm the observations that were made in transfected cells, RET kinase activity and autophosphorylation were investigated in tumor cell lines. A MTC thyroid cancer cell line with a MEN 2A mutation showed significant tyrosine kinase activity and autophosphorylation. The MTC cell line expressed both the normal RET and the MEN 2A–RET protein. A neuroblastoma cell line expressing wild-type RET, on the other hand, did not show any kinase activity or any evidence of autophosphorylation. These observations further supported the hypothesis that MEN 2A mutations result in RET activation via increased tyrosine kinase activity (Santoro et al., 1995).

Wild-type, MEN 2B, and MEN 2A forms of the RET all migrate as 150–160 kDa species on sodium dodecyl sulfate (SDS) polyacrylamide gels. MEN 2A–RET also gave rise to a 300 kDa species, suggestive of homodimeric proteins. Dimeric RET proteins have a kinase activity 10 times that of the monomers. Based on these observations, it was suggested that MEN 2A mutations, all of which result in substitution of cysteine residues, lead to aberrant homodimerization and activation (Santoro et al., 1995). The MEN 2B mutant proteins, on the other hand, did dimerize. To explain how the MEN 2B protein might lead to transformation, Santoro and colleagues (Santoro et al., 1995) investigated the specificity of the RET–MEN 2B kinase in comparison to MEN 2A and a epidermal growth factor receptor/RET fusion protein. They demonstrated that MEN 2B–RET had altered autocatalytic specificity and substrate phosphorylation. Altered activity of MEN 2B–RET is consistent with the more extensive phenotype associated with MEN 2B in comparison with MEN 2A and FMTC. The effects of MEN 2B mutation in multiple tissues could be explained by tissue-specific differences in substrates, each with a unique mitogenic potential.

Songyang and colleagues (1995) provided experimental evidence that MEN 2B results in disease because of altered substrate specificity. Wild-type RET and MEN 2B–RET were shown to have different preferred substrates. In MEN 2B, phosphorylation of inappropriate substrates could result in stimulation of signal transduction pathways that normal (wild-type) RET usually does not act through (Songyang et al., 1995). The mechanism by which the RET protein influences cell division and/or differentiation is unknown.

CLINICAL APPLICATIONS

MTC occurs in sporadic and hereditary clinical settings (Table 1). In sporadic MTC, tumors are usually single and unilateral, and occur without a pattern of familial predisposition. These tumors almost always present as a mass in the neck, and metastases to lymph nodes in the neck are frequently present at the time of diagnosis. A hereditary predisposition to MTC is seen in patients with MEN 2A and 2B, and in patients with the related disorder, FMTC (see Table 1). In these disorders, the tumor occurs as multifocal, bilateral disease, and there is an autosomal dominant pattern of inheritance (Cance and Wells, 1985). In addition to MTC, patients with MEN 2A may develop pheochromocytomas (catecholamine-secreting tumors of the adrenal medulla) and parathyroid hyperplasia (which result in elevated calcium levels). MTCs that develop in patients with MEN 2B are more aggressive and have an earlier age of onset than those seen in patients with sporadic or MEN 2A tumors. MEN 2B patients also develop pheochromocytomas, but they do not have parathyroid hyperplasia. They have multiple mucosal neuromas, ganglioneuromas of the gastrointestinal tract, and a characteristic phenotype (Fig. 2). FMTC is the most indolent form of MTC, and is inherited as an autosomal dominant trait. Patients with FMTC develop only MTC, at a later age than MEN 2A or 2B. MTC is a tumor of the thyroid C-cells, also known as the parafollicular cells. These cells are of neural crest

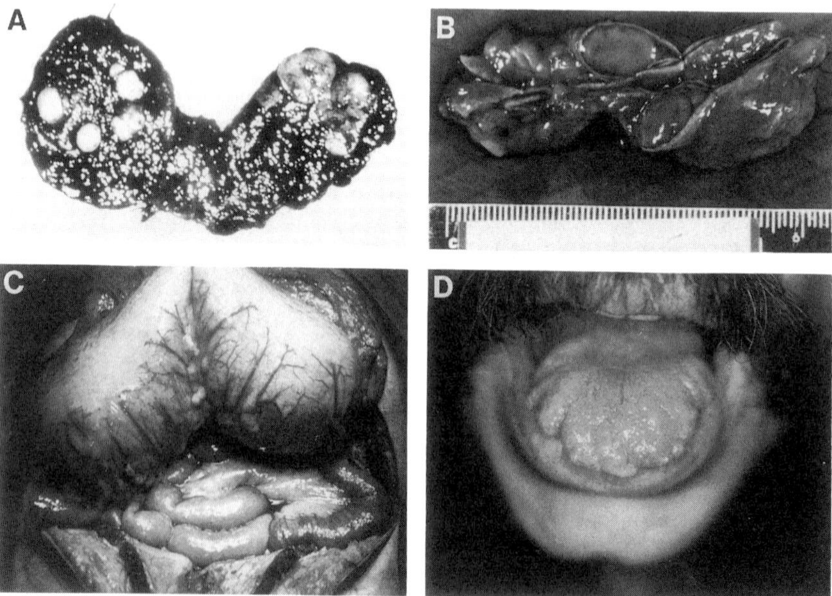

Figure 2. Clinical features of the MEN 2 syndromes. **A**: Thyroid gland from a patient with MEN 2A. This section demonstrates multifocal, bilateral foci of MTC in the upper poles. **B**: Adrenalectomy sample from a patient with MEN 2A. A pheochromocytoma and surrounding medullary hyperplasia are evident. **C**: Dilated colon from a patient with MEN 2B. **D**: Tongue of a patient with MEN 2B. Characteristic plexiform neuromas are seen. [Reprinted from Moley JF (1995): Medullary thyroid cancer. *Endocr Surg* 75: 407. Used with permission.]

derivation and are considered to be part of the amine precursor uptake and decarboxylation (APUD) system of neuroendocrine cells. Other neural crest-derived tumors include melanomas, pheochromocytomas, and neuroblastomas. C-cells are dispersed throughout the thyroid gland, with the highest concentration of cells in the upper poles. C-cells secrete calcitonin, a hormone involved in calcium metabolism. Plasma immunoreactive calcitonin is an excellent marker for the presence of MTC, and it is routinely used in the screening of individuals predisposed to the hereditary forms of the disease and in the follow-up of patients who have been treated. MTC cells express and secrete a variety of substances in addition to calcitonin. Some of these are carcinoembryonic antigen (CEA), histaminase, neuron-specific enolase, calcitonin-gene related peptide, somatostatin, thyroglobulin, thyrotropin-stimulating hormone, adrenocortical-stimulating hormone, gastrin-related peptide, serotonin, chromogranin, substance P, and proopiomelanocortin. MTC is often associated with C-cell hyperplasia, which may be a precursor of MTC.

Of the estimated 13,900 new cases of thyroid cancer in the United States in 1995, approximately 5–10% will be MTC (Wingo et al., 1995). These tumors develop as unifocal or multifocal clonal populations of tumor cells that may spread to perithyroidal lymph nodes, paratracheal lymph nodes, nodes of the jugular chain, and upper mediastinal nodes. MTC may remain confined to the neck for long periods of time. In more advanced stages of disease, metastases are found in liver, lungs, and bone. Histologic features of aggressiveness include vascular invasion, lymphatic invasion, invasion of the thyroid capsule, and extranodal spread of tumor in lymph node metastases. The disease can be locally aggressive, and local structures that may be invaded by tumor include the trachea, jugular vein, strap muscles, posterior cervical fascia, and recurrent laryngeal nerves. Erosion of tumor into the trachea can cause airway obstruction and death.

Screening and follow-up of patients with MTC has been greatly facilitated by the fact that these tumors secrete calcitonin, which is easily measured in the blood. The most sensitive way to test for plasma calcitonin is after the administration of the provocative agents calcium and pentagastrin (Wells et al., 1978). After obtaining basal levels, intravenous calcium (2 mg/kg/per 1 min) is infused, followed immediately by pentagastrin (0.5 μm/kg/5 per sec), and then blood is drawn for measurement of plasma calcitonin levels at 1, 3, and 5 minutes. In families affected by hereditary forms of MTC, the institution of biochemical and genetic screening of at-risk family members has resulted in the routine detection of MTC before a mass is palpable in the neck. In those cases, metastases are rarely present at the time of thyroidectomy. Reliance on the calcium–pentagastrin-stimulated calcitonin test for clinical screening of kindred members at risk for development of MTC has certain drawbacks, however. First, a positive-stimulated calcitonin test usually indicates that cancer has already developed, and there are a few patients who will develop distant metastatic disease, even after thyroidectomy with removal of tiny (1 mm) primary tumors. These patients would benefit from earlier thyroidectomy. Second, because the disease is inherited in an autosomal dominant fashion, 50% of at-risk patients will never develop disease and would be spared the expense and inconvenience of routine scheduled testing if a definitive genetic test were applied. Third, the provocative calcitonin test is unpleasant and uncomfortable and some patients do not keep scheduled testing appointments because of this. In principle, genetic testing, which requires the drawing of blood for extraction of lymphocyte DNA, need be performed only once in an at-risk individual's lifetime. Stimulated calcitonin testing remains an important modality in following patients for recurrent or residual disease after thyroidectomy. Genetic testing, however, has made possible earlier identification of mutant gene carriers, and hopefully this will obviate the need to continue testing those family members who have not inherited the mutation. Control of MTC may be more effective because surgical intervention will be carried out before there is clinical or biochemical evidence of tumor.

A number of techniques have been described to look for RET mutations in the blood or tissue of patients. These include direct DNA sequencing, analysis of restriction sites introduced or deleted by a mutation, or by gel electrophoresis analyses (denaturing gradient gel electrophoresis or single-strand conformation polymorphism analysis) (Donis-Keller et al., 1993; Mulligan et al., 1993a; Wells et al., 1994; Decker and others, in press).

THYROIDECTOMY IN CARRIERS OF RET MUTATIONS

Because patients with MEN 2A and FMTC are virtually certain to develop MTC at some point in their lives, at-risk family members who are found to have inherited the RET gene mutation are candidates for thyroidectomy, regardless of their stimulated plasma calcitonin levels. We recently were involved in a study to evaluate the application of thyroidectomy to patients found to carry mutations in the RET gene (Wells et al., 1994). Of 132 individuals from seven kindreds affected by MEN 2A, 48 had an established diagnosis of MEN 2A and 58 were at 50% risk for inheriting the disease but had no clinical evidence of endocrine neoplasia. All individuals were evaluated by both direct and indirect genetic testing for the presence of RET gene mutations. In patients at 50% risk for disease (having an affected parent or sibling), calcium–pentagastrin-stimulated calcitonin levels were also obtained. PCR- and sequence-based direct mutation analysis of genomic DNA from the 58 individuals at 50% risk for MEN 2A identified 21 individuals who had inherited a RET mutation associated with disease. The other 37 family members had two normal RET alleles. All 26 unaffected control individuals had normal RET alleles. Total thyroidectomy was offered to all 21 individuals who were found to have inherited a RET mutation. In 12 of the 21 patients, the stimulated plasma calcitonin levels were within normal limits. In the remaining nine patients the stimulated plasma calcitonin levels were elevated.

In six genetically positive individuals with normal plasma calcitonin levels, and in seven genetically positive individuals with elevated levels, total thyroidectomy, lymph node dissection, and parathyroid autotransplantation were performed. Of the seven patients whose preoperative plasma calcitonin levels were elevated, each had microscopic evidence of MTC on histologic examination. Two of the seven had macroscopic disease. Of the six patients with normal preoperative plasma calcitonin levels, either macroscopic MTC ($n = 1$), microscopic MTC ($n = 2$), or C-cell hyperplasia only ($n = 3$) were evident. A total of 212 lymph nodes were resected from the central zone of the neck (14.5 per patient) and, on histologic examination, none contained metastases. In each of the 13 patients, the stimulated plasma calcitonin levels were normal after total thyroidectomy.

Lips et al. (1994b), from the Netherlands, described a similar series. In that report, 14 young members of families affected by MEN 2A who had normal calcitonin testing but who were found to be MEN 2A gene carriers by DNA testing were offered thyroidectomy. Thyroidectomy was done on 8 of these 14, and foci of MTC were identified in all 8.

Because cancer was found in the glands of patients with normal stimulated calcitonin testing, there is some urgency to apply this genetic test to other at-risk individuals and performing thyroidectomy on those who test positive genetically. The ideal age for performance of thyroidectomy in these patients has not been determined unequivocally. We recommend that patients who have inherited a MEN 2A or FMTC mutation should undergo thyroidecytomy by age 5 or 6 years. Patients with MEN 2B (in whom the diagnosis is apparent by physical examination) should undergo thyroidectomy during infancy because of the aggressiveness and earlier age of onset of MTC in these patients (O'Riordain et al., 1994).

SUMMARY

In the last 2 years there have been tremendous advances in our understanding of the genetics of the MEN 2 syndromes and FMTC. As discussed in this chapter, the ability to directly detect mutations that result in MEN 2A, MEN 2B, and FMTC has changed how clinicians care for members of families in which these diseases are segregating.

We are just beginning to understand the normal function of RET. The *in vitro* studies of RET that we reviewed in this chapter provide strong evidence that mutant RET can lead to aberrant signal transduction. *In vivo* studies of both normal and mutant forms of RET have provided important clues as to the gene's normal functions. In mouse, *ret* is expressed in the developing nervous and excretory systems (Pachnis et al., 1993). Mice lacking *ret* (deficient in kinase activity) were recently described (Pachnis et al., 1993). They show severe defects in kidney and enteric nervous system development that result in death shortly after birth. The defect in the enteric nervous system is consistent with what is proposed to be an absence of normal RET activity in Hirschsprung's disease in humans (Edery et al., 1994; Romeo et al., 1994). Hirschsprung's patients are characterized by an absence of autonomic ganglion cells in the hindgut.

The identification of the RET ligand and the downstream members of the signal transduction pathway of which RET is a component are necessary to understand how mutant forms of the protein bring about tumor formation. Transgenic mice carrying either the human MEN 2A and MEN 2B mutations or the murine equivalents are equally likely to provide important clues as to how altered RET function contributes to tumorigenesis.

REFERENCES

Brooks-Wilson AR, Lichter JB, Ward DC, Kidd KK, Goodfellow PJ (1993): Genomic and yeast artificial chromosome long-range physical maps linking six loci in 10q11.2 and spanning the multiple endocrine neoplasia type 2A (MEN 2A) region. *Genomics* 17:611–617

Cance WG, Wells SA Jr (1985): Multiple endocrine neoplasia type 2a. *Curr Prob Surg* 21:1

Carlson KM, Dou S, Chi D, Scavarda N, Toshima K, Jackson C, Wells SA Jr, Goodfellow PJ, Donis-Keller H (1994a): Single missense mutation in the tyrosine kinase catalytic domain of the RET protooncogene is associated with multiple endocrine neoplasia type 2B. *Proc Natl Acad Sci USA* 91:1579–1583

Carlson KM, Bracamontes J, Jackson CE, Clark R, Lacroix A, Wells SA, Goodfellow P (1994b): Parent-of-origin effects in multiple endocrine neoplasia type 2B. *Am J Hum Genet* 55:1076–1082

Decker RA, Peacock ML, Borst MJ, Sweet JD, Thompson NW (in press): Progress in genetic screening of MEN 2A: Is calcitonin testing obsolete? *Ann Surg*

Donis-Keller H, Dou S, Chi D (1993): Mutations in the RET proto-oncogene are associated with MEN 2A and FMTC. *Hum Mol Genet* 2:851–856

Edery P, Lyonnet S, Mulligan LM, Pelet A, Dow E, Abel L, Holder S, Nihoul-Fekete C, Ponder BAJ, Munnich A (1994): Mutations of the RET proto-oncogene in Hirschprung's disease. *Nature* 367:378–380

Eng C, Smith DP, Mulligan LM, Nagai MA, Healey CS, Ponder MA, Gardner E, Scheumann GF, Jackson CE, Tunnacliffe A (1994): Point mutation within the tyrosine kinase domain of the RET proto-oncogene in multiple endocrine neoplasia type 2B and related sporadic tumours [published erratum appears in *Hum Mol Genet* 1994; 3(4):686]. *Hum Mol Genet* 3:237–241

Farndon, JR, Leight GS, Dilley WG, Baylin SB, Smallridge RC, Harrison TS, Wells SA (1986): Familial medullary thyroid carcinoma without associated endocrinopathies: a distinct clinical entity. *Br J Surg* 73:278–281

Gardner, E, Papi L, Easton DF, Cummings T, Jackson CE, Kaplan M, Love DR, Mole SE, Moore JK, Mulligan LM, Norum RA, Ponder MA, Reichlen S,

Stall G, Telenius H, Telenius-Berg M, Tunnacliffe A, Ponder BAJ (1993): Genetic linkage studies map the multiple endocrine neoplasia type 2 loci to a small interval on chromosome 10q11.2. *Hum Mol Gen* 2:241–246

Goodfellow PJ, Myers S, Anderson LL, Brooks-Wilson AR, Simpson LL (1990): A new DNA marker (D10S94) very tightly linked to the multiple endocrine neoplasia type 2A (MEN 2A) locus. *Am J Hum Genet* 47:952–956

Hofstra RMW, Landsvater RM, Ceccherini I, Stulp RP, Stelwagen T, Luo Y, Pasini B, Hoppener JWM, Van Amstel HKP, Romeo G, Lips JM, Buys CHCM (1994a): A mutation in the RET protooncogene associated with multiple endocrine neoplasia type 2B and sporadic medullary thyroid carcinoma. *Nature* 367:375–376

Hofstra RM, Landsvater RM, Ceccherini I, Stulp RP, Stelwagen T, Luo Y, Pasini B, Hoppener JW, Amstel Hk van, Romeo G (1994b): A mutation in the RET protooncogene associated with multiple endocrine neoplasia type 2B and sporadic medullary thyroid carcinoma [see comments]. *Nature* 367:375–376

Ishizaka Y, Ushijima T, Sugimura T, Nagao M (1990): cDNA cloning and characterization of ret activated in a human papillary thyroid carcinoma cell line. *Biochem Biophys Res Comm* 168:402–408

Kwok JBJ, Gardner E, Warner JP, Ponder BAJ, Mulligan LM (1993): Structural analysis of the human Ret protooncogene using exon trapping. *Oncogene* 8:2575–2582

Lairmore TC, Howe JR, Korte JA, Dilley WG, Aine L, Aine E, Wells SA, Donis-Keller H (1991): Familial medullary thyroid carcinoma and multiple endocrine neoplasia type 2B map to the same region of chromosome 10 as multiple endocrine neoplasia type 2A. *Genomics* 9:181–192

Lairmore TC, Howe JR, Korte JA (1992): Familial medullary thyroid carcinoma and multiple endocrine neoplasia type 2B map to the same region of chromosome 10 as multiple endocrine neoplasia type 2A. *Genomics* 9:181–192

Lairmore TC, Dou S, Howe JR (1993): A 1.5 megabase contig of yeast artificial chromosome clones containing three loci (RET, D10S94, and D10S102) closely linked to the MEN 2A locus. *Proc Natl Acad Sci USA* 90:492–496

Landsvater RM, Mathew CGP, Smith BA, Marcus EM, Meerman GJ, Lips CJM, Geerdink RA, Nakamura Y, Ponder BAJ, Buys CHCM (1989): Development of multiple endocrine neoplasia type 2A does not involve substantial deletions of chromosome 10. *Genomics* 4:246–250

Lichter JB, Wu J, Brewster S, Brooks-Wilson AR, Goodfellow PJ, Kidd KK. A new polymorphic marker (D10S97) tightly linked to MEN 2A. *Am J Hum Gen* 49:414

Lichter JB, Wu J, Miller D, Goodfellow PJ, Kidd KK (1992). A high-resolution meiotic mapping panel for the pericentromeric region of chromosome 10. *Genomics* 13:607–612

Lips CJM, Landsvater RM, Hoppener JWM (1994a): Clinical screening as compared with DNA analysis in families with multiple endocrine neoplasia type 2A. *N Engl J Med* 331:828–835

Lips CJM, Landsvater RM, Hoppener JWM, Geerdink RA, Blijham G, Jansen-Schillhorn van Veen JM, van Gils APJ, DeWit MJ, Zewald RA, Berends MJH, Beemer FA, Brouwers-Smalbraak J, Jansen RPM, van Amstel HKP, van Vroonhoven TJMV, Vroom TM (1994b): Clinical screening as compared with DNA analysis in families with multiple endocrine neoplasia type 2A. *N Engl J Med* 331:828–835

Mathew CGP, Chin KS, Easton DF (1987a): A linked genetic marker for multiple endocrine neoplasia type 2A on chromosome 10. *Nature* 328:527–528

Mathew CGP, Chin KS, Easton DF, Thorpe K, Carter C, Liou GI, Fong SL, Bridges CDB, Haak H, Nieuwenhuijzen-Kruseman AC (1987b): A linked genetic marker for multiple endocrine neoplasia type 2A on chromosome 10. *Nature* 328:527–528

Mathew CGP, Smith BA, Thorpe K, Wong Z, Royle NJ, Jeffreys AJ, Ponder BAJ (1987c). Deletion of genes on chromosome 1 in endocrine neoplasia. *Nature* 328:524–526

Mathew CPG, Easton DF, Nakamura Y, Ponder BAJ (1991a). Presymptomatic screening for multiple endocrine neoplasia type 2A with linked DNA markers. *Lancet* 337:7–11

Mathew CGP, Easton DF, Nakamura Y, Ponder BAJ (1991b): Presymptomatic screening for multiple endocrine neoplasia type 2A with linked DNA markers. *Lancet* 337:7–11

Mole SE, Mulligan LM, Healey CS, Ponder BAJ, Tunnacliffe A (1993): Localization of the gene for multiple endocrine neoplasia type 2A to a 480 kb region in chromosome band 10q11.2. *Hum. Mol. Gen.* 2:247–252

Mulligan LM, Gardner E, Mole SE, Nakamura Y, Papi L, Telenius H, Ponder BAJ (1991): Is the ret protooncogene a candidate for the MEN2 gene? *Am J Hum Genet* 49:414

Mulligan LM, Kwok JBJ, Healy CS (1993a): Germ-line mutations of the RET protooncogene in multiple endocrine neoplasia type 2A (MEN 2A). *Nature* 363:458–460

Mulligan LM, Gardner E, Smith BA, Mathew CGP, Ponder BAJ (1993b) Genetic events in tumour initiation and progression in multiple endocrine neoplasia type 2. *Genes Chromosomes Cancer* 6:166–167

Mulligan LM, Eng C, Healey CS, Clayton D, Kwok JB, Gardner E, Ponder MA, Frilling A, Jackson CE, Lehnert H (1994): Specific mutations of the RET proto-oncogene are related to disease phenotype in MEN 2A and FMTC. *Nat Genet* 6:70–74

Nelkin BD, Nakamura Y, White RW, de Bustros AC, Herman J, Wells SA, Baylin SB (1989): Low incidence of loss of chromosome 10 in sporadic and hereditary human medullary thyroid carcinoma. *Cancer Res* 49:4114–4119

Norum RA, Lafreniere RG, O'Neal LW (1990a): Linkage of the multiple endocrine neoplasia type 2B gene (MEN 2B) to chromosome 10 markers linked to MEN 2A. *Genomics* 8:313–317

Norum RA, Lafreniere RG, O'Neal LW, Nikolati F, Delaney JP, Sisson JC, Sobol H, Lenoir GM, Ponder BAJ, Willard HF, Jackson CE (1990b): Linkage of the multiple endocrine neoplasia type 2B gene (MEN 2B) to chromosome 10 markers linked to MEN 2A. *Genomics* 8:313–317

O'Riordain D S, O'Brien T, Weaver AL, Gharib H, Hay ID, Grant CS, Heerden Ja van (1994): Medullary thyroid carcinoma in multiple endocrine neoplasia types 2A and 2B. *Surgery* 116:1017–1023

Pachnis V, Mankoo B, Costantini F (1993): Expression of the *c-ret* proto-oncogene during mouse embryogenesis. *Submitted*

Romeo G, Ronchetto P, Luo Y, Barone V, Seri M, Ceccherini I, Pasini B, Bocciardi R, Lerone M, Kaariainen H, Martucciello G. Point mutations affecting the tyrosine kinase domain of the RET proto-oncogene in Hirschprung's disease. *Nature* 367:377–378

Saad MF, Ordonex NG, Rashid RK, Guido JJ, Hill CS, Hickey RC, Samaan NA (1984): Medullary carcinoma of the thyroid: a study of the clinical features and prognostic factors in 161 patients. *Medicine* 63:319–342

Santoro M, Rosati R, Grieco M, Berlingieri MT, D'Amato GL, de Franciscis V, Fusco A (1990): The *ret* proto-oncogene is consistently expressed in human pheochromocytomas and thyroid medullary carcinomas. *Oncogene* 5:1595–1598

Santoro M, Carlomagno F, Romano A, Botttaro DP, Dathan NA, Grieco M, Fusco A, Vecchio G, Matoskova B, Kraus MH, Di Fiore PP (1995): Activation of RET as dominant transforming gene by germline mutations of MEN 2A and MEN 2B. *Science* 267:381–383

Schimke, RN (1984): Genetic aspects of multiple endocrine neoplasia. *Ann Rev Med* 35:25–31

Simpson NE, Kidd KK, Goodfellow PJ (1987): Assignment of multiple endocrine neoplasia type 2A to chromosome 10 by genetic linkage. *Nature* 328:527–528

Songyang Z, Carraway III KL, Eck MJ, Harrison SC, Feldman RA, Mohammadi M, Schlessinger J, Hubbard SR, Smith DP, Eng C, Lorenzo MJ, Ponder BAJ, Mayer BJ, Cantley LC (1995): Catalytic specificity of protein-tyrosine kinases is critical for selective signalling. *Nature* 373:536–539

Takahashi M, Cooper GM (1987): *ret* transforming gene encodes a fusion protein homologous to tyrosine kinases. *Mol Cell Biol* 7:1378–1385

Takahashi M, Buma Y, Iwamoto T, Inaguma Y, Ikeda H, Hiai H (1988): Cloning and expression of the ret proto-oncogene encoding a tyrosine kinase with two potential transmembrane domains. *Oncogene* 3:571–578

Wells SA, Baylin SB, Linehan WM, Farrel RE, Cox EB, Cooper CW (1978): Provocative agents and the diagnosis of medullary carcinoma of the thyroid gland. *Ann Surg* 188:139–141

Wells SA Jr,Chi DD, Toshima K, Dehner LP, Coffin CM, Dowton SB, Ivanovich JL, DeBenedetti MK, Moley JF, Donis-Keller H (1994): Predictive DNA testing and prophylactic thyroidectomy in patients at risk for multiple endocrine neoplasia type 2A. *Ann Surg* 220:237–247

Wingo PA, Tong T and Bolden S (1995): Cancer Statistics, 1995. *CA Cancer J Clin* 45:8–30.

4

The Nuclear Hormone Receptor Superfamily: Structure and Function

XIAOHUA LENG, SOPHIA Y. TSAI AND MING-JER TSAI

INTRODUCTION

Members of the nuclear receptor superfamily are ligand-dependent transcription factors that regulate the expression of target genes via binding to specific *cis*-acting elements (Evans, 1988; Green and Chambon, 1988; Beato, 1989; O'Malley, 1990; Parker, 1990; Wahli and Martinez, 1991; Tsai and O'Malley, 1994). Members of this superfamily respond to endocrine, paracrine, and possibly autocrine signals and therefore modulate diverse aspects of development, differentiation, homeostasis, and behavior in vertebrates. The superfamily consists of receptors for steroid hormones (e.g., estrogens, progestins, androgens, and corticosteroids), steroid derivatives (dihydroxyl vitamin D_3), and nonsteroids (retinoids and thyroid hormone). It also includes a growing number of structurally related proteins for which their ligands have yet to be identified, referred to as "orphan receptors," or members that have lost ligand binding function (e.g., thyroid hormone receptor $\alpha2$ and $\alpha3$; TR$\alpha2$ and TR$\alpha3$). In addition, a group of proteins that regulate a variety of developmental pathways in invertebrates, mostly *Drosophila melanogaster*, also has been classified as members of this superfamily (Table 1) (Amero et al., 1992; Laudet et al., 1992; Parker, 1993; Lutz et al., 1994).

The nuclear receptor superfamily can be divided into subgroups on the basis of either evolutionary relationships, characteristic amino acid sequence motifs, or functional criteria, such as their cognate hormone response elements (HREs). Group A includes all the classical steroid hormone receptors. In the absence of ligands, these receptors are complexed with heat shock proteins and are not able to bind DNA. Binding of hormone

Hormones and Cancer
Wayne V. Vedeckis, Editor

Table 1. Nuclear receptor superfamily

Hormone receptors	Orphan receptors	*Drosphila* proteins
Glucocorticoid	COUP-TFI (EAR3), COUP-TFII (ARP1)	*Seven-up*
	ERR1, ERR2	
Mineralcorticoid	SF-1 (ELP)	E75
	NGFI-B (Nur77)	
Androgen	nurr1 (RNR-1, NOT)	*Fushi-tarazu*
	ROR	
Progesterone	RVR	*DHR3*
	DAX-1, Odr-7	
Estrogen	TR2, TR4 (TR2r1)	*Embryonic gonad*
	HNF-4	
Thyroid hormone	Rev-Erbα (EAR1), BD73	*Knirps*
	EAR2	
Vitamin D_3	LXR	*Knirps*-related
	RLD-1	
All-*trans* retinoic acid	UR (TAK1, OR)	*Ecdysone* receptor
	MB67	
9-*cis* retinoic acid	PPAR, NUC1	*Ultraspiracle*
	RIP14, RIP15	
Melatonin (RZR)	GCNF	*Tailless*
	Tlx	

ARP ApoA1 regulatory protein, EAR, erbA-related; ERR, ER-related; HNF, hepatocyte nuclear factor; SF, steroidogenic factor; NGFI, nerve growth factor–induced factor; RNR, regenerating liver nuclear receptor; GCNF, germ cell nuclear factor; DHR3, *Drosophila* hormone receptor 3.

transforms the receptor into an active form by mediating the dissociation of at least some of the heat shock proteins and allowing the dimerization of receptor proteins (Allan, 1994, and the references therein). The activated receptor is now able to bind the HRE as a homodimer and stimulate, or in some cases repress, the expression of target genes. Group B receptors, which consist of the receptors for thyroid hormone (TR), retinoic acids (RAR and RXR), vitamin D (VDR), peroxisome proliferators (PPAR) and ecdysone (*EcR*), are located in the nucleus and are consistently associated with the chromatin. These receptors can bind their HREs as homo- or heterodimers in the absence of hormone, and at least some members inhibit the basal activity of the *cis*-linked gene (silencing). Hormone binding, however, drastically alters the activity of these receptors. Some of the orphan receptors, such as NGFI-B/Nur77, ELP/SF-1, *FTZ-F1*, and Rev-ErbAα define yet another group of receptors (group C). These receptors can bind as monomers to extended, single PuGG($^T/_A$)CA motifs and may be constitutively active. It is not clear if these receptors can form homodimers or heterodimerize with other nuclear receptors on other DNA elements. Finally, an unexpected variety of isoforms of PR, ER, RAR, RXR, TR, and many orphan receptors such as COUP-TF, *EcR*, PPAR, and *FTZ-F1* have been identified. These isoforms often have distinct spatial and temporal expression patterns, indicating that they may play a variety of physiological roles (Ohno and Petkovich, 1992; Rowe et al., 1991; Ruberte et al., 1991; Bradley et al., 1994; Qiu et al., 1994a,b).

STRUCTURAL AND FUNCTIONAL DOMAINS OF NUCLEAR HORMONE RECEPTORS

Amino acid sequence analysis indicates that members of the nuclear hormone receptor superfamily share common structural motifs, thus suggesting that they are evolutionarily linked and may be derived from a common ancestor receptor. Intermolecular sequence comparison and mutational dissection of these nuclear hormone receptors led to the definition of six regions of the primary structure, commonly referred to as regions A through F (Fig. 1). Each one of these regions may harbor one or several functional domains.

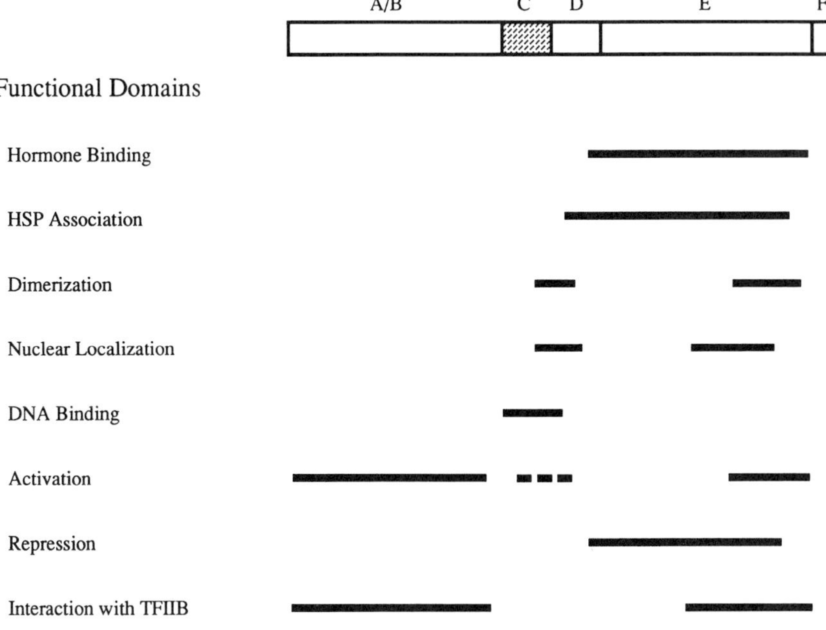

Figure 1. Schematic illustration of the structure-function organization of nuclear hormone receptors. The structure of nuclear receptors can be divided into six regions: A through F. The summary of localized functional domains include DNA binding, hormone binding, heat-shock protein association, nuclear localization, activation, repression, dimerization, and interaction with TFIIB. Not every receptor contains all the listed regions and domains.

A/B Region

The N-terminal half of the receptor contains the A/B region, which is highly variable in sequence and length. It is the major determinant of the size differences among the receptors and ranges from 24 amino acids (aa) in the VDR to 603 aa in the mineralocorticoid receptor (MR). Domain swapping and deletion studies have revealed that the A/B region contains a transferable transactivation function (AF) (Hollenberg and Evans, 1988; Tora et al., 1989; Tasset et al., 1990; Sartorius et al., 1994), which activates the target gene presumably by interacting with components of the core transcriptional machinery, coactivators and other transactivators (Bagchi, 1994, and the references therein). This region may also be important for determining cell-type and receptor isoform-specific activation of target genes (Bocquel et al., 1989; Tora et al., 1989; Folkers et al., 1993).

Regions C and D

Region C includes the conserved DNA-binding domain (DBD). It represents the DNA-binding motif required for sequence specific recognition and binding of the receptor to HREs. This 66-amino acid DBD is the most conserved region, with over half the amino acids being highly conserved among all the receptors. The positions of the Cys residues are absolutely conserved, yielding two type II zinc fingers: CXXC-X(13)-CXXC-X(15–17)-C-X(5)-C-X(12)-C-X(4)-C. It has since become the most easily recognizable and, thus, defining structure of the nuclear receptor superfamily. Many new members were identified on the basis of homology in the DBD and possession of two type II zinc modules, including RAR, v-erbA, and COUP-TF (Debuire et al., 1984; Weinberger et al., 1985, 1986; Giguere et al., 1987; Petkovich et al., 1987; Wang et al., 1989).

The two zinc fingers are distinct from each other both structurally and functionally. The N-terminal zinc motif is involved in DNA sequence recognition by providing the base-specific contacts with the nucleotides in the major groove of the response elements. The C-terminal zinc module mediates high-affinity DNA binding by providing the phosphate backbone contacts and the dimerization interface between the two DBDs bound at the response element (Umesono and Evans, 1989; Schwabe et al., 1990, 1993; Luisi et al., 1991; Lee et al., 1993). This dimerization interface may account for some of the cooperative DNA binding ability of the receptor (Tsai et al., 1988, 1989; Dahlman-Wright et al., 1990; Bradshaw et al., 1991).

In vitro transcription assays demonstrate that the receptor DBD contains a weak transactivation domain (Freedman et al., 1989; Klein-Hitpass et al., 1990). In addition, a nuclear localization signal is also localized to this region (Picard and Yamamoto, 1987). Thus, region C is multifunctional and contains sequence for specific DNA recognization, information for dimerization and cooperative DNA binding, a weak transcriptional activation function and a nuclear localization signal (Fig. 1).

Region D, also called the "hinge-region," is highly variable in length and less conserved in sequence. Some amino acids of region D are required for efficient DNA binding, indicating that flanking amino acid sequences of the DBD are also essential for DNA binding (Mader et al., 1993; unpublished data of X. Leng, B. W. O'Malley, and M.-J. Tsai). Finally, deletion studies have shown that this region is also important, at least in TR and RAR, for hormone binding (Lin et al., 1991; Leng et al., 1993).

Regions E and F

The receptor C-terminus, which is the most complicated portion of the receptor in terms of function and structure, is subdivided into regions E and F (the F region is only present in some receptors, i.e., ER, RAR, and PPAR). Alignment of nuclear hormone receptor superfamily cDNA sequences on the basis of maximum similarity led to the identification of three highly conserved regions: I, II, and III in all receptor members (Wang et al., 1989). Region I resides in the DNA binding motif of region C, whereas regions II and III, each with 43 and 26 amino acids in length, respectively, are localized to the region E (Fig. 2). This strong amino acid conservation indicates that they are likely involved in receptor ligand binding and transactivation functions.

Indeed, the hormone binding function is mapped to the E region. Intermolecular and intramolecular "domain swapping" experiments with GR, TR, and ER demonstrate that this region is a functionally independent region, as it can be transferred to a different part of the receptor and remain entirely functional (Kumar et al., 1987; Freedman et al., 1989; Thompson and Evans, 1989). In addition, the C-termini of various receptors have been fused to a heterologous DBD (i.e., DBD of the yeast Gal4 transcription factor) to create chimeric receptors. All these Gal4-receptor chimeras can thus induce gene expression through a *cis*-linked Gal4 binding site upon addition of the cognate hormone (Webster et al., 1988; Hollenberg and Evans, 1988; Baniahmad et al., 1992a; Leng et al., 1995). Similarly, fusion of the estrogen receptor C-terminus to the oncogene myc led to hormone-dependent cell transformation (Eilers et al., 1989). The borders of the ligand binding domain (LBD) have been precisely defined by utilization of the Gal4-receptor chimeras and more recently by partial proteolytic analysis (Kumar et al., 1986; Leng et al., 1993, 1995). These studies indicate that the minimal LBD comprises almost the entire E region and part of the D region.

In addition to hormone binding function, the E region also contains a transactivation function. For most nuclear hormone receptors, the ligand-dependent transactivation function is mapped within the ligand-binding domain of the receptor (Kumar et al., 1987; Hollenberg and Evans, 1988; Baniahmad et al., 1995). Almost all mutations that compromise the ligand binding ability also impair the transactivation function of the receptor. Thus, it seems that the transactivation function of the C-terminus requires the binding of hormone. Interestingly, for RXR, the LBD and at least one of its transactivation domains can be separated from each other, thus indicating that the E region represents a functional unit with multifunctional domains of the receptor (Leng et al., 1995). In fact, a domain for the hormone-dependent nuclear localization signal (NLS) is located in the E region (Picard et al., 1988; Guiochon-Mantel et al., 1989). This NLS domain shows sequence homology to the NLS of SV40 large T antigen.

The E region also contains a dimerization domain. This dimerization domain is much stronger than the one in the DBD (Kumar and Chambon, 1988; Farwell et al., 1990a). Parker and co-workers have characterized this dimerization domain in ER and identified a leucine-rich fragment that is important for receptor dimerization (Fawell et al., 1990a). Sequence comparison among all nuclear hormone receptors over this leucine-rich fragment reveals a conserved nine heptad–repeat of hydrophobic residues, which suggests that receptor dimerization may involve a coiled-coil type dimerization interphase similar to the leucine zipper of the C/EBP protein (Landschultz et al., 1988; Lees et al., 1990). Deletion analysis with TR and RXR indicates that the ninth "heptad" is important for heterodimer formation and is involved in the repression function of these receptors (Au-Fliegner et al., 1993; Leng et al., 1995).

Finally, a transcriptional inhibitory function is localized in the C-terminus of group B receptors such as TR and RAR (Baniahmad et al., 1992a). Although generally these receptors activate gene transcription in the presence of hormone, in the absence of ligands TR and RAR repress basal promoter activity in the same way as the v-erbA oncogene product (Damm et al., 1989; Sap et al., 1989). Repression of basal promoter activity also has been demonstrated for COUP-TF (Cooney et al., 1993) and recently for RXR mutants (Leng et al., 1995). The repression function of these receptors has been analyzed and a transferable silencing domain is localized in the E region of these receptors. Further studies show that a minimal promoter, which contains only a TATA-box, is a target for repression *in vivo* and *in vitro* (Baniahmad et al., 1992a; Fondell et al., 1993). These results indicate that silencing may involve direct interactions between nuclear hormone receptors and basal transcription factors. Indeed, TR and RAR are able to interact specifically with the basal transcription factors TFIIB and TBP

Figure 2. Alignment of members of the nuclear hormone receptor superfamily using COUP-TFI as the basis for comparison. Three conservative regions (I, II, III) are most recognizable. Region I is the DBD and Regions II and III localize in the LBD. Numbers indicate the percentage of identity.

The Nuclear Hormone Receptor Superfamily

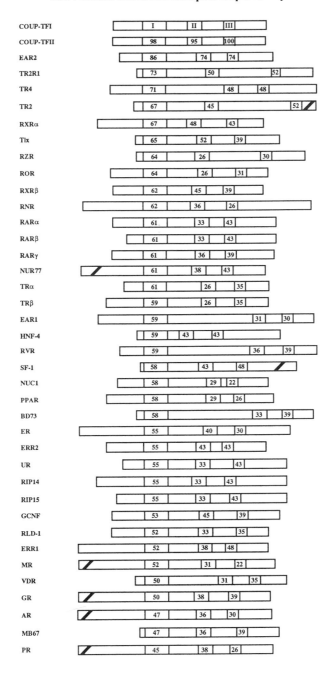

COUP-TFI	I	II	III
COUP-TFII	98	95	100
EAR2	86	74	74
TR2R1	73	50	52
TR4	71	48	48
TR2	67	45	52
RXRα	67	48	43
Tlx	65	52	39
RZR	64	26	30
ROR	64	26	31
RXRβ	62	45	39
RNR	62	36	26
RARα	61	33	43
RARβ	61	33	43
RARγ	61	36	39
NUR77	61	38	43
TRα	61	26	35
TRβ	59	26	35
EAR1	59	31	30
HNF-4	59	43	43
RVR	59	36	39
SF-1	58	43	48
NUC1	58	29	22
PPAR	58	29	26
BD73	58	33	39
ER	55	40	30
ERR2	55	43	43
UR	55	33	43
RIP14	55	33	43
RIP15	55	33	43
GCNF	53	45	39
RLD-1	52	33	35
ERR1	52	38	48
MR	52	31	22
VDR	50	31	35
GR	50	38	39
AR	47	36	30
MB67	47	36	39
PR	45	38	26

(Baniahmad et al., 1993; Fondell et al., 1993; unpublished data of X. Leng, B. W. O'Malley, and M.-J. Tsai). Interestingly, hormone binding interferes with this interaction (Baniahmad et al., 1993). Thus, one possible mechanism of silencing may be due to unliganded receptor sequestering TFIIB in a non-functional configuration (Baniahmad et al., 1993; Fondell et al., 1993).

The F region, located at the C-terminal end, is only present in certain receptors. For ER, deletion studies indicate that the F region is not required for any known function of the receptor (Kumar et al., 1987). The F-region of RAR, however, blocks the silencing function of RAR in a cell specific manner (Baniahmad et al., 1992a).

In summary, the receptor C-terminus harbors multiple functions including, hormone binding, transactivation, dimerization, nuclear localization and, for some receptors, a silencing function. Domains for these functions have been delineated and examined as functional units that can be transferred to heterologous proteins. Thus, analyses of nuclear hormone receptor functional domains adds substantially to our general understanding of protein structure-function relationship. It becomes obvious that all receptors possess this modular structure and these modules still retain their functional properties even when separated from the natural receptor context. One should be cautious, however, because the functional modulation of a given domain by other domains is lost when it is separated from the natural receptor context.

NUCLEAR RECEPTOR DNA INTERACTIONS

In response to inter- or intracellular signals, nuclear receptors selectively regulate a set of gene promoters, thus controlling diverse responses involved in differentiation, development, and homeostasis. One of the key steps in these processes is receptor-DNA interaction. Nuclear receptors mediate transcription by specifically recognizing cognate nucleotide sequences, termed "hormone response elements" (HREs). These HREs provide hormone- and receptor-dependent transcriptional regulation on a gene as enhancer elements. Thus, in general, the specificity of the hormonal regulation is determined by the sequence of HREs and the structure of nuclear receptors.

Formation of Dimers

For most nuclear receptors, one of the distinct features for receptor-DNA interaction is that receptors bind to their target sites as dimers. Dimer formation, in general, enhances the efficiency of receptor-HRE interaction, possibly through cooperative DNA binding. In fact, upon binding ligand, all group A receptors form stable homodimers and two receptor molecules can thus cooperate with each other in recognizing palindromic HREs. No monomer DNA binding has been detected with these receptors. Unlike group A receptors, the homodimer DNA binding of TR, RAR, and VDR is relatively weak (Glass et al., 1989, 1990; Lazar et al., 1991). However, it is known that the binding of TR (Murry and Towle, 1989; Darling et al., 1991; O'Donnell et al., 1991), RAR (Glass et al., 1989), and VDR (Liao et al., 1990) to their cognate response elements can be greatly enhanced by a nuclear factor(s) present in a variety of cells. Cloning of one of these proteins revealed it to be a retinoid X receptor (RXR) (Yu et al., 1991; Leid et al., 1992). It appears that RXR, which binds 9-*cis* retinoic acid, stimulates the binding of TR, RAR, VDR, and PPAR to their corresponding response elements by the formation of heterodimers (Yu et al., 1991; Bugge et al., 1992; Kliewer et al., 1992c, 1992b; Leid et al., 1992; Marks et al., 1992; Zhang et al., 1992). In addition, RXR can also heterodimerize with COUP-TF (Kliewer et al., 1992a; Widom et al., 1992; Cooney et al., 1993). Heterodimers between RXR and TR or RAR can form in solution through an interaction between the LBDs, and heterodimerization is further stabilized by DNA binding (Marks et al., 1992; Kurokawa et al., 1993). Transient transfection studies with wild-type and dominant-negative mutants suggest that heterodimers are more important in transcriptional regulation than homodimers (Durand et al., 1992). Thus, it is proposed that some *in vivo* actions of thyroid hormone, retinoids, and vitamin D_3 may be mediated by heterodimers that contain RXR as a common partner. In view of this promiscuous interaction ability of RXR, one is tempted to speculate that it is also capable of interacting with receptors in group C. In *Drosophila*, *ultraspiracle* (*usp*) appears to be an invertebrate homologue of RXR because it can substitute for RXR in forming the heterodimer with TR, RAR, and VDR. Moreover, it heterodimerizes with another *Drosophila* protein, the *EcR*, and stimulates its DNA binding, ligand binding and transactivation (Yao et al., 1992). In view of the importance of heterodimerization in both vertebrates and invertebrates, Evans and co-workers suggest that RXR/*usp* plays a central role in many hormonal signaling pathways (Yao et al., 1993).

However, RXR is not the only receptor capable of forming heterodimers with other members of group B receptors. TR (Umesono et al., 1988; Schrader et al., 1994b), COUP-TF (Berrodin et al., 1992), and PPAR (Bogazzi et al., 1994) all have been reported to form heterodimers with other receptors. Interestingly, although heterodimerization with RXR usually results in enhanced transactivation, COUP-TF seems to function primarily as a transcriptional repressor (Cooney et al., 1992, 1993; Kliewer et al., 1992a; Tran et al., 1992; Widom et al., 1992). The formation of inactive heterodimers with COUP-TF represents yet another possible regulatory mechanism in nuclear receptor-mediated transcription.

Heterodimerization between individual receptors clearly offers a great potential to generate many different combinations of receptors, especially when we consider that RAR, RXR, TR, COUP-TF, and PPAR are each encoded by multiple genes. The complexity of RAR is further amplified because each gene gives rise to a number of isoforms. As the expression of each isoform depends on cell type and the stage of growth and development, heterodimerization among all these receptors results in the formation of an extremely diverse group of receptors. It is conceivable that the enormous number of heterodimers would cover a wide range of transcriptional activities, thus generating significant diversity in gene regulation.

Hormone Response Elements

Sequence comparison of many HREs reveals one of the most remarkable features of HREs: all nuclear receptor binding sites consist of only two distinct sequence motifs related to either AGAACA or PuGGTCA (for review see Parker, 1993; Tsai and O'Malley, 1994). DNA binding specificity appears to be largely determined by the configurations of each HRE, the nucleotide sequence of the core motifs and the surrounding bases, the spacing between two half-site core motifs, and the orientation of these core motifs. The configurations of HREs not only determine whether a given receptor can bind to the response element, but also if the DNA-bound receptor bestows activation, repression, or no transcriptional activity on the gene. Four types of HREs have been established so far in terms of their distinct configurations.

One type is best depicted by the HREs of steroid receptors that bind as homodimers to inverted repeats (IRs, palindromes). The response element for GR was the first such HRE to be characterized (Yamamoto, 1985). Sequences conferring the glucocorticoid responsiveness were identified from various promoters (Karin et al., 1984; Renkawitz et al., 1984; Jantzen et al., 1987; Strahle et al., 1987). Later, response elements for progesterone, mineralocorticoid, androgen and estradiol were identified and shown to be similar to that of GR (Cato et al., 1986, 1987; Darbre et al., 1986; Arriza et al., 1987; Strahle et al., 1987; Cato and Weinmann, 1988; Ham et al., 1988). Sequence comparison revealed that all these HREs contain inverted repeats of a hexamer sequence separated by a three–base pair nonconserved spacer (IR3). The consensus half-site core motif derived for GRE and ERE are AGAACA and AGGTCA, respectively (for review see Gronemeyer, 1993; Cooney and Tsai, 1994). Interestingly, GR, PR, MR, and AR can activate through the same GRE sequence (von der Ahe et al., 1985; Darbre et al., 1986; Arriza et al., 1987; Cato et al., 1987). It is therefore concluded that group A receptors invariably bind to the IR3 type of HREs. More appealingly, by sharing common response elements, different hormonal signaling pathways can cross-talk with each other. This principle of intracellular cross-talk between different signaling pathways was greatly substantiated when another special type of GRE (a composite GRE) was characterized by its ability to confer both positive and negative regulation on target genes (Diamond et al., 1990; Jonat et al., 1990; Konig et al., 1992). Detailed analysis indicates that one composite GRE can confer a novel regulatory cross-talk between GR and AP-1 activity (Miner and Yamamoto, 1991, and references therein). For example, the ability of GR to stimulate proliferin gene transcription is enhanced by cJun homodimers but repressed by cJun/cFos heterodimers. The effects are mediated through a composite GRE that binds both GR and AP-1 and shares almost no homology with the consensus GRE. The effects cannot be reproduced by replacing GR with MR and appear to involve an interaction between the basic region of AP-1 and the DBD of the receptor (Miner and Yamamoto, 1992). Similarly, the AP-1 family members are also involved in retinoid- and thyroid-specific responses. Since AP-1 activities are regulated by membrane receptors, the interaction of nuclear receptors with AP-1 offers one means by which the signaling pathway of nuclear receptors can be coupled with that of the membrane receptors. More significantly, the nuclear receptors have been found to be coupled directly to a number of extracellular signals, such as forskolin (Denner et al., 1990), dopamine (Power et al., 1991a), insulin-like growth factor (Aronica and Katzenellenbogen, 1993) and epidermal growth factor (Ignar-Trowbridge et al., 1993). These unexpected findings indicate that any given HRE is not specific to a certain hormonal regulation, and it remains to be determined how the correct differential responses dictated by different hormones are specified. One possible explanation is that subtle variations in the HRE sequence may greatly affect the optimum response to different hormone signals *in vivo*.

Extensive work by many laboratories has uncovered a great number of HREs for thyroid hormone, retinoid, and vitamin D_3 receptors. Unlike the steroid receptors, these group B receptors tend to bind a different type of HRE containing direct repeats (DRs) with PuGGTCA consensus motifs. Detailed sequence analysis of these HREs has revealed an amazing 3–4–5 rule (Fig. 3) (Näär et al., 1991; Umesono et al., 1991). The 3–4–5 rule indicates that the specificity of a given group B receptor HRE is not determined by the exact sequence of the half-site core motif, but rather by the length of the spacer that separates the half sites (Fig. 3). Thus, direct repeats with a spacing of three nucleotides (DR3) dictates a vitamin D_3 response; DR4 dictates a thyroid hormone response; and a retinoic acid response is conferred by DR5. Soon after, DR1 was found to be a response element for RXR (Mangelsdorf et al., 1991) and PPAR (Kliewer et al., 1992c) and the 3–4–5 rule was extended to the 1–3–4–5 rule (Fig. 3).

However, this view of HREs may be overly simplified, and it has been challenged consistently since proposed. Chambon and co-workers have found that the selective discrimination of half-site spacing in direct repeats can be more complex. They showed that DR1 and DR2 are responsive to both all-*trans* and 9-*cis* RA and to RAR/RXR heterodimers (Durand et al., 1992). Subsequent work by Perlmann et al. (1993) showed that DR5 elements

Interaction Between Nuclear Hormone Receptors and Direct Repeats

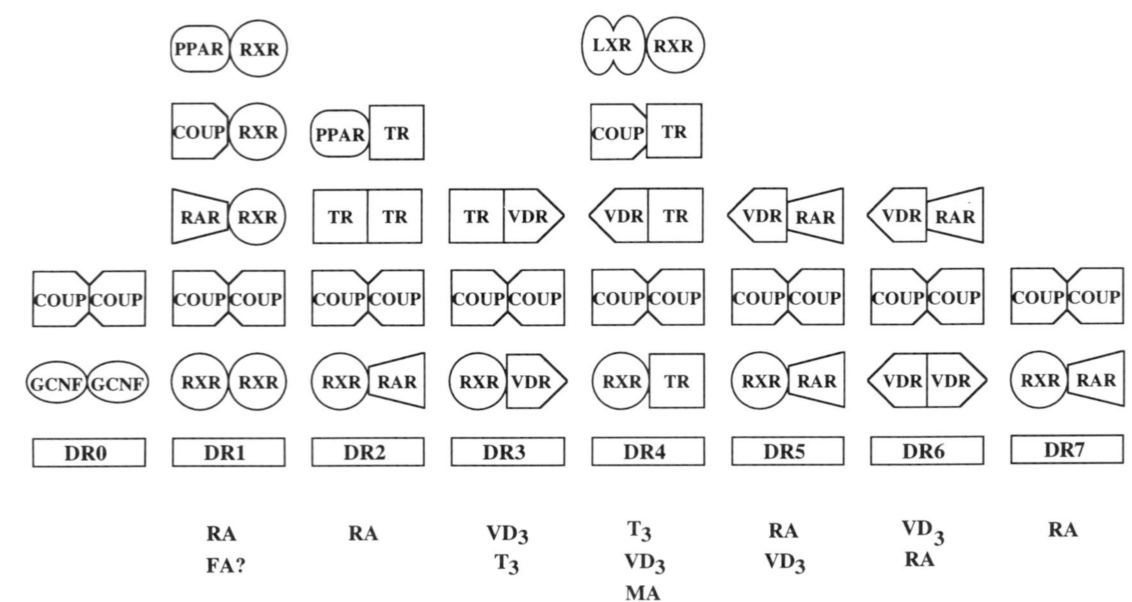

Figure 3. Hormone responsiveness of direct repeats containing PuGGTCA core motif. RA, retinoic acid; FA, fatty acid; MA, methoprenic acid; T_3, thyroid hormone; VD_3, vitamin D_3.

are high-affinity binding sites and DR2 elements are low-affinity binding sites for RAR/RXR heterodimers. DR2 also has been shown to be a negative TRE (Näär et al., 1991). More recently, Carlberg et al. (1993) showed that VDR mediates diverse hormonal responses on DR elements. Although VDR/RXR heterodimers synergistically activate a subset of DR3-dependent gene expression in the presence of both vitamin D_3 and 9-*cis* RA, VDR also activates gene expression as a homodimer through DR6 elements in the presence of vitamin D_3 alone (Carlberg et al., 1993). Thus, the presence of a DR3 or DR6 element in a gene will determine whether its expression is sensitive to both the vitamin D_3 and 9-*cis* RA signaling pathways or to the vitamin D_3 pathway alone. In addition, Carlberg and co-workers also have found that VDR can form heterodimers with TR and RAR (Fig. 3) (Schrader et al., 1994b, 1994a). VDR/TR heterodimers can activate gene expression through some imperfect DR3 or DR4 elements (Schrader et al., 1994b), and these respond to both thyroid hormone and vitamin D_3. More intriguingly, the sensitivity of VDR/TR heterodimers to thyroid hormone or vitamin D_3 depends on the relative position of TR and VDR on the two half-sites of a given response element. The heterodimer is 10 times more sensitive to the ligand for the receptor that occupies the 3′ half-site (Schrader et al., 1994b).

In addition to the spacing between half-sites, the sequence of the spacer and the surrounding bases may also play an important role in determining the relative potency of each TRE, VDRE, and RARE. Evans and co-workers have found that at spacer positions 3 and 4 of DR4 there is strong selection against A and T residues, respectively (Perlmann et al., 1993). Moreover, TR monomers and homodimers have higher affinity and transactivity for their response elements if the two nucleotides 5′ to AGGTCA motifs are TG (Force et al., 1994). Thus, spacers and surrounding sequences are more than inert nucleotide sequences; they influence the activation potential of a given receptor (Leng et al., 1994).

Group B receptors can also recognize HREs containing the hexameric PuGGTCA half-site being arranged as everted repeats (ERs). Natural HREs of such a type have been found in the chicken lysozyme gene promoter (ER6: TGACCC cagctg AGGTCA) (Baniahmad et al., 1990) and in the promoter of the lens specific gene encoding γF-crystallin (ER8) (Tini et al., 1993; see Chapter 5, this volume). In view of these findings, an extensive screening of everted repeats for TR binding reveals that TR can bind to ER4–6 as a homodimer (Kurokawa et al., 1993). More recently, ER7 has been shown to be a response element for VDR/TR heterodimers (Schrader et al., 1994b). This is in contrast to DR4, to which TR preferentially binds as a TR/RXR heterodimer (Kurokawa et al., 1993). Thus, homodimers and various heterodimers of group B receptors can have distinct preferences for their response elements.

Finally, orphan receptors in group C, such as NFGI-B (Hazel et al., 1988, 1991; Milbrandt, 1988; Ryseck et al., 1989) and SF-1 (Lala et al.,

1992), recognize HREs containing only a single PuGGTCA half site. The HRE for NGFI-B/Nur77 (NBRE) has been identified as AAAGGTCA. Two extra 5′ adenosine residues are required for high-affinity NGFI-B binding to DNA (Wilson et al., 1991). Nurrl/RNR-1 can also bind to this NBRE (Law et al., 1992; Scearce et al., 1993). Similar to NGFI-B, orphan receptor SF-1 can bind to TCAAGGTCA as a monomer (Wilson et al., 1993a). It remains unclear if these orphan receptors also can bind to response elements as heterodimers, in which their heterodimeric partner(s) (most likely RXR) makes nonspecific DNA contacts.

The finding of various HREs with distinct configurations reveals the complexity of hormonal regulation of gene expression. Although steroid receptors mostly bind to palindromic HREs as homodimers, other receptors, such as VDR, TR, and RAR, are rather flexible and can bind to inverted, direct, and everted PuGGTCA repeats as either homo- or heterodimers. Furthermore, TR and RAR also have been shown to transactivate through a natural single half-site (Leng et al., 1994). The exact physiological implication of such a flexibility is not known, but it may explain the more diverse roles of this group of receptors in development and differentiation as compared to the more restricted roles of group A receptors.

DBD Subdomains

It is generally accepted that there are a number of functional modules within the DBD that individually detect the sequence, spacing and orientation of a given HRE. To date, five functional "boxes" have been found. These include the P and D boxes (Mader et al., 1989; Umesono and Evans, 1989; Danielian et al., 1992), the T and A boxes (Wilson et al., 1992), and the recently defined DR box (Perlmann et al., 1993). These functional related boxes dictate the specificity of an individual nuclear receptor for its cognate HRE.

The ER/GR domain swapping experiments by Green and Chambon (1987) indicated that the first zinc module of the nuclear receptors can functionally determine sequence specificity of receptor DNA binding. The amino acid residues that discriminate an HRE are located in a cluster of highly conserved amino acids at the C-terminal end of the first zinc module (Fig. 4) (Danielsen et al., 1989; Mader et al., 1989; Umesono and Evans, 1989). Point mutations in this region can alter the DNA recognition specificity. Thus, Evans and co-workers have designated it as P box (Proximal box) (Umesono and Evans, 1989). Using the P box and its cognate HRE half-site sequences as a criterion, at least six distinct groups of receptors can be identified (Table 2) (Forman and Samuels, 1990; Tsai and O'Malley, 1994). For example, group I includes GR, MR, PR, and AR, which has a P box sequence of GSckV recognizing a TGTTCT half-site.

The D box (Distal box) is located between the first and second cysteines of the second zinc finger of the DBD (Fig. 4), and is usually responsible for

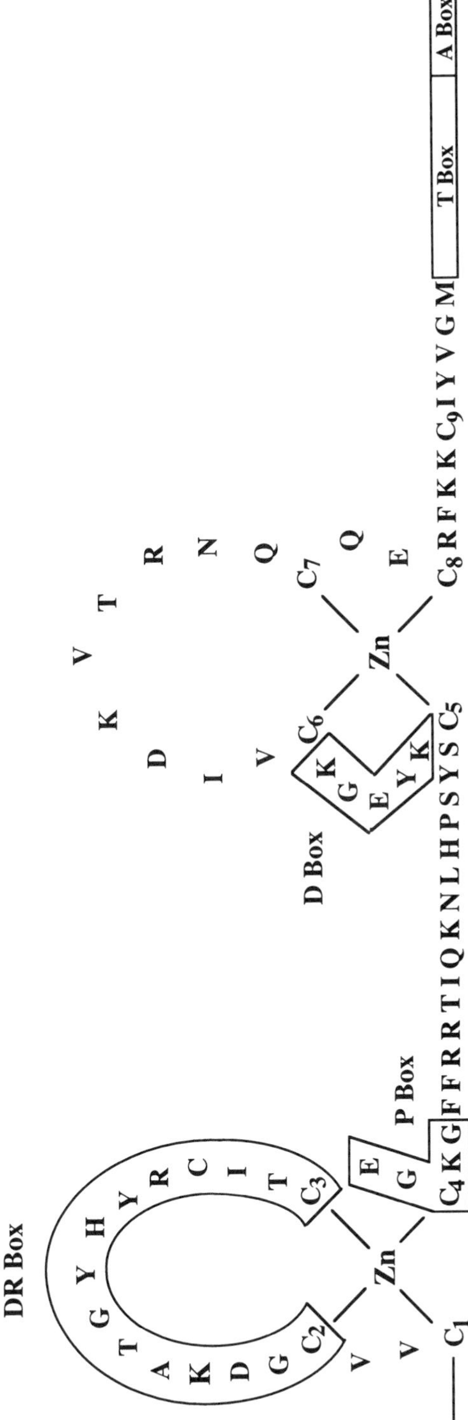

Figure 4. Schematic representation of the functional subdomains within the nuclear receptor DBD. Amino acid sequences of the thyroid hormone receptor β DBD (C region) are shown. A/T box sequences are not conserved among receptors. Zn, zinc ions. The *subscripts* on the cysteine residues refer to their numbering (in the DBD) from the amino terminus.

Table 2. Subgroups of the nuclear receptor superfamily based on the P-box sequence

Group	P-box sequence	Nuclear hormone members
I	GSckV	GR, MR, PR, AR
II	EDckA	ER, ERR1, ERR2
III	EGckG	TR, RAR, RXR, VDR, PPAR, NUC1, NGFI-B (Nur77), RVR, Rev-Erbα (EAR1), Rev-Erbβ, BD73, UR(TAK1, OR), MB67, TR2, TR4 (TR2r1), Nurr1 (RNR-1, NOT), ROR, RIP14, GCNF, RLD-1, *DHR3*, *Usp*, E75, *EcR*
IV	EGckS	v-erbA, EAR2, COUP-TFI (EAR3), COUP-TFII (ARP-1), *Kni*, *Knrl*, *Svp*, *Egon*
V	ESckG	SF-1 (ELP), *FTZ-F1*
VI	DGckG	Tlx, HNF4, *Tll*

dimerization and discriminating the half-site spacing of HREs. For example, the D box sequence AGRND in GR recognizes a three-nucleotide spacing whereas KYEGK of TR recognizes a 0 nucleotide spacing (Umesono and Evans, 1989). In contrast to the rather conserved P box, the D box is widely varying among receptors recognizing the same half-site sequences, which may correlate with the different spacing of their cognate HREs.

Wilson et al. (1992) identified two other important regions downstream of the second zinc finger, designated as the A and T boxes (Fig. 4). Both boxes are required for high-affinity DNA binding of an NGFI-B monomer. The A box consists of seven mostly charged residues, and it makes contact with the two adenosine residues in the minor groove 5′ to the AGGTCA core sequences to stabilize the DNA binding. Adjacent to the A box is the T box, which forms an α-helix to interact with the phosphate backbone of HREs (Lee et al., 1989; Wilson et al., 1993a). X-ray crystallographic and mutation analysis of the T box in RXR indicates that besides protein-DNA interaction, it is also involved in protein–protein dimerization interactions (Lee et al., 1989; Wilson et al., 1992). Thus, in contrast to the DBD of steroid receptors, which dimerizes only via the D-box, the homo- and heterodimers of RXR with other receptors may use both T and D boxes for dimerization to bind HREs with direct repeats (Wilson et al., 1992). Although the exact function of A/T boxes in other receptors is not well defined, it is likely that they are involved in response element selection and receptor heterodimerization.

In addition to the A/T box, the heterodimerization between RXR and TR or RAR also requires another region, the DR box, in the first zinc finger (Fig. 4) (Perlmann et al., 1993). Experimental results using receptor chimeras indicate that heterodimerization of TR/RAR with RXR on direct repeats involves direct interaction between the DR box of TR/RAR and the D box of RXR. The consequence of such an interaction between RXR and TR or RAR is that RXR binds to the 5' half site and TR or RAR binds to the 3′ half site of the response element (Kurokawa et al., 1993; Perlmann et al., 1993). It is also through the DR box/D box interaction that the spacing between the half-site repeats is determined. For example, a spacing of 4 bp permits the TR DR box to interact with the RXR D box, and likewise a spacing of 5 bp favors RAR/RXR interaction. It is still not clear whether the D box-DR box interaction plays any role in heterodimer binding to an everted repeat response element.

NUCLEAR RECEPTOR LIGAND INTERACTION

Hormones for nuclear receptors are crucial for a wide range of essential processes and pathways in higher organisms. These lipophilic hormones pass through the cell membrane and bind to the cognate intracellular receptors with rather high affinity (in the range of 10^{-9}–10^{-11} M). Hormone binding has a drastic effect on the transformation of the receptor from either an inactive molecule or silencer of transcription to a positive transactivator. One of the major questions in this field is the mechanism by which ligands activate their cognate receptors. Although the detailed mechanism is not fully understood yet, significant advances have been made in the past decade. In general, for group A receptors, hormone binding induces heat-shock protein (hsp) dissociation, receptor dimerization, DNA binding, and eventually receptor activation. Since group B receptors do not associate with hsps and can bind to DNA as homodimers or heterodimers in the absence of ligands, the ligand activation is mainly to convert the receptors from inactive molecules or transcriptional repressors to transcriptional activators. Sequence comparison suggests that group C receptors are more closely related to group B receptors than to the steroid receptor subfamily (Wahli and Martinez, 1991), which indicates that they may function similarly to the TR/RAR subfamily members. Detailed analyses reveal that for almost all receptors examined, ligand binding induces phosphorylation and conformational changes (for review, see Tsai and O'Malley, 1994). In addition, in transient transfection assays, the LBDs of various receptors from different subfamilies are all functionally interchangeable (Thompson and Evans, 1989). Therefore, it is possible that upon hormone binding, the basic event converting all receptors to their active forms is the same.

Phosphorylation

Most nuclear receptors are phosphoproteins (for review see Weigel, 1994). For PR, ER, VDR, and RAR, treatment of these receptors with the cognate hormone significantly enhances phosphorylation above basal levels (Pike and Sleator, 1985; Auricchio, 1989; Washburn et al., 1991; Orti et al., 1992; Rochette-Egly et al., 1992). In addition, cellular treatment with known protein kinase A activators dramatically increases receptor-dependent activation of reporter genes (Denner et al., 1990; Somers and DeFranco, 1992; Aronica and Katzenellenbogen, 1993; Cho and Katzenellenbogen, 1993). Theoretically, extra phosphates could be required for efficient hsp dissociation, to stabilize or trigger ligand-induced conformational changes, to promote receptor binding to DNA, to facilitate receptor interactions with coactivators or basal transcription factors, or for receptor recycling (Weigel, 1994, and references therein). Experiments by Chambon and co-workers indicate that mutation of a single phosphorylated serine reduces estrogen-dependent transactivation by ER (Ali et al., 1993). Similarly, phosphorylation in VDR and v-erbA appears to be important for their activity (Glineur et al., 1989; Jurutka et al., 1993). Although the precise role of phosphorylation in steroid receptor transactivation is largely unknown, it is established that the phosphorylation is an integral part of the regulation of receptor function. One should be cautious that in many cases mutation of phosphorylation sites has little effect on receptor function. It should be noted, however, that such mutation usually leads to the emergence of secondary phosphorylation sites, which may compensate for the original ones in receptor function (Weigel, 1994, and references therein).

Conformational Change

Ligand-dependent structural change of nuclear receptors has always been accepted as one of the key steps in receptor activation. Immunorecognition and aqueous two-phase partitioning analyses have suggested that there is hormone-induced conformational change in ER and TR (Allan, 1994, and references therein). Gel-shift assays also have revealed that the mobility of hormone-bound receptor-DNA complexes is different from that of unliganded receptor-DNA complexes on nondenaturing gels (Kumar and Chambon, 1988; Fawell et al., 1990b; Sabbah et al., 1991). More recently, partial proteolytic analysis has convincingly shown that ligands can induce conformational changes in almost all receptors examined (Allan et al., 1992a; Leng et al., 1993; Keidel et al., 1994; Leid, 1994; Zeng et al. 1994; Leng et al., 1995). These experiments indicate that at a gross level, the nature of the conformational change is conserved across the nuclear receptor superfamily. It centers on the receptor LBD and renders the whole LBD more resistant to protease digestion. Native gel electrophoresis indicates that the receptor LBD is folded into a more compact structure (Leng et al., 1993). Since the hormone-induced conformational change in steroid receptors is not affected by hsp dissociation or receptor dimerization, the ligand-dependent conformational change may be required mainly for steps after receptor DNA binding (Allan et al., 1992b; Leng et al., 1993). Given the fact that hormone binding to receptors like TR apparently affects only their ability to transactivate, it is likely that the observed conformational change generates a form of the receptor capable of positively interacting with the transcriptional machinery to turn on target gene expression. One cannot exclude, however, the possibility that for hsp-associated receptors, the conformational change may also play a role in hsp dissociation, receptor dimerization, DNA binding, and phosphorylation because the structural change precedes all these steps.

TRANSCRIPTIONAL REGULATION BY NUCLEAR RECEPTORS

Like many other transcription factors, nuclear receptors modulate gene transcription by interacting with other transcription factors at the promoter region. Their action is also limited by the accessibility of genes in chromatin and mediated by ubiquitous and specific factors interacting with receptors. For example, GR has been shown to act synergistically with a variety of transcriptional factors, including NF-1, Sp1, and CACC box binding factors (Schule et al., 1988a,b; Strahle et al., 1988). More significantly, the hormonal response of MMTV LTR requires two proximal octamer motifs and a binding site for CTF/NF-1 (Buetti et al., 1989; Toohey et al., 1990; Bruggemeier et al., 1991, 1994; Miksicek et al., 1994). As mentioned previously, GR and RAR can also repress AP-1 activation for several promoters (Schule et al., 1990a,b; Mordacq and Linzer, 1994). On the other hand, overexpression of Jun or Fos can also block glucocorticoid responsiveness of target genes (Schule et al., 1990b; Lucibello et al., 1990; Yang-Yen et al., 1990). In general, the outcome of the interaction between nuclear receptors and other factors depends on the cell type and promoter context and can be either potentiation or repression (Shemshedini et al., 1991). Such interactions may be of great physiological significance and they have been detected in several systems.

In addition to the sequence-specific transcription factors, nuclear receptor-regulated gene transcription also requires the general transcription factors of all Pol II genes and co-regulators. Several distinct general transcription factors, namely, TFIIA, TFIIB, TFIID, TFIIE, TFIIF, TFIIH, and RNA polymerase II, have been found to participate in the accurate and efficient assembly of the initiation complex at the basic promoter (Conaway and Conaway, 1993; Zawel and Reinberg, 1993; Buratowski, 1994). In addition, a co-activator associated with the TFIID complex has been identified to be important in ER-dependent transcriptional activation (Jacq et al., 1994). Although initiation complexes composed of different subsets of these general transcription factors may assemble at different promoters (Parvin et al., 1993), they all follow a well-defined scheme of stepwise

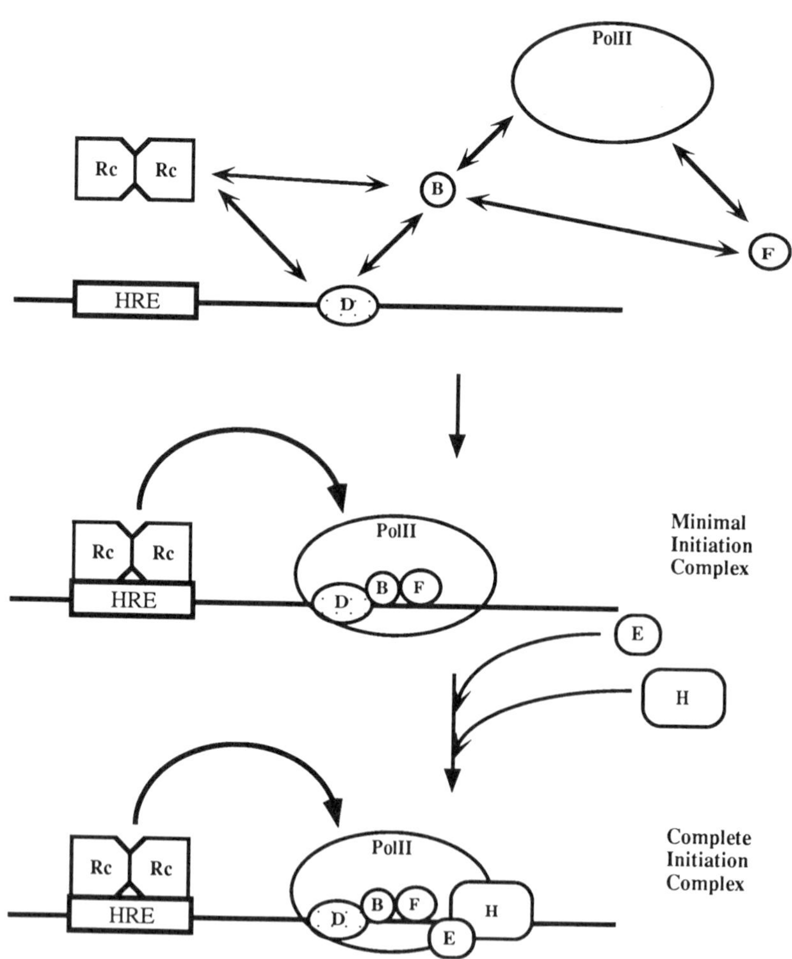

Figure 5. Schematic illustration of transcription preinitiation complex assembly. Proteins that are components of RNA polymerase II transcription factors are indicated by their designated letters. *Double-headed* arrows indicate protein-protein interactions. *Single-headed* arrows indicate that nuclear receptors can stabilize the preinitiation complex.

assembly (Buratowski, 1994, and references therein) (Fig. 5). An initial complex is formed by TFIID (the TATA-binding protein TBP and TBP associated factors, TAFIIs) binding to the TATA element of a promoter. This complex acts as an anchor for subsequent assembly of TFIIA and

TFIIB, which recruit Pol II and TFIIF into the complex. The association of these factors results in a minimal initiation complex, stable in both kinetic assays and native gel electrophoresis. This minimal complex can thus initiate basal level transcription. Subsequently, TFIIE and TFIIH attach to this initiation complex, which may be necessary for the conversion of an initiation complex into an elongation complex. Once the complete initiation complex is assembled, an adenosine triphosphate (ATP)-dependent activation step is necessary for transcription to occur. In a TATA-less promoter, an initiator motif (Inr) serves as the anchor site for the initiation complex formation (Hernandez, 1993, and references therein), and a specific Inr-binding protein (IBP or TFII I) helps to recruit TFIID and the other basal factors to the TATA-less promoter. Although it remains largely unclear how nuclear receptors regulate such a system, *in vitro* transcription experiments indicate that steroid receptors can stabilize the formation of the preinitiation complex on a TATA box (Bagchi et al., 1990b; Elliston et al., 1990; Klein-Hitpass et al., 1990; Tsai et al., 1990).

Gene Activation by Nuclear Receptors

During the past several years, *in vitro* transcription systems have emerged as powerful tools to understand the mechanism of gene regulation by nuclear receptors (Corthesy et al., 1988; Freedman et al., 1989; Bagchi et al., 1990a; Klein-Hitpass et al., 1990). Using baculovirus-expressed GR, PR, and ER, O'Malley and co-workers have developed highly efficient *in vitro* transcription systems (Tsai and O'Malley, 1994, and references therein). More than 30-fold activation by hormone was observed with minimal promoter constructs (Allan et al., 1991). Recently, Fondell et al. (1993) successfully reconstituted target gene repression by unliganded TR *in vitro* using highly purified general transcription factors and bacteria-expressed receptor. Development of these hormone-dependent cell-free transcription systems would enable us to study the role of nuclear receptors in the formation of the preinitiation complex for transcription.

As discussed previously, transcription initiation is largely controlled by the assembly of the initiation complex. The ability of nuclear receptors to activate minimal promoter constructs suggests that nuclear receptors may regulate the rate of transcription initiation by direct interaction with the general transcriptional machinery. Using template challenge assays *in vitro*, O'Malley and co-workers have demonstrated the essential role of PR, ER, and GR in the formation of a stable preinitiation complex (Elliston et al., 1990; Klein-Hitpass et al., 1990; Tsai et al., 1990).

The direct and specific interaction between nuclear receptors and general transcription factors was first demonstrated with COUP-TF, which can bind specifically to TFIIB (Tsai et al., 1987; Ing et al., 1992). Subsequently, PR, ER, TR, and RAR have all been found to interact with TFIIB (Tsai and

O'Malley, 1994, and references therein). Similar results also have been obtained with other activators, such as VP16 (Lin et al., 1991) and FTZ protein (Colgan et al., 1993). Since binding of TFIIB to the TFIID-DNA complex is one of the rate-limiting steps in preinitiation complex formation, it is plausible that by facilitating this interaction nuclear receptors can help to recruit TFIIB into the preinitiation complex (Lin et al., 1991). It is largely unclear if nuclear receptors can interact with other components in the transcriptional machinery (Tsai and O'Malley, 1994).

Gene Repression by Nuclear Receptors

Transcriptional repression plays an important role in prokaryotes and lower eukaryotes. In higher eukaryotes, transcriptional repression is not well understood. However, a combination of both positive and negative regulation is often responsible for the observed gene expression pattern (Levine and Manley, 1989). Members of the nuclear receptor superfamily have been found to be involved in transcriptional repression by one of these means: competitive DNA binding, blocking activation, squelching, or direct silencing (for review see Renkawitz, 1990).

Gene repression through competitive DNA binding is best exemplified in the glucocorticoid-dependent inhibition of human glycoprotein α-subunit gene expression. The inhibition is mediated through the GRE overlapping with a functional cyclic adenosine monophosphate (cAMP)-responsive element (CRE) (Levine and Manley, 1989, and references therein). In transfection assays, deletion of the GR DBD abolishes such a repression, whereas the N-terminal and the C-terminal halves of the receptor are dispensable (Oro et al., 1988). This suggests that the repression results from the DNA-bound GR inhibiting binding of a positive factor to the CRE by steric hindrance. More recently, COUP-TF I (ear3) and II (ARP-1) have been shown to repress HNF4-, TR-, PPAR-, RAR-, RXR-, VDR- or ER-dependent gene transcription via competitive DNA binding (for review see Qiu et al., 1994b).

Nuclear receptors also can block transactivation of other transcription factors; examples of such include the T3- and RA-dependent negative regulation of the epidermal growth factor receptor (EGFR) promoter (Hudson et al., 1990). In this case, TR/RXR or RAR/RXR heterodimers bind to an element downstream of the TATA box, which contains a TRE half site and a binding site for Sp1. Since TR and RAR can bind to DNA in the absence of ligand and yet the repression is ligand-dependent, it is suggested that ligand induces allosteric changes in receptor-DNA complexes, thus resulting in a block of transcription complex formation at the core promoter. GR, TR, and RAR also have been found to block transactivation of AP-1, for example, the glucocorticoid-dependent repression of AP-1 induction of the collagenase or the stromelysin genes (Jonat et al., 1990; Nicholson et al., 1990; Schule et al., 1990a; Yang-Yen et al., 1990). In these cases, GR and AP-1 can form heterodimers in the presence of glucocorticoid, and AP-1/GR heterodimers can still bind to DNA but are unable to transactivate.

Another alternative mechanism for transcriptional repression by nuclear receptors is "squelching" of limiting factors (Gill and Ptashne, 1988; Meyer et al., 1989, 1992). For example, in HeLa cells, but not in lymphoid cells, expression of GR inhibits transactivation by OTF-2 but not OTF-1, which indicates that GR and OTF-2 may share a common co-factor(s) in their transactivation pathways (Weiland et al., 1991). This type of mechanism could be of general significance if these factors are confirmed to play an important role for transactivation by nuclear receptors under physiological conditions (Shemshedini et al., 1992; Truss and Beato, 1993).

Finally, nuclear receptors can mediate direct silencing. Although group A receptors are inert *in vivo* in the absence of ligand, TR/RAR subfamily members can frequently repress basal promoter activity, an activity referred to as silencing. Like activation, silencing activity can be detected on a minimal construct containing only the TATA box and HREs (Baniahmad et al., 1992b), thus suggesting that silencing is a result of direct interaction of the receptor with the basal transcriptional machinery. Indeed, the silencing activity of TR *in vitro* is due to its ability to inhibit the formation of a preinitiation complex, probably by interacting with TFIIB (Baniahmad et al., 1993; Fondell et al., 1993).

More detailed analysis by Baniahmad et al. (1993) indicates that the N-terminal and the C-terminal halves of the receptor can interact with different regions of TFIIB. The N-terminal region of TR interacts with the C-terminal half of TFIIB. This interaction may be important for transactivation because the same region of TFIIB also interacts with the activation domain of VP16 and is required for VP16-dependent activation (Zawel and Reinberg, 1993). In contrast, the C-terminal domain of TR, which confers silencing activity, interacts with the N-terminal putative zinc finger of TFIIB. This region of TFIIB has been suggested to be important for the formation of the preinitiation complex by interacting with the small subunit of TFIIF and the largest subunit of Pol II (Ha et al., 1993). More appealingly, the interaction of the TR C-terminus and TFIIB is sensitive to hormone (Baniahmad et al., 1993). In the presence of near physiological concentration of T_3, the interaction is significantly decreased. Similar results with RAR have also been obtained (X. Leng, B. W. O'Malley, and M.-J. Tsai, unpublished data). These results indicate that the interaction between TFIIB and TR or RAR may be necessary for silencing. We wish to point out, however, that the direct interaction of nuclear receptors with general transcriptional machinery is not sufficient for receptor-mediated silencing and there is likely more than one target involved (Baniahmad et al., 1992b). Recent studies indicate that the silencing activity of TR can be competed with by overexpression of an internal fragment of TR LBD, which does not interact with TFIIB (Baniahmad et al., 1995). These results suggest an involvement of a cellular co-repressor(s).

NUCLEAR RECEPTORS IN TUMORS

Nuclear receptors regulate the growth and differentiation of a wide range of cell types in different tissues, including malignant cells. The mutated forms of some nuclear receptors, such as v-erbA (TR derivative) (Gandrillon et al., 1989) and PML/RARα (see Chapters 19 and 20, this volume), are directly involved in cell transformation and tumorigenesis. Recently, considerable interest has been focused on understanding the role of mutated nuclear receptors in necrosis and oncogenesis, with particular emphasis on mutations that occur in malignant cells or in various pathological states. Examples of such include the many variants of ER that have been identified in breast cancer biopsies (see Chapter 9, this volume), specific point mutations in AR that lead to androgen insensitivity and prostate cancer (see Chapter 16, this volume), mutations in the TR LBD which result in general resistance to thyroid hormone (GRTH) (Usala, 1991), various GR mutants that have been observed in different lymphoma cells (Ip et al., 1993, and references therein), and PML/RARα in acute promyelocytic leukemia (APL) (see Chapter 20, this volume). Studies combining the molecular biology of nuclear receptors and cancer biology can thus offer a solid basis for remedial treatments. Clinical therapies stemming from such studies include the application of retinoids and vitamin D_3 derivatives in cancer treatments. One of the reasons is that retinoids and metabolites of vitamin D_3 can regulate cell proliferation and differentiation. RA, for example, can induce tumor cell differentiation, such as in the human promyelocytic leukemia cell line HL-60 and leukemia cells in patients with APL. Specifically, 13-*cis* RA has been shown to be effective for treating oral leukoplakia, for preventing skin tumors in patients with xeroderma pigmentosum and for treating squamous cell carcinomas of the cervix and of the skin (Rowe and Brickell, 1993, and references therein). All-*trans* RA has been used to induce complete remission in APL patients (see Chapter 19, this volume). Similarly, vitamin D_3 analogs have also been developed for the treatment of psoriasis, breast cancer, and leukemia. In view of these, the correlation between nuclear receptors and neoplastic processes has been one of the most fruitful fields in nuclear receptor research.

ACKNOWLEDGMENTS

This work was supported by NIH Grants DK44988 (to S.Y.T.), HD08188, and DK45641 (to M.J.T.), and a student fellowship from Glaxo Inc. (to X.L.).

REFERENCES

Ali S, Metzger D, Bornert JM, Chambon P (1993): Modulation of transcriptional activation by ligand-dependent phosphorylation of the human oestrogen receptor A/B region. *EMBO J* 12:1153–1160

Allan GF, Ing NH, Tsai SY, Srinivasan G, Weigel NL, Thompson EB, Tsai M-J, O'Malley BW (1991): Synergism between steroid response and promoter elements during cell-free transcription. *J Biol Chem* 266:5905–5910

Allan GF, Leng X, Tsai SY, Weigel NL, Edwards DP, Tsai M-J, O'Malley BW (1992a): Hormone and antihormone induce distinct conformational changes which are central to steroid receptor activation. *J Biol Chem* 267:19513–19520

Allan GF, Tsai SY, Tsai M-J, O'Malley BW (1992b): Hormonal induced conformational changes in the progesterone receptor are required for events following binding to DNA. *Proc Natl Acad Sci USA* 89:11750–11754

Allan GF (1994): Mechanism of ligand activation. In: *Mechanism of Steroid Hormone Regulation of Gene Transcription*, Tsai M-J, O'Malley BW, eds. Austin: R.G. Landes Company

Amero SA, Kretsinger RH, Moncrief MD, Yamamoto KR, Pearson WR (1992): The origin of nuclear hormone receptor proteins: A single precursor distinct from other transcription factors. *Mol Endocrinol* 6:3–7

Aronica SM, Katzenellenbogen BS (1993): Stimulation of estrogen-receptor-mediated transcription and alteration in the phosphorylation state of the rat uterine estrogen receptor by estrogen cyclic adenosine monophosphate and insulin-like growth factor-1. *Mol Endocrinol* 7:743–752

Arriza JL, Weinberger C, Cerelli G, Glaser TM, Handelin BL, Housman DE, Evans RM (1987): Cloning of human mineralocorticoid receptor complementary DNA: structural and functional kinship with the glucocorticoid receptor. *Science* 237:268–275

Au-Fliegner M, Helmer E, Casanova J, Raaka BM, Samuels HH (1993): The conserved ninth C-terminal heptad in thyroid hormone and retinoic acid receptors mediates diverse responses by affecting heterodimer but not homodimer formation. *Mol Cell Biol* 13:5725–5735

Auricchio F (1989): Phosphorylation of steroid receptors. *J Steroid Biochem* 32:613–622

Bagchi M (1994): Mechanisms of target gene activation by steroid hormone receptors: Insights from cell-free transcription system. In: *Mechanism of Steroid Hormone Regulation of Gene Transcription*. Tsai M-J, O'Malley BW, eds. Austin: R.G. Landes Company

Bagchi MK, Tsai SY, Tsai M-J, O'Malley BW (1990a): Identification of a functional intermediate in receptor activation in progesterone-dependent cell-free transcription. *Nature* 345:547–550

Bagchi MK, Tsai SY, Weigel NL, Tsai M-J, O'Malley BW (1990b): Regulation of in vitro transcription by progesterone receptor: characterization and kinetic studies. *J Biol Chem* 265:5129–5134

Baniahmad A, Steiner C, Kohne AC, Renkawitz R (1990): Modular structure of a chicken lysozyme silencer: involvement of an unusual thyroid hormone receptor binding site. *Cell* 61:505–514

Baniahmad A, Kohne AC, Renkawitz R (1992a): A transferable silencing domain is present in the thyroid hormone receptor in the v-erbA oncogene product and in the retinoic acid receptor. *EMBO J* 11:1015–1023

Baniahmad A, Tsai SY, O'Malley BW, Tsai M-J (1992b): Kindred S thyroid hormone receptor is an active and constitutive silencer and a repressor for thyroid hormone and retinoic acid responses. *Proc Natl Acad Sci USA* 89:10633–10637

Baniahmad A, Ha I, Reinberg D, Tsai SY, Tsai M-J, O'Malley BW (1993): Interaction of human thyroid hormone receptor beta with transcription factors TFIIB may mediate target gene derepression and activation by thyroid hormone. *Proc Natl Acad Sci USA* 90:8832–8836

Baniahmad A, Leng X, Burris TP, Tsai SY, Tsai M-J, O'Malley BW (1995): The τ4 activation domain of the thyroid hormone receptor is required to release a corepressor(s) necessary for silencing. *Mol Cell Biol* 15:76–86

Beato M (1989): Gene regulation by steroid hormones. *Cell* 56:335–344

Berrodin TJ, Marks MS, Ozato K, Linney E, Lazar MA (1992): Heterodimerization among thyroid hormone receptor retinoic acid receptor retinoid X receptor chick ovalbumin upstream promoter transcription factor and an endogenous liver protein. *Mol Endocinol* 6:1468–1478

Bocquel MT, Kuman V, Stricker C, Chambon P, Gronemeyer H (1989): The contribution of the N- and C-terminal regions of steroid receptors to activation of transcription is both receptor and cell specific. *Nuc Acids Res* 17:2581–2594

Bogazzi F, Hudson LD, Nikodem VM (1994): A novel heterodimerization partner for thyroid hormone receptor. *J Biol Chem* 269:11683–11686

Bradley DJ, Towle HC, Young III (1994): Alpha and beta thyroid hormone receptor (TR) gene expression during auditory neurogenesis: evidence for TR isoform-specific transcriptional regulation in vivo. *Proc Natl Acad Sci USA* 91:439–443

Bradshaw MS, Tsai SY, Leng X, Dobson ADW, Conneely OM, O'Malley BW, Tsai M-J (1991): Studies on the mechanism of functional cooperativity between progesterone and estrogen receptors. *J Biol Chem* 266:16684–16690

Bruggemeier U, Kalff M, Franke S, Scheidereit C, Beato M (1991): Ubiquitous transcription factor OTF-1 mediates induction of the mouse mammary tumor virus promoter through synergistic interaction with hormone receptors. *Cell* 64:565–572

Bruggemeier U, Rogge L, Winnacker EL, Beato M (1994): Nuclear factor I acts as a transcription factor on the MMTV promoter but competes with steroid hormone receptors for DNA binding. *EMBO J* 9:2233–2239

Buetti E, Kuhnel B, Diggelmann H (1989): Dual function of a nuclear factor I binding site in MMTV transcription regulation. *Nucleic Acids Res* 17:3065–3078

Bugge TH, Pohl J, Lonnoy O, Stunnenberg HG (1992): RXR alpha a promiscuous partner of retinoic acid and thyroid hormone receptors. *EMBO J* 11:1409–1418

Buratowski S (1994): The basics of basal transcription by RNA polymerase II. *Cell* 77:1–3

Carlberg C, Bendik I, Wyss A, Meier E, Sturzenbecker LJ, Grippo JF, Hunziker W (1993): Two nuclear signalling pathways for vitamin D. *Nature* 361:657–660

Cato ACB, Miksicek R, Schutz G, Arnemann J, Beato M (1986): The hormone regulatory element of mouse mammary tumor virus mediates progesterone induction. *EMBO J* 5:2237–2240

Cato ACB, Henderson D, Ponta H (1987): The hormone response element of the mouse mammary tumour virus DNA mediates the progestin and androgen induction of transcription in the proviral long terminal repeat region. *EMBO J* 6:363–368

Cato ACB, Weinmann J (1988): Mineralocorticoid regulation of transfected mouse mammary tumour virus DNA in cultured kidney cells. *J Cell Biol* 106:2119–2125

Cho H, Katzenellenbogen BS (1993): Synergistic activation of estrogen receptor-mediated transcription by estradiol and protein kinase activators. *Mol Endocrinol* 7:441–452

Colgan J, Wampler S, Manley JL (1993): Interaction between a transcriptional activator and transcription factor IIB in vivo. *Nature* 362:549–553

Conaway RC, Conaway JW (1993): General initiation factors for RNA polymerase II. *Annu Rev Biol* 3:760–769

Cooney AJ, Tsai SY, O'Malley BW, Tsai M-J (1992): Chicken ovalbumin upstream promoter transcription factor (COUP-TF) dimer binds to different GGTCA response elements allowing COUP-TF to repress hormonal induction of the vitamin D_3, thyroid hormone and retinoic acid receptors. *Mol Cell Biol* 12:4153–4163

Cooney AJ, Leng X, Tsai SY, O'Malley BW, Tsai M-J (1993): Multiple mechanisms of chicken ovalbumin upstream promoter transcription factor-dependent repression of transactivation by the vitamin D, thyroid hormone, and retinoic acid receptors. *J Biol Chem* 268:4152–4160

Cooney AJ, Tsai SY (1994): Nuclear receptor—DNA interactions. In: *Mechanism of Steroid Hormone Regulation of Gene Transcription*, Tsai M-J, O'Malley BW, eds. Austin: R.G. Landes Company

Corthesy B, Hipskind R, Theulaz I, Wahli W (1988): Estrogen-dependent in vitro transcription from the vitellogenin promoter in liver nuclear extracts. *Science* 239:1137–1139

Dahlman-Wright K, Wright A, Gustafsson JÅ, Carlstedt-Duke J (1990): Interaction of the glucocorticoid receptor DNA-binding domain with DNA as a dimer is mediated by a short segment of five amino acids. *J Biol Chem* 265:14030–14035

Damm K, Thompson CC, Evans RM (1989): Protein encoded by v-erbA functions as a thyroid-hormone receptor antagonist. *Nature* 339:593–597

Danielian PS, White R, Lees JA, Parker MG (1992): Identification of a conserved region required for hormone dependent transcriptional activation by steroid hormone receptors. *EMBO J* 11:1025–1033

Danielsen M, Hinck L, Ringold GM (1989): Two amino acids within the knuckle of the first zinc finger specify DNA response element activation by the glucocorticoid receptor. *Cell* 57:1131–1132

Darbre P, Page M, King RJB (1986): Androgen regulation by the long terminal repeat of mouse mammary tumour virus. *Mol Cell Biol* 6:2847–2854

Darling DS, Beebe JS, Burnside J, Winslow ER, Chin WW (1991): 3,5,3′-triiodothyronine (thyroid hormone) receptor-auxiliary protein (TRAP) binds DNA and forms heterodimers with the thyroid hormone receptor. *Mol Cell Endocrinol* 5:73–84

Debuire B, Henry C, Benaissa M, Biserte G, Claverie JM, Saule S, Martin P, Stehelin D (1984): Sequencing the erbA gene of avian erythroblastosis virus reveals a new type of oncogene. *Science* 224:1456–1459

Denner LA, Weigel NL, Maxwell BL, Schrader WT, O'Malley BW (1990): Regulation of progesterone receptor-mediated transcription by phosphorylation. *Science* 250:1740–1743

Diamond MI, Miner JN, Yoshinaga SK, Yamamoto KR (1990): Transcription factor interactions: Selectors of positive or negative regulation from a single DNA element. *Science* 249:1266–1272

Durand B, Saunders M, Leroy P, Leid M, Chambon P (1992): All-*trans* and 9-*cis* retinoic acid induction of CRABP II transcription is mediated by RAR-RXR heterodimers bound to DR1 and DR2 repeated motifs. *Cell* 71:73–85

Eilers M, Picard D, Yamamoto KR, Bishop JM (1989): Chimaeras of myc oncoprotein and steroid receptors cause hormone-dependent transformation of cells. *Nature* 340:66–68

Elliston JF, Fawell SE, Klein-Hitpass L, Tsai SY, Tsai M-J, Parker MG, O'Malley BW (1990): Mechanism of estrogen receptor-dependent transcription in a cell-free system. *Mol Cell Biol* 10:6607–6612

Evans RM (1988): The steroid and thyroid hormone receptor superfamily. *Science* 240:889–895

Fawell SE, Lees JA, White R, Parker MG (1990a): Characterization and colocalization of steroid binding and dimerization activities in the mouse estrogen receptor. *Cell* 60:953–962

Fawell SE, White R, Hoare S, Sydenham M, Page M, Parker MG (1990b): Inhibition of estrogen receptor-DNA binding by the "pure" antiestrogen ICI 164,384 appears to be mediated by impaired receptor dimerization. *Proc Natl Acad Sci USA* 87:6883–6887

Folkers GE, Vander Leede BM, van der Saag PT (1993): The retinoic acid receptor-beta2 contains two separate cell-specific transactivation domains, at the N-terminus and in the ligand-binding domain. *Mol Endo* 7:616–627

Fondell JD, Roy AL, Roeder RG (1993): Unliganded thyroid hormone receptor inhibits formation of a functional preinitiation complex: implications for active repression. *Genes Dev* 7:1400–1410

Force W, Tillman JB, Sprung CN, Spindler SR (1994): Homodimer and heterodimer DNA binding and transcriptional responsiveness to triiodothyronine (T3) and 9-cis retinoic acid are determined by the number and order of high affinity half-sites in a T3 response element. *J Biol Chem* 269:8863–8871

Forman BM, Samuels HH (1990): Interactions among a subfamily of nuclear hormone receptors: The regulatory zipper model. *Mol Endocrinol* 4:1293–1301

Freedman LP, Yoshinaga SK, Vanderbilt JN, Yamamoto KR (1989): In vitro transcription enhancement by purified derivatives of the glucocorticoid receptor. *Science* 245:298–301

Gandrillon O, Jurdic P, Pain B, Desbois C, Madjar JJ, Moscovici MG, Moscovici C, Samarut J (1989): Expression of the v-erbA product an altered nuclear hormone receptor is sufficient to transform erythrocytic cells in vitro. *Cell* 58:115–121

Giguere V, Ong ES, Segui P, Evans RM (1987): Identification of a receptor for the morphogen retinoic acid. *Nature* 330:624–629

Gill G, Ptashne M (1988): Negative effect of the transcriptional activator GAL4. *Nature* 334:721–724

Glass CK, Lipkin SM, Devary OV, Rosenfeld MG (1989): Positive and negative regulation of gene transcription by a retinoic acid-thyroid hormone receptor heterodimer. *Cell* 59:697–708

Glass CK, Devary OV, Rosenfeld MG (1990): Multiple cell type-specific proteins differentially regulate target sequence recognition by the α retinoic acid receptor. *Cell* 63:729–738

Glineur C, Bailly A, Ghysdael J (1989): The c-erbA alpha-encoded thyroid hormone receptor is phosphorylated in its amino terminal domain by casein kinase II. *Oncogene* 4:1247–1254

Green S, Chambon P (1987): Oestradial induction of a glucocorticoid-responsive gene by a chimaeric receptor. *Nature* 325:75–78

Green S, Chambon P (1988): Nuclear receptors enhance our understanding of transcription regulation. *Trends Genet* 4:309–314

Gronemeyer H (1993): Transcription activation by nuclear receptors. *Receptor Res* 13:667–691

Guiochon-Mantel A, Loosfelt H, Lescop P, Sar S, Atger M, Perrot-Applanat M, Milgrom E (1989): Mechanisms of nuclear localization of the progesterone receptor: evidence for interaction between monomers. *Cell* 57:1147–1154

Ha I, Roberts S, Maldonado E, Sun X, Kim LU, Green M, Reinberg D (1993): Multiple functional domains of human transcription factor IIB distinct interactions with two general transcription factors and RNA polymerase II. *Gene Dev* 7:1021–1032

Ham J, Thomson A, Needham M, Webb P, Parker M (1988): Characterization of response elements for androgen glucocorticoids and progestins in mouse mammary tumor virus. *Nucleic Acids Res* 16:5263–5277

Hazel TG, Nathans D, Lau LF (1988): A gene inducible by serum growth factors encodes a member of the steroid and thyroid hormone receptor superfamily. *Proc Natl Acad Sci USA* 85:8444–8448

Hazel TG, Misra R, Davis IJ, Greenberg ME, Lau LF (1991): NUR77 is differentially modified in PC12 cells upon membrane depolarization and growth factor treatment. *Mol Cell Biol* 11:3239–3246

Hernandez N (1993): TBP, a universal eukaryotic transcription factor? *Genes Dev* 7:1291–1308

Hollenberg SM, Evans RM (1988): Multiple and cooperative trans-activation domains of the human glucocorticoid receptor. *Cell* 55:899–906

Hudson LG, Thompson KL, Xu J, Gill GN (1990): Identification and characterization of a regulated promoter element in the epidermal growth factor receptor gene. *Proc Natl Acad Sci USA* 87:7536–7540

Ignar-Trowbridge DM, Teng CT, Ross KA, Parker MG, Korach KS, McLachlan JA (1993): Peptide growth factors elicit estrogen receptor-dependent transcriptional activation of an estrogen-responsive element. *Mol Endocrinol* 7:992–998

Ing NH, Beekman JM, Tsai SY, Tsai M-J, O'Malley BW (1992): Members of the steroid hormone receptor superfamily interact with TFIIB (S300II). *J Biol Chem* 267:17617–17623

Ip MM, Shea WK, Rowan BG (1993): Mutant glucocorticoid receptors in lymphoma. *Ann NY Acad Sci* 684:94–115

Jacq X, Brou C, Lutz Y, Davidson I, Chambon P, Tora L (1994): Human TAFII30 is present in a distinct TFIID complex and is required for transcriptional activation by the estrogen receptor. *Cell* 79107–17

Jantzen HM, Strahle U, Gloss B, Stewart F, Schmid W, Boshart M, Miksicek R, Schutz G (1987): Cooperativity of glucocorticoid response elements located far upstream of the tyrosine aminotranferase gene. *Cell* 49:29–38

Jonat G, Rahmsdorf HJ, Park KK, Cato ACB, Gebel S, Ponta H, Herrlich P (1990): Antitumor promotion and antiinflammation: down-modulation of AP-1 (Fos/Jun) activity by glucocorticoid hormone. *Cell* 62:1189–1204

Jurutka PW, Hsieh JC, MacDonald PN, Terpening CM, Haussler CA, Haussler MR, Whitfield GK (1993): Phosphorylation of serine 208 in the human vitamin D receptor The predominant amino acid phosphorylated by casein kinase II in vitro and identification as a significant phosphorylation site in intact cells. *J Biol Chem* 268:6791–6799

Karin M, Haslinger A, Holtgreve H, Richards RI, Krauter P, Westphal HM, Beato M (1984): Characterization of DNA sequences through which cadmium and glucocorticoid hormones induce human metallothionein—IIA gene. *Nature* 308:513–519

Keidel S, LeMotte P, Apfel C (1994): Different agonist- and antagonist-induced conformational changes in retinoic acid receptors analyzed by protease mapping. *Mol Cell Biol* 14:287–298

Klein-Hitpass L, Tsai SY, Weigel NL, Allan GF, Riley D, Rodriguez R, Schrader WT, Tsai M-J, O'Malley BW (1990): The progesterone receptor stimulates cell-free transcription by enhancing the formation of a stable preinitiation complex. *Cell* 60:247–257

Kliewer SA, Umesono K, Heyman RA, Mangelsdorf DJ, Dyck JA, Evans RM (1992a): Retinoid X receptor-COUP-TF interactions modulate retinoic acid signalling. *Proc Natl Acad Sci USA* 89:1448–1452

Kliewer SA, Umesono K, Mangelsdorf DJ, Evans RM (1992b): Retinoid X receptor interacts with nuclear receptors in retinoic acid thyroid hormone and vitamin D signalling. *Nature* 355:446–449

Kliewer SA, Umesono K, Noonan DJ, Heyman RA, Evans RM (1992c): Convergence of 9-*cis* retinoic acid and peroxisome proliferator signalling pathways through heterodimer formation of their receptors. *Nature* 358:771–774

Konig H, Ponta H, Rahmsdorf HJ, Herrlich P (1992): Interference between pathway-specific transcription factors: glucocorticoids antagonize phorbol ester-induced AP-1 activity without altering AP-1 site occupation in vivo. *EMBO J* 11:2241–2246

Kumar V, Green S, Staub A, Chambon P (1986): Localization of the oestradiol-binding and putative DNA-binding domains of the human oestrogen receptor. *EMBO J* 5:2231–2236

Kumar V, Green S, Stack G, Berry M, Jin J, Chambon P (1987): Functional domains of the human estrogen receptor. *Cell* 51:941–951

Kumar V, Chambon P, (1988): The estrogen receptor binds tightly to its response element as a ligand-induced homodimer. *Cell* 55:145–156

Kurokawa R, Yu VC, Naar A, Kyakumoto S, Han Z, Silverman S, Rosenfeld MG, Glass CK (1993): Differential orientations of the DNA-binding domain and carboxy-terminal dimerization interface regulatory binding site selection by nuclear receptor heterodimers. *Genes Dev* 7:1423–1435

Lala DS, Rice DA, Parker KL (1992): Steroidogenic factor 1 a key regulator of steroidogenic enzyme expression is the mouse homolog of *fushi tarazu*-factor 1. *Mol Endocrinol* 6:1249–1258

Landschultz WH, Johnson PF, McKnight SL (1988): The leucine zipper: a hypothetical structure common to a new class of DNA binding proteins. *Science* 240:1759–1764

Laudet V, Hanni C, Coll J, Catzeflis F, Stehelin D (1992): Evolution of the nuclear receptor gene superfamily. *EMBO J* 11:1003–1013

Law SW, Conneely OM, DeMayo FJ, O'Malley BW (1992): Identification of a new brain specific transcription factor Nurr1. *Mol Endocrinol* 6:2129–2135

Lazar MA, Berrodin TJ, Harding HP (1991): Differential DNA binding by monomeric homodimeric and potentially heteromeric forms of the thyroid hormone receptor. *Cell Biol* 11:5005–5015

Lee MS, Gippert GP, Soman KV, Case DA, Wright PE (1989): Three dimensional solution structure of a single zinc finger DNA-binding domain. *Science* 245:635–637

Lee MS, Kliewer SA, Provencal J, Wright PE, Evans RM (1993): Structure of the retinoid X receptor alpha DNA binding domain: a helix required for homodimeric DNA binding. *Science* 260:1117–1121

Lees JA, Fawell SE, White R, Parker MG (1990): A 22-amino-acid peptide restores DNA-binding activity to dimerization-defective mutants of the estrogen receptor. *Mol Cell Biol* 10:5529–5531

Leid M, Kastner P, Lyons R, Nakshatri H, Saunders M, Zacharewski T, Chen JY, Staub A, Garnier JM, Chambon P (1992): Purification cloning and RXR identity of the HeLa cell factor with which RAR or TR heterodimerizes to bind target sequences efficiently. *Cell* 68:377–397

Leid M (1994): Ligand-induced alteration of the protease sensitivity of retinoid X receptor alpha. *J Biol Chem* 269:14175–14181

Leng X, Blanco J, Tsai SY, Ozato K, O'Malley BW, Tsai M-J (1994): Mechanisms for synergistic activation of thyroid hormone receptor and retinoid X receptor on different response elements. *J Biol Chem* 269:31436–31442

Leng X, Blanco J, Tsai SY, Ozato K, O'Malley BW, Tsai M-J (1995): Mouse retinoid X receptor contains a separable ligand binding and transactivation domain in its E region. *Mol Cell Biol* 15:255–263

Leng X, Tsai SY, O'Malley BW, Tsai M-J (1993): Ligand-dependent conformational changes in homodimers and heterodimers of the thyroid hormone and retinoic acid receptors. *J Steroid Biochem Mol Biol* 46:643–661

Levine M, Manley JL (1989): Transcriptional repression of eukaryotic promoters. *Cell* 59:405–408

Liao J, Ozono K, Sone T, McDonnell DP, Pike WJ (1990): Vitamin D receptor interaction with specific DNA requires a nuclear protein and 125-dihydroxy vitamin D_3. *Proc Natl Acad Sci USA* 87:9751–9755

Lin K, Parkison C, McPhie P, Cheng S (1991): An essential role of domain D in the hormone-binding activity of human beta1 thyroid hormone nuclear receptor. *Mol Endocrinol* 5:485–492

Lin YS, Ha I, Maldonado E, Reinberg D, Green MR (1991): Binding of general transcription factor TFIIB to an acidic activating region. *Nature* 353: 569–571

Lucibello FC, Slater EP, Jooss KU, Beato M, Muller R (1990): Mutual transrepression of Fos and the glucocorticoid receptor: involvement of a functional domain in Fos which is absent in FosB. *EMBO J* 9:2827–2834

Luisi BF, Xu WX, Otwinowski Z, Freedman LP, Yamamoto KR, Sigler PB (1991): Crystallographic analysis of the interaction of the glucocortioid receptor with DNA. *Nature* 352:497–505

Lutz B, Kuratani S, Thaller C, Eichele G (1994): Nuclear receptors in development and differentiation In: *Mechanism of Steroid Hormone Regulation of Gene Transcription*. Tsai M-J, O'Malley BW, eds. Austin: R.G. Landes Company

Mader S, Kumar V, deVerneuil H, Chambon P (1989): Three amino acids of the oestrogen receptor are essential to its ability to distinguish an oestrogen from a glucocorticoid-responsive element. *Nature* 338:271–274

Mader S, Chambon P, White JH (1993): Defining a minimal estrogen receptor DNA binding domain. *Nucleic Acids Res* 21:1125–1132

Mangelsdorf DJ, Umesono K, Kliewer SA, Borgmeyer U, Ong ES, Evans RM (1991): A direct repeat in the cellular retinol-binding protein type II gene confers differential regulation by RXR and RAR. *Cell* 66:555–561

Marks MS, Hallenbeck PL, Nagata T, Segars JH, Appella E, Nikodem VM, Ozato K (1992): H-2RIIBP [RXR-β] heterodimerization provides a mechanism for combinatorial diversity in the regulation of retinoic acid and thyroid hormone responsive genes. *EMBO J* 11:1419–1435

Meyer ME, Gronemeyer H, Turcotte B, Bocquel MT, Tasset D, Chambon P (1989): Steroid hormone receptors compete for factors that mediate their enhancer function. *Cell* 57:433–442

Meyer ME, Quirin-Stricker C, Lerouge T, Bocquel MT, Gronemeyer H (1992): A limiting factor mediates the differential activation of promoters by the human progesterone receptor isoforms. *J Biol Chem* 267:10882–10887

Miksicek R, Borgmeyer U, Nowock J (1994): Interaction of the TGGCA-binding protein with upstream sequences is required for efficient transcription of mouse mammary tumour virus. *EMBO J* 6:1355–1360

Milbrandt J (1988): Nerve growth factor induces a gene homologous to the glucocorticoid receptor gene. *Neuron* 1:183–188

Miner JN, Yamamoto KR (1991): Regulatory crosstalk at composite response element. *Trends Biochem Sci* 16:423–426

Miner JN, Yamamoto KR (1992): The basic region of AP-1 specifies glucocorticoid receptor activity at a composite response element. *Genes Dev* 6:2491–2501

Mordacq JC, Linzer DIH (1994): Co-localization of elements required for phorbol ester stimulation and glucocorticoid repression of proliferin gene expression. *Genes Dev* 3:760–769

Murry MB, Towle HC (1989): Identification of nuclear factors that enhance binding of the thyroid hormone receptor to a thyroid hormone response element. *Mol Endocrinol* 3:1434–1442

Näär AM, Boutin JM, Lipkin SM, Yu VC, Holloway JM, Glass CK, Rosenfeld MG (1991): The orientation and spacing of core DNA-binding motifs dictate selective transcriptional responses to three nuclear receptors. *Cell* 657: 1267–1279

Nicholson RC, Mader S, Nagpal S, Leid M, Rochette-Egly C, Chambon P (1990): Negative regulation of the rat stromelysin gene promoter by retinoic acid is mediated by an AP1 binding site. *EMBO J* 9:4443–4454

O'Donnell AL, Rosen ED, Darling DS, Koenig RJ (1991): Thyroid hormone receptor mutations that interfere with transcriptional activation also interfere with receptor interaction with a nuclear protein. *Mol Endocrinol* 5:94–99

O'Malley BW (1990): The steroid receptor superfamily: more excitement predicted for the future. *Mol Endocrinol* 4:363–369

Ohno CK, Petkovich M (1992): FTZ-F1b a novel member of the drosophila nuclear receptor family. *Mech Dev* 40:13–24

Oro AE, Hollenberg SM, Evans RM (1988): Transcriptional inhibition by a glucocorticoid receptor-beta-galactosidase fusion protein. *Cell* 55:1109–1114

Orti E, Bodwell JE, Munck A (1992): Phosphorylation of steroid hormone receptors. *Endocrine Rev* 13:105–128

Parker MG (1990): Structure and function of nuclear hormone receptors. *Semin Cancer Biol* 1:81–87

Parker MG (1993): Steroid and related receptors. *Cell Biol* 5:499–504

Parvin JD, Timmers HT, Sharp PA (1993): Promoter specificity of basal transcription factors. *Cell* 68:1135–1144

Perlmann T, Rangarajan PN, Umesono K, Evans RM (1993): Determinants for selective RAR and TR recognition of direct repeat HREs. *Genes Dev* 7:1411–1422

Petkovich M, Brand NJ, Krust A, Chambon P (1987): A human retinoic acid receptor which belongs to the family of nuclear receptors. *Nature* 330:444–450

Picard D, Salser SJ, Yamamoto KR (1988): A movable and regulatable inactivation function within the steroid binding domain of the glucocorticoid receptor. *Cell* 54:1073–1080

Picard D, Yamamoto KR (1987): Two signals mediate hormone-dependent nuclear localization of the glucocorticoid receptor. *EMBO J* 6:3333–3340

Pike WJ, Sleator NM (1985): Hormone-dependent phosphorylation of the 1,25 dihydroxyvitamin D_3 receptor is generated through a hormone-dependent phosphorylation. *Biochem Biophys Res Commun* 131:378–385

Power RF, Lydon JP, Conneely OM, O'Malley BW (1991a): Dopamine activation of an orphan member of the steroid receptor superfamily. *Science* 252:1546–1548

Power RF, Mani SK, Codina J, Conneely OM, O'Malley BW (1991b): Dopaminergic and ligand-independent activation of steroid hormone receptors. *Science* 254: 1636–1639

Qiu YH, Cooney AJ, Kuratani S, DeMayo FJ, Tsai SY, Tsai M-J (1994a): Differential expression of COUP-TFI and II in mouse developing CNS. *Proc Natl Acad Sci USA* 91:4451–4455

Qiu YH, Tsai SY, Tsai M-J (1994b): COUP-TF: An orphan member of the steroid/thyroid hormone receptor superfamily. *Trends Endocrinol Metab* 5:234–239

Renkawitz R, Schutz G, von der Ahe D, Beato M (1984): Sequences in the promoter region of the chicken lysozyme gene required for steroid regulation and receptor binding. *Cell* 37:503–510

Renkawitz R (1990): Transcriptional repression in eukaryotes. *Trends Genet* 6:192–197

Rochette-Egly C, Gaub M, Lutz Y, Ali S, Scheuer I, Chambon P (1992): Retinoic acid receptor-beta: Immunodetection and phosphorylation on tyrosine residues. *Mol Endocrinol* 6:2197–2209

Rowe A, Eager NSC, Brickell PM (1991): A member of the RXR nuclear receptor family is expressed in the neural-crest-derived cells of the developing peripheral nervous system. *Development* 111:771–778

Rowe A, Brickell PM (1993): Current status review: the nuclear retinoid receptors. *Int J Exp Path* 74:117–126

Ruberte E, Kastner P, Dolle P, Krust A, Leroy P, Mendelsohn C, Zelent A, Chambon P (1991): Retinoic acid receptor in the embryo. *Semin Devel Biol* 2:153–159

Ryseck RP, MacDonald-Bravo H, Mattei MG, Ruppert S, Bravo R (1989): Structure mapping and expression of a growth factor inducible gene encoding a putative nuclear hormonal binding receptor. *EMBO J* 8:3327–3335

Sabbah M, Gouilleux F, Sola B, Redeuilh G, Baulieu EE (1991): Structural differences between the hormone and antihormone estrogen receptor complexes bound to the hormone response element. *Proc Natl Acad Sci USA* 88:390–394

Sap J, Munoz A, Schmitt J, Stunnenberg HG, Vennstrom B (1989): Repression of transcription mediated at a thyroid hormone response element by the v-erb-A oncogene product. *Nature* 340:242–244

Sartorius CA, Melville MY, Hovland AR, Tung L, Takimoto GS, Horwitz KB (1994): A third transactivation function (AF3) of human progesterone receptors located in the unique N-terminal segment of the B-isoform. *Mol Endocrinol* 8:1347–1360

Scearce LM, Laz TM, Hazel TG, Lau LF, Taub R (1993): RNR-1 a nuclear receptor in the NGFI-B/Nur77 family that is rapidly induced in degenerating liver. *J Biol Chem* 268:8855–8861

Schrader M, Muller KM, Carlberg C (1994a): Thyroid hormone receptor function as monomeric ligand-induced transcription factor on octameric half-sites. *J Biol Chem* 269:5501–5504

Schrader M, Muller KM, Nayeri S, Kahlen JP, Carlberg C (1994b): Vitamin D_3-thyroid hormone receptor heterodimer polarity directs ligand sensitivity of transactivation. *Nature* 370:382–386

Schule R, Muller E, Kaltschmidt C, Renkawitz R (1988a): Many transcription factors interact synergistically with steroid receptors. *Science* 242:1418–1420

Schule R, Muller E, Otsuka-Murakami H, Renkawitz R (1988b): Cooperativity of the glucocorticoid receptor and the CACCC-box binding factor. *Nature* 332:87–90

Schule R, Rangarajan PN, Kliewer SA, Ransone LJ, Bolado J, Yang N, Verma IM, Evans RM (1990a): Retinoic acid is a negative regulator of AP-1 responsive genes. *Cell* 62:1217–1226

Schule R, Rangarajan PN, Kliewer SA, Ransone LJ, Bolado J, Yang N, Verma IM, Evans RM (1990b): Functional antagonism between oncoprotein c-jun and the glucocorticoid receptor. *Cell* 62:1217–1226

Schwabe JWR, Neuhaus D, Rhodes D (1990): Solution structure of the DNA-binding domain of the oestrogen receptor. *Nature* 340:458–461

Schwabe JWR, Chapman L, Finch JT, Rhodes D (1993): The crystal structure of the complex between the oestrogen receptor DNA-binding domain and DNA at 24A: how receptors discriminate between their response elements. *Cell* 75: 567–578

Shemshedini L, Knauthe R, Sassone-Corsi P, Pornon A, Gronemeyer H (1991): Cell specific inhibitory and stimulatory effects of fos and jun on transcription activation by nuclear receptors. *EMBO J* 10:3839–3849

Somers JP, DeFranco DB (1992): Effects of okadaic acid a protein phosphatase inhibitor on glucocorticoid receptor-mediated enhancement. *Mol Endocrinol* 6:26–34

Strahle U, Klock G, Schutz G (1987): A DNA sequence of 15 base pairs is sufficient to mediate both glucocorticoid and progesterone induction of gene expression. *Proc Natl Acad Sci USA* 84:7871–7875

Strahle U, Schmid W, Schutz G (1988): Synergistic action of the glucocorticoid receptor with transcription factors. *EMBO J* 7:3389–3395

Tasset D, Tora L, Fromental C, Scheer E, Chambon P (1990): Distinct classes of transcriptional activating domains function by different mechanisms. *Cell* 62:1177–1187

Thompson CC, Evans RM (1989): Trans-activation by thyroid hormone receptors: functional parallels with steroid hormone receptors. *Proc Natl Acad Sci USA* 86:3494–3498

Tini M, Otulakowski G, Breitman ML, Tsui LC, Giguere V (1993): An everted repeat mediates retinoic acid induction of the F-crystallin gene: Evidence of a direct role for retinoids in lens development. *Genes Dev* 7:295–307

Toohey MG, Lee JW, Huang M, Peterson DO (1990): Functional elements of the steroid hormone-responsive promoter of mouse mammary tumor virus. *J Virol* 64:4477–4488

Tora L, White JH, Brou C, Tasset DM, Webster NJG, Scheer E, Chambon P (1989): The human estrogen receptor has two independent nonacidic transcriptional activation functions. *Cell* 59:477–487

Tran PBV, Zhang XK, Salbert G, Hermann T, Lehmann JM, Pfahl M (1992): COUP orphan receptors are negative regulators of retinoic acid response pathways. *Mol. Cell Biol* 12:4666–4676

Truss M, Beato M (1993): Steroid hormone receptors: Interaction with deoxyribonucleic acid and transcription factors. *Endocrine Rev* 14:459–479

Tsai M-J, O'Malley BW (1994): Molecular mechanisms of action of steroid/thyroid receptor superfamily members. *Annu Rev Biochem* 63:451–486

Tsai SY, Sagami I, Wang LH, Tsai M-J, O'Malley BW (1987): Interactions between a DNA-binding transcription factor (COUP) and a non-DNA binding factor (S300-11). *Cell* 50:701–709

Tsai SY, Carlstedt-Duke J, Weigel NL, Dahlman K, Gustafsson JÅ, Tsai M-J, O'Malley BW (1988): Molecular interactions of steroid hormone receptor with its enhancer element: evidence for receptor dimer formation. *Cell* 55: 361–369

Tsai SY, Tsai M-J, O'Malley BW (1989): Cooperative binding of steroid hormone receptors contributes to transcriptional synergism at target enhancer elements. *Cell* 57:443–448

Tsai SY, Srinivasan G, Allan GF, Thompson EB, O'Malley BW, Tsai M-J (1990): Recombinant human glucocorticoid receptor induces transcription of hormone response genes in vitro. *J Biol Chem* 265:17055–17061

Umesono K, Giguere V, Glass CK, Rosenfeld MG, Evans RM (1988): Retinoic acid and thyroid hormone induce gene expression through a common responsive element. *Nature* 336:262–265

Umesono K, Murakami KK, Thompson CC, Evans RM (1991): Direct repeats as selective response elements for the thyroid hormone retinoic acid and vitamin D_3 receptors. *Cell* 65:1255–1266

Umesono K, Evans RM (1989): Determinants of target gene specificity for steroid/thyroid hormone receptors. *Cell* 57:1139–1146

Usala SJ (1991): Molecular diagnosis and characterization of thyroid hormone resistance syndromes. *Thyroid* 1:361–367

von der Ahe D, Janich S, Schneider C, Renkawitz R, Schutz G, Beato M (1985): Glucocorticoid and progesterone receptors bind to the same sites in two hormonally regulated promoters. *Nature* 313:706–709

Wahli W, Martinez E (1991): Superfamily of steroid nuclear receptors: positive and negative regulators of gene expression. *FASEB J* 5:2243–2249

Wang LH, Tsai SY, Cook RG, Beattie WG, Tsai M-J, O'Malley BW (1989): COUP transcription factor is a member of the steroid receptor superfamily. *Nature* 340:163–166

Webster NJG, Green S, Jin J, Chambon P (1988): The hormone-binding domains of the estrogen and glucocorticoid receptors contain an inducible transcription activation function. *Cell* 54:199–207

Weigel NL (1994): Receptor phosphorylation. In: *Mechanism of Steroid Hormone Regulation of Gene Transcription*. Tsai M-J, O'Malley BW, eds. Austin: R.G. Landes Company

Weiland S, Dobbeling U, Rusconi S (1991): Interference and synergism of glucocorticoid receptor and octamer factors. *EMBO J* 10:2513–2522

Weinberger C, Hollenberg SM, Rosenfeld MG, Evans RM (1985): Domain structure of human glucocorticoid receptor and its relationship to the V-erbA oncogene product. *Nature* 318:670–672

Weinberger C, Thompson CC, Ong ES, Lebo R, Gruol DJ, Evans RM (1986): The c-erb-A gene encodes a thyroid hormone receptor. *Nature* 324:641–646

Widom RL, Rhee M, Karathanasis SK (1992): Repression by ARP-1 sensitizes apolipoprotein AI gene responsiveness to RXR alpha and retinoic acid. *Mol Cell Biol* 12:3380–3389

Wilson TE, Fahrner TJ, Johnston M, Milbrandt J (1991): Identification of the DNA binding site for NGFI-beta by genetic selection in yeast. *Science* 252:1296–1300

Wilson TE, Paulsen RE, Padgett KA, Milbrandt J (1992): Participation of non-zinc finger residues in DNA binding by two nuclear orphan receptors. *Science* 256:107–110

Wilson TE, Fahrner TJ, Milbrandt J (1993a): The orphan receptors NGFI-B and steroidogenic factor 1 establish monomer binding as a third paradigm of nuclear receptor-DNA interaction. *Mol Cell Biol* 13:5794–5804

Wilson TE, Mouw AR, Weaver CA, Milbrandt J, Parker KL (1993b): The orphan nuclear receptor NGFI-β regulates expression of the gene encoding steroid-21-hydroxylase. *Mol Cell Biol* 13:861–868

Yamamoto KR (1985): Steroid receptor-regulated transcription of specific genes and gene network. *Ann Rev Genet* 19:209–252

Yang-Yen H, Chambard J, Sun Y, Smeal T, Schmidt TJ, Drouin J, Karin M (1990): Transcriptional interference between c-Jun and the glucocorticoid receptor: mutual inhibition of DNA binding due to direct protein-protein interaction. *Cell* 62:1205–1215

Yao TP, Segraves WA, Oro AE, McKeown M, Evans RM (1992): Drosophila ultraspiracle modulates ecdysone receptor function via heterodimer formation. *Cell* 71:63–72

Yao TP, Forman BM, Jiang Z, Cherbas L, Chen JD, Mckeown M, Cherbas P, Evans RM (1993): Functional ecdysone receptor is the product of EcR and Ultraspiracle genes. *Nature* 366:476–479

Yu VC, Delsert C, Andersen B, Holloway JM, Devary OV, Näär AM, Kim SY, Boutin JM, Glass CK, Rosenfeld MG (1991): A coregulator that enhances binding of retinoic acid thyroid hormone and vitamin D receptors to their cognate response elements. *Cell* 67:1251–1266

Zawel L, Reinberg D (1993): Initiation of transcription by RNA polymerase II: A multiple step process. *Progress in Nucleic Acid Res Mol Biol* 44:69–108

Zeng Z, Allan GF, Thaller C, Cooney AJ, Tsai SY, O'Malley BW, Tsai M-J (1994): Detection of potential ligands for nuclear receptors in cellular extracts. *Endocrinol* 135:248–52

Zhang XK, Hoffmann B, Tran PBV, Graupner G, Pfahl M (1992): Retinoid X receptor is an auxiliary protein for thyroid hormone and retinoic acid receptors. *Nature* 355:441–446

5

Retinoids: Concepts for Separation of Desirable and Undesirable Effects in the Treatment or Prevention of Cancer

MAGNUS PFAHL

INTRODUCTION

Since its discovery early this century, vitamin A has been known to play essential biological roles (Lotan, 1981; Roberts and Sporn, 1984; Morris-Kay 1992). It has now been established that the active derivatives of vitamin A such as all-*trans* retinoic acid (ATRA) and 9-*cis* retinoic acid (9-*cis* RA) as well as their synthetic analogs—the retinoids—function as hormone-like signaling molecules that affect important biological processes at the level of gene transcription. The ability of retinoids to regulate growth and development, cell proliferation, and differentiation made them particularly attractive as therapeutic candidates for the treatment of cancer and other proliferative diseases. In fact, 13-*cis* RA and related retinoids are the most effective drugs known today for the treatment of acne vulgaris and other skin diseases (reviewed in Schaefer and Reichert, 1990). In addition, ATRA has recently been approved for the treatment of acute promyelocytic leukemia (APL) where it leads to complete remission (Huang et al., 1988; Castaigne et al., 1990; Chen et al., 1991). Retinoids have also shown promise for the treatment and prevention of many other cancers including cancers of the head and neck, cervical cancer, and most other epithelial tissue-derived cancers (reviewed in Bollag and Holdener, 1992; Smith et al., 1992; Meyskens et al., 1994).

A major problem, however, that prevented the full development of retinoids as anticancer agents has been their undesirable side effects, which can range from mild skin irritations to severe nausea and malformation in fetuses born from mothers taking retinoids during pregnancy. In the last

Hormones and Cancer
Wayne V. Vedeckis, Editor

20 years, a large effort has therefore been devoted to the synthesis of retinoic acid derivatives and analogs, with the aim to separate the beneficial from the undesirable retinoid effects (reviewed in Dawson and Okamura, 1990). Because of the complex biological responses to retinoids and a lack of understanding of mechanisms of retinoid action, the success in development of retinoids with more optimal therapeutic profiles was limited.

More recently, important breakthroughs in the elucidation of the molecular mechanism of retinoid action, combined with the demonstration that retinoids with selective biological activity can be designed, have strongly revived the interest in retinoids as therapeutic agents. This interest is further enhanced by a number of recent clinical studies demonstrating the effectiveness of retinoids in the treatment and prevention of several cancers (Chen et al., 1991; Hong et al., 1991; Meyskens et al., 1994). Since the large majority of the clinical and preclinical studies and treatments have been carried out so far with natural retinoids or closely related compounds and have been well reviewed recently (i.e., see Bollag and Holdener, 1992; Smith et al., 1992), this chapter will focus on the present understanding of retinoid signal transduction by specific receptors and discuss the possibilities that now exist for the development of retinoids with more targeted activities and fewer side effects, and thus higher therapeutic index.

RETINOID SIGNALING MECHANISMS

The Nuclear Receptors

It is now well established that retinoid signals are mediated by specific nuclear receptors. These receptors, the three retinoic acid receptors (RARα, β, γ) and the three retinoid X receptors (RXRα, β, γ), are part of a large family of regulatory proteins that also includes the steroid and thyroid hormone receptors. The relationship of RARs and RXRs to the other receptors as well as their structural domains has been reviewed elsewhere in detail (Pfahl et al., 1994; also see Chapter 4 of this volume). Briefly, the receptors contain a highly conserved DNA binding domain (DBD) that allows specific DNA recognition as well as protein-protein interaction. The second general feature of the receptors is a "ligand binding domain" (LBD) comprising the carboxy terminal half of the receptors. This LBD is less well conserved among the receptors and, besides specific ligand binding, also contains a transcriptional activation function as well as a strong dimerization domain. In addition to the six major retinoid receptor subtypes, isoforms for each of the receptors have been identified that differ in their amino terminal regions, a domain that can also contain a transactivation function. The various receptor isoforms are generated by differential promoter usage and alternative splicing (for reviews see Morris-Kay, 1992; Chambon, 1994; Pfahl et al., 1994).

Although the RARs and RXRs represent the only receptors that to date are known to directly interact with or bind retinoids, an increasing number of other receptors are being identified that interact with the retinoid receptors either by heterodimer formation or by competing for the same specific DNA binding sites—the RA response elements (RAREs). These receptors include the thyroid hormone receptors (TRs), vitamin D_3 receptor (VDR), peroxisome proliferator activator receptor (PPAR), and a number of "orphan receptors," receptors for which specific ligands have not yet been identified. Thus, the retinoid hormone receptor subfamily comprises a complex network of receptors that modulates the retinoid response, and in addition, allows for cross-talk with various other hormones and vitamin derivatives.

Transcriptional Regulation via Response Elements by RXR Containing Hetero- and Homodimers.

Like the steroid hormone receptors, retinoid receptors regulate gene expression by binding as dimeric complexes to specific DNA sites, the RAREs. As observed for the steroid hormone responsive elements, these RAREs are made up of a minimum of two identical or similar half-sites (usually PuGGTCA or a closely related sequence) and are located usually in the promoter or 5′ flanking region of RA responsive genes (Pfahl, 1994). However, whereas steroid hormone response elements are palindromes where the half-sites are separated by a 3-bp spacer (Beato, 1989), RAREs comprise a variety of differently arranged half-sites including direct repeats with 1-, 2-, 4-, 5- or 8-bp spacers, palindromes with no or a 9-bp spacer, and everted repeats with an 8-bp spacer (Fig. 1). This diversity of RAREs may (as discussed in more detail below) allow differential responses to retinoid signals.

In contrast to the steroid hormone receptors that bind as homodimers, RARs were found to interact with their response elements as heterodimers with RXR; RARs or RXRs alone were found to bind poorly to DNA (Yu et al., 1991; Bugge et al., 1992; Kliewer et al., 1992a; Leid et al., 1992; Marks et al., 1992; Zhang et al., 1992a). Since effective DNA binding is required for receptor function, these data suggested that RARs and RXRs mostly function as heterodimers. These conclusions from the *in vitro* data have indeed recently been confirmed by gene "knockout" experiments demonstrating that also *in vivo* the RAR/RXR heterodimer is a transmitter of the retinoid signal (Lohnes et al., 1994; Mendelsohn et al., 1994).

In the presence of ATRA or 9-*cis* RA, the heterodimers act as activators of transcription; however, DNA binding of the heterodimers is independent of the ligand, and this may allow the receptors to function as gene repressors in the absence of ligand, as has also been observed for other members of the nuclear receptor family (Graupner et al., 1989; Tran et al., 1992). Interestingly, RXRs not only enhance RAR DNA binding but are also required

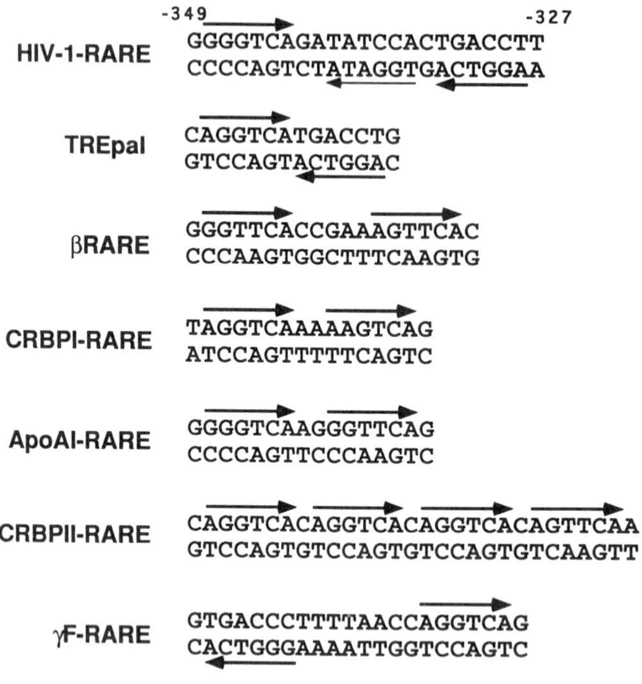

Figure 1. RA response elements. RAREs are composed of two or more half-sites (*arrows*) arranged in various orientations. The half-site is usually AGGTCA or a closely related sequence (for further details see Pfahl et al., 1993).

for efficient DNA binding of TRs, VDR, PPAR (reviewed in Zhang and Pfahl, 1993), and the v-*erb*A oncogene [a mutated form of TR (Hermann et al., 1993)], as well as several orphan receptors (Apfel et al., 1994). These various RXR-containing heterodimers bind to various response elements that can have overlapping specificities. Thus, one class of retinoid receptors, the RXRs, have a central role and function as co-regulators of several other nuclear receptors that are activated by structurally unrelated hormones and signaling molecules (see Fig. 2). This role of RXR appears to have been maintained through evolution because the RXR homologue ultraspiracle of Drosophila was found to heterodimerize with the ecdysone receptor (Yao et al., 1992; Thomas et al., 1993).

Besides their central role as co-receptors, RXRs can also function as homodimers. In the presence of 9-*cis* RA, a natural ligand of RXR, RXR homodimers are induced in solution, allowing for efficient DNA binding of RXR in the absence of RARs or other receptors (Zhang et al., 1992b) (Fig. 2). RXR homodimers mediate a distinct retinoid signaling pathway by interacting selectively with a subset of RAREs, including those present in the promoter regions of the CRBPII and the Apolipoprotein AI genes (Mangelsdorf et al., 1991; Lehmann et al., 1992b; Zhang et al., 1992b). The

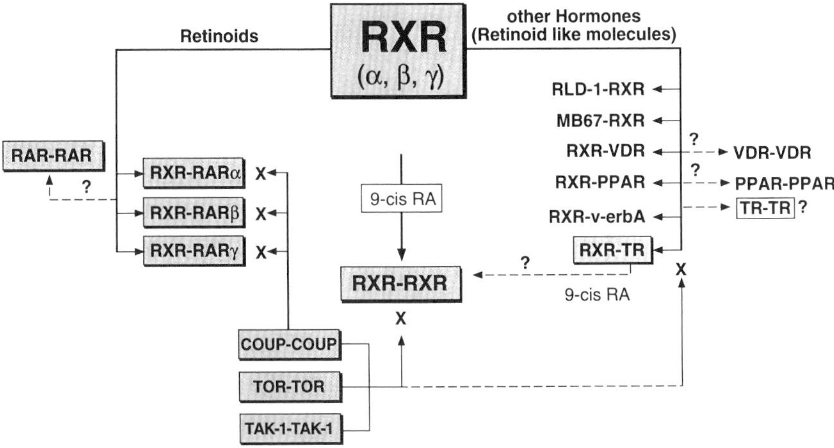

Figure 2. The retinoid receptor network. RXRs form heterodimers with RAR that are activated by retinoids. RXRs also form heterodimers with TRs (including V-*erb*A), VDR, PPAR, and probably other receptors yet to be determined. In the presence of RXR-specific retinoids like 9-*cis* RA, RXR homodimers that recognize a specific subset of RAREs are formed. The presence of ligands that induce the formation of RXR homodimers can inhibit the formation of certain heterodimers; for instance, this can lead to a reduction of expression of T_3-responsive genes, whereas RAR containing heterodimers do not appear to be negatively affected by RXR-specific ligands. COUP and some other receptors form homodimers that bind with high affinity to several RAREs and can repress RAR/RXR heterodimer as well as RXR homodimer activity (Tran et al., 1992). Thus, orphan receptors can restrict retinoid responses to a subset of retinoid-responsive genes. These orphan receptors may also antagonize other receptors. Other orphan receptors, including RLD-1 and MB67, form heterodimers with RXRs that may allow a subclass of RXR ligands to activate from novel response elements. VDR homodimers bind to a subset of VDREs. TR homodimers only bind in the absence of T_3 and can function as ligand-responsive repressors. Other homodimers (RAR, PPAR) may also form on specific response elements. The inhibition of activity of one set of dimers by another is depicted by an "X". (For a more detailed discussion of these mechanisms, see Zhang and Pfahl, 1993.)

unique capability of RXR to form heterodimers with a number of receptors that bind other hormones and signaling molecules, in combination with the ability of RXR to form homodimers in response to 9-*cis* RA and certain synthetic retinoids, allows RXR to mediate cross-talk among retinoids and other regulatory signals and sets the stage for the important physiological role of the retinoids. One obvious mechanism for the cross-talk is the competition of various receptors for RXR.

In addition to the induction of RXR homodimers by specific retinoids like 9-*cis* RA, this binding can shift the equilibrium from heterodimers to RXR homodimers, resulting in the inhibition of heterodimer-mediated signals. When we investigated, for instance, the effect of 9-*cis* RA on TR/RXR heterodimer activity, we observed that thyroid hormone (triiodothyronine; T_3)-induced transcription was strongly repressed by 9-*cis* RA and other synthetic retinoids that lead to RXR homodimer formation (Lehmann et al., 1993). A detailed analysis of the phenomena suggested a mechanism whereby 9-*cis* RA induces the sequestering of RXR molecules from TR-RXR heterodimers into RXR homodimers, thereby leading to a repression of the T_3 response (Lehmann et al., 1993) (see Figs. 1 and 2). The vitamin D_3 response, it appears, can also be inhibited by 9-*cis* RA via this mechanism (MacDonald et al., 1993). Thus, certain retinoids can control the availability of RXR molecules for heterodimerization with other receptors and thereby allow cross-talk between retinoids and other hormone and vitamin signals.

An additional mechanism by which retinoids can control the availability of RXR for other receptors is through the up-regulation of RARs. The RARs belong to the class of so-called master regulators, regulatory proteins that can control their own synthesis. The RARα, β, and γ genes have been shown to contain RAREs in their promoter regions, which allows up-regulation of transcription in response to retinoids (Hoffmann et al., 1990; Leroy et al., 1991; Lehmann et al., 1992a; Chambon, 1994). In particular, RARβ was found to be rapidly increased in lung and other tissues when vitamin A-starved rats were fed ATRA (Haq et al., 1991). Such an RA-induced increase in RAR molecules will necessarily affect the equilibrium of the heterodimers when RXR molecules are limited in the cell and, therefore, this can affect signal responses of other hormone and vitamin receptors that require RXR for DNA binding. Thus, a complex network of receptor interactions has been unraveled that endows retinoids with central biological roles and allows them to influence a network of receptors and thus other hormone and vitamin signals.

The Roles of Orphan Receptors in the Retinoid Response

Although the various heterodimers and ligand-induced RXR homodimers allow for a large degree of diversity and specificity of the retinoid responses, the question of how in certain cell types some retinoid responses can be restricted or modified has not yet been clarified. The orphan receptor COUP and its closely related homologue, ARP-1, are the evolutionary most highly conserved members of the nuclear receptor family. Since their DBDs are closely related to the DBDs of RARs and RXRs, they could recognize similar binding sites or sites overlapping with those of the retinoid receptors. Indeed, COUP-α and -β were found to bind with high affinity to several RAREs but they did not function as activators of these RAREs (Tran et al., 1992). Instead, COUPs were able to inhibit activation of RAREs by RAR/RXR heterodimers and RXR homodimers. Inhibition of retinoid receptors was only observed on those RAREs to which COUP bound strongly as a homodimer including, the palindromic TRE, the CRBPI-RARE, and the RXR homodimer binding ApoA1, CRBPII, and HIV-1-RAREs (Rottman et al., 1991; Tran et al., 1992; Lee et al., 1994). Response elements like the RARβ2-RARE were not recognized by COUP and were also not repressed (Tran et al., 1992). These data and similar results reported by others (Cooney et al., 1992; Kliewer et al., 1992b; Windom et al., 1992) suggest that the COUP receptors are repressors that can restrict RA signaling to certain genes (Fig. 2). A major expression site for COUP are various regions of the developing brain (Lu et al., 1994), where COUP may have important roles in controlling the RA response.

Are there then other orphan receptors in other tissues that could play similar roles as tissue-specific repressors of certain portions of the RA response? Recently two new orphan receptors have been characterized that bind strongly as homodimers to certain RAREs and can repress RA induction of genes containing such elements. One, TAK-I, is expressed in a cell type-specific manner in several tissues, including testis. TAK-I can repress the ATRA response on βRARE type elements and RXR homodimer activity on the CRBPII RARE (Hirose et al., 1995). Another orphan receptor that represses RAREs is TOR. This receptor is expressed almost exclusively in the thymus and mature T cells and represses selectively a subset of DR-4 and DR-5 elements (Ortiz et al., 1995). Thus, a subgroup of orphan receptors exists that can block various RAREs, and these receptors show a tissue-specific expression pattern. The tissue-specific expression of these orphan receptors (repressors) thus allows for a mechanism to restrict the RA response to certain genes in a tissue-specific manner.

Another group of orphan receptors may also modulate the retinoid response and allow for additional more targeted retinoid responses. The orphan receptors RLD-1 and MB67 form heterodimers with RXR that interact with specific DNA sequences, a DR-4-type element in the case of RLD-1/RXR (Apfel et al., 1994) and a DR-5 in the case of MB67/RXR (Baes et al., 1994). Although it was observed that these RXR-containing heterodimer/DNA complexes showed little response to 9-*cis* RA (Apfel et al., 1994; Baes et al., 1994), we have more recently observed that certain DR-4-type elements allow activation by the RLD-1/RXR heterodimer in the presence of 9-*cis* RA. Others have recently reported that a receptor closely related to RLD-1—LXR—also forms heterodimers with RXR on the DR-4 type element. This heterodimer is also activated by 9-*cis* RA (Willy et al., 1995). These observations that RXR ligands can activate an orphan receptor/RXR complex on additional response elements provide an important extension of the concept for signal transduction by heterodimeric receptors and points to novel roles for orphan receptors functioning as auxiliary receptors for RXR. Further research in this area promises to extend the network of the retinoid response pathways even further.

Receptor Interaction with the Activator Protein 1 Transcription Factor

An important alternative mechanism by which retinoid receptors can mediate retinoid signals is by interacting with the transcription factor AP-1 (for a recent review see Pfahl, 1993b). The products of the two proto-oncogenes c-*fos* and c-*jun*, and several related proteins, form a complex in the nucleus termed activator protein 1 (AP-1) that binds specifically to a DNA sequence motif referred to as the AP-1 binding site or TPA-responsive element (TRE) (Angel et al., 1987). AP-1 mediates signals from growth factors, inflammatory peptides, oncogenes, and tumor promoters, usually resulting in cell proliferation (Angel and Karin, 1991). In the presence of their ligands, RARs as well as RXRs can inhibit AP-1 activity via protein/protein interactions. The retinoid receptors can thus directly interfere with many cell proliferation signals when blocking AP-1 activity. In addition, AP-1 can also inhibit RAR activity via the same mechanism (Schüle et al., 1991; Yang-Yen et al., 1991; Salbert et al., 1993). Thus, depending on the concentrations of the AP-1 components and that of the nuclear receptors, AP-1 activity or the activity of the receptors is inhibited, thus allowing for major cross-talk between cellular proliferation and differentiation signals. In Figure 3 a model for the interaction between the retinoid receptors and AP-1 (cJun/cFos) is presented; this has been reviewed in detail elsewhere (Pfahl, 1993b).

SEPARATION OF RETINOID ACTIVITIES

RAR Selective Compounds

With the knowledge of multiple receptors, one way to achieve the long-time goal of chemists in separating the beneficial effects of the retinoids from the

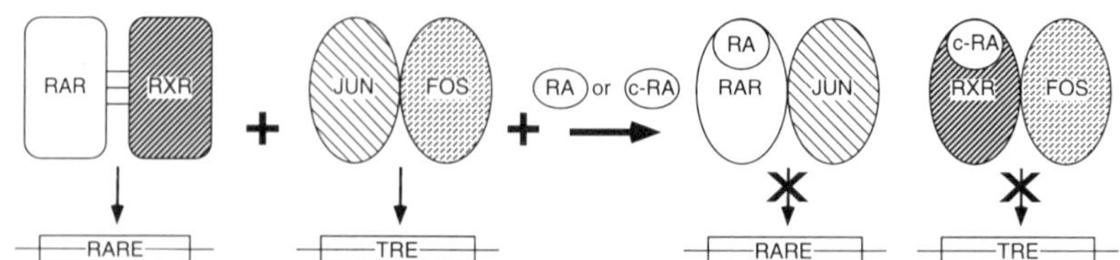

Figure 3. Nuclear receptor/AP-1 interaction. A basic model for cross-talk between retinoid receptors and the transcription factor AP-1 (cJun/cFos) is proposed. RARs and RXRs form heterodimers in solution that bind with high affinity to RAREs. Similarly, the components of AP-1, Jun and Fos, form heterodimers that bind with high affinity to TREs. In the presence of RA or 9-*cis* RA (c-RA), RARs or RXRs undergo a conformational change that allows them to form complexes with Jun and/or Fos. These receptor/Jun or receptor/Fos complexes do not bind to DNA. Thus, when the retinoid receptors are in excess, AP-1 activity is inhibited, while an excess of Jun or Fos can lead to the inhibition of the retinoid receptor activity. (For further details, see Pfahl, 1993b.)

undesirable side effects was to define receptor-selective retinoids that could only activate, for instance, one subtype of receptor. If such receptor-selective retinoids still retained antiproliferative activity, they could be expected to have an overall improved therapeutic index because of their overall reduced alternative biological activities. The natural retinoids are unique among the fat-soluble hormones in that they are quite flexible molecules that might adapt slightly different configurations when interacting with different receptor subtypes. Close examination of the RARs from several different species revealed that amino acid sequence differences among the RAR subtypes were well conserved among species, that is, human RARα was essentially identical in its LBD to mouse RARα whereas both were clearly distinct in their LBDs from mouse and human RARβ and RARγ. This suggested that the evolutionarily conserved differences in the receptor LBDs might serve to allow for differential ligand responses. Indeed, the three RAR subtypes showed differential sensitivity to ATRA when examined in co-transfection assays. Lehmann et al. (1991) and Graupner et al. (1991) were first able to demonstrate that the differential ligand sensitivities of the three RARs could be largely enhanced through the use of conformationally restricted synthetic retinoids. For example, the synthetic retinoids R20 and R22 showed striking differences in their activation capacity of individual RARs, in that they were efficient activators of RARβ and RARγ but poor activators or non-activators of RARα (for retinoid structures, see Fig. 4). The RARβ/γ selectivity of these retinoids may be related to the higher degree of sequence similarity between the ligand-binding domain of RARβ and RARγ compared to the ligand-binding domain of RARα (Giguere et al., 1987, 1980; Petkovich et al., 1987; Benbrook et al., 1988; Brand et al., 1988; Krust et al., 1989). The selectivity of RARβ/γ over RARα of some of the retinoids was also noted in *in vitro* binding assays (Delescluse et al., 1991). In contrast, two retinobenzoic acid derivatives, Am80 and Am580, representatives of a distinct class of all-*trans* RA analogs, showed preferential RARα transcriptional activation (Graupner et al., 1991) combined with higher affinities for RARα (Crettaz et al., 1990; Delescluse et al., 1991; Reichert et al., 1993).

Even more highly selective retinoids that allowed separation of RARβ from RARγ activities were subsequently reported by Bernard et al. (1992). Surprisingly, retinoids exhibiting RARβ selectivity and RARγ selectivity were derived from a common parental compound, indicating that a simple chemical modification can be sufficient to differentiate between the ligand-binding pockets of these two RAR subtypes (for structural formulas see Fig. 4). Taken together, these first studies demonstrated that retinoids with unique RAR subtype activation profiles could be defined. These retinoids now need to be tested for their impact on cellular programs and for therapeutic applications.

RXR Selective Retinoids

The central role of RXR in regulating distinct hormonal response pathways

Retinoid			Retinoid		
Structure	Name	Selectivity	Structure	Name	Selectivity
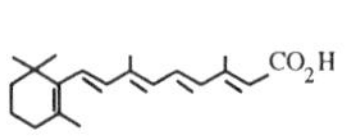	ATRA	RARα, β, γ	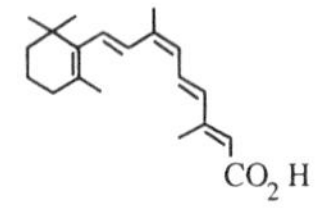	*9-cis* RA	Pan-Agonist
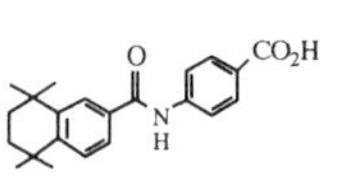	Am-580	RARα*	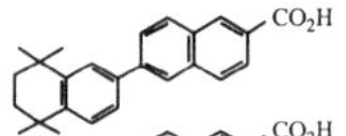 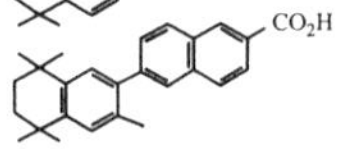	R20 R22	RARγ, β RARβ, γ
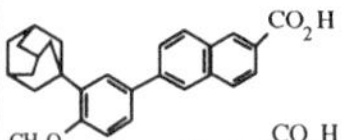 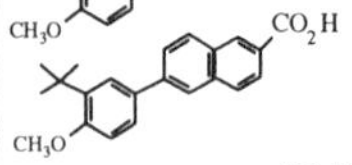 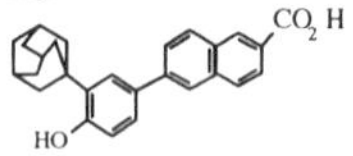	CD271 CD417 CD437	RARβ,γ RARβ RARγ	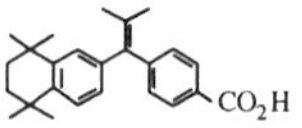 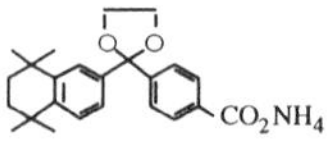	11217 11237	RXRα RXRα
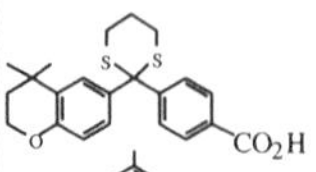 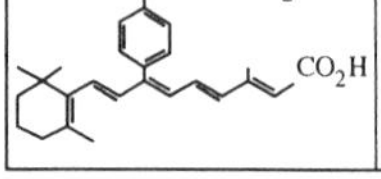	11238 11302	Anti-AP-1 Anti-AP-1	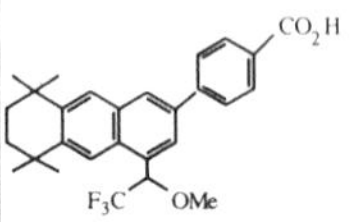	11335	Antagonist

Figure 4. Retinoids with selective activities. Note: Not all retinoids are fully selective; in particular, Am580, CD417, and CD437, at higher concentrations, can also activate other RAR subtypes.

has been addressed above (see Fig. 2). RXR homodimers were shown to activate RAREs different from RAR/RXR heterodimers (Zhang et al., 1992b). However, the only naturally available ligand for RXRs, 9-*cis* RA, also activates RARs (Heyman et al., 1992; Levin et al., 1992). Lehmann et al. (1992b) were the first to demonstrate that conformationally restricted 9-*cis* RA analogs could be used to separate RXR from RAR activities. Several retinoids were identified that activate RXR homodimers but not RAR/RXR heterodimers (Lehmann et al., 1992b). Conformational analyses indicated that the spatial orientations of the lipophilic head and carboxyl termini of these retinoids were similar to those of 9-*cis* RA, and thus activity could be related to the length and volume of the substituent group (Lehmann et al., 1992b; Jong et al., 1993), linking the tetrahydronaphthalene and phenyl ring systems (see Fig. 4). Selectivity of these retinoids for RXR homodimers was shown using co-transfection assays with several different reporter constructs. Similar to 9-*cis* RA, the RXR-selective retinoids SR11217 and

SR11237 were strong activators of the CRBPII-RARE (for response elements, see Fig. 1), which is activated only by RXR homodimers. However, in contrast to 9-*cis* RA, the synthetic retinoids did not induce the rat CRBPI-RARE, which is activated only by RAR/RXR heterodimers (Lehmann et al., 1992b; Husmann et al., 1992).

This group of RXR-selective retinoids has now been enlarged (Jong et al., 1993; Boehm et al., 1994; Dawson and Pfahl, unpublished observations) and some of these compounds are now being evaluated in various biological systems. As expected, they show a much more restricted biological activity spectrum; for instance, they do not induce differentiation in F9 teratocarcinoma cells, although they are activators of the growth hormone gene (Gendimenico et al., 1994; Lotan et al., 1995). Further roles of RXR-selective ligands, including possibly molecules outside the retinoid family like Methoprene acid (Harmon et al., 1995) and other ligands, that might only activate RXR in the context of specific orphan receptors, will deserve special attention.

Anti-AP-1 Selective Retinoids

The antiproliferative activities of retinoids are likely to be linked to their ability to induce differentiation or to their direct interference with cell proliferation signals, in particular their ability to inhibit the transcription factor AP-1. Since the retinoid receptors can down-regulate AP-1 activity without binding to specific DNA sequences, the anti-AP-1 function (Pfahl, 1993b) of the retinoid receptors is clearly mechanistically distinct from their gene activation function and could therefore include different ligand requirements. We have recently investigated this and analyzed whether retinoids can be designed that function primarily in the anti-AP-1 pathway, but not as activators of transcription via hetero- and homodimers. We found that such retinoids can indeed be defined and discovered several retinoids that inhibit only AP-1 activity but do not induce transcriptional activation (Fanjul et al., 1994). Interestingly, these anti-AP-1 selective retinoids (e.g., see Fig. 4) have maintained their antiproliferative activity, and they are potent growth inhibitors of certain cancer cell lines, including lung cancer and breast cancer cells. Importantly, anti-AP-1 selective retinoids do not induce differentiation in F9 embryonal carcinoma cells. This latter observation is consistent with the notion that induction of differentiation in F9 cells by ATRA appears to involve the transcriptional activation of several genes through RAREs (Vasios et al., 1989; Hu and Gudas et al., 1990; Wu et al., 1992).

The anti-AP-1 selective retinoids only affect some of the cellular retinoid response pathways and, in contrast to the natural retinoids ATRA, 9-*cis* RA, and 13-*cis* RA, have a much more restricted spectrum of biological activities. Therefore, they are likely to have fewer side effects. This new class of retinoids are thus good candidates for a new generation of antiproliferative retinoid therapeutics. Although such retinoids have so far been found only among synthetic compounds, it is also conceivable that they (or related molecules with similar specificity) occur naturally as metabolites of β-carotenes and related molecules.

"Gene Selective" Retinoids

The ideal retinoid therapeutic would be targeted to that subset of genes that can positively influence the diseased state by, for instance, interfering with cell proliferation. The receptor selective compounds described above were mainly defined as being specific for one receptor subtype. However, it became subsequently clear that in most cases the receptors function as heterodimers. Part of the diversity of the cellular retinoid responses could therefore be based on the fact that the various RAR/RXR heterodimers formed in specific cell types or tissues can differentially activate genes through heterodimer-specific response elements.

Indeed, in contrast to steroid hormone response elements, RAREs are diverse and include palindromes and direct repeats with various spacers, as well as everted repeats (see Fig. 1). Although no significant differences in RARE binding have been observed so far for different RAR/RXR heterodimers (Hermann et al., 1992; Zhang et al., 1992a), it has been observed that RARγ1 is a poor activator of DR-5 RAREs (direct repeat of the half-site with a 5-bp spacer) (Husmann et al., 1991; Lehmann et al., 1992a) and that a DR-1, the CRBPII RARE, is mostly activated by RXR homodimers (Mangelsdorf et al., 1991; Lehmann et al., 1992b). In addition, Nagpal et al. (1992) have analyzed various RA responsive genes and observed that activation by different receptors can vary.

To obtain further insights into the principles of differential signal transduction by the various retinoid receptor complexes, we have recently compared the transactivation potentials of different RAR/RXR heterodimer complexes on structurally distinct RAREs in a constant promoter context (La Vista-Picard and Pfahl, unpublished observations). Since RAR/RXR heterodimer formation requires interaction of the receptor LBDs, we were particularly interested in whether interaction of the heterodimer with the variously spaced and oriented half-sites of different RAREs could affect the configuration of the LBDs. This could alter the sensitivity of the ligand response and could be exploited with synthetic retinoids having structurally restricted configurations. We made three important observations. First, various retinoid receptor heterodimers show distinct activation patterns with various RAREs. Second, a given heterodimer can be activated at different ATRA concentrations on specific RAREs, allowing for a differential ATRA response for various RAREs and their associated genes. For example, we found that the RARβ/RXRα heterodimer requires approximately 40 nM ATRA for half-maximal activation via the CRBPI-RARE (a DR-2), whereas the same heterodimer requires more than 10-fold higher ATRA concentration for the half-maximal activation via the β-RARE (a DR-5). Third, synthetic retinoids can be identified that preferentially activate a specific RAR/RXR heterodimer on one response element but not on another. Thus, the retinoid response of a given gene is determined by a receptor-DNA complex, and not the receptor alone, and this significantly enhances the possibilities for the selective regulation of subsets of retinoid responsive genes. Conformationally restricted retinoids that can only activate a particular receptor heterodimer/RARE complex could therefore act as "gene-selective" agents with optimal therapeutic index.

PERSPECTIVES

The molecular analysis of the retinoid response has revealed a complex network of signaling pathways that allow retinoids to play a central role in hormonal and nutritional responses. Pro-vitamin A is usually taken up in the form of β-carotenes and related compounds. The active vitamin A derivatives, ATRA and 9-*cis* RA, then essentially function like hormones, interacting with specific nuclear receptors. One class of these receptors, the RXRs, also plays an important role in mediating other vitamin and hormone responses, in particular vitamin D_3 (derived from cholesterol) and thyroid hormone, which requires iodine for its synthesis. Moreover, it is well known that RA is necessary for erythrocyte differentiation, thereby controlling the usage of iron. The nuclear receptors themselves require zinc atoms for their DBDs. Thus, vitamins and micronutrients can directly affect the complex retinoid response network.

Using molecular biology-based analysis systems, it was possible to show that retinoids with receptor- or pathway-selective activities can be designed. These retinoids induce only a limited spectrum of retinoid responses and are thus likely to have fewer side effects. In addition, these retinoids may at times be more effective than the natural retinoids. Using more complex analysis systems, we have now also observed that the various RAREs can influence the activity of selective retinoids, suggesting that "gene-selective retinoids" can be obtained that may even interfere more specifically with defined diseases (LaVista-Picard and Pfahl, unpublished observations). Retinoid antagonists have also been described, one of which is able to inhibit RA induction of a RARE present in the HIV-1 promoter (Lee et al., 1994). HIV and some other viruses have been shown to contain RAREs that allow enhancement of viral transcription in the presence of RA. Therefore, RA antagonists could serve in some cases as antiviral agents. Overall, synthetic retinoids with selective activity show much promise as new therapeutics for a large variety of diseases.

Many clinical and preclinical studies have now shown the importance of retinoids for the treatment of cancer, and this has been discussed in detail elsewhere (Bollag and Holdener, 1992; Hill and Grubbs, 1992; Smith et al., 1992). However, although more than a thousand different retinoids have been synthesized, only a handful have been evaluated in preclinical and clinical studies. In addition, in most clinical studies the (natural) retinoid 13-*cis* RA has been used, and this can easily be converted to ATRA and 9-*cis* RA, molecules with broad activities. Although valuable information has been obtained from these clinical studies, it is now necessary to switch to retinoids with narrower biological activities to reduce their side effects. One important step in this direction is the RARβ, γ-selective retinoid Adapalene (CD271) developed by Galderma (see Fig. 4) that has now been approved for acne treatment in several countries. One other example is the synthetic retinoid N-(4-hydroxyphenyl)retanimide (4HPR), which has shown particular promise for the treatment of breast cancer. However, the molecular mechanism by which 4HPR functions has not yet been defined. From our recently obtained data it appears that 4HPR is also a receptor-selective activator. In conclusion, with the present understanding of retinoid action, it is now possible to develop selective retinoids that are likely to have fewer side effects than the natural compounds and could thus represent more optimal retinoid therapeutic agents in the treatment of various cancers.

REFERENCES

Andrews P (1984): Retinoic acid induces neuronal differentiation of a cloned human embryonal carcinoma cell line *in vitro*. *Dev Biol* 103:285–293

Angel, P, Imagawa, M, Chiu, R, Stein, B, Imbra, RJ, Rahmsdorf, HJ, Jonat, C, Herrlich, P, Karin M (1987): Phorbol ester-inducible genes contain a common cis element recognized by a TPA modulated *trans*-acting factor. *Cell* 49:729–739

Angel P, Karin M (1991): The role of Jun, Fos and the AP-1 complex in cell-proliferation and transformation. *Biochim Biophys Acta* 1072:129–157

Apfel R, Benbrook D, Lernhardt E, Ortiz-Caseda MA, Pfahl M (1994): A novel orphan receptor with a unique ligand binding domain and its interaction with the retinoid/thyroid hormone receptor subfamily. *Mol Cell Biol* 14:7025–7035

Baes M, Gulick T, Choi H-S, Martinoli MG, Simha D, Moore DD (1994): A new orphan member of the nuclear hormone receptor superfamily that interacts with a subset of retinoic acid response elements. *Mol Cell Biol* 14:1544–1552

Beato M (1989): Gene regulation by steroid hormone. *Cell* 56:335–344

Benbrook D, Lernhardt W, Pfahl M (1988): A new retinoic acid receptor identified from a hepatocellular carcinoma. *Nature* 333, 669–672

Bernard BA, Bernardon J-M, Delescluse C, Martin B, Lenoir M-C, Maignan J, Charpentier B, Pilgrim WR, Reichert U, Shroot B (1992): Identification of synthetic retinoids with selectivity for human nuclear retinoic acid receptor γ. *Biochem Biophys Res Commun* 186:977–983

Boehm MF, Zhang L, Badea BA, White SK, Mais DE, Berger E, Suto CM, Goldman ME, Heyman RA (1994): Synthesis and structure-activity relationhips of novel retinoid X receptor-selective retinoids. *J Med Chem* 37:2930–2941

Bollag W, Holdener, EE (1992): Retinoids in cancer prevention and therapy. *Ann Oncol* 3:513–526

Brand N, Petkovich M, Krust A, de Th H, Marchio A, Tiollais P, Dejean D (1988): Identification of a second human retinoic acid receptor. *Nature* 332:850–853

Bugge TH, Pohl J, Lonnoy O, Stunnenberg, HG (1992): RXRa, a promiscuous partner of retinoic acid and thyroid hormone receptors. *EMBO J* 11:1409–1418

Castaigne S, Chomienne C, Daniel MT, Ballerini P, Berger R, Fenaux P, Degos L (1990): All-*trans* retinoic acid as a differentiation therapy for acute promyelocytic leukemia. *Blood* 76:11704–1709

Chambon P (1994): The retinoid signaling pathway: molecular and genetic analyses. *Semin Cell Biol* 5:115–125

Chen, Z-X, Xue, Y-Q, Zhang R, Tao RF, Xia, X-M, Li C, Wange W, Zu W-Y, Yao X- Z, Ling BJ (1991): A clinical and experimental study on all-*trans* retinoic acid-treated acute promyelocytic leukemia patients. *Blood* 78:1413–1419

Chu EW, Malgren RA (1965): An inhibitory effect of vitamin A on the induction of tumors of forestomach and cervix in the Syrian hamster by carcinogenic polycyclic hydrocarbons. *Cancer Res* 25:884–895

Cooney AJ, Tsai SY, O'Malley BW, Tsai MJ (1992): Chicken ovalbumin upstream promoter transcription factor (COUP-TF) dimers bind to different GGTCA response elements allowing COUP-TF to repress hormonal induction of the vitamin D_3, thyroid hormone, and retinoic acid receptors. *Mol Cell Biol* 12:4153–4163

Crettaz M, Baron A, Siegenthaler G, Hunzker W (1990): Ligand specificities of recombinant retinoic acid receptors RARα and RARβ. *Biochem J* 272:391–397

Dawson MI, Okamura WH (1990): *Chemistry and Biology of Synthetic Retinoids*. Boca Raton, FL: CRC Press

Delescluse C, Cavey MT, Martin B, Bernard BA, Reichert U, Maignan J, Darmon M, Shroot B (1991): Selective high affinity retinoic acid receptor a or β-γ ligands. *Mol Pharmacol* 40:556–562

De Thé H, Marchio A, Tiollais P, Dejean A (1989): Differential expression and ligand regulation of the retinoic acid receptor α and β genes. *EMBO J* 8:429–433

Doyle L, Giangiulo D, Hussain A, Park H-J, Chin Yen, R-W, Borges M (1989): Differentiation of human variant small cell cancer cell lines to a classic morphology by retinoic acid. *Cancer Res* 49:6745–6751

Fanjul A, Hobbs P, Graupner G, Zhang, X-K, Dawson MI, Pfahl M (1994): A new class of retinoids with selective anti-AP-1 activity exhibit anti-proliferative activity. *Nature* 372:107–111

Findley H, Steuber C, Tuymann F, McKolanis JR, Williams DL, Ragab AH (1986): Effect of retinoic acid on myeloid antigen expression and clonal growth of leukemic cells from children with acute nonlymphocytic leukeamia—A Pediatric Oncology Group study. *Leuk Res* 10:43–50

Fontana J (1987): Interaction of retinoids and tamoxifen on the inhibition of human mammary carcinoma cell proliferation. *Exp Cell Biol* 55:136–144

Gendimenico GJ, Stim TB, Corbo M, Janssen B, Mezick JA (1994): A pleiotropic response is induced in F9 embryonal carcinoma cells and rhino mouse skin by all-*trans* retinoic acid, a RAR agonist but not by SR11237, a RXR-selective agonist. *J Invest Dermatol* 102:676–680

Giguere V, Ong ES, Seigi P, Evans RM (1987): Identification of a receptor for the morphogen retinoic acid. *Nature* 330:624–629

Giguere V, Shago M, Zirngibl R, Tate P, Rossant J, Varmuza S (1990): Identification of a novel isoform of the retinoic acid receptor γ expressed in the mouse embryo. *Mol Cell Biol* 10:2335–2340

Graupner G, Wills KN, Tzukerman M, Zhang X- K, Pfahl M (1989): Dual regulatory role for thyroid hormone receptors allows control of retinoic-acid receptor activity. *Nature* 340:653–656

Graupner G, Malle G, Maignan J, Lang G, Prunieras M, Pfahl M (1991): 6′-Substituted naphthalene-2-carboxylic acid analogs, a new class of retinoic acid receptor subtype-specific ligand. *Biochem Biophys Res Commun* 179:1554–1561

Griffiths CEM, Elder JT, Bernard BA, Rossio P, Cromie MA, Finkel LJ, Shroot B, Voorhees JJ (1993): Comparison of CD271 (Adapalene) and all-*trans* retinoic acid in human skin: dissociation of epidermal effects and CRABP-II mRNA expression. *J Invest Dermatol* 101:325–328

Grunberg S, Itri L (1987): Phase II study of isotretinoin in treatment of advanced nonsmall cell lung cancer. *Cancer Treat Rep* 71:1097–1098

Halter S, Fraker L, Adcock D, Vick S (1988): Effect of retinoids on xenotransplanted human mammary carcinoma cells in athymic mice Cancer Res 48:3733–3736

Haq R-U, Pfahl M, Chytil F (1991): Retinoic acid affects the expression of nuclear retinoic acid receptors in tissues of retinol deficient rats. *Proc Natl Acad Sci USA* 88:8272–8276

Harmon MA, Boehm MF, Heyman RA, Mangelsdorf DJ (1995): Activation of mammalian retinoid X receptors by the insect growth regulator methoprene. *Proc Natl Acad Sci USA* 92:6157–6160

Hermann T, Hoffmann B, Zhang X-K, Tran P, Pfahl M (1992): Heterodimeric receptor complexes determine 3,5,3′-triiodothyronine and retinoid signaling specificities. *Mol Endocrinol* 6:1153–1162

Hermann T, Hoffmann B, Piedrafita JF, Zhang X-K, Pfahl M (1993): V-*erb*A requires auxiliary proteins for dominant negative activity. *Oncogene* 8:55–65

Heyman RA, Mangelsdorf DJ, Dyck JA, Stein RB, Eichele G, Evans, R M, Thaller C (1992): 9-*cis* retinoic acid is a high affinity ligand for the retinoid X receptor Cell 68:397–406

Hill DL and Grubbs CJ (1992): Retinoids and cancer prevention. *Annu Rev Nutr* 12:161–181

Hirose T, Apfel R, Pfahl M, Jetten AM (1995): The orphan receptor TAK1 acts as a competitive repressor of RAR-, RXR-, and T_3R-mediated signaling pathways. *Biochem Biophys Res Commun*

Hoffmann B, Lehmann JM, Zhang X-K, Hermann T, Graupner G, Pfahl M (1990): A retinoic acid receptor specific element controls the retinoic acid receptor-β promoter. *J Mol Edocrinol* 4:1734–1743

Hong WK, Lippman SM, Itri LM, Karp DD, Lee JS, Byers RM, Schants SP, Kramer AM, Lotan R, Peters IJ, Dimery IW, Brown BW, Goepfert H (1991): Prevention of second primary tumors with isotretinoin in squamous-cell carcinoma of the head and neck. *N Engl J Med* 323:795–801

Hu L, Gudas LJ (1990): Cyclic AMP analogs and retinoic acid influence the expression of retinoic acid receptor α, β, and γ mRNAs in F9 teratocarcinoma cells. *Mol Cell Biol* 10:391–396

Huang M-E, Ye Y-C, Chen S-R, Chai J-R, Lu J-X, Zhoa L, Gu L-J, Wang Z-Y (1988): Use of all-*trans* retinoic acid in the treatment of acute promyelocytic leukemia. *Blood* 72:567–572

Husmann J, Hoffmann B, Stump DG, Chytil F, Pfahl M (1992): A retinoic acid response element from the rat CRBPI promoter is activated by an RAR/RXR heterodimer. *Biochem Biophys Res Commun* 187:1558–1564

Husmann M, Lehmann J, Hoffmann B, Hermann T, Tzuckermann M, Pfahl M (1991): Antagonism between retinoic acid receptors. *Mol Cell Biol* 11:4097–4103

Jong L, Lehmann JM, Hobbs PD, Harlev E, Hoffman JC, Pfahl M, Dawson MI (1993): Conformational effects on retinoid receptor selectivity. 1. Effect of 9-double bond geometry on retinoid X receptor activity. *J Med Chem* 36:2605–2613

Kliewer SA, Umesono K, Mangelsdorf DJ, Evans RM (1992a): Retinoid X receptor interacts with nuclear receptors in retinoic acid, thyroid hormone and vitamin D_3 signaling. *Nature* 355:446–449

Kliewer SA, Umesono K, Heyman RA, Mangelsdorf DJ, Dyck JA, Evans RM (1992b): Retinoid X receptor-COUP-TF interactions modulate retinoic acid signaling. *Proc Natl Acad Sci USA* 89:1448–1452

Krust A, Kastner PH, Petkovich M, Zelent A, Chambon P (1989): A third human retinoic acid receptor, hRAR-γ. *Proc Natl Acad Aci USA* 86:5310–5314

Lee M-O, Hobbs PD, Zhang X-K, Dawson MI, Pfahl M (1994): A new retinoid antagonist inhibits the HIV-1 promoter. *Proc Natl Acad Sci USA* 91:5632–5636

Lehmann JM, Zhang X-K, Pfahl M (1992a): RARγ2 expression is regulated through a retinoic acid response element embedded in Sp1 sites. *Mol Cell Biol* 12:2976–2985

Lehmann JM, Jong L, Fanjul A, Cameron JF, Lu XP, Haefner P, Dawson MI, Pfahl M (1992b): Retinoids selective for retinoid X receptor response pathways. *Science* 258:1944–1946

Lehmann JM, Zhang X-K, Graupner G, Hermann T, Hoffmann B, Pfahl M (1993): Formation of RXR homodimers leads to repression of T_3 response: hormonal cross-talk by ligand induced squelching. *Mol Cell Biol* 13:7698–7707

Lehmann JM, Dawson MI, Hobbs PD, Husmann M, Pfahl M (1991): Identification of retinoids with Nuclear Receptor subtype-selective activities. *Cancer Res* 51:4804–4809

Leid M, Kastner P, Lyons R, Nakshatri H, Saunders M, Zacharewski T, Chen J-Y, Staub A, Garnier J-M, Mader S, Chambon P (1992): Purification, cloning, and RXR identity of the HeLa cell factor with which RAR or TR heterodimerizes to bind target sequences efficiently. *Cell* 68:377–395

Leroy P, Nakshatri H, Chambon P (1991): Mouse retinoic acid receptor α2 isoform is transcribed from a promoter that contains a retinoic acid response element. *Proc Natl Acad Sci USA* 88:10138–10142

Levin AA, Sturzenbecker LJ, Kazmer S, Bosakowski T, Huselton C, Allenby G, Speck J, Kratzeisen C, Rosenberger M, Lovey A, Grippo JF (1992): 9-*cis* retinoic acid stereoisomer binds and activates the nuclear receptor RXRa. *Nature* 355:359–361

Lippman SM, Kavanagh JJ, Paredes-Espinoza M, Delgadillo-Madrueno F, Paredes-Casillas P, Hong WK, Holdener E, Krakoff, IH (1992): 13-*cis*-retinoic acid plus interferon alpha: highly active systemic therapy for squamous cell carcinoma of the cervix. *J Natl Cancer Inst* 84:241–245

Lohnes D, Mark M, Mendelsohn C, Dollé P, Dierich A, Gorry P, Gansmuller A, Chambon P (1994): Function of the retinoic acid receptors (RARs) during development (I) Craniofacial and skeletal abnormalities in RAR double mutants. *Development* 120:2723–2748

Lotan R (1979): Different susceptibilities of human melanoma and breast carcinoma cell lines to retinoic acid-induced growth inhibition. *Cancer Res* 39:1014–1019

Lotan R (1981): Effects of vitamin A and its analogs (retinoids) on normal and neoplastic cells. *Biochim Biophys Acta* 605:33–91

Lu X-P, Salbert G, Pfahl M (1994): An evolutionary conserved COUP-TF binding element in a neural specific gene and COUP-TF expression patterns support a major role for COUP-TF in neural development. *Mol Endocrinol* 8:1774–1788

MacDonald PN, Dowd DR, Nakajima S, Galligan MA, Reeder MC, Haussler CA, Ozato K, Haussler M (1993): Retinoid X receptors stimulate and 9-*cis* retinoic acid inhibits 1,25-dihydroxyvitamin D_3-activated expression of the rat osteocalcin gene. *Mol Cell Biol* 13:5907–5917

Mangelsdorf DJ, Umesono K, Kliewer SA, Borgmeyer U, Ong ES, Evans RM (1991): A direct repeat in the cellular retinol-binding protein type II gene confers differential regulation by RXR and RAR. *Cell* 66:555–561

Marks MS, Hallenbeck PL, Nagata T, Segars JH, Appella E, Nikodem VM, Ozato K (1992): H-2RIIBP (RXRβ) heterodimerization provides a mechanism for combinatorial diversity in the regulation of retinoic acid and thyroid hormone response genes. *EMBO J* 11:1419–1435

Mendelsohn C, Lohnes D, Décimo D, Lufkin T, LeMeur M, Chambon P, Mark M (1994): Function of the retinoic acid receptors (RARs) during development (II) Multiple abnormalities at various stages of organogenesis in RAR double mutants. *Development* 120:2749–2771

Meyskens FL, Surwit E, Moon TE, Childers JM, Davies JR, Door RT, Johnson, C S, Alberts, DS (1994): Enhancement of regression of cervical intraepithelial neoplasia II (moderate dysplasia) with topically applied all-*trans* retinoic acid: a randomized trial. *J Natl Cancer Inst* 86:539–543

Morris-Kay, G (ed) (1992): *Retinoids in Normal Development and Teratogenesis.* Oxford: Oxford Science Publications

Munker M, Munker R, Saxton R, Koeffler HP (1987): Effect of recombinant monokines, lymphokines and other agents on clonal proliferation of human lung cancer cell lines. *Cancer Res* 47:4081–4085

Nagpal S, Saunders M, Kastner P, Durand B, Nakshatri H, Chambon P (1992): Promoter context- and response element-dependent specificity of the transcriptional activation and modulating functions of retinoic acid receptors. *Cell* 70:1007–1019

Ortiz MA, Piedrafita FJ, Pfahl M, Maki R (1995): A new orphan receptor that can modulate the retinoid and thyroid hormone responses in T-cells. *Mol Endocrinol* 9:1679–1691

Petkovich M, Brand NJ, Krust A, Chambon P (1987): Human retinoic acid receptor belongs to the family of nuclear receptors. *Nature* 330:444–450

Pfahl M (1993a): Signal transduction by retinoid receptors. In *Skin*, Merk H, ed. *Pharmacology* 6:8–16

Pfahl M (1993b): Nuclear receptor/AP-1 interaction. *Endocr Rev* 14:651–658

Pfahl M (1994): Vertebrate receptors: Molecular biology, dimerization and response elements. *Semin in Cell Biol.* 5:95–103

Pfahl M, Apfel R, Bendik I, Fanjul A, Graupner G, Lee M-O, La-Vista N, Lu X-P, Piedrafita J, Ortiz-Caseda A, Salbert G, Zhang X-K (1994): Nuclear retinoid receptors and their mechanism of action. In: *Vitamins and Hormones*, Litwack G, ed. Academic Press

Reichert U, Bernardon JM, Charpentier B, Nedoncelle P, Martin B, Bernard BA, Asselineau D, Michel S, Lenoir MC, Delescluse C, Pilgrim WR, Darmon YM, Shroot B (1993): Synthetic retinoids: receptor selectivity and biological activity. In: *Pharmacol Skin*, Bernard BA, Shroot B, eds. Basel: Karger

Reynolds C, Kane D, Einhorn P, Matthay KK, Crouse VL, Wilbur JR, Shunn SB, Seeger RC (1991): Response of neuroblastoma to retinoic acid *in vitro* and *in vivo*. In *Advances in Neuroblastoma Research*, Evans A, D'Angio G, Knudson A, et al, eds. New York: Alan R Liss

Roberts AB, Sporn MB (1984): Cellular biology and biochemistry of the retinoids. In: *The Retinoids*, Sporn MB, Roberts AB, Goodman DS, eds. Orlando: The Academic Press

Rottman JN, Widom RL, Nadal-Ginard B, Mahdavi V, Karathanasis SK (1991): A retinoic acid-responsive element in the apolipoprotein AI gene distinguishes between two different retinoic acid response pathways. *Mol. Cell Biol* 11:3814–3820

Sacks P, Oke V, Amos B, Vasey T, Lotan R (1989): Modulation of growth, differentiation, and glycoprotein synthesis by all-*trans* retinoic acid in a multicellular tumor spheroid model for squamous carcinoma of the head and neck. *Int J. Cancer* 44:926–933

Salbert G, Fanjul A, Piedrafita J Lu, X-P, Kim S-J, Tran P, Pfahl M (1993): Retinoic acid receptors and retinoid X receptor-α down-regulate the transforming growth factor-β_1 promoter by antagonizing AP-1 activity. *Mol Endocrinol* 7:1347–1356

Schaefer H, Reichert U (1990): Retinoids and their perspectives in dermatology. *Nouv Dermatol* 9:3–6

Schliecher R, Moon R, Patel M, Beattie C (1988): Influence of retinoids on growth and metastasis of hamster melanoma in athymic mice. *Cancer Res* 48:1465–1469

Schüle R, Rangarajan P, Yang N, Kliewer S, Ransone LJ, Bolado J, Verma IM, Evans RM (1991): Retinoic acid is a negative regulator of AP-1-responsive genes. *Proc Natl Acad Sci USA* 88:6092–6096

Smith MA, Parkinson DR, Cheson BD, Friedman MA (1992): Retinoids in Cancer Therapy. *J Clin Oncol* 10:839–864

Thein R, Lotan R (1982): Sensitivity of cultured human osteosarcoma and chondrosarcoma cells to retinoic acid. *Cancer Res* 42:4771–4775

Thomas HE, Stunnenberg HG, Stewart AF (1993): Heterodimerization of the Drosophila ecdysone receptor with retinoid X receptor and ultraspiracle. *Nature* 362:471–475

Tran P, Zhang X-K, Salbert G, Hermann T, Lehmann JM, Pfahl M (1992): COUP orphan receptors are negative regulators of retinoic acid response pathways. *Mol. Cell Biol* 12:4666–4676

Vasios GW, Gold JD, Petkovich M, Chambon P, Gudas LJ (1989): A retinoic acid-responsive element is present in the 5′ flanking region of the laminin B1 gene. *Proc Natl Acad Sci USA* 86:9099–9103

Warrell RP, Frankel SR, Miller WH Jr, Scheinberg DA, Itri LM, Hittelman WN, Vyas R, Andreeff M, Tafuri A, Jakubowski A, Gabrilove J, Gordon MS, Dmitrovsky E (1991): Differentiation therapy of acute promyelocytic leukemia with tretinoin (all-*trans* retinoic acid). *N Engl J Med* 324:1385–1393

Willy PJ, Umesono K, Ong ES, Evans RM, Heyman RA, Mangelsdorf DJ (1995): LXR, a nuclear receptor that defines a distinct retinoid response pathway. *Genes Dev* 9:1033–1045

Windom RL, Rhee M, Karathanasis SK (1992): Repression by ARP-1 sensitizes apolipoprotein AI gene responsiveness to RXRa and retinoic acid. *Mol. Cell Biol* 12:3380–3389

Wu T-CJ, Wang L, Wan Y-JY (1992): Retinoic acid regulates gene expression of retinoic acid receptors α, β and γ in F9 mouse teratocarcinoma cells. *Differentiation* 51:219–224

Yang-Yen H-F, Zhang X-K, Graupner G, Tzukerman M, Sakamoto B, Karin M, Pfahl M (1991): Antagonism between retinoic acid receptors and AP-1: implication for tumor promotion and inflammation. *New Biol* 3:1216–1219

Yao T-P, Segraves WA, Oro AE, McKeown M, Evans RM (1992): Drosophila ultraspiracle modulates ecodysone receptor function via heterodimer formation. *Cell* 71:63–72

Yu VC, Delsert C, Andersen B, Holloway JM, Devary OV, Nr AM, Kim SY, Boutin J-M, Glass CK, Rosenfeld MG (1991): RXR β: a coregulator that enhances binding of retinoic acid, thyroid hormone, and vitamin D receptors to their cognate response elements. *Cell* 67:1251–1266

Zhang X-K, Hoffmann B, Tran P, Graupner G, Pfahl M (1992a): Retinoid X receptor is an auxiliary protein for thyroid hormone and retinoic acid receptors. *Nature* (London) 355:441–446

Zhang X-K, Lehmann J, Hoffmann B, Dawson MI, Cameron J, Graupner G, Hermann T, Pfahl M (1992b): Homodimer formation of retinoid X receptor induced by 9-*cis* retinoic acid. *Nature* (London) 358:587–591

Zhang X-K, Pfahl M (1993): Regulation of retinoid and thyroid action through homo- and heterodimeric receptors. *Trends Endocrinol Met* 4:156–162

Part II

Breast Cancer

Women face a one in eight chance of developing breast cancer in their lifetime. In the United States, nearly 200,000 women will be diagnosed with breast cancer this year, and almost 50,000 will die as a result of this disease. Besides the life-threatening nature of breast cancer, the emotional and psychological toll of surgical disfigurement in its treamnent is enormous. The ongoing controversy on the effect of postmenopausal estrogen replacement therapy on breast cancer risk, and the intense public interest in this debate, emphasizes the immense impact that breast cancer has on the overall human condition.

Chapter 6 provides a thorough discussion of breast cancer and defines the multiplicity of pathologies that comprise this condition. A strong emphasis is placed on the modalities of treatment used, specific approaches used in certain conditions (e.g., premenopausal v. postmenopausal), and expected outcomes. Of great importance are the results obtained from clinical trials, and these are covered extensively. After all, the patient benefit derived from alternative treatment protocols is the true endpoint of all biomedical research. The use of neoadjuvant therapy, an important clinical topic, is also addressed.

Chapter 7 reviews the molecular genetics of breast cancer. With the recent spectacular success in isolating the BRCA1 and BRCA2 genes, this area of study has achieved a top priority in the study of breast cancer. As more is learned, we will finally begin to develop a coherent picture of the molecular pathways that are involved in neoplastic transformation of mammary tissue. This may lead to new translational research, by indicating which biochemical pathways may be particulary sensitive to hormonal and biochemical intervention.

Chapter 8 provides an extensive overview of the complex and interwoven roles of numerous growth factors and their receptors in breast cancer. Of particular interest is the determination of which growth factors may modify the response rate of human breast tumors to therapy. The parallel discussion of animal model system results and human clinical studies highlights the way in which basic and clinical science cross-fertilize to foster medical advances.

The role of estrogens in stimulating the growth of hormone-dependent breast tumors is well known. Chapter 9 outlines the molecular mechanisms involved in this process. It presents data that may give us a clue as to why some breast cancers are (or become) hormone-resistant. Mutations in the estrogen receptor can lead to such resistance. Although this represents only one of many potential ways in which a breast tumor can progress from hormone dependence to hormone resistance, it begins to dissect the molecular pathway in progression to the hormone-independent state.

Expression of the progesterone receptor has been used diagnostically to assess the potential hormone-dependence of tumors, but little is known about the effects of progestins on breast cancer. Chapter 10 provides surprising and important data suggesting that the progesterone receptor may be active in controlling breast tissue proliferation. Of particular significance is the fact that one cell signaling pathway (that involving cyclic AMP) actually can cause an antiprogestin to acquire agonist activity. These studies have major clinical implications, as a controversy exists about whether estrogens alone or estrogens plus progestins should be used in hormone replacement therapy, and which of these approaches might increase breast cancer risk more.

Chapters 11 and 12 deal with the impressive results that have been obtained using therapy directed at the estrogenic pathway. The success of tamoxifen treatment (although not complete) indicates the power of this expreimental approach. A complementary tact, the targeting of the enzyme systems that synthesize estrogens, is also bearing clinical fruit. Both chapters point to future directions in which basic science knowledge can be translated into clinical treatment modalities.

This section, then, presents a fairly complete picture of breast cancer, the genes and growth factors that control its development and progression, the role of steroid receptors in endocrine-responsive tumors, and the hormonal and chemical approaches to treatment that are currently employed.

6

Clinical Aspects of Breast Cancer

WILLIAM J. GRADISHAR AND MONICA MORROW

OVERVIEW OF BREAST CANCER INCIDENCE AND MORTALITY

Except for skin cancer, breast cancer is the most common malignancy in women. It accounts for approximately one-third of cancers in females in the United States. In 1994, the American Cancer Society estimated that 182,000 new cases of breast cancer would be diagnosed in women and 46,000 women would die of the disease. In contrast, breast cancer in males is rare, with only 1000 new diagnoses and 300 cancer-related deaths expected in 1994 (Boring et al., 1994).

Most breast cancer cases (78%) occur in women above the age of 50 years, whereas only 6.5% of breast cancer is diagnosed in women younger than 40 years (Fig. 1). Tumor registries established on both the state and national levels have observed an uninterrupted 1% per year increase in breast cancer incidence until the 1980s, when the slope of the incidence curve increased dramatically (Fig. 2) (Miller et al., 1994). Several hypotheses have been proposed to explain this trend. Some epidemiologists believe the slow, long-term increase in incidence before 1980 is attributable to life style factors such as widespread oral contraceptive use, exogenous estrogen replacement therapy following menopause, or the earlier onset of menarche in recent decades (Holford et al., 1991). The recent, dramatic increase in breast cancer incidence during the 1980s has been studied by many investigators and most analyses suggest that the introduction of widespread mammography screening programs during the 1980s explains the increased detection of early breast cancers (Miller et al., 1994). Interestingly, the largest increase in the incidence of breast cancer has been identified in women 60 to 75 years of age who have historically been underutilizers of mammography screening compared to younger women. One group suggests that lead time,

Hormones and Cancer
Wayne V. Vedeckis, Editor

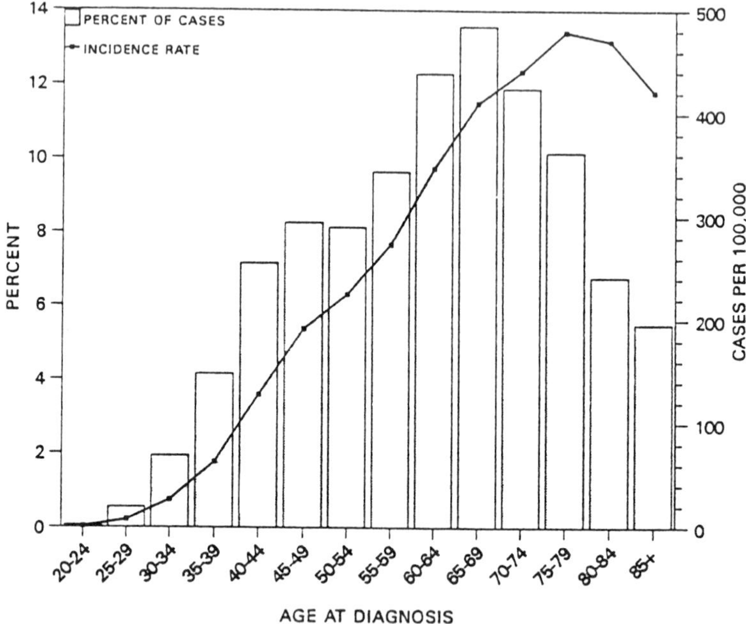

Figure 1. Percent of incident cases (*bars*) and incident rate (*line*) by age at diagnosis for invasive breast cancer diagnosed between 1986 and 1990. (Reprinted with permission of the J.B. Lippincott Company, from Miller et al., 1994).

defined as the point in time that the diagnosis of breast cancer has been advanced by mammography, may differ between younger and older age groups (Feuer and Wun, 1992; Swanson et al., 1993). If the lead time for women over the age of 70 is 5 years and for women 40 to 59 is 2 years, then the incidence data are completely consistent with the increased use of mammography.

Although the media have focused significant attention on breast cancer in younger women, a careful analysis of United States population statistics and national cancer databases reveal that the incidence rates for breast cancer in women under the age of 40 has remained stable for 20 years, but the number of women in the United States in the 20 to 39 year age group over the same time period has increased. As a result, the absolute number of young patients diagnosed with breast cancer has increased (Miller et al., 1994). Several interesting similarities and differences have been detected in cancer statistics when analyzed by race. Just as in whites, mammography screening programs have resulted in an overall increased incidence of both invasive and noninvasive breast cancers in blacks. In contrast, there are subtle differences in breast cancer incidence and mortality patterns for blacks and whites. For the time period 1973–1989, the incidence of invasive breast cancer in blacks is only slightly greater than in whites, and the difference appears to be stable. The

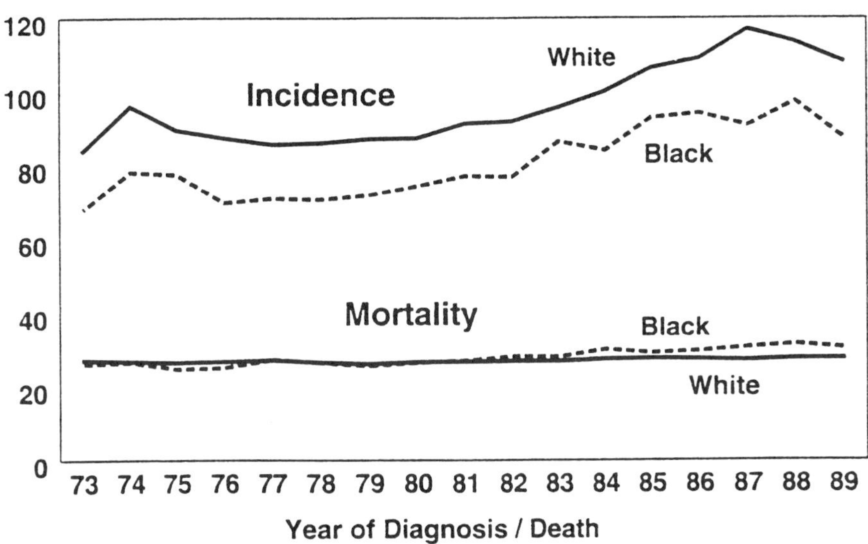

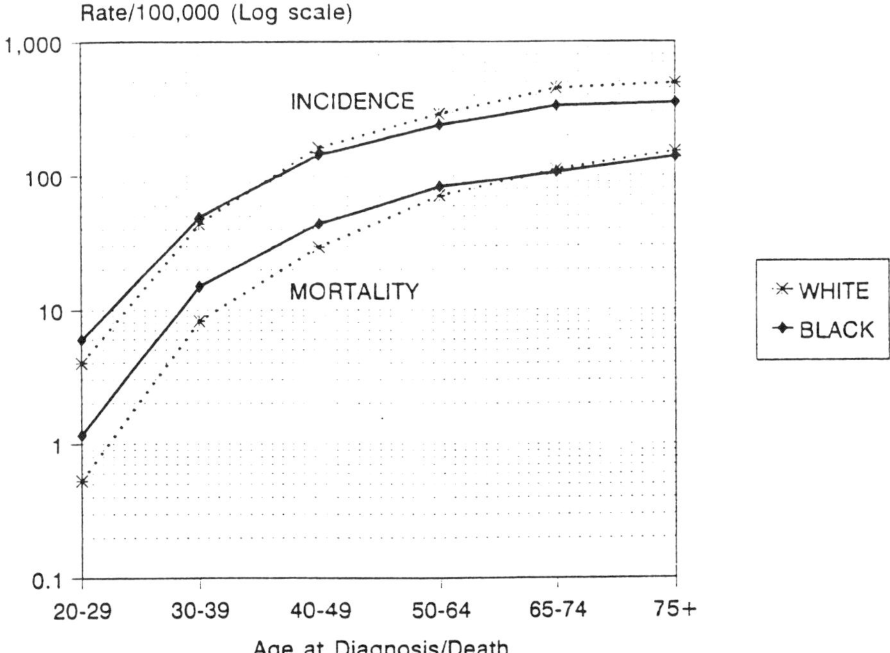

Figure 2. **A**: Breast cancer incidence and mortality rates by race, SEER, 1973–1989. (Reprinted with permission of the J.B. Lippincott Company, from Garfinkel et al., 1994). **B**: Age-specific female invasive breast cancer incidence rates and age-specific female breast cancer mortality rates by race (Hankey et al., 1994).

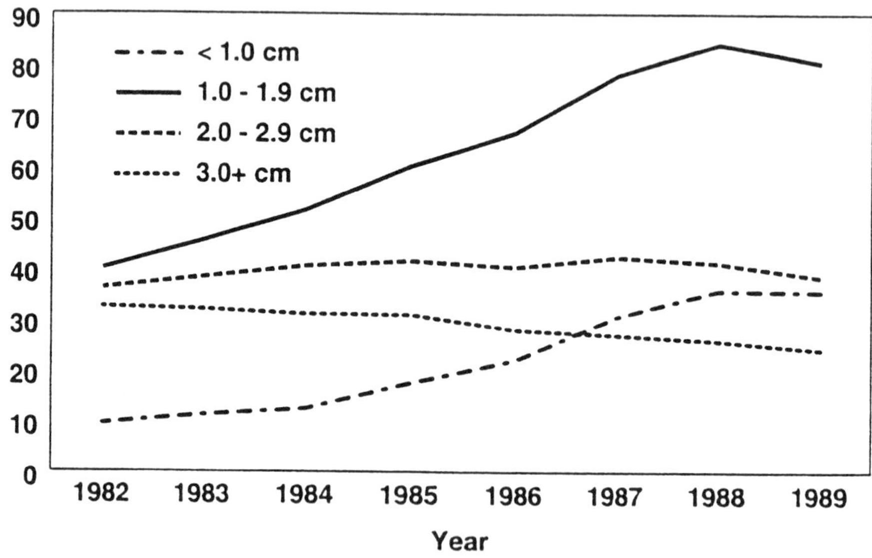

Figure 3. Breast cancer incidence rates by size at diagnosis, SEER, 1982–1989. (Reprinted with permission of the J.B. Lippincott Company, from Garfinkel et al., 1994).

difference in breast cancer mortality between blacks and whites is more significant and has been steadily increasing (Fig. 2) (Swanson et al., 1993; Hankey et al., 1994). In age-cohort analysis, the incidence of breast cancer is slightly higher in blacks up to age 40 years, and breast cancer mortality is higher in blacks up to age 65 years (Swanson et al., 1993; Hankey et al., 1994). These observations are not easily explained, but there is some evidence that breast cancer presenting in black women may possess biological characteristics imparting a more aggressive behavior (Elledge et al., 1994).

The widespread use of screening mammography has resulted in a significant increase in the incidence of ductal carcinoma *in situ* and axillary node-negative, invasive breast cancer (Garfinkel et al., 1994; Hankey et al., 1994). The incidence of primary tumors measuring ≤ 2 cm steadily increased from 1982 to 1989, whereas the incidence of larger primary tumors decreased (Fig. 3). Similarly, significant decreases in incidence were observed for: (1) invasive breast cancers ≥ 2 cm with axillary lymph node involvement among women 50 to 70 years of age at diagnosis, (2) invasive breast cancers 1 to 1.9 cm with distant metastases at diagnosis among women 50 to 59 years of age; (3) invasive breast cancer ≥ 2 cm, with distant metastases, in every age category (Garfinkel et al., 1994; Hankey et al., 1994).

RISK FACTORS

Identification of risk factors for breast cancer is important for at least two reasons: (1) closely monitoring patients thought to be at highest risk may lead to earlier diagnosis and a better prognosis, and (2) understanding the pathophysiology of the risk factor's association with breast cancer development may lead to ways of modifying the risk factor (prevention) and/or developing more effective means of treating the disease. There are several established and probable risk factors for breast cancer (Table 1), the most strongly associated being age and gender. Breast cancer is a rare disease in males and breast cancer incidence dramatically increases with increasing age in women (see Fig. 1). A family history of first-degree relatives with breast cancer (Sattin et al., 1990), early menarche (Colditz, 1993), delayed first pregnancy and late menopause (Colditz, 1993) are also associated with an increased risk of breast cancer. The risk associated with family history is greatest when a mother or sister is identified with breast cancer before the age of 50 years and/or if the disease was bilateral (Lynch et al., 1988; Claus et al., 1990). Women who undergo breast biopsy and are determined to have benign, proliferative changes, particularly atypical hyperplasia, are at increased risk of developing breast cancer (Dupont and Page, 1985; Dupont et al., 1993). Other potential risk factors have been extensively studied, but their contribution to a patient's risk profile remains controversial. Specifically, fat intake, oral contraceptives, exogenous estrogen, previous abortions and alcohol intake have all been studied in large populations and their contribution to breast cancer risk remains inconclusive and controversial (Harris et al., 1992; Daling et al., 1994; Rosenberg, 1994; Velentgas and Daling, 1994).

Table 1. Risk factors for breast cancer

Risk factor	Relative risk
Age >50	++
Early age at menarche	+
Late age at menopause	+
Late age at first birth	+
Benign breast disease	
Proliferative without atypical hyperplasia	+
Proliferative with atypical hyperplasia	++
BRCA1	+++
Family history of breast cancer	++
Oral contraceptives	?
Exogenous estrogens	?
Dietary fat	?
Alcohol	?
Induced abortion	?

Hereditary breast cancer (see Chapter 7, this volume) accounts for a small fraction (~5%) of all newly diagnosed breast cancers (Hall et al., 1990; Goldgar et al., 1994). In 1994, after almost a decade of work, the gene on chromosome 17 (BRCA1) responsible for some hereditary breast cancer was identified (Miki et al., 1994). Other genes that contribute to breast cancer risk have already been identified (Vogelstein and Kinzler, 1994). Studies of families with inherited alterations in BRCA1 suggest that more than 50% of the women who carry a cancer-associated mutation in the gene will be diagnosed with breast cancer by age 50 years, and more than 85% by the age of 70 years (Hall et al., 1990; Goldgar et al., 1994). Ultimately, an understanding of the function of BRCA1 may provide insights into malignant transformation in the breast and possibly new prevention and treatment strategies (Vogelstein and Kinzler, 1994). At present, the vast majority of women with breast cancer have no identifiable risk factors.

SIGNS AND SYMPTOMS

Most breast cancers present as an asymptomatic lump. The upper outer quadrant of the breast is the most common site of breast carcinoma, with 38% of tumors in a series of 1007 patients found in this location (Haagensen, 1971). The central region (within 1 cm of the areola) was the site of 29% of tumors, and 8.8% were found in the lower outer quadrant, 6.6% in the lower inner quadrant and, 15.3% in the upper inner quadrant. An additional 1.8% arose from multiple areas within the breast.

Breast pain is an uncommon presenting symptom of breast carcinoma, but was the only breast complaint in 7% of 240 women with carcinoma reported by Preece et al. (1982). Nipple discharge is another presentation of breast carcinoma, although most nipple discharges are due to benign breast disease. Nipple discharges that are suspicious for malignancy include those that are unilateral, spontaneous, or contain blood (Morrow, 1991). Only 4–20% of discharges that are felt to be worrisome enough to biopsy are due to malignant disease (Leis et al., 1967; Urban and Egeli, 1978; Morrow, 1991), although the risk of malignancy increases with increasing patient age. Seltzer et al. (1970) reported a 32% incidence of carcinoma in women over age 60 years presenting with discharge and no palpable breast mass compared with a 7% incidence in women under 60 years of age.

Eczematous changes of the nipple, often accompanied by itching, and erythema are an uncommon presentation of primary breast cancer. These symptoms are characteristic of Paget's disease, a rare form of breast cancer (Ashikari et al., 1970). Breast cancer may also present as axillary adenopathy in the absence of clinical or mammographic evidence of a tumor in the breast. This presentation was seen in fewer than 0.5% of cases in two large series (Vezzoni et al., 1979; Barone et al., 1990), and microscopic carcinoma is usually identified in the breast after surgical therapy. Patients with more advanced disease may have skin changes over the breast such as erythema and skin warmth, edema or peau d'orange, retraction, or frank ulceration of the breast. Some patients may not present with symptoms referable to the breast, but may complain of symptoms suggesting metastatic disease (i.e., bone pain, neurological symptoms).

A diagnostic mammogram is a routine part of the evaluation of clinical breast abnormalities. The mammographic features of malignancy include irregular or spiculated masses (Fig. 4), clustered microcalcifications (Fig. 5), architectural distortion, and skin thickening or retraction. In the presence of a clinical abnormality, mammography is not done to determine if a biopsy is necessary. Although a mammogram may reinforce the clinical suspicion of cancer, a negative study cannot be used to exclude cancer because mammography is known to have a false-negative rate of approximately 15% (Edeiken, 1988). The purpose of mammography is to determine the extent of clinically occult cancer within the breast, an important factor in selecting local therapy, and to identify occult carcinomas in the contralateral breast. In a report from Guy's Hospital, the use of routine preoperative mammography resulted in a 2.4% incidence of synchronous breast cancer detection, a fivefold increase over prior clinical detection rates (Chaudry et al., 1984).

BREAST CANCER DETECTED MAMMOGRAPHICALLY

The impact that mammography has made on the incidence of breast cancer has been discussed. The primary purpose of screening is the detection of an abnormality rather than diagnosis. A screening mammogram consists of one or two views of each breast with a goal of high image quality and low radiation exposure.

Several nonrandomized studies done in the 1960s and 1970s demonstrated that mammography is able to detect nonpalpable breast lesions, primarily masses less than 1 cm in size and microcalcifications, leading to an earlier diagnosis of cancer (Morrison, 1993). Subsequent randomized trials and case control or cohort studies in which women underwent regular mammographic screening or routine clinical follow-up have demonstrated a reduction in breast cancer mortality in women over the age of 50 years undergoing mammographic screening (Fletcher et al., 1993; Kerlikowske et al., 1995). A recent meta-analysis of 13 screening studies demonstrated a 26% reduction (95% confidence interval, 17–34%) in breast cancer mortality in screened women aged 50 to 74 years (Kerlikowske et al., 1995). No mortality reduction was noted in women younger than age 50 years, and the benefit of mammography screening to women less than age 50 years remains a source of controversy (Dupont, 1994; Kaluzny et al., 1994; Kopans et al., 1994; Smart, 1994). The finding that screening reduces breast cancer mortality indicates that some breast cancers are not systemic from their inception and provides

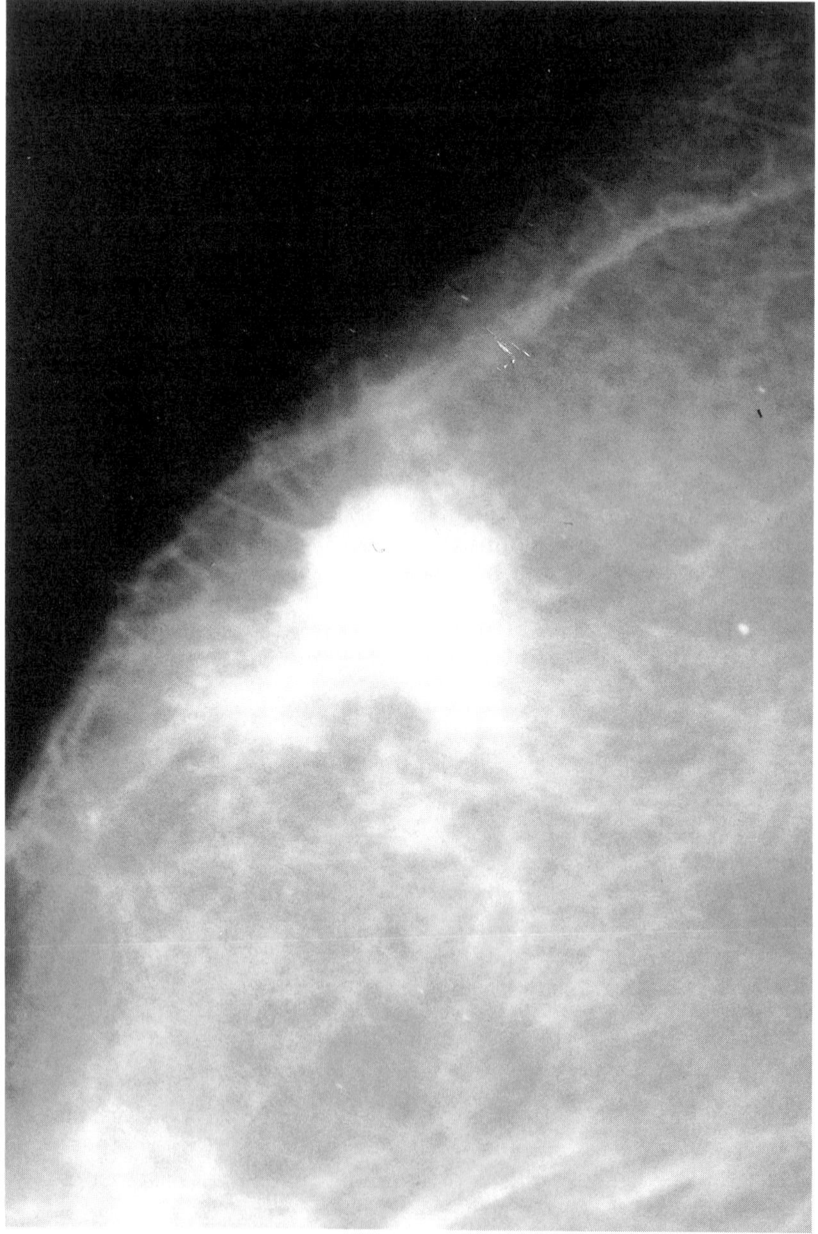

Figure 4. Mammogram showing a spiculated mass.

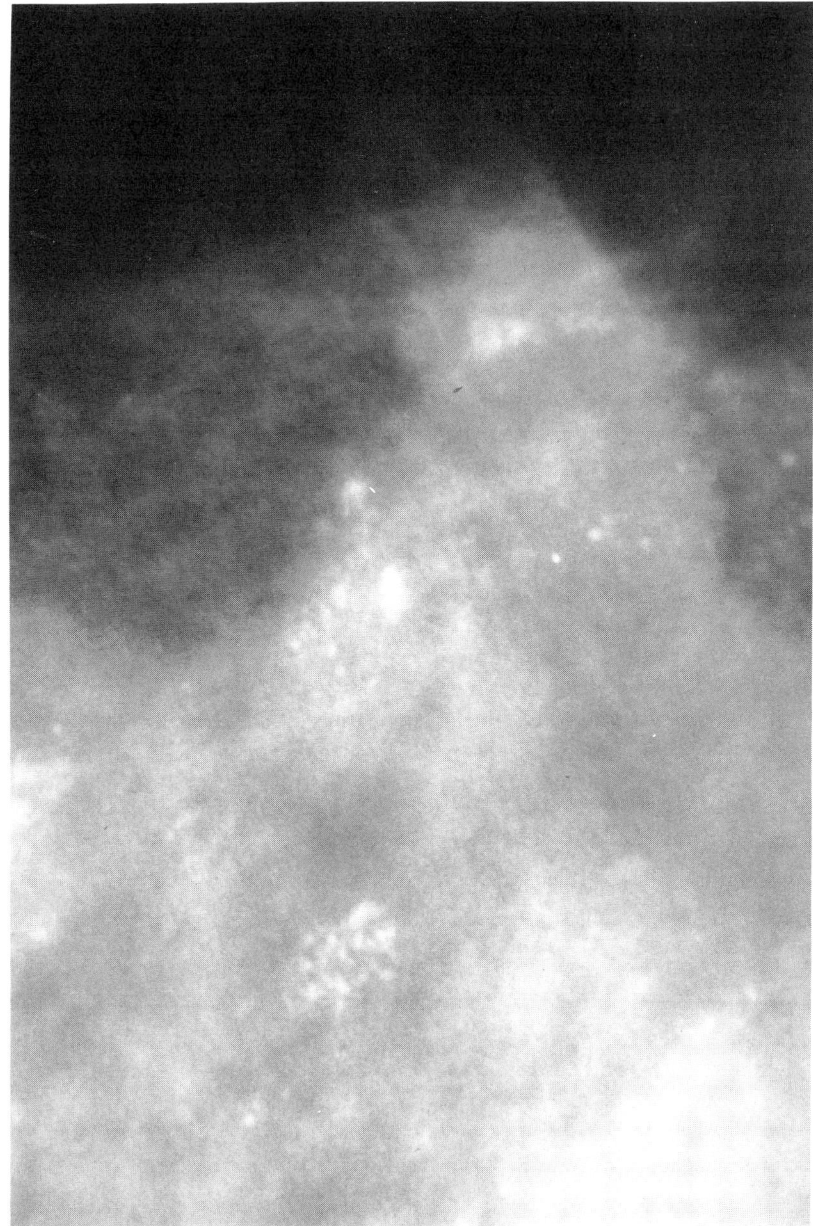

Figure 5. Mammogram showing malignant-appearing microcalcifications.

important evidence that early diagnosis and treatment will prevent the development of metastases.

The American Cancer Society recommends that screening should begin at the age of 40 years and consist of an annual clinical examination and a screening mammogram performed at 1- to 2-year intervals. At the age of 50 years, both the clinical examination and screening mammogram should be performed annually. Women who have a very strong family history of breast cancer or other risk factors that may predispose them to an elevated risk of breast cancer may need closer and more frequent surveillance.

BREAST BIOPSY TECHNIQUES

Four biopsy techniques are available for the diagnosis of breast masses: fine-needle aspiration (FNA), core-cutting needle biopsy, and incisional and excisional biopsy (Harris et al., 1993; Morrow, 1990). FNA is an outpatient procedure that is quick and relatively painless. The requirement of an experienced cytopathologist, a documented false-negative rate of 10% (Bell et al., 1983), and the inability to distinguish *in situ* from invasive carcinoma are the main drawbacks of the procedure (Strawbridge et al., 1984). A core-cutting needle biopsy has many of the advantages of an FNA, but since a core of tissue is obtained more details of tumor structure are available and ductal carcinoma *in situ* can be identified. The choice of FNA or core-cutting needle biopsy depends on the availability of an experienced cytopathologist.

Excisional biopsy has the advantage of allowing a complete evaluation of the tumor size and its histologic characteristics before instituting definitive treatment. If the mass is excised as a single piece of tissue with a surrounding margin of normal breast tissue and the margins are painted with ink, the diagnostic biopsy will often serve as the definitive lumpectomy if breast-conserving therapy is chosen (Kearney and Morrow, 1995). Hormone receptor evaluation is a routine part of the pathological evaluation of breast malignancies. Incisional biopsies are usually used to establish a diagnosis of breast cancer in masses too large to excise completely, or in patients with metastatic or locally advanced disease where systemic therapy will be administered as the initial treatment modality.

Biopsy of nonpalpable, mammographic abnormalities poses a special problem. Blind excision of the suspected site of the abnormality is inadequate. Localization of the abnormality can be accomplished using straight needles, hook wires, or dye injection along a needle tract. The most important factor in the success of localization is how close the wire is placed to the mammographic abnormality. Specimen radiography should be carried out on all biopsies of nonpalpable lesions to confirm that they are present in the specimen.

STAGING

Staging is a systematic way of grouping patients according to the extent of disease (American Joint Committee on Cancer, 1987). The TNM staging system as outlined by the American Joint Committee on Cancer is described in Table 2. The system is based on the clinical features of the primary tumor (T), lymph nodes (N), and the presence or absence of metastasis (M). Initially, a patient is staged according to the clinical evaluation, which includes a history, physical examination, chest x-ray, and routine blood tests. Clinical assessment of the axillary nodes has both a false-positive and false-negative rate of approximately 30%, so an accurate determination of stage occurs only after the pathological evaluation of the surgical specimen (e.g., pathological stage).

Fewer than 5–10% of newly diagnosed breast cancer patients present with grossly detectable metastatic disease. Therefore, outside the context of an investigational clinical trial, additional radiological investigation (i.e., CT scans, bone scans, etc.) should be obtained only if there is a strong suspicion of metastasis based on abnormalities detected in the preliminary evaluation. As an example, the positive yield of a bone scan in an asymptomatic patient is extremely low. In the National Surgical Adjuvant Breast and Bowel Project (NSABP), only 52 of 7984 (0.06%) bone scans detected bone metastases in asymptomatic patients (Wickerham et al., 1984). Furthermore, the predictive value of a positive bone scan in asymptomatic patients with Stage I or II disease is only about 11–12%, meaning that only about 1 in 9 patients with a positive bone scan will have confirmed bone disease. Interestingly, tumor cells have been detected in conventional bone marrow aspirates in up to 33% of breast cancer patients, but most studies did not distinguish between patients with metastatic and nonmetastatic disease (Mendoza et al., 1969; Mansi et al., 1989). Some investigators believe the presence of tumor cells in the bone marrow is a transient event occurring around the time of surgery. These tumor cells may lack metastatic potential and, as such, are unimportant clinically.

PATHOLOGY

Breast cancer is divided into two broad categories. Noninvasive cancers, or carcinoma *in situ*, remain within the ducts or lobules without any microscopic invasion into the surrounding breast stroma, and they lack the ability to metastasize. In contrast, invasive breast cancer is defined by the presence of tumor cells in the breast stroma.

Noninvasive Carcinoma

Noninvasive carcinomas can be separated into ductal carcinoma *in situ* (DCIS), also known as intraductal carcinoma, and lobular carcinoma *in situ* (LCIS). DCIS and LCIS have distinct clinical features, morphologic appearances, and biological characteristics (Table 3).

Table 2. Staging of primary breast cancer

Stage groupings

	T	N	M
Stage 0	T_{is}	N_0	M_0
Stage 1	T_1	N_0	M_0
Stage IIA	T_0	N_1	M_0
	T_1	N_1	M_0
	T_2	N_0	M_0
Stage IIB	T_2	N_1	M_0
	T_3	N_0	M_0
Stage IIIA	T_0	N_2	M_0
	T_1	N_2	M_0
	T_2	N_2	M_0
	T_3	N_1,N_2	M_0
Stage IIIB	T_4	Any N	M_0
	Any T	N_3	M_0
Stage IV	Any T	Any N	M_1

TNM nomenclature for breast cancer

Tumor (T)

T_0 No evidence of primary tumor
T_{is} Carcinoma *in situ*
$T_{1a} \leq 0.5\,cm$
$T_{1b} \leq 0.5\,cm–1\,cm$
$T_{1c} \leq 1\,cm–2\,cm$
T_2 >2 cm–5 cm
T_3 >5 cm
T_4 Any size, with direct extension to chest wall or skin (excluding pectoral muscle)
T_{4a} Extension to chest wall
T_{4b} Edema or ulceration of skin or presence of satellite nodules
T_{4c} Both T_{4a} and T_{4b}
T_{4d} Inflammatory carcinoma

Nodes (N)

N_0 No regional lymph node metastasis
N_1 Metastasis to movable ipsilateral axillary lymph node or nodes
N_2 Metastasis to ipsilateral axillary lymph node or nodes fixed to one another or to other structures
N_3 Metastasis to ipsilateral internal mammary lymph node or nodes

Metastasis (M)

M_0 No distant metastases
M_1 Distant metastasis, including metastasis to ipsilateral supraclavicular lymph node or nodes

Table 3. Distinguishing features of *in situ* carcinomas

	LCIS	DCIS
Age at diagnosis	Premenopausal	Pre- and postmenopausal
Physical findings	None	Mass, nipple discharge
Mammogram	None	Microcalcifications or mass
Axillary nodal involvement	None	<2%
Malignant potential	Bilateral	Ipsilateral

LCIS, lobular carcinoma *in situ*; DCIS, ductal carcinoma *in situ*.

LCIS was first described in 1941, and was thought to be the anatomic precursor of infiltrating carcinoma (Foote and Stewart, 1941). Mastectomy was the recommended treatment. LCIS is typically multicentric and bilateral. Its true incidence is unknown because it lacks both clinical and mammographic signs and is an incidental finding in breast biopsies done for another indication (Morrow and Schnitt, 1995). Reported incidences of LCIS range from 0.5% (Page et al., 1991) to 3.6% (Haagensen et al., 1981) of breast biopsies.

Over time it has become apparent that not all cases of LCIS progress to invasive carcinoma, and the major issue in the management of LCIS is the risk of invasive carcinoma following the diagnosis. Five studies (Wheeler et al., 1974; Andersen, 1977; Rosen et al., 1978; Haagensen et al., 1981; Page et al., 1991) with long-term follow-up of women with LCIS treated with biopsy alone demonstrate that the risk of breast cancer development is 7 to 10 times that of the index population. This is equal to a risk of cancer development of about 1% per year. Interestingly, most carcinomas that develop in women with LCIS are infiltrating ductal, not infiltrating lobular carcinomas (Wheeler et al., 1974; Rosen et al., 1978; Haagensen et al., 1981). This, coupled with the observation that the incidence of infiltrating carcinoma is equal in both breasts, regardless of whether LCIS is unilateral or bilateral, has led to the current belief (Rosen et al., 1978; Haagensen et al., 1981; Page et al., 1991; Morrow and Schnitt, 1996) that LCIS is a marker for cancer development rather than a premalignant lesion. One management option for the woman with LCIS is careful observation, as would be carried out for any woman known to be at increased risk for breast cancer development. The alternative, for women unwilling to accept the cancer risk associated with this policy, is bilateral total (simple) mastectomy, usually with immediate reconstruction. This is prophylactic surgery, and is viewed by many as unnecessarily radical in this era of breast-conserving surgery for invasive carcinoma. Treatment strategies addressing one breast, such as unilateral simple mastectomy with contralateral biopsy would not be logical because the risk of LCIS is bilateral, regardless of the findings of the contralateral biopsy.

In contrast to LCIS, DCIS has a variety of clinical presentations. In the past, most DCIS was "gross" or palpable. This form of DCIS was uncommon, and accounted for only about 2% of palpable breast carcinomas (Rosner et al., 1980). DCIS may also present as nipple discharge, or be identified as an incidental finding. Today, an abnormal mammogram is the most common presentation of DCIS (Morrow et al., 1996), and in many reports of mammographically directed biopsies DCIS accounts for 50% or more of the malignancies that are identified (Silverstein et al., 1989; Morrow et al., 1994).

In the past, DCIS was generally classified according to architecture into comedo type (associated with central necrosis) and noncomedo types (solid, cribriform, micropapillary). The comedo type of DCIS is associated with several biologically aggressive characteristics compared to the noncomedo types of DCIS: high-grade morphology, high proliferative index, aneuploidy, overexpression of the HER-2/*neu* oncogene, and an increased frequency of microinvasion (Morrow et al., 1996). Because of the observation that 50% of cases of DCIS have more than one morphologic type of DCIS in the lesion (Lennington et al., 1994), newer classification schemes have been proposed that group lesions according to grade rather than architecture (Holland et al., 1994).

The major issue in the management of DCIS is its risk of progression to invasive carcinoma. A minimal amount of clinically relevant data is available to address this issue, primarily because DCIS has traditionally been treated with mastectomy. In addition, most of the patients with DCIS for whom long-term follow-up is available had gross DCIS, which may not be equivalent to mammographic DCIS, which is more commonly diagnosed today. Studies of women found to have DCIS on review of biopsy specimens originally classified as benign indicate that some, but not all, cases of DCIS progress to invasive carcinoma. Page et al. (1982) observed a 28% incidence of invasive carcinoma at a mean of 6.1 years postbiopsy, a finding similar to that of Rosen et al. (1982). It is important to note that most cases included in these studies were low grade noncomedo lesions, representing one extreme of the histologic spectrum of DCIS. In contrast to the bilateral risk of carcinoma seen with LCIS, invasive cancers in women with DCIS undergoing biopsy alone occur in the same breast, and usually the same quadrant as the DCIS, suggesting that this lesion is the anatomic precursor to invasive carcinoma.

Uncertainty regarding the natural history of DCIS has resulted in a wide range of treatment recommendations for the disease. Mastectomy is a curative treatment for approximately 98% of patients with DCIS, whether gross or mammographic (Morrow et al., 1996). Although highly effective, mastectomy is a radical approach to a lesion that may not progress to invasive carcinoma during the patient's lifetime.

A number of investigators have examined the use of excision alone as a treatment for DCIS. In general, patients treated with this approach have been highly selected, usually on the basis of low histologic grade and/or small tumor size. Schwartz et al. (1991) reported 72 cases of DCIS treated with excision alone. Two-thirds of the cases were detected as mammographic calcifications and one-third as incidental findings. At a mean follow-up of 49 months, 15% of patients have recurred locally. Other studies report local failure rates ranging from 8–43% (Hetelekidis et al., 1995; Morrow et al., 1996), depending on the criteria used for patient selection and the length of follow-up. The time course to local failure is quite prolonged, with studies having longer follow-up showing higher local failure rates. When local failure occurs after excision alone, DCIS is documented in about half of the cases, and invasive carcinoma in the other half (Schwartz et al., 1992; Hetelkidis et al., 1995; Morrow et al., 1996).

Radiation therapy has been combined with excision in an attempt to improve local control in women with DCIS treated with a breast-conserving approach. Solin et al. (1991) reported the results of 259 women with DCIS treated with excision and irradiation at 9 institutions in Europe and the United States. The 10-year actuarial rate of local failure was 16%, and half of the failures were invasive carcinoma. Other studies of excision and radiation (McCormick et al., 1991; Silverstein et al., 1992; Morrow et al., 1996; Hetelekidis et al., 1995) also suggest that the local recurrence rate after excision and radiation is approximately 50% less than that seen after excision alone. The pattern of recurrence is similar, with invasive carcinoma seen in one half the recurrences, and most recurrences occur at or near the site of the primary tumor.

The NSABP has reported the initial results of a prospective trial designed to evaluate the role of radiation therapy in DCIS (Fisher et al., 1993). In this study, 818 women were randomized to excision alone or excision plus radiotherapy to the breast. At a median follow-up of 43 months, event-free survival was significantly higher in the women receiving irradiation (84% vs. 74%; $p=0.01$). This difference was due to a 58.8% reduction in the annual incidence of ipsilateral breast recurrence in the irradiated group. Although the incidence of both invasive and intraductal recurrences was reduced, the benefit of radiation was most evident for invasive recurrences.

The development of an invasive breast recurrence after a diagnosis of DCIS has the potential to decrease survival. High nuclear grade and comedo histology have been found to be predictors of local recurrence after excision and irradiation (Solin et al., 1991; Schwartz et al., 1992; Silverstein et al., 1992), yet most patients with these tumor characteristics do not have recurrences. The currently available information on DCIS suggests that although all patients can be treated with mastectomy, many are candidates for treatment with excision and irradiation. A smaller group, as yet poorly defined, may be candidates for excision alone. After a diagnosis of DCIS, treatment options and their risks and benefits should be discussed in detail. It is becoming increasingly clear that DCIS represents a spectrum of diseases of differing biological potential (Morrow et al., 1996). The final answers on the risks of breast-conserving therapy for DCIS will be obtained only after studies of patients with more precisely characterized tumors and long-term follow-up are available. The role of tamoxifen in reducing local failure is currently under evaluation, but no data are available. At present, there is no indication for the use of tamoxifen in DCIS outside of a clinical trial.

Invasive Carcinoma

Infiltrating ductal carcinoma is the most common invasive mammary tumor, accounting for 65–80% of mammary carcinomas (Rosen, 1996). A review of the SEER data from 1983 to 1987 found infiltrating ductal carcinoma to account for 75% of invasive tumors in white women and 61% in black women (Berg and Hutter, 1995). The median age at diagnosis was 62 years for white women and approximately 57 years for black women. Many infiltrating ductal carcinomas have some histologic features of other specific types of breast carcinoma, but unless these features are extensive, the tumors are classified as infiltrating ductal carcinoma not otherwise specified (NOS). Infiltrating ductal carcinoma classically forms a firm, irregular lump and a stellate mass, often with microcalcifications, is seen on a mammogram. Approximately 30% of infiltrating ductal carcinomas have extensive intraductal carcinoma (EIC) in and around the primary tumor (Vicini et al., 1992).

Infiltrating lobular carcinoma is the second most common type of breast cancer, accounting for 5–10% of cases. The classic microscopic appearance of infiltrating lobular carcinoma is single files of cytologically uniform cells. This growth pattern makes the clinical detection of lobular carcinoma difficult in some cases because tumors may present as vague thickening of the breast rather than a well defined mass. On a mammogram, infiltrating lobular carcinoma may result in only increased density of the involved area, and calcifications are uncommon with this lesion. Despite this, the stage distribution for infiltrating lobular carcinoma was similar to that of infiltrating ductal carcinoma in the SEER database (Berg and Hutter, 1995), although survival was better for infiltrating lobular carcinoma. When it metastasizes, infiltrating lobular carcinoma involves the intra-abdominal organs and peritoneal surface more often than other types of breast cancer (Harris et al., 1984; Rosen, 1991). The incidence of bilateral breast cancer, both synchronous and metachronous, is higher in women with infiltrating lobular carcinomas than in those with ductal tumors (Lesser et al., 1982), although the inclusion of cases of LCIS in these calculations artificially elevates the incidence of bilaterality.

The remaining fraction of invasive breast cancers are referred to as special-type carcinomas (Table 4). These include medullary, tubular, and mucinous carcinomas. Since mammography screening has come into widespread use, special-type carcinomas are being diagnosed more frequently. These special-type carcinomas tend to have a better prognosis than the

Table 4. Frequency of histologic types of invasive breast cancer

Type	%
Infiltrating ductal	75
Infiltrating lobular	10
Medullary	9
Mucinous	2
Tubular	2
Other	2

more common types of invasive breast cancer, and they are less likely to metastasize to the axillary nodes. Rosen et al. (1993) reported a large series of women with axillary node-negative breast cancer who were treated with surgery alone and followed for 30 years. Approximately 30% of the patients in the study had a survival similar to the general population. Included in this group were women with tumors of special histologic type and infiltrating ductal carcinoma <1 cm in diameter.

PRIMARY SURGICAL MANAGEMENT

The local therapy of breast cancer has evolved over many decades. During much of this century it was generally accepted that breast cancer spread in an orderly and time-dependent fashion from the primary tumor to the regional nodes and then to distant sites (Halstedian theory). The radical mastectomy, an *en bloc* resection of the breast, the overlying skin, the pectoral muscles, and the entirety of the axillary contents was developed in response to this theory of cancer spread. Unfortunately, although most women with breast cancer were technically able to undergo radical mastectomy, few were cured by the procedure. Of 1458 women treated with radical mastectomy at Memorial Hospital between 1940 and 1943, only 13% survived 30 years free of cancer (Adair et al., 1974). In the last several decades, as knowledge of tumor biology has increased, an alternative paradigm of breast cancer biology has evolved that suggests that treatment failure after breast cancer surgery is determined by the intrinsic biological characteristics of the tumor rather than the extent of the operative procedure on the breast (Fisher, 1980). This change in our understanding of breast cancer had led to less extensive surgical treatment of the disease.

Modified radical mastectomy is now the most common operative treatment for patients with breast cancer in the United States (Nattinger et al., 1992; Samet et al., 1994). The term "modified radical mastectomy" is used to describe a variety of surgical procedures, but all involve complete removal of the breast and the removal of some of the axillary nodes. Although the modified radical mastectomy may not seem to differ significantly from the radical mastectomy, it represented a major departure from the principles of *en bloc* cancer surgery exemplified by the radical mastectomy. Two prospective randomized trials (Turner et al., 1981; Maddox et al., 1983) demonstrating no difference in survival between women undergoing modified radical and radical mastectomy supported this change in philosophy regarding the local therapy of breast cancer.

An extension of the principles that led to the use of the modified radical mastectomy, coupled with the increasingly frequent identification of small breast cancers by mammography, and the success of moderate doses of radiotherapy (RT) in eliminating subclinical foci of breast cancer after mastectomy, have contributed to the development of breast-conserving therapy (lumpectomy and radiotherapy). An initial objection to the use of breast-conserving treatment was the known multicentricity of breast carcinoma. The reported incidence of multicentricity ranged from 9–75%, depending on the definition employed and the technique of pathologic examination used (Qualheim and Gall, 1957; Rosen et al., 1975). Considerable clarification of this issue has occurred as a result of the work of Holland and co-authors, who demonstrated that multifocality (foci of cancer in the vicinity of the primary tumor) is common, but true multicentricity is rare (Holland et al., 1985). Using detailed pathological and radiographic assessment of mastectomy specimens, they mapped tumor distribution in 264 cases with apparently localized tumor. Only 39% of specimens showed no evidence of cancer beyond the primary tumor, and in an additional 20% of specimens all residual carcinoma was within 2 cm of the primary tumor. Although 41% of specimens had tumor 2 cm or more from the primary site, only 11% had additional foci of tumor at distances greater than 4 cm from the index tumor. Of the 41% of cases with residual cancer more than 2 cm from the reference tumor, two-thirds had pure intraductal carcinoma. The percent of cases with residual cancer more than 2 cm from the reference tumor corresponds well to the rate of local failure reported in patients treated with excision of the primary tumor alone (Freeman et al., 1981; Lagios et al., 1983; Fisher and Redmond et al., 1989). Thus, the strategy behind breast-conserving treatment is to remove the bulk of the tumor surgically, recognizing that microscopic residual cancer remains in a significant number of women, and to use radiation to eradicate any residual disease.

The general acceptance of the Halstedian theory of cancer spread necessitated a relatively large number of randomized trials of conservative surgery (CS) and radiotherapy (RT) before this approach was accepted in clinical practice. Since 1970, six prospective randomized trials using modern radiation techniques have compared CS and RT and mastectomy (Fisher et al., 1989c; Sarrazin et al., 1989; Veronesi et al., 1990; Blichert-Toft et al., 1992; Lichter et al., 1992; Van Dongen et al., 1992). The results of these trials are summarized in Table 5. No significant survival differences between surgical treatments were observed in any of these studies. In the randomized studies presented, with widely varying surgical and RT techniques,

Table 5. Randomized trials comparing breast conserving therapy to mastectomy

Trial	Treatment	Results at years	Local relapse (%)	Survival (%)
Institut Gustave Roussy[a]	MRM	10	9	79
	LE + RT		7	78
Milan[b]	RM	13	2	69
	QUART		4	71
NSABP BO6[c]	MRM	8	8	76
	LE + RT		16	71
NCI[d]	MRM	8	6	85
	LE + RT		20	89
EORTC[e]	MRM	8	9	75
	LE + RT		13	75
Danish[f]	MRM	6	4	82
	LE + RT		3	79

MRM, modified radical mastectomy; RM, radical mastectomy; RT, radiation therapy; QUART, quadrantectomy, axillary dissection, radiation therapy; LE, excision of the primary tumor with a variable amount of normal breast tissue; NSABP, National Surgical Adjuvant Breast and Bowel Project; NCI, National Cancer Institute; EORTC, European Organization for Research and Treatment of Cancer.
[a]Sarrazin et al., 1989.
[b]Veronesi et al., 1990.
[c]Fisher et al., 1989.
[d]Lichter et al., 1992.
[e]Van Dongen et al., 1992.
[f]Blichert-Toft et al., 1992.

the rates of tumor recurrence in the breast ranged from 4–20%. A great deal of emphasis has been placed on the problem of recurrence in the breast after breast-conserving treatment. However, in these studies 2–9% of the patients treated with mastectomy developed local recurrence, emphasizing that mastectomy does not guarantee freedom from chest wall failure, even in women with Stage I and II breast carcinoma.

The goal of breast-conserving surgery is to maintain local tumor control in the breast while preserving a good cosmetic appearance. The natural history of local recurrence after CS and RT, as well as factors that influence the risk of local recurrence has been reviewed in detail elsewhere (Harris and Morrow, 1996). The risk of local recurrence after CS and RT varies with patient factors such as young age (Matthews et al., 1988; Fourquet et al., 1989), tumor factors such as the presence of EIC associated with the invasive tumor (Jacquemier et al., 1990; Vicini et al., 1992), and treatment factors such as the extent of surgical resection (Veronesi et al., 1990) and the radiation dose employed (Osborne et al., 1984). Some of these factors are interrelated. For example, cancers with an EIC are more common in premenopausal than in postmenopausal patients. Both EIC and young patient age are associated with an increased risk of local failure, suggesting a possible hormonal influence. As more information on local recurrence after CS and RT is accumulated, it appears that there may be two types of local recurrence. The first reflects an inherent biological aggressiveness of the tumor, which places the patient at high risk for treatment failure whether CS and RT or mastectomy is undertaken as local therapy. The other type is related to the local manifestations of the cancer (e.g., the extent of the disease in the breast), and probably is not a reflection of poor prognosis. Further information is needed to clarify the significance of local recurrence. Approximately 90% of women treated with CS and RT report an excellent or good cosmetic outcome 5 years after treatment (Rose et al., 1989). A combination of patient, tumor, and treatment factors also influence the cosmetic outcome of breast-conserving surgery. However, the amount of breast tissue resected appears to be the major determinant of outcome (Veronesi et al., 1990b).

Based on the information available from prospective and retrospective studies, representatives from the American College of Surgeons, the American College of Radiology, the College of American Pathologists, and the Society of Surgical Oncology have developed standards of care for breast-conserving treatment (Winchester and Cox, 1992). Absolute and relative contraindications to CS and RT, as determined by the panel, are listed in Table 6. Since the development of this statement in 1991, further experience has demonstrated that large breast size and tumor beneath the nipple are rarely contradictions to CS and RT. Although studies (Morrow et al., 1994; Foster et al., 1995) have indicated that 75–80% of women with Stage I and II breast cancer are medical candidates for CS and RT, fewer than 50% of women in the United States receive this type of therapy at present (Nattinger et al., 1992; Samet et al., 1994).

Breast-conserving therapy and modified radical mastectomy are equivalent options in the management of Stage I and II breast carcinoma. Breast reconstruction, either done at the time of mastectomy or as a separate procedure, is also an option. Although mastectomy with and without reconstruction has never been studied in a randomized trial, retrospective studies (Noone et al., 1994; Petit et al., 1994) do not demonstrate any difference in the incidence of local or distant disease recurrence in women undergoing

Table 6. Contraindications to breast-conserving therapy

Absolute	Relative
First, second trimester of pregnancy	Large primary tumor in small breast
Multiple primary tumors	History of collagen vascular disease
Prior breast irradiation	Tumor beneath nipple
Diffuse indeterminate or suspicious microcalcifications	Very large breasts

breast reconstruction. Patient selection for local treatment is based on a detailed evaluation by mammography, physical examination, review of the pathology specimen, and an assessment of the patient's wishes. In the absence of medical contraindications to a method of local therapy, the choice of local treatment rests with the patient.

The issue of whether there is a subset of patients with invasive breast cancer in whom breast irradiation is not needed for breast-conserving therapy has been investigated in a number of studies (Freeman et al., 1981; Lagios et al., 1983; Fisher et al., 1989c; Gelber and Goldhirsch, 1994; Liljegren et al., 1994). Using a variety of selection criteria, including patients with very small primary tumors, all patient subsets have an improvement in local control with the addition of breast irradiation to surgical excision. At the present time, breast irradiation should be recommended to all patients with invasive breast cancer who undergo breast-conserving therapy.

Management of the Regional Nodes

The most common sites of regional node involvement in breast cancer are the axillary, supraclavicular, and internal mammary node regions. The axillary nodes are the primary site of lymphatic drainage from all quadrants of the breast, and approximately 40% of patients have spread to the axillary nodes at presentation (Carter et al., 1989). A variety of surgical procedures have been described for the removal of the axillary nodes, ranging from an anatomic dissection of all nodal tissue to an attempt to identify and remove individual axillary nodes. The complete removal of the axillary nodes was an important part of the Halstedian concept of the surgical cure of breast cancer. As axillary nodes have come to be regarded as markers of the risk of metastatic disease rather than the cause of metastatic disease, axillary dissection is primarily considered a staging procedure. The failure of axillary dissection to impact on survival was most convincingly demonstrated in the NSABP B04 trial (Fisher et al., 1985). In this trial, patients with clinically negative axillary nodes were randomized to radical mastectomy, total mastectomy with observation of the axillary nodes and a delayed dissection if positive nodes appeared, or total mastectomy and irradiation of the regional lymphatics. No statistically significant differences in survival were observed at 10 years, even though approximately 40% of women undergoing axillary dissection had positive nodes and a similar percentage were presumed to have positive nodes in the observation-only group.

Current techniques of axillary dissection are based on the removal of an anatomically defined volume of tissue rather than a specific number of lymph nodes because nodes of normal size are not readily identifiable at the time of surgery. To limit axillary dissection to patients with involved lymph nodes who will benefit from the procedure, Giuliano et al. (1994) have studied a biopsy technique of the primary draining lymph node, or "sentinel node," as a staging procedure. Initial results indicated accurate staging of 95% of patients, but further follow-up is needed before this technique can be considered as a substitute for axillary dissection.

RATIONALE FOR ADJUVANT SYSTEMIC THERAPY

Clinical trials conducted in the 1960s and 1970s, in which patients were treated with radical mastectomy, support the paradigm that breast cancer is a systemic disease in many patients even when the disease is diagnosed at an early clinical stage. For patients diagnosed with axillary node-negative breast cancer, the disease-free survival and overall survival at 10 years is 76% and 65%, respectively (Nemoto et al., 1980). For patients with axillary node-positive breast cancer, the disease-free survival and overall survival is 24% and 25%, respectively (Nemoto et al., 1980). These data suggested that axillary lymph node status was of paramount importance in determining a patient's overall prognosis. The probability of axillary lymph node involvement is closely correlated with the size of the primary tumor. In a study of 24,740 breast cancer cases, Carter et al. (1989) found axillary lymph node involvement in ~20% of cases where the primary tumor measured <0.5 cm compared to 27% in tumors up to 1 cm and 80% in tumors ≥10 cm in size.

Multiple reports have demonstrated that the absolute number of involved axillary nodes is precisely correlated with the risk of recurrence and death. In a study of 20,547 cases of breast cancer analyzed by the American College of Surgeons, recurrences at 5 years increased from 19.4% when no lymph nodes were involved to 82% when 21 or more lymph nodes contained tumor metastases (Nemoto et al., 1980). Survival declined from 72% when no lymph nodes were involved to ~50% when 4 lymph nodes were involved to

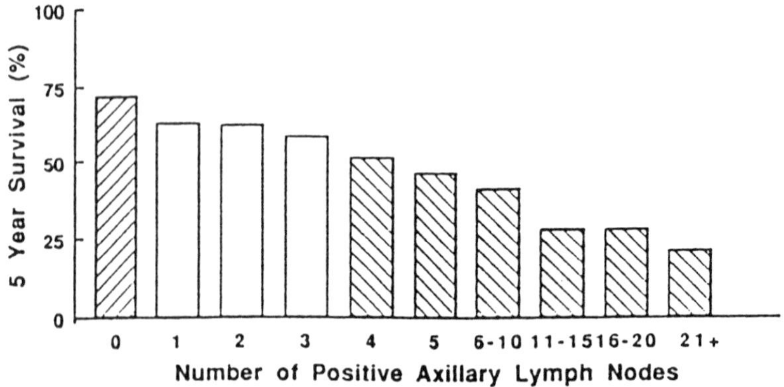

Figure 6. Survival of 20,547 women with breast cancer according to the number of histologically involved axillary nodes (Nemoto et al., 1980). The *unshaded bars* indicate the three categories of involvement most often reported (0, 1–3+, and >3+). (Reprinted with permission of the J.B. Lippincott Company, from Nemoto et al., 1980).

22% when 21 or more lymph nodes were involved (Fig. 6) (Nemoto et al., 1980). The observation that lymph node involvement directly correlated with recurrence and decreased survival suggested that micrometastases were established at the time of diagnosis. This clinical observation served as the rationale for the initiation of adjuvant systemic chemotherapy trials in an effort to determine if the micrometastases could be eradicated and if the natural history of axillary node-positive disease could be altered.

In 1958, the NSABP initiated the first adjuvant trial in which women underwent a radical mastectomy and were randomized to receive postoperative Thio-tepa (2 days) or placebo. The results were first reported by Fisher et al. (1968) and showed a significant improvement in survival for women receiving chemotherapy that persisted at 10 years, primarily in the subset of patients with 4 involved lymph nodes (Fisher et al., 1975a). A subsequent trial conducted by the NSABP and the Eastern Cooperative Oncology Group (ECOG) randomized axillary node-positive patients to receive L-phenylalanine mustard (L-PAM) or placebo for 24 months following radical mastectomy (Fisher et al., 1975). With follow-up of greater than 10 years, the premenopausal women who received L-PAM had a significant prolongation of both disease-free and overall survival (Fisher et al., 1986). Clinical trials in patients with advanced disease demonstrated the superiority of combination chemotherapy over single agents, so subsequent adjuvant chemotherapy trials investigated various combinations of chemotherapy drugs.

The Milan trials used a combination of cyclophosphamide, methotrexate, and 5-fluorouracil (CMF) (Bonadonna, 1992). In the first Milan trial, initiated in 1973, patients received 12 months of CMF or no postoperative adjuvant treatment. With follow-up of greater than 10 years, significant prolongation of relapse-free survival and overall survival in axillary node-positive, premenopausal women (but not in postmenopausal women) was observed. The second Milan trial compared 12 months of CMF to 6 months of CMF. With follow-up of greater than 10 years, the relapse-free survival and overall survival are equivalent between the 2 treatment arms and the survival benefit is equal in both pre- and postmenopausal women. The findings from these trials demonstrate that combination chemotherapy (CMF) altered the natural history of premenopausal node-positive disease, and that fewer cycles of chemotherapy produced the same effect as treatment of longer duration. Numerous clinical trials investigating variations of the CMF program and alternative programs utilizing doxorubicin have shown similar results.

In the 1960s, the steroid hormone receptors were discovered, including the estrogen receptor (ER). In 1971 Jensen et al. (Jensen et al., 1971) first reported that measurement of ER content in tumor-biopsy material predicted the response to adrenalectomy. Other investigators determined that the ER content in excised breast tumors or metastatic lesions predicted the likelihood of response to endocrine therapy (oophorectomy, adrenalectomy, hypophysectomy, etc.) for most breast cancer patients (McGuire et al., 1975; Byar et al., 1979). Approximately 70–80% of patients with breast cancer have estrogen-receptor-positive tumors. Postmenopausal women are more likely to have ER positive tumors compared to their premenopausal counterparts. Fifty percent of ER-positive tumors respond to hormone therapy compared to 10% of ER-negative tumors (McGuire et al., 1975; Byar et al., 1979).

During the 1970s, evaluation of the antiestrogen, tamoxifen, was initiated in patients with metastatic breast cancer. Approximately 50% of ER-positive, metastatic breast cancer patients responded to tamoxifen therapy (Legha et al., 1978; Santen et al., 1990). As a result of this observation, the Nolvadex Adjuvant Trial Organization initiated a clinical trial in 1977 randomizing 1131 patients with axillary node-positive (pre- and postmenopausal) or node-negative (postmenopausal) breast cancer to 2 years of tamoxifen or no treatment (Analysis at Six Years by Nolvadex Adjuvant Trial Organization, 1985). This trial and others that followed showed a modest but significant survival benefit for patients treated with tamoxifen, particularly the patients who were ER-positive and node-positive (Legha et al., 1978; Analysis at Six Years by Nolvadex Adjuvant Trial Organization, 1985; Breast Cancer Trials Committee, 1987; Baum et al., 1990; Santen et al., 1990).

In 1985, the NCI convened a Consensus Conference to synthesize the observations made from the numerous adjuvant chemotherapy and hormonal trials conducted over the previous decade (Consensus Conference, 1985). Outside the setting of a clinical trial, the conference recommended that: (1) node-negative, premenopausal women should not receive adjuvant therapy, (2) node-positive, premenopausal women, regardless of hormone receptor status, should receive adjuvant chemotherapy, (3) node-positive, ER-positive postmenopausal women should receive tamoxifen, (4) node-positive, ER-negative postmenopausal women could be *considered* for chemotherapy, and (5) node-negative, postmenopausal women should not receive adjuvant treatment.

Since the 1985 NCI Consensus Conference, a major change has occurred in the approach to women with node-negative breast cancer. Although most axillary node-negative breast cancer patients will be cured following local therapy, 25% of such patients will develop a recurrence and ultimately die of breast cancer at 10 years. Clinicians have used a variety of prognostic factors to identify patients at the greatest risk for disease recurrence (Table 7) (McGuire et al., 1992). Although new prognostic factors are reported almost weekly, only a few of them are widely accepted and have undergone prospective validation. The clinical challenge is to identify the axillary node-negative patients at greatest risk of disease recurrence in whom adjuvant therapy is potentially of greatest benefit, and to avoid administering adjuvant therapy to patients who are already cured of breast cancer and will only experience the toxicity of therapy.

Tumor size and axillary nodal status are established prognostic factors in assessing the risk for recurrence and metastatic disease. The 5-year

Table 7. Factors that predict prognosis for disease recurrence in axillary node-negative breast cancer

Factor	Favorable association
Primary tumor size	<1 cm
DNA content of tumor	Diploid
Proliferative index	Low S-phase fraction
Nuclear grade	Low
Hormone receptor status	High

recurrence-free survival in patients with axillary node-negative breast cancer ranges from 89% for tumors < 1 cm to 86% for tumors 1.1 to 2.0 cm to 79% for tumors 2.1 to 5 cm in diameter (Nemoto et al., 1980; Carter et al., 1989; Rosen et al., 1993). Ten-year survival rates in relation to tumor size in axillary node-negative disease are shown in Table 8 (Schottenfeld et al., 1976). The increasing risk of recurrence (either locally or distantly) and decreasing survival as the primary tumor size increases suggests that the risk of micrometastatic disease increases with increasing tumor size.

Histologic grading systems have been developed that consider the architectural arrangement of cells (degree of tubule formation), the degree of nuclear differentiation (size and shape), and the mitotic rate (Fisher et al., 1988). The NSABP demonstrated that patients with good nuclear grade tumors (well-differentiated) have a 5-year survival of 93% compared to 79% for patients with poor nuclear grade tumors (poorly differentiated) (Fisher et al., 1988). One problem that has been identified with this type of assessment is interobserver variability between different pathologists. The presence of ERs in a tumor is an excellent marker of differentiation and, as such, identifies a subset of patients who have a modestly improved prognosis. The 5-year disease-free survival difference between tumors with and without estrogen receptors is 8–9%. In addition, the presence of ERs predicts a response to hormonal therapy.

Proliferative indices, such as the S-phase fraction, are used to measure the DNA synthetic activity of breast tumors. Flow cytometry is a tool that can be used to measure the percentage of cells entering the S-phase of the cell cycle and the DNA content of the cells. Tumors that have an abnormal

Table 8. 10-year survival in relation to tumor size in patients with axillary node-negative disease[a]

Size (cm)	10-year survival (%)
<2	82%
2–5	65%
>5	44%

[a]Schottenfeld et al., 1976.

amount of DNA content (aneuploid) have been associated with a worse prognosis compared to tumors containing a normal amount of DNA (diploid). However, not all studies have confirmed this observation. S-phase appears to correlate well with both the ER and ploidy status of tumors. Tumors with a low S-phase are more commonly diploid and ER-positive than tumors with a high S-phase. In a study by Clark et al. (1992), the difference in disease-free survival in patients with axillary node-negative disease by S-phase status is highly significant (Fig. 7).

Numerous other prognostic factors have been proposed for patients with axillary node-negative disease, including presence or absence of amplification of the HER-2/*neu* oncogene, expression of p53, levels of the protease cathepsin D, measures of angiogenesis within the primary tumor, presence or absence of nm23, etc. None of these prognostic factors has an established role for assigning patients into risk groups or for determining the likely

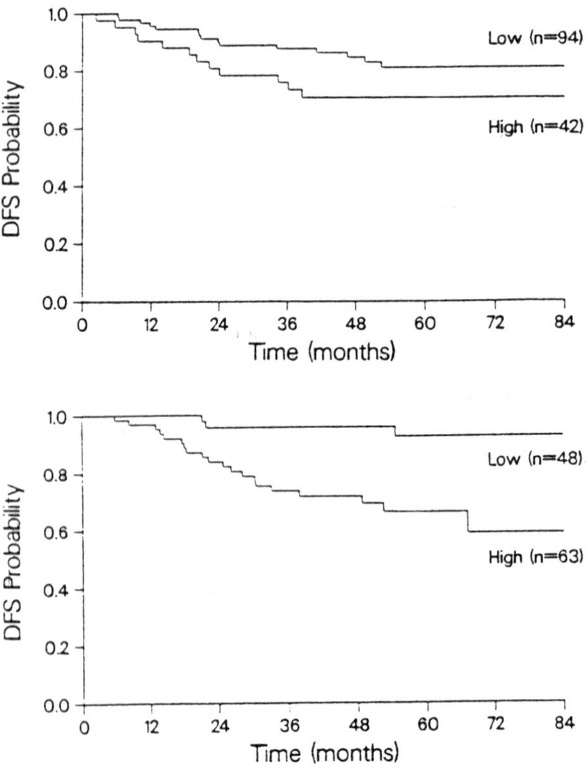

Figure 7. **A**: Disease-free survival by S-phase fraction (SPF) for patients with diploid tumors. High SPF is defined as >4.4%. $p = 30.053$. **B**: Disease-free survival by S-phase fraction (SPF) for patients with aneuploid tumors. High SPF is defined as >7.0%. $p = 30.0012$. (Reprinted with permission of the W.B. Saunders Company, from Clark et al., 1992).

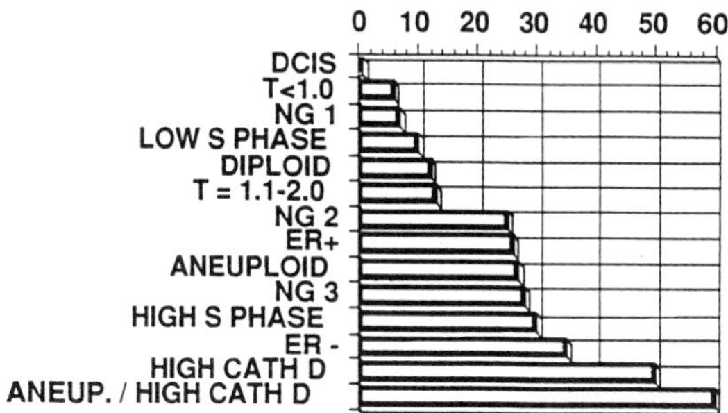

Figure 8. Probability of recurrence at 5 years based on tumor characteristics in axillary node-negative breast cancer. (Reprinted with permission of the J.B. Lippincott Company, from McGuire et al., 1992). (*DCIS*, ductal carcinoma *in situ*; *T*, tumor diameter; *NG*, nuclear grade; *ER*, estrogen receptor; *CATH D*, cathepsin D).

benefit of adjuvant therapy. At present no single prognostic factor that accurately differentiates high risk and low risk has been identified. Risk assessment involves integrating all known information and arriving at an estimate for disease recurrence. McGuire et al. (1992) calculated the probability of recurrence based on specific prognostic factors (Fig. 8), but the integration of multiple prognostic variables in an individual patient requires further study.

Before 1980, few patients with node-negative breast cancer were included in adjuvant trials and there was little enthusiasm for recommending treatment to such patients at the 1985 NCI Consensus Conference (Consensus Conference, 1985). The failure of prognostic factors to identify high-risk, node-negative breast cancer patients resulted in a series of trials to determine if adjuvant therapy was beneficial in node-negative breast cancer. In 1981, the NSABP initiated a clinical trial in ER-negative, node-negative breast cancer patients. Patients received sequential antimetabolite therapy (methotrexate → fluorouracil) or no postoperative therapy (Fisher et al., 1989). At 5 years treatment failures were reduced by 50% in patients older than 50 years and by 20% in patients younger than 50 years (Fisher et al., 1989). A significant survival benefit was detected in the older group of patients. At the same time, the NSABP initiated another trial in "good" prognosis, ER-positive, node-negative patients in which randomization was between postoperative tamoxifen or placebo (Fisher et al., 1989). This study demonstrated a highly significant improvement in disease-free survival for all patients receiving tamoxifen ($\leq$49, $\geq$50 years). There was a reduction in both ipsilateral breast recurrences and distant metastasis in patients receiving tamoxifen compared to the placebo group. In addition, this study also demonstrated a reduction in contralateral breast cancers in patients receiving tamoxifen.

Large clinical trials conducted by the Intergroup (Mansour et al., 1989) and the Ludwig Breast Cancer Study Group (The Ludwig Breast Cancer

Study Group, 1989) (LBCSG) also demonstrated a significant improvement in disease-free survival in node-negative breast cancer patients receiving adjuvant chemotherapy. The Intergroup study showed a significant improvement in disease-free survival in poor prognosis, node-negative patients (ER-negative or ER-positive, T > 3 cm) who received 6 cycles of adjuvant cyclophosphamide, methotrexate, 5-fluorouracil and prednisone (CMFP) compared to patients who received no adjuvant therapy (Mansour et al., 1989). The LBSCG study randomized patients to a single postoperative course of cyclophosphamide, methotrexate, 5-fluorouracil, and leucovorin or no therapy. This study demonstrated a modest improvement in disease-free survival for all patients receiving therapy, but the greatest effect was demonstrated in patients with ER-poor tumors (The Ludwig Breast Cancer Study Group, 1989). At that time, only the Milan IV trial (Bonadonna et al., 1986) was able to demonstrate an actual improvement in overall survival. In this trial, 89 patients were randomized to receive either 9 months of intravenous CMF or no adjuvant therapy. In an early follow-up report, 93% of the CMF-treated patients were alive and free of disease recurrence compared to 50% of the untreated patients. At 3 years, all CMF-treated patients were alive compared to 75% of the control patients.

In 1990, the NCI convened another Consensus Conference on Early Stage Breast Cancer (Consensus Statement, 1990). This conference cautiously reviewed the data from the trials cited above as well as others, but the recommendations were not a full endorsement of the use of adjuvant therapy in axillary node-negative breast cancer patients. However, the consensus statement did suggest that the natural history of axillary node-negative disease was perturbed by adjuvant therapy. As a result, clinical practice quickly changed with more medical oncologists recommending adjuvant therapy to patients with axillary node-negative breast cancer. Because of the small size of individual trials and short follow-up of patients on these trials, the consensus statement focused on what had been observed in previous trials rather than providing clear practice recommendations. A critical analysis of the available studies demonstrated great variability in the quality of trial design and as a result the data and conclusions generated from many of these studies were questionable. Many of the studies lacked adequate statistical power to reach the stated conclusions because too few patients were enrolled on the study. Because of many studies that have been conducted, both the possibility of a positive outcome and magnitude of benefit may be dictated by chance alone. Data can be referenced from the medical literature, citing individual trials, that support or refute the benefit of adjuvant therapy.

To analyze the results of adjuvant clinical trials more critically, meta-analysis methodology has recently come into widespread use. This technique gives the greatest weight to results from the largest studies and corrects false-negative results produced by small randomized trials. As a result, a meta-analysis is helpful in detecting a modest but real difference between two treatments that may not be detected in a single randomized trial. Results from a meta-analyses are commonly expressed in terms of the percentage reduction in the odds of an event occurring (i.e., recurrence or death), in contrast to the absolute benefit of treatment. This provides a way of expressing the percentage of patients destined to have a disease recurrence or die from the disease in whom treatment can be expected to delay or avoid these events. For instance, in a group of patients with only a 10% mortality rate without treatment (i.e., axillary node-negative, tumor <1 cm), a 50% proportional reduction in mortality would lower the mortality rate to about 5%. In contrast, in a group of high-risk patients with a mortality risk of 50% without treatment (i.e., >10 involved axillary nodes), a 50% proportional reduction in mortality would lower the mortality rate to about 25%. The biological effect of this hypothetical treatment is the same in the low- and high-risk group (e.g., 50% proportional reduction in mortality), but the absolute effect of therapy is greatest in the group of patients at highest risk. Another way of describing the results above is: the absolute benefit would be 5 additional patients out of every 100 treated alive at the end of a defined time period in the low-risk group; the absolute benefit would be an additional 25 patients alive out of every 100 treated at the end of the same time period in the high-risk group. Obviously, the number of patients who will not benefit from the treatment and who will be exposed to the toxicity of treatment is greatest in the low-risk group of patients.

In a worldwide collaborative effort, the Early Breast Cancer Trialists' Group conducted a meta-analysis of mortality and recurrence for more than 75,000 women enrolled in 131 randomized trials of systemic adjuvant therapy before 1985 (Early Breast Cancer Trialists' Collaborative Group, 1992). The randomized treatment groups included 30,000 women in tamoxifen trials, 3000 women in ovarian ablation trials, 26,000 women in chemotherapy trials, and 6000 women in immunotherapy trials. The analyses involved all deaths and recurrences. The following information was obtained on all patients: age, menopausal status, nodal status, type of local treatment, hormone receptor measurements, first contralateral, second primary breast cancer, first local recurrence, first distant recurrence, and death.

The meta-analyses can be summarized in the following manner. Among premenopausal women with positive nodes, chemotherapy will reduce the annual odds of recurrence by 36% and the annual odds of death by 25%. At 10 years the difference between treated and control patients was 12% for disease-free survival and 10% for overall survival. Chemotherapy produced a statistically significant delay in recurrence in node-positive, postmenopausal patients 50 to 59 years of age, a marginal delay in death in women 50 to 59 years of age, and a reduction of borderline significance in mortality in women 60 to 69 years of age. For postmenopausal women with positive nodes, tamoxifen reduced the odds of death by at least 20%. At 10 years, the difference between treated and control patients was 9% for disease-free survival and 7% for overall survival. Even women under the age of 50 years had a statistically significant reduction in the odds of recurrence when treated with tamoxifen (12%), but the magnitude of benefit is significantly less than in older women (28%). The meta-analyses did not show a significant tamoxifen treatment effect on mortality in women under 50 years of age. Cumulative differences with tamoxifen and combination chemotherapy were larger at 10 years of follow-up than at 5 years (Fig. 9A). In the group of patients at lower risk of disease recurrence, axillary node-negative disease, a more modest benefit from adjuvant therapy was identified. Combination chemotherapy reduced the annual odds of recurrence by 29% and the annual odds of death by 16%. At the end of 10 years, the difference between treated and control groups was 5% for disease-free survival and 3.5% for overall survival (Fig. 9B).

Additional insight can be gained regarding the benefits of adjuvant therapy by subset analysis (Gelber et al., 1993). Most patients treated with tamoxifen received treatment for 2 years or less. Patients treated with tamoxifen for more than 2 years had a greater reduction in the annual odds of recurrence compared to patients treated with tamoxifen for less than 2 years (38% vs. 16%). A similar effect on the annual odds of death was observed with longer tamoxifen therapy (24% vs. 11%). The benefit derived from tamoxifen on both the annual odds of recurrence and death was greatest in patients with ER-positive tumors, with increasing benefit correlating with increasing ER levels. Not surprisingly, most postmenopausal patients have ER-positive tumors, the subset of patients where tamoxifen has its greatest benefit.

The subset analysis of chemotherapy trials also suggests that multiple cycles of chemotherapy are superior to a single course of perioperative chemotherapy (Gelber et al., 1993). The optimal number of chemotherapy treatment cycles has not been defined, but 4 to 6 months of treatment has been accepted as equal to longer durations of therapy. In addition, combination chemotherapy is superior to single-agent therapy. Whether chemotherapy regimens containing doxorubicin are superior to other regimens like CMF was not definitively answered by the overview analysis and remains controversial.

Active areas of clinical investigation include trials exploring whether chemotherapy plus tamoxifen is superior to chemotherapy alone in premenopausal women with ER-positive tumors. Other trials are attempting to determine if chemohormonal therapy provides more benefit than tamoxifen alone in postmenopausal women. New agents such as paclitaxel and docetaxel are being introduced into adjuvant chemotherapy regimens. New antiestrogens and tamoxifen/retinoid analogue combinations are under investigation. For patients at high risk of relapse (involvement >10 axillary lymph nodes), NCI-sponsored clinical trials comparing high-dose chemotherapy and peripheral stem cell transplantation to conventional dose chemotherapy are actively accruing patients.

A reasonable synthesis of available information (Early Breast Cancer Trialists' Collaborative Group, 1992; Glick et al., 1992; Gelber et al., 1993) would lead to recommendations for the use of adjuvant therapy in routine practice, as outlined in Table 9.

TREATMENT OF ADVANCED BREAST CANCER

Advanced breast cancer can be divided into those patients with locally advanced disease (Stage III) and patients with metastatic disease (Stage IV). Approximately 20% of all patients presenting with breast cancer have locally advanced disease (Nemoto et al., 1980; Carter et al., 1989). Stage III disease represents a heterogeneous group of clinical presentations, including patients with very large tumors (>5 cm), matted axillary lymph nodes, tumors extending to the chest wall or skin, and inflammatory breast cancers (see Table 2). The approach to this hetergeneous group of disorders is always individualized, but multimodality therapy involving surgery, radiation therapy and chemotherapy is accepted as standard practice because it appears that both disease-free survival and local-regional tumor control are optimized by integrating all 3 treatment modalities. The controversy regarding management centers on the optimal sequence of integrating the different treatment modalities.

Neoadjuvant or "Upfront" Chemotherapy

The rationale for the use of neoadjuvant therapy is based on the ability of chemotherapy to reduce the tumor burden before definitive local therapy, to facilitate the treatment of micrometastases, and to allow a direct assessment of the response of the primary tumor to chemotherapy. Tumor regression (clinical partial remission + complete remission) is noted in 60–90% of patients treated with induction chemotherapy and approximately 10% of patients attain a pathologic complete response (Morrow et al., 1986; Jacquiilat et al., 1988; Bonadonna et al., 1990; Singeltary et al., 1992). The extent of the pathological response may be predictive of long term outcome. Feldman et al. (1986) compared survival on the basis of the pathological response in 90 women receiving neoadjuvant chemotherapy. In the 15 patients who had only microscopic residual cancer or a complete pathological response, the mean disease-free survival was more than 61 months, compared to a mean of 22 months in the 75 women with gross residual tumor, regardless of the amount of shrinkage of the tumor.

Two randomized trials (DeLena et al., 1981; Perloff et al., 1988) have compared the effectiveness of surgery and radiotherapy following chemotherapy for locally advanced breast cancer. No differences in local control or survival between the two therapies were noted in either study. No data from prospective, randomized trials comparing the combination of surgery and irradiation to either modality alone are available, but nonrandomized studies suggest that the combination may be more effective.

A

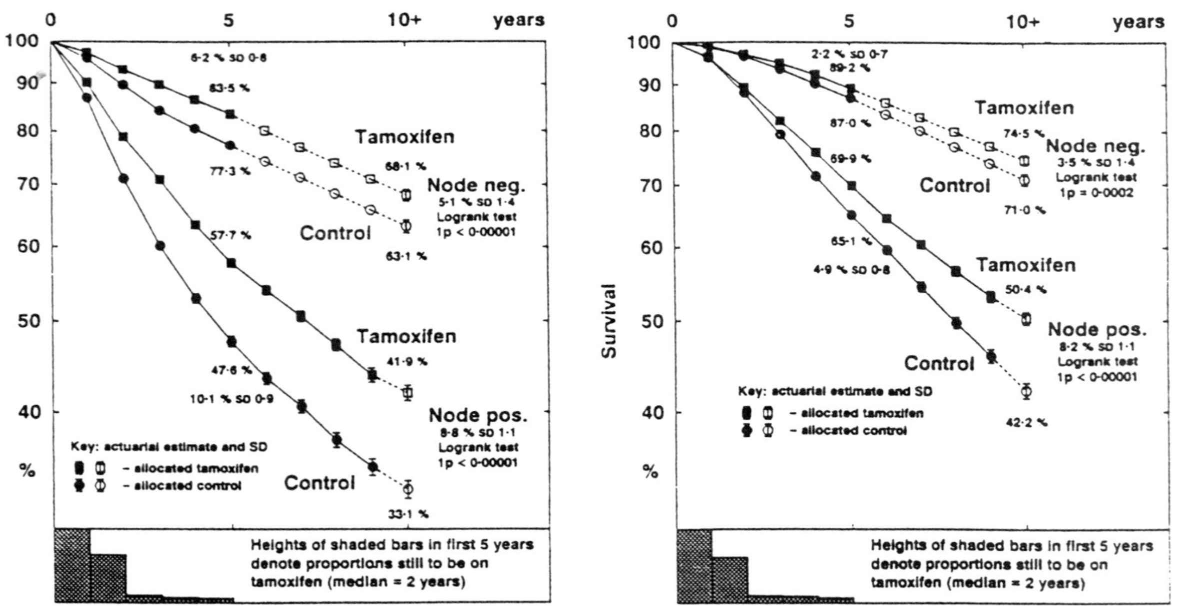

Figure 9. **A**: Ten-year outcome in tamoxifen trials, subdivided by nodal status (Early Breast Cancer Trialists' Collaborative Group, 1992). **B**: Ten-year outcome in polychemotherapy trials, subdivided by nodal status (Early Breast Cancer Trialists' Collaborative Group, 1992) (*left panels*: recurrence-free survival; *right panels*: overall survival).

B

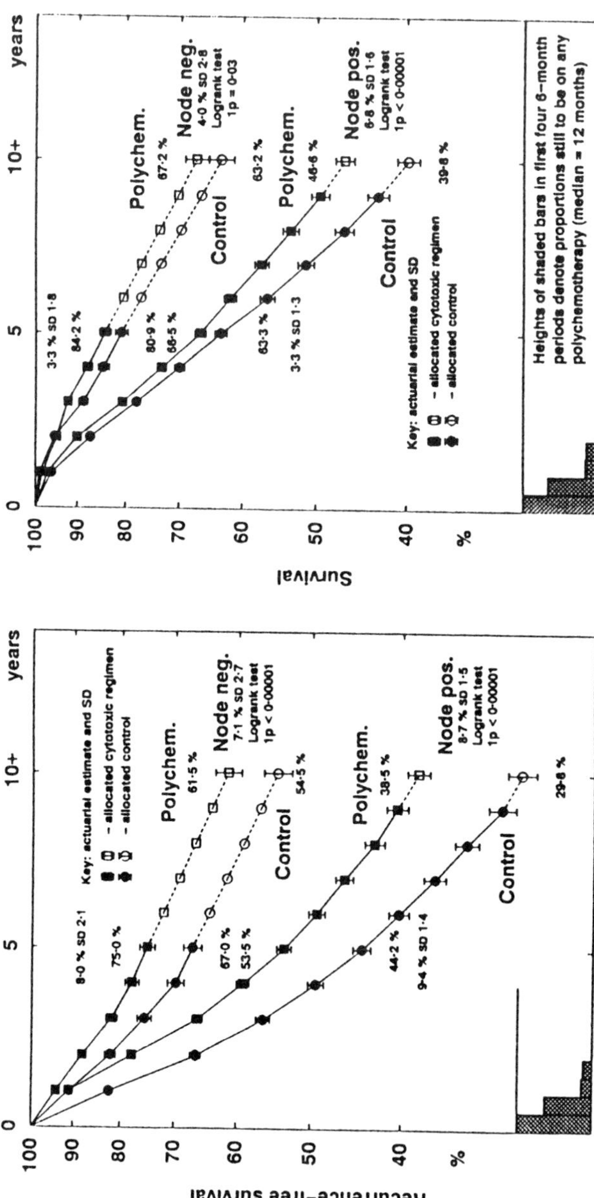

Figure 9. Continued.

Table 9. Recommendations for the use of adjuvant systemic therapy in routine practice

Nodal Status	Premenopausal	Postmenopausal
Positive		
Hormone-receptor positive	CMF × 6 or AC × 4 tamoxifen 2–5 yrs	Tamoxifen 2–5 yrs chemotherapy
Hormone-receptor negative	CMF × 6 or AC × 4	CMF × 6 or AC × 4 or tamoxifen based on clinical judgment
Negative (T > 1 cm)		
Hormone-receptor positive	CMF × 6 or AC × 4 ±tamoxifen 2–5 yrs	Tamoxifen 2–5 yrs
Hormone-receptor negative	CMF × 6 or AC × 4	CMF × 6 or AC × 4 or tamoxifen 2–5 yrs
Negative (T ≤ 1 cm)	No treatment	No treatment

CMF, cyclophosphamide, methotrexate, 5-fluorouracil administered for 6 cycles; AC, Adriamycin (doxorubicin), cyclophosphamide administered for 4 cycles.

More recently, breast-conserving treatments have been attempted following neoadjuvant therapy (Jacquiilat et al., 1988; Bonadonna et al., 1990; Singeltary et al., 1992). With short-term follow-up, local failure rates have not been significantly different from those observed after breast conservation for early-stage disease, but further follow-up is necessary.

Inflammatory breast cancer (IBC) is an uncommon form of breast cancer, accounting for only 1–4% of newly diagnosed cases (Jaiyesimi et al., 1992). IBC is a subset of Stage III disease (T_{4d}) characterized by a high rate of local-regional and metastatic failures that tend to occur very early (Jaiyesimi et al., 1992; Attia-Sobol et al., 1993). Clinical features of IBC include skin erythema, edema and warmth over the breast, but not necessarily a discrete mass. The pathological feature of this type of breast cancer is dermal lymphatic invasion of tumor cells. Because of the gross and subclinical involvement of skin, adequate surgical margins are often difficult, if not impossible, to obtain. As a result, initial surgical treatment results in a high incidence of local recurrence and is contraindicated.

Clinical management of IBC has evolved over the last several decades because of the observation that any single treatment modality is largely ineffective in controlling the disease. Even with aggressive surgical and radiation techniques, the overall survival may not exceed 5% at 5 years. IBC is now approached as a systemic disease from the time of diagnosis and neoadjuvant therapy is the initial treatment. Using this approach, patients are treated with 2 to 4 cycles of combination chemotherapy (Jaiyesimi et al., 1992). If a good response is observed, then local-regional therapy is initiated, usually mastectomy and axillary lymph node dissection. Surgery is usually followed by chest wall irradiation, and additional adjuvant chemotherapy is administered as a final treatment step.

Investigational approaches under evaluation for the treatment of IBC include new neoadjuvant chemotherapy regimens (i.e., paclitaxel, docetaxel) and adjuvant, high-dose chemotherapy (HDC) programs administered to patients responding to neoadjuvant chemotherapy. HDC uses chemotherapy at doses 5 to 30 times higher than those conventionally administered. The rationale for using high-dose chemotherapy is based on laboratory experiments demonstrating that certain chemotherapy drugs, particularly alkylating agents, exhibit a steep *in vitro*, dose-response curve that is log-linear in nature (Henderson et al., 1988; Teicher et al., 1990). This means that a two-fold increase in drug dose results in a 10-fold increase in tumor cell kill. In the laboratory, this response has been observed through multiple logs of tumor cell kill. Applying this concept to the clinic was previously impossible because of the myeloablative effect of HDC. Conventional dose chemotherapy is myelosuppressive, but bone marrow function generally normalizes predictably. With the introduction of autologous bone marrow and peripheral stem cell transplantation and hematopoietic growth factors, HDC can be administered more safely and with shortened hospital stays. The dose-limiting toxicities of the chemotherapy drugs are now nonhematopoietic in nature (gastrointestinal, neurologic, cardiac, etc.). Results from pilot studies using HDC in IBC are sparse and will be described below.

For the 10% of patients who present with metastatic disease (Stage IV), the prognosis remains poor. The median survival of patients with stage IV disease receiving systemic therapy is approximately 2 years (Garber and Henderson, 1989; Henderson et al., 1990). The natural history of patients with metastatic disease is heterogeneous: some patients follow a very indolent course (bone disease), whereas others tend to develop rapid disease dissemination and a precipitous clinical decline, resulting in death within months of diagnosis (Garber and Henderson, 1989; Henderson et al., 1990). Since the disease is not localized, systemic therapy consisting of combination chemotherapy and/or endocrine therapy is considered the primary treatment modality (Gregory et al., 1993). Surgery and radiation therapy remain important in specific situations. Patients who develop symptomatic bone metastases or early spinal cord compression are often treated with local radiation therapy. Similarly, patients with a large, necrotic breast mass may be palliated by a "toilet" mastectomy (surgical removal of gross tumor) and spared the discomfort of tumor necrosis on the chest wall.

Treatment decisions for patients with metastatic disease need to be individualized and based on both patient and tumor characteristics (Henderson, 1991). For elderly patients or patients debilitated from other medical conditions, chemotherapy may cause an irreversible decline in performance status. In these situations, particularly when the tumor is ER-positive, tamoxifen or other hormonal therapies may be considered. Alternatively, supportive care alone may be a rational recommendation. As previously discussed, the hormone receptor content of a breast tumor will predict responsiveness to hormonal manipulation. Breast tumors in patients older than 50 years more commonly express hormone receptors [ER positive—80%, progesterone receptor (PgR) positive—54%] compared to younger patients (ER positive—62%, PgR positive—53%). Multiple studies have shown that hormonal treatments result in response rates exceeding 50% in patients with ER+ tumors compared to <10% in patients with ER− tumors (Clark et al., 1984). In the past, surgical ablative therapies such as oophorectomy, adrenalectomy and hypophysectomy have proven effective (Santen et al., 1990). Alternatively, hormonal additive therapies such as estrogens, androgens and steroids have activity in metastatic disease. Most of these approaches have been abandoned for treatments that achieve the same physiological effect with less treatment-related toxicity. Synthetic antiestrogens (i.e., tamoxifen), aromatase inhibitors (i.e., aminoglutethimide), progestins, and inhibitors of pituitary function such as luteinizing hormone-releasing hormone analogues (i.e., leuprolide, buserelin, goserelin) are in widespread use (Santen et al., 1990).

Older patients or patients with well-differentiated tumors, a long disease-free interval following treatment for early stage disease, and metastases predominantly in soft tissue and bone are most likely to respond to hormonal therapy (Clark et al., 1984; Henderson, 1991). In contrast, patients with rapidly disseminating disease or disease involving the liver are less likely to benefit from hormonal therapies. The synthetic antiestrogen, tamoxifen, is the most commonly used first-line hormonal therapy. Its efficacy is equal to other hormonal agents and it has a superior toxicity profile. Tamoxifen is well tolerated when given in the standard oral dose of 20 mg/day. Higher daily doses have not proven to be more effective and may be associated with more side effects (Henderson, 1991). The median duration of response to tamoxifen is approximately 12 months, which compares favorably to other hormonal therapies. Although there is little evidence from clinical trials to support the use of combinations of hormonal agents, patients responding to one hormonal agent have a good chance of responding to a second- and third-line hormonal therapy when disease progression is detected (Henderson, 1991). Ultimately, all patients responding to hormonal therapies will become refractory to hormonal manipulation.

Combination chemotherapy is considered standard treatment for patients with hormone receptor-negative disease and in patients with rapidly progressing disease (Henderson, 1991; Pfieffer et al., 1992). Combination chemotherapy is able to achieve response rates of 30–80%, with 10–25% of responding patients attaining a complete clinical remission. The median survival after combination chemotherapy has been reported to be between 15 and 24 months (Garber and Henderson, 1989; Henderson et al., 1990; Henderson, 1991; Gregory et al., 1993). The average duration of response for metastatic breast cancer treated with combination chemotherapy is 7 to 14 months, whereas patients attaining a complete remission have more durable responses lasting 12 to 18 months. A small fraction of patients (10%) attaining a complete response will remain disease-free for periods exceeding 15 years (Hortobagyi et al., 1988). The most commonly used regimens are CMF and CAF (cyclophosphamide, doxorubicin, 5-fluorouracil) or variations of these regimens. In a large series of 758 consecutive patients, the response rate to first-line therapy was 34%, but the response rate dropped to 16% in the 249 patients who went on to receive second-line therapy (Gregory et al., 1993).

In the last several years several new chemotherapy drugs have entered clinical trials and several show encouraging activity in metastatic disease, particularly as first-line agents. The taxanes, paclitaxel and docetaxel, have achieved response rates >50% (Reichman et al., 1993). Vinorelbine tartrate (Fumoleau et al., 1993) and edatrexate (Vandenberg et al., 1993) have also produced response rates of ~40%. As might be expected, incorporation of these new agents into standard regimens has produced higher response rates, but the durability of the responses do not appear to have increased. Another area of active clinical investigation is the use of HDC and autologous bone marrow or peripheral stem cell transplantation (discussed above) (Eddy, 1992; Ghalie et al., 1994; Somlo et al., 1994). The efficacy of this approach in comparison to conventional dose chemotherapy cannot be judged at this time because of lack of randomized trial data and the limited information and follow-up from pilot studies. Information from the pilot studies suggests that higher response rates are achieved compared to conventional dose chemotherapy and a higher fraction of patients attain a complete clinical remission, but response durability continues to be <1 year for most patients. The results of the NCI-sponsored randomized trials comparing conventional dose chemotherapy to HDC with peripheral stem cell transplantion are eagerly awaited.

Breast cancer represents a significant health-care problem. Better screening has resulted in the identification of most breast cancers at an early stage. Breast-conserving therapies allow most women to retain their breast without compromising outcome. Advances in adjuvant chemotherapy and hormonal therapies have decreased the risk of recurrence and improved survival. Ultimately a better understanding of the biology of breast cancer will lead to improved treatment for patients with the disease.

REFERENCES

Adair F, Berg J, Joubert L, Robbins JF (1974): Long term follow-up of breast cancer patients. The 30 year report. *Cancer* 33:1145–1150

American Joint Committee on Cancer (1988): Breast cancer. In: *Manual for Staging of Cancer*, Beahrs OH, Henson DE, Hutter RVP, Meyers MH, eds. Philadelphia: JB Lippincott, 3rd Ed

Analysis at six years by Nolvadex Adjuvant Trial Organization (1985): Controlled trial of tamoxifen as single adjuvant agent in management of early breast cancer. *Lancet* 1:836–839

Andersen JA (1977): Lobular carcinoma in situ of the breast. Long term follow-up. *Cancer* 39:2597–2602

Ashikari R, Park K, Huvos AG, Urban JA (1970): Paget's disease of the breast. *Cancer* 26:680–686

Attia-Sobol J, Ferriere JP, Cure H, Kwiatkowski F, Achard JL, Verrelle P, Feillel V, DeLatour M, Lafaye C, Deloach C, Dauplat J, Dely A, Rozan R, Chollet P (1993): Treatment results, survival and prognositic factors in 109 inflammatory breast cancers: univariate and multivariate analyses. *Eur J Cancer* 29A:1081–1088

Barone P, Moore M, Kinne D (1990): Occult primary cancer presenting with axillary metastases. *Arch Surg* 125:210–214

Baum M, Ebb S, Brooks M (1990): Biological fall out from trials of adjuvant tamoxifen in early breast cancer. In: *Adjuvant Therapy of Cancer* VI, Salmon SE, ed. Philadelphia: WB Saunders

Bell D, Hadju S, Urban J (1983): The role of aspiration cytology in the diagnosis and management of mammary lesions in office practice. *Cancer* 51:1182–1189

Berg JW, Hutter RVP (1995): Breast Cancer. *Cancer Suppl* 75:257–269

Blichert-Toft M, Rose C, Andersen J, Overgaard M, Axelsson CK, Andersen KW, Mouridsen HT (1992): Danish randomized trial comparing breast conservation therapy with mastectomy: six years of life table analysis. *J Natl Cancer Inst Monogr* 11:19–25

Bonadonna G (1992): Evolving concepts in the systemic adjuvant treatment of breast cancer. *Cancer Res* 52:2127–2137

Bonadonna G, Valagussa P, Tancini G, Rossi A, Brambilla C, Zambetti M, Bignami P, Di Fronzo G, Silvestrini R (1986): Current status of Milan adjuvant chemotherapy trials for node-positive and node-negative breast cancer. *Natl Cancer Inst Monogr* 1:45–49

Bonadonna G, Veronesi U, Brambilla C, Ferrari L, Luini A, Greco M, Bortoli C, deYoldi C, Zucali R, Rilke F, Andreola S, Silvestrini R, DiFronzo G, Valagussa P (1990): Primary chemotherapy to avoid mastectomies in tumor with diameters of three centimeters or more. *J Natl Cancer Inst* 82:1539–1545

Boring CC, Squires TS, Tong T, Montgomery S (1994): Cancer statistics 1994. *Cancer* 44:7–26

Breast Cancer Trials Committee (1987): Adjuvant tamoxifen in the management of operable breast cancer: the Scottish trial. *Lancet* 2:171–175

Byar DP, Sears ME, McGuire WL (1979): Relationship between estrogen receptor values and clinical data in predicting the response to endocrine therapy for patients with advanced breast cancer. *Eur J Cancer* 15:299–310

Carter C, Allen C, Henson D (1989): Relation of tumor size, lymph node status and survival in 24,740 breast cancer cases. *Cancer* 64:181–187

Chaudry M, Millis R, Hoskins E (1984): Bilateral primary breast cancer: a prospective study of disease incidence. *Br J Surg* 71:711–714

Clark G, Osborne CK, McGuire W (1984): Correlations between estrogen receptor, progesterone receptor, and patient characteristics in human breast cancer. *J Clin Oncol* 2:1102–1109

Clark GM, Mathieu MC, Owens MA, Dressler LG, Eudey L, Tormey DC, Osborne CK, Gilchrist KW, Mansour EG, Abeloff MD, McGuire WL (1992): Prognostic significance of S-phase fraction in good-risk, node-negative breast cancer patients. *J Clin Oncol* 10:428–432

Claus EB, Risch NJ, Thompson WD (1990): Age at onset as an indicator of familial risk of breast cancer. *Am J Epidemiol* 131:961–972

Colditz GA (1993): Epidemiology of breast cancer: findings form the Nurses Health Study. *Cancer* 71:1480–1489

Consensus Conference (1985): Adjuvant chemotherapy for breast cancer. *JAMA* 254:3461–3463

Consensus Statement (1990): Early stage breast cancer. NIH Consensus Development Conference. June 18–21, Vol. 8:6

Daling JR, Malone KE, Voight LF, White E, Weiss NS (1994): Risk of breast cancer among young women: realtionship to induced abortion. *J Natl Cancer Inst* 86:1584–1592

DeLena M, Varini M, Zucali R, Rovini D, Viganotti G, Valagussa P, Veronesi U, Bonadonna G (1981): Multimodality treatment for locally advanced breast cancer. Results of chemotherapy-radiotherapy versus chemotherapy-surgery. *Cancer Clin Trials* 4:229–236

Dupont WD (1994): Evidence of efficacy of mammographic screening for women in their forties. *Cancer* 74:1204–1206

Dupont WD, Page DL (1985): Risk factors for breast cancer in women with proliferative breast disease. *N Engl J Med* 312:146–151

Dupont WD, Parl FF, Hartmann WH, Brinton LA, Winfield AC, Worrell JA, Schuyler PA, Plummer WD (1993): Breast cancer risk associated with proliferative breast disease and atypical hyperplasia. *Cancer* 71:1258–1265

Early Breast Cancer Trialists' Collaborative Group (1992): Systemic treatment of early breast cancer by hormonal, cytotoxic, or immune therapy: 133 randomized trials involving 31,000 recurrences and 24,000 deaths among 75,000 women. *Lancet* 339: 1–15,71–85

Eddy DM (1992): High-dose chemotherapy with autologous bone marrow transplantation for the treatment of metastatic breast cancer. *J Clin Oncol* 10:657–670

Edeiken S (1988): Mammography and palpable cancer of the breast. *Cancer* 61:263–265

Elledge RM, Clark GM, Chamness GC, Osborne CK (1994): Tumor biologic factors and breast cancer prognosis among white, hispanic and black women in the U.S. *J Natl Cancer Inst* 86:705–712

Feldman LD, Hortobagyi GN, Buzdar AU, Ames FC, Blumenschein GR (1986): Pathological assessment of response to induction chemotherapy in breast cancer. *Cancer Res* 46:2578–2581

Feuer EJ, Wun LM (1992): How much of the recent rise in breast cancer incidence can be explained by an increase in mammography utilization? *Am J Epidemiol* 136:1423–1436

Fisher B (1980): Laboratory and clinical research in breast cancer; a personal adventure: the David Karnofsky Memorial Lecture. *Cancer Res* 40:3863–3874

Fisher B, Ravdin RG, Ausman RK, Slack NH, Moore GE, Noer RJ (1968): Surgical adjuvant chemotherapy in cancer of the breast: results of a decade of cooperative investigation. *Ann Surg* 168:337–356

Fisher B, Carbone PP, Economou SG, Frelick R, Glass A, Lerner H, Redmond C, Zelen M, Bond P, Katrych DL, Wolmark N, Fisher ER (1975a): L-phenylalanine mustard (L-PAM) in the management of primary breast cancer: a report of early findings. *N Engl J Med* 292:117–122

Fisher B, Slack N, Katrych D, Wolmark N (1975a): Ten-year follow-up of patients with carcinoma of the breast in a cooperative clinical trial evaluating surgical chemotherapy. *Surg Gynecol Obstet* 140:528–534

Fisher B, Redmond C, Fisher E, Bauer M, Wolmark N, Wickerham DL, Deutsch M, Montague E, Margolese R, Foster R (1985): Ten year results of a randomized clinical trial comparing radical mastectomy and total mastectomy with or without irradiation. *N Engl J Med* 311:674–681

Fisher B, Redmond C, Fisher ER, Wolmark N (1986): Systemic adjuvant therapy in treatment of primary operable breast cancer: National Surgical Adjuvant Breast and Bowel Project experience. Adjuvant chemotherapy and endocrine therapy for breast cancer. *Natl Cancer Inst Monogr* 1:35–43

Fisher B, Redmond C, Fisher ER, Caplan R (1988): Relative worth of estrogen and progesterone receptor and pathologic characteristics of differentiation as indicators of prognosis in node-negative breast cancer patients: findings from National Surgical Adjuvant Breast and Bowel Project Protocol B-06. *J Clin Oncol* 6:1076–1087

Fisher B, Redmond C, Dimitrov NV, Bowman D, Legault-Poisson S, Wickerham DL, Wolmark N, Fisher ER, Margolese R, Sutherland C, Glass A, Foster R, Caplan R (1989a): A randomized clinical trial evaluating sequential methotrexate and fluorouracil in the treatment of patients with node-negative breast cancer who have estrogen-receptor-negative tumors. *N Engl J Med* 320:473–478

Fisher B, Constantino J, Redmond C, Poisson R, Bowman D, Couture J, Dimitrov NV, Wolmark N, Wickerham DL, Fisher ER, Margolese R, Robidoux A, Shibata H, Terz J, Paterson AHG, Feldman MI, Farrar W, Evans J, Lickley HL, Ketne M (1989b): A randomized clinical trial evaluating tamoxifen in the treatment of patients with node-negative breast cancer who have estrogen-receptor-positive tumors. *N Engl J Med* 320:479–484

Fisher B, Redmond C, Poisson R, Margolese R, Wolmark N, Wickerham L, Fisher E, Deutsch M, Caplan R, Pilch Y, Glass A, Shibata H, Lerner H, Terz J, Sidorovich L (1989c): Eight-year results of a randomized clinical trial comparing total mastectomy and lumpectomy with or without irradiation in the treatment of breast cancer. *N Engl J Med* 320:822–828

Fisher B, Constantino J, Redmond C, Fisher E, Margolese R, Dimitrov N, Wolmark N, Wickerham DL, Deutsch M, Ore L, Mamounes E, Poller E, Kavanah M (1993): Lumpectomy compared with lumpectomy and radiation for the treatment of intraductal breast cancer. *N Engl J Med* 328:1581–1586

Fletcher SW, Black W, Harris R, Rimer BK, Shapiro S (1993): Report of the international workshop on screening for breast cancer. *J Natl Cancer Inst* 85: 1644–1656

Foote FW, Stewart F (1941): Lobular carcinoma in situ: a rare form of mammary carcinoma. *Am J Pathol* 17:491–496

Foster RS, Farwell M, Costanza M (1995): Breast conserving surgery for invasive breast cancer: patterns of care in a geographic region and estimate of potential applicability. *Ann Surg Oncol* (in press)

Fourquet A, Campana F, Zafrani B, Mosseri V, Vielh P, Durand JC, Vilcoq JR (1989): Prognostic factors of breast recurrence in the conservative management of early breast cancer: a 25 year follow-up. *Int J Rad Oncol Biol Phys* 17:719–725

Freeman C, Belliveau N, Kim T, Boivin JF (1981): Limited surgery with or without radiotherapy for early breast carcinoma. *J Can Assoc Radiol* 32:125–128

Fumoleau P, Delgado FM, Delozier T, Monnier MA, Delgado G, Kerbrat P, Garcia-Giralt E, Keiling R, Namer M, Closon MT, Goudier MJ, Chollet P, Lecourt L, Monteuquet P (1993): Phase II trial of weekly vinorelbine in first-line advanced breast cancer chemotherapy. *J Clin Oncol* 11:1245–1252

Garber JE, Henderson IC (1989): The use of chemotherapy in metastatic breast cancer. *Hematol Oncol Clin North Am* 3:807–821

Garfinkel L, Boring CC, Heath CW (1994): Changing Trends. An overview of breast cancer incidence and mortality. *Cancer* 74:222–227

Gelber RD, Goldhirsch A (1994): Radiotherapy to the conserved breast: is it avoidable if the cancer is small? *J Natl Cancer Inst* 86:652–653

Gelber RD, Goldhirsch A, Coates AS (1993): Adjuvant therapy for breast cancer: understanding the overview. *J Clin Oncol* 11:580–585

Ghalie R, Richman CM, Adler SS, Cobleigh MA, Korenblit AD, Manson SD, McCleod BC, Taylor SG, Valentino LA, Wolter J, Kaizer H (1994): Treatment of metastatic breast cancer with a split-course high dose chemotherapy regimen and autologous bone marrow transplantation. *J Clin Oncol* 12:342–346

Giuliano A, Kirgan D, Guenther J, Morton DL (1994): Lymphatic mapping and sentinel lymphadenectomy for breast cancer. *Ann Surg* 220:391–401

Glick JH, Gelber RD, Goldhirsch A, Senn HJ (1992): Meeting highlights: adjuvant therapy for primary breast cancer. *J Natl Cancer Inst* 84:1479–1485

Goldgar DE, Fields P, Lewis CM, Tran TD, Cannon-Albright LA, Ward JH, Swensen J, Skolnick MH (1994): A large kindred with 17q-linked breast and ovarian cancer: genetic, phenotypic, and genealogical analysis. *J Natl Cancer Inst* 86:200–209

Gregory WM, Smith P, Richards MA, Twelves CJ, Knight RK, Rubens RD (1993): Chemotherapy of advanced breast cancer: outcome and prognostic factors. *Br J Cancer* 68:1988–1995

Haagensen CD (1971): The natural history of breast carcinoma. In: *Diseases of the Breast*, Philadelphia: WB Saunders

Haagensen CD, Bodian C, Haagensen DE (1981): Lobular neoplasia (lobular carcinoma in situ). In: *Breast Carcinoma: Risk and Detection*. Philadelphia: WB Saunders

Hall JM, Lee MK, Morrow J, Anderson L, Huey B, King MC (1990): Linkage of early-onset familial breast cancer to chromosome 17q21. *Science* 250:1684–1689

Hankey BF, Miller B, Curtis R, Kosary C (1994): Trends in breast cancer in younger women in contrast to older women. *J Natl Cancer Inst Monogr* 16:7–14

Harris M, Howell A, Chrissohou M, Swindell R, Hudson M, Sellwood R (1984): A comparison of the metastatic pattern of infiltrating lobular and infiltrating duct carcinoma of the breast. *Br J Cancer* 50:23–30

Harris JR, Lippman ME, Veronesi U, Willet W (1992): Breast Cancer. *N Engl J Med* 327:319–328

Harris JR, Morrow M, Bonadonna G (1993): Cancer of the Breast. In: *Cancer. Principles and Practice of Oncology*, DeVita VT, Hellman S, Rosenberg SA, eds. Philadelphia: JB Lippincott

Harris JR, Morrow M (1996): Local management of invasive breast cancer. In: *Diseases of the Breast*, Harris JR, Lippman ME, Morrow M, Hellman S, eds. Philadelphia: Lippincott Raven

Henderson IC (1991): Principles in the mangement of metastatic breast cancer. In: *Breast Diseases*, 2nd ed, Harris JR, Hellman S, Henderson IC, Kinne DW, eds. Philadelphia: JB Lippincott

Henderson IC, Hayes DF, Gelman R (1988): Dose-response in the treatment of breast cancer: a critical review. *J Clin Oncol* 6:1501–1515

Henderson IC, Garber JE, Breitmeyer JB (1990): Comprehensive management of disseminated breast cancer. *Cancer* 66:1439–1448

Hetelekidis S, Schnitt S, Morrow M, Harris JR (1995): Management of ductal carcinoma in situ. *CA Cancer J Clinicians* 45:244–253

Holford TR, Roush GC, McKay LA (1991): Trends in female breast cancer in Connecticut and the United States. *J Clin Epidemiol* 44:29–39

Holland R, Veling S, Mravunac M, Hendriks J (1985): Histologic multifocality of Tis, T1–2 breast carcinomas: implications for clinical trials of breast conserving treatment. *Cancer* 56:979–990

Holland R, Peterse JL, Millis RR, Eusebi V, Faverly D, van de Vijver MJ, Zafrani B (1994): Ductal carcinoma in situ: a proposal for a new classification. *Sem Diag Pathol* 11:167–180

Hortobagyi GN, Frye D, Buzdar AU, Hug V, Fraschini G (1988): Complete remissions in metastatic breast cancer: a thirteen year follow-up report. *Proc Am Soc Clin Oncol* 7:37

Jacquemier J, Kurtz J, Amalric R, Brandone H, Ayme Y, Spitalier J (1990): An assessment of extensive intraductal component as a risk factor for local recurrence after breast conserving therapy. *Br J Cancer* 61:873–876

Jacquiilat C, Baillat F, Weil M, Auclerc G, Housset M, Aucler MF, Sellami M, Jindani A, Thill L, Soubrane C, Khayat D (1988): Results of a conservative treatment combining induction (neoadjuvant) and consolidation chemotherapy, hormonotherapy, and external and interstitial irradiation in 98 patients with locally advanced breast cancer (IIIa-IIIb). *Cancer* 61:1977–1982

Jaiyesimi IA, Buzdar AU, Hortobagyi G (1992): Inflammatory breast cancer: a review. *J Clin Oncol* 10:1014–1024

Jensen EV, Block GE, Smith S, Kyser K, DeSombre ER (1971): Estrogen receptors and breast cancer response to adrenalectomy. *Natl Cancer Inst Monogr* 34:55–70

Kaluzny AD, Rimer B, Harris R (1994): The National Cancer Institute and Guideline Development: lessons from the breast cancer screening controversy. *J Natl Cancer Inst* 86:901–905

Kearney T, Morrow M (1995): Effect of re-excision on the success of breast conserving therapy. *Ann Surg Oncol* 2:303–307

Kerlikowske K, Grady D, Rubin SM, Sandrock C, Ernster VL (1995): Efficacy of screening mammography. A meta-analysis. *JAMA* 273:149–154

Kopans DB, Halpern E, Hulka CA (1994): Statistical power in breast cancer screening trials and mortality reduction among women 40–49 years of age with particular emphasis on the National Breast Screening Study of Canada. *Cancer* 74:1196–1204

Lagios M, Richards V, Rose M, Yee E (1983): Segmental mastectomy without radiotherapy. Short-term follow-up. *Cancer* 52:2173–2179

Legha SS, Powell K, Davis HL, Muggia FM (1978): Hormonal therapy of breast cancer: new approaches and concepts. *Ann Inter Med* 88:69–77

Leis HP, Dursi J, Mersheimer WL (1967): Nipple discharge. Significance and treatment. *NY State J Med* 67:3105–3110

Lennington WJ, Jensen RA, Dalton LW, Page DL (1994): Ductal carcinoma in-situ of the breast: heterogeneity of individual lesions. *Cancer* 73:118–124

Lesser ML, Rosen PP, Kinne DW (1982): Multicentricity and bilaterality in invasive breast carcinoma. *Surgery* 91:234–240

Lichter A, Lippman M, Danforth D, d'Angelo T, Steinberg SM, deMoss E, McDonald HD, Reichert CM, Merino M, Swain SM, Cowan K, Gerber LH, Bader LH, Findlay PA, Schain W, Gorrell C, Straus K, Rosenberg S, Glatstein E (1992): Mastectomy versus breast conserving therapy in the treatment of Stage I and II carcinoma of the breast. A randomized trial at the National Cancer Institute. *J Clin Oncol* 10:976–983

Liljegren G, Holmberg L Adami O, Westman G, Graffman S, Bergh J, Uppsala-Örebro Breast Cancer Study Group (1994): Sector resection with or without post-operative radiotherapy for stage I breast cancer: five-year results of a randomized trial. *J Natl Cancer Inst* 86:717–722

Lynch HT, Watson P, Conway T, Fitzsimmons ML, Lynch J (1988): Breast cancer family history as a risk factor for breast cancer. *Breast Cancer Res Treat* 11:263–267

Maddox W, Carpenter J, Laws H, Soong SJ, Cloud G, Urist MM, Balch CM (1983): A randomized prospective trial of radical (Halsted) mastectomy versus modified radical mastectomy in 311 breast cancer patients. *Ann Surg* 198:207–212

Mansi JL, Berger U, McDonnel T, Pople A, Rayter Z, Gazet JC, Coombes RC (1989): The fate of bone marrow of bone marrow micrometastases in patients with primary breast cancer. *J Clin Oncol* 7:445–449

Mansour EG, Gray R, Shatila AH, Osborne CK, Tormey DC, Gilchrist KW, Cooper MR, Falkson G (1989): Efficacy of adjuvant chemotherapy in high-risk, node-negative breast cancer: an intergroup study. *N Engl J Med* 320:485–490

Matthews R, McNeese M, Montague E, Oswald MJ (1988): Prognostic implications of age in breast cancer patients treated with tumorectomy and irradiation or with mastectomy. *Int J Rad Oncol Biol Phys* 14:659–663

McCormick B, Rosen PP, Kinne DW, Cox L, Yahalom J (1991): Ductal carcinoma in situ of the breast: an analysis of local control after conservation surgery and radiotherapy. *Int J Rad Oncol Biol Phys* 21:289–292

McGuire WL, Vollmer EP, Carbone PP (1975): *Estrogen Receptors in Human Breast Cancer*. New York, Raven Press

McGuire WL, Tandon AK, Allred DC, Chamness GC, Ravdin PM, Clark GM (1992): Prognosis and treatment decisions in patients with breast cancer without axillary node involvement. *Cancer* 70: 1755–1781

Mendoza CB, Moore GE, Crosswhite LH, Sanberg AA, Watne AL (1969): Prognostic significance of tumor cells in the bone marrow. *Surg Gynecol Obstet* 129:483–488

Miki Y, Swensen J, Shattuck-Eidens D, Futreal PA, Harshman K, Tavtigian S, Liu Q, Cochran C, Bennett LM, Ding W, Bell R, Rosenthal J, Hussey C, Tran T, McClure M, Frye C, Hattier T, Phelps R, Haugen-Strano A, Katcher H, Yakumo K, Gholami Z, Shaffer D, Stone S, Bayer S, Wray C, Bogden R, Dayananth P, Ward J, Tonin P, Narod S, Bristow PK, Norris FH, Helvering L, Morrison P, Rosteck P, Lai M, Barrett JC, Lewis C, Neuhausen S, Cannon-Albright L, Goldgar D, Wiseman R, Kamb A, Skolnick MH (1994): A strong candidate for the breast and ovarian cancer susceptibility gene BRCA1. *Science* 266:66–71

Miller BA, Feuer EJ, Hankey BF (1994): The significance of rising incidence of breast cancer in the United States. In: *Important Advances in Oncology*, DeVita VT, Hellman S, Rosenberg SA, eds. Philadelphia: JB Lippincott

Morrison AS (1993): Screening for breast cancer. *Epidemiol Rev* 15:244–255

Morrow M (1990): Management of nonpalpable breast lesions. In: *PPO Updates*, DeVita VT, Hellman S, Rosenberg SA, eds. Philadelphia: JB Lippincott

Morrow M (1991): Nipple discharge. In: *Breast Diseases*, Harris JR, Hellman S, Henderson IC, Kinne DW, eds. Philadelphia: JB Lippincott

Morrow M, Braverman A, Thelmo W, Sohn CK, Sand J, Mora M, Forlenza T, Marti J (1986): Multimodal therapy for locally advanced breast cancer. *Arch Surg* 121:1291–1296

Morrow M, Quiet C, Hellman S, Hassett C, Schmidt R, Ewing C (1994a): Treatment selection in breast cancer: are our biases correct? *Proc Am Soc Clin Oncol* 13:99

Morrow M, Schmidt R, Cregger B, Hassett C, Cox S (1994b): Preoperative evaluation of abnormal mammographic findings to avoid unnecessary breast biopsy. *Arch Surg* 129:1091–1096

Morrow M, Schnitt S (1996): Lobular carcinoma in situ. In: *Diseases of the Breast*, Harris JR, Lippman ME, Morrow M, Hellman S, eds. Philadelphia: Lippincott Raven

Morrow M, Harris JR, Schnitt S (1996): Ductal carcinoma in situ. In: *Diseases of the Breast*, Harris JR, Lippman ME, Morrow M, Hellman S, eds. Philadelphia: Lippincott Raven

Nattinger A, Gottleib M, Verum J, Yahnke D, Goodwin JS (1992): Geographic variation in the use of breast conserving treatment for breast cancer. *N Engl J Med* 326:1102–1107

Nemoto T, Vana J, Bedwani RN, Baker HW, McGregor FH, Murph GP (1980): Management and survival of female breast cancer: results of a national survey by the American College of Surgeons. *Cancer* 45:2917–2924

Noone RB, Frazier TG, Noone GC (1994): Recurrence of breast carcinoma following immediate reconstruction: a 13-year review. *Plast Reconst Surg* 93:96–106

Osborne M, Ormiston N, Harmer C, McKinna J, Baker J, Greening W (1984): Breast conservation in the treatment of early breast cancer. A 20 year follow-up. *Cancer* 53:349–352

Page DL, Dupont WD, Rogers LW, Landenberger M (1982): Intraductal carcinoma of the breast. Follow-up after biopsy only. *Cancer* 49:751–758

Page DL, Kidd TE Jr, Dupont WD, Simpson JF, Rodgers LW (1991): Lobular neoplasia of the breast: high risk for subsequent invasive cancer predicted by more extensive disease. *Hum Pathol* 22:1232–1239

Perloff M, Lesnick GJ, Korzun A, Chu F, Holland JF, Thirwell MP, Ellison RR, Carey RW, Leone L, Weinberg V, Rice MA, Wood WC (1988): Combination chemotherapy with mastectomy or radiotherapy for Stage III breast cancer: a Cancer and Leukemia Group B Study. *J Clin Oncol* 6:261–269

Petit JY, Le MG, Mouriesse H, Rietjens M, Gill P, Contesso G, Lehmann A (1994): Can breast reconstruction with gel-filled silicone implants increase the risk of death and second primary cancer in patients treated by mastectomy for breast cancer? *Plast Reconst Surg* 94:115–119

Pfieffer P, Cold S, Rose C (1992): Cytotoxic treatment of metastatic breast cancer. Which drugs and drug combinations? *Acta Oncol* 31:219–224

Preece P, Baum M, Mansel R, Webster D, Fortt R, Gravelle I, Hughes I (1982): Importance of mastalgia in operable breast cancer. *Br Med J* 284:1299–1300

Qualheim R, Gall E (1957): Breast carcinoma with multiple sites of origin. *Cancer* 10:460–468

Reichman BS, Seidman AD, Crown JPA, Heelan R, Hakes TB, Lebwohl DE, Gilewski TA, Surbone A, Currie V, Hudis CA, Yao TJ, Klecker R, Jamis-Dow C, Collins J, Quinlivan S, Berkery R, Toomasi F, Canetta R, Fisherman J, Arbuck S, Norton L (1993): Paclitaxel and recombinant human granulocyte colony-stimulating factor as initial chemotherapy for metastatic breast cancer. *J Clin Oncol* 11:1943–1951

Rose M, Olivotto I, Cady B, Koufman C, Osteen R, Silver B, Recht A, Harris JR (1989): The long-term results of conservative surgery and radiation therapy for early breast cancer. *Arch Surg* 124:153–157

Rosen P, Fracchia A, Urban J, Schottenfeld D, Robbins GF (1975): "Residual" mammary carcinoma following simulated partial mastectomy. *Cancer* 35:739–747

Rosen PP, Lieberman PH, Braun DW Jr, Kosloff C, Adair F (1978): Lobular carcinoma in situ of the breast. *Am J Surg Pathol* 2:225–261

Rosen PP (1996): The pathology of invasive breast cancer. In: *Diseases of the Breast*, Harris JR, Lippman ME, Morrow M, Hellman S, eds. Philadelphia: Lippincott Raven

Rosen PP, Braun D, Kinne D (1982): The clinical significance of pre-invasive breast carcinoma. *Cancer* 49:751–758

Rosen PP, Groshen S, Kinne DW, Hellman S (1993): Factors influencing prognosis in node-negative breast carcinoma: an analysis of 767 T1N0M0/T2N0M0 patients with long-term follow-up. *J Clin Oncol* 11:2090–2100

Rosenberg L (1994): Induced abortion and breast cancer: more scientific data are needed. *J Natl Cancer Inst* 86:1569–1570

Rosner D, Bedwani RN, Vana J, Baker HW, Murphy GP (1980): Noninvasive breast carcinoma. Results of a national survey by the American College of Surgeons. *Ann Surg* 192:139–147

Samet J, Hunt W, Farrow D (1994): Determinants of receiving breast conserving surgery. The Surveillance, Epidemiology and End Results Program, 1983–1986. *Cancer* 73:2344–2351

Santen RJ, Manni A, Harvey H, Redmond C (1990): Endocrine treatment of breast cancer in women. *Endocr Rev* 11:221–263

Sarrazin D, Le M, Arriagada R, Contesso G, Fontaine F, Spielman M, Rochard F, LeChevalier T, Lacour J (1989): Ten-year results of a randomized trial comparing a conservative treatment to mastectomy in early breast cancer. *Radiother Oncol* 14:177–184.

Sattin RW, Rubin GL, Webster LA, Huezo CM, Wingo PA, Ory HW, Layde PM (1990): Family history and the risk of breast cancer. *JAMA* 253:1908–1913

Schottenfeld D, Nash AG, Robbins GF, Beattie EJ (1976): Ten-year results of the treatment of primary operable breast carcinoma. A summary of 304 patients evaluated by the TNM system. *Cancer* 38:1001–1007

Schwartz GF, Finkel GC, Garcia JC, Patchetsky AS (1992): Sub-clinical ductal carcinoma in situ of the breast. Treatment by local excision and surveillance alone. *Cancer* 70:2468–2472

Seltzer MH, Perloff LJ, Kelley RI, Fitts WT (1970): The significance of age in patients with nipple discharge. *Surg Gynecol Obstet* 131:519–522

Silverstein M, Gamagami P, Colburn W, Colburn WJ, Gierson ED, Rosser RJ, Handel H, Waisman JR (1989): Non-palpable breast lesions: diagnosis with slightly over penetrated screen-film mammography and hook wire directed biopsy in 1014 cases. *Radiology* 171:633–638

Silverstein JM, Cohlan BF, Gierson ED, Furmanski MJ, Gamagami P, Colburn WJ, Lewinsky BS, Waisman JR (1992): Ductal carcinoma in situ: 227 cases without microinvasion. *Eur J Cancer* 28:630–634

Singeltary SE, McNeese MD, Hortobagyi GN (1992): Feasibility of breast conservation surgery after induction chemotherapy for locally advanced breast cancer. *Cancer* 69:2849–2852

Smart CR (1994): Breast cancer screening in women 40–49 years of age: data and 14-year followup of the Breast Cancer detection Demonstration Project. *Breast Dis* 5:20–25

Solin LJ, Recht A, Fourquet A, Kurtz J, Kuske R, McNeese M, McCormick B, Cross M, Schultz D, Bornstein B, Spitalier J, Vilcoq J, Fowble B, Harris J, Goodman R (1991): Ten year results of breast conserving therapy and definitive irradiation for intraductal carcinoma (ductal carcinoma in situ) of the breast. *Cancer* 68:2337–2344

Somlo G, Doroshow JH, Forman SJ, Leong LA, Margolin KA, Morgan RJ, Raschko JW, Akman SA, Ahn C, Sniecinski I (1994): High-dose doxorubicin, etoposide and cyclophosphamide with stem cell reinfusion in patients with metastatic or high risk primary breast cancer. *Cancer* 73:1842–1848

Strawbridge H, Hassett A, Foldes I (1984): Role of cytology in management of lesions of the breast. *Surg Gynecol Obstet* 159:130–132

Swanson GM, Raghab NE, Lin CS, Hankey BF, Miller B, Horn-Ross P, White E, Liff JM, Harlan LC, McWhorter WP, Mullan PB, Key CR (1993): Breast cancer among black and white women in the 1980s. Changing patterns in the United States by race, age and extent of disease. *Cancer* 72:788–798

Teicher BA, Holden SA, Eder JP, Herman TS, Antman KH, Frei E (1990): Preclinical studies relating to the use of thiotepa in the high dose setting alone and in combination. *Semin Oncol* (suppl 3) 17:18–32

The Ludwig Breast Cancer Study Group (1989): Prolonged disease-free survival after one course of perioperative adjuvant chemotherapy for node-negative breast cancer. *N Engl J Med* 320:491–496

Turner L, Swindell R, Bell W (1981): Radical versus modified radical mastectomy for breast cancer. *Ann R Coll Surg Engl* 63:239–243

Urban J, Egeli R (1978): Non lactational nipple discharge. *Cancer* 28:3–13

Van Dongen J, Bartelink H, Fentiman I, Lerut T, Mignolet F, Olthuis G, van der Schueren E, Sylvester R, Winter J, van Zijl K (1992): Randomized clinical trial to assess the value of breast conserving therapy in Stage I and II breast cancer, EORTC 10801 trial. *J Natl Cancer Inst Monogr* 11:15–18

Vandenberg TA, Pritchard KI, Eisenhauer EA, Trudeau ME, Norris BD, Lopez P, Verma SS, Buckman RA, Muldal A (1993): Phase II study of weekly edatrexate as first-lie chemotherapy for metastatic breast cancer: a National Cancer Institute of Canada Clinical Trials Group study. *J Clin Oncol* 11:1241–1244

Velentgas P, Daling JR (1994): Risk factors for breast cancer in younger women. *Monogr Natl Cancer Inst* 16:15–22

Veronesi U, Banfi A, Salvadori B, Luini A, Saccozzi R, Zucali R, Marubini E, Del Vecchio M, Boracchi P, Marchini S, Merson M, Sacchini V, Riboldi G, Santoro

G (1990a): Breast conservation is the treatment of choice in small breast cancer. Long term results of a randomized clinical trial. *Eur J Cancer* 26:668–670

Veronesi U, Volterrani F, Luini A, Saccozzi R, Del Vecchio M, Zucali R, Galimbert V, Rasponi A, Di Re E, Squicciarini P, Salvadori B (1990b): Quadrantectomy versus lumpectomy for small size breast cancer. *Eur J Cancer* 26:671–673

Vezzoni P, Balestrazzi A, Bignami P (1979): Axillary lymph node metastases from occult carcinoma of the breast. *Tumori* 65:87–91

Vicini F, Recht A, Abner A, Boyages J, Cady B, Connolly JL, Gelman R, Osteen RT, Schnitt SJ, Silen W (1992): Recurrence in the breast following conservative surgery and radiation therapy for early-stage breast cancer. *J Natl Cancer Inst Monogr* 11:33–39

Vogelstein B, Kinzler KW (1994): Has the breast cancer gene been found? *Cell* 79:1–3

Wheeler JE, Enterline HT, Roseman JM, Tomasulo JP, McIlvaine CH, Fitts WT (1974): Lobular carcinoma in situ of the breast. Long term follow-up. *Cancer* 34:554–563

Wickerham L, Fisher B, Cronin W (1984): The efficacy of bone scanning in the follow-up of patients with operable breast cancer. *Breast Cancer Res Treat* 4:303–307

Winchester D, Cox J (1992): Standards for breast conserving treatment, *Cancer* 42:134–162

7

Genetics of Breast Cancer

STEVEN M. HILL, DIANE M. KLOTZ AND CLAUDIA S. COHN

For women in the United States, the risk of developing breast cancer by age 85 years is currently about one in eight. It is clear that this disease arises from multiple causes, many of which can even occur within the same extended family. Etiological factors of breast cancer such as social class, body mass, and age at menarche appear to be interrelated, and to occur within certain families without necessarily having a genetic basis. In addition, even when the pattern of breast cancer in a given family strongly suggests that the trait for breast cancer susceptibility is inherited, the apparent mode of inheritance frequently does not conform to the principles of mendelian genetics. Determining the heritability of the trait within a family is often complicated by incomplete penetrance, i.e., by the presence of individuals who appear to have inherited susceptibility to breast cancer without developing the disease themselves. The presence of some individuals in which the cancer may have occurred sporadically rather than genetically is also a complicating factor.

Of all the factors contributing to the risk of breast cancer, a strong family history of the disease is the most powerful. Familial clustering of breast cancer was first described in ancient Rome (Lynch and Guirgis, 1981), but formal documentation was not published until 1866 when the French surgeon, Paul Broca, reported 10 cases of breast cancer in four generations of his wife's family, a total of 24 women. Several other individuals in this family died as a result of hepatic tumors (Broca, 1866). Broca concluded that the clustering of breast cancer in this family could not reasonably be attributed to chance. With the recent strong influence of molecular genetics in medicine, there has been a tendency to assume that familial clustering of breast cancer results from a genetically inherited predisposition. However, other possible explanations for familial clustering of breast cancer include

Hormones and Cancer
Wayne V. Vedeckis, Editor

(1) environmental exposure to carcinogens in a particular geographical area that might affect a family living within the region, and (2) culturally motivated or socioeconomic behavior that may alter a family's risk profile, such as age at first live birth or dietary intake. Although noninherited factors can play a role in the clustering of breast cancer in families, recent reports have provided unequivocal evidence for the presence of breast cancer susceptibility genes that are responsible for 5–10% of all breast cancer (Claus et al., 1991). Clearly, there is still much to understand concerning the heritable factors involved in the development of breast cancer, and the prevalence of genetic mutations within these genes in the population.

EPIDEMIOLOGIC STUDIES

Data from the initial studies to determine the influence of family history on the risk of breast cancer were published in the early part of this century (Lane-Clayton, 1926; Wainwright, 1931; Wassink, 1935). These studies compared the incidence of breast cancer in relatives of known breast cancer patients to those of healthy women. Familial clustering was observed in these early studies; however, these studies were often flawed by the lack of rigorously defined control groups, the absence of adjustments for family size, and, most importantly, by an unverified diagnosis. Also, the age of onset in breast cancer cases or the presence of other cancers in the family were not taken into consideration. A number of ensuing investigations used control groups that were more carefully defined, often using, as control subjects, individuals from the general population from which the families were derived (Martynova, 1937; Crabtree, 1941; Jacobsen, 1946; Penrose et al., 1948; Smithers, 1948; Bucalossi et al., 1954; Anderson et al., 1950; Macklin, 1959; Papadrianos et al., 1967; Tulinius et al., 1982). These studies consistently demonstrated a two- to three-fold increase in breast cancer risk in first degree relatives (mothers and sisters) of breast cancer patients. These results correlate well with data from recent studies (Colditz et al., 1933; Sattin et al., 1985; Slattery and Kerber, 1993).

The largest population-based study using modern epidemiologic methodologies to estimate breast cancer risk associated with a positive family history was conducted in Sweden. This study included 1330 women who had a confirmed diagnosis of breast cancer and 1330 age-matched controls (Adami et al., 1981). In this study, breast cancer in a first-degree relative was reported by 11.2% of breast cancer patients but in only 6.7% of controls ($p < .01$), yielding a standardized relative risk factor of 1.7. Extension of this study to include breast cancer in second-degree relatives yielded a standardized risk factor of 1.6, with 19.8% of breast cancer patients and 12.9% of control women reporting an affected first- or second-degree relative. This study, unlike others, failed to find an association between the increased incidence of bilateral disease and a younger age of onset in patients with a family history of breast cancer. This is in contrast to earlier as well as subsequent studies (Anderson, 1971; Sakamoto et al., 1978; Schneider et al., 1983; Anderson and Badzioch, 1985; Claus et al., 1990; Mettlin et al., 1990). Relative risks of a similar magnitude were found in a Canadian population-based study (Lubin et al., 1982) and in the U.S. Nurses Health Study (Bain et al., 1980). Higher risks were reported in the Breast Cancer Detection Demonstration Project (Brinton et al., 1979) and in the American Cancer Society cohort (Brinton et al., 1979; Seidman et al., 1982).

Large population-based studies often dilute out the effects of a very strong association in a small subset of the study population, but they have the advantage of avoiding selection bias in determining the contribution of risk factors to the occurrence of a specific disease. In large studies where most breast cancer cases are multifactorial in origin, analysis might obscure a small subset of families at high risk of developing breast cancer as the result of genetic defects inherited in classical mendelian fashion. Anderson and colleagues (1971) were among the first to suggest that breast cancer was not a homogeneous disease and, as such, the occurrence of breast cancer may not be influenced by genetic factors in a uniform manner. This group (1971) chose to emphasize the hereditary component, if one were present, by assembling a database enriched for kindred with a family history of breast cancer (Anderson, 1971, 1972, 1974, 1976, 1977). These studies were the first to demonstrate the heterogeneity of risk among breast cancer families. The primary factors that increased risk within families were premenopausal status at the time of diagnosis and bilateral disease in the primary proband. In addition, first-degree relatives of the primary proband were found to be at higher risk than second-degree relatives. Data from these studies (Anderson, 1971, 1972, 1974, 1976, 1977; Anderson and Badzioch, 1985a;b, 1989) are summarized in Table 1.

Risk calculation in first-degree paired relatives (mother–daughter or sister–sister) from the Anderson studies (1971, 1972, 1974, 1976, 1977) yielded estimates of 50% for lifetime breast cancer risk in the presence of both premenopausal diagnosis and bilateral disease. This is in contrast to the 7% calculated risk of breast cancer in first degree relatives of women with postmenopausal diagnosis and unilateral disease (Anderson and Badzioch, 1985b), which does not differ significantly from the risk in the general population. Ottman and colleagues (1986) later confirmed these

Table 1. Relative risks for first-degree relatives of women with breast cancer

Characteristics of affected mother/sister	Relative risk
Premenopausal diagnosis	3.0
Bilateral disease	5.0
Bilateral disease and premenopausal diagnosis	9.0
Postmenopausal diagnosis	1.5

From Anderson, 1972, and Anderson and Badzioch, 1985a.

data by reporting a relative risk of 5 for sisters of patients with bilateral breast cancer diagnosed before age 50 years, a relative risk of 10.5 for sisters of patients with bilateral breast cancer diagnosed before age 40 years, and a relative risk of 2.5 for sisters of patients with unilateral breast cancer diagnosed before age 40 years.

The first evidence for the presence of inherited factors responsible for familial clustering of breast cancer and an autosomal dominant breast cancer susceptibility gene was generated by a Danish study in 1984 (Williams and Anderson, 1994). In this study, Williams and Anderson examined 200 Danish pedigrees obtained from the Danish Cancer Registry and identified that in 95% of the cases, inherited factors appeared to be responsible for the clustering of breast cancer in these families. The results of this study were supported by Newman et al. (1988), who examined 1579 nuclear families of breast cancer probands diagnosed before age 55 years and found the existence of a highly penetrant breast cancer susceptibility gene transmitted as an autosomal dominant trait. The recent identification of the BRCA1, BRCA2, and p53 genes as breast cancer susceptibility genes provide the best explanation for the high incidence of breast cancer in these families (Williams and Anderson, 1984; Hall et al., 1990; Malkin et al., 1990; Wooster et al., 1994) and validate this hypothesis.

Frequency of Inherited Breast Cancer

Numerous recent studies have focused on determining the frequency of breast cancer-related gene mutations in the general population and, thus, the percentage of breast cancer in the general population that is directly attributable to inherited genetic factors. Unfortunately, available data addressing this question are limited. In a study conducted at the Creighton University Oncology Clinic, among 225 breast cancer patients seen, only 5% had family histories consistent with inherited forms of breast cancer (Lynch et al., 1976). A percentage as high as 11.4% was noted in families with a strong history of breast cancer diagnosed before age 50 years. In a study by the Centers for Disease Control (CDC), the Cancer and Steroid Hormone Study (CASH), where family history was gathered on 4730 histologically confirmed breast cancer patients between the ages of 20 and 54 years, and 4688 age-matched controls from 8 geographic regions of the United States, 11% of breast cancer cases reported a first-degree relative with breast cancer as compared to 5% of controls (Sattin et al., 1985), and less than 1% of the cases reported both a mother and sister with breast cancer. It must be recognized that some of these kindred are likely to represent chance clustering of sporadic tumors. Analysis of the CASH data provides further evidence for the existence of a breast cancer-related gene alteration in approximately 0.33% of the population (Claus et al., 1991). In this study, Claus estimates that approximately 36% of breast cancer cases among 20- to 29-year-old women may be attributable to the presence of an autosomal dominant susceptibility gene, with the involvement of this gene in breast cancer decreasing to 1% by age 80. Thus, the effects of these mutations on breast cancer risk appears to be a function of age.

The correlation between an autosomal dominant breast cancer susceptibility gene and the risk of breast cancer was also examined in two studies in 1993. The Nurses Health Study analyzed by Colditz et al. (1993) suggested that breast cancer occurred as a result of inherited genetic factors in 6% of the cases, whereas an estimate of 17–19% was noted using the Utah Population Database (Slattery and Kerban, 1993). The variance between these two studies may be attributable to an under-estimation in the Nurses Health Study as a result of small family size and absence of paternal data, and an over-estimation in the Utah Population Database as a result of the unique composition and large size of the families (Weber and Garber, 1993). In general, current estimates suggest that between 5% and 10% of breast cancer cases are directly attributable to inherited factors. Based on current investigations, it is anticipated that the identification of the major breast cancer susceptibility genes will allow a direct measurement of the percentage of breast cancers resulting from inherited mutations in these genes.

GENETICS OF INHERITED BREAST CANCER

Linkage Analysis

Genetic linkage analysis has been used successfully in recent years to identify a number of important disease-related genes such as those causing cystic fibrosis, hereditary nonpolyposis colon cancer, Huntington's disease, polycystic kidney disease, as well as the breast cancer susceptibility genes BRCA1 and BRCA2 (Wooster et al., 1994). The underlying principle of linkage analysis was clearly set out by Yates and Connor (1986); the key to exploiting this technique is to have large numbers of highly informative genetic markers with known locations widely distributed over the chromosome set. Linkage analysis relies on the fact that genes or genetic markers that lie near one another on a chromosome will be inherited together, depending on the distance between the two loci, because the likelihood of two segments of DNA that are immediately adjacent to one another being separated during meiotic crossover is very low.

Linkage analysis can be a useful genetic tool, providing (1) the ability to trace the chromosome carrying the mutant gene through several generations, (2) the ability to detect a meiotic recombination event within the family that allows narrowing of the interval thought to contain the gene, and (3) the ability to predict which currently unaffected members of the family are at high risk of developing breast and/or ovarian cancer in the future, and which members of the family are not at increased risk as compared to the general population (Biesecker et al., 1974; Lynch et al., 1993). Linkage analysis is still the only means of identifying BRCA2 mutation carriers or other breast cancer-associated genes, whereas BRCA1 mutations can now be detected directly because of the recent cloning of this gene (Miki et al., 1994). Linkage analysis will also be essential for determining the location of other breast cancer-related genes.

CHROMOSOMES AND GENES ASSOCIATED WITH HEREDITARY BREAST CANCER

Genetic factors are unequivocally important in the development of human breast cancer. However, to provide additional insight into the mechanism(s) by which genetic mutations result in the development of breast cancer, a more detailed analysis of chromosomes and genes known to be associated with breast cancer and other clinical cancer syndromes is necessary. The major chromosomes and genes linked with various familial syndromes that are also associated with breast cancer are shown in Table 2. Slack and colleagues (1991) have observed that within these familial syndromes, "core" pedigrees may include multiple cases of breast cancer, but as the pedigree is extended, an elevated frequency of other malignancies (upper and lower gastrointestinal tract, pancreas, ovary, endometrium, and skin) becomes apparent. Consistent with the studies of Lynch and Guirgis (1981), it appears that within these multicancer families there may be branches in which one form of cancer (e.g., ovarian, colon, etc.) predominates. As the size of the pedigree becomes sufficiently large, the heterogeneity of tumor expression becomes obvious, with patients being identified who have had two, or even more, different primary tumors (Table 2).

The availability of genetic markers for breast cancer, such as BRCA1, p53, and HER-2/*neu*, has been important in expanding our understanding of the mechanisms that cause cancer, as well as serving as potential diagnostic tools. The further study of chromosomes and genes involved in, or associated with, an increased incidence of breast cancer will expand our current understanding and will provide insight into individual mechanisms by which genetic mutations result in the development of breast cancer. Breast cancer occurrence has been associated with defects on a number of chromosomes, and in some cases with mutations in specific genes located within these chromosomes—most notably chromosomes 1p, 3p, 11p, 13, 17p, and 22 (Devilee et al., 1989). The genes whose primary phenotypic expression results in nonmalignant syndromes, such as Cowden's and Muir's syndromes, and ataxia telangiectasia (Swift et al., 1986) also may be involved in the development of breast cancer. It also may be that a separate gene important in breast cancer is located on the chromosome near the genes for these syndromes.

Table 2. Patterns of hereditary breast cancer

Chromosome	Gene/syndrome	Identifying characteristics
Chromosome 5	MSH2/MLH1 genes Muir's syndrome	A variant of Lynch type II syndrome. Results from germline mutations in the MSH2/MLH1 loci (hereditary nonpolyposis colon cancer genes). Increased incidence of benign and malignant tumors of the GI and GU tracts. The key mechanism may be defects in the DNA repair.
Chromosomes 11 and 22	11q:22q Chromosomal translocations	Translocation occurs between the long arms of chromosomes 11 and 22. Carriers of this translocation display an elevated risk for developing breast cancer.
	11q15 and 11q23	Loss of heterozygosity at regions q15 and q23 of chromosome 11 are frequently associated with an increased risk for the development of breast cancer and are also predictive of more aggressive postmetastatic disease and decreased survival.
	11q23 Ataxia-tangiectasia	Autosomal recessive trait, mapped to chromosome 11q. Progressive cerebellar ataxia, and telangiectasia, especially of conjunctiva. Increased risk of malignancy. Female heterozygotes display an elevated risk of developing breast cancer. The key mechanism is believed to be a defect in DNA repair mechanisms.
Chromosome 13	13q12,13 BRCA2	An early-onset breast cancer-associated gene localized to chromosome 13q12,13. Segregation analysis indicates a risk of breast cancer development of 87% by age 80 years. In addition, a small but significant risk of male breast cancer is associated with this gene as well as a lower risk of ovarian cancer as compared to BRCA1.
Chromosome 17	Breast/ovarian syndrome	Autosomal dominant pattern of inheritance. Strong association between breast and ovarian cancer. Although not definitively proven, there is a strong correlation between the segregation of this syndrome and the susceptibility gene for early-onset breast cancer, BRCA1.
	p53-17p13.2 Li-Fraumeni syndrome	Autosomal dominant pattern of transmission associated with germline mutations of p53. Characterized by the development of soft tissue sarcomas or adrenal carcinomas. Family members show an increased incidence of other malignancies. Appears to be involved in only a small percentage (1%) of early; onset breast cancers.
	17q21 BRCA1	A tumor-suppressor early-onset breast cancer-associated gene localized to chromosome 17q21. Germline mutations in this gene inactivate tumor-suppressor function and account for appearance of breast and ovarian cancer. May account for only 2–5% of all breast cancer cases, but may approach 30% in breast cancer patients under age 35 years.

Tumor Suppressor Genes

Building on the retinoblastoma paradigm (Knudson, 1985), the favored view of familial breast cancer is that a suppressor gene inactivated by a small deletion or point mutation is the heritable component. Some tumor-suppressor genes appear to be important in cell cycle regulation, normally functioning as checks on cell growth, whereas others appear to be involved in DNA repair, preventing the propagation of mutations in other critical genes (see Chapter 1, this volume). Tumor-suppressor genes are thought to lose these regulatory functions as a result of acquired mutations, which then lead to malignant transformation of the cell. In many cases, inactivated tumor-suppressor genes appear to be responsible for familial cancer syndromes. It was hypothesized that genes existed whose normal function resulted in suppression of malignant cell growth; when the activity of these "tumor-suppressor genes" was lost, a clone of cells capable of malignant growth emerged and caused the development of a clinically detectable cancer (Harris, 1969; Stanbridge et al., 1982; Klein, 1987). All individuals are born with duplicate copies of every gene, and the tumor-suppressor gene's behavior is recessive at the cellular level, i.e., the residual normal copy of the gene compensates for the defect that can therefore persist in the population. An explanation was thus needed for the development of cancer in many individuals who had only a single inherited mutation in a tumor-suppressor gene, with residual function due to the other copy. The "two-hit" hypothesis was proposed in 1971 by Knudson, who suggested that cancer is a result of two genetic events occurring in the same cell, inactivating both copies of a given tumor-suppressor gene. The likelihood that two events would occur in the same cell is quite low in the case of sporadic, noninherited cancer. However, individuals from "cancer families" inherit an inactivating mutation in one allele of the implicated tumor-suppressor gene in all cells (i.e., a germline mutation); therefore, only one somatic (noninherited) event is required to inactivate the single remaining copy, making the development of cancer a much more common event than in individuals born without the "first hit."

The retinoblastoma gene (Rb) was the first tumor-suppressor gene linked to the pathogenesis of human cancers (Lee et al., 1987; Hollstein et al., 1991). It has subsequently been shown that all patients with hereditary retinoblastoma carry a germline mutation in one allele of the Rb gene. The "two-hit" hypothesis postulated the presence of a defect or mutation in the normal or "wild type" allele arising in the tumor. This hypothesis has proven to be correct in the case of the Rb gene. Loss of the wild-type allele (loss of heterozygosity, LOH) results in the inactivation of both Rb alleles: the first one by the inherited germline mutation, and the second one by a somatic event. Thus, germline mutations in tumor-suppressor genes create highly susceptible groups of individuals in whom only a single somatic event is required to inactivate tumor-suppressor activity at a given locus.

Besides the Rb gene, several other tumor-suppressor genes appear to be important in breast cancer, including p53, BRCA1, and, presumably, BRCA2.

Chromosome 5

Recently, it was found that germline mutations in genes on chromosome 5q21 that are involved in DNA repair at the MSH2/MLH1 loci (genes responsible for hereditary nonpolyposis colon cancer) are also responsible for development of the Muir's syndrome (Anderson, 1980; Groden et al., 1991; Kinzler et al., 1991; Bronner et al., 1994; Hall et al., 1994; Papadopoulos et al., 1994). An increased risk for breast cancer has been observed in females, particularly postmenopausal women, from families affected by this syndrome (Muir et al., 1967). This syndrome, a variant of the Lynch type II syndrome, has an autosomal dominant pattern of inheritance, with high penetrance (Borrensen et al., 1992), and also displays an association between multiple skin tumors and tumors of the upper and lower gastrointestinal and genitourinary tracts (Muir et al., 1967; Hall et al., 1994).

Chromosome 11

Among the genomic regions commonly undergoing LOH in breast tumors is chromosome 11. In particular, regions 11p15 and 11q23 appear to be the most highly affected. Deletions of 11p15, manifested as a LOH, have been observed in a variety of different cancers, including breast cancer, suggesting the presence of either a cluster of tumor-suppressor-type genes or a single gene with pleiotropic effects. In addition to being associated with the development of breast cancer and other malignancies, LOH of the 11q23 regions, either alone or in conjunction with LOH of 11p15, in the primary tumor was found to be highly predictive of an aggressive postmetastatic disease course, with substantially reduced survival (Winqvist et al., 1995).

A translocation between the long arm of chromosome 11 and the long arm of chromosome 22 is the most frequently reported heritable balanced translocation in humans, and has now been described in more than 100 families (Fraccaro et al., 1980; Zackai and Emmanuel, 1990; Iselius et al., 1993). An elevated incidence of cancer when compared to the general population (Lindblom et al., 1994) was noted in a review of eight Swedish families containing 22 known carriers. At least one case of breast cancer per family was observed in five of the eight Swedish families. Based on the breast cancer incidence in the Swedish population as a whole, the number of breast cancer cases observed in this cohort was significantly greater than expected, showing a standardized relative risk of 9.8. These data suggest the presence of a breast cancer-associated gene on chromosome 11 or 22, although no such gene has yet been identified. However, efforts in several laboratories are underway to isolate the genes on both chromosomes affected by the translocation, and to perform linkage studies of breast cancer families.

Chromosome 13

In addition to BRCA1, several other genes have been noted to confer susceptibility to breast cancer. Recently, Wooster and colleagues (1994) used genetic linkage analysis to study 15 families that had multiple cases of early-onset breast cancer that were not linked to BRCA1. This analysis localized a second breast cancer susceptibility gene locus, BRCA2, to a 6-centimorgan interval on chromosome 13q12–13. Like BRCA1, BRCA2 appears to confer a high risk of early-onset breast cancer in females; previous segregation analysis of the largest BRCA2-linked family indicated a risk of 87% by age 80 years (Wooster et al., 1992), which is comparable to the penetrance observed for BRCA1. Some clear differences do exist between the BRCA2 and the BRCA1 phenotypes. For example, a significantly lower risk of ovarian cancer is noted with BRCA2 as compared to BRCA1 (Wooster et al., 1994). In addition, a small but elevated risk of male breast cancer has been attributed to BRCA2 (Wooster et al., 1994). Although BRCA1 and BRCA2 appear to be key genes in the development of early-onset breast cancer, it is likely that these genes still do not account for all breast cancer caused by high-risk susceptibility genes.

Chromosome 17

The ultimate goal of gene mapping of breast cancer or other forms of cancer is to progress from a known chromosomal location to the identification of the involved gene or genes, and the characterization of their alterations. Numerous studies have demonstrated that chromosome 17 is the locale for a number of genes linked to the development of cancer in general, and to breast cancer specifically. Included among these gene candidates are *her2/neu* (*erbB2*), which encodes a protein related to the human epidermal growth factor receptor (Coussens et al., 1985; Di Fiore et al., 1987; Popescu et al., 1989); the gene for estradiol-17β dehydrogenase, an enzyme that catalyzes the conversion of estrone to estradiol; *nm23*, a gene associated with lymph node metastasis in primary breast cancer (Steeg et al., 1993); *wnt3*, a gene homologous to the *Drosphila* wingless locus and that is one of the integration sites for the mouse mammary tumor virus (Roelink et al., 1993); *p53*, a tumor-suppressor gene associated with the development of various malignancies including breast, ovarian, and prostate cancer (Nigro et al. 1989); and *BRCA1*, another recently identified tumor-suppressor gene associated with the development of early-onset familial breast cancer (Miki et al., 1994).

p53

The product of the p53 gene is a nuclear phosphoprotein that is believed to play a role in cell cycle regulation (Mercer et al., 1984; Shohat et al., 1987; see Chapter 1, this volume). p53 is a sequence-specific DNA-binding protein that acts as a tetramer to activate the transcription of responsive genes (Stenger et al., 1992; Vogelstein and Kinzler, 1992). Mutations in the p53 gene could lead to dysfunction or aberrant expression of the protein, ultimately resulting in uncontrolled cell proliferation. In fact, genetic alterations in the p53 gene are perhaps the most frequently occurring mutations in sporadic human cancer (Hollstein et al., 1991; Levine et al., 1991). Such alterations include deletion of one p53 allele and subsequent or concomitant alteration of the retained allele. It has been suggested that mutant p53 can act in a dominant negative fashion to prevent proper functioning of the wild-type protein (Farmer et al., 1992; Kern et al., 1992). Other mutants of p53 have been shown to bind heat shock proteins, forming a stable complex with a long p53 half-life (Sturzbecher et al., 1987). The wild-type p53 protein is naturally unstable, with a half-life of only several minutes (Finlay et al., 1988), and increasing the stability of a mutant form of p53 could provide a mechanism for dysregulation of its function. Thus, mutations in the p53 gene are very common, are highly associated with breast cancer, and are indicative of a much poorer prognosis.

Overexpression of p53 has been demonstrated by immunochemical methods in anywhere from 25–50% of primary breast tumors examined (Davidoff 1991a,b; Moll et al., 1992). Wild-type p53 is expressed at low levels in normal cells; however, because of its instability, it is virtually undetectable by most methods. Thus, overexpression of p53 in breast tumors is correlated with mutations that increase the stability of the protein. When such tumors were examined for p53 gene alterations, overexpression of nuclear p53 was always associated with mutation of the gene (Davidoff et al., 1991a;b; Moll et al., 1992). p53 can only act as a tumor suppressor when localized to the nucleus. So, whereas overexpression of mutant p53 in the nucleus may provide a mechanism to overcome normal regulation of proliferation, some tumors may develop another mechanism by sequestering wild-type p53 in the cytoplasm, where it is unable to exert its effects (Shaullsky et al., 1991).

The gene encoding p53 has been identified on 17p13.1 (van Tuinen et al., 1988). Deletions of the short arm of chromosome 17 have been reported to occur frequently in a variety of cancers (Mikkelsen and Cavenee, 1990), including breast cancer (Mackay et al., 1988). It has also been suggested that the presence of a deletion on chromosome 17p is predictive of a mutation in the p53 allele in colorectal and other cancers (Nigro et al., 1989). When this possibility was examined in human breast tumors, however, it was determined that allelic loss at chromosome 17p was not always associated with a mutation in the nearby p53 allele (Davidoff et al., 1991b; Mazars et al., 1992).

In examining mutations in the p53 gene, it is important to consider its structure. The gene is comprised of 11 exons, of which exons 2 and 5–9 have been highly conserved throughout evolution (Soussi et al., 1990). Mutational "hot spots" have been identified throughout exons 5–9 (Hollstein et al., 1991). Studies examining the p53 gene for mutations have thus far concentrated on this region of the gene. When breast cancer patients were screened for germline mutations in the p53 gene, a direct association between the presence of germline p53 mutations and breast cancer was observed. Studies that examined tumor p53 DNA and peripheral blood DNA from the same patients indicated that the only cases in which germline p53 mutations were associated with early-onset breast cancer were in those cases where there was a familial history of cancer, suggestive of Li-Fraumeni syndrome (Srivastava et al., 1990; Borrensen et al., 1992; Sidransky et al., 1992). Li-Fraumeni syndrome is characterized by one inherited mutant p53 allele followed by a somatic mutation in the second allele, resulting in predisposition for a variety of cancers, including breast cancer (Srivastava et al., 1991; Li et al., 1992).

As with any cancer, it is important in breast cancer to determine the point in tumorigenesis at which promotive or progressive events occur. Thus, efforts have been made to correlate p53 mutations with phase of progression of the disease. In a study examining the status of p53 expression and mutation in various stages of breast cancer (*in situ*, invasive, and metastatic), results indicated that mutations in p53 leading to overexpression of the protein can occur in the earliest stages of breast cancer as often as in later stages (Davidoff et al., 1991a). This same study was able to show that specific alterations are maintained during progression from intraductal to infiltrating carcinoma, as well as during metastasis.

Efforts have also been made to correlate p53 mutations and overexpression with estrogen receptor (ER)/progesterone receptor (PgR) status. Such studies indicate a correlation between ER-negative/PgR-negative or ER-negative/growth factor receptor-positive tumors and p53 mutations (Cattoretti et al., 1988; Mazars et al., 1992). However, since p53 is not directly linked to steroid receptor pathways, and steroid receptor-negative tumors are generally considered to have a poor prognosis, this correlation most likely links p53 mutations to aggressive breast cancer.

Although tumors provide a naturally occurring system in which to study p53 status, they do not allow for ongoing functional studies. To assess functionally the role of p53 in human breast cancer, researchers must rely on breast tumor cell lines. Studies have been performed in which breast cancer cell lines that overexpress mutant p53 were transfected with exogenous wild-type p53. Some cell lines were growth arrested by the constitutive expression of exogenous wild-type p53, whereas others continued to grow unaffected (Casey et al., 1991). Even though these data are *in vitro* analogies to a naturally occurring system, they do suggest that wild-type p53 may act to suppress cell growth in some breast cancers even though the protein may be ineffective in others, further indicating that p53 does not play a causative role in all breast cancers.

More recently, investigators have begun to examine the exact molecular pathway(s) through which p53 exerts its effects. A target gene for p53 that is highly involved in cell cycle regulation, WAF1/Cip1, has been identified (El-Deiry et al., 1993; Harper et al., 1993; Xiong et al., 1993; Noda et al., 1994). Wild-type p53 induces transcription of this gene, leading to cell cycle arrest at the G1/S transition. *In vitro* studies suggest that breast cancer cell lines expressing high levels of mutant p53 express extremely low levels of the WAF1/Cip1 gene product, possibly preventing the cells from arresting at G1/S (Sheikh et al., 1994). So even though it appears from retrospective studies that the evidence implicating p53 in the development of human breast cancer is, at best, equivocal, studies such as these should continue to provide insight into direct mechanisms of regulation of breast cancer cell growth by p53, ultimately allowing for the dissecting of individual breast cancer situations and the development of better individual treatment strategies.

Li and Fraumeni (1969) were the first to identify and describe, as a syndrome, the occurrence of excessive cancers including childhood soft tissue sarcomas, breast cancer, and other neoplasms in four kindreds. The existence of this syndrome, the Li-Fraumeni syndrome (LFS), has been confirmed by subsequent epidemiologic studies that have expanded the list of the major component neoplasms to include breast cancer, soft tissue and osteosarcomas, brain tumors, leukemias, and adrenocortical carcinomas, with several additional tumor types likely to merit inclusion (Li et al., 1988; Strong et al., 1992). An autosomal dominant pattern of transmission for this syndrome was determined by segregation analysis of families containing a family member with a sarcoma (Williams and Strong, 1985). Further analysis of these families with LFS identified an age-specific penetrance estimated to reach 90% by age 70 years; however, 30% of tumors in reported families occur before age 15 years (Strong et al., 1992).

The gene associated with this syndrome was identified in affected members of LFS families, and proved to be the p53 tumor-suppressor gene (Malkin et al., 1990). A report the following year by Srivastava et al. (1991) confirmed that the germline transmission of a mutated p53 gene was the common genetic factor in cancer-prone families with the LFS. In subsequent analysis of p53 germline alterations in families meeting the classic criteria for the clinical syndrome of LFS, approximately half show alterations in the p53 gene. Although mutations in the p53 gene are most frequently clustered in "hot spots" within the conserved sequences, exons 5–9, they have also been reported to occur throughout the gene. p53 genes ostensibly normal by sequencing but with abnormal functional assays or expression have also been observed (Barnes et al., 1992; Frebourg et al., 1992a,b; Toguchida et al., 1992).

The prevalence of germline p53 alterations among women with diagnosed breast cancer before age 40 years has been estimated at approximately

1% (Borresen et al., 1992; Sidransky et al., 1992). It is therefore not a common explanation for breast cancer occurrence in the population; nonetheless, p53 alterations have permitted the first predisposition testing programs for breast cancer susceptibility to be developed. However, the technological difficulties of p53 analysis, and the low number of mutations in the general population, have kept this genetic test from widespread application.

BRCA1

Lynch and Krush (1971) were the first to report an association between breast and ovarian cancer. In an analysis of 12 pedigrees, Lynch reported an estimated cumulative risk of 46% for breast or ovarian cancer in daughters of affected mothers. These numbers are consistent with the transmission of an autosomal dominant trait with high penetrance. This has been confirmed by subsequent reports by other investigators (Malkin et al., 1990; Hall et al., 1994) who have localized a susceptibility gene for early-onset breast cancer, now termed BRCA1, on chromosome 17q21 (Miki et al., 1994). Narod and colleagues (1991) have subsequently demonstrated unequivocal linkage between the genetic marker, D17S74, on 17q21 and the appearance of ovarian cancer in several large kindreds, providing strong evidence that germline mutations in BRCA1 account for the clinical appearance of breast/ovarian cancer syndrome. BRCA1 is a novel gene with homology to the zinc finger motif near the 5' end of the gene, suggesting that it may function as a transcription factor. However, the BRCA1 protein does not display homology to any other known motif or cloned gene. The mRNA that encodes BRCA1 is extremely large, approximately 7.8 kb in length, and is spliced from 24 exons that span almost 100 kb of genomic DNA. The BRCA1 protein is also quite large, being comprised of 1863 amino acids (Fig. 1). Preliminary studies indicate that mutations are found throughout the coding region, with two "hot spots" at codon 24 and codon 1756, but additional work is needed to define the spectrum of mutations and their functional significance.

Families in which breast and ovarian cancer are thought to result from BRCA1 mutations have now been identified. Inheritance in these families follows the classic mendelian pattern of autosomal dominant transmission, with 50% of the children carriers inheriting BRCA1 mutations. This inheritance pattern, as well as LOH studies in tumors from affected members of BRCA1-linked families, supports the hypothesis that BRCA1 is a classic tumor-suppressor gene, with loss of the normal or wild-type allele in the tumors of all informative cases (Chamberlain et al., 1993; Easton et al., 1993; Merajver et al., 1995). Female mutation carriers are estimated to have an 85% lifetime risk of developing breast cancer (Easton et al., 1994) and a 20–50% risk of developing ovarian cancer (Fraccaro et al., 1980).

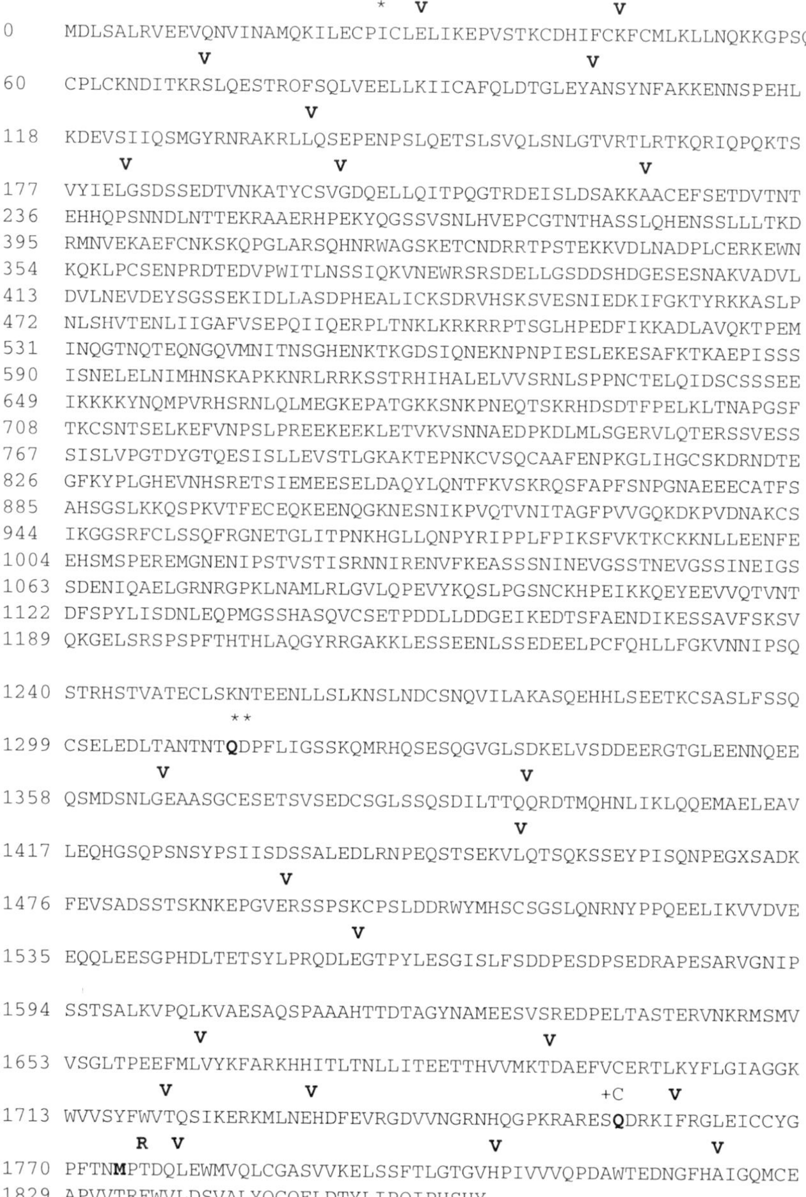

Figure 1. Predicted amino acid sequences for BRCA1 (Miki et al., 1994). Conceptual translation of the BRCA1 open reading frame, indicating the approximate positions on introns (Vs above sequence) and the locations of germline mutations (bold face residues). An 11-bp deletion found in one kindred is shown by an asterisk; the non-sense mutation found in another kindred is designated by a double asterisk; other mutations are identified by "+C", and "R."

BRCA1 mutation carriers also have an increased incidence of bilateral breast cancer. It is estimated that 45% of families with apparent autosomal dominant transmission of breast cancer susceptibility, and approximately 90% of families with dominant inheritance of both breast and ovarian cancer, harbor BRCA1 germline mutations. The percentage of breast cancer-only families that are attributed to BRCA1 mutations rises to almost 70% if the median age of onset of breast cancer in the families is less than 45 years (Easton et al., 1994).

A recent study conducted by the Breast Cancer Linkage Consortium (Easton et al., 1994) examined 33 families with evidence of germline mutations in BRCA1. Families included in these studies contained at least four cases of either breast or ovarian cancer diagnosed before age 60 years. Families from six different countries were included in these studies and they identified a risk heterogeneity among families carrying BRCA1 germline mutations, suggesting that different mutations in the BRCA1 gene could confer different degrees of risk. These data also suggest that the risk estimate is a function of age, and that mutation carriers that live to age 70 years have a 65% cumulative risk of developing a second breast cancer. Thus, even though BRCA1 germline mutations may account for as much as 30% of the breast tumors for women under age 35 years, such mutations account for only 5–10% of breast cancers in the general population and less than 1% of all breast cancers in elderly women (Claus et al., 1991; Easton et al., 1994).

Lynch et al. (1994) analyzed 180 tumors from hereditary breast/ovarian or site-specific breast cancer families to determine whether tumors that may arise as a result of BRCA1 mutations have clinical and pathological characteristics that differ from sporadic tumors. The presence of ovarian cancer in another family member, as well as the results of known linkage, segregated 98 of the 180 tumors into a subset that was more likely to be the result of BRCA1 mutations. The BRCA1 group was found to have more aneuploid and higher S-phase tumors than the other group. Tubular and lobular cancers were more common in the group in whom the presence of BRCA1 mutations was less certain. Both subgroups were significantly younger than the populaton average. Despite the findings of more ominous pathological characteristics in the BRCA1 group, disease-free survival was longer in this group than in the group thought less likely to contain BRCA1 mutations. These investigators suggest that BRCA1 mutations may result in tumors with adverse pathology indicators but, paradoxically, better survival. This study was performed before it was possible to determine which tumors harbored BRCA1 mutations. The isolation of BRCA1 now makes this possible, and the collection of data on the clinical and pathological characteristics of malignancies in BRCA1 mutation carriers will allow verification and refinement of this work. Although estimates of the frequency of BRCA1 germline mutations in the general population will be difficult to determine with accuracy until population-based studies are completed, it is possible that as many as 1 in 500 women in the United States may harbor germline mutations in breast cancer-susceptibility genes with the associated increased risk of developing neoplastic disease, approximately half of which may be accounted for by BRCA1.

The germline mutations in BRCA1 appear to account for only 3–5% of all breast cancers. However, the identification of BRCA1 will likely have far-reaching consequences. Its recent isolation makes possible the studies needed to elucidate its function. A current realistic option for women with two or more affected first-degree relatives is bilateral mastectomy in early adulthood. However, as many as 50% of these women may be at no greater risk than the general population. Although estimates of the percentage of breast cancer cases that occur as a direct result of germline mutations in BRCA1 vary from 5–25% (Colditz et al., 1993; Slattery and Kerber, 1993), what appears certain is that the estimate is a function of age. Thus, although less than 1% of all breast cancers in elderly women are likely to result from BRCA1 germline mutations, this fraction may approach 30% for women diagnosed with breast cancer under the age of 35 years (Claus et al., 1990, 1991; Easton et al., 1993). Additional studies will eventually allow presymptomatic testing for any woman desiring this information. The advisability of and approach to this type of genetic counseling is currently an active area of discussion.

Unlike female breast cancer, male breast cancer does not appear to be a component of this syndrome. Strong evidence against linkage between BRCA1 and male breast cancer comes from a lod score of −16.63 in a recent analysis of 22 pedigrees with a dominant inheritance pattern for female breast cancer and, at least, one case of male breast cancer (Stratton et al., 1994). The results indicated that there is a gene or genes other than BRCA1 that predisposes women to early-onset breast cancer, and that confers an increased risk of male breast cancer. The presence of such a gene is now confirmed by the finding of BRCA2 on chromosome 13 (Wooster et al., 1992).

BRCA2

As noted above, several studies have demonstrated that BRCA1 accounts for a large proportion of the families with multiple cases of both early-onset breast and ovarian cancer and approximately 45% of the families with only breast cancer. However, BRCA1 does not appear to be associated with many families demonstrating both male and female breast cancer (Easton et al., 1994). Wooster et al. (1994), using families demonstrating cases of both female and male breast cancer, have been able to localize a second breast cancer-associated gene (BRCA2) to a 6-centimorgan interval on chromosome 13q12-13. Like BRCA1, this new breast cancer-associated gene appears to confer a high risk of early-onset breast cancer in females.

Using the largest BRCA2-linked family from the Utah database, a risk of 87% for the development of breast cancer by age 80 years was noted, demonstrating comparable penetrance to BRCA1. Preliminary evidence suggests that BRCA2, like BRCA1, confers a significantly elevated risk of breast cancer but, unlike BRCA1, does not appear to confer a significantly elevated risk of ovarian cancer. It is probable that additional high-risk susceptibility genes will be identified that will account for breast cancer, because the evidence for genetic heterogeneity and the analysis of haplotypes in individual families indicate that additional breast cancer susceptibility genes remain to be identified.

Familial Male Breast Cancer

The study of familial male breast cancer is even more difficult than that of female breast cancer because of the infrequent occurrence of any breast cancer in men and the even smaller number of cases that occur in first- or second-degree male relatives. The initial reported case of male breast cancer within a nuclear family appeared in 1889 (Williams, 1889) and described both a father and son diagnosed with breast cancer; no other cases of cancer, including breast cancer, were known in other family members. Since then, only 16 additional families with two or more cases of male breast cancer have been reported (Kozak et al., 1986; Hauser et al., 1992). Analysis of these families suggests that familial male breast cancer is likely to be as heterogeneous as familial female breast cancer. In addition, more than half these families with male breast cancer describe first-degree relatives with other types of cancers including oropharyngeal, ovarian, and female breast cancer (Kozak et al., 1986). Nine of the families on which data were available reported cases of female breast cancer (Hauser et al., 1992). Average age at diagnosis of breast cancer in female relatives (where data were available) was 46 years, considerably younger than the average age at diagnosis in the general population (57 years) (Nemoto et al., 1980) and consistent with studies of familial breast cancer in females.

Epidemiologic case control studies have also documented an increased risk of breast cancer in male relatives of affected males (relative risk 6.1) and affected females (relative risk 2.2), with risk to male relatives increasing with decreasing age at diagnosis of the affected relative (Cassagrande et al., 1988; Rosenblatt et al., 1991). Female relatives of affected males were similarly at increased risk, again suggesting an inherited component to male breast cancer. In 1992, a study was published that provided evidence that inherited male breast cancer in one family was associated with germline mutations in the androgen receptor gene; this was subsequently confirmed in a second family (Wooster et al., 1992; Lobaccaro et al., 1993). This finding was of interest, not only as the first direct evidence that susceptibility to male breast cancer could be inherited, but also because altered hormonal metabolism has been suggested by many investigators as a risk factor for male breast cancer. Indirect evidence for this hypothesis came from reports of an increased incidence of gynecomastia, exogenous estrogen usage, schistosomal liver fibrosis, orchitis, testicular atrophy, obesity, and Klinefelter's syndrome among men diagnosed with breast cancer (for review see Hauser et al., 1992). Evidence for the presence of inherited male breast cancer associated with a defective androgen receptor strengthens this hypothesis.

Future of Molecular Genetics in the Study of Breast Cancer

Based on the evidence presented, the greatest problem confronting the genetic study of breast cancer is the possibility of genetic heterogeneity: there are many genes that are involved, any of which may predispose an individual to this disease. This scenario would make gene localization, via genetic linkage analysis, a quagmire requiring considerably more data and information than can be generated in a single laboratory. It is certain that more "breast cancer-associated genes" in addition to BRCA1 and BRCA2 will be identified in the coming years. However, whether it is a few or many, it will have a major impact on breast cancer prediction as individuals at risk will become identifiable.

As a result of public awareness of the increasing incidence of breast cancer in Western women, combined with media coverage of recent advances in the genetics of breast cancer, women are increasingly concerned about their individual risk of developing breast cancer. Recognition of the contribution of heritable factors to the development of breast cancer has resulted in a role for genetic counseling in the care of women with a family history of the disease. However, it is clear that epidemiologic studies do not always provide accurate risk assessments for individual women. The complexities of genetic counseling for breast cancer risk have been well described (Biesecker et al., 1974). The major categories of effort in providing risk assessment for women and families concerned about breast cancer include the collection of a detailed medical history of the extended family, interpretation of this history in light of available information to assess individual risk, and clear communication of this information to patient and family.

Family history information must include details of all cancers among blood relatives as well as the positions of these relatives within the family pedigree. Unaffected relatives should be included because the distribution of normal and affected relatives within a pedigree is essential to the assessment of potential inheritance patterns. Ideally, family history data should be confirmed with medical and/or death records to ensure accuracy. Details such as age at diagnosis, tumor histologies, laterality of breast tumors, and sites of primary tumors versus metastatic disease should be attained and verified whenever possible.

Recent advances in understanding the molecular genetics of inherited breast cancer, specifically the isolation of BRCA1 and localization of BRCA2, have allowed accurate determination of risk for a few individuals and made risk counseling for most women more complex. Within families that can be directly studied for mutations in BRCA1, estimates of breast cancer risk can be made with a high level of accuracy. Previously, many women with a family history of breast cancer suggestive of dominant inheritance of susceptibility were told that they had a 50% lifetime risk of developing breast cancer, based on the 50:50 chance of inheriting the mutant allele of such a gene. In fact, individuals from these striking families either carry a mutant BRCA1 allele or they do not. Mutation carriers have an 85% lifetime risk of developing breast cancer; individuals with two normal copies of BRCA1 have a breast cancer risk equivalent to the general population and are essentially unaffected by their family history. Although the number of families in which mutations in BRCA1 can be identified with a high degree of certainty is small, these families have access to the most accurate breast cancer risk assessment currently available. Finally, because all available data on families linked to BRCA1 are from a highly selected group of families, generally large families with at least four affected individuals, little is known about the contribution of this or other breast cancer susceptibility genes to breast cancer in smaller families with fewer breast cancer cases.

The basic mechanisms involved in the genesis of breast cancer and numerous potential applications clearly await further study of BRCA1, BRCA2, and other breast cancer-susceptibility genes. Targeted genetic interventions designed to reduce or eliminate risk might be developed for patients carrying germline mutations in such genes. Future studies will address the question of genetic heterogeneity: Do the various hereditary and nonhereditary forms of the disease involve the same or different loci? The mechanisms of the development of associated cancers in the various clinical syndromes associated with familial breast cancer also may be clarified. Finally, the importance of environmental influences may be addressed. The coming years should bring exciting developments in genetic epidemiology and molecular biology that may revolutionize our view of the role of hereditary factors in breast cancer and provide new diagnostic tools for the treatment of women with both inherited and sporadic forms of breast cancer.

REFERENCES

Adami HO, Hansen J, Jung B, Rimsten A (1981): Characteristics of familial breast cancer in Sweden: absence of relation to age and unilateral versus bilateral disease. *Cancer* 48:1688–1695

Anderson DE (1971): Some characteristics of familial breast cancer. *Cancer* 28:1500–1504

Anderson DE (1972): A genetic study of human breast cancer. *J Natl Cancer Inst* 48:1029–1034

Anderson DE (1974): Genetic study of breast cancer: identification of a high risk group. *Cancer* 34:1090–1097

Anderson DE (1976): Genetic predisposition to breast cancer. *Rec Results Cancer Res* 57:10–20

Anderson DE (1977): Breast cancer in families. *Cancer* 40:1855–1860

Anderson DE (1980): An inherited form of large bowel cancer. *Cancer* 45:1103–1107

Anderson DE, Goodman HO, Reed SC, eds. (1950): *Variables Related to Human Breast Cancer*. Minneapolis, University of Minnesota Press

Anderson DE, Badzioch MD (1985a): Risk of familial breast cancer. *Cancer* 56:383–387

Anderson DE, Badzioch MD (1985b): Bilaterality in familial breast cancer patients. *Cancer* 56(8):2092–2098

Anderson DE, Badzioch MD (1989): Combined effect of family history and reproductive factors on breast cancer risk. *Cancer* 63:349–353

Bain C, Speizer FE, Rosner B et al. (1980): Family history of breast cancer as a risk indicator for the disease. *Am J Epidemiol* 111:301–308

Barnes DM, Hanby AM, Gillett CE, Mohammed S, Hodgson S, Bobrow LG, Leigh IM, Purkis T, MacGeoch C, Spurr NK, Bartek J, Vojtesek B, Picksley SM, Lane DP (1992): Abnormal expression of wild type p53 protein in normal cells of a cancer family patient. *Lancet* 340:259–263

Biesecker BB, Boehnke M, Calzone K et al. (1970–1974): Genetic counseling for families with inherited susceptibility to breast and ovarian cancer. *JAMA* 1993:269

Borresen AL, Andersen TI, Garber J, Barbier-Piraux N, Thorlacius S, Eyfjord J, Ottestad L, Smith-Sorensen B, Hovig E, Malkin D, Friend SH (1992a): Screening for germ line p53 mutations in breast cancer patients. *Cancer Res* 52:3234–3236

Borresen A-L, Andersen TI, Garber J, Barbier-Piraux N, Thorlacius S, Eyfjord J, Ottestad L, Smith-Sorensen B, Hovig E, Malkin D, Friend SH (1992b): Screening for germ type TP53 mutations in breast cancer patients. *Cancer Res* 52:3234–3236

Brinton LA, Williams RR, Hoover RN, Stegens NL, Feinleib M, Fraumeni, JF Jr. (1979): Breast cancer risk factors among screening program participants. *J Natl Cancer Inst* 62:37–44

Broca P (1866): *Traite de tumeurs*. Paris: Asselin

Bronner EC, Baker SM, Morrison, PT et al. (1994): Mutation in the DNA mismatch repair gene homologue hMLH1 is associated with hereditary non-polyposis colon cancer. *Nature* 368:258–261

Brownstein MH, Wolf M, Bikowski JB (1978): Cowden's disease: a cutaneous marker of breast cancer. *Cancer* 41:2393–2398

Bucalossi P, Veronesi, Pandolfi A (1954): Il problema dell'ereditarieta neoplastica nell'uomo. Il cancro della mammela. *Tumori* 40:365–402

Casey G, Lo-Hsueh M, Lopez ME, Vogelstein B, Stanbridge EJ (1991): Growth suppression of human breast cancer cells by the introduction of a wild-type p53 gene. *Oncogene* 6:1791–1797

Cassagrande JT, Hanish RT, Pike MC, Ross RK, Brown JB, Henderson BE (1988): A case-control study of male breast cancer. *Cancer Res* 48:1326–1330

Cattoretti G, Rilke F, Andreola S, D'Amato L, Delia D (1988): p53 expression in breast cancer. *Int J Cancer* 41: 178–183

Chamberlain JS, Boehnke M, Frank TS, Kiousis, S Xu, J Gou, S-W, Hauser ER, Norum RA, Helmbold EM, Markel DS, Keshavarzi SM, Jackson CE, Calzone K, Garber J, Collins FS, Weber BL (1993): BRCA1 maps proximal to D17S579 on chromosome 17q21 by genetic analysis. *Am J Hum Genet* 52:792–798

Claus EB, Risch NJ, Thompson WD (1990): Age at onset as an indicator of familial risk of breast cancer. *Am J Epidemiol* 131:961–72

Claus EB, Risch NJ, Thompson WD (1991): Genetic analysis of breast cancer in the cancer and steroid hormone study. *Am J Hum Genet* 48:232–242

Colditz GA, Willett WC, Hunter DJ, Stampfer MJ, Manson JE, Hennekens CH, Rosner BA (1933): Family history, age and risk of breast cancer. *JAMA* 270:338–343

Crabtree JA (1941): Observations on the familial incidence of cancer. *Am J Public Health* 31:49–56

Collins N, McManus R, Wooster R, Mangion J, Seal S, Lakhani SR, Ormiston W, Daly PA, Ford D, Easton DF (1995): Consistent loss of the wild-type allele in breast cancers from a family linked to the BRCA2 gene on chromosome 13q12–13

Coussens L, Yang-Feng TL, Liao YC, Chen E, Gray A, McGrath J, Seeburg PH, Libermann TA, Schlessinger J, Francke, U.(1985): Tyrosine kinase receptor with extensive homology to EGF receptor shares chromosomal location with neu oncogene.. *Science* 230:1132–1139

Davidoff AM, Humphrey PA, Inglehart JD, Marks JR (1991a): Genetic basis for p53 overexpression in human breast cancer. *Proc Natl Acad Sci USA* 88:5006–5010

Davidoff AM, Kerns BJM, Inglehart JD, Marks JR (1991b): Maintenance of p53 alterations throughout breast cancer progression. *Cancer Res* 51:2605–2610

Devilee P, Van den Broek M, Kuipers-Dijkshoorn N, Kolluri R, Meera Khan PM, Pearson PL, Cornelisse CJ (1989): At least four different chromosomal regions are involved in loss of heterozygosity in human breast cancer. *Genomics* 5:554–560

DiFiore PP, Pierce JH, Kraus MH, Segatto O, King CR, Aaronson S. (1987): ErbB-2 is a potent oncogene when overexpressed in NIH/3T3 cells. *Science* 237:178–182

Easton DF, Bishop DT, Ford D, Crockford GP, Breast Cancer Linkage Consortium (1993): Genetic linkage analysis in familial breast and ovarian cancer results from 214 families. *Am J Hum Genet* 52:678–701

Easton DF, Ford D, Bishop, DT and the Breast Cancer Linkage Consortium (1994): Breast and ovarian cancer incidence in BRCA1 mutation carriers. *Lancet* 343:692–695

El-Deiry WS, Tokino T, Veculescu VE, Levy DB, Parsons R, Trent MJ, Lin D, Mercer ,WE, Kinzler KW, Vogelstein B (1993): WAF1, a potential mediator of p53 tumor suppression. *Cell* 75: 817–825

Farmer G, Bargonetti, J Zhu H, Friedman P, Prywes R, Prives C (1992): Wild-type p53 activates transcription *in vitro*. *Nature* 358:83–86

Ferrell RE, Anderson DE, Chidambaram A et al. (1989): A genetic linkage study of familial breast-ovarian cancer. *Cancer Genet Cytogenet* 38:241–248

Finlay CA, Hinds, PW Tan TH, Eliyahu D, Oren M, Levine AJ (1988): Activating mutations for transformation by p53 produce a gene product that forms an hsc 70-p53 complex with an altered half-life. *Mol Cell Biol* 8:531–539

Fishel R, Lescoe, MK Rao, MRS et al. (1993): The human mutator gene homolog MSH2 and its association with hereditary nonpolyposis colon cancer. *Cell* 75:1027–1038

Fraccaro M, Lindsten J, Ford CE (1980): The 11q;22q translocation: a European collaborative analysis of 43 cases. *Hum Genet* 156:21–51

Fraumeni, JF Jr, Grundy GW, Creagan ET et al. (1975): Six families prone to ovarian cancer. *Cancer* 36:364–369

Frebourg T, Barbier N, Kassel J, Ng Y-S, Romero P, Friend SH (1992a): A functional screen for germ line p53 mutations based on transcriptional activation. *Cancer Res* 52:6976–6978

Frebourg T, Kassel, J Lam KT, Gryka MA, Barbier N, Andersen TI, Borresen A-L, Friend SH (1992b): Germ-line mutations of the p53 tumor suppressor gene in patients with high risk for cancer inactivate the p53 protein. *Proc Natl Acad Sci* 89:6413–6417

Friend SH, Bernards R, Rogelj S et al. (1986): A human DNA segment with properties of the gene that predisposes to retinoblastoma and osteosarcoma. *Nature* 323:643–646

Gelehrter, TD and Collins FS (1990): Principles of Medical Genetics. Baltimore: Williams and Wilkins

Groden J, Thliveris A, Samowitz ??? et al. (1991): Identification and characterization of the familial adenomatous polyposis coli gene. *Cell* 66: 589–600

Hall, JM, Lee MK, Newman B, Morrow JE, Anderson LA, Huey B, King M-C (1990): Linkage of early onset breast cancer to chromosome 17q21. *Science* 250:1684–1689

Hall NR, Williams AT, Murday VA, Newton JA, Bishop DT (1994): Muir-Torre syndrome: a variant of the cancer family syndrome. *J Med Genet* 31:627–631

Harper JW, Adami, GR Wei N, Keyomarsi K, Elledge SJ (1993): The p21 cdk-interacting protein Cip1 is a potent inhibitor of G1 cyclin-dependent kinases. *Cell* 75:805–816

Harris H (1969): Suppression of malignancy by cell fusion. *Nature* 223:363–368

Hauser AR, Lerner IJ, King RA (1992): Familial male breast cancer. *Am J Med Genet* 44:839–840

Hollstein M, Sidransky D, Vogelstein B, Harris CC (1991): p53 mutations in human cancers. *Science* 253: 49–53

Iselius L, Lindsten J, Aurias A, Fraccara M, Bastard C, Bottelli, AM, Bui TH et al. (1983): The 11q;22q translocation: a collaborative study of 20 new cases and analyses of 110 families. *Hum Genet* 64:343–355

Itoh H, Houlston RS, Harocofos C, Slack J (1991): Risk of cancer death in first degree relatives with hereditary non-polyposis cancer syndrome (Lynch type II): a study of 130 kindreds in the United Kingdom. *Br J Surg* 56:1221–1229

Jacobsen O (1946): *Heredity in Breast Cancer: A Genetic and Clinical Study of Two Hundred Probands*. London, HK Lewis & Co

Kern SE, Pietenpol JA, Thiagalingam S, Seymour A, Kinzler KW, Vogelstein B (1992): Oncogenic forms of p53 inhibit p53-regulated gene expression. *Science* 256: 827–830

Kinzler KW, Nilber MC, Su LK et al. (1991): Identification of FAP locus genes from chromosome 5q21. *Science* 253:661–665

Klein G (1987): The approaching era of tumor suppressor genes. *Science* 238:1539–1545

Knudson AG (1971): Mutation and cancer: statistical study of retinoblastoma. *Proc Natl Acad Sci USA* 68:820–823

Knudson AG (1985): Herediatry cancer, oncogenes and anti-oncogenes. *Cancer Res* 45:1437–1443

Kozak FK, Hall JG, Baird PA (1986): Familial breast cancer in males. A case report and a review of the literature. *Cancer* 12:2736–2739

Lane-Clayton JE (1926): A further report on cancer of the breast, with special reference to its associated antecedent conditions. In: *Rep Min Health No. 32*, London, HM Stationery Office

Leach FS, Nicolaides N, Papadopoulos N et al. (1993): Mutations of a mutS homolog in hereditary nonpolyposis colon cancer. *Cell* 75:1215–1225

Lee WH, Bookstein R, Hong F et al. (1987): Human retinoblastoma susceptibility gene: cloning, identification, and sequence. *Science* 235:1394–1399

Levine AJ, Momand J, Finlay CA (1991): The p53 tumor suppressor gene. *Nature* 351: 453–456

Li FP, Fraumeni, JF Jr (1969): Soft-tissue sarcomas, breast cancer, and other neoplasms: a familial syndrome? *Ann Intern Med* 71:747–752

Li FP, Fraumeni JF, Mulvihill JJ, Blattner WA, Dreyfus MG, Tucker MA, Miller RW (1988): A cancer family syndrome in 24 kindreds. *Cancer Res* 48:5358–5362

Li FP, Garber JE, Friend SH (1992): Recommendations on predictive testing for germline p53 mutations among cancer-prone individuals. *J Natl Cancer Inst* 84:1156–1160

Lindblom A, Sandelin K, Iselius L, Dumanski J, White I, Nordenskjold, Larsson C (1994): Predisposition for breast cancer in carriers of constitutional translocation 11q;22q. *Am J Hum Genet* 54:871–876

Lobaccaro, J-M et al. (1993): Male breast cancer and the androgen receptor gene. *Nature Genet* 5:109–110

Lubin JH, Burns PE, Blot WJ et al. (1982): Risk factors for breast cancer in women in Northern Alberta, Canada, as related to age at diagnosis. *J Natl Cancer Inst* 68:211–217

Lynch HT, Guirgis HA (1981a): Introduction to Breast Cancer Genetics. In: *Genetics and Breast Cancer*, Lynch HT (ed). New York, Van Nostrand Reinhold

Lynch HT, Guirgis HA (1981b): Genetics and breast cancer. New York: Van Nostrand Reinhold

Lynch HT, Guirgis HA, Brodkey F et al. (1976): Genetic heterogeneity and familial carcinoma of the breast. *Surg Gynecol Obstet* 142:693–699

Lynch HT, Watson P, Conway TA et al. (1993): DNA screening for breast/ovarian cancer susceptibility based on linked markers. *Arch Intern Med* 153: 1979–1987

Lynch HT, Marcus J, Watson P, Page D (1994): Distinctive clinicopathologic features of BRCA1-linked hereditary breast cancer. *Proc ASCO* 13:56

Mackay J, Steel CM, Elder PA, Forrest APM, Evans HJ (1988): Allele loss on short arm of chromosome 17 in breast cancers. *Lancet* 2:1384–1385

Macklin MT (1959): Comparison of the number of breast-cancer deaths observed in relatives of breast-cancer patients, and the number expected on the basis of mortality rates. *J Natl Cancer Inst* 22:927–951

Malkin, D Li, FP, Strong LC, Fraumeni, JF Jr, Nelson, CE Kim DH, Kassel J, Gryka MA, Bischoff FZ, Tainsky MA et al. (1990): Germ line p53 mutations in a familial syndrome of breast cancer, sarcomas, and other neoplasms. *Science* 250(4985):1233–8

Martynova RP (1937): Studies in the genetics of human neoplasms. Cancer of the breast, based on 201 family histories. *Am J Cancer* 29:530–540

Mazars R, Spinardi L, BenCheikh M, Simony-Lafontaine J, Jeanteur P, Theillet C (1992): p53 mutations occur in aggressive breast cancer. *Cancer Res* 52:3918–3923

Merajver SD, Frank TS, Xu J, Calzone KA, Bennett-Baker P, Chamberlain J, Garber JE, Collins FS, Weber, BL. Loss of the wild type allele within the BRCA1 candidate region in tumors from early-onset breast/ovarian cancer families. (Submitted)

Mercer WE, Avignolo C, Baserga R (1984): Role of the p53 protein in cell proliferation as studied by microinjection of monoclonal antibodies. *Mol Cell Biol* 4:276–281

Mettlin C, Croghan I, Natarajan N, Lane W (1990): The association of age and familial risk in a case-control study of breast cancer. *Am J Epidemiol* 131:973–83

Miki Y, Swensen J, Shattuck-Eidens D, Futreal PA, Harshman K, Tavtigian S, Liu Q, Cochran C, Bennett LM, Ding W, Bell R, Rosenthal J, Hussey C, Tran T, McClure M, Frye C, Hattier T, Phelps R, Haugen-Strano A, Katcher H, Yakumo K, Gholami Z, Shaffer D, Stone S, Bayer S, Wray C, Bogden R, Dayanaanth P, Ward J, Tonin P, Narod S, Bristow PK, Norris FH, Helvering L, Morrison P, Rosteck P, Lai M, Barrett JC, Lewis C, Neuhaausen S, Cannon-Albright L, Goldgar D, Wiseman R, Kamb A, Skolnick M (1994): A strong candidate for the breast and ovarian cancer susceptibility gene BRCA1. *Science* 266:66–71

Mikkelsen T, Cavenee WK (1990): Suppressors of the malignant phenotype. *Cell Growth Differ* 1:201–207

Moll UM, Riou G, Levine AJ (1992): Two distinct mechanisms alter p53 in breast cancer: mutation and nuclear exclusion. *Proc Natl Acad Sci USA* 89:7262–7266

Morrell D, Cromartie E, Swift M (1986): Mortality and cancer incidence in 263 patients with ataxia-telangectasia. *J Natl Cancer Inst* 77:89–92

Morton NE (1955): Sequential tests for the detection of linkage. *Am J Hum Genet* 7:277–318

Muir EG, Yates-Bell AJ, Barlow KA (1967): Multiple primary carcinomata of the colon, duodenum, and larynx associated with keratoacanthomata of the face. *Br J Surg* 54:191–195

Narod SA, Feuteun J, Lynch HT, Watson P, Conway T, Lynch J, Lenoir GM (1991): Familial breast-ovarian cancer locus on chromosome 17q12-23. *Lancet* 338:82–83

Nemoto T, Vana J, Bedwani RN, Baker HW, McGregor FH, Murphy GP (1980): Management and survival of breast cancer: results of a national survey by the Americam College of Surgeons. *Cancer* 45:2917–2924

Newman B, Austin, MA Lee M, King MC (1988) Inheritence of breast cancer: evidence for autosomal dominant transmission in high risk families. *Proc Natl Acad Sci. USA* 85:3044–3048

Nigro JM, Baker SJ, Preisinger AC, Jessup JM, Hostetter R, Cleary K, Bigner SH, Davidson N, Bayulin S, Devilee P, Glover T, Collins FS, Weston A, Modali R, Harris CC, Vogelstein B (1989): Mutations in the p53 gene occur in diverse human tumor types. *Nature* 342:705–708

Noda A, Ning Y, Venable SF, Pereira-Smith OM, Smith JR (1994): Cloning of senescent cell-derived inhibitors of DNA synthesis using an expression screen. *Exp Cell Res* 211:90–98

Ottman R, Pike MC, King MC, Casagrande JT, Henderson BE (1986): Familial breast cancer in a population-based series. *Am J Epidemiol 123:15–21*

Papadopoulos N, Nicolaides, NC, Wei Y-F, Ruben SM, Carter KC, Roxen CA, Haseltine WA, Fleischmann RD, Fraser SM, Adams MD, Venter JC, Hamilton,SR, Peterson GM, Watson P, Lynch HT, Peltomaki P, Mecklin J-P, de la Chapelle A, Kinzler KW, Vogelstein B (1994): Mutation of a mutL homolog in hereditary colon cancer. *Science* 263:1625–1629

Papadrianos E, Haagensen CD, Cooley E (1967): Cancer of the breast as a familial disease. *Ann Surg* 165:10–19

Penrose LS, MacKenzie HJ, Karn MNA (1948): Genetical study of human mammary cancer. *Ann Eug* 14:234–266

Popescu NC, King CR, Kraus MH (1989): Localization of the human erbB-2 gene on normal and rearranged chromosomes 17 to bands q12-21.32. *Genomics* 4:362–366

Roelink H, Wang J, Black DM, Solomon E, Nusse R (1993): Molecular cloning and chromosomal location of the *wnt3* gene. *Genomics* 17:760–792

Rosenblatt KA, Thomas DB, McTiernan A, Austin MA, Stalsberg H, Stemhagen A, Thompson WD, Curen MGM, Satariano W, Austin DF, Isacson P, Greenberg, RS Key C, Kolonel L, West D (1991): Breast cancer in men: aspects of familial aggregation. *J Natl Cancer Institute* 83:849–854

Sakamoto G, Sugano H, Kasumi F (1978): Bilateral breast cancer and familial aggregations. *Prev Med* 2:225–229

Sattin RW, Rubin GL, Webster LA, Huezo CM, Wingo PA, Ory HW et al. (1985): Family history and the risk of breast cancer. *JAMA* 253:1908–1913

Schneider NR, Chaganti SR, German J, Chaganti RS (1983): Familial predisposition to cancer and age at onset of disease in randomly selected cancer patients. *Am J Hum Genet* 35:454–467

Seidman H, Stellman SD, Mushinski MH (1982): A different perspective on breast cancer risk factors: some implications of the nonattributable risk. *CA Cancer J Clin* 32:301–13

Shaullsky G, Goldfinger N, Peled A, Rotter V (1991): Involvement of wild-type p53 protein in the cell cycle requires nuclear localization. *Cell Growth Differ* 2:661–667

Sheikh MS, Li XS, Chen JC, Shao ZM, Ordonez JV, Fontana JA (1994): Mechanisms of regulation of WAF1/Cip1 gene expression in human breast carcinoma: role of p53-dependent and independent signal transduction pathway. *Oncogene* 9:3407–3415

Shohat O, Greenberg M, Reisman D, Oren M, Rotter V (1987): Inhibition of cell growth mediated by plasmids encoding p53 antisense. *Oncogene* 1:277–278

Sidransky D, Tokino T, Helzlsouer K, Zehnbauer B, Rausch G, Shelton B, Prestgiacomo L, Vogelstein B, Davidson N (1992): Inherited p53 gene mutations in breast cancer. *Cancer Res* 52:2984–2986

Slattery ML, Kerber RA (1993): A comprehensive evaluation of family history and breast cancer risk: the Utah Population Database. *JAMA* 270:1563–1568

Smith SA, Easton DF, Evans DGR, Ponder BAJ (1991): Allele losses in the region 17q12-21 in familial breast and ovarian cancer involve the wild-type chromosome. *Nature Genet* 2:128–131

Smithers DW (1948): Family histories of 459 patients with cancer of the breast. *Br J Cancer* 2:163–167

Soussi T, Caron de Fromentel, C May P (1990): Structural aspects of the p53 protein in relation to gene evolution. *Oncogene* 5: 945–952

Srivastava, S Zou ZQ, Pirollo D, Blattner W, Chang EH (1990): Germ-line transmission of a mutated p53 gene in a cancer-prone family with Li-Fraumeni syndrome. *Nature* 348:681–682

Srivastava, S Zou Z, Pirollo K, Blattner W, Chang EH (1991): Germline transmission of a mutated p53 gene in a cancer-prone family with Li-Fraumeni syndrome. *Nature* 348:747–479

Stanbridge EJ et al. (1982): Human cell hybrids: analysis of transformation and tumor genecity. *Science* 215:252–259

Starink TM (1984): Cowden's disease: analysis of fourteen new cases. *J Am Acad Dermatol* 11:1127–1141

Steeg PS, de la Rosa A, Fatowll A, MacDonald NJ, Benedict M, Leone A. (1993): Nm23 and breast cancer metastasis. *Breast Cancer Res. Treat.* 25:175–187

Stenger JE, Mayr GA, Mann K, Tegtmeyer P (1992): Formation of stable p53 homotetramers and multiples of tetramers. *Mol Carcinog* 5:102–106

Stratton MR, Ford D, Neuhausen S et al. (1994): Familial male breast cancer is not linked to the BRCA1 locus on chromosome 17q. *Nature Genet* 7:103

Strong LC, Williams WR, Tainsky MA (1992): The Li-Fraumeni syndrome: from clinical epidemiology to molecular genetics. *Am J Epidemiol* 135:190–199

Sturzbecher HW, Chumakov P, Welch WJ, Jenkins JR (1987): Mutant p53 proteins bind hsp 72/73 cellular heat shock-related proteins in SV40-transformed monkey cells. *Oncogene* 1: 201–211

Swift M, Morrell D, Cromartie E, Chamberlin AR, Skolnick MH, Bishop DT (1986): The incidence and gene frequency of ataxia-telangiectasia in the United States. *Am J Hum Genet* 39:573–83

Toguchida J, Yamaguchi T, Dayton SH, Beauchamp RL, Herrera GE, Ishizaki K, Yamamuro T, Meyers PA, Little JB, Sasaki MS, Weichselbaum RR, Yandell DW (1992): Prevalence and spectrum of germline mutations of the p53 gene among patients with sarcoma. *N Engl J Med.* 326:1301–1308

Tulinius, H Day NE, Bjarnason, O et al. (1982): Familial breast cancer in Iceland. *Int J Cancer* 29:365–371

van Tuinen P, Dobyns,WB, Rich DC, Summers KM, Robinson TJ, Nakamura Y, Ledbetter DH (1988): Molecular detection of microscopic and submicroscopic deletions associated with Miller Dieker syndrome. *Am J Hum Genet* 43:587–596

Vogelstein B, Kinzler KW (1992): p53 function and dysfunction. *Cell* 70:523–526

Wassink WF (1935): Cancer et heredite. *Genetika* 17:103–144

Wainwright JM (1931): Comparison of conditions associated with breast cancer in Great Britain and America. *Am J Cancer* 15:2610–2645, 1931

Weber BL, Garber JE (1993): Family history and breast cancer: probabilities and possibilities. *JAMA* 270:1602–1603

Williams WR (1889): Cancer of the male breast: based on the records of 100 cases. *Lancet* 2:261

Williams WR, Anderson DE (1984): Genetic epidemiology of breast cancer: segregation analysis of 200 Danish pedigrees. *Genet Epidemiol* 1:7–20

Williams WR, Strong LC (1985): Genetic epidemiology of soft tissue sarcomas in children. In:Muller, HR and Weber W (Eds.). Familial cancer. First international research conference. Basel: S. Karger

Winqvist R, Hampton GM, Mannermaa A, Blanco G, Alavaikko M, Kiviniemi H, Taskinen PJ, Evans GA, Wright FA, Newsham I, Cavenee WK (1995): Loss of heterozygosity for chromosome 11 is primary breast tumors is associated with poor survival after metastasis. *Cancer Res* 55:2660–2664

Wood DA, Darling HH (1943): A cancer family manifesting multiple occurrences of bilateral carcinoma of the breast. *Cancer Res* 3:509–514

Wooster R et al. (1992): A germline mutation in the androgen receptor in two brothers with breast cancer and Reifenstein syndrome. *Nature Genet* 2:132–143

Wooster R, Neuhausen S, Mangion J, Quirk Y, Ford D, Collins N (1994): Localization of a breast cancer susceptibility gene, BRCA2, to chromosome 13q12-13. *Science* 265:2088–2090

Xiong Y, Hannon GJ, Zhang H, Casso D, Kobayashi R, Beach D (1993): p21 is a universal inhibitor of cyclin kinases. *Nature* 366:701–704

Yates JRW, Connor JM (1986): Genetic linkage, Br J Hosp Med 33:133–136

Zackai EH, Emmanuel BS (1980): Site-specific reciprocal translocation t(11;22)(q23;q11) in several unrelated families with 3:1 meiotic disjunction. *Am J Med Genet* 7:507–521.

8

Growth Factors and Modulation of Endocrine Response in Breast Cancer

ROBERT I. NICHOLSON AND JULIA M. W. GEE

INTRODUCTION

Breast cancer is the most prevalent of all cancer diseases in women throughout the world. In high incidence areas, as many as one woman in eight may be expected to develop the disease. Studies on factors affecting the growth and development of breast cancer have identified steroid hormones to be of importance, with perturbation of the hormone environment of breast cancer cells delaying the appearance of the disease and promoting extensive tumor remissions in established tumors (Nicholson et al., 1992). Since hormone manipulations are relatively nontoxic when compared with other therapies such as cytotoxic chemotherapy, they have become widely-established as a preferred first-line therapeutic choice in many breast cancer management programs (Nicholson, 1993; Nicholson et al., 1993) and have been proposed as potential chemopreventative agents in women identified as being at high risk of developing the disease (Costa. 1993).

Unfortunately, endocrine therapy is not effective in all patients and even in initially responsive disease, patients will sooner or later relapse and die from progressive disease (Patterson, 1981). The transitions that result in the development of primary hormone insensitivity (*de novo* endocrine resistance) and acquired resistance to endocrine measures, therefore, remain a major clinical health-care problem in women suffering from breast cancer.

Importantly, a number of polypeptide growth factors can influence the growth of the breast and can impinge on steroid hormone-driven growth signaling through autocrine, paracrine, and endocrine loops (Brunner et al., 1989; Kern et al., 1990; King, 1990; Clarke et al.,1992). The acquisition

Hormones and Cancer
Wayne V. Vedeckis, Editor

of the ability to constitutively express growth factors, or their receptors, or to circumvent their cellular functions through postreceptor modifications in signal transduction, therefore provides a paradigm for the loss of hormone sensitivity in breast cancer and the development of endocrine resistance. These aspects are the focus of this chapter.

Emphasis is placed on four growth factor-signaling pathways mediated via the receptors for the *erbB* family, insulin-like growth factors, fibroblast growth factors, and transforming growth factor beta (TGF-β) and on the cellular actions of estrogens. This is not to preclude a similar future importance for other growth factors and steroid hormones, but merely reflects the current literature on endocrine sensitivity in breast cancer. Similarly, with the exception of the epidermal growth factor receptor (EGF-R) and transforming growth factor alpha (TGF-α), it has been necessary to acquire the majority of information primarily from experimental studies rather than from clinical data.

THE *erbB* FAMILY AND ASSOCIATED PATHWAYS

Epidermal Growth Factor Receptor (EGF-R)

The EGF-R is a member of the tyrosine kinase family of cell membrane spanning receptors (Type I) that transmit mitogenic signals via a series of phosphorylation steps. EGF-R is highly homologous to the v-*erbB* oncogene product, which encodes a truncated receptor lacking the ligand binding domain, and consequently has the potential to remain constitutively activated in the presence or absence of ligands (Hayman and Enrietto, 1991). The EGF-R is expressed at high levels in a number of cancer types and appears to be associated with aggressive tumor behavior (Neal et al., 1985; Yasui et al., 1988; Klijn et al., 1992). Several ligands for the EGF-R have been identified, including epidermal growth factor (EGF), TGF-α (the tumor homolog of EGF), amphiregulin, and cripto (Prigent and Lemoine, 1992). Characteristically, the cellular actions of ligand-activated EGF-R are mediated following the formation of homodimers. The EGF-R can, however, form heterodimers with the c-*erbB*-2 oncoprotein, a member of the same family of growth factor receptors (Spivak-Kroizman et al., 1992). Such heterodimers are similarly believed to be capable of transmitting growth signals to responsive cells.

Experimental Studies. EGF and TGF-α promote the growth of both normal mammary epithelium and human breast cancer cells *in vitro* (Osborne et al., 1980; Fitzpatrick et al., 1984). Receptors for EGF have been demonstrated on several breast cancer cell lines, especially within estrogen receptor (ER)-negative tumor cells (Kudlow et al., 1985; Davidson et al., 1987). In ER-positive breast cancer cell lines, the EGF-R gene is under a degree of steroid hormone control, with estrogen inducing slight increases in EGF-R levels with a concomitant decrease in ER (Bertois et al., 1989; Chrysogelos et al., 1994). Although EGF-R levels are frequently low in such cells, they nevertheless retain sensitivity to the mitogenic effects of the ligands EGF (Davidson et al., 1987) and TGF-α (Stewart et al., 1992).

Several studies have indicated that the acquisition of endocrine insensitivity is associated with an increase in EGF-R expression. In a series of 26 serially transplanted mouse mammary tumors, EGF-R was maximal in those tumors that gained hormone insensitivity and contained low levels of ER (Kienhuis et al., 1993). Similarly, an increase in EGF-R has also been reported in ZR-75–1 human breast cancer cells that have gained tamoxifen resistance (ZR-75–9al) and estrogen independence (ZR-PR-LT) (Long et al., 1992). Transfection and overexpression of the EGF-R in ER-positive ZR-75–1 human breast cancer cells has also been shown to induce steroid hormone independency (Van Agthoven et al., 1992). This, however, requires the presence of exogenous EGF. Importantly, although such cells are initially growth inhibited by antiestrogenic drugs, prolonged culture in the presence of 4-hydroxytamoxifen results in the outgrowth of antihormone-resistant subclones that lack ER. Similar conditions failed to induce proliferating subclones of the parental ZR-75–1 cells. These data infer a causative role for the EGF-R in the development of estrogen independency, and this may represent an important step in the acquisition of antihormone resistance (Van Agthoven et al., 1992). Other studies, however, have been unable to demonstrate a relationship between increased expression of EGF-R and the loss of hormone sensitivity in transfected MCF-7 breast cancer cells (Kern et al., 1994).

Clinical Studies. The analysis of EGF-R in breast carcinomas has been undertaken by a variety of methods and has produced a spectrum of reported positivity rates. In a recent review of more than 5000 patients (Klijn et al., 1992), approximately 45% were classified as positive and in a number of studies this has been associated with a poorly differentiated phenotype (Lewis et al., 1990; Klijn et al., 1992). Detection of the EGF-R in breast cancer specimens is generally due to increased gene transcription and occasionally (3%) as a consequence of gene amplification (Gullick, 1991). Although the EGF-R can be observed in normal breast tissue using immunohistochemistry, it is primarily localized in the myoepithelial layer toward the basement membrane, with relatively low levels of cell membrane immunostaining being observed in epithelial cells (Walker et al., 1991; Nicholson, unpublished results). This contrasts with the immunostaining seen in breast cancers, where EGF-R positivity occurs both in myoepithelial cells and within the tumor epithelial cell population (Nicholson et al., 1993).

The overexpression of EGF-R is usually inversely associated with ER levels (Klijn 1992). Given this association, it is not surprising that its presence in breast tumors has been linked to endocrine insensitivity (Nicholson et al., 1989, 1990, 1993, 1994a; McClelland, 1994). This is especially evident in ER negative disease, where 95% of patients with EGF-R-positive tumors fail to benefit from endocrine measures, and these patients have extremely poor survival characteristics (Nicholson et al., 1989, 1994a; McClelland et al., 1994a).

In contrast to these observations, patients with EGF-R-negative tumors frequently respond to endocrine therapies, show lower rates of cell proliferation, and have a much more favorable outlook (Nicholson et al., 1989, 1994a; McClelland et al., 1994a). Indeed, in ER-positive disease almost 80% of patients with the EGF-R-negative phenotype showed some degree of response in our own study (Nicholson et al., 1994a; McClelland et al., 1994a), with 43% of patients obtaining complete or partial remissions. Although the proportion of patients benefiting from endocrine therapy is higher than was reported in the study of Nicholson et al. (1989), where an objective response rate of 29% was achieved, nevertheless, the overall biological behavior of the EGF-R-negative cancers clearly distinguishes them from ER-negative/EGF-R-positive disease.

Despite the inverse relationship observed between ER and EGF-R, a small group of patients present with double positive tumors, and these show a response rate to endocrine measures that is intermediary between ER-positive/EGF-R-negative and ER-negative/EGF-R-positive disease (Nicholson et al., 1994a; McClelland et al., 1994a). Of particular significance to this observation is a recent study using a dual staining procedure to investigate whether ER and EGF-R are ever co-expressed within individual cells from clinical breast cancer specimens (Sharma et al., 1994). No such phenotype has been identified, inferring that cellular ER and EGF-R overexpression is mutually exclusive *in vivo*, and that tumor remissions following endocrine therapy in ER-positive/EGF-R-positive disease probably stem from the ER-positive/EGF-R-negative cells. These data therefore suggest that either ER-negative/EGF-R-positive cells are capable of responding to endocrine therapy, possibly through paracrine influences, or that in some instances this phenotype is reversible. Certainly, reversal of the resistance of ZR-75–9al human breast cancer cells to tamoxifen is accompanied by a return of ER and a fall in EGF-R levels (Long et al., 1992).

Although few patients with tumors wholly comprising ER-negative/EGF-R-negative cells have been reported (<10%; Nicholson et al., 1989, 1994a), this cellular phenotype is not uncommon within breast cancer and is observed in both ER-positive/EGF-R-positive and ER-positive/EGF-R-negative tumors. Indeed, such cells are frequently seen in normal breast tissue, where only approximately 15% of the luminal epithelial cells of both ducts and lobules are ER-positive (Walker et al., 1992) and where the cells are often negative using the EGF-R1 antibody raised to the external domain of EGF-R (Walker et al., 1991). Since normal breast is still hormone-responsive, it is perhaps not surprising that tumor remissions have been observed in cancers displaying this phenotype (Nicholson et al., 1994a). Currently, it is unclear whether this phenotype represents true ER and EGF-R negativity or, as is more likely, that the receptor levels fall below the sensitivity of the assays (Nicholson, 1992). Certainly, it is our experience that ER-negative/EGF-R-negative tumors are frequently negative for other biological endpoints, including Ki-67, a cell proliferation marker (Nicholson, unpublished results). Failure to detect hormone and growth factor receptors, therefore, may represent their down-regulation in quiescent cell populations (Nicholson, 1992).

Importantly, although EGF-R expression correlated with a failure to respond to endocrine treatments, it did not appear to influence the quality of remission observed in treated patients (Nicholson et al., 1994a). Thus, in patients who responded to endocrine measures, the proportion of women with complete, partial and static responses was not changed by either EGF-R status of the breast cancer or EGF-R level. Previously, we have suggested that the quality of tumor remissions observed in ER-positive tumors may be influenced, in part, by the prevailing rate of cell proliferation or growth fraction, with indolent cancers most frequently responding well to endocrine therapies (Nicholson et al., 1991). This influence was shown to be independent of EGF-R expression (McClelland et al., 1994a).

c-*erbB*-2 *and its Interactions with the EGF-R*

The c-*erbB*-2 gene encodes a 185-kDa transmembrane receptor protein (p185) with over 50% homology to the EGF-R, and it was first isolated in chemically induced neuroglioblastomas (and hence also termed *neu*). It was shown to transform NIH-3T3 cells when expressed at levels 100-fold greater than normal (Bargmann et al., 1986). Until recently, the ligand for the c-*erbB*-2 receptor has remained elusive (Holmes et al., 1992; Peles et al., 1992; Wen et al., 1992). Following an external stimulus, a series of events is initiated beginning with the autophosphorylation of the c-*erbB*-2 receptor protein. Neither EGF nor TGF-α bind to or directly activate p185 (Yarden and Weinberg 1989).

Experimental Studies. Several papers have reported structural interactions between the EGF-R and the c-*erbB*-2 proteins in normal and transformed rodent and human cell lines (Akiyama et al., 1986; Stern et al., 1986; Kadowaki et al., 1987; King et al., 1988; Kokai et al., 1988; Stern and Kamps, 1988). These include the formation of heterodimers with increased intracellular activities, and EGF-induced tyrosine phosphorylation of normal *neu* (the c-*erbB*-2 equivalent gene product in the rat), oncogenic *neu*, and the human c-*erbB*-2 products. Indeed, Kokai et al. (1989) have demonstrated a synergism between *neu* and EGF-R in the transformation of rodent fibroblasts.

Transfection of a full-length HER2/*neu* cDNA into ER-positive MCF-7 cells (MCF-7/HER2–18) reduces their sensitivity to tamoxifen *in vitro* and, furthermore, can promote complete tamoxifen resistance *in vivo* (Benz et al., 1993). These cells retain sensitivity to estrogen *in vivo*, suggesting that loss of hormone sensitivity may be acquired secondarily both to p185 overexpression and to decreased antiestrogen responsiveness. Other studies have noted some effects of c-*erbB*-2 overexpression on the hormone-dependent growth phenotype, but failed to find that overexpression conferred the ability to form progressive tumors in either ovariectomized nude mice or in tamoxifen-treated animals (Kern et al., 1994). Although some overexpressing cell lines acquired the ability to form small tumor nodules, they eventually regressed. These studies suggest that signal transduction mediated by the c-*erbB*-2 tyrosine kinase can only partially override the estrogen dependence of ER-positive breast cancer cells for growth and that additional genetic alterations are necessary to obtain the fully endocrine resistant phenotype (Kern et al., 1994). Indeed, it has been suggested that an additional factor is also necessary for c-*erbB*-2 to produce the fully-invasive phenotype. Thus, mice transgenic for the activated, point-mutated oncogene rapidly develop metastatic disease, whereas transgenic mice overexpressing the *neu* proto-oncogene develop metastatic tumors with a much longer latency period (Trimble et al., 1993).

Clinical Studies. Several techniques have been exploited to measure alterations in the c-*erbB*-2 proto-oncogene. Southern blot analysis has revealed that gene amplification, with a concomitant overexpression of c-*erbB*-2 mRNA and protein, is a relatively frequent event, occurring in approximately 20% of breast cancers (Slamon et al., 1987). Amplification of c-*erbB*-2 is associated with intense cell membrane immunostaining and may be readily distinguished from the weak cytoplasmic immunostaining that is present in many breast cancers and within normal tissue (Barnes et al., 1988; Lovekin et al., 1991).

In clinical breast cancer specimens, elevated c-*erbB*-2 levels and gene amplification appear to indicate the presence of lymph-node metastases and a poor survival (Slamon et al., 1987) and are associated with a decreased responsiveness to endocrine measures (Nicholson et al., 1990, 1993). In each instance, however, the relationship is relatively weak, with our own study showing 50% and 25% of patients gaining benefit from endocrine treatment in c-*erbB*-2-negative and -positive disease, respectively (Nicholson et al., 1993). Although we recorded no statistically significant difference between the survival curves of women with c-*erbB*-2-positive and -negative disease, the median survival time of women with c-*erbB*-2 protein-negative cancers was 11 months longer than patients with positive tumors (28 and 17 months, respectively).

The co-expression of EGF-R and c-*erbB*-2 in approximately 10% of human breast cancers correlates with a poorer survival than is seen with the expression of either protein alone (Nicholson et al., 1990).

Interestingly, this also appears to apply to their endocrine responsiveness, with patients whose tumors show overexpression of EGF-R and c-*erbB*-2 proteins having a very low likelihood of benefiting from such measures. In contrast, c-*erbB*-2 expression did not influence the proportion of responding patients in an EGF-R-negative group (Nicholson et al., 1993). In our own study, the influence of c-*erbB*-2 protein on endocrine sensitivity in EGF-R-positive tumors was restricted to women with moderately EGF-R-positive disease (Nicholson et al., 1993). Patients with highly EGF-R-positive tumors had largely hormone independent neoplasms and an unfavorable outlook, which was independent of c-*erbB*-2 protein staining. The prognosis of women with moderately EGF-R-positive/c-*erbB*-2-positive tumors was similar to that observed for patients with highly EGF-R-positive disease. Any influence, therefore, of c-*erbB*-2 signaling on the hormone sensitivity of breast cancer appears to require the activation of additional pathways and broadly corresponds to studies performed at an experimental level.

c-erbB-3 *and* c-*erbB*-4

c-*erbB*-3 is a recently identified member of the gene family coding for type I growth factor receptors. It is located on chromosome 12q13 and codes for a 180-kDa glycoprotein. The protein product shows considerable sequence homology with both EGF-R and c-*erbB*-2, especially in the tyrosine kinase domain (Lemoine et al., 1992). Heregulin binds to the c-*erbB*-3 receptor (also binding to c-*erbB*-2) but instigates little or no resultant tyrosine kinase activity (Carraway and Cantley, 1994).

Immunohistochemically detected membrane staining for the c-*erbB*-3 receptor is rare in primary and advanced breast cancer specimens and most staining is cytoplasmic (Lemoine et al., 1992). Strong immunostaining has been recognized in approximately one-third of clinical breast cancer specimens, but staining shows no correlation with established prognostic indicators or survival (Travis et al., in press). The receptor is, however, capable of forming heterodimers with c-*erbB*-2 (Sliwkowski et al., 1994). This generates a high affinity binding site whose activation by heregulin produces unique patterns of phosphorylation on tyrosine residues. This is in contrast to the interaction and complex formation between the c-*erbB*-2 and the fourth member of the family, c-*erbB*-4, where both receptors have active kinase components that are capable of autophosphorylation (Plowman et al., 1993). The potential for type I tyrosine kinase receptors to produce different combinations of heregulin-stimulated heterodimeric complexes could explain some of the myriad biological activities that have been demonstrated for this group of receptors (Carraway and Cantley 1994) that may impinge on the hormonal regulation of breast cancer growth.

Transforming Growth Factor Alpha (TGF-α)

Although a number of ligands for the EGF-R have been identified and characterized, only the actions of EGF and TGF-α have been studied in detail in breast cancer. In most instances, the effects of EGF and TGF-α appear similar and each has been shown to be a potent mitogen for a number of human breast cancer cell lines (Osborne et al., 1980; Bates et al., 1988). Enhanced production of TGF-α has been observed in transformed rodent and human fibroblast and epithelial cells, where it may function as a down-stream intermediary in the transformation pathway by oncogenes (Salomon et al., 1990). It has been suggested that TGF-α may act to induce hyperplastic responses in transformed breast cells and, thereby, act as a promotional agent in combination with a normal background of mutational events (Dixon et al., 1990; Matsui et al., 1990; Sandgren et al., 1990). Antibodies directed against TGF-α or its receptor transiently suppress both the anchorage-dependent and -independent growth of MCF-7 cells (Bates et al., 1988), as well as the anchorage-dependent proliferation of the estrogen-independent MDA-MB-468 human breast cancer cell line (Bates et al., 1988).

Experimental Studies. In human estrogen-responsive breast cancer cells *in vitro*, TGF-α is frequently expressed and secreted from both ER-positive and -negative cells (Dixon et al., 1986). However, in ER-positive cells the baseline expression of TGF-α mRNA is low, and may be induced approximately five-fold by estradiol. This suggests that part of the estrogenic growth stimulus may be mediated through autocrine TGF-α secretion (Dixon et al., 1986; Bates et al., 1988). Studies with pure antiestrogens may be interpreted as supporting such a role, with ICI 182780 preventing the steroid associated rise in TGF-α protein levels and decreasing, in parallel, tumor cell growth rates (Nicholson et al., 1995a). However, TGF-α transfection experiments infer that elevated TGF-α levels *in vitro* and *in vivo* Clarke et al., 1989) are not sufficient to override the growth-promoting activity of estradiol alone or in combination with overexpression of EGF-R (Kern et al., 1984). Indeed, the human breast cancer cell line CAMA-1 responds to estrogen but does not possess EGF-R (Leung et al., 1991).

Importantly, recent cell and molecular biology studies have identified significant interactions between estrogens and TGF-α (Ignar-Trowbridge et al., 1992) with, for example, the growth factor aiding ER-mediated transcriptional events, and with ER improving the efficiency of TGF-α-directed transcription from genes containing AP-1 binding sites in their promoters (Philips et al., 1993). Such activities imply a central importance for the ER in growth signaling, with estrogen and growth factors interacting to improve mitogenic signaling in breast cancer cells.

A number of studies have directly examined the relationship between TGF-α expression and the development of estrogen-directed growth independency in experimental models of human breast cancer. These have been largely performed *in vitro* following the prolonged withdrawal of estrogen from breast cancer cell lines (Katzenellenbogen et al., 1987; Darbre and King, 1988). Such cells often show increased basal growth rates and a reduced requirement for serum, added insulin, and growth factors. For two breast cancer cell lines, derived from both MCF-7 (K3 cells, Katzenellenbogen et al., 1987) and T47D cells (Darbre and King, 1988), TGF-α expression is raised and the growth factor shows a reduced sensitivity to estrogen stimulation (Nicholson, unpublished results). In T47D cells, this is associated with an increased sensitivity to the mitogenic actions of TGF-α (Darbre and King, 1988). Each of these derived cell lines may be growth-inhibited by antiestrogens (Katzenellenbogen et al., 1987; Reese and Katzenellenbogen et al., 1987; Darbre and King 1988; Clarke et al., 1989), with pure antiestrogens being more effective than tamoxifen in reducing the cellular levels of TGF-α in K3 cells (Nicholson et al., 1995a). This parallels their relative abilities to inhibit breast cancer cell growth (Clarke et al., 1989; Nicholson et al., 1994b; 1995a,b).

Although the above data imply a causative autocrine link between overexpression of TGF-α and increased growth rates with subsequent loss of estrogen-induction of growth, as stated above, transfection of a TGF-α cDNA into MCF-7 cells does not fully prevent the mitogenic actions of estradiol (Clarke et al., 1989). In this respect it may be of significance that transfection of MCF-7 cells with v-H-*ras* (the oncogene of Harvey sarcoma virus) increases TGF-α, TGF-β and IGF-1 expression and secretion and produces a corresponding loss of estrogen and antiestrogen sensitivity (Ciardiello et al., 1991). Indeed, in estrogen growth-independent T47D cells, loss of estrogen growth dependency is associated with an increased expression of both TGF-α and β (Darbre and King, 1988). Altered cellular phenotype, therefore, may result from changes in the pattern of production and secretion of a spectrum of factors whose expression is dependent upon an as-yet-unidentified control point(s).

Clinical Studies. The localization of TGF-α in 50–70% of primary breast cancer has recently been reported by several groups (Ciardiello et al., 1991; Lundy et al., 1991; Umekita et al., 1992; McClelland et al., 1994b; Nicholson et al., 1994c). In our own study, TGF-α was more frequently expressed than EGF, but the two were related in a statistically significant positive manner (McClelland et al., 1994b). Although a wide range of TGF-α positivity rates has been reported in breast cancer specimens, its detection is associated with ER expression (McClelland et al., 1994b) and, in one instance, well-differentiated tumor types (Lundy et al., 1991). The association with ER is, however, relatively weak, and heterogeneity of TGF-α expression exists within ER-positive and ER-negative tumors.

As *in vitro*, tamoxifen may partially down-regulate TGF-α in ER-positive disease (Noguchi et al., 1993). Thus, in ER- and PR-positive tumors, short courses of tamoxifen therapy (approximately 11 days; 20 mg twice daily) reduced TGF-α levels in solubilized breast cancer preparations. The effect was deemed to be specific because it was not observed in ER- and PR-negative tumors. These data, which were similar to an earlier study on much smaller numbers (Gregory et al., 1989), contrast with a lack of activity of tamoxifen and a pure antiestrogen, ICI 182780, on immunohistochemically detected TGF-α in primary breast cancer patients administered the drugs for short periods before mastectomy (Gee and Nicholson, unpublished results).

In only one instance has TGF-α expression been related to the responsiveness of breast cancer to endocrine measures (Nicholson et al., 1994c). In that study high levels of TGF-α were observed in 65% of tumors and showed no relationship with tumor EGF-R overexpression. TGF-α levels did, however, relate to the endocrine sensitivity of the disease, with unresponsive tumors frequently showing high levels of TGF-α immunoreactivity. Indeed, in ER-positive tumors, although low TGF-α values were often linked to objective tumor remissions, high values were most often associated with either static or progressive disease.

Significantly, elevated tumor TGF-α and EGF-R levels have recently been linked to increased expression of the nuclear transcription factor Fos in women failing to respond to endocrine therapy (Gee et al., 1995). This observation is of interest because the Fos protein is believed to represent a point of cross-talk between steroid and growth factor pathways and may provide a mechanism for aberrations in growth factor signaling to influence the cellular hormonal requirements. In as much as the relationship was seen in ER-positive disease, growth factor driven Fos expression may indeed be involved in a departure from estrogenic control.

INSULIN-LIKE GROWTH FACTOR (IGF) FAMILY

Most breast carcinomas contain membrane-bound receptors for insulin-like growth factors (IGFRs; Cullen et al., 1990; Krywicki and Yee, 1992), and their ligands IGF-I and IGF-II are generally more potent mitogens for human breast cancer cell lines than either TGF-α or EGF (Wosikowski et al., 1992; Musgrove and Sutherland, 1993). Like TGF-α and EGF, these growth factors can drive T47D cells back into the cell cycle through their induction of cyclin D1 following serum deprivation (Sutherland et al., 1993) and produce growth rates comparable to those observed with calf serum addition (Musgrove and Sutherland, 1993). Their proliferative activity is believed to be heavily influenced by the presence or absence of six specific binding proteins (IGFBPs) known to sequester IGFs (Figueroa and Yee, 1992).

Insulin-like Growth Factor Receptors (IGFR)

Two well-characterized IGFRs exist (Rosen et al., 1991; Krywicki and Yee, 1992). The IGFR-I is a heterotetrameric structure analogous to the insulin receptor. It has 50–60% homology with the insulin receptor and many important features are maintained (Krywicki and Yee 1992). These include cysteine-rich domains and tyrosine kinase activity (Krywicki and Yee 1992). IGFR-I binds both IGF-I and IGF-II with approximately equal affinities, whereas the affinity for insulin is much lower (Cullen et al., 1990; Krywicki and Yee, 1992; Peyrat and Bonneterre, 1992). IGFR-II is a single peptide chain that has no intrinsic tyrosine kinase activity (Kywicki and Yee, 1992). It is identical to the mannose-6-phosphate (M-6-P) receptor and binds IGF-II with much higher affinity than IGF-I (Krywicki and Yee, 1992). Approximately 40% of breast cancers are IGFR-I positive (Railo et al., 1994) and receptor expression associates with ER positivity (Foekens et al., 1989; Peyrat and Bonneterre, 1992). Although an elevated IGFR-I content has been reported by some groups to be prognostically favorable (Bonneterre et al., 1990; Peyrat and Bonneterre 1992; Papa et al., 1993), this is controversial; a further study has demonstrated that IGFR-I expression associates with a shortened disease-free survival, although the effect was most obvious in ER-negative disease (Railo et al., 1994).

Insulin-like Growth Factor-I (IGF-I)

Experimental Studies. IGF-I (somatomedin C) is a 70-amino acid peptide (Krywicki and Yee 1992). A significant role for IGF-I in mediating and modulating the effects of endocrine therapy on breast cancer is likely because it amplifies estrogen-induced growth responses in breast cancer cells (Wakeling et al., 1991; Nicholson, 1992; Stuart et al., 1992). Thus, although both estradiol and IGF-I alone stimulate the growth of breast cancer cells *in vitro*, together they have a powerful synergistic action. This may be aided by the ability of IGF-I to induce enzymes associated with estrogen synthesis and, hence, increase local production of estrogen (Purohit et al., 1992).

Additionally, estradiol can stimulate IGF-I binding to MCF-7 cells (Freiss et al., 1990), and estradiol and IGF-I (as well as TGF-α) can regulate the expression of common genes such as the progesterone receptor (Katzenellenbogen and Norman 1990; Cho et al., 1994), cathepsin D lysosomal enzyme (Cavailles et al., 1989), pS2 (Chalbos et al., 1993), and the nuclear transcription factor, Fos (Wosikowski et al., 1992). Further interactions between Fos and other IGF-I-induced transcription factors (e.g., Jun) efficiently promote TPA-dependent transcriptional activation from AP-1 sites, an important step in growth responses (Schuchard et al.,1993). Importantly, blockade of IGFR-1 by a monoclonal antibody (αIR3) has been used as a strategy to demonstrate the significance of the IGF pathway (Arteaga 1992). Although αIR3 could not block serum-free growth of breast cancer cell lines, it inhibited growth in the presence of whole serum (containing IGF-I and estrogen). Combination of αIR3 with an anti-EGF-R antibody caused significantly greater inhibition of growth than either antibody alone (Arteaga 1992), again illustrating the spectrum of signal transduction pathways used by the breast cancer cell.

The contribution made to growth responses by estradiol can be partially mimicked by the estrogen-like activity of tamoxifen, leading to incomplete growth arrest by the antiestrogen. Indeed, amplification of the agonistic activity of tamoxifen can give rise to cells that are no longer growth inhibited (to any degree) by tamoxifen and that have elevated growth rates and IGF-I binding (Wiseman et al., 1992). Such cells, which have acquired tamoxifen resistance, may be growth arrested by antibodies that block the IGF-I receptor and by pure antiestrogens that do not possess estrogen-like activity (Wiseman et al., 1992). In this model, tamoxifen fails to show any estrogen agonistic activity for the induction of a series of estrogen-regulated mRNAs, suggesting that its increased agonistic activity may be restricted to key components involved in the proliferative response (Wiseman et al., 1992).

Clinical Studies. IGF-I expression is found in many human breast cancer specimens (Foekens et al., 1989) and the major cellular source of IGF-I mRNA is the stroma associated with areas of pathologically normal and benign breast tissue (Cullen et al., 1992a; Paik 1992). No IGF-I mRNA appears to originate from the malignant epithelial tissue or its associated stroma. This is in agreement with studies performed on experimental human breast cancer cells *in vitro* demonstrating the absence of any significant production of IGF-I mRNA or protein (Cullen et al., 1992a,b). It is thus possible that IGF-I is a paracrine regulator of breast cancer cell growth (Rosen et al., 1991), acting via the IGFR-I present in the epithelial component of breast cancer samples. This receptor appears elevated relative to normal breast tissue (Gammes et al., 1992; Peyrat and Bonneterre, 1992), but overexpression is rarely the result of gene amplification, which can be detected in only 2% of cases (Berns et al., 1992). Although preliminary data indicate a correlation between IGFR-1 expression and better clinical outcome (Bonneterre et al., 1990; Peyrat and Bonneterre, 1992; Papa et al., 1993), this is not a universal finding (Railo et al., 1994). Little is known about the IGF family and the clinical responsiveness of breast cancer to endocrine treatments. However, tamoxifen can reduce the serum concentrations of IGF-I (Pollak et al., 1992; Reed et al., 1992), possibly by inhibiting growth hormone secretion (Malaab et al., 1992). A reduction in the tumor cell availability of IGF-I would be expected to contribute to the direct antitumor activity of tamoxifen (Pollak et al., 1992) in ER-positive patients, and it might also be responsible for any inhibitory effects of the drug in ER-negative disease. Unfortunately, no data exist that examine the relative reduction of IGF-I levels in individual patients and the primary response to tamoxifen. Similarly, no study has examined changes in circulating IGF-I levels in patients whose disease has become resistant to tamoxifen therapy. However, in one study, IGF-I levels were reduced in women treated with tamoxifen for 2 years (Friedl et al., 1993). Such an interval would normally encompass the time taken to develop acquired resistance to this antiestrogenic drug. Importantly, the estrogen-like activity of tamoxifen is now thought to be a contributory factor in the clinical development of tamoxifen resistance, with patients obtaining a further remission of their disease on tamoxifen withdrawal (Howell et al., 1992). Whether this is due to improved interactions between tamoxifen and IGF-I signaling remains to be elucidated.

Insulin-like Growth Factor-II (IGF-II)

Experimental Studies. IGF-II is a 67-amino acid peptide showing a 62% homology to IGF-I (Krywicki and Yee, 1992). Investigation of the expression of IGF-II in breast cancer was stimulated by the finding that it is a potent mitogen for breast cancer cells *in vitro* (Osborne et al., 1989; Cullen et al., 1991, 1992a,b). In T47D and MCF-7 cells, production of IGF-II mRNA and protein was induced by estradiol (Brunner et al., 1992, 1993; Cullen et al., 1992a,b) and, in contrast, inhibited by antiestrogens (Osborne et al., 1989). Estradiol-induced secretion of IGF-II from estrogen-sensitive MCF-7 cells, however, is lower than the constitutively expressed levels seen in estrogen-unresponsive MDA-231 and HBL-100 human breast cancer cells (Brunner et al., 1992, 1993). Experiments with MCF-7 cells in nude mice support the notion that the regulation of IGF-II expression by estradiol may be important for overall tumor growth. These cells require estradiol supplements to grow and show low background levels of IGF-II in the absence of the steroid (Brunner et al., 1991, 1993). On estradiol treatment, IGF-II mRNA expression increases. In contrast, T61 breast cancer cell xenografts, which are normally growth inhibited by physiological doses of estrogens, have readily detectable IGF-II mRNA in the absence of estradiol treatment (Brunner et al., 1992, 1993). When nude mice bearing these tumors are treated with estradiol, IGF-II mRNA disappears within 72 hours, closely followed by tumor regression (Brunner et al., 1992, 1993). Furthermore, in this model, tamoxifen also down-regulates IGF-II mRNA levels. In addition to the above, prolonged estrogen withdrawal results in increased IGF-II levels in ZR-75–1 cells. This is accompanied by an elevated basal growth rate of the cells and a loss of estrogen-growth responsiveness (Darbre and Daly, 1990). Although the effect appears clonal, it occurs at a high frequency, suggesting a mechanism that affects a wide proportion of the cell population synchronously. This does not appear to involve any loss of ER or the ability of the ER to regulate a number of molecular markers of ER action (e.g., pS2, PR). A potential role for IGF-II in influencing the endocrine sensitivity of breast cancer has recently been demonstrated following the transfection of MCF-7 cells with either a retroviral vector containing the coding sequence for the IGF-II preprohormone or an inducible IGF-II expression system (Daly et al., 1991; Cullen et al., 1992,b). These cells, which do not normally express either IGF-I or large amounts of IGF-II, became partially estrogen-insensitive for cell growth *in vitro* (Daly et al., 1991; Cullen et al., 1992a,b). Interestingly, the overexpression of IGF-II is also associated with a number of changes in cell phenotype that have been linked to malignant progression (Cullen et al., 1992b) and increased positive growth responses to both tamoxifen and 4-hydroxytamoxifen (Daly et al., 1991). Changes in IGF-II did not, however, confer estrogen insensitivity *in vivo* (Cullen et al., 1992a,b).

Clinical Studies. IGF-II mRNA is expressed in a large proportion of clinical breast cancer specimens and, like IGF-I, is most frequently seen in stroma (Cullen et al., 1992a). Unlike IGF-I, however, IGF-II mRNA is observed in the stroma from both malignant and nonmalignant areas of the tumor (Cullen et al., 1992a). It is also present in the epithelium of some tumors, as observed in cultured breast cancer cells (Cullen et al., 1992a); thus, this growth factor potentially may exert both paracrine and autocrine influences on breast cancer cells (Rosen et al.,1991; Cullen et al., 1992a). No clinical studies exist on the relationship between IGF-II expression and the endocrine responsiveness of breast cancer, although, in contrast to IGF-I, it has been reported that plasma levels of IGF-II remain unaltered after tamoxifen therapy (Reed et al., 1992).

FIBROBLAST GROWTH FACTOR (FGF) FAMILY

To date, nine members of the FGF family [including both acidic (a) and basic (b) forms] have been detected, sharing an overall structural homology of 40–50% (Biard and Klagsbrun, 1991; Tanaka et al., 1992; Miyamoto et al., 1993). Four homologous transmembrane tyrosine kinase receptors for FGFs have also been identified (Lehtola et al., 1992). Considerable variability in the relative affinities of these fibroblast growth factor receptors (FGFR) for the individual ligands has been observed (Dionne et al., 1990; Keegan et al., 1991; Partanen et al., 1991) and heterodimeric forms of the receptors may exist (Bellot et al., 1991; Shi et al., 1993). FGFs are transforming in NIH-3T3 cells (Biard and Klagsbrun 1991), and have been implicated in both tumorigenesis and metastasis of experimental mammary tumors (Dixon et al., 1984; Kern et al., 1994).

Experimental Studies. FGFs, particularly basic FGF, are potent mitogens for several breast cancer cell lines (Peyrat et al., 1992), having the ability to drive growth-inhibited T47D cells back into the cell cycle by their induction of cyclin D1 during the G1 phase (Musgrove and Sutherland 1993; Sutherland et al., 1993). With the exception of FGF-5, Kern et al.

(1994) failed to find elevated expression of any of the FGF mRNAs in various breast cancer cell lines. However, all the cell lines expressed mRNAs for at least one FGFR. ER-positive MDA-MB-134 human breast cancer cells expressed very high levels of mRNA for FGFR-1 (with a parallel amplification of the FGFR-1 gene) and also elevated mRNA for FGFR-4 (McLeskey et al., 1994). Although the FGFRs do not appear to be estrogen regulated in either MCF-7 or ZR-75–1 cells (Lehtola et al., 1992), synergism exists between estradiol and FGFs on the growth of ZR-75–1 cells (Stewart et al., 1992). These effects exceed the activity of both TGF-α and IGF-I.

Transfection of MCF-7 cells with an expression vector for FGF-4 conferred an ability on these cells to form progressively growing, metastatic tumors in ovariectomized, athymic, nude mice without estrogen supplementation (McLeskey et al., 1993). In addition, these tumors were growth stimulated by tamoxifen treatment of the mice and growth inhibited by estradiol (McLeskey et al., 1993; Kern et al., 1994). These data suggest a possible role for FGFs in the progression of breast tumors to an estrogen-independent, antiestrogen-resistant, metastatic phenotype. Importantly, when a cloned population of c-*erbB*-2-overexpressing cells that were themselves incapable of forming progressive tumors were co-injected with the FGF-4-overproducing MCF-7 cells, significantly larger tumors were obtained without estrogen supplementation, compared to co-injection of the FGF-4 producing cells with either parental or IGF-II-overexpressing MCF-7 cells (Kern et al., 1994). These data raise the possibility of synergistic interactions between FGF and c-*erbB*-2 signal transduction pathways.

Clinical Studies. FGFs may also be important mitogens for breast tumors *in vivo* (McLeskey et al., 1993). Transcripts or immunoreactivity for various FGFs have been identified in clinical material (Theillet et al., 1993; Kern et al., 1994), albeit invariably at reduced levels compared with the normal breast myoepithelial cells (Luqmani et al., 1992; Ke et al., 1993). These include the mRNAs for FGF-1 and, to a lesser extent, FGF-2 and 7 (Gomm et al., 1991; Ding et al., 1992; Kern et al., 1994). Approximately 15% of human primary breast cancers appear to exhibit amplification of the 11q13 amplicon (Lafage et al., 1990), which contains the *int*-2/*fgf*-3 and HST1 proto-oncogenes, which code for members of the FGF family and have been implicated in the progression of breast cancer. Amplification and overexpression of the genes encoding FGFR-1/FLG, FGFR-2/BEK, and FGFR-4 have also been recorded in 10–12% of breast cancers, with both FLG and FGFR-4 gene amplification being prevalent in steroid hormone receptor positive-tumors. This correlates with lymph nodal involvement (Adnane et al., 1991; Jaakkola et al., 1993; Theillet et al., 1993), implying that mitogenic effects of FGFs may be of particular importance in such tumors.

TRANSFORMING GROWTH FACTOR BETA (TGF-β)

The 25-kDa TGF-β peptides belong to a family of pluripotent growth and differentiation factors (Gorsch et al., 1992; Dickens and Colletta 1993; Murray et al., 1993). Three members have been identified in mammals, TGF-β1 to 3 (McCune et al., 1992). They are normally secreted in a biologically inactive, latent form that can be activated by acid pH, urea, or proteases, including plasmin and cathepsin-D (Massague, 1987). Cross-linking studies of iodinated TGF-β1 to cell surface proteins has revealed three binding proteins or receptors, with the type I and type II transmembrane receptors probably mediating the cellular actions of TGF-β1, and the type III and type IV (endoglin) receptors primarily aiding the delivery of the ligand to these receptors (Cheifetz et al., 1992).

Experimental Studies. TGF-βs may play a crucial role in tumor biology, having been implicated in tumor progression, invasion, angiogenesis, immune function, and extracellular matrix formation (Gorsch et al., 1992; Dickens and Colletta 1993; Walker et al., 1994). Although TGF-β proteins are pleiotropic in their actions, TGF-β1 and TGF-β2 are equipotent growth inhibitors of ER-positive (Manni et al., 1991) and -negative (Arteaga et al., 1988) human breast cancer cell lines *in vitro*. Low levels of TGF-β have been detected in conditioned medium from human breast cancer cells (HS578T and MDA-231) and this medium inhibits DNA synthesis and cell proliferation when added to subconfluent cells, the inhibition being reversed by anti-TGF-β antibodies (Arteaga et al., 1990). Interestingly, in this study the addition of anti-TGF-β antibodies to quiescent breast cancer cells induced a dose-dependent increase in DNA synthesis and stimulated cell proliferation in serum-free medium and anchorage-independent growth. In MCF-7 cells, the secretion of TGF-β proteins is induced by growth-inhibitory doses of antiestrogens and is inhibited by estrogen *in vitro* (Knabbe et al., 1987, 1991) and *in vivo* (Thompson et al., 1991); these results have led to the concept that TGF-βs have an autocrine inhibitory function. This, however, is controversial (Arteaga et al., 1988; Ji et al., 1994). Indeed, in contrast, stimulatory effects of TGF-β1 have been reported in estrogen-dependent MCF-7 cells *in vitro* (Croxtall et al., 1992), albeit under serum-free conditions.

Importantly, the induction of TGF-β by antiestrogens is not found in the antiestrogen-resistant MCF-7 variant LY-2, even though this cell line can be inhibited by the administration of TGF-β (Knabbe et al., 1987). A lack of induction of TGF-β by antiestrogen was originally suggested to contribute to the antiestrogen-resistant phenotype. However, cell hybrids between MCF-7 and LY-2 cells, which regain antiestrogen sensitivity, do not manifest a corresponding gain in TGF-β induction by antiestrogens (Park et al., 1994). Indeed, a failure to induce TGF-β is not necessarily a general condition for endocrine resistance, because several models of estrogen-resistance produce elevated levels of TGF-β proteins. These include v-H-*ras* transfected MCF-7 cells, where increased expression of TGF-β mRNA (together with increased TGF-α and IGF-I) was observed (Ciardiello et al., 1991), and also estrogen growth-independent T47D cells, where the transition toward loss of estrogen sensitivity is accompanied by an up-regulation of TGF-β mRNA and a loss of growth response to insulin and basic FGF (Daly et al., 1990). Autocrine inhibitory functions are presumably lost in these models, giving rise to TGF-β-induced growth (King et al., 1989; Daly et al., 1990). Additional support for such a concept has arisen from studies involving the transfection of estrogen growth-dependent MCF-7 cells with TGF-β1 cDNA (Arteaga et al., 1993a,b). This resulted in the cells gaining an ability to form tumors in ovariectomized athymic mice in the absence of estrogen supplementation, and this was abrogated by the neutralizing anti-TGF-β 2G7 IgG2B antibody (Arteaga et al., 1993a,b). Similarly, administration of TGF-β1 transiently supported estrogen-independent growth of the parent MCF-7 cells in castrated nude mice (Arteaga et al., 1993a,b). Such cells normally depend on an exogenous source of estrogen for their growth. This indicates that the overexpression of TGF-β by breast cancer cells or their overexposure to the protein can contribute to their escape from steroid hormone dependency.

Clinical Studies. In clinical breast cancer, TGF-β proteins (Walker and Dearing 1992) or mRNAs (MacCullum et al., 1994) are present in most samples examined, usually at higher levels than observed in the normal breast (Travers et al., 1988). These levels, however, are highly variable, and patterns of expression of TGF-β1 to 3 also vary (MacCullum et al., 1994). Several studies have indicated a positive relationship between TGF-β1 and both disease progression (Gorsch et al., 1992) and lymph node metastasis (Walker and Dearing 1991), with TGF-β-1 localizing to the advancing epithelial edge of primary tumors and lymph node metastases (Dalal et al., 1983). Similarly, the detection of all three isoforms of TGF-β mRNA in breast cancer specimens is associated with lymph node involvement (MacCullum et al., 1994).

Although the relationship between TGF-β and the endocrine sensitivity of breast cancer has not been studied in great depth in clinical breast cancer, a pilot study has been performed in 11 patients who had received tamoxifen for 3 to 6 months before surgery (Thompson et al., 1981). Unexpectedly high levels of TGF-β1 mRNA were found in patients whose tumors increased in size and were unresponsive to the antiestrogen. The authors speculated that the progression during tamoxifen therapy may have been due to a failure of the autocrine inhibitory functions of TGF-β1 alone or in combination with a paracrine stimulation of stromal cells or angiogenesis. Certainly tamoxifen has been reported to consistently induce stromal TGF-β in breast cancer biopsies, compared with matched pretreatment samples from the same patient (Mirza 1991; Butta et al., 1992). The induced TGF-β was localized between and around stromal fibroblasts and appeared to be derived from these cells. Lower levels of TGF-β1 to 3 were seen within tumor epithelial cells, and these were not influenced by tamoxifen.

ORIGINS OF GROWTH-FACTOR MEDIATED SUBVERSION OF ENDOCRINE RESPONSIVENESS

It is evident from the above that considerable data exist linking growth factor signaling to the normal cellular actions of estrogen, and that alterations in growth factor signaling may play an important role in the loss of endocrine sensitivity. These data are summarized in Table 1. Before considering how this information may be clinically exploited, however, it is necessary to understand the processes that lead to altered growth factor signaling and how they are coupled to the clinical course of the disease. Three distinct pathways may be envisaged, and these are outlined in Figure 1.

Cellular Origin of Cancer Cells

A proportion of endocrine-insensitive tumors may arise from normal breast cell types, which themselves show a reduced sensitivity to alterations in their steroid hormonal environment. Since it is likely that normal breast development may involve the transition of undifferentiated stem cells to fully differentiated, hormone-sensitive phenotypes, initiating events along this pathway could potentially promote cancers with a spectrum of steroid hormone and growth factor requirements. Studies examining the rat breast have identified that the highly proliferative stem cell population, which gives rise to both the epithelial and myoepithelial cell lineages, fails to express detectable quantities of steroid hormone receptors (Daniel et al., 1987), but is highly responsive to growth factors (Coleman et al., 1988). In contrast, significant quantities of steroid receptors are present in the more differentiated epithelial population (Daniel et al., 1987), and these cells may be growth inhibited by antihormones. Such cell types have their equivalent in human breast, where ER-negative cell populations expressing growth factor receptors have been identified (Walker et al., 1991). Initiating events in these cells might favor the selective outgrowth of an ER-negative/EGF-R-positive phenotype and may be significant in the establishment of *de novo* endocrine insensitivity in breast cancer. It is not envisaged that the development of such tumors could be delayed or halted by antihormonal measures, because they lack the receptor machinery necessary to respond.

Genetic Alterations

Breast cancer cells, in common with other tumor types, are subject to genetic alterations that can activate/suppress components of the pathways that

Table 1

Endocrine sensitivity: experimental

1. Estrogen-induced growth factors can act as mitogens for breast cancer cells (TGF-α, IGF-II). Receptors are present for these and other mitogenic growth factors
2. Estrogen suppression of growth-inhibitory growth factors (TGF-β) can contribute to estrogen-induced growth responses
3. Estrogen-induced growth of breast cancer cells can be amplified by growth factors (predominantly IGF-I and FGFs)
4. Cross-talk between steroid hormones and growth factors can aid mitogenic signaling in breast cancer cells (TGF-α, IGF-I)
5. Antiestrogens can prevent many aspects of the above, contributing to their growth-inhibitory properties

Endocrine insensitivity and resistance: experimental

1. Estrogen growth-independence and tamoxifen resistance can occur by either increasing positive growth factor production or signaling (TGF-α, EGF-R/c-*erbB*-2, IGFs, and FGFs), through a decreased production of a negative growth regulator (TGF-β), or through a change in a cellular growth response pathway (i.e., conversion of negative growth regulation to a positive signal; TGF-β)
2. Changes that increase the cellular expression or signaling of several growth factors/receptors appear more effective than single changes
3. Alterations in growth factor signaling may decrease the sensitivity of breast cancer cells to the mitogenic effects of estrogen, or reduce their sensitivity to the growth-inhibitory properties of antihormones
4. Growth factors can increase the sensitivity of cells to ER-mediated signaling and thereby allow them to survive and proliferate in a low estrogenic environment or hijack the estrogen-like properties of antiestrogens
5. Loss of growth responses to estrogen and antiestrogens does not necessarily result from a loss of the ER or a loss of all ER-directed cellular functions
6. Changes in estrogen and antiestrogen sensitivity can occur independently of each other

Endocrine sensitivity/resistance: clinical

1. Several growth factors are available to tumor cells and are produced by autocrine, paracrine and endocrine mechanisms
2. Breast cancer cells express receptors for growth factors
3. Antiestrogens can reduce the tissue and circulating levels of mitogenic growth factors
4. Alterations in the cellular levels of growth factors and their receptors may be coupled to tumor progression during endocrine treatments

mediate the effects of growth factors on cell proliferation. The presence of mutations or overexpressed oncogenes may hence allow breast cancer cells to bypass estrogen-dependent proliferation and endocrine-mediated growth control.

To date, several activated oncogenes have been identified in breast cancer, together with the loss of two suppressor gene activities (Walker and Varley,

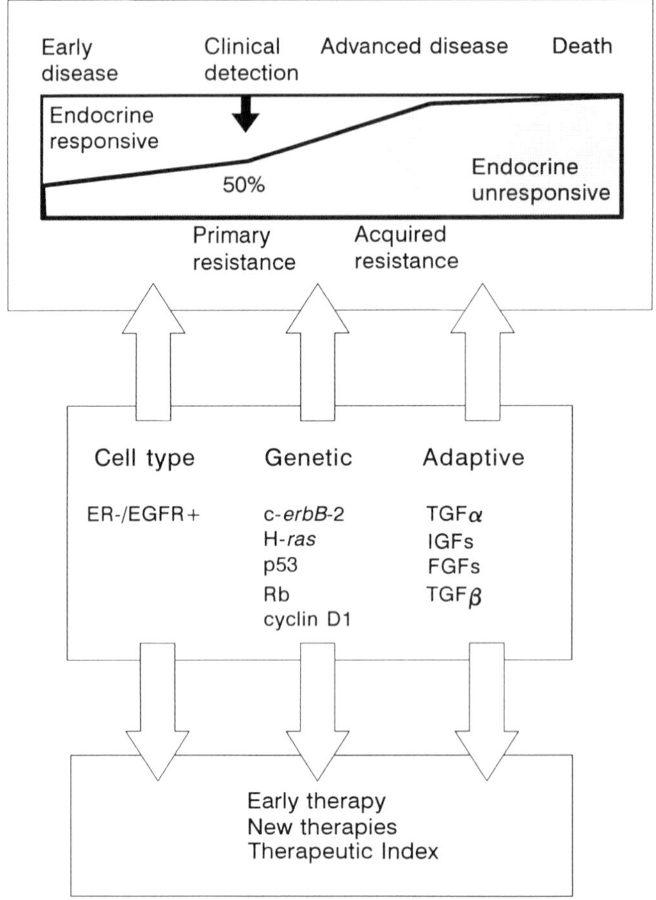

Figure 1. In this model, areas of overlap between the various pathways are not shown. It is noteworthy, however, that a number of the factors associated with primary resistance to endocrine measures (e.g. elevated growth rates and TGF-α expression) also characterize acquired resistance in tumor models. If these factors are causative in the loss of endocrine sensitivity, then primary and acquired resistance may occur on a similar developmental pathway. Similarly, although a clinical scale has been applied to the model (i.e., early disease to death), no time scale has been used. This would depend on the rate of genetic and phenotypic alterations and the aggressive nature of the disease.

1993). These include an amplification of the c-*erbB*-2 oncogene, which potentially directly alters growth factor signaling and steroid hormone sensitivity, and elevated expression of the H-*ras* oncogene (Walker and Varley, 1993), which in culture can increase the cellular output of several autocrine growth factors (Ciardiello et al., 1991). The expression of an altered form of the p53 protein (the defective product of the mutated p53 tumor-suppressor gene) has been associated both with EGF-R expression and elevated cell proliferation in clinical breast cancer specimens (Poller et al., 1992). Retinoblastoma (Rb) gene inactivation is also found in 20–50% of breast cancers (Fung and T'Ang, 1992); and loss of this second tumor-suppressor gene has been linked to an increased cellular importance of the c-*erbB* protein (Yu et al., 1992; Martin and Hung, 1994). Interestingly, cyclin D1, a cell cycle regulator essential for G1 phase progression and a candidate oncogene (Sutherland et al., 1993; Musgrove et al., 1994), is amplified in approximately 20–50% of human breast cancers (Buckley et al., 1993) and is frequently overexpressed as part of the q13 region of chromosome 11 in these tumors (Lebwohl et al., 1994). This amplification has been associated with poor prognosis (Bartkova et al., 1994). A role for cyclin D1 in mediating the mitogenic effects of steroid hormones and growth factors has been proposed, with amplification of the cyclin D1 gene potentially overriding the cellular need for each of these classes of mitogenic agents (Buckley et al., 1993; Sutherland et al., 1993).

A number of the above genetic changes are evident in primary breast cancer and have, moreover, been reported in non-malignant *in situ* carcinomas (Walker and Varley, 1993). They are thus likely to contribute to *de novo* endocrine insensitivity in ER-positive disease, becoming progressively more important as multiple aberrations occur in the advancing disease.

Adaptive changes

The environmental pressures that are placed on the developing breast cancer cells, both through fluctuations in their hormonal milieu during the menstrual cycle and menopause, and through the use of antihormonal procedures and drugs, may promote adaptive changes leading to endocrine insensitivity and resistance. Such changes undoubtedly occur more frequently in the cancer cell than in the normal breast as a consequence of existing genetic aberrations. However, the speed of emergence of endocrine-insensitive and -resistant states suggests that they are unlikely to be due to selection of a chance mutation, but are more likely to represent a high rate of phenotypic modulation (Daly and Darbre, 1990). In at least one instance, such alterations have been shown to be recessive, with hybrids between MCF-7 cells and an LY-2 antiestrogen-resistant variant regaining antiestrogen-induced growth inhibition (Park et al., 1994).

Potential sites for adaptive changes include alterations in "cross-talk" between steroid hormone and growth factor signaling pathways. Adaptive changes in the contribution made by growth factors to such pathways might reduce the cellular threshold necessary for estrogen-directed cell growth, allowing cells to survive and ultimately proliferate in the presence of antihormonal drugs. In this light, it is noteworthy that in a number of models of estrogen growth-independence and tamoxifen resistance, the ER is retained. Moreover, it is of central importance to the growth of such cells, because they may subsequently be growth arrested by pure antiestrogens that bind to the ER (Nicholson et al., 1994). At first sight, although these results appear to represent a paradox (Nicholson et al., 1995a,b), the cellular actions of pure antiestrogens include an ability to reduce the cellular concentrations of the ER (Nicholson et al., 1994b) and interfere with growth factor signaling through AP-1 sites (Philips et al., 1993). Such actions would block both growth factor signaling through the ER and the ability of the ER to aid growth factor-driven transcriptional events. Both IGF-I and TGF-α have been shown to cross-talk with ER signaling pathways and to involve the activation of intracellular protein kinases. For TGF-α, any contribution made to the development of estrogen-growth independence in K3/MCF-7 cells appears to occur distal to alterations in TGF-α-directed tyrosine kinase activity, because an EGF-R specific tyrosine kinase inhibitor appears less active in these cells versus the parent cell line (Nicholson et al., 1995a,b).

Adaptive changes undoubtedly contribute to *de novo* endocrine insensitivity in ER-positive disease, with the selective influence of antihormonal procedures being a major stimulus to their development.

FUTURE THERAPEUTIC INITIATIVES

An essential feature of the cancer phenotype is that growth is less tightly controlled than in normal cells. This is a progressive event, with those aberrations that favor continued growth being selected over those that do not. Thus, although considerable evidence exists to implicate steroid hormones as major factors both in the control of cell proliferation in the normal breast and in providing a permissive environment for initiating events, such initial hormone sensitivity is invariably lost as the breast cancer develops and progresses, and endocrine-based therapies eventually become ineffective. From the data presented in this chapter, it is apparent that growth factor signaling contributes to this progression and represents an essential feature dictating poor prognosis in breast cancer patients.

Based on the progressive and cumulative nature of these events, several important concepts for the future management of breast cancer can be established:

First, the therapy of breast cancer patients with antihormonal measures should be initiated as early as is feasible in the life history of the disease in order to act on tumor cells, or their premalignant counterparts, before the occurrence of multiple alterations in the growth factor signaling pathways. Such an approach should maximize the therapeutic benefit gained from any endocrine therapies. Clear success in this area has recently been demonstrated in a series of trials employing tamoxifen as an adjuvant to mastectomy in primary breast cancer patients, where the drug has improved 10-year survival rates by approximately 20% over treatment of women with more advanced forms of the disease (Early Breast Cancer Trialists'

Collaborative Group, 1992). A similar result has also been observed in premenopausal women with primary breast cancer following the removal of their ovaries or after ovarian irradiation (Early Breast Cancer Trialists' Collaborative Group, 1992). Since such trials have also demonstrated that early therapy can prevent the development of contralateral breast cancer, it is envisaged that the extension of this approach to women who are at high risk of developing breast cancer will aim antihormonal measures at a more susceptible population of premalignant cells.

Second, the targeting of aberrant growth factor signaling by approaches that encompass a range of developing technologies may provide a means of preventing tumor progression and resultant acquired autonomous growth. These techniques may potentially include gene therapy directed against those oncogenes that directly or indirectly modify growth factor signaling; the synthesis of competitive growth factor analogues; the production of growth factor pathway-directed antibodies, and signal transduction modifying drugs (Arteaga, 1992). Indeed, it may even be envisaged that such treatments in tumors that have gained endocrine resistance might restore hormone sensitivity. Although data are sparse on the effectiveness of such measures, exposure of cells overexpressing c-*erbB*-2 to high concentrations of gp30 (a ligand for the c-*erbB*-2 protein) can growth inhibit cells, which then express a more differentiated phenotype (Staebler et al., 1994). It is noteworthy that more than one growth factor pathway may be involved in tumor progression, and that simultaneous targeting may be necessary to maximize tumor inhibition (Arteaga, 1992).

Third, antihormonal drugs that inhibit interactions between steroid hormone and growth factor signaling pathways should be more effective than those that fail to do so. Of particular interest are the pure antiestrogens, which are more effective than either tamoxifen or estrogen withdrawal at decreasing cross-talk between steroid and growth factor pathways, and which are more effective at inhibiting the growth of ER-positive breast cancer cells *in vitro* and *in vivo* (Nicholson et al., 1994). It is also significant that pure antiestrogens are more effective than tamoxifen or estrogen withdrawal in delaying the development of antiestrogen resistance in ER-positive cells (Osborne et al., 1995) and have been successfully used in breast cancer patients who have previously developed an insensitivity to tamoxifen therapy (Howell et al., 1995). A natural progression from the use of antihormones as single agents would be to combine them with novel agents directed against specific growth factor signaling pathways, including tyrosine kinase inhibitors to further inhibit cross-talk. Initial studies with an EGF-R-specific tyrosine kinase inhibitor have shown it to be as effective as pure antiestrogens in reducing the growth of MCF-7 cells (Nicholson et al., 1994b). No published data exist on the combination of these treatments.

Finally, it is evident that a knowledge of the active components of growth factor signaling pathways in breast cancer specimens may help us to stratify patients for the above therapeutic measures. In this light, we are currently beginning to use growth factor/receptor measurements to predict more accurately which patients should/should not receive endocrine therapy (Nicholson et al., 1992). Selection of candidates suitable for anti-growth factor regimes is logically the next phase.

ACKNOWLEDGMENTS

The authors wish to thank the Tenovus Organisation and the Association for International Cancer Research for their generous financial support.

REFERENCES

Adnane J, Gaudray P, Dionne CA, Crumley G, Jaye M, Schlessinger J, Jeanteur P, Birnbaum D, Theillet C (1991): BEK and FLG, two receptors for members of the FGF family, are amplified in subsets of human breast cancers. *Oncogene* 6:659–663

Akiyama T, Sudo C, Ogawara H, Toyoshima K, Yamamoto T (1986): The product of the c-erbB-2 gene: a 185KD glycoprotein with tyrosine kinase activity. *Science* 232:1644–1646

Arteaga CL, Coffey RJ, Dugger TC, McCutchen CM, Moses HL, Lyons RM (1990): Growth stimulation of human breast cancer cells with anti-transforming growth factor beta antibodies: evidence for negative autocrine regulation by transforming growth factor beta. *Cell Growth Differ* 1:367–374

Arteaga CL, Carty-Dugger T, Moses H, L, Hurd SD, Pietenpol JA (1993a): Transforming growth factor beta-1 can induce oestrogen-independent tumourigenesis in human breast cancer cells in athymic mice. *Cell Growth Differ* 4:193–201

Arteaga CL, Carty-Dugger TC, Winnier AR, Forbes JT (1993b): Evidence for a positive role of transforming growth factor-beta in human breast cancer cell tumourigenesis *J Cell Biochem* 17G:187–193

Arteaga, CL (1992): Interference of the IGF system as a strategy to inhibit breast cancer growth. *Breast Cancer Res Treat* 22:101–106

Arteaga CL, Tandon AK, Von Hoff DD, Osborne CK (1988): Transforming growth factor beta: potential autocrine growth inhibitor of oestrogen receptor-negative human breast cancer cells. *Cancer Res* 48:3898–3904

Bargmann C, Hung MC, Weinberg RA (1986): Multiple independent activations of the neu oncogene by a point mutation altering the transmembrane domain of p185. *Cell* 45:649–657

Barnes DM, Lammie GA, Millis RR, Gullick W L, Allen DS, Altman DG (1988): An immunohistochemical evaluation of c-erbB-2 expression in human breast carcinoma. *Br J Cancer* 58:448–452

Bartkova J, Lukas J, Muller H, Lutzhoft D, Strauss M, Bartek J (1994): Cyclin D1 protein expression and function in human breast cancer. *Int J Cancer* 57:353–361

Bates SE, Davidson NE, Valverius EM, Freter CE, Dickson RB, Tam JP, Kudlow JE, Lippman ME, Salomon DS (1988): Expression of transforming growth factor alpha and its messenger ribonucleic acid in human breast cancer: its regulation by oestrogen and its possible functional significance. *Mol Endocrinol* 2:543–555

Bellot F, Crumley G, Kaplow JM, Schlessinger J, Jaye M, Dionne CA (1991): Ligand-induced transphosphorylation between different FGF receptors. *EMBO J* 10:2849–2854

Benz CC, Scott GK, Sarup JC, Johnson RM, Tripathy D, Coronado E, Shepard HM, Osborne CK (1993): Oestrogen-dependent, tamoxifen-resistant tumourigenic growth of MCF-7 cells transfected with HER2/neu. *Breast Cancer Res Treat* 24:85–95

Berns EM, Klijn JG, van Staveren IL, Portengen H, Foekens JA (1992): Sporadic amplification of the insulin-like growth factor I receptor gene in human breast tumours. *Cancer Res* 52:1036–1039

Bertois Y, Dong XF, Martin PM (1989): Regulation of epidermal growth factor-receptor by oestrogen and antioestrogen in the human breast cancer cell-line MCF-7. *Biochem Biophys Res Commun* 159:126–131

Biard A, Klagsbrun M (1991): The fibroblast growth factor family Cancer. *Cells* 3:230–243

Bonneterre J, Peyrat FP, Beuscart R, Demaille A (1990): Prognostic significance of insulin-like growth factor I receptors in human breast cancer. *Cancer Res* 50:6931–6935

Brunner N, Zugmaier G, Bano M, Ennis BW, Clarke R, Cullen KJ, Kern G, Dickson RB, Lippman ME (1989): Endocrine therapy of human breast cancer cells: the role of secreted polypeptide growth factors Cancer. *Cells* 1:81–87

Brunner N, Moser C, Clarke R, Cullen K (1992): IGF-I and IGF-II expression in human breast cancer xenografts: relationship to hormone independence. *Breast Cancer Res Treat* 22:39–45

Brunner N, Yee D, Kern FG, Spang-Thomsen, M Lippman ME, Cullen KJ (1993): Effect of endocrine therapy on growth of T61 human breast cancer xenografts is directly correlated to a specific down-regulation of insulin-like growth factor II (IGF-II). *Eur J Cancer* 29:562–569

Buckley MF, Sweeney KJ, Hamilton JA, Sini RL, Manning DL, Nicholson RI, deFazio A, Watts, C K Musgrove EA, Sutherland, R L(1993): Expression and amplification of cyclin genes in human breast cancer. *Oncogene* 8:2127–2133

Butta A, MacLennan K, Flanders KC, Sacks NP, Smith I, McKinna A, Dowsett M, Wakefield LM, Sporn MB, Baum M (1992): Induction of transforming growth factor beta 1 in human breast cancer *in vivo* following tamoxifen treatment. *Cancer Res* 52:4261–4264

Carraway KL, Cantley LC (1994): A new acquaintance for erbB-3 and erbB-4: a role for receptor heterodimerisation in growth signaling. *Cell* 78:5–8

Cavailles V, Garcia M, Rochefort H (1989): Regulation of cathepsin-D and pS2 gene expression by growth factors in MCF-7 human breast cancer cells. *Mol Endocrinol* 3:552–558

Chalbos D, Philips A, Galtier F, Rochefort H (1993): Synthetic antioestrogens modulate induction of pS2 and cathepsin D messenger ribonucleic acid by growth factors and adenosine 3′,5′-monophosphate in MCF-7 cells. *Endocrinolology* 133:571–576

Cheifetz S, Bellon T, Cales C, Vera S, Bernabeu C, Massague J, Letarte M (1992): Endoglin is a component of the transforming growth factor-beta receptor system in human endothelial cells. *J Biol Chem* 267:19027–19030

Cho H, Aronica SM, Katzenellenbogen BS (1994): Regulation of progesterone receptor gene espression in MCF-7 breast cancer cells: a comparison of the effects of cyclic adenosine $3', 5'$-monophosphate, estradiol, insulin-like growth factor-I and serum factors. *Endocrinology* 134:658–664

Chrysogelos SA, Yarden R I, Lauber AH, Murphy JM (1994): Mechanisms of EGF receptor regulation in breast cancer cells. *Breast Cancer Res Treat* 31:227–236

Ciardiello F, Kim N, McGeady ML, Liscia DS, Saeki T, Bianco C, Salomon DS (1991): Expression of transforming growth factor alpha (TGF alpha) in breast cancer. *Ann Oncol* 2:169–182

Clarke R, Brunner N, Katz D, Glanz P, Dickson RB, Lippman ME, Kern FG (1989a): The effects of a constitutive expression of transforming growth factor alpha on the growth of MCF-7 human breast cancer cells *in vitro* and *in vivo*. *Mol Endocrinol* 3:372–380

Clarke R, Brunner N, Katzenellenbogen BS, Thompson EW, Norman MJ, Koppi C, Paik S, Lippman ME, Dickson RB (1989b): Progression of human breast cancer cells from hormone-dependent to hormone independent growth both *in vitro* and *in vivo*. *Proc Natl Acad Sci USA* 86:3649–3653

Clarke R, Dickson RB, Lippman ME (1992): Hormonal aspects of breast cancer: growth factors, drugs and stromal interactions. *Crit Rev Oncol-Hematol* 12:1–23

Coleman S, Silberstein GB, Daniel CW (1988): Ductal morphogenesis in the mouse mammary gland: evidence supporting a role for epidermal growth factor. *Dev Biol* 127:304–315

Costa A (1993): Breast cancer chemoprevention. *Eur J Cancer* 29A:589–592

Croxtall, JD, Jamil A, Ayub M, Colletta AA, White JO (1992): TGF-beta stimulation of endometrial and breast-cancer cell growth. *Int J Cancer* 50:822–827

Cullen KJ, Yee D, Sly WS, Perdue J, Hampton B, Lippman ME, Rosen N (1990): Insulin-like growth factor receptor expression and function in human breast cancer. *Cancer Res* 50:48–53

Cullen KJ, Allison A, Martire I, Ellis M, Singer C (1992a): Insulin-like growth factor expression in breast cancer epithelium and stroma. *Breast Cancer Res Treat* 22:21–29

Cullen KJ, Lipmann ME, Chow D, Hill S, Rosen N, Zweibel JA (1992b): Insulin-like growth factor-II overexpression in MCF-7 cells induces phenotypic changes associated with malignant progression. *Mol Endocrinol* 6:91–100

Dalal BI, Keown PA, Greenberg AH (1993): Immunocytochemical localisation of secreted transforming growth factor-beta to the advancing edges of primary tumours and to lymph node metastases of human mammary carcinoma. *Am J Pathol* 143:381–389

Daly RJ, Darbre PD (1990). Cellular and molecular events in loss of oestrogen sensitivity in ZR-75–1 and T-47-D human breast cancer cells. *Cancer Res* 50:5868–5875

Daly RJ, King RJ, Darbre PD (1990): Interaction of growth factors during progression towards steroid independence in T-47-D human breast cancer cells. *J Cell. Biochem* 43:199–211

Daly RJ, Harris WH, Wang EDY, Darbre PD (1991): Autocrine production of insulin-like growth factor II using an inducible expression system results in reduced oestrogen sensitivity of MCF-7 human breast cancer cells. *Cell Growth Differ* 2:457–464

Daniel CW, Silberstein GB, Strickland P (1987): Direct action of 17beta-estradiol on mouse mammary ducts analyzed by sustained release implants and steroid autoradiography. *Cancer Res* 47:6052–6057

Darbre PD, Daly RJ (1990): Transition of human breast cancer cells from an oestrogen responsive to unresponsive state. *J Steroid. Biochem Mol Biol* 37:753–763

Darbre PD, King RJ (1988): Steroid hormone regulation of cultured breast cancer cells. *Cancer Treat Res* 40:307–341

Davidson NE, Gelmann EP, Lippman ME, Dickson RB (1987): Epidermal growth factor receptor gene expression in oestrogen receptor-positive and negative human breast cancer cell lines. *Mol Endocrinol* 1:216–223

Dickens TA, Colletta AA (1993): The pharmacological manipulation of members of the transforming growth factor beta family in the chemoprevention of breast cancer. *Bioess* 15:71–74

Dickson C, Smith R, Brookes S, Peters G (1984): Tumourigenesis by mouse mammary tumour virus: proviral activation of a cellular gene in the common integration region int-2. *Cell* 37:529–536

Dickson RB (1990): Stimulatory and inhibitory growth factors and breast cancer. *J Steroid Biochem Mol Biol* 37:795–803

Dickson RB, Bates SE, McManaway M, Lippman ME (1986): Characterisation of oestrogen response transforming activity in human breast cancer cell lines. *Cancer Res* 46:1707–1713

Ding IYF, McLeskey SW, Chang K, Fu YM, Acol JC, Shon MT, Alitalo, K Kern FG (1992): Expression of fibroblast growth factors (FGFs) and receptors in human breast carcinomas. *Proc Am Assoc. Cancer Res* 33:269

Dionne CA, Crumley G, Bellot F, Kaplow JM, Searfoss G, Ruta M, Burgess WH, Jaye M, Schlessinger J (1990): Cloning and expression of two distinct high-affinity receptors cross-reacting with acidic and basic fibroblast growth factors. *EMBO J* 9:2658–2692

Early Breast Cancer Trialists' Collaborative Group (1992): Systemic treatment of early breast cancer by hormonal, cytotoxic or immune therapy. *Lancet* 339:1–15

Figueroa JA, Yee D (1992): The insulin-like growth factor binding proteins (IGFBPs) in human breast cancer. *Breast Cancer Res Treat* 22:81–90

Fitzpatrick SL, LaChance MP, Schultz GS (1984): Characterization of epidermal growth factor receptor and action on human breast cancer cells in culture. *Cancer Res* 44:3442–3447

Foekens JA, Portengen H, Janssen M, Klijn JG (1989): Insulin-like growth factor-I receptors and insulin-like growth factor-I-like activity in human primary breast cancer. *Cancer* 63:2139–2141

Freiss G, Prebois C, Rochefort H, Vignon F (1990): Anti-steroidal and anti-growth factor activities of anti-oestrogens. *J Steroid Biochem Mol Biol* 37:777–781

Friedl A, Jordan VC, Pollak M (1993): Suppression of serum insulin-like growth factor-1 levels in breast cancer patients during adjuvant tamoxifen therapy. *Eur J Cancer* 29:1368–1372

Fung YK, T'Ang A (1992): The role of the retinoblastoma gene in breast cancer development. *Cancer Treat Res* 61:59–68

Gee JMW, Ellis IO, Robertson JFR, Willsher P, McClelland RA, Hewitt KN, Blamey RW, Nicholson RI (1995): Immunocytochemical localization of Fos protein in human breast cancers and its relationship to a series of prognostic markers and response to endocrine therapy. *Int J Cancer (Pred Oncol)* 64:269–273

Gomm JJ, Smith J, Ryall GK, Baillie R, Turnbull L, Coombes RC (1991): Localisation of fibroblast growth factor and transforming growth factor β1 in human mammary gland. *Cancer Res* 51:4685–4692

Gorsch SM, Memoli, VA, Stukel TA, Gold LI, Arrick BA (1992): Immunohistochemical staining for transforming growth factor beta 1 associates with disease progression in human breast cancer. *Cancer Res* 52:6949–6952

Gregory H, Thomas CE, Willshire IR, Yorg JA, Anderson H, Baildam A, Howell A (1989): Epidermal and transforming growth factor alpha in patients with breast tumours. *Br J Cancer* 59:605–609

Gullick WL (1991): Prevalence of aberrant expression of epidermal growth factor receptor in human cancers. *Br Med Bull* 47:87–89

Hayman MJ, Enrietto PJ (1991): Cell transformation by the epidermal growth factor receptor and v-erbB. *Cancer Cells* 3:302–307

Holmes WE, Sliwkowski MX, Akita RW, et al (1992): Identification of Heregulin, a specific activator of p185erbB-2. *Science* 256:1205–1210

Howell A, DeFriend D, Robertson J, Blamey R, Walton P (1995): Response to a specific antioestrogen (ICI 182780) in tamoxifen-resistant breast cancer. *Lancet* 345:29–30

Howell A, Dodwell DJ, Anderson E, Redford J (1992): Response after withdrawal of tamoxifen and progestogen in advanced breast cancer. *Ann Oncol* 3:611–617

Ignar-Trowbridge D, M, Nelson KG, Bidwell MC, Curtis SW, Washburn TF, McLachlan JA, Korach KS (1992): Coupling of dual signaling pathways: epidermal growth factor action involves the oestrogen receptor. *Proc Natl Acad Sci USA* 89:4658–4662

Jaakkola S, Salmikangas P, Nylund S, Partanen J, Armstrong E, Pyrhonen S, Lehtovirta P, Nevanlinna H (1993): Amplification of FGFR4 gene in human breast and gynaecological cancers. *Int J Cancer* 54:378–382

Jammes H, Peyrat JP, Ban E, Vilain MD, Haor F, Djiane J, Bonneterre J (1992): Insulin-like growth factor I receptors in human breast tumour: localisation and quantification by histo-autoradiographic analysis. *Br J Cancer* 66:248–253

Ji H, Stout LE, Zhang Q, Leung HT, Leung BS (1994): Absence of transforming growth factor-beta responsiveness in the tamoxifen growth-inhibited human breast cancer cell line CAMA-1. *J Cell Biochem* 54:332–342

Kadowaki T, Kasuga M, Tobe K et al. (1987): A Mr = 190 000 glycoprotein phosphorylated on tyrosine residues in epidermal growth factor stimulated cells is the product of the c-erbB-2 gene. *Biochem Biophys Res Commun* 144:699–704

Katzenellenbogen BS, Kendra KL, Norman MJ, Berthois Y (1987): Proliferation, hormone responsiveness and oestrogen receptor content of MCF-7 human breast cancer cells grown in short-term and long-term absence of oestrogens. *Cancer Res* 47:4355–4360

Katzenellenbogen K, Norman MJ (1990): Multihormonal regulation of the progesterone receptor in MCF-7 human breast cancer cells: interrelationships among insulin/insulin-like growth factor-I, serum and oestrogen. *Endocrinology* 126:891–898

Ke Y, Fernig DG, Wilkinson MC, Winstanley JH, Smith JA, Rudland PS, Barraclough R (1993): The expression of basic fibroblast growth factor and its receptor in cell lines derived from normal human mammary gland and a benign mammary lesion. *J Cell Sci* 106:135–143

Keegan K, Johnson DE, Williams LT, Hayman MJ (1991): Isolation of an additional member of the fibroblast growth factor receptor family, FGFR-3. *Proc Natl Acad Sci USA* 88:1095–1099

Kern F, Cheville AL, Liu Y (1990): Growth factor receptors and the progression of breast cancer. *Sem Cancer Biol* 1:317–328

Kern FG, McLeskey SW, Zhang L, Kurebayashi J, Lui Y, Ding IYF, Kharbanda D, Miller D, Cullen K, Paik S, Dickson RB (1994): Transfected MCF-7 cells as a model for breast cancer progression. *Breast Cancer Res Treat* 31:153–165

Kienhuis CB, Sluyer M, Goeij CC, Koenders PG, Benraad TJ (1993): Epidermal growth factor receptor levels increase but epidermal growth factor ligand levels decrease in mouse mammary tumours during the progression from hormone dependence to hormone independence. *Breast Cancer Res Treat* 26:289–295

King CR, Borrello I, Bellot F, Comoglio P, Schlessinger J (1988): EGF binding to its receptor triggers a rapid tyrosine phosphorylation of erbB-2 protein in the mammary cell line SK-BR-3. *EMBO J* 7:1645–1651

King RJ, Wang DY, Daly RJ, Darbre PD (1989): Approaches to studying the role of growth factors in the progression of breast tumours from the steroid sensitive to insensitive state. *J Steroid Biochem* 34:133–138

King RJB (1990). Receptors, growth factors and steroid insensitivity of tumours. *J Endocrinol* 124:179–181

Klijn JGM, Berns PM, Schmitz PI, Foekens JA (1992): The clinical significance of epidermal growth factor receptor in human breast cancer: a review on 5232 patients. *Endocr Rev* 13:3–17

Knabbe C, Lippman ME, Wakefield LM, Flanders KC, Kasid A, Derynck R, Dickson RB (1987): Evidence that transforming growth factor-beta is a hormonally regulated negative growth factor in human breast cancer cells. *Cell* 48:417–428

Knabbe C, Zugmaier G, Schmahl M, Dietel M, Lippman ME, Dickson RB (1991): Induction of transforming growth factor beta by the antioestrogens droloxifene, tamoxifen and toremifene in MCF-7 cells. *Am J Clin Oncol* 14:S15–20

Kokai Y, Dobashi K, Weiner DB, Myers JN, Nowell PC, Greene MI (1988): Phosphorylation process induced by epidermal growth factor alters the oncogenic and cellular neu (NGL) gene products. *Proc Natl Acad Sci USA* 85:5389–5393

Kokai Y, Myers JN, Wada T et al. (1989): Synergistic interactions of p185 c-neu and the EGF receptor leads to transformation of rodent fibroblasts. *Cell* 58:287–292

Krywicki, R F and Yee D (1992): The insulin-like growth factor family of ligands, receptors and binding proteins. *Breast Cancer Res Treat* 22:7–19

Kudlow JE, Cheng CYM, Bjorge JD (1985): Epidermal growth factor stimulates the synthesis of its own receptor in a human breast cancer cell line. *J Biol Chem* 261:4134–4138

Lafage M, Nguyen C, Szepetowski P, Pebusque MJ, Simonetti J, Courtois G, Gaudray P, deLapeyriere O, Jordan B, Birnbaum D (1990): The 11q13 amplicon of a mammary carcinoma cell line. *Genes Chrom Cancer* 2:171–181

Lebwohl DE, Muise-Helmericks R, Sepp-Lorenzino L, Serve S, Timaul M, Bol R, Borgen P, Rosen N (1994): A truncated cyclin D1 gene encodes a stable mRNA in a human breast cancer cell line. *Oncogene* 9:1925–1929

Lehtola L, Partanen J, Sistonen L, Korhonen J, Warri A, Harkonen P, Clarke R, Alitalo K (1992): Analysis of tyrosine kinase mRNAs including four FGF receptor mRNAs expressed in MCF-7 cells. *Int J Cancer* 50:598–603

Lemoine NR, Barnes DM, Hollywood DP et al. (1992): Expression of erbB-3 gene product in breast cancer. *Br J Cancer* 66:1116–1121

Leung BS, Stout L, Zhou L, Ji HJ, Zang QQ, Leung HT (1991): Evidence of an EGF/TGFalpha-independent pathway for oestrogen-regulated cell proliferation. *J Cell Biochem* 46:125–133

Lewis S, Locker A, Todd JH, Bell JA, Nicholson RI, Elston CW, Blamey RW, Ellis IO (1990): Expression of epidermal growth factor receptor in breast cancer. *J Clin Pathol* 43:385–389

Long B, McKibben BM, Lynch M, Van den Berg HW (1992): Changes in epidermal growth factor receptor expression and response to ligand associated with acquired tamoxifen resistance or oestrogen independence in the ZR-75-1 human breast cancer cell line. *Br J Cancer* 65:865–869

Lovekin C, Ellis IO, Locker A, Robertson JFR, Bell J, Nicholson RI, Gullick WV, Elston CW, Blamey RW (1991): c-erb-B2 oncoprotein expression in primary and advanced breast cancer. *Br J Cancer* 63:439–443

Lundy J, Schuss A, Stanick D, McCormack ES, Kramer S, Sorvillo JM (1991): Expression of neu protein, epidermal growth factor receptor, and transforming growth factor alpha in breast cancer Correlation with clinicopathologic parameters. *Am J Pathol* 138:1527–1534

Luqmani YA, Graham M, Coombes RC (1992): Expression of basic fibroblast growth factor, FGFR1 and FGFR2 in normal and malignant human breast, and comparison with other normal tissues. *Br J Cancer* 66:273–280

MacCullum J, Bartlett JM, Thompson AM, Keen JC, Dixon JM, Miller WR (1994): Expression of transforming growth factor beta mRNA isoforms in human breast cancer. *Br J Cancer* 69:1006–1009

Malaab SA, Pollak MN, Goodyer CG (1992): Direct effects of tamoxifen on growth hormone secretion by pituitary cells *in vitro*. *Eur J Cancer* 28:788–793

Manni A, Wright C, Buck H (1991): Growth factor involvement in the multihormonal regulation of MCF-7 breast cancer cell growth in soft agar. *Breast Cancer Res Treat* 20:43–52

Massague J (1987): The TGF-beta family of growth and differentiation factors. *Cell* 49:437–438

Matin A, Hung MC (1994): The retinoblastoma gene product, Rb, represses neu expression through two regions within the neu regulatory sequence. *Oncogene* 9:1333–1339

Matsui Y, Halter SA, Holt JT, Hogan BLM, Coffey R (1990): Development of mammary hyperplasia and neoplasia in MMTV-TGFalpha transgenic mice. *Cell* 61:1147–1155

McClelland RA, Finlay P, Gee JMW, Manning DL, Hoyle H, Ellis IO, Blamey RW, Nicholson RI (1994a): Immunocytochemically-localized epidermal growth factor receptor and oestrogen receptor in breast cancer: relationship to endocrine therapy. *Oncol (Life Sci Adv)* 12:143–155

McClelland RA, Finlay P, Gee JMW, Manning DL, Hoyle H B, Ellis I O, Blamey RW, Nicholson RI (1994b): An immunocytochemical analysis of EGF, TGFalpha, amphiregulin and phosphotyrosine in human breast cancer. *Oncol (Life Sci Adv)* 13:91–105

McCune BK, Mullin BR, Flanders KC, Jaffurs WJ, Mullen LT, Sporn MB (1992): Localization of transforming growth factor-beta isotypes in lesions of the human breast. *Hum Pathol* 23:13–20

McLeskey SW, Kurebayashi J, Honig SF, Zwiebel J, Lippman ME, Dickson RB, Kern FG (1993): Fibroblast growth factor-4 transfection of MCF-7 cells produces cell lines that are tumourigenic and metastatic in ovariectomized or tamoxifen-treated athymic nude mice. *Cancer Res* 53:2168–2177

McLeskey SW, Ding TY, Lipmann ME, Kern FG (1994): MDA-MB-134 breast carcinoma cells overexpress fibroblast growth factor (FGF) receptors and are growth-inhibited by FGF ligands. *Cancer Res* 54:523–530

Mirza MR (1991): Anti-oestrogen induced synthesis of transforming growth factor-beta in breast cancer patients. *Cancer Treat Rev* 18:145–148

Miyamoto K, Naruo K, Seko C, Matsumoto K, Kondo T, Kurokawa T (1993): Molecular cloning of a novel cytokine cDNA encoding the ninth member of the fibroblast growth factor family, which has a unique secretion property. *Mol Cell Biol* 13:4251–4259

Murray PA, Barrett-Lee P, Travers M, Luqmani Y, Powles T, Coombes RC (1993): The prognostic significance of transforming growth factors in human breast cancer. *Br J Cancer* 67:1408–1412

Musgrove EA, Sutherland RL (1993): Acute effects of growth factors on T-47D breast cancer cell cycle progression. *Eur J Cancer* 29A:2273–2279

Musgrove EA, Lee CS, Buckley MF, Sutherland RL (1994): Cyclin D1 induction in breast cancer cells shortens G1 and is sufficient for cells arrested in G1 to complete the cell cycle. *Proc Natl Acad Sci USA* 91:8022–8026

Neal DE, Bennett MK, Hall RR, Marsh C, Abel PD, Sainsbury JRC, Harris AL (1985): Epidermal-growth-factor receptors in human bladder cancer: comparison of invasive and superficial tumours. *Lancet* 1:366–368

Nicholson RI (1992): Pure antioestrogens and breast cancer. In: *Current Directions in Cancer Chemotherapy*. London, UK: IBC Technical Services

Nicholson RI (1992): Why ER level may not reflect endocrine responsiveness in breast cancer. *Rev Endocr-Rel Cancer* 40:25–28

Nicholson RI (1993): Recent advances in the anti-hormonal therapy of breast cancer. *Curr Opin Invest Drugs* 2:1259–1268

Nicholson S, Halcrow P, Fardon JR, Sainsbury JRC, Chambers P, Harris AL (1989): Expression of epidermal growth factor receptors associated with lack of response to endocrine therapy in recurrent breast cancer. *Lancet* i:182–185

Nicholson S, Wright C, Sainsbury JRC, Fardon JR, Harris AL (1990): Epidermal growth factor receptor as a marker of poor prognosis in node-negative breast cancer patients: neu and tamoxifen failure. *J Steroid Biochem Mol Biol* 37:811–814

Nicholson RI, Bouzubar N, Walker KJ, McClelland RA, Dixon AR, Robertson JFR, Ellis IO, Blamey RW (1991): Hormone sensitivity in breast cancer: influence of heterogeneity of oestrogen receptor expression and cell proliferation. *Eur J Cancer* 27:908–913

Nicholson RI, Eaton CL, Manning DL (1992): New developments in the endocrine management of breast cancer. In: *Recent Advances in Endocrinology and Metabolism*, Edwares CRW, Lincoln DW eds. UK: Churchill Livingstone

Nicholson RI, Gee JMW, Manning DL (1993a): New anti-hormonal approaches to breast cancer therapy. *Drugs of Today* 29:363–372

Nicholson RI, McClelland RA, Finlay P, Eaton, CL, Gullick WJ, Dixon AR, Robertson JFR, Ellis IO, Blamey RW (1993b): Relationship between EGF-R, c-erbB-2 protein expression and Ki-67 immunostaining in breast cancer and hormone sensitivity. *Eur J Cancer,* 29A:1018–1023

Nicholson RI, McClelland RA, Gee JMW, Manning DL, Cannon P, Robertson JFR, Ellis IO, Blamey RW (1994a): Epidermal growth factor receptor expression in breast cancer: association with response to endocrine therapy. *Breast Cancer Res Treat* 29:117–125

Nicholson RI, Gee JMW, Eaton CL, Manning DL, Mansel RE, Sharma AK, Douglas-Jones A, Price-Thomas M, Howell A, DeFriend DJ, Bundred NJ, Anderson E, Robertson JFR, Blamey RW, Dowsett M, Baum, Walton P, Wakeling AE (1994b): Pure antioestrogens in breast cancer: experimental and clinical observations. In: *Sex Hormones and Antihormones in Endocrine Dependent Pathology: Basic and Clinical Aspects*, Motta M, Serio M, eds. Amsterdam: Elsevier

Nicholson RI, McClelland RA, Gee JMW, Manning DL, Cannon P, Robertson JFR, Ellis I O, Blamey RW (1994c): Transforming growth factor-alpha and endocrine sensitivity in breast cancer. *Cancer Res* 54:1684–1689

Nicholson RI, Francis AB, McClelland RA, Manning DL, Gee JMW (1994d): Pure anti-oestrogen (ICI 164384 and ICI 182780) and breast cancer: is the attainment of complete oestrogen withdrawal worthwhile? *Endocr-Rel Cancer* 1:5–17

Nicholson RI, Gee JMW, Manning DL, Wakeling AE, Montano MM, Katzenellenbogen BS (1995a): Responses to pure antioestrogens (ICI 164384, ICI 182780) in oestrogen-sensitive and -resistant experimental and clinical breast cancer. *Ann NY Acad Sci* 761:148–163

Nicholson RI, Gee JMW, Francis AB, Manning DL, Wakeling AE, Katzenellenbogen BS (1995b): Observations arising from the use of pure antioestrogens on oestrogen-responsive (MCF-7) and oestrogen growth-independent (K3) human breast cancer cells. *Endocr-Rel Cancer* 2:115–121

Noguchi S, Motomura K, Inaji H, Imaoka S, Koyama H (1993): Down regulation of transforming growth factor alpha by tamoxifen in human breast cancer. *Cancer* 72:131–136

Osborne CK, Hamilton B, Titus G, Livingston RB (1980): Epidermal growth factor stimulation on human breast cancer cells in culture. *Cancer Res* 40:2361–2366

Osborne CK, Coronado EB, Kitten LJ, Arteaga CI, Fuqua SA, Ramasharma K, Marshall M, Li CH (1989): Insulin-like growth factor-II: a potential autocrine/paracrine growth factor for human breast cancer acting via the IGF-I receptor. *Mol Endocrinol* 3:1701–1709

Osborne CK, Coronado-Heinsohn EB, Hilsenbeck SG, McCue BL, Wakeling AE, McClelland RA, Manning DL, Nicholson RI (1995): Pure steroidal anti-oestrogens are superior to tamoxifen in a model of human breast cancer. *J Natl Cancer Inst* 87:746–750

Paik S (1992): Expression of IGF-I and IGF-II mRNA in breast tissue. *Breast Cancer Res Treat* 22:31–38

Papa V, Gliozzo B, Clark GM, McGuire Wl, Moore D, Fujita-Yamaguchi Y, Vigneri R, Goldfine ID, Pezzino V (1993): Insulin-like growth factor-I receptors are overexpressed and predict low risk in human breast cancer. *Cancer Res* 53:3736–3740

Park S, Hartmann DP, Dickson RB, Lipmann ME (1994): Antioestrogen resistance in ER-positive breast cancer cells. *Breast Cancer Res Treat* 31:301–307

Partanen J, Makela TP, Korhonen J, Hirvonen H, Claesson-Welsh L, Alitalo K (1991): FGFR-4, a novel acidic fibroblast growth factor receptor with a distinct expression pattern. *EMBO J* 10:1347–1354

Patterson, JS (1981): Clinical aspects and development of antioestrogen therapy: a review of the endocrine effects of tamoxifen in animals and man. *J Endocrinol* 89:67P-75P

Peles E, Bacus SS, Koski RA et al. (1992): Isolation of the neu/HER-2 stimulatory ligand: a 44kd glycoprotein that induces differentiation of mammary tumour cells. *Cell* 69:205–216

Peyrat J, Bonneterre J (1992): IGF receptor in human breast diseases. *Breast Cancer Res Treat* 22:59–67

Peyrat JP, Hondermarck H, Hecquet B, Adenis A, Bonneterre J (1992): FGFB binding sites in cancers of the human breast. *Bulletin du Cancer* 79:251–260

Philips A, Chalbos D, Rochefort H (1993): Estradiol increases and anti-oestrogens antagonize the growth factor-induced activator protein-1 activity in MCF-7 breast cancer cells without affecting c-fos and c-jun synthesis. *J Biol Chem* 268:14103–14108

Plowman GD, Green JM, Culouscou JM, Carlton GW, Rothwell VM, Buckley S (1993): Heregulin induces tyrosine phosphorylation of HER4/p180erbB-4. *Nature* 366:473–475

Pollak MN, Huynh HT, LeFebvre SP (1992): Tamoxifen reduces serum insulin-like growth factor I (IGF-I). *Breast Cancer Res Treat* 22:91–100

Poller DN, Hutchings CE, Galea M, Bell JA, Nicholson RI, Elston CW, Blamey RW, Ellis IO (1992): p53 protein expression in human breast carcinoma: relationship to expression of epidermal growth factor receptor, c-erbB-2 protein overexpression, and oestrogen receptor. *Br J Cancer* 66:583–588

Prigent SA, Lemoine NR (1992): The type 1 (EGFR-related) family of growth factor receptors and their ligands. *Prog Growth Factor Res* 4:1–24

Purohit A, Chapman O, Duncan L, Reed MJ (1992): Modulation of oestrone sulphatase activity in breast cancer cell lines by growth factors. *J Steroid Biochem Mol Biol* 41:563–566

Railo MJ, von Smitten K, Pekonen F (1994): The prognostic value of insulin-like growth factor-I in breast cancer patients Results of a follow-up study on 126 patients. *Eur J Cancer* 30:307–311

Reed MJ, Christodoulides A, Koistinen R, Teale JD, Ghilchick, MW (1992): The effect of endocrine therapy with medroxyprogesterone acetate, 4-hydroxyandrostenedione or tamoxifen on plasma concentrations of insulin-like growth factor (IGF)-I, IGF-II and IGFBP-1 in women with advanced breast cancer. *Int J Cancer* 52:208–212

Reese JC, Katzenellenbogen BS (1992): Examination of the DNA-binding of oestrogen receptor in whole cells: implications for hormone independent transactivation and the actions of antioestrogens. *Mol Cell Biol* 12:4531–4538

Rosen N, Yee D, Lippman ME, Paik S, Cullen KJ (1991): Insulin-like growth factors in human breast cancer. *Breast Cancer Res Treat* 18:S55-S62

Salomon DS, Kim N, Saeki T, Ciardiello F (1990): Transforming growth factor alpha: an oncodevelopmental growth factor. *Cancer Cells* 2:389–397

Sandgren EP, Luetteke NC, Palmiter RD, Brinsten RL, Lee DC (1990): Overexpression of TGFalpha in transgenic mice: induction of epithelial hyperplasia, pancreatic metaplasia and carcinoma of the breast. *Cell* 61:1121–1135

Schuchard M, Landers JP, Sandhu NP, Spelsberg TC (1993): Steroid hormone regulation of nuclear protooncogenes. *Endocr Rev* 14:659–669

Sharma AK, Horgan K, Douglas-Jones A, McClelland R, Gee J, Nicholson RI (1994): Dual immunocytochemical analysis of oestrogen and epidermal growth factor receptors in human breast cancer. *Br J Cancer* 69:1032–1037

Shi E, Kan M, Xu J, Wang F, Hou J, McKeehan WL (1993): Control of fibroblast growth factor receptor kinase signal transduction by heterodimerisation of combinatorial splice variants. *Mol Cell Biol* 13:3907–3918

Slamon D, J, Clark GM, Wong SG et al. (1987): Human breast cancer: correlation of relapse and survival with amplification of the HER-2/neu oncogene. *Science* 235:177–182

Sliwkowski MX, Schaefer G, Akita RW et al. (1994): Co-expression of erbB-2 and erbB-3 proteins reconstitutes a high affinity receptor for Heregulin. *J Biol Chem* 269:14661–14665

Spivak-Kroizman T, Rotin D, Pinchasi D, Ullrich A, Schlessinger J, Lax I (1992): Heterodimerization of c-erbB-2 with different epidermal growth factor receptor mutants elicits stimulatory or inhibitory responses. *J Biol Chem* 267:8056–8063

Staebler A, Sommer C, Mueller SC, Byers S, Thompson EW, Lupu R (1994): Modulation of breast cancer progression and differentiation by gp30/heregulin. *Breast Cancer Res Treat* 31:175–182

Stern DF, Kamps MP (1988): EGF-stimulated tyrosine phosphorylation of p185 neu: a potential model for receptor interactions. *EMBO J* 7:995–1001

Stern, Heffernan PA, Weinberg RA (1986): p185 a product of the neu protooncogene, is a receptor-like protein associated with tyrosine kinase activity. *Molec Cell Biol* 6:1729–1740

Stewart AJ, Westley BR, May FEB (1992): Modulation of the proliferative response of breast cancer cells to growth factors by oestrogen. *Br J Cancer* 66:640–648

Sutherland RL, Watts CK, Musgrove EA (1993): Cyclin gene expression and growth control in normal and neoplastic human breast epithelium. *J Steroid Biochem Mol Biol* 47:99–106

Tanaka A, Miyamoto K, Minamino N, Takeda M, Sato B, Matsuo H, Matsumoto K (1992): Cloning and characterisation of an androgen-induced growth factor essential for the androgen-dependent growth of mouse mammary carcinoma cells. *Proc Soc Natl Acad Sci USA* 89:8928–8932

Theillet C, Adelaide J, Louason G, Bonnet-Dorion F, Jacquemier J, Adnane J, Longy M, Katsaros D, Sismondi P, Gaudray, P et al (1993): FGFRI and PLAT genes and DNA amplification at 8p12 in breast and ovarian cancers. *Genes Chrom Cancer* 7:219–226

Thompson AM, Kerr DJ, Steel CM (1991): Transforming growth factor beta 1 is implicated in the failure of tamoxifen therapy in human breast cancer. *Br J Cancer* 63:609–614

Travers MT, Barrett-Lee PJ, Berger U, Luqmani YA, Gazet JC, Powles TJ, Coombes RC (1988): Growth factor expression in normal, benign, and malignant breast tissue. *Br Med J Clin Res* 296:1621–1624

Travis A, Bell JA, Wencyk P, Robertson JF, Poller DN, Gullick WJ, Nicholson RI, Blamey RW, Elston CW, Ellis IO (in press): c-erbB-3 in human breast cancer-expression and relation to prognosis and established prognostic factors

Trimble MS, Xin JH, Guy CT, Muller WJ, Hassell JA (1993): PEA3 is overexpressed in mouse metastatic mammary adenocarcinomas. *Oncogene* 8:3037–3042

Umekita Y, Enokizono N, Sagara Y, Kuriwaki K, Takasaki T, Yoshida A, Yoshida H (1992): Immunohistochemical studies on oncogene products (EGF-R, c-erbB-2) and growth factors (EGF, TGF-alpha) in human breast cancer: their relationship to oestrogen receptor status, histological grade, mitotic index and nodal status. *Virchows Archiv A Pathol Anat* 420:345–351

Van Agthoven T, Van Agthoven TLA, Portengen H, Foekens JA, Dorssers LCJ (1992): Ectopic expression of epidermal growth factor receptors induces hormone independence in ZR-75–1 human breast cancer cells. *Cancer Res* 52:5082–5088

Wakeling AE, Dukes M, Bowler J (1991): A potent specific pure antioestrogen with clinical potential. *Cancer Res* 51:3867–3873

Walker KJ, Price-Thomas JM, Candlish W, Nicholson RI (1991): Influence of the antioestrogen tamoxifen on normal breast tissue. *Br J Cancer* 64:764–768

Walker KJ, McClelland RA, Candlish W, Blamey RW, Nicholson RI (1992): Heterogeneity of oestrogen receptor expression in normal and malignant breast tissue. *Eur J Cancer* 28:34–37

Walker RA, Dearing SJ (1992): Transforming growth factor beta 1 in ductal carcinoma in situ and invasive carcinomas of the breast. *Eur J Cancer* 28:641–644

Walker RA, Varley JM (1993): The molecular pathology of human breast cancer. *Cancer Surv* 16:31–57

Walker RA, Dearing SJ, Gallacher B (1994): Relationship of transforming growth factor beta 1 to extracellular matrix and stromal infiltrates in invasive breast carcinoma. *Br J Cancer* 69:1160–1165

Wen D, Peles E, Cupples R et al. (1992): Neu differentiation factor: a transmembrane glycoprotein containing an EGF domain and an immunoglobulin homology unit. *Cell* 69:559–572

Wiseman LR, Johnson MD, Wakeling AE, Lykkesfeld AE, May FE, Westley BR (1993): Type I IGF receptor and acquired tamoxifen resistance in oestrogen-responsive human breast cancer cells. *Eur J Cancer* 29:2256–2264

Wosikowski K, Eppenberger U, Kung W, Nagamine Y, Mueller H (1992): c-fos, c-jun and c-myc expressions are not growth rate limiting for the human MCF-7 breast cancer cells. *Biochem Biophys Res Commun* 188:1067–1076

Yarden Y, Weinberg RA (1989): Experimental approaches to hypothetical hormones: detection of a candidate ligand of the neu protooncogene. *Proc Natl Acad Sci (USA)* 86:3179–3183

Yasui W, Hata J, Yokozaki H, Nakatani H, Ochiai A, Ito H, Tahara E (1988): Interaction between epidermal growth factor and its receptor in progression of human gastric cancer. *Int J Cancer* 41:211–217

Yu D, Matin A, Hung, MC (1992): The retinoblastoma gene product suppresses neu oncogene-induced transformation via transcriptional repression of neu. *J Biol Chem* 267:10203–10206

9

Role of Altered Estrogen Receptors in Breast Cancer

Sami G. Diab, Carl G. Castles and Suzanne A.W. Fuqua

INTRODUCTION

Breast cancer is the most common malignancy in women in the United States, and is predicted to account for about 32% of new cancer cases in 1995 (Wingo et al., 1995). One of the first pieces of evidence to suggest that breast tumors depend on estrogen for growth comes from the observation, made more than a century ago, that oophorectomy can cause tumor shrinkage in women with metastatic breast cancer (Beatson, 1896). The discovery of the estrogen receptor (ER) as a mediator of estrogen's biological effects revolutionized our understanding of breast cancer biology and eventually led to the classification of breast cancer into two major groups: those tumors that express the ER (ER-positive), and those tumors with undetectable levels of the ER (ER-negative). We now appreciate that each of these groups has its own distinct biological and clinical features.

The expression of ER in breast cancer cells has important clinical implications. First, the absence of ER is associated with a more aggressive disease. Patients with ER-positive tumors tend to be older, and have a more indolent cancer with longer disease-free survivals and a better overall survival compared to patients with ER-negative tumors (Knight et al., 1977; Osborne et al., 1980; Stewart et al., 1981). This clinical phenotype is a reflection of the general biological features of these tumors, as they tend to be well differentiated histologically, are diploid, and exhibit a lower % S-phase. The presence of ER is also a major predictor of response to endocrine therapy with roughly two-thirds of ER-positive tumors responding to the antiestrogen tamoxifen (TAM). Moreover, high ER levels and the presence of

Hormones and Cancer
Wayne V. Vedeckis, Editor

a functional ER signaling pathway, as indicated by the co-expression of ER-regulated proteins such as progesterone receptor (PgR) and pS2, identify women who will most likely respond to TAM (Osborne et al., 1980).

Estrogen has also been implicated in the promotion of mammary neoplasms. This is suggested by the observation that women have a much higher incidence of breast cancer than men. This is further supported by the elevated incidence of breast cancer in young women who have used oral contraceptives with higher estrogen doses than those currently being used (Meirik et al., 1986). Also, oophorectomy before age 50 years decreases the risk of breast cancer; the earlier the surgery is performed, the more risk reduction is observed (Trichopoulos et al., 1972). These findings, and others, have led to a nationwide breast cancer prevention trial using TAM in women at high risk for the disease.

Estrogen is a major growth factor for breast cancer, so alterations in its signal transduction pathway might have significant clinical implications. The presence in a tumor of altered forms of the estrogen receptor with aberrant functions might provide biological explanations for several clinical observations that might seem paradoxical at first. These observations include the 5–10% of ER-negative tumors that behave as if they are ER-positive, expressing estrogen-regulated proteins PgR and pS2 (Horwitz et al., 1985) and responding to hormonal therapy (Osborne et al., 1980). On the other hand, some ER-positive tumors fail to respond to endocrine therapy, and many also fail to express PgR and pS2, suggesting a dysfunctional ER signaling pathway. In this chapter we will first discuss the structure and function of the normal ER and then review specific ER alterations in breast tumors with emphasis on their clinical implications.

ESTROGEN RECEPTOR STRUCTURE AND FUNCTION

General Organization

The ER belongs to the steroid/thyroid nuclear receptor superfamily, which includes the steroid, thyroid, vitamin D, retinoic acid, and other orphan receptors whose ligands are not yet identified (Evans, 1988; see Chapter 4, this volume). These receptors are all part of the zinc finger family of transcription factors. The nuclear receptors share common structural organizations with separable domains with distinct functions. These domains include the hormone binding domain (HBD), which mediates the interaction between the receptor and its ligand, and the DNA binding domain (DBD), which mediates the binding of the receptor to the DNA upon ligand activation (Kumar et al., 1987; Webster et al., 1988). When activated, these nuclear receptors function as transcription factors and through their various transactivation domains modulate the expression of other transcription factors. Other domains such as those required for nuclear localization

(Guiochon-Mantel et al., 1989), dimerization (Fawell et al., 1990), and binding of regulatory heat shock proteins (Chambraud et al., 1990) are also characteristic of many members of the nuclear receptor superfamily.

The ER gene is located on chromosome 6, extends over 140 kbp, and consists of 8 exons (Gosden et al., 1986). The gene is under the control of two promoters that give rise to two mRNA transcripts that differ only in their 5′ ends (Keaveney et al., 1991). These two ER mRNAs encode an identical 595–amino acid protein with a molecular weight of approximately 65 kDa. By convention, the domains of the ER are classically identified by the letters A through F (Kumar et al., 1987) (Fig. 1).

Transactivation

The ER has three transactivation domains, known as activation functions 1, 2, and 2a (AF-1, AF-2, and AF-2a, respectively), that are responsible for receptor transcriptional activity. The AF-2 domain is located in the carboxy-terminal region of the ER. The activity of AF-2 depends on estrogen and requires the presence of the entire HBD (Webster et al., 1989). Site-directed mutagenesis studies have shown that mutations in only a few amino acids in the AF-2 domain will abolish its activity without affecting receptor hormone binding function (Danielian et al., 1992). The AF-1 domain is located at the amino-terminal region of the receptor within the A/B domain of the receptor (Metzger et al., 1988). AF-1's function is constitutively active in the absence of estrogen. A third domain, AF-2a, containing a hormone-independent transcriptional activation function, has recently been localized to the N-terminal portion of AF-2 (Pierrat et al., 1994).

We now know that the various functions of the activation domains are manifested in both a cellular and promoter-specific manner (Tasset et al., 1990; Tzukerman et al., 1994), which can be modulated by proteins that interact with the ER (McDonnell et al., 1992). Halachmi et al. (1994) have identified a 160-kDa ER-associated protein (ERAP 160) that interacts with the HBD/AF-2 domain. The ability of ER to activate transcription in the presence of estrogen parallels its ability to bind ERAP 160, suggesting that this protein is involved in AF-2 estrogen-dependent transcriptional activity. Cavailles et al. (1994) have also identified several proteins that interact with the wild-type ER HBD/AF-2 domain, but not with a mutated HBD lacking AF-2, again suggesting that these proteins are somehow involved in mediating AF-2 activity. McDonnell et al. (1992) have also identified a yeast-repressor protein, called SSN6, that suppresses AF-1 hormone-independent transcriptional activity. A protein homologue to SSN6 may exist in mammalian cells, but it has yet to be identified. It is obvious from these and other studies that the function of the activation domains is complex, and that their ultimate biological effects depend on the environment in which the receptor exists.

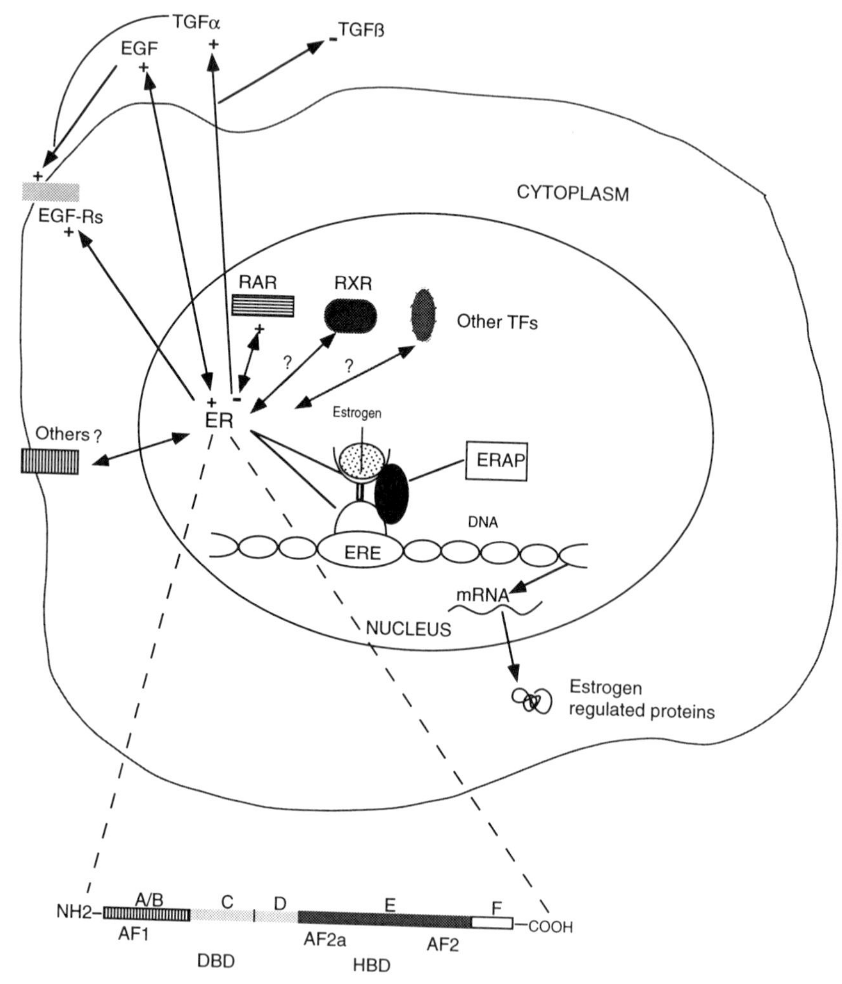

Figure 1. A simplified model for the ER signaling pathway and the interaction of ER with other modulators of cellular growth. Upon binding of hormone, the ER binds as a homodimer to its ERE in the upstream promoter regions of ER-regulated genes, modulating their transcription. Estrogen is able to increase the production of growth factors and their receptors such as TGF-α and EGF-R, initiating a loop that augments the effect of estrogen on cellular proliferation. In breast tumor cells, estrogen is also able to decrease the expression of negative growth regulators such as TGF-β (Arrick et al., 1990). Some estrogen-induced proteins such as RARα can have an inhibitory effect on ER, establishing a negative feedback loop. It is clear that the interaction of these different pathways is essential in maintaining controlled cellular growth and differentiation. *ERAP*, estrogen receptor associated protein; *TF*, transcription factor; *RXR*, retinoid X receptor.

Ligand Binding

The HBD is a large region of approximately 250 amino acids encoded by exons 4 through 8. It is located at the carboxy-terminus of the receptor (region E) and is believed to exist as a hydrophobic pocket that binds estradiol. Several studies have investigated whether this domain plays a direct role in the development of estrogen resistance in mammalian cells. However, before further discussing this possibility, the concept of hormone independence (as opposed to the related, but different, phenomenon of TAM resistance) must be clarified. Hormone independence is defined as the state when a malignant cell no longer depends on estrogen or its receptor for growth. Some ER-negative breast tumors, or tumor-derived cell lines such as MDA-MB-231 that do not respond to various antiestrogenic agents and grow in media depleted of estrogen, are examples of hormone independence. In this regard, it is interesting that the introduction of ER into MDA-MB-231 cells leads to cell death in the presence of estrogen (Jiang and Jordan, 1992). The exact mechanism of this phenomenon is unknown, but it is tempting to speculate that ER-negative tumors might suppress ER expression to escape an ER-mediated death pathway, thereby acquiring a survival advantage. In contrast, TAM-resistant and/or TAM-stimulated growth infers that the tumor has developed a mechanism (or mechanisms) to overcome the antiestrogenic effects of the drug, but this does not necessarily imply that the tumor is truly hormone independent in the sense that its growth can still be stimulated by hormone. This is supported by evidence that TAM-resistant breast tumor growth can often be inhibited by other pure antiestrogens such as ICI 182780 (Wakeling, 1993). The mechanisms underlying these two breast tumor phenotypes (hormone-independent and TAM-resistant) are probably dissimilar, and both are discussed in more detail later.

The HBD is necessary for binding of both estrogen and TAM as demonstrated by affinity labeling experiments (Harlow et al., 1989; Ratajczak et al., 1989), and numerous studies have investigated whether mutations in this domain lead to hormone resistance in mammalian cells. Several mutagenesis studies have revealed that the stretches of amino acids responsible for estrogen and hydroxytamoxifen (4-OHT) binding are distinct, even though they are partially overlapping. Mutation of the lysines at positions 529 and 531 to glutamines in the human ER reduces its affinity for estradiol 5- to 10-fold, but has little effect on 4-OHT binding (Pakdel and Katzenellenbogen, 1992). Several mutations in the HBD (at amino acids 521, 522, and 525) that are capable of reducing estrogen binding activity about 1000-fold without affecting the sensitivity to 4-OHT have similarly been identified (Danielian et al., 1993). Mutation of a valine for glycine at codon 400 in the HBD of human ER leads to enhanced estrogenic activity of 4-OHT in stable ER transfectants of the breast cancer cell line MDA-MB-231 (Jiang et al., 1992). Whether such point mutations exist in human breast cancers, and whether they play a significant role in the development of hormone independence, remains to be seen. At the present time, no mutations within the HBD that reduce TAM binding without affecting estrogen binding have been reported in human breast tumors. Such mutations, if found, could have important clinical significance if they are found to be associated with the development of TAM resistance.

DNA Binding

Alterations in other regions of the ER, such as the two dimerization domains, could also have important consequences in ER-regulated cellular functions. The ER dimerization domain, which mediates most of the interaction between the two ER dimer molecules, is located at the far carboxy-terminus of the HBD (Fawell et al., 1990). A second, weaker dimerization domain is located in the DBD, possibly on the second zinc finger (Kumar and Chambon, 1988). Upon ligand activation, it is believed that the ER normally forms a homodimer, and functions as a transcription factor by binding to estrogen response elements (EREs) in the upstream promoter region of target genes (see Fig. 1). The consensus ERE is a 13-bp palindromic element consisting of two 5-bp half sites (GGTCA) separated by a 3-bp spacer (Klein-Hitpass et al., 1986). Recently, Dana et al. (1994) have shown that there are other EREs, including some that are direct repeats of the GGTCA half site, and others that have different half-site sequences. Some of these novel EREs are fully functional *in vivo*, yet they are capable of binding ER only weakly *in vitro*, indicating that other cellular elements are most probably required for full receptor activity. The identification of novel EREs is clearly important progress toward the identification of new estrogen-regulated target genes. It is somewhat surprising that our current knowledge of the important genes regulated by ER is limited to only a few genes. The identification of new ER-regulated genes may open the door to our understanding of the complex role played by estrogen in growth control and cellular differentiation.

Finally, it should be emphasized that growth control is the ultimate result of the maintenance of cellular homeostasis. This is achieved, in part, by balanced biofeedback loops (see Fig. 1) between the different cellular receptors (both nuclear and membrane receptors). It has been well established that dysregulation of these loops will trigger the process of carcinogenesis. Estrogen, for instance, can induce the expression of retinoic acid receptor alpha (RARα) in the T47D breast cancer cell line by an ER-dependent pathway, as evidenced by the ability of TAM to inhibit this induction (Roman et al., 1992). Conversely, retinoic acid will inhibit estradiol-induced transcription of PgR and pS2 (Balaguer et al., 1991), suggesting that RAR may exert a negative regulatory effect on the ER-mediated pathway. Estrogen is also capable of inducing both the epidermal growth factor receptor (EGF-R) (Chrysogelos and Dickson, 1994) and its ligand TGF-α (Murphy and Dotzlaw, 1989), while decreasing the expression of TGF-β in a number of breast cancer cell lines (Arrick et al., 1990), all of which have been shown to enhance estrogen-stimulated growth of breast tumor cells. Epidermal growth factor and TGF-α may also activate ER in an estrogen-independent fashion (Ignar-Trowbridge et al., 1993), thus establishing a positive feedback loop between the ER and these growth factors. Therefore, it is apparent that estrogen regulation of many complex cellular growth pathways plays a major role in the proliferative ability of breast tissue, and that alterations in the ER could potentially contribute to the dysregulation of cellular growth, which is a hallmark of malignancy.

THE ROLE OF THE ESTROGEN RECEPTOR IN NORMAL BREAST TISSUES AND PREMALIGNANT BREAST LESIONS

An important concept that is currently the subject of much exploration is whether specific ER alterations are causally involved in the malignant evolution of breast lesions, or whether such alterations are a later epiphenomenon associated with the known genetic instability inherent during breast tumor progression. The detection of ER alterations in premalignant breast lesions would be an important first step in linking such changes to the process of breast carcinogenesis.

Histologically, breast lesions can be separated into four categories (Fig. 2), as defined by microscopic architecture of the lesion according to specific nuclear and cytoplasmic features (Fearon and Vogelstein, 1990; O'Connell et al., 1994; see Chapter 6, this volume, for an in-depth discussion of breast disease). The lesion considered most benign, and perhaps the "earliest" in the scale, is termed "proliferative disease without atypia" (PDWA). If cellular atypia is present, then the lesion is termed an "atypical ductal hyperplasia" (ADH). When the cell morphology is markedly disturbed to meet the criteria of malignancy, but the basement membrane remains intact, the lesion is called a "ductal carcinoma *in situ*" (DCIS). DCIS is further divided into two subtypes, called comedo or noncomedo DCIS, according to the presence or absence of cellular necrosis, respectively.

Currently, our knowledge of the ER status of normal breast tissues is limited to a few small studies. Peterson et al. (1987) examined 18 normal breast tissues obtained from women undergoing reduction mammoplasty using ER immunohistochemistry (IHC), and showed that only 7% of normal breast epithelial cells stained positive for ER. In another study, Ricketts et al. (1991), also using IHC, found that 16% of normal breast samples obtained by fine needle aspirate were positive for ER. One problem with these and other similar studies examining ER expression in normal breast tissues is that different cut-points of positivity are used to define ER states. In addition, the "normal" tissue evaluated in many studies was obtained from breasts that were resected because of the presence of breast cancer, and the issue of contamination of normal adjacent breast epithelium

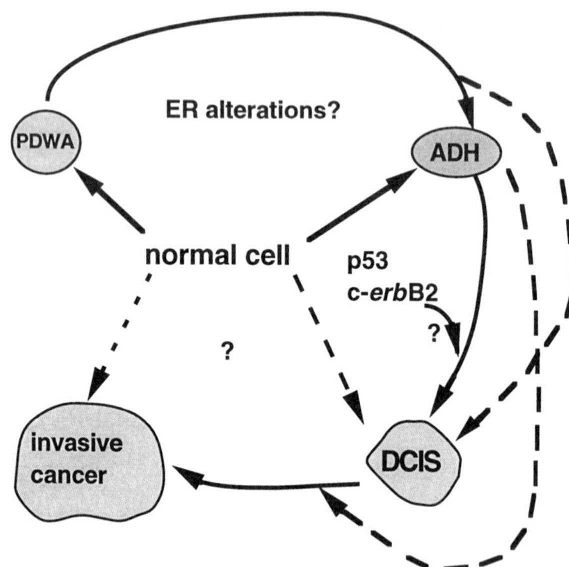

Figure 2. A model for breast cancer evolution. In the above model, a normal breast epithelial cell acquires a genetic defect and begins proliferating (proliferative disease without atypia; PDWA), then accumulates additional genetic defects (atypical ductal hyperplasia; ADH) while progressing to a malignant lesion (ductal carcinoma *in situ*; DCIS), and ultimately becomes an invasive tumor. Based on the genetic alterations acquired, a lesion can bypass any of these stages and go directly to subsequent steps (*dashed lines*). Our knowledge of the genetic alterations responsible for progression is limited. Alterations in some factors involved in cellular regulation, such as p53 and c-*erb*B-2, have been identified in ductal carcinoma *in situ* lesions but not atypical ductal hyperplasia or proliferative disease without atypia. ER alterations may be present at earlier stages such as PDWA.

with cancer cells cannot be excluded in these analyses. Despite these problems, most studies agree that ER levels are generally low in normal breast epithelium (Jacquemier et al., 1982, 1990; Petersen et al., 1987; Khan et al., 1994).

In a comparison of normal breast tissues with ADH, Jacquemier et al. (1990) found, again using IHC, that ADH lesions express higher levels of the ER. In a study of DCIS, Poller et al. (1993b) showed that about one-third of noncomedo DCIS lesions stained positive for ER. This was in contrast to comedo DCIS, where ER expression was observed in only 16% of these lesions. Finally, Petrangeli et al. (1994), using enzyme immunoassay, have shown that "normal" breast tissues have a significantly lower level of ER expression as compared to adjacent malignant lesions in the breast. In addition, they have reported that the ER gene is frequently hypomethylated in neoplastic as compared to preneoplastic lesions, and they suggest that increased transcription of the ER gene as a result of hypomethylation is responsible for this difference in ER expression. In summary, these studies suggest that ER expression increases as one evaluates cells along the proposed evolutionary pathway from normal, to premalignant, and finally to the malignant phenotype. This increase in ER expression might lead to an amplification in the signals generated by estrogen, leading to an increase in expression of a number of growth factors or other proteins that drive breast mitogenesis. Obviously, this model is by no means exclusive, and it may only apply to the carcinogenic process in, at best, a subset of breast tumors.

In addition to altered levels of ER in premalignant breast lesions, the presence of mutant ERs with altered functions represents another potential mechanism accelerating a cell's progression to malignancy. The detection of ER mutations and their correlation with clinical follow-up could help determine which premalignant lesions might be at high risk of progression to a malignant phenotype, and thus require early medical intervention. Furthermore, such ER alterations could potentially serve as intermediate endpoints in chemoprevention studies. A point mutation in the ER has recently been detected in PDWA lesions, and characterization of its functional significance in these lesions is currently underway in our laboratory. Preliminary results suggest that this ER alteration might confer increased estrogen sensitivity to breast tumor cells (Wiltschke and Fuqua, unpublished results).

However, since carcinogenesis is obviously a complex and heterogeneous process, alterations in growth factors such as the TGF-α (Matsui et al., 1990), oncogenes such as c-*erbB*-2 (Gusterson et al., 1988, Tsutsumi et al., 1990), and tumor-suppressor genes such as p53 (Davidoff et al., 1991; Poller et al., 1993a; Allred et al., 1994) play important roles along with the ER during the process of breast cancer evolution (see Fig. 2). It has been suggested that the presence of more than one molecular alteration is stochastic and required for progression toward an invasive malignancy (Smith et al., 1993). Clearly, an understanding of the genetic changes involved in each step of tumor evolution would guide the development of screening methods for early detection and intervention.

ER ALTERATIONS IDENTIFIED IN MALIGNANT BREAST LESIONS

The determination of tumor ER status is one of the most important tests performed in the clinical evaluation of patients with breast cancer. As previously mentioned, the ER status of a tumor can influence the aggressiveness of the disease (McGuire et al., 1975; Clark et al., 1984) and can predict the probability of a response to endocrine therapy (McGuire, 1978; Osborne et al., 1980). Two methods are routinely used to measure ER levels in breast tumors: the ligand binding assay and antibody-based assays. The first method relies on the detection of unoccupied ER with a radiolabeled estrogen. The limitation of the ligand binding assay is its inability to detect an occupied receptor or an altered ER that is unable to bind estrogen (Fuqua et al., 1991). In these situations, this assay will give false-negative results. The second ER detection method relies on ER-specific monoclonal antibodies used either for enzyme-linked immunosorbent assay (ELISA) or IHC (Greene et al., 1984; Andersen and Poulsen, 1989). Since antibodies have been raised to epitopes other than the HBD, such methods might offer greater sensitivity, allowing the detection of occupied and/or altered receptors (Allred, 1993; Tesch et al., 1993). Another means of defining a tumor's ER status or an intact estrogen response pathway is by evaluation of estrogen-regulated products such as PgR (Horwitz et al., 1975; Horwitz and McGuire, 1978). The inclusion of other functional assays such as this may provide better insights into the true functional capability of the ER, or lack thereof, of a particular breast tumor, and this might confer a greater definition of a tumor's hormone dependence. Thus, tumors that are ER-negative/PgR-negative and negative for other estrogen-regulated proteins are those tumors most likely to be truly negative for ER expression.

ER-negative/PgR-negative Breast Tumor Phenotype

The absence of ER expression in breast tumors is poorly understood at present. Those who have studied ER-negative/PgR-negative breast tumors initially explored the fundamental question of whether ER deletions or rearrangements are responsible for this phenotype. However, there is little evidence to suggest that this is indeed the case. Hill et al. (1989) examined 65 ER-negative breast tumors by Southern blot hybridization and showed that the ER gene was not deleted in any of these cases. Yaich et al. (1992) also did not detect any ER gene deletion events in 12 ER-negative breast tumors they examined. In a recent study by Roodi et al. (1995), 70 ER-negative tumors were evaluated by single-strand conformational polymorphism and DNA sequencing. None of these tumors exhibited an ER gene deletion. Therefore, all these studies suggest that the absence of ER expression in tumors is not the result of genomic deletion but perhaps rather transcriptional inactivation of the ER gene. This hypothesis is further supported by the inability to detect ER mRNA in ER-negative tumors using very sensitive molecular techniques (Weigel and deConinck, 1993)

One possible explanation for the absence of ER transcription in ER-negative tumors is the methylation of specific regulatory regions of the ER gene. Recently, it has been shown in ER-negative breast cancer cell lines that there is an extensive methylation in the cytosine-rich "CpG islands" within the 5′ promoter region of the ER gene (Ottaviano et al., 1994). Furthermore, demethylation of the ER gene, using methylation inhibitors such as 5-azacytidine or 5-aza-2-deoxycytidine, was found to induce ER expression in MDA-MB-231 cells (Ferguson et al., 1995). This result further supports the hypothesis that ER gene methylation may be an important factor in transcriptional regulation or suppression of the ER in cells. Clearly, delineating the exact mechanisms underlying this suppression could have clinical implications by offering new opportunities to design treatment strategies based on reactivation of ER expression, in an attempt to convert tumors to a less aggressive and hormone-dependent ER-positive phenotype.

One of the most critical aspects of the absence or loss of ER expression in breast tumors is the relationship of the ER-negative phenotype with the acquisition of hormone independence. As expected, ER-negative tumors do not respond to the growth-stimulatory effects of estrogen, and are not inhibited by antiestrogens such as TAM. One creative research approach has explored the possibility of converting ER-negative cells to a hormone-dependent state by the introduction of wild-type ER (wtER) (Jiang and Jordan, 1992). These investigators stably introduced wtER into MDA-MB-231 cells and demonstrated that the ER was functional, as evidenced by induction of PgR after administration of estrogen. Paradoxically, however, they found that estrogen inhibited, rather than stimulated, the growth of these transfected cells. This growth suppression phenomenon by estrogen has also been reported using other non-breast cancer cell lines transfected with the wtER (Wattset al., 1989; Kushner et al., 1990). The clinical observation that high dose estrogens can also inhibit the growth of some breast tumors suggests that this inhibitory effect can also occur *in vivo*. Observations such as this raise an important question. What are the cellular conditions or milieu that determine whether ER is a stimulator, or an inhibitor, of tumor growth? Although this question remains to be answered, the answer may foster the development of models to explain the pathogenesis of ER-negative tumors. For example, if the ER inhibits the growth of tumors in some cellular contexts, then a cell that can suppress ER expression would escape this inhibitory effect. Reactivation of ER expression in these tumors would conceivably reverse aberrant, uncontrolled proliferation. At the present time there is little experimental evidence to support or refute this model. However, one line of evidence that can partially support this model is that comedo DCIS, which may be a precursor of aggressive ER-negative tumors, is also ER-negative in most cases (Bobrow et al., 1994). This suggests that ER gene inactivation occurs at the stage of DCIS or earlier lesions of carcinogenesis. Further examination of ER inhibition of growth is clearly warranted, and could enhance our understanding of the biology of ER.

ER-negative/PgR-positive Breast Tumor Phenotype

The observation that some tumors determined to be ER-negative by ligand binding assay still express ER-induced genes, such as PgR and pS2, has led to the search for altered ERs that are functionally constitutively active

(Fuqua et al., 1991). The hypothesis is that expression of such an altered ER that is capable of binding to DNA, but is lacking the ability to bind to estrogen, would appear as an ER-negative tumor in a ligand binding assay. Using reverse transcriptase-polymerase chain reaction (RT-PCR) analysis, we have identified an alternatively spliced ER mRNA variant with a precise deletion of exon 5 (Δ5 variant) from ER-negative/PgR-positive breast tumors (Fuqua et al., 1991). This variant encodes a truncated protein predicted to be lacking a large portion of the HBD, but retaining the AF-1, DBD, and AF-2a domains.

Before expanding on the functional significance of the Δ5 variant, it is important first to emphasize a major technical limitation of many of the reported ER studies using clinical samples. Most altered ERs that have been detected were found by evaluating not the receptor protein itself, but its mRNA. Although it is well documented that the level of ER mRNA expression generally correlates with ER protein levels (Piva et al., 1988), the detection of altered ER mRNA is not necessarily indicative of the expression of an altered ER protein. We would like to suggest that the ultimate clinical significance of these altered ER mRNAs, such as the Δ5 ER variant, will not be determined until studies evaluate the ER alterations at the protein level, a technically challenging task at present.

The Δ5 ER variant is transcriptionally active in the absence of estrogen, and its relative activity compared to the activity of wtER ranges from 15–45%, depending on the cell in which it is evaluated (Fuqua et al., 1991; Castles et al., 1993). Although the Δ5 variant was initially cloned from ER-negative/PgR-positive breast tumors, it was later found to be present in most ER-positive/PgR-positive and ER-positive/PgR-negative tumors (Zhang et al., 1993). This variant has also been reported to be expressed in the breast cancer cell lines BT-20 (Castles et al., 1993), MCF-7 (Fuqua et al., 1991), T47D, ZR-75–1 (Zhang et al., 1993), and MDA-MB-330 (Klotz et al., 1993). Furthermore, it has been determined that the Δ5 variant is often expressed at two- to threefold higher levels than wtER in many tumors and cell lines based on RNase protection analysis (Zhang et al., 1993). However, we have used semiquantitative RT-PCR (Fuqua et al., 1990) to simultaneously measure the Δ5 variant and wtERs in three human breast tumors and have compared this result to that measured by RNase protection assay. Interestingly, discordant reults were obtained (Fuqua et al., 1995). Thus, we believe that one should be cautious in interpreting quantitative results of ER mRNA variants based solely on mRNA measurements. The development of a specific antibody to the Δ5 ER variant would help alleviate this problem. One conclusion that can be made from the existing studies is that the Δ5 variant and wtER are almost always co-expressed in the same tumor. Recently, a study of Δ5 variant expression in 120 breast tumors confirmed that the Δ5 variant was co-expressed along with wtER (Daffada et al., 1995). In this study, about two-thirds of these tumors expressed wtER transcripts as detected by RT-PCR, and most of these patients (88%) co-expressed the Δ5 variant. No patients expressed Δ5 variant mRNA in the absence of wtER mRNA. Another important conclusion from this study was that ER-negative/PgR-positive/pS2-positive tumors expressed significantly higher levels of the variant compared to ER-negative/PgR-negative/pS2-negative tumors (Daffada et al., 1995), supporting our hypothesis that the Δ5 variant may be an important determinant factor in the discordant receptor phenotype (Fuqua et al., 1991).

Although the Δ5 ER variant is co-expressed with wtER in tumors, it is unknown whether these proteins are co-expressed in the same cell, or if tumors are composed of heterogeneous subpopulations, with some clones expressing the variant while others express the wild-type receptor. This is an important question to address because the clinical implications are most probably different for these two possibilities. If the Δ5 variant and wtERs are expressed in different cells of the same tumor, blocking estrogen's growth effects with TAM or another antiestrogen might eliminate only those cells carrying wtER, but not those cells expressing the constitutively active variant. This might eventually lead to the selection of clones expressing the Δ5 variant and the development of clinical hormone independence in a tumor that was initially responsive to growth suppression. In contrast, if the same cell simultaneously expresses both the variant and the wtER, one could envision several possible outcomes based on either the cellular environment or the abundance of the variant versus the wtER protein. For instance, the wild-type and variant receptors might compete for other ER-associated proteins (repressors or activators), or they might compete for binding to estrogen-responsive elements in the DNA. The final outcome could be a gradual conversion from growth control mediated by the wtER pathway to constitutive growth stimulus from the Δ5 variant, with different cell subclones perhaps transitioned at various stages. One possible trigger for such a shift to hormone resistance could be TAM administration during the course of treatment.

Recently, our lab investigated the role of the Δ5 ER variant by stably transfecting it into the MCF-7 breast cancer cell line, which expresses high levels of wtER and which is normally growth-inhibited by TAM. Interestingly, the growth of these transfected cells was not affected by either TAM or 4-hydroxy-TAM administration. However, the growth of both untransfected and Δ5 variant–transfected cells was inhibited by the pure steroidal antiestrogen ICI 164,384, indicating that the Δ5 variant conferred a TAM-resistant phenotype to these cells. This situation mimics the emerging clinical scenario in which breast cancer patients receiving TAM eventually develop resistance to the drug, yet some of these patients will respond when treated with a pure steroidal antiestrogen (DeFriend et al., 1993). We would like to propose a model that might explain some forms of TAM resistance. We know that the bulk of ER-positive tumors will initially respond to TAM, but almost all of them eventually become resistant. If indeed most ER-positive tumors also express the Δ5 variant even before being exposed to TAM (Zhang et al., 1993; Daffada et al., 1995), we would like to suggest that these tumors are destined to become TAM-resistant, and that TAM exposure would trigger the emergence of TAM-resistance. One can envision several mechanisms by which this could occur. For instance, TAM could induce the expression of a repressor protein that associates with wtER, but not the Δ5 variant, thus inhibiting wtER function. In summary, the Δ5 variant may play a role in the development of clinical TAM resistance, and further study of this variant in human breast tumor samples is currently underway in a number of laboratories.

ER-positive/PgR-negative Breast Tumor Phenotype

The ER-positive/PgR-negative receptor phenotype is observed in about 15–20% of breast tumors (Osborne et al., 1980). Collectively, this discordant group of tumors falls in between the ER-positive/PgR-positive and ER-negative/PgR-negative groups in terms of their TAM responsiveness (Osborne et al., 1980; Ravdin et al., 1992). This group is also biologically heterogeneous, with some tumors expressing estrogen-regulated proteins such as pS2 but not PgR, whereas others fail to express either protein. Tumors that are ER-positive and PgR/pS2-negative are those most likely to have a defective estrogen response pathway. One hypothesis is that these tumors contain altered ERs that are unable to bind to DNA and are thus nonfunctional, but are still able to bind estrogen. To investigate this possibility, sequence analysis of ER mRNA obtained from tumors with this phenotype was performed (Fuqua et al., 1992, 1993). We identified two ER variants with a precise deletion of either exon 7 (Δ7) or exon 3 (Δ3), and a functional analysis in a yeast expression system revealed that the Δ7 variant was able to function as a dominant-negative receptor, interfering with the activity of the wild-type receptor while failing to induce transcription itself (Fuqua et al., 1992). The Δ3 variant, which would express a receptor protein missing the second zinc finger and would therefore be expected to lack DNA-binding ability, was not functional in this assay system (Fuqua et al., 1993). These same two ER variants were also cloned from the T47D breast cancer cell line (Wang and Miksicek, 1991), and in functional studies in HeLa cells the Δ3, but not the Δ7 variant, acted as a dominant-negative receptor, contrary to our observations in the yeast expression vector system. This discrepancy in results is not understood at present.

The ultimate functional consequences of potential dominant-negative ER variants will depend on the mechanism by which the altered receptor interferes with wtER function. One question is whether or not the variant affects the DNA binding of wtER. This was the case for the Δ7 ER variant in the yeast system, where binding of wtER to consensus ERE was markedly reduced when wtER was co-expressed along with the Δ7 variant (Fuqua et al., 1992). Thus, it appears that the Δ7 ER variant functions by forming an inactive heterodimer complex with wtER. Squelching of other modulating proteins (such as low abundance transcription factors) by the ER variant is another potential mechanism by which a dominant-negative variant could interfere with wtER function. This may also help explain the discrepancies in the function of the Δ7 ER variant in different assay systems. It is also likely that the relative expression levels of potential dominant-negative variants as compared to the levels of wtER will influence the estrogen response pathway of a tumor. Finally, a dominant-negative variant might play a role in the progression of hormone-independent breast tumor cells by blocking the estrogen response pathway at the transcriptional level. In summary, although altered ERs having a dominant-negative function may play a role in many ER-positive/PgR-negative tumors, further work is required to evaluate their clinical significance and their potential role in the development of hormone independence.

CONCLUSIONS

Our knowledge of the structure/function relationship of the ER and the presence of altered forms of the ER has been greatly expanded on in the last few years. The isolation of several altered forms of the ER in breast tumor samples has led to the generation of several hypotheses that attempt to explain how a tumor may evolve from dependence on estrogen for growth to a state of hormone insensitivity. These hypotheses, if proven correct, may open the door to the design of new therapies to reverse or prevent the emergence of this more aggressive phenotype. However, it is not until the tools become available to evaluate ER variants at the protein level that we will be able to determine clearly their role in clinical breast cancer and the TAM-resistant breast tumor phenotype.

At present, one promising new therapeutic approach in clinical oncology is chemoprevention. For this strategy to be successful, delineation of the molecular evolution of tumors and the pathways involved is essential. Unfortunately, our knowledge of the underlying early genetic events in breast cancer evolution is rudimentary at best. The role of the ER, its various forms, and the dynamic interaction among different cellular receptors in the process of carcinogenesis is an active area of research, which hopefully will enhance our understanding of this process, opening the way to the design of new approaches in the prevention of breast cancer.

ACKNOWLEDGMENT

The unpublished work presented in this chapter was supported by NIH CA52351 to Suzanne A.W. Fuqua.

REFERENCES

Allred DC (1993): Should immunohistochemical examination replace biochemical hormone receptor assays in breast cancer? *Am J Clin Pathol* 99:1–3

Allred DC, O'Connell P, Fuqua SA, Osborne CK (1994): Immunohistochemical studies of early breast cancer evolution. *Breast Cancer Res Treat* 32:13–18

Andersen J, Poulsen HS (1989): Immunohistochemical estrogen receptor determination in paraffin-embedded tissue. Prediction of response to hormonal treatment in advanced breast cancer. *Cancer* 64:1901–1908

Arrick BA, Korc M, Derynck R (1990): Differential regulation of expression of three transforming growth factor β species in human breast cancer cell lines by estradiol. *Cancer Res* 50:299–303

Balaguer P, Demirpence E, Pons M, Gagne D, Bocquel MT, Gronemeyer H, Nicolas JC (1991): Retinoic acid has an antiestrogenic effect on different regulated estrogen genes in different cellular types. *C R Seances Soc Biol Fil* 185:434–443

Beatson GT (1896): On the treatment of inoperable cases of carcinogen of the mamma: suggestions for a new method of treatment with illustrative cases. *Lancet* 2:104–107, 162–167

Bobrow LG, Happerfield LC, Gregory WM, Springall RD, Millis RR (1994): The classification of ductal carcinoma *in situ* and its association with biological markers. *Semin Diagn Pathol* 11:199–207

Castles CG, Fuqua SAW, Klotz DM, Hill SM (1993): Expression of a constitutively active estrogen receptor variant in the estrogen receptor-negative BT-20 human breast cancer cell line. *Cancer Res* 53:5934–5939

Cavailles V, Dauvois S, Danielian PS, Parker MG (1994): Interaction of proteins with transcriptionally active estrogen receptors. *Proc Natl Acad Sci USA* 91:10009–10013

Chambraud B, Berry M, Redeuilh G, Chambon P, and Balieu EE (1990): Several regions of human estrogen receptor are involved in the formation of receptor-heat shock protein 90 complexes. *J Biol Chem* 265:20686–20691

Chrysogelos SA, Dickson RB (1994): EGF receptor expression, regulation, and function in breast cancer. *Breast Cancer Res Treat* 29:29–40

Clark GM, Osborne CK, McGuire WL (1984): Correlations between estrogen receptor, progesterone receptor, and patient characteristics in human breast cancer. *J Clin Oncol* 2:1102–1109

Daffada AA, Johnston SR, Smith IE, Detre S, King N, Dowsett M (1995): Exon 5 deletion variant estrogen receptor messenger RNA expression in relation to tamoxifen resistance and progesterone receptor/pS2 status in human breast cancer. *Cancer Res* 55:288–293

Dana SL, Hoener PA, Wheeler DA, Lawrence CB, McDonnell DP (1994): Novel estrogen response elements identified by genetic selection in yeast are differentially responsive to estrogens and antiestrogens in mammalian cells. *Mol Endocrinol* 8:1193–1207

Danielian PS, White R, Lees JA, Parker MG (1992): Identification of a conserved region required for hormone dependent transcriptional activation by steroid hormone receptors. *EMBO J* 11:1025–1033

Danielian PS, White, R, Hoare SA, Fawell SE, Parker MG (1993): Identification of residues in the estrogen receptor that confer differential sensitivity to estrogen and hydroxytamoxifen. Mol Endocrinol 7:232–240

Davidoff AM, Kerns BJ, Iglehart JD, Marks JR (1991): Maintenance of p53 alterations throughout breast cancer progression. *Cancer Res* 51:2605–2610

DeFriend DJ, Blamey RE, Robertson JF, Walton P, Howell, A (1993): Response to the pure antiestrogen ICI 182780 after tamoxifen failure in advanced breast cancer: 16th annual San Antonio breast cancer symposium (abstract). *Breast Cancer Res Treat* 27:136

Evans RM (1988): The steroid and thyroid hormone receptor superfamily. *Science* 240:889–895

Fawell SE, Lees JA, White R, Parker MG (1990): Characterization and co-localization of steroid binding and dimerization activities in the mouse estrogen receptor. *Cell* 60:953–962

Fearon ER and Vogelstein B (1990): A genetic model for colorectal tumorigenesis. *Cell* 61:759–767

Ferguson AT, Lapidus RG, Davidson NE (1995): Demethylation of the estrogen receptor gene in estrogen receptor-negative breast cancer can reactivate estrogen receptor gene expression. In Proceedings of 86th Ann Meeting of Am Assoc Cancer Res, Toronto, Canada, 269.

Fuqua SAW, Fitzgerald SD, McGuire WL (1990): A simple polymerase chain reaction method for detection and cloning of low-abundance transcripts. *BioTechniques* 9:206–211

Fuqua SAW, Fitzgerald SD, Chamness GC, Tandon AK, McDonnell DP, Nawaz Z, O'Malley BW, McGuire WL (1991): Variant human breast tumor estrogen receptor with constitutive transcriptional activity. *Cancer Res* 51:105–109

Fuqua SAW, Fitzgerald SD, Allred DC, Elledge RM, Nawaz Z, McDonnell DP, O'Malley BW, Greene GL, McGuire WL (1992): Inhibition of estrogen receptor action by a naturally occurring variant in human breast tumors. *Cancer Res* 52:483–486

Fuqua SAW, Allred DC, Elledge RM, Krieg SL, Benedix MG, Nawaz Z, O'Malley BW, Greene GL, McGuire WL (1993): The ER-positive/PgR-negative breast cancer phenotype is not associated with mutations within the DNA binding domain. *Breast Cancer Res Treat* 26:191–202

Fuqua SAW, Wiltschke, C, Castles C, Wolf D, and Allred DC (1995): A role for estrogen-receptor variants in endocrine resistance. *Endocrine-Related Cancer* 2:19–25

Gosden JR, Middleton PG, Rout D (1986): Localization of the human oestrogen receptor gene to the chromosome 6q24 → q27 by *in situ* hybridization. *Cytogenet Cell Genet* 43:218–220

Greene GL, Sobel N, King WJ, Jensen EJ (1984): Immunochemical studies of estrogen receptors. *J Steroid Biochem* 20:51–56

Guiochon-Mantel A, Loosfelt H, Lescop P, Sar S, Atger M, Perrot-Applanat M, Milgrom, E (1989): Mechanisms of nuclear localization of the progesterone receptor: evidence for interaction between monomers. *Cell* 57:1147–1154

Gusterson BA, Machin LG, Gullick WJ, Gibbs NM, Powles TJ, Elliott C, Ashley S, Monaghan P, Harrison, S (1988): c-*erb*-B2 expression in benign and malignant breast disease. *Br J Cancer* 58:453–457

Halachmi S, Marden E, Martin G, MacKay H, Abbondanza C, Brown, M (1994): Estrogen receptor-associated proteins: possible mediators of hormone-induced transcription. *Science* 264:1455–1458

Harlow KW, Smith DN, Katzenellenbogen JA, Green GL, Katzenellenbogen BS (1989): Identification of cysteine 530 as the covalent attachment site of an affinity-labeling estrogen (ketononestrol aziridine) and antiestrogen (tamoxifen aziridine) in the human estrogen receptor. *J Biol Chem* 164:17476–17485

Hill SM, Fuqua SAW, Chamness GC, Greene GL, McGuire WL (1989): Estrogen receptor expression in human breast cancer associated with an estrogen receptor gene restriction fragment length polymorphism. *Cancer Res* 49:145–148

Horwitz KB, McGuire WL (1978): Estrogen control of progesterone receptor in human breast cancer. *J Biol Chem* 253:2223–2228

Horwitz KB, McGuire WL, Pearson OH, Segaloff A (1975): Predicting response to endocrine therapy in human breast cancer: a hypothesis. *Science* 189:726–727

Horwitz KB, Wei LL, Sedlacek SM, D'Arville CN (1985): Progestin action and progesterone receptor structure in human breast cancer: a review. *Rec Prog Horm Res* 41:249–316

Ignar-Trowbridge DM, Teng CT, Ross KA, Parker MG, Korach KS, McLachlan JA (1993): Peptide growth factors elicit estrogen receptor-dependent transcriptional activation of an estrogen-responsive element. *Mol Endocrinol* 7:992–998

Jacquemier JD, Rolland PH, Vague D, Lieutaud R, Spitalier JM, Martin PM (1982): Relationships between steroid receptor and epithelial cell proliferation in benign fibrocystic disease of the breast. *Cancer* 49:2534–2536

Jacquemier JD, Hassoun J, Torrente M, Martin PM (1990): Distribution of estrogen and progesterone receptors in healthy tissue adjacent to breast lesions at various stages—immunohistochemical study of 107 cases. *Breast Cancer Res Treat* 15: 109–117

Jiang S-Y, Jordan VC (1992): Growth regulation of estrogen receptor-negative breast cancer cells transfected with complementary DNAs for estrogen receptor. *J Natl Cancer Inst* 84:580–591

Jiang S-Y, Langan-Fahey SM, Stella AL, McCague R, Jordan VC (1992): Point mutation of estrogen receptor (ER) in the ligand binding domain changes the pharmacology of antiestrogens in ER-negative breast cancer cells stably expressing cDNA's for ER. *Mol Endocrinol* 6:2167–2174

Keaveney M, Klug J, Dawson M, Nestor PV, Neilan JG, Forde RC, Gannon F (1991): Evidence of a previously unidentified upstream exon in the human oestrogen receptor gene. *J Molec Endocrinol* 6:111–115

Khan SA, Rogers MA, Obando JA, Tamsen A (1994): Estrogen receptor expression of benign breast epithelium and its association with breast cancer. *Cancer Res* 54:993–997

Klein-Hitpass L, Schorpp M, Wagner U, Ryffel GU (1986): An estrogen-responsive element derived from the 5' flanking region of the Xenopus vitellogenin A2 gene functions in transfected human cells. *Cell* 46:1053–1061

Klotz DK, Castles CG, Hill SM (1993): Variant ER mRNAs are expressed in the MDA-MB-330 and other human breast tumor cell lines. In Proceedings of 75th Annual Meeting Endocrine Society, Las Vegas NV, 515.

Knight WAI, Livingston RB, Gregory EJ, McGuire WL (1977): Estrogen receptor is an independent prognostic factor for early recurrence in breast cancer. *Cancer Res* 37:4669–4671

Kumar V, Chambon P (1988): The estrogen receptor binds tightly to its responsive element as a ligand-induced homodimer. *Cell* 55:145–156

Kumar V, Green S, Stack G, Berry M, Jin J, Chambon, P (1987): Functional domains of the human estrogen receptor. *Cell* 51:941–951

Kushner PJ, Hort E, Shine J, Baxter JD, Greene GL (1990): Construction of cell lines that express high levels of the human estrogen receptor and are killed by estrogens. *Mol Endocrinol* 4:1465–1473

Matsui Y, Halter SA, Holt JT, Hogan BL, Coffey RJ (1990): Development of mammary hyperplasia and neoplasia in MMTV-TGFα transgenic mice. *Cell* 61:1147–1155

McDonnell DP, Vegeto E, O'Malley BW (1992): Identification of a negative regulatory function for steroid receptors. *Proc Natl Acad Sci USA* 89:10563–10567

McGuire WL (1978): Hormone receptors: their role in predicting prognosis and response to endocrine therapy. *Semin Oncol* 5:428–433

McGuire WL, Carbone PP, Vollmer EP (1975): *Estrogen Receptors in Human Breast Cancer*. New York: Raven Press

Meirik O, Lund E, Adami H (1986): Oral contraceptives and breast cancer in young women. *Lancet* 2:650–653

Metzger D, White JH, Chambon, P (1988): The human oestrogen receptor functions in yeast. *Nature* 334:31–36

Murphy LC, Dotzlaw, H (1989): Regulation of transforming growth factor α and transforming growth factor β messenger ribonucleic acid abundance in T-47D, human breast cancer cells. *Mol Endocrinol* 3:611–617

O'Connell P, Pekkel V, Fuqua S, Osborne CK, Allred DC (1994): Molecular genetic studies of early breast cancer evolution. *Breast Cancer Res Treat* 32:5–12

Osborne CK, Yochmowitz MG, Knight WA, McGuire WL (1980): The value of estrogen and progesterone receptors in the treatment of breast cancer. *Cancer* 46:2884–2888

Ottaviano YL, Issa JP, Parl FF, Smith HS, Baylin SB, Davidson NE (1994): Methylation of the estrogen receptor gene CpG island marks loss of estrogen receptor expression in human breast cancer cells. *Cancer Res* 54:2552–2555

Pakdel F, Katzenellenbogen BS (1992): Human estrogen receptor mutants with altered estrogen and antiestrogen ligand discrimination. *J Biol Chem* 267: 3429–3437

Petersen OW, Hoyer PE, van Deurs, B (1987): Frequency and distribution of estrogen receptor-positive cells in normal, nonlactating human breast tissue. *Cancer Res* 47:5748–5751

Petrangeli E, Lubrano C, Ortolani F, Ravenna L, Vacca A, Sciacchitano S, Frati L, Gulino, A (1994): Estrogen receptors: new perspectives in breast cancer management. *J Steroid Biochem Mol Biol* 49:327–331

Pierrat B, Heery DM, Chambon P, Losson, R (1994): A highly conserved region in the hormone-binding domain of the human estrogen receptor functions as an efficient transactivation domain in yeast. *Gene* 143:193–200

Piva R, Bianchini E, Kumar VL, Chambon P, Del Senno, L (1988): Estrogen induced increase of estrogen receptor RNA in human breast cancer cells. *Biochem Biophys Res Commun* 155:943–949

Poller DN, Roberts EC, Bell JA, Elston CW, Blamey RW, Ellis IO (1993a): p53 protein expression in mammary ductal carcinoma *in situ*: relationship to immunohistochemical expression of estrogen receptor and c-erbB-2 protein. *Hum Pathol* 24:463–468

Poller DN, Snead DRJ, Roberts EC, Galea M, Bell JA, Gilmour A, Elston CW, Blamey RW, Ellis IO (1993b): Oestrogen receptor expression in ductal carcinoma in-situ of the breast: relationship to flow cytometric analysis of DNA and expression of the c-erb-2 oncoprotein. *Br J Cancer* 68:156–161

Ratajczak T, Wilkinson SP, Brockway MJ, Hahnel R, Moritz RL, Begg GS, Simpson RJ (1989): The interaction site for tamoxifen aziridine with the bovine estrogen receptor. *J Biol Chem* 264:13453–13459

Ravdin PM, Green S, Dorr TM, McGuire WL, Fabian C, Pugh RP, Carter RD, Rivkin SE, Borst JR, Belt RJ, Metch B, Osborne CK (1992): Prognostic significance of progesterone receptor levels in estrogen receptor-positive patients with metastatic breast cancer treated with tamoxifen: results of a prospective Southwest Oncology Group study. *J Clin Oncol* 10:1284–1291

Ricketts D, Turnbull L, Ryall G, Bakhshi R, Rawson NSB, Gazet J-C, Nolan C, Coombes RC (1991): Estrogen and progesterone receptors in the normal female breast. *Cancer Res* 51:1817–1822

Roman SD, Clarke CL, Hall RE, Alexander IE, Sutherland RL (1992): Expression and regulation of retinoic acid receptors in human breast cancer cells. *Cancer Res* 52:2236–2242

Roodi N, Bailey LR, Kao WY, Verrier CS, Yee CJ, Dupont WD, Parl FF (1995): Estrogen receptor gene analysis in estrogen receptor-positive and receptor-negative primary breast cancer. *J Natl Cancer Inst* 87:446–451

Smith HS, Lu Y, Deng G, Martinez O, Krams S, Ljung BM, Thor A, Lagios M (1993): Molecular aspects of early stages of breast cancer progression. *J Cell Biochem Suppl* 17G:144–152

Stewart JF, King, R JB, Sexton SA, Millis RR, Rubens RD, Hayward JL (1981): Oestrogen receptors, sites of metastatic disease and survival in recurrent breast cancer. *Eur J Cancer* 17:449–453

Tasset D, Tora L, Fromental C, Scheer E, Chambon, P (1990): Distinct classes of transcriptional activating domains function by different mechanisms. *Cell* 62: 1177–1178

Tesch M, Shawwa A, Henderson, R (1993): Immunohistochemical determination of estrogen and progesterone receptor status in breast cancer. *Am J Clin Pathol* 99:8–12

Trichopoulos D, MacNahon B, Cole, P (1972): Menopause and breast cancer risk. *J Natl Cancer Inst* 48:605–613

Tsutsumi Y, Naber SP, DeLellis RA, Wolfe HJ, Marks PJ, McKenzie SJ, Yin S (1990): neu oncogene protein and epidermal growth factor receptor are independently expressed in benign and malignant breast tissues. *Hum pathol* 21:750–758.

Tzukerman MT, Esty A, Santisomere D, Danielian P, Parker MG, Stein RB, Pike JW, McDonnell DP (1994): Human estrogen receptor transactivational capacity is determined by both cellular and promoter context and mediated by two functionally distinct intramolecular regions. *Mol Endocrinol* 8:21–30

Wakeling AE (1993): Are breast tumours resistant to tamoxifen also resistant to pure antiestrogens? *J Steroid Biochem Molec Biol* 47:107–114

Wang Y, Miksicek RJ (1991): Identification of a dominant negative form of the human estrogen receptor. *Mol Endocrinol* 5:1707–1715

Watts CKW, Parker MG, King RJB (1989): Stable transfection of the oestrogen receptor gene into a human osteosarcoma cell line. *J Steroid Biochem* 34:483–490

Webster NJG, Green S, Jin J-R, Chambon, P (1988): The hormone-binding domains of the estrogen and glucocorticoid receptors contain an inducible transcription activation function. *Cell* 54:199–207

Webster NJG, Green S, Tasset D, Ponglikitmongkol M, Chambon, P (1989): The transcription activation function located in the hormone-binding domain of the human oestrogen receptor is not encoded by a single exon. *EMBO J* 8: 1441–1446

Weigel RJ, deConinck EC (1993): Transcriptional control of estrogen receptor in estrogen receptor-negative breast carcinoma. *Cancer Res* 53:3472–3474

Wingo PA, Tong T, Bolden S (1995): Cancer Statistics, 1995. *CA* 45:8–31

Yaich LE, Dupont WD, Cavener DR, Parl FF (1992): Analysis of the PvuII restriction fragment polymorphism and exon structure of the estrogen receptor gene in breast cancer and peripheral blood. *Cancer Res* 52:77–83

Zhang, Q-X, Borg Å, Fuqua SAW (1993): An exon 5 deletion variant of the estrogen receptor frequently coexpressed with wild-type estrogen receptor in human breast cancer. *Cancer Res* 53:5882–5884

10

Progestins, Progesterone Receptors, and Breast Cancer

KATHRYN B. HORWITZ, LIN TUNG AND GLENN S. TAKIMOTO

Endocrine therapy used either prophylactically or therapeutically for the treatment of locally advanced or metastatic breast cancers offers many advantages to patients whose tumors contain functional estrogen (ER) and progesterone (PR) receptors (Horwitz et al., 1975). The range of treatments defined as endocrine include surgical ablation of endocrine glands, administration of pharmacologic doses of steroid hormones, chemical blockade of steroid hormone biosynthesis, and inhibition of endogenous steroid hormone action at the tumor with synthetic antagonists. The last of these approaches is the most widely used, making the antiestrogen tamoxifen the preferred first-line therapeutic agent for treatment of hormone-dependent metastatic breast cancer. The widespread use of tamoxifen reflects its efficacy and low toxicity, and the fact that it makes good physiological sense to block the local proliferative effects of estrogens directly at the breast. But are estrogens the only hormones with a proliferative impact on the breast and on breast cancers? This chapter focuses on evidence that progesterone also has proliferative actions in the breast; on preliminary data showing that progesterone antagonists may be new tools for the management of metastatic breast cancer; and on recent data suggesting that antiprogestin-occupied PRs have novel mechanisms of action that bear on tissue specificity and development of hormone resistance.

PROGESTERONE AND THE NORMAL BREAST

Conventional wisdom holds that the mechanisms by which estradiol and progesterone regulate the proliferation and differentiation of uterine

Hormones and Cancer
Wayne V. Vedeckis, Editor

epithelial cells apply equally to the breast. This is probably inaccurate (Anderson et al., 1987, 1989; Going et al., 1988). In the uterus, estrogens are mitogenic, and addition of progesterone to the estrogenized endometrium leads to the appearance of a secretory pattern characterized by cells engaged in protein synthesis rather than cell division. Thus, in the uterus, estradiol is a proliferative hormone, whereas progesterone is a differentiating hormone. For this reason the unopposed actions of estradiol are tumorigenic in the uterus, whereas the risk of endometrial hyperplasia and cancer is lowered when progestins are added to the estrogens. In fact, the combined regimen may even be protective because a decrease in endometrial cancers has been reported in women prescribed combined estrogens and progestins, compared to women receiving no treatment (Henderson et al., 1988; Clarke and Sutherland, 1990, and references therein).

However, considerable data suggest that in breast epithelia, progesterone, like estradiol, has a strong proliferative effect. Studies in support of this come from experimental models and from normal cycling women. Both the proliferation of normal mammary epithelium in virgin mice, and the lobular-alveolar development of mammary tissues in pregnant mice, require progesterone (Imagawa et al., 1985; Haslam, 1988). A fundamental difference in the actions of estradiol and progesterone in the breast is that the latter stimulates DNA synthesis, not only in the epithelium of the terminal bud, but also in the ductal epithelium (Bresciani, 1971). The stimulating effects of progesterone on the development of mammary gland buds can be inhibited by progesterone antagonists (Michna et al., 1991).

Studies of the mitotic rate in breast epithelial cells during the normal menstrual cycle and in women taking oral contraceptives also support a proliferative role for progesterone. They show that the highest thymidine labeling indices in the breast occur during the progestin-dominated, secretory phase of the menstrual cycle. Both the estrogenic and the progestational components of oral contraceptives increase the thymidine labeling index with progestin-only formulations exhibiting high activity (Going et al., 1988). Investigators conclude that it is difficult to sustain the idea that progestins are protective in the breast (Anderson et al., 1989). Clearly, more work must be done to understand the actions of progestins in the normal breast, but clinical decisions based on inappropriate uterine models are flawed.

PROGESTERONE AND BREAST CANCER

A discussion of the role of progestins in breast cancer must distinguish between their effects on carcinogenesis and their role in regulating proliferation of established cancers.

Progestin Agonists and Tumor Induction

Progestin agonists have been shown to be carcinogenic or to increase the incidence of spontaneous mammary tumors in dogs and mice. In mice, results vary with the strain tested, suggesting the contribution of a genetic component; however, tumorigenic effects of progestins have been observed whether or not the strain harbors the mouse mammary tumor virus (MMTV). The importance of progesterone in carcinogen-induced rat mammary cancers is documented by the early reports (Huggins et al., 1962; Huggins and Yang, 1962; Huggins, 1965) that showed that pregnancy promotes the growth of dimethylbenzanthracene (DMBA)-induced mammary tumors, and that administration of progesterone together with the carcinogen to intact rats accelerates the appearance of tumors, increases the number of tumors, and augments the growth rate of established tumors. The relationship between progestins and carcinogenesis is temporally complex. In general, progesterone administered simultaneously with, or after the carcinogen enhances tumorigenesis, whereas progesterone administered before the carcinogen inhibits tumorigenesis (Welsch, 1985, and references therein). Thus, the high progesterone levels associated with pregnancy can be protective if they precede the administration of the carcinogen (Russo et al., 1989). Extrapolation of these experimental models to human disease is difficult because the only data available for the latter are epidemiologic in nature and relate hormone use, particularly oral contraceptive use, to the risk of breast cancer. The trend toward increased risk with increased duration of hormone use appears repeatedly (Meirik et al., 1986; Hulka,1990), so that an adverse effect of progestins is likely (Ewertz, 1988; Bergkvist, 1989). This is discouraging when taken together with the probability that in the breast, unlike the uterus, progestins enhance proliferation.

Progestin Agonists and Tumor Growth

Carcinogen-induced rat mammary tumors are a major model for *in vivo* studies of progestin-regulated growth. Following ovariectomy, progesterone alone is usually unsuccessful in preventing regression of established tumors. The rapid decrease of PR levels due to estrogen withdrawal is probably a critical factor (Horwitz and McGuire, 1977). In intact animals, which more closely mimic the clinical situation, progestin agonists at moderate doses promote tumor growth and reverse the antitumor effects of tamoxifen (Robinson and Jordan, 1987). Thus, there exists the possibility that endogenous circulating progesterone enhances breast cancer growth. Enigmatically, progestins at higher pharmacologic doses are growth inhibitory. The molecular mechanisms responsible for the opposing actions of physiological and high-dose progestins remain unclear.

In vitro cell culture models designed to assess the role of progestin agonists in tumor cell proliferation have generated contradictory results. Experiments can be cited in support of the fact that progestins stimulate, inhibit, or have no effect on growth (Horwitz, 1992, and references therein). Explanations for the lack of a consensus are as varied as the results. Responses of cells in culture are critically dependent on the conditions in which they are grown. In a rich medium, where growth is optimized, further growth enhancement is difficult to demonstrate, whereas inhibitory stimuli may be exaggerated. In a deprived medium the reverse is true, although here key co-factors may be lacking. There is no simple solution to these inherent problems. Couple this generic uncertainty with other variables including the use of different cell lines, heterogeneity and genetic instability even within the same cell lines, a burgeoning list of factors besides estradiol and progestins that directly or indirectly modulate progestin sensitivity through regulation of PR levels, the possibility that progestin-sensitive cells can generate resistant subpopulations, and the lack of consensus is not surprising. The timing of events is also critical since recent data demonstrate a biphasic effect of progestins, in which a transient proliferative burst is followed by chronic growth suppression (Musgrove and Sutherland, 1993).

Where does this leave us on the critical issue of the use of progestin agonists in breast cancer treatment? Interestingly, here there is more agreement, but the data contradict the conclusion that physiological levels of progestins are growth stimulatory. Especially at high doses, progestins appear to be antiproliferative in breast cancers. A comprehensive review of the clinical literature shows that synthetic progestins, used at pharmacological doses for first- or second-line therapy, are as effective as tamoxifen in the treatment of advanced breast cancer (Sedlacek and Horwitz, 1984). That is, in patients whose tumors are not screened for steroid receptors, approximately 30% have an objective, positive response. Since, in addition, progestins are well tolerated and have a relatively low toxicity, their use in the treatment of advanced breast cancer is experiencing a resurgence. However, the mechanisms underlying the actions of intermediate and high doses of progestin agonists in breast cancer regression remain unclear, when compared to their proliferative actions at physiological doses. Some studies suggest that PR-negative tumors respond just as well as do PR-positive tumors (implying that PRs are not involved), but others suggest that methodological problems produce false PR-negative values in responders, and that PRs are indeed required to obtain a response to progestins. An interesting study, in which tamoxifen therapy was compared to therapy in which tamoxifen was alternated with medroxyprogesterone acetate (MPA) in ER-positive patients, showed a 40% response to tamoxifen alone versus a 62% response with the alternating treatment (Gundersen et al., 1990). It is postulated that when tamoxifen is cycled, its agonist properties predominate, which increases PR levels (Horwitz and McGuire, 1978), thereby enhancing the efficacy of MPA. The same argument is made for enhancing the therapeutic efficacy of antiprogestins (Canobbio et al., 1987). In general, tamoxifen inducibility of PR is considered to be a good indicator for a positive response to hormone therapy (Howell et al., 1987). Thus, although definitive data are still lacking, it is likely that responses to progestin therapy in breast cancer are mediated by their PR.

PROGESTERONE ANTAGONISTS AND BREAST CANCER

Animal Models

The antiproliferative properties of progesterone antagonists are well documented in animal models of hormone-dependent mammary cancer. These include rats bearing DMBA-induced or nitrosomethylurea (NMU)-induced tumors, and mice bearing the transplantable MXT tumor line. Growth of these tumors is inhibited by ovariectomy and maintained by physiological doses of estrogens (Welsch, 1985). Treatment of rats with progestins at the time of DMBA administration accelerates tumor formation. In contrast, prophylactic treatment of rats with RU486 (an antiprogestin) at the time of DMBA administration delays the initial appearance of tumors from an average of 39 days to more than 80 days (Bakker et al., 1987). RU486 also blocks progesterone-induced proliferation in tamoxifen suppressed tumors (Robinson and Jordan, 1987), a fact that implicates PR in the RU486 mechanism.

Treatment of established DMBA tumors with RU486 prevents their further enlargement, analogous to the effect obtained with tamoxifen (Bakker et al., 1989; Klijn et al., 1989). When the two antagonists are combined, the inhibition is additive, leading to tumor remission similar to that induced by ovariectomy (Bakker et al., 1990). This effect of combined treatment with an antiprogestin and antiestrogen is extremely exciting and has considerable therapeutic promise. The mechanisms underlying their cooperativity are unclear, but several proposals have surfaced. First, tamoxifen can have agonist actions, among which one is the induction of PR (Horwitz and McGuire, 1978). A tumor with increased, or restored, PR may have greater or more sustained sensitivity to RU486. This hypothesis could be tested by the use of an antiestrogen having no agonist activity. Second, among the physiological effects seen in RU486-treated intact female rats are increased plasma levels of LH, prolactin, estradiol, and progesterone, as well as the persistence of numerous and actively secretory corpora lutea associated with hypertrophic pituitaries (Bakker et al., 1989, 1990; Michna et al., 1989a,b; Schneider et al., 1989). It has therefore been proposed that the efficacy of simultaneous tamoxifen results from its ability to counteract the proliferative effects of the high estrogen levels induced by RU486.

Several newer antiprogestins, ORG31710 and ORG31806 (Bakker et al., 1990), and ZK98299 and ZK112993, have equal or greater antiproliferative actions than RU486 (Michna et al., 1989a,b). In the hormone-dependent MXT-transplantable tumor model, treatment with ZK98299 or RU486 starting one day after transplantation led to an almost complete inhibition of tumor growth. Their effect on established tumors was equivalent to that of ovariectomy (Michna et al., 1989a; Schneider et al., 1989). In this model, the potent antiproliferative actions of the antiprogestins completely counteracted the growth-stimulatory actions of estradiol, or of approximately equimolar doses of MPA; however, at higher MPA doses, the agonist actions of the progestin prevailed (Michna et al., 1989b). It appears that antiprogestins inhibit growth by direct antagonism of progesterone action at the tumor, probably mediated by PR. This conclusion is bolstered by the fact that the hormone-independent MXT tumor is antiprogestin-resistant (Michna et al., 1989a).

In DMBA-induced tumors, ZK98299 was more potent than an equal concentration of RU486. It produced tumor regression analogous to that of ovariectomy, rather than the tumor stasis observed with RU486. A similar trend was observed with NMU-induced rat mammary tumors (Michna et al., 1989a,b; Schneider et al., 1989). However, lack of comparative metabolic and pharmacokinetic data on the two antiprogestins in rats and mice makes these quantitative differences uninterpretable at present. Also of interest is the finding that strong antitumor activity was noted at 20% of the doses needed to obtain abortifacient actions in these rodent systems. This is important because by use of lower doses of antiprogestins, their antiglucocorticoid effects may be mitigated. After treatment with the antiprogestins, the morphology of the hormone-dependent MXT and DMBA tumors showed signs of differentiation of the mitotically active polygonal epithelial tumor cells toward the nonproliferating glandular secretory pattern, with formation of acini and evidence of secretory activity (Michna et al., 1989a). Based on this, it is suggested that the antiproliferative efficacy of the antiprogestins is related to their ability to induce terminal differentiation, perhaps by blocking tumor cells in G0/G1 (Michna et al., 1990). Note that an antiproliferative mechanism based on induction of terminal differentiation is fundamentally different from a mechanism involving tumor cell death, which are features of ovariectomy-induced regression. The mechanisms underlying the antitumor effects of antiprogestational agents require further study, especially in human tissues and cells. Additionally, little is known about the chronic endocrinologic effects of different antiprogestins in humans, especially on the pituitary-gonadal axis.

Human Clinical Trials

The promise of progestin antagonists to treat breast cancer remains largely unexplored in clinical practice. Only two small clinical trials using RU486 have been reported. The first involved a series from France (Maudelonde et al., 1987) of 22 oophorectomized or postmenopausal patients in whom chemotherapy, radiotherapy, or tamoxifen and other hormonal therapy had already been used. RU486 at 200 mg/day led to partial regression or stabilization of lesions in 12 of 22 (53%) women following 4 to 6 weeks of treatment. The response rate at 3 months had dropped to 18%. It is important to note that for ethical reasons, this untried therapy was used only in patients with advanced breast cancers in whom other treatment modalities had already failed.

The second trial, from the Netherlands (Michna et al., 1989b; Bakker et al., 1990), involved 11 postmenopausal patients with metastatic breast cancer who were treated with 200 to 400 mg of RU486 for 3 to 34 weeks as second-line therapy after first-line treatment with tamoxifen, irrespective of the response to tamoxifen. After RU486 treatment, 6 of 11 patients had a short-term (3–8 months) stabilization of disease and one had an objective response lasting 5 months. Several newer trials of RU486, onapristone (ZK98299), and others are ongoing; however, none is being carried out, to our knowledge, in which the antiprogestins are tested as first-line therapy alone or in combination with tamoxifen.

PROGESTERONE RECEPTORS AND RESISTANCE TO PROGESTERONE ANTAGONISTS

The emergence of hormone-resistant cells eventually reduces the effectiveness of all endocrine therapies in advanced breast cancer, and progestin agonists or antagonists are unlikely to be exceptions. In general, resistance is defined as tumor progression while therapy is ongoing, and results either from loss of responsiveness to the hormone, or alternatively to a switch in responsiveness to the hormone. Antagonists behaving like agonists would represent such a switch. The mechanisms by which steroid hormone antagonists produce unexpected agonist-like effects, or different tissue-specific effects, are unknown, but have important clinical implications. For example, tamoxifen, the estrogen antagonist used widely to treat breast cancers, is an agonist in bone and uterus and has estrogenic effects on lipid and lipoprotein levels. Tamoxifen can even be an agonist in breast cancers, producing undesirable side effects that exacerbate the disease. Thus, at the start of tamoxifen therapy, patients often experience an estrogenic tumor flare; tamoxifen induces tumor PR; and after long-term tamoxifen therapy inappropriate proliferative effects camouflage as "resistance." A number of different mechanisms probably account for these agonist-like effects of tamoxifen and some may be activated concomitant with development of resistance (reviewed in Horwitz, 1995).

Although antiprogestins will undoubtedly also prove to be useful hormonal agents for the treatment of breast cancer, agonist-like proliferative effects have been reported with RU486 in cultured breast cancer cell lines and in postmenopausal women, under conditions in which inhibition would be

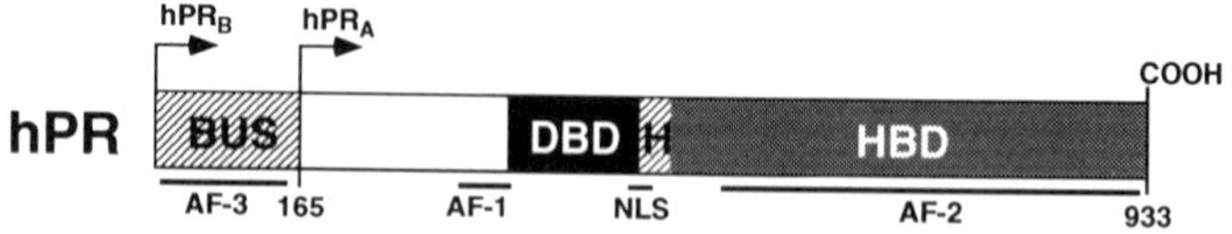

Figure 1. Model of human progesterone receptors (hPR) showing origin of B-receptors (hPR_B) and A-receptors (hPR_A). *BUS*, B-upstream segment; *DBD*, DNA binding domain; *H*, hinge region; *HBD*, hormone binding domain; *NLS*, nuclear localization signal; *AF*, activation function.

expected. Thus, these agents will not be exempt from the resistance issue. Recent studies in our laboratory have addressed molecular mechanisms by which these unexpected actions of antiprogestins occur. The studies focus on the two natural isoforms of human (h)PR: B-receptors (hPR_B), which are 933 amino acids in length, and-A-receptors (hPR_A), which lack 164 amino acids at the N-terminus but are otherwise identical to B-receptors. When A- and B-isoforms are present in equimolar amounts in wild-type PR-positive cells or are transiently co-expressed in PR-negative cells, they dimerize and bind DNA as three species: A/A and B/B homodimers, and A/B heterodimers. This heterogeneity has complicated the study of antiprogestins because each isoform and dimeric species has a unique ligand response profile that varies with the gene and cell in question (Horwitz, 1992).

The structure of PR and other members of the steroid receptor family of proteins has been extensively reviewed (see Truss and Beato, 1993, or Horwitz, 1992, or Chapter 4, this volume, and references therein). Briefly, these receptors are intranuclear transcription factors whose DNA binding capacity is activated by ligand occupancy. Receptor proteins bind as dimers to specific transcriptional enhancers called hormone response elements (HRE), and regulate gene expression. DNA binding is effected through a DNA binding domain (DBD) in the protein that is composed of two zinc fingers. The DBD also has weak dimerization functions and may be involved in other protein-protein interactions. Upstream of the DBD is an N-terminal region that usually contains at least one activation function (AF-1). Downstream of the DBD is a nuclear localization sequence (NLS), followed by a C-terminal region that contains the hormone binding domain (HBD), a strong dimerization domain, a second transcriptional activation function (AF-2), and other regions involved in protein-protein interactions. The 164-amino acid hPR B-receptor upstream segment (BUS) in the far N-terminus contains a third activation function (AF-3) (Fig. 1).

Conventional Inhibitory Actions of Antiprogestins

Two fundamentally different mechanisms underly the actions of antagonist-hPR complexes. First is the classical effect of antagonists; namely, their ability to directly inhibit agonist actions (Horwitz, 1992). In this scenario, agonist-occupied hPR regulate transcription by binding as dimers to PREs present on the promoter of the regulated gene. Antagonist-occupied hPR complexes also bind to PREs but are transcriptionally inert. Thus, by this mechanism, antagonist inhibition involves competition between the two ligands, agonist versus antagonist, for hPR occupancy, followed by competition between the two ligand-hPR classes for binding to PREs. With agonists, DNA binding leads to a specific transcriptional response, whereas with antagonists, the DNA binding is nonproductive (Fig. 2Ia). It follows that the nonproductive or inhibitory potency of an antagonist is controlled by numerous factors, which include its affinity for the receptors, the affinity of antagonist-occupied hPR complexes for PREs, the number and occupancy of PREs on a promoter, and probably other factors. As is discussed below, the two hPR isoforms often have unequal inhibitory potency. In general, we find that, at equimolar concentrations, antagonist-occupied A-receptors are stronger transcriptional inhibitors and proliferation inhibitors than B-receptors.

Non-conventional Inhibitory Actions of Antiprogestins

In addition to, or complementing, their inhibitory effects through direct competition with agonist-occupied hPR at PREs, recent evidence suggests that antiprogestin-occupied hPR can be inhibitory without direct binding to DNA at PREs. Truss et al. (1994) recently showed by *in vivo* footprinting of the MMTV promoter that under conditions in which antiprogestin-hPR complexes inhibit transcription by PRE-bound agonist-occupied hPRs, the antagonist-hPR complexes are not bound to the PREs. We have shown, in unpublished studies, that antiprogestin-occupied hPR mutants whose DNA binding specificity has been switched from a PRE to an estrogen response element (ERE) can nevertheless still inhibit transcription by a constitutively active PRE-bound hPR. The mechanisms underlying this inhibition by a receptor that cannot bind a PRE are unknown but could, in theory, be due to sequestration by soluble antagonist-occupied hPRs of an accessory factor required by the PRE-bound agonist-occupied hPR (Fig. 2Ib). Again in our hands, A-receptors are more potent inhibitors than B-receptors through this pathway.

Additionally, it is now becoming evident that antagonist-occupied A-receptors, but not B-receptors, have more general inhibitory properties than previously thought. As discussed further below, antagonist-occupied A-receptors can inhibit transactivation by ER from an ERE-reporter construct—inhibition that, again, does not require hPR_A binding to PREs. The mechanisms are unknown, but squelching of one or more transcriptional regulatory proteins common to the function of several steroid receptors can be envisioned (Fig. 2Ic).

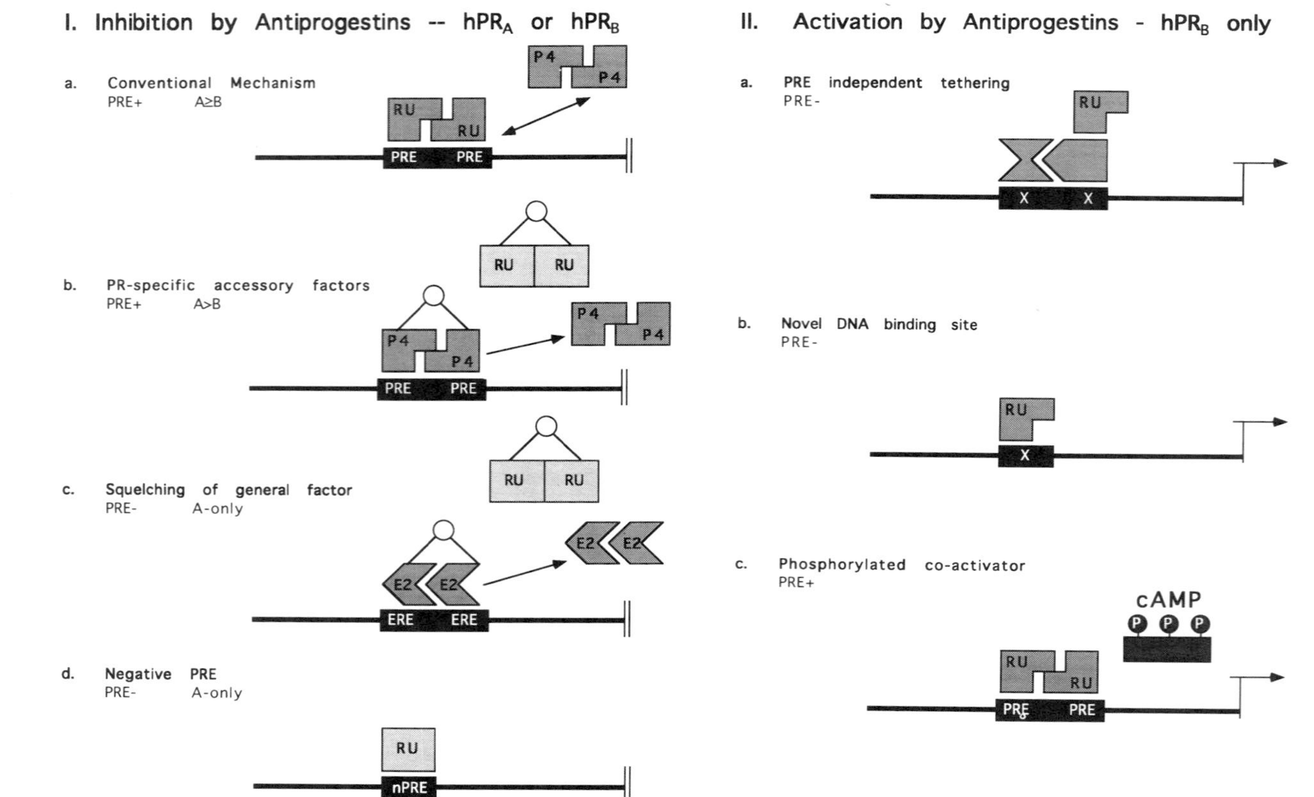

Figure 2. Models of antiprogestin action described in the text. A-receptors are depicted as *squares*; B-receptors are *L-shaped*. *RU*, RU486; *P4*; progesterone; *E2*, estradiol bound to estrogen receptors.

Finally, another mechanism for transcriptional inhibition—once more restricted to A-receptors—is through noncanonical DNA binding sites (hPREs) analogous to the negative glucocorticoid response elements (nGREs). One such nPRE has been described in the bovine prolactin promoter (Cairns et al., 1993). Our unpublished studies show that anti-progestin-occupied A-receptors, but not B-receptors, inhibit constitutive transactivation through this promoter (Fig. 2Id).

Novel Transactivation Mechanisms of Antagonist-occupied hPR_B

In addition to novel mechanisms by which antagonist-occupied hPR can be inhibitory, we have been exploring conditions under which antiprogestins can have agonist-like effects. We find that regardless of the model system in which such agonist-like effects can be elicited, it is a property restricted to PR B-receptors (Fig. 2II). In some conditions, we find that antagonist-occupied hPR_B have inadvertent transcriptional stimulatory actions through DNA-binding sites or DNA-binding proteins that do not involve the canonical PREs. These novel agonist-like effects could, in theory, affect not only genes that contain PREs, but even genes that were never meant to be regulated by hPR and on which agonists have no effects. Such mechanisms could explain how an antiprogestin can inadvertently regulate genes that are not normally targets of progesterone control. There are several experimental models that demonstrate these unusual mechanisms.

PRE-INDEPENDENT TRANSACTIVATION OF THE THYMIDINE KINASE PROMOTER. We have studied the transcriptional activity of antagonist-occupied hPRs using a chloramphenicol acetyltransferase (CAT) reporter driven by a PRE cloned upstream of the thymidine kinase (*tk*) gene promoter (Tung et al., 1993). Transient transfection of HeLa cervicocarcinoma cells with an hPR_A expression vector and treatment with the progestin agonist R5020 leads to a 20-fold increase in CAT transcription compared to basal levels. As expected, none of three antiprogestins, RU486, ZK112993, or ZK98299, stimulate transcription. Instead, the antiprogestins typically suppress basal levels of transcription. However, in cells expressing hPR_B, not only the agonist, but all three antagonists, strongly stimulate CAT transcription. We were surprised that ZK98299 was a transcriptional activator on hPR_B because, in our hands, receptors occupied by this antagonist do not bind to DNA *in vitro* or *in vivo*. This suggested that transcriptional activation by antagonist-occupied hPR_B was independent of PRE binding. We therefore removed the PRE from the *tk* promoter-reporter constructs to test this theory. As expected, agonist-dependent transcription was abolished when the PRE was removed, but to our surprise, the anomalous antagonist-dependent transactivation was retained. Similar results were observed with a DNA-binding domain (DBD) B-receptor mutant whose specificity was altered so that it would no longer recognize a PRE but would instead bind an ERE. When occupied by antiprogestins, these mutant B-receptors still activated transcription of the PRE-containing reporter.

Recent data show that other members of the steroid receptor superfamily can have effects independent of the canonical HREs. Potential mechanisms fall into two broad categories. Either the receptors bind to novel DNA sites that differ substantially from the consensus HREs, or the receptors do not bind DNA at all, but interact with other DNA-binding proteins instead (Oro et al., 1988; Sakai et al., 1988; Diamond et al., 1990; Jonat et al., 1990; Schüle et al., 1990; Yang-Yen et al., 1990; Miner and Yamamoto, 1991; Kutoh et al., 1992). By the latter mechanism, termed factor tethering, two factors establish protein-protein contacts on the DNA, but only one of the two actually binds DNA. However, both the DNA-bound protein and its tethered partner contain a DBD. This model is of particular significance for antagonist-occupied hPR_B-mediated transcription, because we also find a requirement for an intact DBD. Thus, an hPR_B mutant lacking an ordered first zinc finger fails to stimulate transcription when occupied by RU486. In addition to its DNA-binding function, the DBD of steroid receptors is implicated in mediating protein-protein interactions (Diamond et al., 1990; Schüle et al., 1990; Yang-Yen et al., 1990), perhaps through conserved surfaces that face away from the DNA. Indeed, several recent studies show that glucocorticoid receptors (GR) and cJun repress one another's activity by protein-protein binding mechanisms that are independent of DNA binding. Nevertheless, to produce repression, an intact GR DBD is required. Additionally, a dimerization function has been assigned to the second zinc finger (Härd et al., 1990), providing further evidence that the DBD mediates protein interactions. We speculate that induction of transcription by antagonist-occupied hPR_B can proceed through a mechanism in which the receptors are tethered to a DNA-bound protein partner, but do not bind DNA themselves (Fig. 2IIa). Alternatively, hPR_B could function by linking an activator protein bound to the *tk* promoter, to the basal transcriptional machinery, or by binding to novel, as yet undefined, DNA elements (Fig. 2IIb).

ANTAGONIST-OCCUPIED HUMAN hPR_B ARE SWITCHED TO AGONISTS BY cAMP. By contrast, we have described an entirely different antagonist-mediated activation mechanism, in which hPR does have to be bound to DNA (Sartorius et al., 1993). Because we have a specific interest in the actions of steroid antagonists in breast cancer, we studied antiprogestins in a derivative of T47D human breast cancer cells that express high endogenous levels of hPR_B and hPR_A and that stably express the MMTV promoter cloned upstream of the CAT gene. Treatment of these cells with the agonist R5020 produces high levels of CAT. When tested alone, the three antiprogestins, RU486, ZK98299, and ZK112993, are unable to stimulate transcription, and all three inhibit R5020-mediated transcription. Thus, in this model, all three antiprogestins are good antagonists, presumably through the conventional pathway shown in Fig. 2Ia.

However, when cellular cyclic adenosine monophosphate (cAMP) levels are raised, two of the antagonists demonstrate a surprisingly strong agonist activity: when present alone, RU486 and ZK112993 are transcriptionally inactive, but in the presence of 8-Br-cAMP, their transcriptional effect is agonist-like. Of interest is the fact that ZK98299 is entirely different, and despite elevated cAMP levels, this antagonist does not function as an agonist. Recall that ZK98299-occupied hPRs either do not bind to DNA at PREs or have anomalous DNA-binding properties. From this and other controls we deduce that in order for antagonist-occupied hPRs to become transcriptional activators under cAMP control, the receptors have to be bound to DNA, presumably at the PREs.

The amplification of steroid-mediated responses in the presence of cAMP is not limited to hPRs. Even though PR levels are overexpressed in T47D cells, the levels of GRs, androgen (AR), and ERs are extremely low. In addition to hPRs, the PREs of the MMTV promoter can be regulated by ARs and GRs (Cato et al., 1987). However, the MMTV promoter lacks an estrogen response element, and is not regulated by ERs. In T47D cells expressing the MMTV-CAT reporter, neither dexamethasone nor dihydrotestosterone stimulate CAT transcription, suggesting that GR and AR levels are too low in these cells to activate this promoter in the absence of other influences. However, when cAMP levels are raised, the cells acquire sensitivity to the steroid hormones, resulting in strong transcription. Thus, 8-Br-cAMP sensitizes the MMTV promoter to the actions of glucocorticoids and androgens. In contrast, no transcriptional amplification is seen with estradiol, consistent with the inability of ERs to bind the MMTV promoter. Since the MMTV promoter lacks an ERE, this again suggests that the cooperative effects of 8-Br-cAMP require that the receptors be bound to DNA. In fact, when ER are tested on an ERE-containing promoter-reporter in the presence of elevated cellular levels of cAMP, the antiestrogen tamoxifen also becomes a strong transactivator. Thus, cAMP-induced amplification of transcription is a general property of antagonist-occupied steroid receptors.

Signal transduction pathways ultimately converge at the level of transcription to produce patterns of gene regulation that are specific to the gene and cell in question. Composite promoters may be regulated by multiple independent and interacting factors. In extreme cases, a transcription factor can yield opposite regulatory effects from one DNA-binding site because of modulation by a second factor. A case in point are GRs, which regulate proliferin gene transcription either positively or negatively. The direction of transcription by glucocorticoids is selected by DNA-bound Jun and Fos, which are postulated to interact with GRs at composite GREs. cAMP-responsive signal transduction pathways are often involved in such cooperative interactions. These models suggest that on complex promoters, nonreceptor factors, among which are cAMP-regulated proteins, can interact with steroid receptors to select the direction of transcription (Gruol et al., 1986; Diamond et al., 1990).

Our studies demonstrate that cAMP can amplify the transcriptional signals of agonist-occupied steroid receptors and can switch the transcriptional direction of some antiprogestins to render them potent agonists—an effect that can have unintended clinical consequences. We believe that this functional reversal requires that hPRs bind to DNA, but that it is not due to ligand-independent phosphorylation of the receptor or direct activation of the receptors by protein kinase A-dependent pathways. We find that elevated cAMP levels do not enhance phosphorylation of hPRs in breast cancer cells and do not modulate the hormone-dependent phosphorylation induced by progestins, and we therefore conclude that cAMP does not directly influence hPR activity by phosphorylating the receptors (Sartorius et al., 1993). Instead, our data are consistent with a model in which the direction of transcription by DNA-bound hPRs is indirectly regulated by "co-activator" proteins whose activity is perhaps controlled by cAMP-dependent phosphorylation (Fig. 2IIc). This cooperativity between two signal transduction pathways, one involving steroid receptors, another involving cAMP-regulated proteins, requires that the steroid receptors bind to DNA. It therefore does not occur on the MMTV promoter with ERs, or when hPRs are occupied by ZK98299. However, it does occur with many steroid receptors when they are bound to their cognate HREs, suggesting that the "co-activator" is a general mediator of steroid hormone-induced transcription. As we show again below, PR A-receptors do *not* activate transcription by this route, suggesting that there are fundamental differences between the mechanisms of action of the two PR isoforms.

New T47D Breast Cancer Cell Lines for the Independent Study of Progesterone B- and A-receptors: Only Antiprogestin-occupied hPR_B are Switched to Transcriptional Agonists with cAMP. The studies with wild-type T47D cells described above do not permit analysis of the relative contributions of hPR_B and hPR_A to the synergism observed with cAMP because these cells contain mixtures of the two receptors. However, their constitutive high-level production of PRs have made T47D cells the major model in which to study the actions of progesterone in human breast cancer cells, unencumbered by the need for estradiol priming. Because of several special phenotypic properties of T47D cells and because factors other than receptors may be missing in persistently receptor-negative cells, we thought it prudent to retain the T47D cellular milieu in developing new models to study the independent actions of the two PR isoforms (Sartorius et al., 1994a). First we needed a PR-negative T47D subline. We developed a monoclonal PR-negative cell line, called T47D-Y, by selecting a PR-negative subpopulation from a parental T47D line that contained mixed PR-positive/PR-negative cells identified by flow cytometry. T47D-Y cells are PR-negative immunologically and by ligand binding assays, by growth resistance to progestins, by failure to bind a PRE *in vitro*, and by failure to transactivate PRE-regulated promoters.

T47D-Y cells were then stably transfected with expression vectors encoding one or the other hPR isoform, and two monoclonal cell lines were selected that express only B-receptors (called T47D-YB) or only A-receptors (called T47D-YA). The ectopically expressed receptors are properly phosphorylated, and like endogenously expressed receptors, they undergo ligand-dependent down-regulation. The expected B/B or A/A homodimers are present in cell extracts from each cell line, but A/B heterodimers are missing in both (Sartorius et al., 1994a).

To study isoform-specific transcriptional effects of agonists and antagonists when cAMP levels are raised, YA or YB cells were transiently transfected with the MMTV-CAT reporter and treated with R5020 or the three antiprogestins in the presence or absence of 8-Br-cAMP. 8-Br-cAMP alone does not stimulate CAT synthesis in either cell line. In YA cells, R5020 alone moderately stimulated CAT transcription from MMTV-CAT, and the agonist effect was synergistically enhanced by raising cAMP levels. Thus, agonist-occupied A-receptors are relatively weak transactivators, the activity of which is strongly enhanced by cAMP. When only A-receptors are available, as they are in YA cells, the three antiprogestins RU486, ZK98299, or ZK112993 have no intrinsic agonist-like activity and 8-Br-cAMP does not alter this.

However, the agonists and antagonists have quite different effects in the B-receptor-containing YB cells. R5020-regulated transcription from the MMTV-CAT reporter is very strong in these cells, with little further cAMP amplification. In contrast, 8-Br-cAMP strongly enhances the transcriptional phenotype of two of the antagonists, RU486 and ZK112993. Both of these antiprogestins are weak agonists on the MMTV-CAT promoter, but become strong agonists when 8-Br-cAMP is added. The antagonist ZK98299 is entirely different because it has no intrinsic agonist activity alone and no enhancement is produced by 8-Br-cAMP. This resistance of ZK98299 in YB cells to the activating effects of cAMP is similar to the one we described in wild-type T47D cells that contain the natural mixture of both receptors.

These studies using our new stable cell lines show that the two PR isoforms behave differently in their cooperativity with cAMP. With regard to R5020, the synergism between cAMP and agonist-occupied receptors is most pronounced in YA cells. We speculate that in YA cells, cAMP sensitizes the MMTV promoter to the weak signal transmitted by R5020-occupied A-receptors. This is similar to the manner in which cAMP amplifies the weak signals transmitted by hormone-occupied GR and AR in wild-type T47D cells. Since in YB cells agonist-occupied B-receptors are already strong transactivators, cAMP has only modest further effects on this isoform.

With regard to progesterone antagonists, the isoform specificity of the cAMP effect is even more interesting. We find absolutely no effect of cAMP in YA cells, perhaps because the antagonists (specifically RU486 and ZK112993) exhibit no agonist-like activity on A-receptors. Is there no minimal signal for cAMP to amplify? In contrast, the two antagonists appear to have some weak agonist-like activity in YB cells; hence, cAMP strongly amplifies this signal, converting the antagonist-occupied B-receptors to potent transactivators. Therefore, it is significant that B-receptors occupied by the antiprogestin ZK98299 are not subject to this functional modulation by cAMP. We speculate that ZK98299-occupied PRs are physically removed from cAMP control by their failure to bind DNA. This again implies that cooperativity between DNA-bound PRs and a cAMP-regulated coactivator accounts for the transcriptional synergism depicted in Fig. 2IIc.

In additional studies, the details of which will be reported elsewhere, we find that cell growth regulation by progestins is also modulated by cellular cAMP levels. In the first 48 hours of treatment, R5020 stimulates proliferation, and the antiprogestins inhibit the agonist. However, when cAMP levels are raised the antiprogestins have agonist-like proliferative effects—but only in YB cells. We believe that the ratio of B- to A-receptors in tumors, and perhaps also in normal progesterone target tissues, controls the response to progestational agents, and that in the future not only their PR status, but also their isoform distribution, will have to be defined.

A-RECEPTORS ARE TRANSDOMINANT REPRESSORS OF B-RECEPTORS WITHOUT BINDING A PRE. If B- and A-receptors are so different, what happens when the two are mixed? To determine the effects of A-receptors on antagonist-stimulated transcription by B-receptors, expression vectors encoding hPR_B and increasing levels of hPR_A were co-transfected into HeLa cells together with the PRE-*tk*-CAT reporter, and the cells were treated with either R5020 or RU486. hPR_B alone stimulate CAT transcription in this model whether the receptors are occupied by agonist or antagonist, whereas hPR_A alone are stimulatory only when they are agonist-occupied. In fact, when RU486 is bound to hPR_A, transcription is always suppressed below basal levels. When the two receptor isotypes are co-expressed, strong transcription is maintained in the presence of the agonist, regardless of the hPR_B to hPR_A ratio. However, in the presence of the antagonist and at approximately equimolar amounts of the two receptors, the transcriptional phenotype of hPR_A predominates, so that hPR_B-stimulated transcription is almost entirely extinguished (Tung et al., 1993).

When hPR_A and hPR_B are equimolar, a 1:2:1 ratio of A/A, A/B, and B/B dimers is expected. The extensive inhibition by A-receptors suggested that A/B heterodimers have the same inhibitory transcriptional activity as A/A homodimers, and that only B/B homodimers are stimulatory. However, presence of the two competing homodimeric species complicates functional analysis of the heterodimers and B/B homodimers probably account for the incomplete suppression of transcription seen when A- and B-receptors are co-expressed. We therefore decided to construct receptors in which the heterodimeric species was the only class present.

When they are mixed, cJun and cFos preferentially form heterodimers over homodimers by at least 1000-fold (O'Shea et al., 1989). Therefore, to force heterodimerization of hPR, the leucine zippers of cFos or cJun were fused to the C-terminus of hPR_A or hPR_B (Mohamed et al., 1994). These chimeric hPRs retain agonist and antagonist binding capacity, and agonist or antagonist-occupied hPR_A-Jun and hPR_B-Fos, when each is expressed alone has the same transcriptional phenotype as the wild-type receptors. However, when the two are co-transfected, the weak residual transcription seen with wild-type RU486-occupied B/A receptor mixtures is entirely eliminated. Thus, CAT levels are reproducibly below control values with B-Fos/A-Jun. These data confirm the A-dominance hypothesis and show that antagonist-occupied pure A/B heterodimers exhibit exclusively the inhibitory transcriptional phenotype of antagonist-occupied A/A homodimers.

The dominance of A-receptors is observed even when the antagonist used is ZK98299. The strong PRE binding-independent transcriptional stimulation imparted by ZK98299-occupied hPR_B in the *tk* promoter model is 80% suppressed by approximately equimolar concentrations of hPR_A and fully suppressed by a two-fold molar excess of hPR_A. Because ZK98299-occupied hPR_A do not bind to a PRE, these data imply that the inhibitory effects of antagonist-occupied hPR_A, like the stimulatory effects of antagonist-occupied hPR_A, are mediated by novel non-PRE-dependent mechanisms. This was confirmed by experiments in which the antagonist-occupied A-receptor DBD specificity mutant, which cannot bind a PRE, was used as the competing receptor species. On PRE-*tk*-CAT, activation of CAT transcription by RU486-occupied wild-type hPR_B was completely inhibited by the antagonist-occupied hPR_A-DBD specificity mutant.

Our studies demonstrate that A-receptors can inhibit the activity of B-receptors (Tung et al., 1993). In related studies, it was shown that A-receptors inhibit not only their B-receptor partners, but also the activities of other members of the steroid receptor family, including ERs (Vegeto et al., 1993; McDonnell and Goldman, 1994). Thus, the dominant inhibitory effects of A-receptors are extensive, and may explain some of the "anti-estrogenic" actions reported for antiprogestins. The mechanisms underlying these "trans" inhibitory effects are unknown. Meyer et al. (1989) demonstrated several years ago that transcription by PRs is inhibited by co-expressed ERs, and that both PR and GR expression inhibits activation by ERs. They suggested that steroid hormone receptors compete for limiting transcription factors that they all use in common. Since ER and PR bind to different DNA response elements, their mutual inhibition appears to occur without the direct DNA binding of the interfering receptor. Thus, when PRs interfere with ER action, the gene being suppressed need not contain a PRE or be otherwise progestin-regulated, as modeled in Fig. 2Ic.

A Third Transactivation Function (AF-3) of Human Progesterone Receptors Located in the Unique N-terminal Segment of the hPR_B—the B-upstream Segment (BUS). Why do B-receptors differ from A-receptors? We postulated that the unique 164 amino acid BUS is in part responsible for the functional differences between the two isoforms, and we constructed a series of hPR expression vectors encoding BUS fused to individual downstream functional domains of the receptors (Sartorius et al., 1994b). These include the two transactivation domains, AF-1 located in a 90-amino acid segment just upstream of the DBD and nuclear localization signal (NLS); and AF-2 located in the HBD. BUS is a highly phosphorylated domain and contains the serine residues responsible for the hPR_B triplet protein structure seen on SDS-PAGE. The construct containing BUS-DBD-NLS binds tightly to DNA when aided by accessory nuclear factors. In HeLa cells, BUS-DBD-NLS strongly and constitutively activates CAT transcription from a promoter containing two progesterone response elements (PRE_2-$TATA_{tk}$-CAT) to levels comparable to those of hormone-activated, full-length B-receptors. Thus, we conclude that this construct contains an autonomous third transactivation function, AF-3.

In HeLa cells, transcription levels with BUS-DBD-NLS are equivalent to those seen with full-length hPR_B, and are higher than those seen with hPR_A. Additional studies show that BUS specifically requires an intact hPR DBD in order to be transcriptionally active. DBD mutants that cannot bind DNA, or whose DNA binding specificity has been altered, cannot cooperate in BUS transcriptional activity. This suggests that the autonomous AF-3 activity resides in a discontinuous domain formed from BUS and the hPR DBD, or that the AF-3 domain in BUS must be brought to the DNA to activate the transcription apparatus. We also find that the autonomous function of BUS-DBD-NLS is promoter and cell-specific. BUS-DBD-NLS does not transactivate MMTV-CAT in HeLa cells, and poorly transactivates PRE_2-$TATA_{tk}$-CAT in the PR-negative T47D-Y breast cancer cells. In the latter case, however, transcription can be restored either by elevating cellular levels of cAMP or by linking BUS to AF-1 or AF-2, each of which alone is also inactive in T47D-Y cells. Thus, whereas in T47D-Y cells each AF alone is inactive, when AF-3 is linked to either of the other two AFs (AF-3 + AF-1 or AF-3 + AF-2), strong transcriptional activation is regenerated, which is approximately equal to that obtained with B-receptors. These data suggest that in the appropriate cell or promoter context, BUS can supply an important transactivation function in two different ways: either by autonomously activating transcription in the absence of the other two AFs, as it does in HeLa cells on PRE_2-$TATA_{tk}$-CAT, or by synergizing with the other AFs on the hPR molecule, as it does in T47D-Y cells on PRE_2-$TATA_{tk}$-CAT (Sartorius et al., 1994b). Is it the autonomous function of BUS that produces agonist-like effects from antagonist-occupied B-receptors?

SUMMARY

This year represents the 21st anniversary of our first demonstration that human breast cancers contain PRs (Horwitz et al., 1975). These receptors are now routinely measured in tumors as markers of hormone dependence and disease prognosis. Theoretically, their central function in breast cancers should not be as markers, however, but as effectors of the proliferative signals of endogenous progesterone in premenopausal women and as targets for the therapeutic effects of progestins and antiprogestins. At present, PRs are rarely measured for these functional purposes. As the aforegoing shows, the actions of PRs are complex, and responsiveness to progestin agonists or antagonists will depend on the gene whose activity is being measured, the peculiarities of the cell and tissue under study, and, most importantly, the PR isoform that predominates in a tissue or tumor. Thus, although this chapter has focused on progesterone actions in the breast and breast cancer, the principles described here will undoubtedly also apply to other progesterone target tissues, including the female reproductive tract, the ovaries, the brain, and bone. The differential expression of PR isoforms, perhaps under developmental and hormonal control, serves, we believe, to fine-tune responsiveness to this important reproductive hormone. It follows that knowledge, not just of the PR content of a tissue, but of the expression of B- versus A-receptors in that tissue, will be vital to understanding the effects of progestins therein.

Additionally, recent studies with PRs have forced us to revise the standard model of steroid receptor action. The conventional model, which depicts receptors as ligand-activated proteins that bind to specific DNA sequences at "consensus" hormone response elements and activate transcription, is not incorrect. It is, however, oversimplified, as studies with PRs demonstrate. This should not have been surprising given the complex regulatory demands on these receptors. These demands include requirements for both positive and negative transcriptional regulation, for tissue specificity of action, and for regulation of composite and simple gene promoters. It should also not have been surprising given the complex structural organization of these proteins. This includes multiple covalent modifications by phosphorylation, and multiple functional domains that control intramolecular contacts, intermolecular protein-protein interactions, and DNA binding. Finally, because steroid antagonists are synthetic rather than natural hormones, it is perhaps not surprising that their binding produces structural alterations in the receptors that unveil additional novel interactive capabilities. Thus, whereas antiprogestins can indeed competitively inhibit agonists by forming nonproductive receptor-DNA complexes, this is not their sole mechanism of action. Depending on the promoter and cell regulated, antiprogestin effects may also be mediated by receptor interactions with co-activators whose function is in turn controlled by nonsteroidal signals. Therefore, when two different signaling pathways are simultaneously activated, they can cooperate to produce unintended effects. Additionally, it seems clear from several studies that antagonist-occupied receptors can act without binding to canonical PREs, or without binding to DNA at all, relying perhaps on tethering proteins. This may be a consequence of the unusual allosteric structure imparted on the receptors by synthetic ligands. For some of these unusual actions, the receptors may be monomeric rather than dimeric. Because of these, and undoubtedly other, mechanisms yet to be discovered, the most serious mistake that investigators can make when studying antiprogestins is to assume that a specific mechanism is operating. It is our contention that these novel actions begin to explain two properties of steroid antagonists that have puzzled investigators. One is the common observation that antagonists are agonists in some normal tissues. The other, which is an extension of the first, is that in malignant cells, antagonists can acquire agonist-like properties as tumors progress, leading to treatment failure. Although such tumors are called "resistant," they may in fact be responding quite well to the antagonist!

With respect to receptor protein structure, we are only beginning to appreciate its complexity. For example, it appeared at first that the structural independence of functional domains permitted the analysis of receptor fragments by fusing them to heterologous proteins. However, we now know that important functional domains can overlap, that other functional domains may be discontinuous, and that one domain can modulate the activity of another. This means that analysis of receptor fragments in chimeras is an incomplete test of domain function, and that we need innovative experimental strategies to understand this intramolecular cross-talk. Finally, what could be more unexpected than finding that one receptor isoform can inhibit not just its partner, but even distantly related receptor cousins! More surprises can be expected from this fascinating protein family.

ACKNOWLEDGMENTS

The studies described herein were generously supported by the NIH through CA26869, CA55595, and DK48238, by the National Foundation for Cancer Research, and by the Johnson and Johnson Focused Giving Program. Versions of this chapter are in press in *Acta Oncologica* and *The Breast*, based on lectures presented at the European Breast Cancer Consortium Meeting, Lillehammer, Norway, 1995, and the ESO Woekshop on Endocrine Therapy, Paris, France, 1995.

REFERENCES

Anderson TJ, Howell A, King RJB (1987): Comment on progesterone effects in breast tissue. *Breast Cancer Res Treat* 10:65–66

Anderson TJ, Battersby S, King RJB, McPherson K, Going JJ (1989): Oral contraceptive use influences resting breast proliferation. *Hum Pathol* 20:1139–1144

Bakker GH, Setyono-Han B, Henkelman MS, De Jong FH, Lamberts SWJ, van der Schoot P, Klijn JGM (1987): Comparison of the actions of the antiprogestin mifepristone (RU486), the progestin megestrol acetate, the LHRH analog buserelin, and ovariectomy in treatment of rat mammary tumors. *Cancer Treat Rep* 71:1021–1027

Bakker GH, Setyono-Han B, Portengen H, De Jong FH, Foekens JA, Klijn JGM (1989): Endocrine and antitumor effects of combined treatment with an antiprogestin and antiestrogen or luteinizing hormone-releasing hormone agonist in female rats bearing mammary tumors. *Endocrinology* 125:1593–1598

Bakker GH, Setyono-Han B, Portengen H, De Jong FH, Foekens JA, Klijn JGM (1990): Treatment of breast cancer with different antiprogestins: preclinical and clinical studies. *J Steroid Biochem Mol Biol* 37:789–794

Bergkvist L, Adami HO, Persson I, Hoover R, Schairer C (1989): The risk of breast cancer after estrogen and estrogen-progestin replacement. *N Engl J Med* 321: 293–297

Bresciani F (1971): Ovarian steroid control of cell proliferation in the mammary gland and cancer. In: *Basic Actions of Sex Steroids on Target Organs*. Basel: Karger Publishing Company

Cairns C, Cairns W, Okret S (1993): Inhibition of gene expression by steroid hormone receptors via a negative glucocorticoid response element. *DNA Cell Biol* 12:695–702

Canobbio L, Galligioni E, Gasparini G, Fassio T, Crivellari D, Villalta D, Santini G, Monfardini S, Boccardo F (1987): Alternating tamoxifen and medroxyprogesterone acetate in postmenopausal advanced breast cancer patients—short and long term endocrine effects. *Breast Cancer Res Treat* 10:201–204

Cato ACB, Henderson D, Ponta H (1987): The hormone response element of the mouse mammary tumor virus DNA mediates the progestin and androgen induction of transcription in the proviral long terminal repeat region. *EMBO J* 6:363–368

Clarke CL, Sutherland RL (1990): Progestin regulation of cellular proliferation. *Endocr Rev* 11:266–301

Diamond MI, Miner JN, Yoshinaga SK, Yamamoto KR (1990): Transcription factor interactions: selectors of positive or negative regulation from a single DNA element. *Science* 249:1266–1272

Ewertz M (1988): Influence of non-contraceptive exogenous and endogenous sex hormones on breast cancer risk in Denmark. *Int J Cancer* 42:832–838

Going JJ, Anderson TJ, Battersby S, MacIntyre CCA (1988): Proliferative and secretory activity in human breast during natural and artificial menstrual cycles. *Am J Path* 130:193–204

Gruol DJ, Campbell NF, Bourgeois S (1986): Cyclic AMP-dependent protein kinase promotes glucocorticoid receptor function. *J Biol Chem* 261:4909–4914

Gundersen S, Kvinnsland S, Lundgren S, Klepp O, Lund E, Bormer O, Host H (1990): Cyclical use of tamoxifen and high-dose medroxyprogesterone acetate in advanced estrogen receptor positive breast cancer. *Breast Cancer Res Treat* 17:45–50

Härd T, Kellenbach E, Boelens R, Maler BA, Dahlman K, Freedman LP, Carstedt-Duke J, Yamamoto KR, Gustafsson J-AÅ (1990): Solution structure of the glucocorticoid receptor DNA-binding domain. *Science* 249:157–160

Haslam SZ (1988): Progesterone effects on deoxyribonucleic acid synthesis in normal mouse mammary glands. *Endocrinology* 122:464–470

Henderson BE, Ross R, Bernstein L (1988): Estrogens as a cause of human cancer. The Richard and Hinda Rosenthal Foundation Award Lecture. *Cancer Res* 48:246–253

Horwitz KB, McGuire WL, Pearson OH, Segaloff A (1975): Predicting response to endocrine therapy: a hypothesis. *Science* 189:726–727

Horwitz KB, McGuire WL (1977): Progesterone and progesterone receptors in experimental breast cancer. *Cancer Res* 37:1733–1738

Horwitz KB, McGuire WL (1978): Estrogen control of progesterone receptor in human breast cancer. Correlation with nuclear processing of estrogen receptors. *J Biol Chem* 253:2223–2228

Horwitz KB (1992): The molecular biology of RU486. Is there a role for antiprogestins in the treatment of breast cancer? *Endocr Rev* 13:146–163

Horwitz KB (1995): Editorial: When tamoxifen turns bad. *Endocrinology* 136:821–823.

Howell A, Harland RNL, Barnes DM, Baildam AD, Wilkinson MJS, Hayward E, Swindell R, Sellwood RA (1987): Endocrine therapy for advanced carcinoma of the breast: relationship between the effect of tamoxifen upon concentrations of progesterone receptor and subsequent response to treatment. *Cancer Res* 47: 300–304

Huggins C, Moon RC, Morii S (1962): Extinction of experimental mammary cancer. I. Estradiol-17ß and progesterone. *Proc Natl Acad Sci USA* 48:379–386.

Huggins C, Yang NC (1962): Induction and extinction of mammary cancer. *Science* 137:257–262.

Huggins C (1965): Two principles in endocrine therapy of cancers: hormone deprival and hormone interference. *Cancer Res* 25:1163–1167

Hulka BS (1990): Hormone-replacement therapy and the risk of breast cancer. *CA* 40:289

Imagawa W, Tomooka Y, Hamamoto S, Nandi S (1985): Stimulation of mammary epithelial cell growth *in vitro*: Interaction of epidermal growth factor and mammogenic hormones. *Endocrinology* 116:1514

Jonat C, Rahmsdorf HJ, Park K-K, Cato AC, Gebel S, Ponta H, Herrlich P (1990): Antitumor promotion and antiinflammation: down-modulation of AP-1 (fos/jun) activity by glucocorticoid hormone. *Cell* 62:1189–1204

Klijn JGM, De Jong FH, Bakker GH, Lamberts SWJ, Rodenburg CJ, Alexieva-Figusch J (1989): Antiprogestins, a new form of endocrine therapy for human breast cancer. *Cancer Res* 49:2851–2856

Kutoh E, Stromstedt P-E, Poellinger L (1992): Functional interference between the ubiquitous and constitutive octamer transcription factor 1 (OTF-1) and the glucocorticoid receptor by direct protein-protein interaction involving the homeo subdomain of OTF-1. *Mol Cell Biol* 12:4960–4969

Maudelonde T, Romieu G, Ulmann A, Pujol H, Grenier J, Khalaf S, Cavalie G, Rochefort H (1987): First clinical trial on the use of the antiprogestin RU486 in advanced breast cancer. In: *Hormonal Manipulation of Cancer: Peptides, Growth Factors and New (Anti-)Steroidal Agents*, Klijn JGM, Paridaens R, Foekens JA, eds. New York: Raven Press

McDonnell DP, Goldman ME (1994): RU486 exerts antiestrogenic activities through a novel progesterone receptor A form-mediated mechanism. *J Biol Chem* 269: 11945–11949

Meirik O, Lund E, Hans-Olov A, Bergström R, Christoffersen T, Bergsjö P (1986): Oral contraceptive use and breast cancer in young women. *Lancet* 2:650–654

Meyer ME, Gronemeyer H, Turcotte B, Bocquel M-T, Tasset D, Chambon P (1989): Steroid hormone receptors compete for factors that mediate their enhancer function. *Cell* 57:433–442

Michna H, Schneider MR, Nishino Y, El Etreby FM (1989a): Antitumor activity of the antiprogestins ZK98299 and RU38486 in hormone dependent rat and mouse mammary tumors: mechanistic studies. *Breast Cancer Res Treat* 14:275–288

Michna H, Schneider MR, Nishino Y, El Etreby MF (1989b): The antitumor mechanism of progesterone antagonists is a receptor mediated antiproliferative effect by induction of terminal cell death. *J Steroid Biochem* 34:447–453

Michna H, Schneider M, Nishino Y, El Etreby MF, McGuire WL (1990): Progesterone antagonists block the growth of experimental mammary tumors in G_0/G_1. *Breast Can Res Treat* 17:155–156

Michna H, Nishino Y, Schneider MR, Louton T, El Etreby MF (1991): A bioassay for the evaluation of antiproliferative potencies of progesterone antagonists. *J Steroid Biochem Molec Biol* 38:359–365

Miner JN, Yamamoto KR (1991): Regulatory cross-talk at composite response elements. *Trends Biol Sci* 16:423–426

Mohamed KM, Tung L, Takimoto GS, Horwitz KB (1994): The leucine zippers of c-Fos and c-Jun for progesterone receptors dimerization: A-dominance in the A/B heterodimer. *J Steroid Biochem Mol Biol* 51:241–250

Musgrove, EA and Sutherland, RC (1993): Effects of the progestin antagonist RU486 on T47D breast cancer cell cycle kinetics and cell cycle regulatory genes. *Biochem Biophys Res Commun* 195:1184–1190

O'Shea EK, Rutkowski R, Stafford WF III, Kim PS (1989): Preferential heterodimer formation by isolated leucine zippers from Fos and Jun. *Science* 245:646–648

Oro AE, Hollenberg SM, Evans RM (1988): Transcriptional inhibition by a glucocorticoid receptor-β-galactosidase fusion protein. *Cell* 55:1109–1114

Robinson SP, Jordan VC (1987): Reversal of the antitumor effects of tamoxifen by progesterone in the 7,12-dimethylbenzanthrcene-induced rat mammary carcinoma model. *Cancer Res* 47:5386–5390

Russo IH, Gimotty P, Dupuis M, Russo J (1989): Effect of medroxyprogesterone acetate on the response of the rat mammary gland to carcinogenesis. *Br J Cancer* 59:210–216

Sakai DD, Helms S, Carlstedt-Duke J, Gustafsson JAÅ, Rottman FM, Yamamoto KR (1988): Hormone-mediated repression of transcription: a negative glucocorticoid response element from the bovine prolactin gene. *Genes Dev* 2:1144–1154

Sartorius CA, Tung L, Takimoto GS, Horwitz KB (1993): Antagonist-occupied human progesterone receptors bound to DNA are functionally switched to transcriptional agonists by cAMP. *J Biol Chem* 5:9262–9266

Sartorius CA, Groshong SD, Miller LA, Powell RL, Tung L, Takimoto GS, Horwitz KB (1994a): New T47D breast cancer cell lines for the independent study of progesterone B- and A-receptors: only antiprogestin-occupied B-receptors are switched to transcriptional agonists by cAMP. *Cancer Res* 54: 3868–3877

Sartorius CA, Melville MY, Hovland AR, Tung L, Takimoto GS, Horwitz KB (1994b): A third transactivation function (AF3) human progesterone receptors located in the unique, N-terminal segment of the B-isoform. *Mol Endocrinol* 8:1347–1360

Schneider MR, Michna H, Nishino Y, El Etreby FM (1989): Antitumor activity of the progesterone antagonists ZK 98299 and RU 38486 in the hormone-dependent MXT mammary tumor model of the mouse and the DMBA- and the MNU-induced mammary tumor models of the rat. *Eur J Cancer Clin Oncol* 25:691–701

Schüle R, Rangarajan P, Kliewer S, Ransone LJ, Bolado J, Yang N, Verma IM, Evans RM (1990): Functional antagonism between oncoprotein c-Jun and the glucocorticoid receptor. *Cell* 62:1217–1226

Sedlacek SM, Horwitz KB (1984): The role of progestins and progesterone receptors in the treatment of breast cancer. *Steroids* 44:467–484

Truss M, Beato M (1993): Steroid hormone receptors: interaction with deoxyribonucleic acid and transcription factors. *Endocr Rev* 14:459–479

Truss M, Bartsch J, Beato M (1994): Antiprogestins prevent progesterone receptor binding to hormone responsive elements *in vivo*. *Proc Natl Acad Sci USA* 91:11333–11337

Tung L, Mohamed KM, Hoeffler JP, Takimoto GS, Horwitz KB (1993): Antagonist-occupied human progesterone B-receptors activate transcription without binding to progesterone response elements, and are dominantly inhibited by A-receptors. *Mol Endocrinol* 7:1256–1265

Vegeto E, Shahbaz MM, Wen DX, Goldman ME, O'Malley BW, McDonnell DP (1993): Human progesterone receptor A form is a cell- and promoter-specific repressor of human progesterone receptor B function. *Mol Endocrinol* 7:1244–1255

Welsch CW (1985): Host factors affecting the growth of carcinogen-induced rat mammary carcinomas: a review and tribute to Charles Brenton Huggins. *Cancer Res* 45:3415–3443

Yang-Yen H-F, Chambard J-C, Sun Y-L, Smeal T, Schmidt TJ, Drouin J, Karin M (1990): Transcriptional interference between c-Jun and the glucocorticoid receptor: mutual inhibition of DNA binding due to direct protein-protein interaction. *Cell* 62:1205–1215

11

Molecular, Cellular, and Systemic Mechanisms of Antiestrogen Action

William H. Catherino and V. Craig Jordan

INTRODUCTION

Tamoxifen provides a survival advantage for women with breast cancer (Early Breast Trialists Collaborative Group, 1992). Because of its proven clinical efficacy, there are approximately 2 to 3 million women receiving treatment in more than 70 countries. With such widespread use, it is imperative that the scientific community develop an understanding of tamoxifen's molecular mechanism of action so as to exploit the antitumor effects while minimizing side effects. By studying this effective cancer therapeutic, investigators will develop a better understanding of breast carcinogenesis, which may translate to superior treatment regimens.

In this chapter we review some of the most intensive areas of antiestogen research. Tamoxifen clearly provides a therapeutic benefit by blocking estrogen-mediated proliferation, but we will also examine a growing body of literature suggesting that tamoxifen may have antitumor activity via non-estrogen receptor-mediated mechanisms. Furthermore, we will examine studies that suggest that tamoxifen not only produces a direct inhibition of malignant cell proliferation, but also interacts within surrounding tissue to create a hostile environment for tumor growth.

We also review the clinical history of tamoxifen development, with particular emphasis on the advantages and drawbacks of tamoxifen therapy. Current models of tamoxifen treatment recommend 5-year (Breast Cancer Trials Committee, 1987; Fisher et al., 1989) or even indefinite therapy (Falkson et al., 1990; Tormey et al., 1992). Indeed, there are ongoing efforts to use tamoxifen as a preventative for breast cancer development in healthy

Hormones and Cancer
Wayne V. Vedeckis, Editor

women who are at high risk (Powles et al., 1989; Fisher, 1992; Jordan, 1993). Although the benefits of tamoxifen therapy for women who have had breast cancer far outweigh the potential side effects, healthy women who take part in the preventative trials may be exposed to accumulated long-term side effects. Toxicity associated with tamoxifen therapy, as well as the threat of tumor resistance to tamoxifen, has resulted in multiple efforts to develop novel antihormonal therapies. The new antiestrogens we discuss have similar antiestrogenic actions as tamoxifen, but do not have some of tamoxifen's undesirable toxicology.

Although described as an antiestrogen, tamoxifen does have estrogenic actions. The drug can prevent breast cancer proliferation, but it acts as an estrogen on the endometrium, bone, and cardiovascular system. Such estrogenic action provides a protective effect against osteoporosis and cardiovascular disease, but may stimulate endometrial cancer proliferation. We review the current literature on each subject so as to tie the laboratory findings with the clinical observations.

The multiplicity of tamoxifen action in biological systems and the success of tamoxifen as a therapeutic agent has resulted in an intensive effort to understand its mechanism of action. The studies described herein provide a basic understanding of how this drug prevents tumor growth by blocking cellular processes. The future holds great promise for the development of novel anticancer therapeutics that will build on the laboratory and clinical studies with tamoxifen.

ANTITUMOR EFFECTS OF TAMOXIFEN

Tamoxifen has proven activity at various steps along the carcinogenic pathway, from blocking the intracellular signaling for cancer cell proliferation to limiting metastatic potential Tamoxifen's antiestrogenic action is a cornerstone of its antitumor activity, but this drug is capable of inhibiting cancer cell proliferation via mechanisms that do not involve the estrogen receptor (ER). By clarifying tamoxifen action on both the cancerous cell and the surrounding tissue, investigators provide insight into the carcinogenic process as a whole.

Intracellular Tamoxifen Action

ESTROGEN RECEPTOR-MEDIATED ACTION OF TAMOXIFEN. Tamoxifen's designation as an antiestrogen resulted from its ability to block estradiol-mediated vaginal cornification and uterine growth in rats (Harper and Walpole, 1966). Furthermore, treatment in mice produced a vaginal epithelium refractory to the actions of estradiol (Emmens, 1971; Jordan, 1975). During the same time period, investigators recognized that radiolabeled estradiol was selectively retained in the uterus of both rats and mice, tissues responsive to estradiol (Jensen and Jacobson, 1962; Stone, 1963). Tamoxifen was shown to prevent estradiol retention by the mouse uterus (Terenius, 1971) and block estradiol binding to a macromolecule (Skidmore et al., 1972) later identified as the ER.

In the following sections, we discuss how the interaction of tamoxifen with the ER results in antitumor action. The ER is a DNA binding protein that is activated by estradiol, and once activated it stimulates gene transcription (Green and Chambon, 1991; see Chapter 4, this volume). By blocking estradiol-ER interaction, tamoxifen can inhibit estradiol-mediated gene transcription. Regulation of gene transcription represents a likely means of tamoxifen's antiproliferative action.

Interaction of tamoxifen and metabolites with the estrogen receptor. Various biochemical studies suggest that tamoxifen's antiestrogenic action results from its ability to compete with estradiol for binding to the ER (Skidmore et al., 1972; Jordan and Koerner, 1975; Jordan and Dowse, 1976). Because the ER is a ligand-dependent transcription factor (Green and Chambon, 1991), tamoxifen theoretically prevents estradiol-mediated ER activation, while possessing no activity itself. However, tamoxifen can act as a weak estrogen in certain systems (Harper and Walpole, 1966; Terenius, 1971). The pharmacological properties of tamoxifen are complex; however, study of tamoxifen metabolism has provided clues on the chemical structures important for tamoxifen's estrogenic and antiestrogenic action.

Tamoxifen metabolites (Fig. 1) have different affinities for the ER. Those metabolites with greater affinity have higher potency as antiestrogens *in vitro*, and the activity of these metabolites have established the pharmacological principles for all the novel compounds under development (described later). Oral dosing with tamoxifen at 20 mg daily results in blood levels of 50–500 ng/ml (Langan-Fahey et al., 1990). High drug levels are required because tamoxifen has a binding affinity for the ER that is 1% that of estradiol's (Murphy et al., 1990). Tamoxifen's major metabolite, *N*-desmethyltamoxifen, has a serum concentration and ER affinity in the same range as the parent compound in humans (Wakeling and Slater, 1980). In contrast, a minor metabolite, 4-hydroxytamoxifen, has a serum concentration of approximately 1–5 ng/ml, but has an affinity for the ER similar to estradiol (Jordan et al., 1977). 3,4-Dihydroxytamoxifen is a minor metabolite that also has high affinity for the ER, but it has weak estrogenic activity in rats and mice (Jordan et al., 1977, 1978). The hydroxyl group associated with 4-hydroxy- and 3,4-dihydroxytamoxifen probably increases the affinity of these metabolites for the ER by anchoring the ligand at the binding site of the ER normally used by the 3-phenolic group of estradiol (Jordan et al., 1977; Jordan, 1984).

Although the hydroxyl group influences affinity for the ER, the antiestrogenic activity of tamoxifen depends on the aminoethoxy side chain.

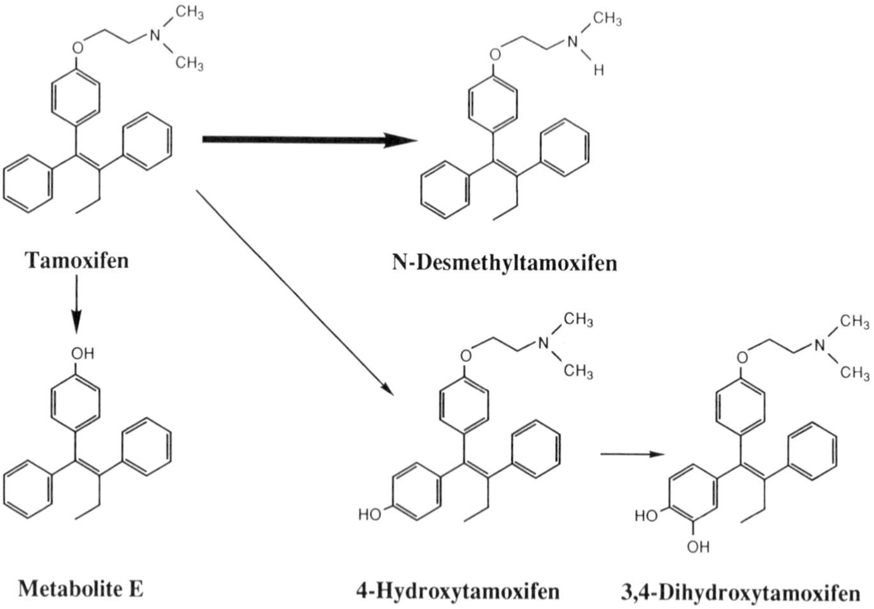

Tamoxifen **N-Desmethyltamoxifen**

Metabolite E **4-Hydroxytamoxifen** **3,4-Dihydroxytamoxifen**

Figure 1. Tamoxifen and major metabolites. At equilibrium, N-desmethyltamoxifen represents the major metabolite at 150% of tamoxifen concentration, whereas 4-hydroxytamoxifen is only 2% of tamoxifen concentration (Langan-Fahey et al., 1994). Hydroxylation of 4-hydroxytamoxifen to 3,4-dihydroxytamoxifen produces a very rare metabolite that possesses estrogenic activity.

Removal of this side chain results in a compound (metabolite E) that acts as a weak estrogen (Jordan, 1984; Murphy and Jordan, 1989; Murphy et al., 1990). This metabolite could potentially stimulate estrogen-dependent breast cancer, but it has only been detected in dogs and rats (Fromson, 1973a,b). Addition of the phenyl ring substituted with a *p*-dimethyamino-ethoxy side chain of tamoxifen to estradiol at the 11β position results in the compound RU 39411 (Fig. 2), an antiestrogen with partial agonistic activity (Gottardis et al., 1989a, 1990; Robinson and Jordan, 1989; Jordan and Koch, 1989). The addition of lipophilic side chains longer than the aminoethoxy group in the 7α or 11β position of estradiol produces anti-estrogens (ICI 164,384; ICI 182,780; RU 58664) that lack estrogenic action (Fig. 2; described later). It is thought that the side chain disrupts proper ER folding required for effective transcriptional activation (Spona et al., 1980; Fauque et al., 1985; Katzenellenbogen et al., 1985; Hansen and Gorski, 1986).

Structure-function studies of tamoxifen and its metabolites have provided useful information on the "estradiol binding site" of the ER. However, recent evidence suggests that there may also be an alternate site for tamoxifen binding. Various laboratories (Martin et al., 1988; Gottardis et al., 1989b; Berthois et al., 1994; Hedden et al., 1995) have shown that exposing cell lysates to

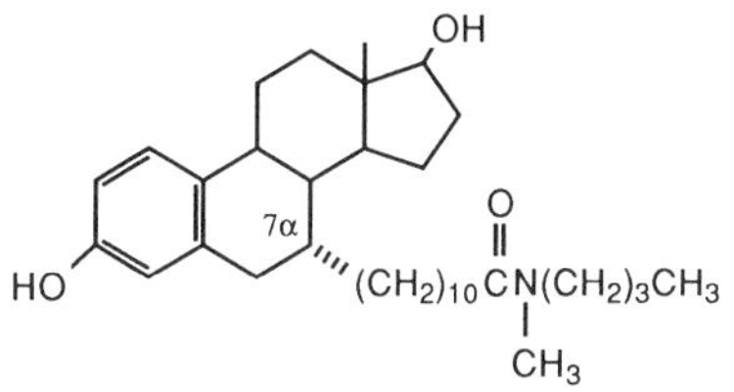

Figure 2. Estradiol-based antiestrogens. This generation of compounds represents high-affinity estradiol analogues with lipophilic side chains either at the 7α or 11β positions. RU39411 most resembles tamoxifen in that the 11β side chain is a *p*-phenylaminoethoxy structure. RU39411 has partial agonistic activity, whereas the other antiestrogens have longer lipophilic side chains and are devoid of estrogenic action.

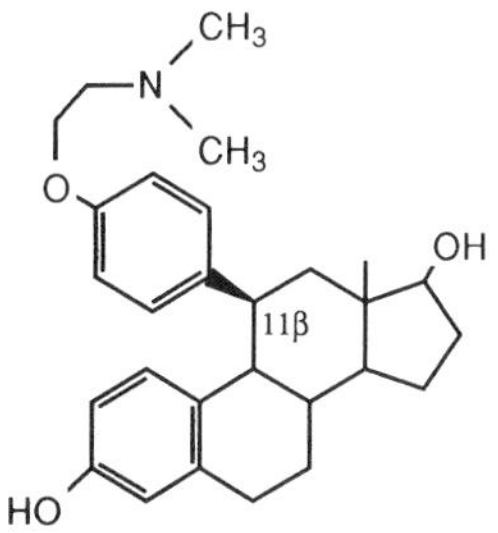

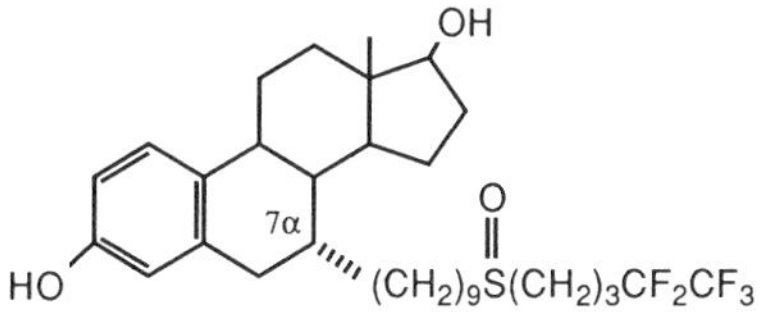

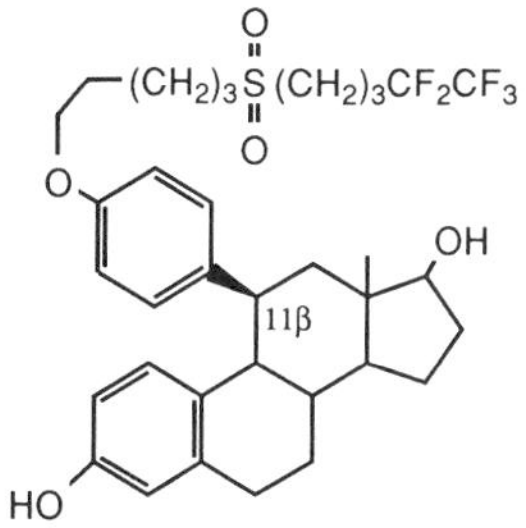

tamoxifen increases the number of potential binding sites for anti-ER antibodies. Even when the ER is occupied with estradiol, low concentrations of antiestrogen result in greater anti-ER antibody binding (Berthois et al., 1994). Perhaps tamoxifen binds to an alternate site on the ER, resulting in an additional conformational change and leading to greater epitope exposure (and therefore greater antibody binding). Furthermore, ER sedimentation studies reveal a greater hydroxytamoxifen binding to the ER compared to estradiol (Hedden et al., 1995). In tissue culture studies, tamoxifen and analogues display agonistic action at low concentration, but at higher concentrations, this agonistic activity is lost (Murphy and Jordan, 1989). Perhaps tamoxifen acts as a weak estrogen when bound to the estradiol binding site, but at higher concentrations, it binds to a secondary site, resulting in ER inactivation (Hedden et al., 1995). Studies of this secondary site using tamoxifen metabolites may not only better define the molecular structures required for high affinity binding, but also may direct the discovery of novel compounds that distinguish between ER binding sites and thereby vary in agonist/antagonist properties.

Interaction of the ER with the preinitiation complex

The net result of estradiol/tamoxifen competition binding is a change in the transcriptional activity of the ER. When bound to either estradiol of 4-hydroxytamoxifen, the ER can bind to DNA (Klinge et al., 1992) at a site known as the estrogen response element (ERE). However, estradiol activates the ER to become a functional transcriptional protein, whereas tamoxifen and tamoxifen metabolites often do not. It has been suggested (Spona et al., 1980; Fauque et al., 1985; Katzenellenbogen et al., 1985; Hansen and Gorski, 1986) that the ER undergoes different conformational changes when bound to tamoxifen, thereby preventing preinitiation complex formation with the ER at sites on the protein known as transcription activation functions (TAFs). Tamoxifen may exert its agonist/antagonist activity by altering the ER conformation, resulting in decreased transcriptional protein interaction at one TAF site located in the ligand binding domain, while protein interaction at a separate TAF in the N-terminal region of the ER may still occur (Gronemeyer et al., 1992). The agonist/antagonist balance of tamoxifen action may therefore be determined by the array of proteins that interact with the liganded ER. Recent data support such a hypothesis (Brown and Sharp, 1990; Halachmi et al., 1993; Cavailles et al., 1994). These investigators demonstrate that when the ER is bound to estradiol it can interact with several specific proteins, and that the protein interactions differ when tamoxifen or 4-hydroxytamoxifen is the ligand. It has become a priority to identify the proteins and understand their interactions for transcription initiation. Perhaps such studies will explain the differences in the pharmacology of antiestrogens at different sites.

Transcriptional control by tamoxifen. Gene regulation by estradiol and tamoxifen has been carefully studied to determine the mechanism of estradiol-mediated cell proliferation. EREs for many genes have been identified (Kumar et al., 1983; Jost et al., 1984; Walker et al., 1984; Klein-Hitpass et al., 1986, 1988; van het Schip et al., 1986; Weisz et al., 1986; Maurer and Notides, 1987; Shyu et al., 1987; Berry et al., 1989; Weisz and Rosales, 1990; Burch and Fischer, 1990; Adan et al., 1991; Darwish et al., 1991; Shupnick and Rosenzweig, 1991; van Dijk and Verhoeven, 1992), and antiestrogens have been used in many cases to confirm that a given DNA sequence indeed interacts with an estradiol-bound ER. Estradiol increases the transcription of early or intermediate-early genes such as *myc*, *fos*, *jun*, and the second messenger *ras* in ER-positive cells, whereas tamoxifen blocks this stimulation (Dubik et al., 1987; Murayama et al., 1988; Santos et al., 1988; Lau et al., 1991; LeRoy et al., 1991; Hyder et al., 1992; Pellerin et al., 1992; Sakakibrar et al., 1992; Nephew et al., 1993; Wosikowski et al., 1993; Bhattacharyya et al., 1994; Hyder and Stancel,. 1994). However, several investigators have not demonstrated tamoxifen inhibition of *fos* and *jun* mRNA expression (Sakakibara et al., 1992; Kirkland et al., 1993; Nephew et al., 1993; Philips et al., 1993; Wosikowski et al., 1993). Tamoxifen also decreases the transcription of mitogenic autocrine/paracrine growth factors while increasing the transcription of inhibitory factors (discussed later). Tamoxifen may therefore inhibit cell proliferation by blocking both the production of proteins involved in the cell cycle cascade and the stimulatory autocrine/paracrine loops.

The partial agonistic properties of tamoxifen that are observed on growth (Katzenellenbogen et al., 1987; Cormier and Jordan, 1989; Poulin et al., 1989) and progesterone receptor expression (Campen et al., 1985) may also exist in part from the ER-DNA interaction. Tzuckerman and colleagues (1994) have hypothesized that the estrogen-like actions of tamoxifen on some estrogen-receptive genes, but not all, result from unknown differences in the promoter regions. We have recently suggested a more specific hypothesis to explain tamoxifen-stimulated gene transcription (Catherino and Jordan, 1995a). We have noted that tamoxifen has no estrogenic activity on a luciferase reporter construct containing one ERE when transiently transfected into the MCF-7 human breast cancer cell line. However, if two or more EREs are present, tamoxifen is able to stimulate gene transcription in a concentration-dependent manner. Perhaps tamoxifen can stimulate progesterone receptor gene transcription because this gene contains several EREs in its promoter region (Savouret et al., 1991; Krause et al., 1993). Tamoxifen may exert its partial agonistic activity only on genes that have multiple EREs. As we better understand the genes transcribed by tamoxifen, we will hopefully gain new insights to the partial agonistic activity of this drug.

TAMOXIFEN ACTION IN THE ABSENCE OF ESTROGEN RRECEPTOR. Tamoxifen has a profound effect on ER-positive breast cancer, but it can also inhibit proliferation of approximately 10% of ER-negative tumors (Early Breast Cancer Trialists Collaborative Group, 1992). Although a portion of these tumors may have been misclassified as ER negative when they were actually ER positive, there is a growing literature describing the control of ER-negative cells by tamoxifen (Grenman et al., 1990; Pollack et al., 1990; Maenpaa et al., 1993). Typically, the mechanisms described require dramatically higher tamoxifen concentrations for inhibition of ER-negative cells (micromolar region), but the concentration of tamoxifen at the tumor site may be high enough to support the proposed mechanisms (Johnston et al., 1993).

Antiestrogen binding sites. More than a decade ago, it was suggested that tamoxifen regulates proliferation by mechanisms that do not require ER (Murphy and Sutherland, 1983). This concept was supported by the identification of a high affinity tamoxifen binding site distinct from the ER

(Reddel et al., 1983; Winneker and Clark, 1983; Winneker et al., 1983). This protein-bound tamoxifen with an affinity comparable to the ER-tamoxifen interaction, and tamoxifen binding could not be blocked with increasing concentrations of estradiol (Chouvet and Saez, 1984; Mehta et al., 1984). The protein was therefore designated an antiestrogen binding site (AEBS). The AEBS is localized in the microsomes (Reddel et al., 1983; Chouvet and Saez, 1984) and has a molecular weight of 265 kDa (Lazier and Bapat, 1988). Although the precise function of AEBS remains to be elucidated, it has been suggested that the AEBS acts as a histamine receptor (Brandes et al., 1885, 1986, 1987; Kroeger and Brandes, 1985; Brandes and Bogdanovic, 1986), a metabolic enzyme for cholesterol (Cypriani et al., 1988; Lazier and Breckenridge, 1990), or a tamoxifen-binding protein that can inhibit lactation (Biswas and Vonderhaar, 1989, 1991).

To dissect the activity of AEBS on tamoxifen-mediated tumor inhibition from ER-mediated tamoxifen action, several laboratories have attempted to construct ligands specific to the AEBS that do not interact with the ER. Sheen et al. (1985) found that an AEBS-specific compound does not possess the antiproliferative activity of tamoxifen. However, Fargin et al (1988) and Teo et al. (1992) have identified compounds that may inhibit cell proliferation. Despite these findings, several lines of evidence suggest that the AEBS is not involved in tamoxifen-mediated inhibition of growth. Cell lines that express AEBS are not necessarily growth inhibited by tamoxifen (Hayashida et al., 1987; Teske et al., 1987). Furthermore, both tamoxifen-inhibited and tamoxifen-resistant cells lines have no differences in AEBS expression (Miller et al., 1984), suggesting that AEBS overexpression is not involved in tamoxifen-resistant cancer cell proliferation. Further study is necessary to clarify the role of AEBS in tamoxifen action.

Tamoxifen interaction with membrane and membrane proteins. A clinically valuable observation is the ability of tamoxifen to block P-glycoprotein activity and thus inhibit multidrug resistance (MDR) (Chatterjee and Harris, 1990; Berman et al., 1991; Hu et al., 1991; Ramu et al., 1991). Although the exact the mechanism remains unclear, recent studies suggest that tamoxifen competes with other chemotherapeutic agents for the binding sites of P-glycoprotein (Safa et al., 1994). Since tamoxifen can be given in short concentrated bursts during cytotoxic treatment with minimal side effects (Millward et al., 1992; Stuart et al., 1992; Trump et al., 1992), it represents a superior choice over verapamil (a commonly used P-glycoprotein blocker) for MDR inhibition. Verapamil in high doses can lead to potentially fatal cardiac arrythmias, whereas tamoxifen doses 15 to 30 times greater than the dose given to control breast cancer have side effects that are minor compared to those associated with the cytotoxic agent used for therapy.

Another membrane-associated protein inhibited by tamoxifen is protein kinase C (PKC) (O'Brian et al., 1985). Tamoxifen may inhibit PKC action by interacting with the phospholipid binding domain (O'Brian et al., 1988a,b;

Issandou et al., 1990). Indeed, the interaction between the substrate protein for PKC and the membrane is disrupted by tamoxifen (Edashige et al., 1991). Although tamoxifen has proved to be a valuable laboratory tool to study the activity of PKC, current evidence suggests that PKC does not play a role in tamoxifen inhibition of malignant growth (Issandrou et al., 1990). PKC *activation* inhibits MCF-7 human breast cancer cell proliferation (Osborne et al., 1981; Darbon et al., 1986; Roos et al., 1986; Issandrou and Darbon, 1988; Issandrou et al., 1988), suggesting that tamoxifen-mediated inhibition of PKC and inhibition of MCF-7 cell proliferation are not directly related.

Tamoxifen also acts as an antioxidant that can block lipid peroxidation (Wiseman et al., 1990a,b; Custodio et al., 1994; Thangaraju et al., 1994). Tamoxifen, like cholesterol, is able to insert into the plasma membrane and can scavenge free radicals within the lipid bilayer. Furthermore, the ability of tamoxifen to decrease peroxidation in cardiac microsomes (Wiseman et al., 1993a) and/or protect low-density lipoproteins from oxidative damage (Wiseman et al., 1993b) may represent one mechanism for tamoxifen's cardioprotective action (discussed later).

Regulation of Growth Factors

Tamoxifen can inhibit the production of various mitogenic autocrine/paracrine factors such as insulin-like growth factors and transforming growth factor-α while increasing the production of the various transforming growth factor-β's, which inhibit proliferation (Fig. 3). These proteins cause changes in proliferation by interacting with their membrane-bound receptor and initiating a second messenger cascade. By regulating the absolute levels of these autocrine/paracrine factors, their receptors, or serum binding proteins, tamoxifen can alter cell proliferation.

INSULIN-LIKE GROWTH FACTORS. Insulin-like growth factors (IGF)-I and II are potent mitogens in many breast cancer cell lines (Huff et al., 1986; Karey and Sirbasku, 1988; Osborne et al., 1989). Although IGF-II messenger RNA is actively produced in various breast tumors (Yee et al., 1988), IGF-I is produced mainly by stromal cells rather than by malignant cells (Yee et al., 1989). However, most experimental and clinical breast cancer cells contain receptors for both IGF-I and -II (DeLeon et al., 1988; Peyrat et al., 1988; Cullin et al., 1990), suggesting that both IGFs can play a role in an autocrine/paracrine mechanism of tumor stimulation. Tamoxifen decreases plasma levels of IGF-I (Colletti et al., 1989; Pollak et al., 1990; Lien et al., 1992; Lonning et al., 1992; Malaab et al., 1992; Fornander et al., 1993; Friedl et al., 1993). Intracellularly, IGF-I down-regulation by tamoxifen occurs via an ER-mediated mechanism because ER-negative cells do not decrease IGF production in the presence of tamoxifen (Vignon et al., 1987). Many studies discussed below examine IGF activity, rather than

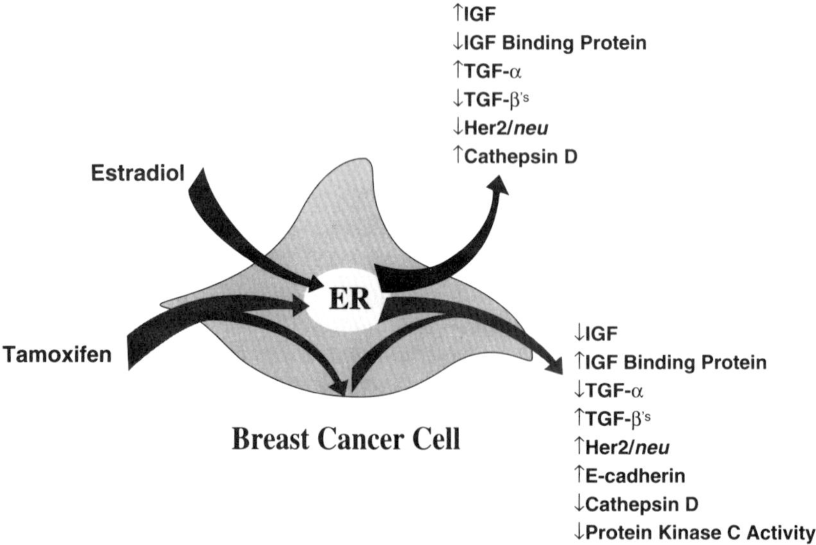

Figure 3. Proliferative regulation by estradiol and tamoxifen. Estradiol is presumed to act via estrogen receptor (ER) regulation. Tamoxifen not only prevents estradiol from interacting with the ER, but also interacts with membrane proteins to influence cell homeostasis.

differentiating the action of IGF-I and IGF-II. However, since both IGF proteins are mitogens, the findings of these studies remain useful for determining the relation of insulin-like growth factors and tamoxifen action.

The ability of tamoxifen to block IGF production via the ER pathway suggests that estradiol-mediated proliferation may include the IGF signaling pathway. Indeed, estradiol stimulation can be attenuated by antibodies directed at either IGF-I directly, or the receptor for IGF-I (Ernst et al., 1989; Manni et al., 1991). These findings support efforts to develop new therapies that can block estradiol-mediated proliferation by competing with IGF signaling.

Since IGF-I regulation is controlled by growth hormones (Clemmons et al., 1981; Clemmons, 1984; Mathews et al., 1986; Roberts et al., 1986), which can be influenced by tamoxifen (Paterson et al., 1983; Malaab et al., 1992; Tannenbaum et al., 1992), this was thought to be the mechanism of tamoxifen control. We now know that this is not the only mechanism of tamoxifen-regulated IGF production. Tamoxifen can block estradiol-stimulated IGF production in human breast cancer cell lines (Berthois et al., 1989; Murphy and Dotzlaw, 1989a; Poulin et al., 1989; Wakeling et al., 1989), whereas it increases IGF in the rat uterus, where tamoxifen acts as a partial agonist for proliferation (Huynh and Pollak, 1993). Such results suggest a role for IGFs in the mechanism of both breast and endometrial cancer proliferation, and demonstrate that systemic IGF levels may not provide an accurate measure of IGF stimulation at particular sites.

IGF binding proteins interact with and prolong the activity of IGFs (Guler et al., 1989), but a subset [binding protein-3(BP3)] may attenuate the IGF function (Knaur and Smith, 1980; Rechler and Brown, 1992; Cohen et al., 1993; Pratt and Pollak, 1993). Because nearly 90% of IGF binds to BP3 (Baxter and Martin, 1989), regulation of this subset has dramatic effects on IGF activity. Estradiol down-regulates BP3 (Krywicki et al., 1993), whereas tamoxifen has no effect, or increases BP3 (Pratt and Pollak, 1993, 1994) in breast and ovarian cancer cell lines. In the uterus, both estradiol and tamoxifen decrease BP3 (Huynh and Pollak, 1994), further suggesting a role for IGFs in estradiol- and tamoxifen-mediated uterine proliferation.

TRANSFORMING GROWTH FACTORS. Transforming growth factors (TGFs) were identified in the supernatant of various transformed cells and are produced by breast cancer cells (Salomon et al., 1989). TGF-α is a potent mitogen that may either be secreted and bind to the epidermal growth factor receptor (EGF-R), or in its membrane-bound precursor form may interact with and stimulate adjacent cells (juxtacrine stimulation) (Wong et al., 1989). Estradiol stimulates (Bates et al., 1986, 1988; Lippman et al., 1986) and tamoxifen inhibits (Murphy and Jordan, 1989; Murphy and Dotzlaw, 1989b; Wakeling, 1989; Noguchi et al., 1993) TGF-α production and/or activity, but the role of TGF-α in the pathway of estrogen stimulation is still controversial. Recent evidence using antisense mRNA to block TGF-α production suggests that TGF-α is required for estradiol-mediated proliferation because antisense TGF-α mRNA inhibits the ability of estradiol to increase proliferation (Reddy et al., 1994). However, there is a mammary carcinoma cell line (CAMA-I) that is estrogen-stimulated but EGF-R negative (Leung et al., 1991), and anti-EGF-R antibodies do not block estradiol stimulation in other cells (Arteaga et al, 1988a; Arteaga and Osborne, 1989), suggesting that TGF-α stimulation via the EGF-R is not central in the pathway of estradiol-mediated proliferation. Furthermore, there are TGF-α-secreting cells that cannot support the growth of estrogen-responsive cells (Osborne et al., 1988), and MCF-7 human breast cancer cells transfected with TGF-α cDNA still require estradiol to proliferate (Clarke et al., 1989). TGF-α expression alone apparently cannot provide an autocrine/paracrine stimulation large enough to replace estradiol dependency.

The family of proteins referred to as transforming growth factors β (TGF-β 1–3) inhibit the proliferation of several transformed cells (Tucker et al., 1984; Roberts et al., 1985; Arteaga et al., 1988b). Estradiol decreases TGF-β expression (Knabbe et al., 1987), whereas antiestrogens either have no effect (Murphy and Dotzlaw, 1989b) or increase (Knabbe et al., 1987; Cohen et al., 1990; Butta et al., 1992) TGF-β messenger RNA's and protein levels. Stromal TGF-β expression is induced by tamoxifen (Butta et al., 1992), suggesting a potential paracrine mechanism for tamoxifen control of ER-negative cancer cells. In fact, increased TGF-β expression by antiestrogens can control the proliferation of ER-negative cell line in a paracrine fashion, and anti-TGF-β antibodies prevent this antiestrogen-mediated inhibition (Knabbe et al., 1987), supporting a role for TGF-β in the mechanism of antiestrogen inhibition. TGF-β regulation by tamoxifen in the MCF-7 cell line may represent one of the mechanisms by which tamoxifen inhibits breast cancer proliferation, because added TGF-β1 and -β2 inhibit proliferation (Arrick et al., 1990). However, the proliferation of T47D cells (another human breast cancer cell line) is inhibited by tamoxifen but are not influenced by TGF-β1 and -β2 (Arrick et al., 1990). TGF-β and growth inhibition are therefore not necessarily tied in all breast cancer cells.

Although TGF-β is growth-inhibitory under certain instances, it may actually support the growth of more advanced breast cancer. Loss of hormonal dependency is correlated with increased TGF-β mRNA and protein expression (Dickson et al., 1987; Kasid et al., 1987), and overexpression of TGF-β can promote the escape of breast cancer cells from hormone dependency to hormone independence (Arteaga et al., 1993). These findings are supported by clinical data suggesting that tamoxifen-resistant tumors are unaffected by high levels of TGF-β (Thompson et al., 1991). Indeed, highly tumorigenic cell lines secrete high levels of TGF-β (Dickson et al., 1986; Arteaga et al., 1988b). Approximately three of four ER- and progesterone receptor-negative cells (which are a highly malignant subtype of breast cancer) overexpress TGF-β, whereas less than 15% of cells expressing these steroid receptors have high TGF-β expression (King et al., 1989). TGF-β expression may even increase the ability of malignant cells to invade the basement membrane and metastasize (Welch et al., 1990). By increasing TGF-β expression, tamoxifen treatment may result in the clonal selection of a metastatic subpopulation. If this is true, it is essential to develop therapeutic modalities to inhibit cancer cell proliferation by TGF-β.

HER2/*neu* (*ErbB*-2). HER2/*neu* is overexpressed in approximately 20% of human breast cancers and is a marker that indicates poor prognosis (Gusterson et al., 1992). It is a transmembrane receptor with protein tyrosine kinase activity and high homology to the EGF-R (Coussens et al., 1985; Bargmann et al., 1986; Yamamoto et al., 1986). Although the ligand for this receptor remains controversial, it is clear that HER2/*neu* does not use the same ligands as EGF-R (Bacus et al., 1994). Estrogen treatment decreases HER2/*neu* expression, both at the mRNA and protein levels (LeRoy et al., 1991; Warri et al., 1991), whereas tamoxifen both blocks the estrogen-mediated decrease and stimulates HER2/*neu* induction (Antoniotti et al., 1992; DeBortoli et al., 1992). Recent studies suggest that estradiol and tamoxifen may bind directly to HER2/*neu* (Matsuda et al., 1993), providing a potential mechanism for estradiol-mediated signaling aside from the ER.

Tamoxifen-mediated HER2/*neu* overexpression may represent a mechanism of breast cancer proliferation during tamoxifen treatment. Overexpression of HER2/*neu in vitro* results in a cell line resistant to the growth-inhibitory activity of tamoxifen (Benz et al., 1993). Furthermore, the clinical impact of HER2/*neu* as an indicator of prognosis is greatest after tumor recurrence

(Perren, 1991), and is most relevant for tamoxifen-resistant breast tumors (Wright et al., 1992). Perhaps future therapeutic efforts will combine tamoxifen treatment with new treatments directed at early control of breast cancer cell clones that use HER2/*neu* as a stimulus for proliferation.

Systemic Effects

Tamoxifen has antiproliferative activity on breast cancer, but therapy is currently initiated after the tumor has reached a clinically detectable size. Tamoxifen treatment is therefore directed at the control of micrometastases. Such cells would already possess the genetic changes necessary for metastatic spread, and ideal control would therefore not only limit proliferation of these cells, but would also prevent these cells from spreading throughout the body. Such control would include disruption of tumor cell tissue invasion and immune-mediated destruction of metastatic foci. Although control of metastatic spread on a systemic scale was not an expected tamoxifen action, the studies described below suggest that tamoxifen may act in this fashion. Perhaps these findings will provide a framework for future targets of antitumor therapeutics.

ANGIOGENESIS AND METASTASIS. Recent studies indicate that tamoxifen can inhibit angiogenesis (Tanaka et al., 1991; Gagliardi and Collins, 1993). Although the mechanism is still not clear, it does not involve an ER-mediated pathway, as excess estradiol cannot overcome the tamoxifen blockade (Gagliardi and Collins, 1993). Since tumor size is limited by the blood supply that supports it, early inhibition of angiogenesis by tamoxifen may result in a substantial delay of tumor growth even after tamoxifen withdrawal. The findings from an overview analysis of tamoxifen action on clinical breast cancer support this hypothesis. Tamoxifen provides a survival advantage that increases over time, despite fairly short treatment regimens (Early Breast Cancer Trialists Collaborative Group, 1992). The time-dependent increase in survival advantage, even after withdrawal (Early Breast Cancer Trialists Collaborative Group, 1992), suggests a mechanism of tumor control beyond the defined cytostatic action of tamoxifen (Robinson et al., 1989).

Control of the metastatic process by tamoxifen could also provide an increased survival advantage. By limiting the ability of malignant epithelial cells to degrade the basement membrane and invade the stroma, tamoxifen may prevent the spread of localized disease. Tamoxifen has been found to increase expression of E-cadherin (Bracke et al., 1994), an adhesion protein whose loss has been correlated with increased invasiveness (Frixen et al., 1991; Vleminckx et al., 1991; Gamallo et al, 1993; Oka et al., 1993; D'Souza and Taylor-Papadimitriou, 1994). Furthermore, expression of cathepsin D [a lysosomal protease that has been correlated with relapse and metastases (Rochefort, 1992)] is stimulated by estrogen, and this stimulation can be blocked by antiestrogens (Westley and Rochefort, 1980). Cathepsin D may be associated with degradation of and migration through extracellular matrix (Montcourrier et al., 1990), as evidenced by an MCF-7 human breast cancer cell line that overexpresses cathepsin D and can metastasize in nude mice (Thompson et al., 1993). Tamoxifen's ability to control this and other as yet uncharacterized proteases (as well as antiproteases) may result in further antimetastatic activity.

IMMUNE MECHANISMS. Although the ultimate impact of tamoxifen on a woman's immune system remains poorly understood, ideally tamoxifen action would synergize with immune-mediated cancer cell eradication. To test this hypothesis, we have examined combination therapy with tamoxifen and β-interferon (an immune cell cytokine). We have found that this combination can block the growth of estradiol-stimulated laboratory tumors *in vivo*, even after treatment withdrawal (Gibson et al., 1993). Furthermore, Campisi and colleagues (1993) have demonstrated that this combination can also induce the regression of bone metastases, suggesting that tamoxifen's antitumor activity can be amplified by an activated immune system. Tamoxifen can also potentiate that activity of tumor necrosis factor (Tiwari et al., 1991; Matsuo et al., 1992; Teodorczyk et al., 1993). However, whereas immune-mediated cancer cell destruction may be amplified by tamoxifen, the tumor inhibitory effect of this drug clearly does not require a functioning immune system, as this drug can inhibit estradiol-stimulated tumor growth in athymic animals (Jordan et al., 1989).

CLINICAL EXPERIENCE WITH TAMOXIFEN

The development of tamoxifen therapy has involved a close interaction between the laboratory and the clinic. Laboratory efforts have provided a mechanistic understanding of antiestrogen action that has been exploited to target tamoxifen therapy to those women with breast cancer who are most likely to respond. Furthermore, clinical findings have provided clues to direct the study of tamoxifen in the laboratory. This section reviews the history of clinical studies with tamoxifen, with particular emphasis on tamoxifen-resistant breast cancer growth. The clinical findings have described the benefits and well as the limitations of tamoxifen therapy. Laboratory and clinical studies on tamoxifen therapy have resulted in a conceptual evolution from antiestrogen use only in advanced disease to use at strategic stages of disease, starting before the development of clinically detectable disease and throughout disease progression.

Evolution of Tamoxifen as a Treatment Modality

Endocrine therapy for breast cancer originated at the beginning of this century (Beatson, 1896; Boyd, 1990). Oophorectomy was the treatment regimen, providing tumor control in approximately 30% of treated women. It was not until the mid-1940s before an effective medicinal therapy, diethylstilbesterol (DES), was developed that could be used to treat postmenopausal patients (Haddow, 1944). Unfortunately, DES has substantial toxicity. Development of nonsteroidal antiestrogens (Lerner et al., 1958; Harper and Walpole, 1967), particularly tamoxifen, in the 1960s provided a medical therapy not limited by unacceptable side effects.

Antiestrogens, like other endocrine treatments, produce an objective response in approximately one-third of unselected women with advanced breast cancer, but the discovery of the ER (Jensen and Jacobson, 1962; Toft et al., 1967) and the finding that this protein is present in a subset of breast cancers (Jensen et al., 1971; McGuire and Chamness, 1973) suggested a predictive test to select which patients will or will not respond to tamoxifen therapy. Following the discovery of the ER, our laboratory demonstrated that tamoxifen could interact with the ER and prevent estradiol binding (Hunter and Jordan, 1975; Jordan and Koerner, 1975a,b; Jordan and Prestwich, 1977; Jordan and Dix, 1979). A potential mechanism of antitumor activity involving competition for the ER was proposed, and this theory directed the use of ER measurements in clinical studies for women with breast cancer (Mouridsen et al., 1978; Fisher et al., 1981, 1983, 1986). Approximately 50–60% of patients with ER-positive tumors respond to tamoxifen therapy. Identification of patients with ER-positive tumors therefore facilitates the clinician's ability to tailor therapy and improve outcomes for the breast cancer patient.

The ER has proven an excellent predictor for response over the past two decades. Furthermore, by measuring progesterone receptor levels, response to endocrine treatment can be further defined (Horwitz et al., 1975). Progesterone receptor is induced by ER when it is activated by estradiol. Therefore, it serves as a marker for the activity of the ER as a functional transcriptional activator protein. Women with ER- and progesterone receptor-positive tumors have a 70% probability of response, whereas only 10% of women lacking these markers will respond to tamoxifen (Deschens, 1991).

Tamoxifen was originally used for the treatment of advanced disease, but laboratory studies demonstrated that longer term tamoxifen treatment of subclinical disease would improve outcome. When the carcinogen dimethylbenzathracene (DMBA) is used in rats, mammary tumors develop within 6 months in all animals (Jordan et al., 1979). Tamoxifen treatment delays tumor occurrence, and this delay is increased with longer tamoxifen exposure (Jordan et al., 1979). When tamoxifen is withdrawn, the tumors begin to grow (Robinson et al., 1988), suggesting that tamoxifen is cytostatic rather than cytotoxic. Such results provided the impetus to use tamoxifen as a treatment adjuvant after surgical removal of the tumor, rather than withholding treatment until tumor relapse. Indeed, clinical studies demonstrate a benefit for patients who receive adjuvant tamoxifen after tumor removal compared to those who receive tamoxifen after recurrence (Breast Cancer Trials Committee, 1987).

Major clinical trials of tamoxifen began in the late 1970s, with treatment times limited to 1 year (Ribeiro and Palmer, 1983; Ludwig Breast Cancer Study Group, 1984; Rose et al., 1985; Ribeiro and Swindell, 1985). Tamoxifen proved beneficial for breast cancer control, but no survival advantages were noted. Nevertheless, the side effects of short courses of tamoxifen were mild compared to the surgical or chemotherapeutic alternatives. Studies that followed examined 2-year (Nolvadex Adjuvant Tamoxifen Organization, 1985), 3-year (Delozier et al., 1986), and 5-year (Tormey and Jordan, 1984; Tormey et al., 1987; Breast Cancer Trials Committee, 1987) tamoxifen therapy and have demonstrated that this drug provides a survival advantage for women with breast cancer. In other words, longer treatment times are correlated with longer survival. Furthermore, an overview analysis (Early Breast Cancer Trialists Committee, 1992) of many tamoxifen trials suggests that it provides benefit to women with breast cancer regardless of nodal or menopausal status.

Current Status of Tamoxifen Therapy

To capitalize on the finding that longer tamoxifen treatment is better for the patient, there are current protocols examining indefinite tamoxifen treatment until tumor relapse (Falkson et al., 1990; Tormey et al., 1992). Trials are also underway to examine the ability of tamoxifen to act as a preventative in women at high risk for breast cancer (Powles et al., 1989, Fisher, 1992; Jordan, 1993).

The basis for preventative trials using tamoxifen results from studies demonstrating the inhibitory effect of the drug on the development of tumors in animals and humans. Tamoxifen treatment prevents the development of mammary tumors in rats and mice (Jordan, 1976, 1978; Jordan et al., 1979, 1980, 1991b; Gottardis and Jordan, 1987), and several clinical studies (Cuzick and Baum, 1985; McDonald and Stewart, 1991) found a reduction in contralateral breast cancer associated with treatment (for review, see Nayfield et al., 1991; Early Breast Cancer Trialists Collaborative Group, 1992; Catherino and Jordan, 1993). If tamoxifen can inhibit preclinical tumor growth in the contralateral breast, it may also prevent primary tumor growth. This, however, can only be established by random clinical trials.

Mechanisms of Tamoxifen Resistance

Tamoxifen does provide a survival advantage for women with breast cancer, but many women have disease recurrence (Early Breast Cancer Trialists

Table 1. Investigated mechanisms of tamoxifen resistance

- Loss of estrogen receptor expression or activity
- Systemic metabolism to estrogenic compounds
- Autocrine/paracrine growth factor stimulation
- Tumor efflux of tamoxifen
- Intracellular metabolism to estrogenic compounds
- Estrogen receptor mutation resulting in activation
- Estrogen response element variations

Collaborative Group, 1992). With ever-increasing exposure times, it should be expected that recurring tumors have developed mechanisms to overcome the tamoxifen blockade. Table 1 lists various mechanisms that could result in tamoxifen resistance. The most obvious mechanism of tamoxifen resistance is the loss of ER with concomitant estrogen-independent proliferation, and although this mechanism does occur, many tumor recurrences are tamoxifen resistant but ER positive (Encarnacion et al., 1993). Because ER-positive tumors that have recurred after tamoxifen therapy may remain susceptible to hormonal therapies (Wilson, 1983; Stoll, 1988; Iveson et al., 1993), a better understanding of this population may provide directions for the development of second-line therapies.

An obvious difference between the breast tumors that initially respond to tamoxifen therapy and those that recur is the time of tamoxifen exposure. Perhaps long-term tamoxifen therapy results in changes in a woman's metabolic profile, resulting in the production of estrogenic tamoxifen metabolites. We have found (Langan-Fahey et al., 1990) that this hypothesis is unlikely to be a major cause of tamoxifen resistance, as the relative levels of tamoxifen metabolites do not change dramatically over a 10-year period.

Osborne and colleagues have suggested two other hypotheses for tamoxifen-resistant tumor growth. They have developed a model system in which the MCF-7 human breast cancer cell line is implanted into athymic nude mice, and these mice are then treated with tamoxifen (Osborne et al., 1987). This model mimics the clinical situation in which tamoxifen is given to patients to control the growth of micrometastases. After a period of 2 to 6 months, many of the tumors begin to grow despite the continued presence of tamoxifen. These investigators found that the tumors may either selectively efflux tamoxifen (thereby lowering the competition at the ER) or metabolize tamoxifen intracellularly to estrogenic compounds (Osborne et al., 1991, 1992; Wiebe et al., 1992). We have reproduced these experiments using a tamoxifen analogue that *cannot* be metabolized to estrogenic compounds, and have shown that tamoxifen-stimulated tumor growth still occurs (Wolf et al., 1993). Based on further study, there appears to be little evidence to support intracellular tamoxifen metabolism as a mechanism of tamoxifen-resistant tumor growth (Osborne et al., 1994).

MCF-7 tumors that become resistant to tamoxifen may actually use tamoxifen as a mitogenic stimulus. There is clinical precedence for tamoxifen-stimulated tumors; women who had tumor recurrence while on tamoxifen therapy can have tumor regression when tamoxifen is withdrawn (Legault et al., 1979; Canney et al., 1987; Belani et al., 1989; Howell et al., 1992). One explanation for this phenomenon is the occurrence of a defective signaling pathway through a mutant receptor. Our laboratory has recently identified a naturally occurring ER mutation in a tamoxifen-stimulated human breast cancer tumor line that can increase the estrogenicity of tamoxifen (Wolf and Jordan, 1995; Catherino et al., 1995; Catherino and Jordan, 1995b). Other mutations of the ER may occur in breast tumors that would allow cells to use tamoxifen as a stimulus for proliferation. However, not all tamoxifen-stimulated tumors have mutant receptors (Wolf and Jordan, 1994). We have proposed an alternative mechanism that depends on an amplification of EREs in the promoter region of a gene, such that tamoxifen acts as an estrogen when bound to a wild-type receptor (Catherino and Jordan, 1995a).

Changes in ER or enhancer sequences that bind ER suggest a means of altering the transcriptional activity of tamoxifen, but the important step for tumor control lies in the identification of the transcribed genes that provide the stimulus for proliferation. There is an ever-increasing body of literature suggesting a role for various growth factors in tamoxifen-resistant and tamoxifen-stimulated tumor growth. Toi and colleagues (1993) have found that generation of tamoxifen-resistant tumors by random cDNA transfections results in cells that secrete factors that can support the proliferation of untransfected cells. Other investigators have identified tamoxifen-stimulated cancer cell proliferation when these cells overexpress IGF-I receptor (Wiseman et al., 1993c), EGF-R (Long et al., 1992), Her2/*neu* (Borg et al., 1994), or fibroblast growth factor 4 (McLeskey et al., 1993). Regulation of growth factor stimuli by tamoxifen (discussed earlier) provides a plausible means of cancer cell regulation. By clarifying the association of tamoxifen with this means of proliferation, the clinician will hopefully gain new means of combating breast cancer recurrences.

NEW ANTIESTROGENS

Tamoxifen has revolutionized the treatment of breast cancer therapy over the past 20 years; however, improving efficacy and decreasing potential side effects remains a priority. Several novel antiestrogens resemble tamoxifen structurally, whereas the others have an estradiol nucleus with a side chain that perhaps disrupts ER conformation (Figs. 2 and 4). Many of these antiestrogens are either in the preclinical or early clinical stages, but they have potential to be useful for breast cancer treatment in the near future.

Toremifene

Toremifene is a tamoxifen analogue (Fig. 4) developed in 1981. It is less active than tamoxifen both in the laboratory (Robinson et al., 1988; Valavaara and Kangas, 1988; DeGregorio et al., 1989; Kangas et al., 1989; DiSalle et al., 1990; Grenman et al., 1991) and the clinic (Gundersen, 1990; Pyrhonen, 1990; Valavaara, 1990; Homesley et al., 1993), but it does not have the genotoxic activity associated with tamoxifen (Hard et al., 1993; Styles et al., 1994). Tamoxifen can cause protein and DNA adduct formation as well as hepatocarcinogenesis in the rat (Mani and Kupfer, 1991; Ching et al., 1992; Han and Liehr, 1992; White et al., 1992; Hirsimaki et al., 1993; Williams et al., 1993; Pathak and Bodell, 1994; Sargent et al., 1994). Although a case can be made against the relevance of the laboratory studies in human disease

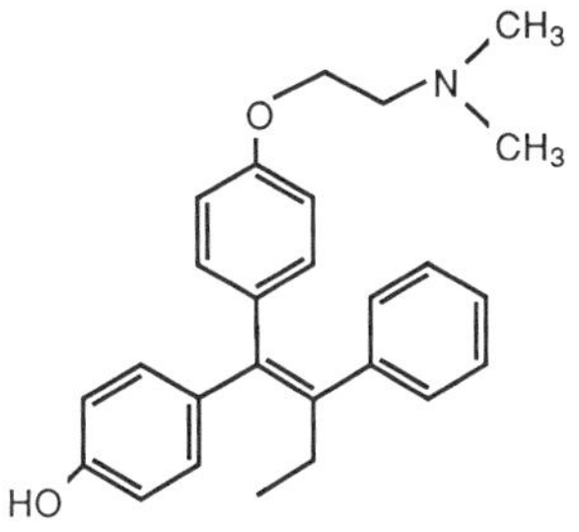

4-Hydoxytamoxifen 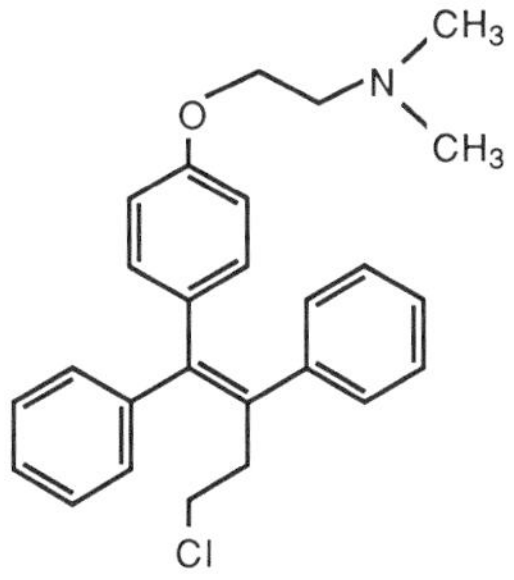 **Toremifene**

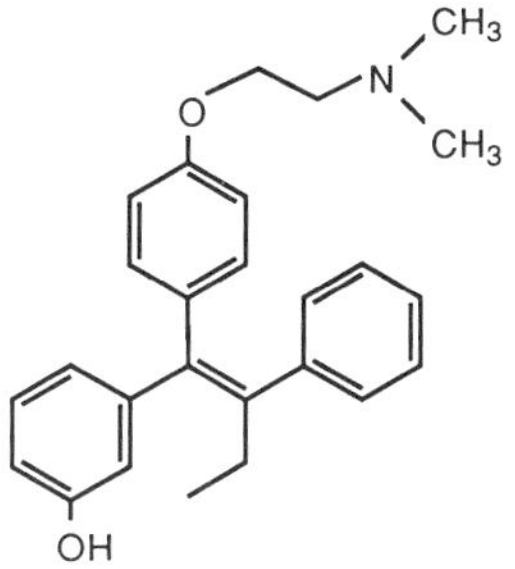

Droloxifene 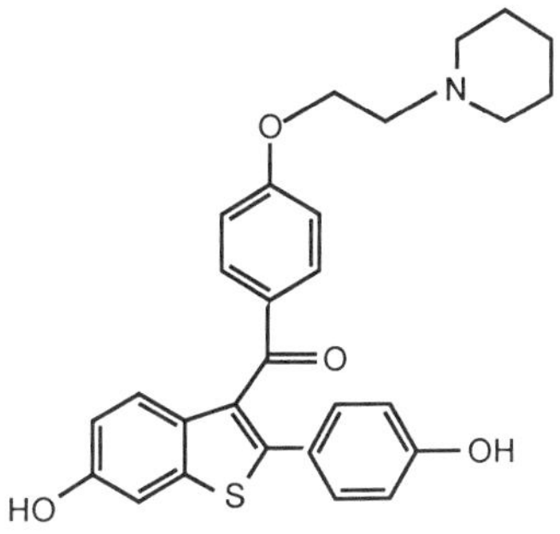 **Raloxifene / Keoxifene**

Figure 4. Tamoxifen-based antiestrogens. 4-hydroxytamoxifen is the most active antiestrogenic metabolite of tamoxifen. Toremifene and droloxifene are tamoxifen analogues that have efficacy against clinical breast cancer. Keoxifene/raloxifene has some structureal similarity and is currently under trial as a therapeutic for osteoporosis.

(Jordan and Morrow, 1994), one concern for the use of tamoxifen in the prevention trials (Powles et al., 19; Fisher, 1992; Jordan, 1993) is the risk of hepatotoxicity. There have been only two reported cases of women with hepatocarcinoma (Fornander et al., 1989) while taking tamoxifen (in comparison with more than 6 million woman-years of tamoxifen exposure), but the use of tamoxifen in preventative trials puts healthy women at risk for this malignancy. Toremifene does not modify bases in DNA (Hard et al., 1993), nor does it cause hepatocarcinoma, even at high doses. Although it is not clear how hepatotoxicity is avoided with toremifene, current hypotheses include differential metabolism of toremifene and tamoxifen via either oxidative metabolism (Potter et al., 1994) or by regulation of glutathione-S-transferase and the hexose monophosphate shunt (Ahotupa et al., 1994), which may result in reactive tamoxifen metabolites. The preventative trials have been organized to carefully monitor hepatotoxicity, and in the event that tamoxifen results in unacceptable liver damage, toremifene may provide a reasonable and effective alternative after long-term treatment trials have been completed to accurately assess its toxicological profile.

Droloxifene

Droloxifene, or 3-hydroxytamoxifen (Fig. 4), has about 10 to 60 times the affinity for the ER as tamoxifen (Roos et al., 1983, Löser et al., 1985a,b; Hasmann et al., 1994), an affinity comparable to 4-hydroxytamoxifen (Jordan et al., 1977). It is superior to tamoxifen on some laboratory tumors (Kawamura et al., 1991; Winterfeld et al., 1992), showing a greater activity on both tissue culture mammary cancer cell lines and tumors in rats and mice (Hasmann et al., 1994). Furthermore, doloxifene is capable of increasing TGF-β and blocking IGF-I activity, as well as blocking estradiol-mediated c-*myc* transcription, far more effectively than tamoxifen (Hasmann et al., 1994). Droloxifene is also comparable to tamoxifen in the treatment of advanced disease in postmenopausal patients, although higher doses have to be used (Bruning, 1992). It appears to have decreased estrogen-like activity compared to tamoxifen (Abe, 1991; Kawamura et al., 1993; Hasmann et al., 1994), and lacks the hepatotoxic effects of tamoxifen in rats (Hasmann et al., 1994). If this drug is found to provide benefit comparable to tamoxifen in several large clinical trials, it may represent a good alternative to tamoxifen. Furthermore, droloxifene may become clinically useful if antiestrogens with less estrogen-like activities are desired.

ICI 164,384 and ICI 182,780

The pure antiestrogens differ from the antiestrogens described previously because they are based on estradiol with a long lipophilic side chain at the

7α position (see Fig. 2). Bucourt and colleagues (1978) identified that various lipophilic side chains attached to the 7α position of estradiol retained high affinity for the ER. The compounds were used for ER affinity chromatography. Variations of these compounds were later tested as potential antiestrogens. This generation of antiestrogens possesses none of the partial agonistic activities associated with tamoxifen or other nonsteroidal antiestrogens (Wakeling and Bowler, 1987; Wakeling et al., 1991). ICI 164,384, the prototypical 7α substituted antiestrogens, binds to the ER, resulting in decreased ER-DNA interaction (Fawell et al., 1990) by inducing the rapid destruction of ER (Gibson et al., 1991). There is no evidence that prolonged ICI 164,384 exposure results in resistant breast cancer cell lines (Wakeling, 1990) and various tamoxifen-resistant cell and tumor lines are inhibited by ICI 164,384 and ICI 182,780 (Gottardis et al., 1989a, 1990; Brunner et al., 1993; Hu et al., 1993; Osborne, 1993; Wakeling, 1993; Coopman et al., 1994; Lykkesfeldt et al., 1994; Osborne et al., 1994), suggesting that this compound may be a valuable treatment after tamoxifen failure. In fact, there is only one published example of an ER-positive, tamoxifen-resistant cell line that is also resistant to ICI 164,384 (Jiang et al., 1992). ICI 182,780 is similar to ICI 164,384, except that its antiestrogenic potency is high enough to warrant clinical study (Wakeling et al., 1991). An initial Phase I study suggests that ICI 182,780 is well tolerated (DeFriend et al., 1994) in short-term exposure, and a small clinical trial suggests that ICI 182,780 has activity on tamoxifen-resistant human tumors (Howell et al., 1995; Dowsett et al., 1995). These results provide compelling evidence that this generation of antiestrogens may provide a better antiestrogenic alternative than tamoxifen in breast cancer regulation, but there is evidence that these drugs also result in bone loss (Gallagher et al., 1993), perhaps because of their lack of estrogen-like action. Until more is known about the activity of these drugs, it would be prudent to maintain this therapeutic option as a secondary treatment after tamoxifen failure.

RU 39411 and RU 58668

Based on the finding that the aminoethoxy side chain of tamoxifen determines the antiestrogenic activity of this drug (for a review, see Jordan, 1984), rational design of high affinity antiestrogens dictates that estradiol with a side chain in the analogous location as tamoxifen's aminoethoxy side chain (the 11β position) should produce a high affinity antiestrogen. This hypothesis was supported by the finding that compounds with 11β substitutions on estradiol bind with high affinity to the ER (Belanger et al., 1980). RU 39411 (see Fig. 2) was the prototype of such antiestrogens. In this compound, the 11β side chain is a phenyl ring with an attached *p*-dimethylaminoethoxy side chain. This compound has greater antiestrogenic activity that 4-hydroxytamoxifen *in vitro* (Robinson and Jordan, 1989;

Jordan and Koch, 1989) and is a potent antiestrogen *in vivo* (Gottardis et al., 1989a, 1990). However, this antiestrogen also expresses partial agonistic activity comparable to tamoxifen in these studies. Subsequent drug development focused on 11β-substituted estradiol with longer lipophilic side chains. This work has led to several pure antiestrogens (Claussner et al., 1992; Nique et al., 1994), including the potent and pure antiestrogen RU 58668 (Van de Velde, 1994). Such compounds block the activity of estradiol both *in vitro* and *in vivo* without displaying any agonistic activity associated with tamoxifen. As pure antiestrogens, they may prove valuable as second-line treatment for tumors that recur after tamoxifen therapy.

ESTROGEN-LIKE EFFECTS OF TAMOXIFEN

Because of the agonist/antagonist activity of tamoxifen, study of this drug in the laboratory and the clinics has provided exciting clues on ER action. Although we have described several potential mechanisms by which tamoxifen can exert agonistic action, it remains unclear how tamoxifen acts as an estrogen at some targets and as an antiestrogen at others. However, this complex pharmacology has provided both benefits and drawbacks when used in clinical treatment. We have previously discussed tamoxifen's stimulatory action on the uterus. As a result, if endometrial cells become neoplastic, they may also be stimulated by tamoxifen. However, tamoxifen's estrogenic action extends beyond uterine carcinogenesis, but most importantly provides beneficial effects on estrogen-regulated systems throughout the body. Physiological control of osteoporosis and cardiovascular disease appear estrogen dependent in women, and we will discuss the value that tamoxifen treatment may provide for protection against these diseases. In the following sections, we review the current literature on the influence of tamoxifen on endometrium, bone, and cardiovascular disease so as to define systemic effects of tamoxifen that should be considered with tamoxifen therapy.

Endometrial Cancer

Over the past 10 years, there has been concern about tamoxifen's potential role in stimulating endometrial carcinoma proliferation during breast cancer palliation. Satyaswaroop and colleagues (1984) first developed a laboratory model system to study the influence of tamoxifen on human endometrial carcinoma implanted in athymic mice. They found that the endometrial cancer grows more rapidly when treated with tamoxifen than controls. We extended these studies by comparing an endometrial tumor line (EnCa101) with the human breast cancer tumor line MCF-7 when implanted in the same athymic animal (Gottardis et al., 1988; Jordan et al., 1901a). When treated with tamoxifen, the endometrial carcinoma grows while the breast tumor remains static, suggesting that EnCa101 stimulation is due to local rather than systemic factors. Numerous breast cancer clinical trials (for review, see Assikis and Jordan, 1995) have now identified an increased incidence of endometrial carcinoma detected in tamoxifen-treated patients. Although several tamoxifen side effects (bleeding after menopause and spotting, for example) may heighten suspicion of endometrial cancer in these patients and therefore result in greater detection (Horowitz and Feinstein, 1986), the consistency of the clinical data (reviewed in Nayfield et al., 1991; Catherino and Jordan, 1993; Jordan and Assikis, 1995) as well as the support by laboratory models suggest a stimulatory role for tamoxifen.

With this information in mind, some perspective must be developed to optimize patient care. Tamoxifen therapy for women with breast cancer results in a survival advantage that far outweighs the risk of endometrial cancer (Catherino and Jordan, 1993; Jordan and Assikis, 1995). However, healthy women who are using tamoxifen as a preventative for breast cancer may experience an increased incidence of endometrial cancer. These women must understand the signs and symptoms of endometrial cancer, and they must be carefully monitored to decrease the probability of developing advanced uterine malignancy.

Bone

Although unrelated to cancer, osteoporosis nonetheless is a significant cause of morbidity and mortality (Peck et al., 1988). Osteoporosis is a disease resulting in part from estrogen decline; it is confined to postmenopausal women and can be controlled by treatment with exogenous estrogens. There has been concern in the laboratory and clinical communities about the effect of long-term use of an antiestrogen on the rate of bone loss in postmenopausal women. Fortunately, tamoxifen is capable of inhibiting parathyroid hormone-mediated bone resorption (Stewart and Stern, 1986) as well as inhibiting osteoclast activity and bone loss in the rat (Jordan et al., 1987; Turner et al., 1988). Human studies suggest that tamoxifen either has no impact on bone loss or indeed slows the rate of bone resorption in postmenopausal women (Love et al., 1988, 1992; Fentiman et al., 1989; Powles et al., 1989; Fornander et al., 1990; Ward et al., 1993; Kristensen et al., 1994). The mechanism of tamoxifen-mediated bone maintenance in postmenopausal women is not clear, but typical indices for bone loss such as serum osteocalcin, alkaline phosphatase, phosphate, calcium and urinary hydroxyproline support an estrogen-like action of tamoxifen (Ward et al., 1993; Kristensen et al., 1994).

An exciting clinical development that has resulted from this finding is the development on keoxifene (now renamed raloxifene) as a treatment for osteoporosis (see Fig. 4). Keoxifene (LY 156,758) is an antiestrogen with high affinity for the ER (Black et al., 1983). It proved less effective than tamoxifen as an antiproliferative agent on mammary tumors in rats (Gottardis and Jordan, 1987) and was not an effective second-line therapy for treatment of tamoxifen-resistant tumors (Gottardis et al., 1990). However, these studies demonstrated a partial agonistic activity similar to tamoxifen, and keoxifene proved comparable to tamoxifen in preventing bone loss (Jordan et al., 1987). Studies with raloxifene (which is the same antiestrogenic chemical in keoxifene complexed as a hydrochloride salt) has supported these findings (Black et al., 1994; Evans et al., 19; Sato et al., 1994). High-dose raloxifene therapy is currently under evaluation as a treatment for osteoporosis as well as for the treatment of advanced breast cancer.

Cardiovascular Disease

Cardiovascular disease is an estrogen-controlled pathology in women. Again, although there was concern that the antiestrogenic activity of tamoxifen would result in greater morbidity and mortality because of cardiovascular disease, tamoxifen decreases the incidence of fatal myocardial infarction and hospital visits for cardiac conditions (McDonald and Stewart, 1991; Rutqvist and Mattsson, 1993). Tamoxifen lowers serum cholesterol by an average of 13%, mostly due to a 19% decrease in low density lipoprotein (Rossner and Wallgren, 1984; Bertelli et al., 1988; Bruning et al., 1988;

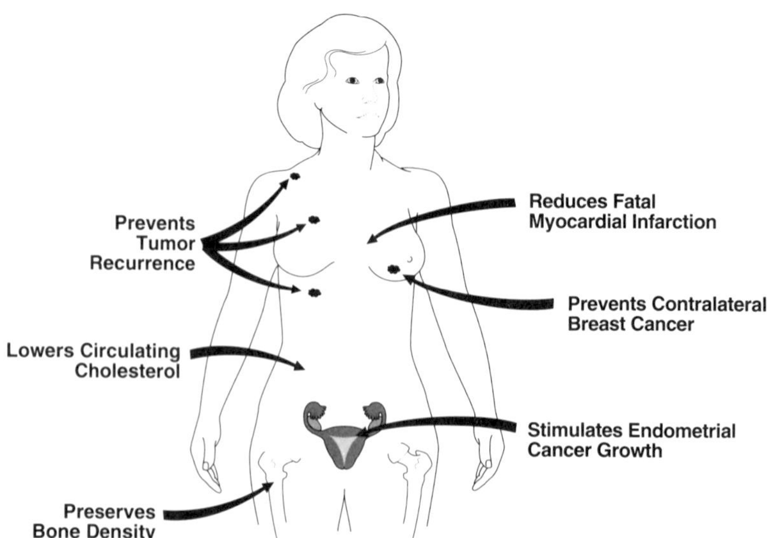

Figure 5. Summary of the target site specific effects of tamoxifen. Although tamoxifen protects against breast cancer recurrence, it also acts on various organ systems throughout the body.

Bagdade et al., 1990; Ingram, 1990; Love et al., 1990, 1991, 1994; Powles et al., 1990; Cuzick et al., 1992; Dnistrian et al., 1993; Thangaraju et al., 1994; Bilimoria et al., 1995 for review). Castelli (1988) has estimated that a 1% decrease in total cholesterol translates to a 2% decrease in coronary heart disease, and therefore tamoxifen may increase survival by providing a substantial estrogen-like effect on cardiovascular disease. Indeed, cardiovascular protection associated with tamoxifen is comparable to that seen with estrogen supplements (Love et al., 1994).

Aside from the partial agonistic action of tamoxifen on cardiovascular disease, this drug may provide further benefit via non-ER-mediated mechanisms. By blocking lipid peroxidation, protecting low density lipoproteins from oxidative damage (Wiseman et al., 1993a), and interacting with the antiestrogen binding site (a potential cholesterol metabolic enzyme) (Cypriani et al., 1988; Lazier and Breckenridge, 1990) tamoxifen may improve cholesterol and lipid clearance. Because cardiovascular disease represents the greatest cause of death in postmenopausal women, the estrogen-like action of tamoxifen may extend survival above and beyond its influence on breast cancer.

SUMMARY

Over the past three decades, the concept of antiestrogen therapy for breast cancer has moved from the laboratory to become a clinical reality, and it represents the most effective treatment strategy for breast cancer to date. The development of tamoxifen highlights the value of close collaboration between the laboratory and the clinics, as laboratory findings can direct clinical use to optimize treatment and clinical discovery can be used to design laboratory experimentation (Jordan, 1994). This drug is now considered a long-term therapeutic option for both breast cancer treatment and prevention. As a result, potential long-term side effects of tamoxifen therapy identified in the laboratory and clinic, such as endometrial cancer stimulation and hepatotoxicity, must be monitored closely to maximize the benefits of tamoxifen while minimizing the risk. Newer antiestrogens developed from the understanding of tamoxifen action may avoid the more serious potential side effects of tamoxifen therapy and may ultimately replace tamoxifen for long-term endocrine treatment. If other side effects of tamoxifen, such as hot flashes, could also be minimized in novel antiestrogens, these compounds may improve therapy by increasing compliance.

Although tamoxifen has provided a better understanding of breast cancer treatment, the results have suggested new applications for antiestrogen therapy other than breast cancer (Lerner and Jordan, 1990). Target antiestrogens are being tested for the prevention of osteoporosis, and the use of these compounds to control cardiovascular disease is certainly a worthy goal for further study. If the drug can also decrease uterine stimulation by having an antiestrogenic effect on that target organ, then there would be a decrease in the risks of endometrial carcinoma. Long-term use of these antiestrogens in postmenopausal women would be expected to have an inhibitory effect on breast carcinogenesis. The general use of targeted antiestrogens by postmenopausal women would provide all the benefits of hormone replacement therapy but with the added advantage of preventing breast and endometrial cancer. New innovations in therapeutics could potentially reduce the death rate of women for four diseases with a single drug.

REFERENCES

Abe O (1991): Japanese early phase II study of droloxifene in the treatment of advanced breast cancer. Preliminary dose-finding study. *Am J Clin Oncol* 14:S40–S45

Adan RAH, Walther N, Cox JJ, Ivell R, Burbach JPH (1991): Comparison of the estrogen responsiveness of the rat and bovine oxytocin gene promotors. *Biochem Biophys Res Commun* 175:117–122

Ahotupa M, Hirsimaki P, Parssinen R, Mantyla E (1994): Alterations of drug metabolizing and antioxidant enzyme activities during tamoxifen-induced hepatocarcinogenesis in the rat. *Carcinogenesis* 15:863–868

Andersson M, Storm HH, Mouridsen HT (1991): Incidence of new primary cancers after adjuvant therapy and radiotherapy for early breast cancer. *J Natl Cancer Inst* 83:1013–1017

Antoniotti S, Maggiora P, Dati C, DeBortoli M (1992): Tamoxifen up-regulates c-erbB-2 expression in oestrogen-responsive breast cancer cells in vitro. *Eur J Cancer* 28:318–321

Arrick BA, Korc M, Derynck R (1990): Differential regulation of expression of three transforming growth factor beta species in human breast cancer cell lines by estradiol. *Cancer Res* 50:299–303

Arteaga CL, Coronado E, Osborne CK (1988a): Blockade of the epidermal growth factor receptor inhibits transforming growth factor alpha-induced but not estrogen-induced growth of hormone-dependent human breast cancer. *Mol Endocrinol* 2:1064–1069

Arteaga CL, Osborne CK (1989): Growth inhibition of breast cancer cells in vitro with an antibody against the type I somatomedin receptor. *Cancer Res* 49:6237–6241

Arteaga CL, Tandon AK, VonHoff DD, Osborne CK (1988b): Transforming growth factor beta: potential autocrine growth inhibitor of estrogen receptor-negative human breast cancer cells. *Cancer Res* 48:3898–3904

Arteaga CL, Carty DT, Moses HL, Hurd SD, Pietenpol JA (1993): Transforming growth factor beta 1 can induce estrogen-independent tumorigenicity of human breast cancer cells in athymic mice. *Cell Growth Diff* 4:193–201

Assikis VJ, Jordan VC (1995): Gynecological effects of tamoxifen and the association with endometrial cancer. *Int J Gynecol Oncol* 49:241–257

Bacus SS, Zelnick CR, Plowman G, Yarden Y (1994): Expression of the erbB-2 family of growth factor receptors and their ligands in breast cancers. Implication for tumor biology and clinical behavior. *Am J Clin Pathol* 102:S13–S24

Bagdade JD, Wolter J, Subbaiah PV, Ryan W (1990): Effects of tamoxifen treatment on plasma lipids and lipoprotein lipid composition. *J Clin Endocrinol Metab* 70:1132–1135

Bargmann CI, Hung MC, Weinberg RA (1986): The neu oncogene encodes an epidermal growth factor receptor-related protein. *Nature* 319:226–230

Bates SE, McManaway ME, Lippman ME, Dickson RB (1986): Characterization of estrogen responsive transforming activity in human breast cancer cell lines. *Mol Endocrinol* 46:1707–1713

Bates SE, Davidson NE, Valverius EM, Freter CE, Dickson RB, Tam JP, Kudlow JE, Lippman ME, Salomon DS (1988): Expression of transforming growth factor alpha and its messenger ribonucleic acid in human breast cancer: its regulation by estrogen and its possible functional significance. *Mol Endocrinol* 2:543–555

Baxter RC, Martin JL (1989): Structure of the Mr 140,000 growth hormone-dependent insulin-like growth factor binding protein complex: determination by reconstitution and affinity-labelling. *Proc Natl Acad Sci USA* 86:6898–6902

Beatson GT (1896): On the treatment of inoperable cases of carcinoma of the mamma: suggestions for a new method of treatment with illustrative cases. *Lancet* 2:104–107

Belanger A, Philibert D, Teutsch G (1980): Regio and stereospecific synthesis of 11β-substituted 19-norsteroids. *Steroids* 37:361–382

Belani CP, Pearl P, Whitley NO, Aisner J (1989): Tamoxifen withdrawal response. Report of a case. *Arch Intern Med* 149:449–450

Benz CC, Scott GK, Sarup JC, Johnson RM, Tripathy D, Coronado E, Shepard HM, Osborne CK (1993): Estrogen-dependent, tamoxifen-resistant tumorigenic growth of MCF-7 cells transfected with HER2/neu. *Breast Cancer Res Treat* 24:85–95

Berman E, Adams M, Duigou OR, Godfrey L, Clarkson B, Andreeff M (1991): Effect of tamoxifen on cell lines displaying the multidrug-resistant phenotype. *Blood* 77:818–825

Berry M, Nunez AM, Chambon P (1989): Estrogen-responsive element of the human pS2 gene is an imperfectly palindromic sequence. *Proc Natl Acad Sci USA* 86:1218–1222

Bertelli G, Pronzato P, Amoroso D, Cusimano MP, Conte PF, Montagna G, Bertolini S, Rosso R (1988): Adjuvant tamoxifen in primary breast cancer: influence on plasma lipids and antithrombin III levels. *Breast Cancer Res Treat* 12:307–310

Berthois Y, Dong XF, Martin PM (1989): Regulation of epidermal growth factor receptor by estrogen and antiestrogens in the human breast cancer cell line MCF-7. *Biochem. Biophys. Res. Commun.* 159:126–131

Berthois Y, Pons M, Dussert C, Crastes de Paulet A, Martin PM (1994): Agonist-antagonist activity of anti-estrogens in the human breast cancer cell line MCF-7: an hypothesis for the interaction with a site distinct from the estrogen binding site. *Mol Cell Endocrinol* 99:259–268

Bhattacharyya N, Ramsammy R, Eatman E, Hollis VW, Anderson WA (1994): Protooncogene, growth factor, growth factor receptor, and estrogen and progesterone receptor gene expression in the immature rat uterus after treatment with estrogen and tamoxifen. *J Submicroscopic Cyt Path* 26:147–162

Bilimoria MM, Assikis VJ, Jordan VC (1995): Should adjuvant tamoxifen therapy be stopped at 5 years? *Cancer Journal* (in press)

Biswas R, Vonderhaar BK (1989): Antiestrogen inhibition of prolactin-induced growth of the Nb2 rat lymphoma cell line. *Cancer Res* 49:6295–6299

Biswas R, Vonderhaar BK (1991): Tamoxifen inhibition of prolactin action in the mouse mammary gland. Endocrinology 128:532–538

Black LJ, Jones CD, Falcone JF (1983): Antagonism of estrogen action with a new benzothiopene-derived antiestrogen. *Life Sci* 32:1031–1036

Black LJ, Sato M, Rowley ER, Magee DE, Bekele A, Williams DC, Cullinan GJ, Bendele R, Kauffman RF, Bensch WR, Frolik CA, Termine JD, Bryant HD (1994): Raloxifene (LY139481 HCl) prevents bone loss and reduces serum cholesterol without causing uterine hypertrophy in ovariectomized rats. *J Clin Invest* 93:63–69

Borg A, Baldetorp B, Ferno M, Killander D, Olsson H, Ryden S, Sigurdsson H (1994): ERBB2 amplification is associated with tamoxifen resistance in steroid-receptor positive breast cancer. *Cancer Lett* 81:137–144

Boyd S (1900): On oophorectomy in cancer of the breast. *Brit Med J* 2:1161–1167

Bracke ME, Charlier C, Bruyneel EA, Labik C, Mareel MM, Castronovo V (1994): Tamoxifen restores the E-cadherin function in human breast cancer MCF-7/6 cells and suppresses their invasive phenotype. *Cancer Res* 54:4607–4609

Brandes LJ, Bogdanovic RP (1986): New evidence that the antiestrogen binding site may be a novel growth-promoting histamine receptor (?H3) which mediates the antiestrogenic and antiproliferative effects of tamoxifen. *Biochem Biophys Res Commun* 134:601–608

Brandes LJ, Bogdanovic RP, Cawker MD, Bose R (1986): The antiproliferative properties of tamoxifen and phenothiazines may be mediated by a unique histamine receptor (?H3) distinct from the calmodulin-binding site. *Cancer Chem Pharmacol* 18:21–23

Brandes LJ, Macdonald LM, Bogdanovic RP (1985): Evidence that the antiestrogen binding site is a histamine or histamine-like receptor. *Biochem Biophys Res Commun* 126:905–910

Brandes LJ, Bogdanovic RP, Cawker MD, LaBella FS (1987): Histamine and growth: interaction of antiestrogen binding site ligands with a novel histamine site that may be associated with calcium channels. *Cancer Res* 47:4025–4031

Breast Cancer Trials Committee, Scottish Cancer Trials Office (1987): Adjuvant tamoxifen in the management of operable breast cancer: the Scottish trial. *Lancet* 2:171–175

Brown M, Sharp PA (1990): Human estrogen receptor forms multiple protein-DNA complexes. *J Biol Chem* 265:11238–43

Bruning PF (1992): Droloxifene, a new anti-oestrogen in postmenopausal advanced breast cancer: preliminary results of a double-blind dose-finding phase II trial. *Eur J Cancer* 28A:1404–1407

Bruning PF, Bonfrer JM, Hart AA, de J, Bakker M, Linders D, van LJ, Nooyen WJ (1988): Tamoxifen, serum lipoproteins and cardiovascular risk. *Br J Cancer* 58:497–499

Brunner N, Frandsen TL, Holst HC, Bei M, Thompson EW, Wakeling AE, Lippman ME, Clarke R (1993): MCF7/LCC2: a 4-hydroxytamoxifen resistant human breast cancer variant that retains sensitivity to the steroidal antiestrogen ICI 182,780. *Cancer Res* 53:3229–3232

Bucourt R, Vignau M, Torelli V, Richard-Foy H, Geynet C, Secco-Millet C, Redeuilh G, Baulieu EE (1978): New biospecific adsorbents for the purification of estradiol receptor. *J Biol Chem* 253:8221–8228

Burch JBE, Fischer AH (1990): Chromatin studies reveal that an ERE is located far upstream of a vitellogenin gene and that a distal tissue specific hypersensitivity site is conserved for two coordinately regulated vitellogenin genes. *Nucleic Acids Res* 18:4157–4165

Butta A, MacLennan K, Flanders KC, Sacks NP, Smith I, McKinna A, Dowsett M, Wakefield LM, Sporn MB, Baum M, Colletta AA (1992): Induction of transforming growth factor beta 1 in human breast cancer in vivo following tamoxifen treatment. *Cancer Res* 52:4261–4264

Campen CA, Jordan VC, Gorski J (1985): Opposing biological actions of antiestrogens in vitro and in vivo: induction of progesterone receptor in the rat and mouse uterus. Endocrinology 116:2327–2336

Campisi C, Terenzi S, Bandierini A, Ferrone L, Mozzicafreddo A, Raffaele M, Zappala A, Tomao S (1993): Complete resolution of breast cancer bone metastasis through the use of beta-interferon and tamoxifen. *Eur J Gynaecol Oncol* 14:479–483

Canney PA, Griffiths T, Latief TN, Priestman TJ (1987): Clinical significance of tamoxifen withdrawal response. *Lancet* 1:36

Castelli WP (1988): Cholesterol and lipids in the risk of coronary artery disease—the Framingham Heart Study. *Canadian J Cardiol* 4:5A–10A

Catherino WH, Jordan VC (1993): A risk-benefit assessment of tamoxifen therapy. *Drug Safety* 8:381–397

Catherino WH, Jordan VC (1995a): Increasing the number of estrogen response elements increases the estrogenic activity of a tamoxifen analogue. *Cancer Lett* 92:39–47

Catherino WH, Jordan VC (1995b): The biological action of cDNA's from mutated estrogen receptors transfected into breast cancer cells. *Cancer Lett* 90:35–42

Catherino WH, Wolf DM, Jordan VC (1995): A naturally occuring estrogen receptor mutation results in increased estrogenicity of a tamoxifen analogue. *Mol Endocinol* 9:153–163

Cavailles V, Dauvois S, Danielian PS, Parker MG (1994): Interaction of proteins with transcriptionally active estrogen receptors. *Proc Natl Acad Sci USA* 91:10009–10013

Chatterjee M, Harris AL (1990): Reversal of acquired resistance to adriamycin in CHO cells by tamoxifen and 4-hydroxy tamoxifen: role of drug interaction with alpha 1 acid glycoprotein. *Br J Cancer* 62:712–717

Ching CK, Smith PG, Long RG (1992): Tamoxifen-associated hepatocellular damage and agranulocytosis [letter]. *Lancet* 339:940

Chouvet C, Saez S (1984): High affinity cytosol binding site(s) for antiestrogens in two human breast cancer cell lines and in biopsy specimens devoid of estrogen receptors. *J Steroid Biochem* 21:755–761

Clarke R, Brunner N, Katy D, Glanz P, Dickson RB, Lippman ME, Kern FG (1989): The effect of a constitutive expression of transforming growth factor alpha on the growth of MCF-7 human breast cancer cells in vitro and in vivo. *Mol Endocrinol* 3:373–380

Claussner A, Nique NF, Philibert D, Teutsch G, Van de Velde P (1992): 11β-amidoalkyl estradiols, a new series of pure antiestrogens. *J Steroid Biochem Mol Biol* 41:609–614

Clemmons DR (1984): Multiple hormones stimulate the production of somatomedin by cultured human fibroblasts. J Clin Endocrinol Metab 58:850–856

Clemmons DR, Underwood LE, VanWyk JJ (1981): Hormonal control of immunoreactive somatomedin production by cultured fibroblasts. J Clin Invest 67:10–19

Cohen FJ, Manni A, Glikman P, Bartholomew M, Demers L (1990): Interactions between growth factor secretion and polyamines in MCF-7 breast cancer cells. *Eur J Cancer* 26:603–608

Cohen P, Lamson G, Okajima T, Rosenfeld RG (1993): Transfection of the human insulin-like growth factor binding protein-3 gene into Balb/c fibroblasts inhibits cellular growth. *Mol Endocrinol* 7:380–386

Colletti RB, Roberts JD, Devlin JT, Copeland KC (1989): Effect of tamoxifen on plasma insulin-like growth factor I in patients with breast cancer. *Cancer Res* 49:1882–1884

Coopman P, Garcia M, Brunner N, Derocq D, Clarke R, Rochefort H (1994): Antiproliferative and anti-estrogenic effects of ICI 164,384 and ICI 182,780 in 4-OH-tamoxifen-resistant human breast-cancer cells. *Int J Cancer* 56:295–300

Cormier EM, Jordan VC (1989): Contrasting ability of antiestrogens to inhibit MCF-7 growth stimulated by estradiol or epidermal growth factor. *Eur J Cancer Clin Oncol* 25:57–63

Coussens L, Yang-Feng TL, Liao YC, Chen E, Gray A, McGrath J, Seeburg PH, Libermann TA, Schlessinger J, Francke U, Levinson A, Ullrich A (1985): Tyrosine kinase receptor with extensive homolgy to EGF receptor shares chromosomal location with neu oncogene. *Science* 230:1132–1139

Cullin KJ, Yee D, Sly WS, Perdue J, Hampton B, Lippman ME, Rosen N (1990): Insulin-like growth factor receptor expression and function in human breast cancer. *Cancer Res* 50:48–53

Custodio JB, Dinis TC, Almeida LM, Madeira VM (1994): Tamoxifen and hydroxytamoxifen as intramembraneous inhibitors of lipid peroxidation. Evidence for peroxyl radical scavenging activity. *Biochem Phearmacol* 47:1989–1998

Cuzick J, Baum M (1985): Tamoxifen and contralateral breast cancer. *Lancet* 2: 282

Cuzick J, Allen D, Baum M, Barrett J, Clark G, Kakkar V, Melissari E, Moniz C, Moore J, Parson V, Pemberton K, Pitt P, Richmond W, Houghton J, Riley D (1992): Long term effects of tamoxifen. Biological effects of tamoxifen working party. *Eur J Cancer* 29A:15–21

Cypriani B, Tabacik C, Descomps B, Crastes de, Paulet A (1988): Role of estrogen receptors and antiestrogen binding sites in an early effect of antiestrogens, the inhibition of cholesterol biosynthesis *J Steroid Biochem* 31:763–771

D'souza B, Taylor-Papadimitriou J (1994): Overexpression of ERBB2 in human mammary epithelial cells signals inhibition of transcription of the E-cadherin gene. *Proc Natl Acad Sci USA* 91:7202–7206

Darbon JM, Valette A, Bayard F (1986): Phorbol esters inhibit the proliferation of MCF-7 cells: possible implications of protein kinase C. *Biochem Pharmacol* 35:2683–2686

Darwish H, Krisinger J, Furlow JD, Smith C, Murdoch FE, DeLuca HF (1991): An estrogen-responsive element mediates the transcriptional regulation of calbindin D-9K gene in rat uterus. *J Biol Chem,* 266:551–558

DeBortoli BM, Dati C, Antoniotti S, Maggiora P, Sapei ML (1992): Hormonal regulation of c-erbB-2 oncogene expression in breast cancer cells. *J Steroid Biochem Mol Biol* 43:21–25

DeFriend DJ, Howell A, Nicholson RI, Anderson E, Dowsett M, Mansel RE, Blamey RW, Bundred NJ, Robertson JF, Saunders C, Baum M, Walton P, Sutcliffe F, Wakeling AE (1994): Investigation of a new pure antiestrogen (ICI 182780) in women with primary breast cancer. *Cancer Res* 54:408–414

DeGregorio MW, Ford JM, Benz CC, Wiebe VJ (1989): Toremifene: pharmacologic and pharmacokinetic basis of reversing multidrug resistance. *J Clin Oncol* 7:1359–1364

DeLeon DD, Bakker B, Wilson DM, Hintz RL, Rosenfeld RG (1988): Demonstration of insulin-like growth factor (IGF-I and -II) receptors and binding protein in human breast cancer cell lines. *Biochem Biophys Res Commun* 152:398–405

Delozier T, Julien JP, Juret P, Veyret C, Couette JE, Graic Y, Ollivier JM, de Ranieri E (1986): Adjuvant tamoxifen in postmenopausal breast cancer: preliminary results of a randomized trial. Breast Cancer Res Treat 7:105–119

Deschens L (1991): Droloxifene, a new antiestrogen, in advanced breast cancer. A double blind dose finding study. *Am J Clin Oncol* 14:s52–s55

DiSalle SE, Zaccheo T, Ornati G (1990): Antiestrogenic and antitumor properties of the new triphenylethylene derivative toremifene in the rat. *J Steroid Biochem* 36:203–206

Dickson RB, Bates SE, McManaway ME, Lippman ME (1986): Characterization of estrogen responsive transforming activity in human breast cancer cell lines. *Proc Natl Acad Sci USA* 46:1707–1713

Dickson RB, Kasid A, Huff KK, Bates SE, Knabbe C, Bronzert D, Gelman EP, Lippman ME (1987): Activation of growth factor secretion in tumorigenic states of breast cancer induced by 17 beta-estradiol or v-Ha-ras oncogene. *Proc Natl Acad Sci USA* 84:837–841

Dnistrian AM, Schwartz MK, Greenberg EJ, Smith CA, Schwartz DC (1993): Effect of tamoxifen on serum cholesterol and lipoproteins during chemohormonal therapy. *Clin Chim Acta* 223:43–52

Dowsett M, Johnston SRD, Iveson TJ, Smith IE (1995): Response to specific anti-oestrogen (ICI182780) in tamoxifen-resistant breast cancer. *Lancet* 1:525

Dubik D, Dembinski TC, Shiu RP (1987): Stimulation of c-myc oncogene expression associated with estrogen-induced proliferation of human breast cancer cells. *Cancer Res* 47:6517–6521

Early Breast Cancer Trialists Collaborative Group, (1992): Systemic treatment of early breast cancer by hormonal, cytotoxic, or immune therapy: 133 randomized trials involving 31000 recurrences and 24000 deaths among 75000 women. *Lancet* 339:1–15, 71–85

Edashige K, Sato EF, Akimaru K, Yoshioka T, Utsumi K (1991): Nonsteroidal anti-estrogen suppresses protein kinase C—its inhibitory effect on interaction of substrate protein with membrane. *Cell Struct Funct* 16:273–281

Emmens CW (1971): Compounds exhibiting prolonged anti-oestrogenic and anti-fertiliy activity in rats and mice. *J Reprod Fertil* 26:175–182

Encarnacion CA, Ciocca DR, McGuire WL, Clark GM, Fuqua SA, Osborne CK (1993): Measurement of steroid hormone receptors in breast cancer patients on tamoxifen. *Breast Cancer Res Treat* 26:237–246

Ernst M, Heath JK, Rodan GA (1989): Estradiol effects on proliferation, messenger ribonucleic acid for collagen and insulin-like growth factor-I, and parathyroid hormone-stimulated adenylate cyclase activity in osteoblastic cells from calvariae and long bones. *Endocrinology* 125:825–833

Evans G, Bryant HU, Magee D, Sato M, Turner RT (1994): The effects of raloxifene on tibia histomorphometry in ovariectomized rats. *Endocrinology* 134:2283–2288

Falkson HC, Gray R, Wolberg WH, Gillchrist KW, Harris JE, Tormey DC, Falkson G (1990): Adjuvant trial of 12 cycles of CMFPT followed by observation or continuous tamoxifen versus four cycles of CMFPT in postmenopausal women with breast cancer: an eastern cooperative oncology group phase III study. *J Clin Oncol* 8:599–607

Fargin A, Bayard F, Faye JC, Traore M, Poirot M, Klaebe A, Perie JJ (1988): Further evidence for a biological role of anti-estrogen-binding sites in mediating the growth inhibitory action of diphenylmethane derivatives. *Chem Biol Interact* 66:101–109

Fauque J, Borgna JL, Rochefort H (1985): A monoclonal antibody to the estrogen receptor inhibits in vitro criteria of receptor activation by an estrogen and an anti-estrogen. *J Biol Chem* 260:15547–15553

Fawell SE, White R, Hoare S, Sydenham M, Page M, Parker MG (1990): Inhibition of estrogen receptor-DNA binding by the "pure" antiestrogen ICI 164,384 appears to be mediated by impaired receptor dimerization. *Proc Natl Acad Sci USA* 87:6883–6887

Fentiman IS, Caleffi M, Rodin A, Murby B, Fogelman I (1989): Bone mineral content of women receiving tamoxifen for mastalgia. *Br J Cancer* 60:262–264

Fisher B, Redmond C, Brown A, Wolmark N, Wittliff J, Fisher ER, Plotkin D, Sachs S, Wolter J, Frelick R, Desser R, DiCalzi N, Geggie P, Campbell T, Elias EG, Prager D, Koontz P, Volk H, Dimitrov N, Gardner B, Lerner H, Shibata H, and other NSABP investigators (1981): Treatment of primary breast cancer with chemotherapy and tamoxifen. *N Engl J Med* 305:1–6

Fisher B, Redmond C, Brown A, Wickerham DL, Wolmark N, Allegra J, Escher G, Lippman M, Savlov E, Wittliff J, Fisher ER, and other NSABP investigators (1983): Influence of tumor estrogen and progesterone receptor levels on the response to tamoxifen and chemotherapy in primary breast cancer. *J Clin Oncol* 1:227–241

Fisher B, Redmond C, Brown A, Fisher ER, Wolmark N, Bowman D, Plotkin D, Wolter J, Bornstein R, Legault-Poisson S, Saffer EA, and other NSABP investigators (1986): Adjuvant chemotherapy with and without tamoxifen in the tratment of primary breast cancer: 5-year results from the National Surgical Adjuvant Breast and Bowel Project trial. *J Clin Oncol* 4:459–471

Fisher B, Redmond C, Wickerham L, Wolmark N, Bowman D, Couture J, Dimitrov NV, Margolese R, Legault-Poisson S, Robidoux A (1989): Systemic therapy in patients with node-negative breast cancer. A commentary based on two National Surgical Adjuvant Breast and Bowel Project (NSABP) clinical trials. *Ann Intern Med* 111:703–712

Fisher B (1992): The evolution of paradigms for management of breast cancer: a personal perspective. *Cancer Res* 52:2371–2383

Fornander T, Rutqvist LE, Cedermark B, Glas U, Mattsson A, Silfversward C, Skoog L, Somell A, Theve T, Wilking N, Askergren J, Hjalmar ML (1989): Adjuvant tamoxifen in early breast cancer: occurrence of new primary cancers. *Lancet* 1:117–120

Fornander T, Rutqvist LE, Sjoberg HE, Blomqvist L, Mattsson A, Glas U (1990): Long-term adjuvant tamoxifen in early breast cancer: effect on bone mineral density in postmenopausal women. *J Clin Oncol* 8:1019–1024

Fornander T, Rutqvist LE, Wilking N, Carlstrom K, von Schoultz SB (1993): Oestrogenic effects of adjuvant tamoxifen in postmenopausal breast cancer. *Eur J Cancer* 29A:497–500

Friedl A, Jordan VC, Pollak M (1993): Suppression of serum insulin-like growth factor-1 levels in breast cancer patients during adjuvant tamoxifen therapy. *Eur J Cancer* 29A:1368–1372

Frixen UH, Behrens J, Sachs M, Eberle G, Voss B, Warda A, Lochner D, Birchmeier W (1991): E-cadherin-mediated cell-cell adhesion prevents invasiveness of human carcinoma cells. *J Cell Biol* 113:173–185

Fromson JM, Pearson S, Bramah S (1973a): The metabolism of tamoxifen (ICI 46,474). Part I. In laboratory animals. *Xenobiotica* 3:693–709

Fromson JM, Pearson S, Bramah S (1973b): The metabolism of tamoxifen (ICI 46,474). Part II. In female patients. *Xenobiotica* 3:711–703

Gagliardi A, Collins DC (1993): Inhibition of angiogenesis by antiestrogens. *Cancer Res* 53:533–535

Gallagher A, Chambers TJ, Tobias JH (1993): The estrogen antagonist ICI 182,780 reduces cancellous bone volume in female rats. *Endocrinology* 133:2787–2791

Gamallo C, Palacios J, Suarez A, Pizarro A, Navarro P, Quintanilla M, Cano A (1993): Correlation of E-cadherin expression with differentiation grade and histological type in breast carcinoma. *Am J Pathol* 142:987–993

Gibson MK, Nemmers LA, Beckman WJ, Davis VL, Curtis SW, Korach KS (1991): The mechanism of ICI 164,384 antiestrogenicity involves rapid loss of estrogen receptor in uterine tissue. *Endocrinology* 129:2000–2010

Gibson DF, Johnson DA, Goldstein D, Langan-Fahey SM, Borden EC, Jordan VC (1993): Human recombinant interferon-beta SER and tamoxifen: growth suppressive effects for the human breast carcinoma MCF-7 grown in the athymic mouse. *Breast Cancer Res Treat* 25:141–150

Gottardis MM, Jordan VC (1987): Antitumor actions of keoxifene and tamoxifen in the N-nitrosomethylurea-induced rat mammary carcinoma model. *Cancer Res* 47:4020–4024

Gottardis MM, Robinson SP, Satyaswaroop PG, Jordan VC (1988): Contrasting actions of tamoxifen on endometrial and breast tumor growth in the athymic mouse. *Cancer Res* 48:812–815

Gottardis MM, Jiang SY, Jeng MH, Jordan VC (1989a): Inhibition of tamoxifen-stimulated growth of an MCF-7 tumor variant in athymic mice by novel steroidal antiestrogens. *Cancer Res* 49:4090–4093

Gottardis MM, Wagner RJ, Borden EC, Jordan VC (1989b): Differential ability of antiestrogens to stimulate breast cancer cell (MCF-7) growth in vivo and in vitro. *Cancer Res* 49:4765–4769

Gottardis MM, Ricchio ME, Satyaswaroop PG, Jordan VC (1990): Effect of steroidal and nonsteroidal antiestrogens on the growth of a tamoxifen-stimulated human endometrial carcinoma (EnCa101) in athymic mice. *Cancer Res* 50:3189–3192

Green S, Chambon P (1991): The oestrogen receptor: from perception to mechanism. In: *Nuclear Hormone Receptors: Molecular mechanisms, cellular functions, clinical abnormalities*, Parker MG, ed. New York: Academic Press

Grenman SE, Worsham MJ, Van DD, England B, McClatchey KD, Babu VR, Roberts JA, Maenpaa J, Carey TE (1990): Establishment and characterization of UM-EC-2, a tamoxifen-sensitive, estrogen receptor-negative human endometrial carcinoma cell line. *Gynecol Oncol.* 37:188–199

Grenman R, Laine KM, Klemi PJ, Grenman S, Hayashida DJ, Joensuu H (1991): Effects of the antiestrogen toremifene on growth of the human mammary carcinoma cell line MCF-7. *J Cancer Res Clin Oncol* 117:223–226

Gronemeyer H, Benhamou B, Berry M, Bocquel MT, Gofflo D, Garcia T, Lerouge T, Metzger D, Meyer ME, Tora L, Vergezac A, Chambon P (1992): Mechanisms of antihormone action. *J Steroid Biochem Mol Biol* 41:217–221

Guler HP, Zapf J, Schmid C, Froesch ER (1989): Insulin-like growth factors I and II in healthy man: estrimations of half-lives and production rates. *Acta Endocrinol (Copenh)* 121:753–758

Gundersen S (1990): Toremifene, a new antiestrogenic compound in the treatment of metastatic mammary cancer. A phase II study. *J Steroid Biochem* 36:233–234

Gusterson BA, Gelber RD, Goldhirsch A, Price KN, Save SJ, Anbazhagan R, Styles J, Rudenstam CM, Golouh R, Reed R, Martinez-Tello F, Tiltman A, Torhorst J, Grigolato P, Bettelheim R, Neville AM, Brki K, Castiglione M, Collins J, Lindtner J, Senn HJ (1992): Prognostic importance of c-erbB-2 expression in breast cancer. International (Ludwig) Breast Cancer Study Group. *J Clin Oncol* 10:1049–1056

Haddow A, Watkinson JM, Paterson E, Koller PC (1944): Influence of synthetic oestrogens upon advanced malignant disease. *Br Med J* ii:393–398

Halachmi S, Marden E, Martin G, MacKay H, Abbondanza C, Brown M (1993): Estrogen receptor-associated proteins: possible mediators of hormone-induced transcription. 264:1455–1458

Han XL, Liehr JG (1992): Induction of covalent DNA adducts in rodents by tamoxifen. *Cancer Res* 52:1360–1363

Hansen JC, Gorski J (1986): Conformational transitions of the estrogen receptor monomer. Effects of estrogens, antiestrogen, and temperature. *J Biol Chem* 261:13990–13996

Hard GC, Iatropoulos MJ, Jordan K, Radi L, Kaltenberg OP, Imondi AR, Williams GM (1993): Major difference in the hepatocarcinogenicity and DNA adduct forming ability between toremifene and tamoxifen in female Crl:CD(BR) rats. *Cancer Res* 53:4534–4541

Harper MJK, Walpole AL (1966): Contrasting endocrine activity of cis and trans isomers in a series of substituted triphenylethylenes. *Nature* 212:87

Harper MJK, Walpole AL (1967): A new derivative of triphenylethylene: effect on implantation and mode of action in rats. *J Reprod Fertil* 13:101–119

Hasmann M, Rattel B, Löser R (1994): Preclinical data for droloxifene. *Cancer Lett* 84:101–116

Hayashida M, Terakawa N, Shimizu I, Ikegami H, Wakimoto H, Aono T, Tanizawa O, Matsumoto K (1987): Roles of antiestrogen binding sites in human endometrial cancer cells. *J Steroid Biochem* 26:705–711

Hedden A, Müller V, Jensen EV (1995): A new interpretation of antiestrogen action. *Ann NY Acad Sci* (in press)

Hirsimaki P, Hirsimaki Y, Nieminen L, Payne BJ (1993): Tamoxifen induces hepatocellular carcinoma in rat liver: a 1-year study with two antiestrogens. *Arch Tox* 67:49–54

Homesley HD, Shemano I, Gams RA, Harry DS, Hickox PG, Rebar RW, Bump RC, Mullin TJ, Wentz AC, O'Toole RV, Lovelace JV, Lyden C (1993): Antiestrogenic potency of toremifene and tamoxifen in postmenopausal women. *Am J Clin Oncol* 16:117–122

Horwitz RI, Feinstein AR (1986): Estrogens and endometrial cancer. Responses to arguments and current status of an epidemiologic controversy. *Am J Med* 81:503–507

Horwitz KB, McGuire WL, Pearson OH, Segaloff A (1975): Predicting response to endocrine therapy in human breast cancer: a hypothesis. *Science* 189:726–727

Howell A, Dodwell DJ, Anderson H, Redford J (1992): Response after withdrawal of tamoxifen and progestogens in advanced breast cancer. *Ann Oncol* 3:611–617

Howell A, DeFriend D, Robertson J, Blamey R, Walton P (1995): Response to a specific antioestrogen (ICI 182780) in tamoxifen-resistant breast cancer. *Lancet* 1:29–30

Hu XF, Nadalin G De, LM, Martin TJ, Wakeling A, Huggins R, Zalcberg JR (1991): Circumvention of doxorubicin resistance in multi-drug resistant human leukaemia and lung cancer cells by the pure antioestrogen ICI 164384. *Eur J Cancer* 27:773–777

Hu XF, Veroni M, DeLuise M, Wakeling A, Sutherland R, Watts CK, Zalcberg JR (1993): Circumvention of tamoxifen resistance by the pure anti-estrogen ICI 182,780. *Int J Cancer* 55:873–876

Huff KK, Kaufman D, Gabbay KH, Spencer EM, Lippman ME, Dickson RB (1986): Secretion of an insulin-like growth factor I related protein by human breast cancer cells. *Cancer Res* 46:4613–4619

Hunter RE, Jordan VC (1975): Detection of the 8S oestrogen binding component in human uterine endometrium during the menstrual cycle. *J Endocrinol* 65:457–458

Huynh HT, Pollak M (1993): Insulin-like growth factor I gene expression in the uterus is stimulated by tamoxifen and inhibited by the pure antiestrogen ICI 182780. *Cancer Res* 53:5585–5588

Huynh H, Pollak M (1994): Uterotrophic actions of estradiol and tamoxifen are associated with inhibition of uterine insulin-like growth factor binding protein 3 gene expression. *Cancer Res* 54:3115–3119

Hyder SM, Stancel GM, Nawaz Z, McDonnell DP, Loose MD (1992): Identification of an estrogen response element in the 3′-flanking region of the murine c-fos protooncogene. *J Biol Chem* 267:18047–18054

Hyder SM, Stancel GM (1994): In vitro interaction of uterine estrogen receptor with the estrogen response element present in the 3′-flanking region of the murine c-fos protooncogene. *J Steroid Biochem Mol Biol* 48:69–79

Ingram D (1990): Tamoxifen use, oestrogen binding and serum lipids in post-menopausal women with breast cancer. *Austr NZ J Surg* 60:673–675

Issandou M, Darbon JM (1985): 1,2-dioctanoylglycerol induces a discrete but transient translocation of protein kinase C as well as the inhibition of MCF-7 cell proliferation. *Biochem Biophys Res Commun* 151:458–465

Issandou M, Bayard F, Darbon JM (1988): Inhibition of MCF-7 cell growth by 12-o-tetradecanoylphorbol-13-acetate and 1,2-dioctanoyl-sn-glycerol: distinct effects on protein kinase C activity. *Cancer Res* 48:6943–6950

Issandou M, Faucher C, Bayard F, Darbon JM (1990): Opposite effects of tamoxifen on in vitro protein kinase C activity and endogenous protein phosphorylation in intact MCF-7 cells. *Cancer Res* 50:5845–50

Iveson TJ, Ahern J, Smith IE (1993): Response to third-line endocrine treatment for advanced breast cancer. *Eur J Cancer* 29A:572–574

Jensen EV, Jacobson HI (1962): Basic guides to the mechanism of estrogen action. *Rec Prog Horm Res* 18:387–414

Jensen EV, Block GE, Smith S, Kyser K, De Sombre ER (1971): Estrogen receptors and breast cancer response to adrenalectomy. *NCI Monogr* 34:55–70

Jiang SY, Wolf DM, Yingling JM, Chang C, Jordan VC (1992): An estrogen receptor positive MCF-7 clone that is resistant to antiestrogens and estradiol. *Mol Cell Endocrinol* 90:77–86

Johnston SR, Haynes BP, Smith IE, Jarman M, Sacks NP, Ebbs SR, Dowsett M (1993): Acquired tamoxifen resistance in human breast cancer and reduced intra-tumoral drug concentration. *Lancet* 342:1521–1522

Jordan VC (1975): Prolonged antioestrogenic activity of ICI 46,474 in the ovariectomized mouse. *J Reprod Fertil* 42:251–258

Jordan VC, Koerner S (1975a): Tamoxifen (ICI 46,474) and the human carcinoma 8S oestrogen receptor. *Eur J Cancer* 11:205–206

Jordan VC, Koerner S (1975b): Inhibition of oestradiol binding to mouse uterine and vaginal oestrogen receptors by triphenylethylenes. *J Endocrinol* 64:193–194

Jordan VC (1976): Effect of tamoxifen (ICI 46,474) on initiation and growth of DMBA-induced rat mammary carcinomata. *Eur J Cancer* 12:419–424

Jordan VC, Dowse LJ (1976): Tamoxifen as an anti-tumour agent: effect on oestrogen binding. *J Endocrinol* 68:297–303

Jordan VC, Collins MM, Rowsby L, Prestwich G (1977): A monohydroxylated metabolite of tamoxifen with potent antioestrogenic activity. *J Endocrinol* 75:305–316

Jordan VC, Prestwich G (1977): Binding of (^{3}H) tamoxifen in rat uterine cytosols: a comparison of swinging bucket and vertical tube rotor sucrose density gradient analysis. *Mol Cell Endocrinol* 8:179–180

Jordan VC (1978): Use of the DMBA-induced rat mammary carcinoma system for the evaluation of tamoxifen as a potential adjuvant therapy. *Rev Endocr Rel Cancer* October Suppliment:49–55

Jordan VC, Dix CJ, Naylor KE, Prestwich G, Rowsby L (1978): Non-steroidal antioestrogens: their biological effects and potential mechanisms of action. *J Tox Environ Health* 4:364–390

Jordan VC, Dix CJ (1979): Effect of oestradiol benzoate, tamoxifen and monohydroxytamoxifen on immature rat uterine progesterone receptor synthesis and endometrial cell division. *J Steroid Biochem* 11:285–291

Jordan VC, Dix CJ, Allen KE (1979): The effectiveness of long-term treatment in a laboratory model fro adjuvant hormone therapy of breast cancer. In: *Adjuvant Therapy of Cancer*, vol. 2, Salmon SE, Jones SE, eds. New York: Grune and Stratton

Jordan VC, Allen KE, Dix CJ (1980): Pharmacology of tamoxifen in laboratory animals. *Cancer Treat Rep* 64:745–759

Jordan VC (1984): Biochemical pharmacology of antiestrogenic action. *Pharmacol Rev* 36:245–276

Jordan VC, Phelps E, Lindgren JU (1987): Effects of anti-estrogens on bone in castrated and intact female rats. *Breast Cancer Res Treat* 10:31–35

Jordan VC, Gottardis MM, Robinson SP, Friedl A (1989): Immune-deficient animals to study "hormone-dependent" breast and endometrial cancer. *J Steroid Biochem* 34:169–176

Jordan VC, Koch R (1989): Regulation of prolactin synthesis in vitro by estrogenic and antiestrogenic derivatives of estradiol and estrone. *Endocrinology* 124: 1717–1726

Jordan VC, Gottardis MM, Satyaswaroop PG (1991a): Tamoxifen-stimulated growth of human endometrial carcinoma. *Ann NY Acad Sci* 622:439–446

Jordan VC, Lababidi MK, Langan-Fahey S (1991b): The suppression of mouse mammary tumorigenesis by long-term tamoxifen therapy. *J Natl Cancer Inst* 83:492–496

Jordan VC (1993): A current view of tamoxifen for treatment and prevention of breast cancer. *Br J Pharmacol* 110:507–517

Jordan VC (1994): The development of tamoxifen for breast cancer therapy. In: *Long-term Tamoxifen Treatment for Breast Cancer*, Jordan VC, ed. Madison: University of Wisconsin Press, pp 3–26

Jordan VC, Morrow M (1994): Should clinicians be concerned about the carcinogenic potential of tamoxifen? *Eur J Cancer* 30A:1714–1721

Jordan VC, Assikis VJ (1995): Tamoxifen and endometrial cancer: clearing up a controversy. *Clin Cancer Res* 1:467–472

Jost JP, Seldran M, Geiser M (1984): Preferential binding of estrogen-receptor mediated complex to a region containing the estrogen-dependent hypomethylation site preceding the chicken vitellogenin II gene. *Proc Natl Acad Sci USA,* 81:429–433

Kangas L, Haaparanta M, Paul R, Roeda D, Sipila H (1989): Biodistribution and scintigraphy of ^{11}C-toremifene in rats bearing DMBA-induced mammary carcinoma. Pharm Tox 64:373–377

Karey KP, Sirbasku DA (1988): Differential responsiveness of human breast cancer cell lines MCF-7 and T47D to growth factors and 17-beta-estradiol. *Cancer Res* 48:4083–4092

Kasid A, Knabbe C, Lippman ME (1987): Effet of v-rasH oncogene transfection on estrogen-independent tumorigenicity of estrogen-dependent human breast cancer cells. *Cancer Res* 47:5733–5738

Katzenellenbogen BS, Miller MA, Mullick A, Sheen YY (1985): Antiestrogen action in breast cancer cells: modulation of proliferation and protein synthesis, and interaction with estrogen receptors and additional antiestrogen binding sites. Breast Cancer Res Treat 5:231–243

Katzenellenbogen BS, Kendra KL, Norma MJ, Berthois Y (1987): Proliferation, hormonal responsiveness, and estrogen receptor content of MCF-7 human breast cancer cells grown in the short-term and long-term absence of estrogens. *Cancer Res* 47:4355–4360

Kawamura I, Mizota T, Kondo N, Shimomura K, Kohsaka M (1991): Antitumor effects of droloxifene, a new antiestrogen drug, against 7,12-dimethylbenz(a)-anthracene-induced mammary tumors in rats. Jap J Pharmacol 57:215–224

Kawamura I, Mizota T, Lacey E, Tanaka Y, Manda T, Shimomura K, Kohsaka M (1993): The estrogenic and antiestrogenic activities of droloxifene in human breast cancers. Jap J Pharmacol 63:27–34

King RJB, Wang DY, Daly RJ, Darbre PD (1989): Approaches to studying the role of growth factors in the progression of breast tumors from the steroid sensitive to insensitive state *J Steroid Biochem* 34:133–138

Kirkland JL, Murthy L, Stancel GM (1993): Tamoxifen stimulates expression of the c-fos proto-oncogene in rodent uterus. Mol Pharmacol 43:709–714

Klein-Hitpass L, Schorpp M, Wagner U, Ryffel GU (1986): An estrogen-responsive element derived from the 5′ flanking region of the Xenopus vitellogenin A2 gene functions in transfected human cells. *Cell,* 46:1053–1061

Klein-Hitpass L, Ryffel GU, Heitlinger E, Cato ACB (1988): A 13 bp palindrome is a functional estrogen responsive element and interacts specifically with estrogen receptor. *Nucleic Acids Res*, 16:647–663

Klinge CM, Bambara RA, Hilf R (1992): What differentiates antiestrogen-liganded vs estradiol-liganded estrogen receptor action? *Oncol Res* 4:137–144

Knabbe C, Lippman ME, Wakefield LM, Flanders KC, Kasid A, Derynck R, Dickson RB (1987): Evidence that transforming growth factor-beta is a hormonally regulated negative growth factor in human breast cancer cells. *Cell* 48:417–428

Knauer DJ, Smith GL (1980): Inhibition of biological activity of multiplication-stimulating activity by binding to its carrier protein. *Proc Natl Acad Sci USA* 77:7252–7256

Kraus WL, Montano MM, Katzenellenbogen BS (1993): Cloning of the rat progesterone receptor gene 5′-region and identification of two functionally distinct promotors. *Mol Endocrinol* 7:1603–1616

Kristensen B, Ejlertsen B, Dalgaard P, Larsen L, Holmegaard SN, Transbol I, Mouridsen HT (1994): Tamoxifen and bone metabolism in postmenopausal low-risk breast cancer patients: a randomized study. *J Clin Oncol* 12:992–997

Kroeger EA, Brandes LJ (1985): Evidence that tamoxifen is a histamine antagonist. *Biochem Biophys Res Commun* 131:750–755

Krywicki RF, Figueroa JA, Jackson JG, Kozelsky TW, Shimasaki S, Von HD Yee D (1993): Regulation of insulin-like growth factor binding proteins in ovarian cancer cells by oestrogen. *Eur J Cancer* 29A:2015–2019

Kumar SA, Beach TA, Dickerman HW (1983): Oligodeoxynucleotide base recognition by steroid hormone receptors. J Cell Biochem, 21:19–27

Langan-Fahey SM, Tormey DC, Jordan VC (1990): Tamoxifen metabolites in patients on long-term adjuvant therapy for breast cancer. *Eur J Cancer* 26:883–888

Langan-Fahey SM, Jordan VC, Fritz NF, Robinson SP, Waters D, Tormey DC (1994): Clinical pharmacology and endocrinology of long-term tamoxifen therapy. In: *Long-term Tamoxifen Treatment for Breast Cancer*, Jordan VC, ed. Madison: University of Wisconsin Press, pp 27–56

Lau CK, Subramaniam M, Rasmussen K, Spelsberg TC (1991): Rapid induction of the c-jun protooncogene in the avian oviduct by the antiestrogen tamoxifen. *Proc Natl Acad Sci USA* 88:829–833

Lazier CB, Bapat BV (1988): Antiestrogen binding sites: general and comparative properties *J Steroid Biochem* 31:665–669

Lazier CB, Breckenridge WC (1990): Comparison of the effects of tamoxifen and of a tamoxifen analogue that does not bind the estrogen receptor on serum lipid profiles in the cockerel. *Biochem Cell Biol* 68:210–217

Legault PS, Jolivet J, Poisson R, Beretta PM, Band PR (1979): Tamoxifen-induced tumor stimulation and withdrawal response. *Cancer Treat Rep* 63:1839–1841

Lerner HJ, Holthaus JF, Thompson CR (1958): A nonsteroidal estrogen antagonist 1-(p-2-diethylaminoethoxyphenyl)-1-phenyl-2-p-methoxy ethanol. *Endocrinology* 63:295–318

Lerner LJ, Jordan VC (1990): Development of antiestrogen and their use in breast cancer: Eighth Cain Memorial Award Lecture. *Cancer Res* 50:4177–4189

LeRoy X, Escot C, Brouillet JP, Theillet C, Maudelonde T, Simony LJ, Pujol H, Rochefort H (1991): Decrease of c-erbB-2 and c-myc RNA levels in tamoxifen-treated breast cancer. *Oncogene* 6:431–437

Leung BS, Stout L, Zhou L Ji, HJ, Zhang QQ, Leung HT (1991): Evidence of an EGF/TGF α-independent pathway for estrogen-regulated cell proliferation. *J Cell Biochem* 46:125–133

Lien EA, Johannessen DC, Aakvaag A, Lonning PE (1992): Influence of tamoxifen, aminoglutethimide and goserelin on human plasma IGF-I levels in breast cancer patients. *J Steroid Biochem Mol Biol* 41:541–543

Lippman ME, Dickson RB, Bates S, Knabbe C, Huff K, Swain S, McMacaway M, Bronzert D, Kasid A, Gelmann EP (1986): Autocrine and paracrine growth regulation of human breast cancer. *Breast Cancer Res Treat* 7:59–70

Long B, McKibben BM, Lynch M van, den, Berg HW (1992): Changes in epidermal growth factor receptor expression and response to ligand associated with acquired tamoxifen resistance or oestrogen independence in the ZR-75-1 human breast cancer cell line. *Br J Cancer* 65:865–869

Lonning PE, Hall K, Aakvaag A, Lien EA (1992): Influence of tamoxifen on plasma levels of insulin-like growth factor I and insulin-like growth factor binding protein I in breast cancer patients. *Cancer Res* 52:4719–4723

Löser R, Seibel K, Eppenberger U (1985a): No loss of estrogenic or anti-estrogenic activity after demethylation of droloxifene (3-OH-tamoxifen). Int J Cancer 36:701–703

Löser R, Seibel K, Roos W, Eppenberger U (1985b): In vivo and in vitro antiestrogenic action of 3-hydroxytamoxifen, tamoxifen and 4-hydroxytamoxifen. *Eur J Cancer Clin Oncol* 21:985–990

Love RR, Mazess RB, Tormey DC, Barden HS, Newcomb PA, Jordan VC (1988): Bone mineral density in women with breast cancer treated with adjuvant tamoxifen for at least two years. *Breast Cancer Res Treat* 12:297–302

Love RR, Newcomb PA, Wiebe DA, Surawicz TS, Jordan VC, Carbone PP, DeMets DL (1990): Effects of tamoxifen therapy on lipid and lipoprotein levels in postmenopausal patients with node-negative breast cancer. *J Natl Cancer Inst* 82:1327–1332

Love RR, Wiebe DA, Newcomb PA, Cameron L, Leventhal H, Jordan VC, Feyzi J, DeMets DL (1991): Effects of tamoxifen on cardiovascular risk factors in postmenopausal women. *Ann Intern Med* 115:860–864

Love RR, Mazess RB, Barden HS, Epstein S, Newcomb PA, Jordan VC, Carbone PP, DeMets DL (1992): Effects of tamoxifen on bone mineral density in postmenopausal women with breast cancer. *N Engl J Med* 326:852–856

Love RR, Wiebe DA, Feyzi JM, Newcomb PA, Chappell R (1994): Effects of tamoxifen on cardiovascular risk factors in postmenopausal women after 5 years of treatment. *J Natl Cancer Inst* 86:1534–1539

Ludwig Breast Cancer Study Group (1984): Randomized trial of chemoendocrin therapy, endocrine therapy and mastectomy alone in postmenopausal patients with operable breast cancer and axillary node metastases. *Lancet* 1:1256–1260

Lykkesfeldt AE, Madsen MW, Briand P (1994): Altered expression of estrogen-regulated genes in a tamoxifen-resistant and ICI 164,384 and ICI 182,780 sensitive human breast cancer cell line, MCF-7/TAMR-1. *Cancer Res* 54:1587–1595

Maenpaa J, Wiebe V, Koester S, Wurz G, Emshoff V, Seymour R, Sipila P, DeGregorio M (1993): Tamoxifen stimulates in vivo growth of drug-resistant estrogen receptor-negative breast cancer. *Cancer Chem Pharm* 32:396–398

Malaab SA, Pollak MN, Goodyer CG (1992): Direct effects of tamoxifen on growth hormone secretion by pituitary cells in vitro. *Eur J Cancer* 28A:788–793

Mani C, Kupfer D (1991): Cytochrome P-450-mediated activation and irreversible binding of the antiestrogen tamoxifen to proteins in rat and human liver: possible involvement of flavin-containing monooxygenases in tamoxifen activation. *Cancer Res* 51:6052–6058

Manni A, Wright C, Buck H (1991): Growth factor involvement in the multihormonal regulation of MCF-7 breast cancer cell growth in soft agar. *Breast Cancer Res Treat* 20:43–52

Martin PM, Berthois Y, Jensen EV (1988): Binding of antiestrogens exposes an occult antigenic determinant in the human estrogen receptor. *Proc Natl Acad Sci USA* 85:2533–2537

Mathews LS, Norstedt G, Palmiter RD (1986): Regulation of insulin-like growth factor I gene expression by growth hormone. *Proc Natl Acad Sci USA* 83:9343–9347

Matsuda S, Kadowaki Y, Ichino M, Akiyama T, Toyoshima K, Yamamoto T (1993): 17 beta-estradiol mimics ligand activity of the c-erbB2 protooncogene product. *Proc Natl Acad Sci USA* 90:10803–10807

Matsuo S, Takano S, Yamashita J, Ogawa M (1992): Synergistic cytotoxic effects of tumor necrosis factor, interferon-gamma and tamoxifen on breast cancer cell lines. *Anticancer Res* 12:1575–1579

Maurer RA, Notides AC (1987): Identification of an estrogen-responsive element from the 5′-flanking region of the rat prolactin gene. *Mol Cell Biol,* 7:4247–4254

McDonald CC, Stewart HJ (1991): Fatal myocardial infarction in the Scottish adjuvant tamoxifen trial. The Scottish Breast Cancer Committee. *Br Med J* 303:435–437

McGuire WL, Chamness GC (1973): Studies on the estrogen receptor in breast cancer. In: *Receptors for Reproductive Hormones*, O'Malley BW, Means AR, eds. New York: Plenum Publishing Corp, pp 113–136

McLeskey SW, Kurebayashi J, Honig SF, Zwiebel J, Lippman ME, Dickson RB, Kern FG (1993): Fibroblast growth factor 4 transfection of MCF-7 cells produces cell lines that are tumorigenic and metastatic in ovariectomized or tamoxifen-treated athymic nude mice. *Cancer Res* 53:2168–2177

Mehta RG, Cerny WL, Moon RC (1984): Distribution of antiestrogen-specific binding sites in normal and neoplastic mammary gland. *Oncology* 41:387–392

Miller MA, Lippman ME, Katzenellenbogen BS (1984): Anti-estrogen growth-resistant estrogen-responsive clonal variants of MCF-7 human breast cancer cells. *Cancer Res* 44:5038–5045

Millward MJ, Cantwell BM, Lien EA, Carmichael J, Harris AL (1992): Intermittent high-dose tamoxifen as a potential modifier of multidrug resistance. *Eur J Cancer* 28A:805–810

Montcourrier P, Mangeat PH, Salazar G, Morisset M, Sahuguet A, Rochefort H (1990): Cathepsin D in breast cancer cells can digest extracellular matrix in large acidic vesicles. *Cancer Res* 50:6045–6054

Mouridsen H, Palshof T, Patterson J, Battersby L (1978): Tamoxifen in advanced breast cancer. *Cancer Treat Rev* 5:131–141

Murayama Y, Kurata S, Mishim Y (1988): Regulation of human estrogen receptor gene, epidermal growth factor receptor gene, and oncogenes by estrogen and antiestrogen in MCF-7 breast cancer cells. *Cancer Det Prev* 13:103–107

Murphy CS, Jordan VC (1989): Structural components necessary for the anti-estrogenic activity of tamoxifen. *J Steroid Biochem* 34:407–411

Murphy CS, Langan FS, McCague R, Jordan VC (1990): Structure-function relationships of hydroxylated metabolites of tamoxifen that control the proliferation of estrogen-responsive T47D breast cancer cells in vitro. *Mol Pharmacol* 38:737–743

Murphy LC, Dotzlaw H (1989a): Endogenous growth factor expression in T-47D, human breast cancer cells, associated with reduced sensitivity to antiproliferative effects of progestins and antiestrogens. *Cancer Res* 49:599–604

Murphy LC, Dotzlaw H (1989b): Regulation of transforming growth factor alpha and transforming growth factor beta messenger ribonucleic acid abundance in T-47D, human breast cancer cells. *Mol Endocrinol* 3:611–617

Murphy LC, Sutherland RL (1983): Antitumor activity of clomiphene analogs in vitro: relationship to affinity for the estrogen receptor and another high affinity antiestrogen-binding site. *J Clin Endocrinol Metab* 57:373–379

Nayfield SG, Karp JE, Ford LG, Dorr FA, Kramer BS (1991): Potential role of tamoxifen in prevention of breast cancer. *J Natl Cancer Inst* 83:1450–1459

Nephew KP, Polek TC, Akcali KC, Khan SA (1993): The antiestrogen tamoxifen induces c-fos and jun-B, but not c-jun or jun-D, protooncogenes in the rat uterus. *Endocrinology* 133:419–422

Nique F, Van de Velde P, Hardy M, Philibert D, Teutsch G (1994): 11β-Amidoalkoxyphenyl estradiols, a new series of pure antiestrogens. *J Steroid Biochem Mol Biol* 50:21–29

Noguchi S, Motomura K, Inaji H, Imaoka S, Koyama H (1993): Down-regulation of transforming growth factor-alpha by tamoxifen in human breast cancer. *Cancer* 72:131–136

Nolvadex Adjuvant Trial Organisation (1985): Controlled trial of tamoxifen as a single adjuvant agent in the management of early breast cancer. *Lancet* 1:836–840

O'Brian CA, Liskamp RM, Solomon DH, Weinstein IB (1985): Inhibition of protein kinase C by tamoxifen. *Cancer Res* 45:2462–2465

O'Brian CA, Housey GM, Weinstein IB (1988a): Specific and direct binding of protein kinase C to an immobilized tamoxifen analogue. *Cancer Res* 48:3626–3629

O'Brian CA, Ward NE, Anderson BW (1988b): Role of specific interactions between protein kinase C and triphenylethylenes in inhibition of the enzyme. *J Natl Cancer Inst* 80:1628–1633

Oka H, Shiozaki H, Kobahashi K, Inoue M, Tahara H, Kobayashi T, Takatsuka Y, Matsuyoshi N, Hirano S, Takeichi M, Mori T (1993): Expression of E-cadherin cell adhasion molecules in human breast cancer tissues and its relationship to metastases. *Cancer Res* 53:1696–1701

Osborne CK, Hamilton B, Nover M, Ziegler J (1981): Antagonism between epidermal growth factor and phorbol ester tumor promoters in human breast cancer cells. *J Clin Invest* 67:943–951

Osborne CK, Hobbs K, Clark GM (1985): Effect of estrogens and antiestrogens on growth of human breast cancer cells in athymic nude mice. *Cancer Res* 45:584–590

Osborne CK, Coronado EB, Robinson JP (1987): Human breast cancer in the athymic nude mouse: cytostatic effects of long-term antiestrogen therapy. *Eur J Cancer Clin Oncol* 23:1189–1196

Osborne CK, Ross CR, Coronado EB, Fuqua SAW, Kitten LJ (1988): Secreted growth factors from estrogen receptor-negative human breast cancer do not support growth of estrogen receptor-positive breast cancer in the nude mouse model. *Breast Cancer Res Treat* 11:211–219

Osborne CK, Coronado EB, Kitten LJ, Arteaga CI, Fuqua SAW, Ramasharma K, Marshall M Li CH (1989): Insulin-like growth factor-II (IGF-II): a potential autocrine/paracrine growth factor for human breast cancer acting via the IGF-I receptor. *Mol Endocrinol* 3:1701–1709

Osborne CK, Coronado E, Allred DC, Wiebe V, DeGregorio M (1991): Acquired tamoxifen (TAM) resistance: correlation with reduced breast tumor levels of tamoxifen and isomerization of trans-4-hydroxytamoxifen. *J Natl Cancer Inst* 83:1477–1482

Osborne CK (1993): Mechanisms for tamoxifen resistance in breast cancer: possible role of tamoxifen metabolism. *J Steroid Biochem Mol Biol* 47:83–89

Osborne CK, Jarman M, McCague R, Coronado EB, Hilsenbeck SG, Wakeling AE (1994): The importance of tamoxifen metabolism in tamoxifen-stimulated breast tumor growth. *Cancer Chemother Pharm* 34:89–95

Oxenhandler RW, McCune R, Subtelney A, Truelove C, Tyrer HW (1984): Flow cytometry determination of estrogen receptor in intact cells. *Cancer Res* 44:2516–2523

Paterson AG, Turkes A, Groom GV, Webster DJ (1983): The effect of tamoxifen on plasma growth hormone and prolactin in postmenopausal women with advanced breast cancer. *Eur J Cancer Clin Oncol* 19:919–922

Pathak DN, Bodell WJ (1994): DNA adduct formation by tamoxifen with rat and human liver microsomal activation systems. *Carcinogenesis* 15:529–532

Peck WA, Riggs BL, Bell NH, Wallace RB, Jonston CC, Gordon SL, Shulman LE (1988): Research directions in osteoporosis. *Am J Med* 84:275–280

Pellerin I, Vuillermoz C, Jouvenot M, Royez M, Ordener C, Marechal G, Adessi G (1992): Superinduction of c-fos gene expression by estrogen in cultured guinea-pig endometrial cells requires priming by a cycloheximide-dependent mechanism. *Endocrinology* 131:1094–1100

Perren TJ (1991): c-erbB-2 oncogene as a prognostic marker in breast cancer. *Br J Cancer* 63:328–332

Peyrat JP, Bonneterre J, Beuscart R, Djiane J, Demaille A (1988): Insulin-like growth factor I receptors in human breast cancer and their relation to estradiol and progesterone receptors. *Cancer Res* 48:6429–6433

Philips A, Chalbos D, Rochefort H (1993): Estradiol increases and anti-estrogens antagonize the growth factor-induced activator protein-1 activity in MCF7 breast cancer cells without affecting c-fos and c-jun synthesis [published erratum appears in J Biol Chem 1993 Dec 5;268(34):26032]. *J Biol Chem* 268:14103–14108

Pollack IF, Randall MS, Kristofik MP, Kelly RH, Selker RG, Vertosick FJ (1990): Effect of tamoxifen on DNA synthesis and proliferation of human malignant glioma lines in vitro. *Cancer Res* 50:7134–7138

Pollak M, Costantino J, Polychronakos C, Blauer SA, Guyda H, Redmond C, Fisher B, Margolese R (1990): Effect of tamoxifen on serum insulin-like growth factor I leves in stage I breast cancer patients. *Cancer Res* 82:1693–1697

Potter GA, McCague R, Jarman M (1994): A mechanistic hypothesis for DNA adduct formation by tamoxifen following hepatic oxidative metabolism. *Carcinogenesis* 15:439–442

Poulin R, Dofour JM, Labrie F (1989a): Progestin inhibition of estrogen-dependent proliferation in ZR-75–1 human breast cancer cells: antagonism by insulin. *Breast Cancer Res Treat* 13:265–276

Poulin R, Merand Y, Poirer D, Levesque C, Dufour JM, Labrie F (1989b): Antiestrogenic properties of keoxifene, trans-4-hydroxytamoxifen, and ICI 164384, a new steroidal antiestrogen, in ZR-75-1 human breast cancer cells. *Breast Cancer Res Treat* 14:65–76

Powles TJ, Hardy JR, Ashley SE, Farrington GM, Cosgrove D, Davey JB, Dowsett M, McKinna JA, Nash AG, Sinnett HD, Tillyer CR, Treleaven J (1989): A pilot trial to evaluate the acute toxicity and feasibility of tamoxifen for prevention of breast cancer. *Br J Cancer* 60:126–131

Powles TJ, Tillyer CR, Jones AL, Ashley SE, Treleaven J, Davey JB, McKinna JA (1990): Prevention of breast cancer with tamoxifen—an update on the Royal Marsden Hospital pilot programme. *Eur J Cancer* 26:680–684

Pratt SE, Pollak MN (1993): Estrogen and antiestrogen modulation of MCF7 human breast cancer cell proliferation is associated with specific alterations in accumulation of insulin-like growth factor-binding proteins in conditioned media. *Cancer Res* 53:5193–5198

Pratt S, Pollak M (1994): Insulin-like growth factor binding protein 3 inhibits estrogen-stimulated breast cancer cell proliferation. *Biochem Biophys Res Commun* 198:292–297

Pyrhonen SO (1990): Phase III studies of toremifene in metastatic breast cancer. *Breast Cancer Res Treat* 16:541–546

Ramu A, Ramu N, Rosario LM (1991): Circumvention of multidrug-resistance in P388 cells is associated with a rise in the cellular content of phosphatidylcholine. *Biochem Phearmacol* 41:1455–1461

Rechler MM, Brown AL (1992): Insulin-like growth factors and their binding proteins. *Growth Regul* 2:55–68

Reddel RR, Murphy LC, Sutherland RL (1983): Effects of biologically active metabolites of tamoxifen on the proliferation kinetics of MCF-7 human breast cancer cells in vitro. *Cancer Res* 43:4618–4624

Reddy KB, Yee D, Hilsenbeck SG, Coffey RJ, Osborne CK (1994): Inhibition of estrogen-induced breast cancer cell proliferation by reduction in autocrine transforming growth factor alpha expression. *Cell Growth Diff* 5:1275–1282

Ribiero G, Palmer MK (1983): Adjuvant tamoxifen for operable carcinoma of the breast: report of a clinical trial by the Christie Hospital and Holt Radium Institute. *Br Med J* 286:827–830

Ribiero G, Swindell R (1985): The Christie Hospital tamoxifen (Nolvadex) adjuvant trial for operable breast carcinoma—seven-year results. *Eur J Cancer Clin Oncol* 21:1817–1821

Roberts AB, Anzano MA, Wakefield LM, Roche NS, Stern DF, Sporn MB (1985): Type beta transforming growth factor: a bifunctional regulator of cellular growth. *Proc Natl Acad Sci USA* 82:119–123

Roberts CT, Brown AL, Graham DE, Seelig S, Berry S, Gabbay KH, Rechler MM (1986): Growth hormone regulates the abundance of insulin-like growth factor I RNA in adult rat liver. *J Biol Chem* 261:10025–10028

Robinson SP, Jordan VC (1989): The paracrine stimulation of MCF-7 cells by MDA-MB-231 cells: possible role in antiestrogen failure. *Eur J Cancer Clin Oncol* 25:293–297

Robinson SP, Mauel DA, Jordan VC (1989): Antitumor actions of toremifene in the 7,12-dimethylbenzanthracene (DMBA)-induced rat mammary tumor model. *Eur J Cancer Clin Oncol* 24:1817–1821

Rochefort H (1992): Cathepsin D in breast cancer: a tissue marker associated with metastasis Cancer Cells 28A:1780–1783

Rochefort H, Capony F, Garcia M (1990): Cathepsin D in breast cancer: from molecular and cellular biology to clinical applications. *Eur J Cancer* 2:383–388

Roos W, Oeze L, Löser R, Eppenberger U (1983): Antiestrogenic action of 3-hydroxytamoxifen in the human breast cancer cell line MCF-7. *J Natl Cancer Inst* 71:55–59

Roos W, Fabbro D, Kung W, Costa SD, Eppenberger U (1983): Correlation between hormone dependency and the regulation of epidermal growth factor receptor by tumor promoters in human mammary carcinoma cells. *Proc Natl Acad Sci USA* 83:991–995

Rose C, Thorpe SM, Andersen KW, Pedersen BV, Mouridsen HT, Blichert-Toft M, Rasmussen BB (1985): Beneficial effect of adjuvant tamoxifen theapy in primary breast cancer patients with high oestrogen receptor values. *Lancet* 1:16–19

Rossner S, Wallgren A (1984): Serum lipoproteins and proteins after breast cancer surgery and effects of tamoxifen. *Atherosclerosis* 52:339–346

Rutqvist, LE, Mattsson A (1993): Cardiac and thromboembolic morbidity among postmenopausal women with early-stage breast cancer in a randomized trial of adjuvant tamoxifen The Stockholm breast cancer group. *J Natl Cancer Inst* 85:1398–1406

Safa AR, Roberts S, Agresti M, Fine RL (1994): Tamoxifen aziridine, a novel affinity probe for P-glycoprotein in multidrug resistant cells. *Biochim Biophys Res Commun* 202:606–612

Sakakibara K, Kan NC, Satyaswaroop PG (1992): Both 17 beta-estradiol and tamoxifen induce c-fos messenger ribonucleic acid expression in human endometrial carcinoma grown in nude mice [see comments]. *Am J Obstet Gynecol* 166:206–212

Salomon DS, Ciardiello F, Valverius E, Saeki T Kim N (1989): Transforming growth factors in human breast cancer. *Biomed Pharmaco* 43:661–667

Santos GF, Scott GK, Lee WM, Liu E, Benz C (1988): Estrogen-induced post-transcriptional modulation of c-myc proto-oncogene expression in human breast cancer cells. *J Biol Chem* 263:9565–9568

Sargent LM, Dragan YP, Bahnub N, Wiley JE, Sattler CA, Schroeder P, Sattler GL, Jordan VC, Pitot HC (1994): Tamoxifen induces hepatic aneuploidy and mitotic spindle disruption after a single in vivo administration to female Sprague-Dawley rats. *Cancer Res* 54:3357–3360

Sato M, McClintock C, Kim J, Turner CH, Bryant HU, Magee D, Slemenda CW (1994): Dual-energy x-ray absorptiometry of raloxifene effects on the lumbar vertebrae and femora of ovariectomized rats. *J Bone Mineral Res* 9:715–724

Satyaswaroop PG, Zaino RJ, Mortel R (1984): Estrogen-like effects of tamoxifen on human endometrial carcinoma transplanted into nude mice. *Cancer Res* 44:4006–4010

Savouret JF, Bailly A, Misrahi M, Rauch C, Redeuilh G, Chauchereau A, Milgrom E (1991): Characterization of the hormone responsive element involved in the regulation of the progesterone receptor gene. *EMBO J* 10:1875–1883

Sheen YY, Simpson DM, Katzenellenbogen BS (1985): An evaluation of the role of antiestrogen-binding sites in mediating the growth modulatory effects of antiestrogens: studies using t-butylphenoxyethyl diethylamine, a compound lacking affinity for the estrogen receptor. *Endocrinology* 117:561–564

Shupnik MA, Rosenzweig BA (1991): Identification of an estrogen-responsive element in the rat LHb gene. *J Biol Chem,* 266:17084–17091

Shyu AB, Blumenthal T, Raff RA (1987): A single gene encoding vitellogenin in the sea urchin Strongylocentrotus purpuratus: Sequence at the 5′ end. *Nucleic Acids Res* 15:10405–10417

Skidmore J, Walpole AL, Woodburn J (1972): Effect of some triphenylethylenes on oestradiol binding in vitro to macromolecules from uterus and anteroir pituitary. *J Endocrinol* 52:289–298

Spona J, Bieglmayer C, Leibl H (1980): Estrogen interaction with the anterior pituitary of female rats: Differential cytosol binding, nuclear translocation and stimulation of RNA synthesis by 17 beta-estradiol and tamoxifen. *Biochim Biophys Act* 633:361–375

Stewart PJ, Stern PH (1986): Effects of the antiestrogens tamoxifen and clomiphene on bone resorption in vitro. *Endocrinology* 118:125–131

Stoll BA (1988): Second endocrine responses in breast, prostatic and endometrial cancers. *Rev Endocr Rel Cancer* 30:19–25

Stone GM (1963): The uptake of tritiated oestrogens by various organs of the ovariectomized mouse following subcutaneous administration. *J Endocrinol* 27:281–288

Stuart NS, Philip P, Harris AL, Tonkin K, Houlbrook S, Kirk J, Lien EA, Carmichael J (1992): High-dose tamoxifen as an enhancer of etoposide cytotoxicity. Clinical effects and in vitro assessment in p-glycoprotein expressing cell lines. *Br J Cancer* 66:833–839

Styles JA, Davies A, Lim CK, De MF, Stanley LA, White IN, Yuan ZX, Smith LL (1994): Genotoxicity of tamoxifen, tamoxifen epoxide and toremifene in human lymphoblastoid cells containing human cytochrome P450s. *Carcinogenesis* 15:5–9

Tanaka NG, Sakamoto N, Korenaga H, Inoue K, Ogawa H, Osada Y (1991): The combination of a bacterial polysaccharide and tamoxifen inhibits angiogenesis and tumour growth. *Int J Rad Biol* 60:79–83

Tannenbaum GS, Gurd W, Lapointe M, Pollak M (1992): Tamoxifen attenuates pulsatile growth hormone secretion: mediation in part by somatostatin. *Endocrinology* 130:3395–3401

Teo CC, Kon OL, Sim KY, Ng SC (1992): Synthesis of 2-(p-chlorobenzyl)-3-aryl-6-methoxybenzofurans as selective ligands for antiestrogen-binding sites Effects on cell proliferation and cholesterol synthesis. *J Med Chem* 35:1330–1339

Teodorczyk IJ, Cembrzynska NM, Lalani S, Kellen JA (1993): Modulation of biological responses of normal human mononuclear cells by antiestrogens. *Anticancer Res* 13:279–283

Terenius L (1971): Structure-activity relationships of antioestrogens with regard to interaction with 17 beta-estradiol in the mouse uterus and vagina. *Acta Endocrinol* 66:431–447

Teske E, Besselink CM, Blankenstein MA, Rutteman GR, Misdorp W (1987): The occurrence of estrogen and progestin receptors and anti-estrogen binding sites (AEBS) in canine non-Hodgkin's lymphomas. *Anticancer Res* 7:857–860

Thangaraju M, Vijayalakshmi T, Sachdanandam P (1994): Effect of tamoxifen on lipid peroxide and antioxidative system in postmenopausal women with breast cancer. *Cancer* 74:78–82

Thompson AM, Kerr DJ, Steel CM (1991): Transforming growth factor beta 1 is implicated in the failure of tamoxifen therapy in human breast cancer. *Br J Cancer* 63:609–614

Thompson EW, Brunner N, Torri J, Johnson MD, Boulay V, Wright A, Lippman ME, Steeg PS, Clarke R (1993): The invasive and metastatic properties of hormone-independent but hormone-responsive variants of MCF-7 human breast cancer cells. *Clin Exp Metastasis* 11:15–26

Tiwari RK, Wong GY, Liu J, Miller D, Osborne MP (1991): Augmentation of cytotoxicity using combinations of interferons (types I and II), tumor necrosis factor-alpha, and tamoxifen in MCF-7 cells. *Cancer Lett* 61:45–52

Toft D, Shyamala G, Gorski J (1967): A receptor molecule for estrogens: studies using a cell free system. *Proc Natl Acad Sci USA* 57:1740–1743

Toi M, Harris AL, Bicknell R (1993): cDNA transfection followed by the isolation of a MCF-7 breast cell line resistant to tamoxifen in vitro and in vivo. *Br J Cancer* 68:1088–1096

Tormey DC, Jordan VC (1984): Long-term tamoxifen adjuvant therapy in node-positive breast cancer: a metabolic and pilot clinical study. *Breast Cancer Res Treat* 4:297–302

Tormey DC, Rasmussen P, Jordan VC (1987): Long-term adjuvant tamoxifen study: clinical update [Letter]. *Breast Cancer Res Treat* 9:157–158

Tormey DC, Gray R, Abeloff MD, Roseman DL, Gilchrist KW, Barylak EJ, Stott P, Falkson G (1992): Adjuvant therapy with a doxorubicin regimen and long-term tamoxifen in postmenopausal breast cancer patients: an eastern oncology cooperative group trial. *J Clin Oncol* 10:1848–1856

Trump DL, Smith DC, Ellis PG, Rogers MP, Schold SC, Winer EP, Panella TJ, Jordan VC, Fine RL (1992): High-dose oral tamoxifen, a potential multidrug-resistance-reversal agent: phase I trial in combination with vinblastine. *J Natl Cancer Inst* 84:1811–1816

Tucker RF, Shipley GD, Moses HL, Holley RW (1984): Growth inhibitor from BSC-1 cells closely related to type beta transforming growth factor. *Science* 226:705–707

Turner RT, Wakley GK, Hannon KS, Bell NH (1988): Tamoxifen inhibits osteoclast-mediated resorption of trabecular bone in ovarian hormone-deficient rats. *Endocrinology* 122:1146–1150

Tzuokerman MT, Esty A, Santiso MD, Danielian P, Parker MG, Stein RB, Pike JW, McDonnell DP (1994): Human estrogen receptor transactivational capacity is determined by both cellular and promoter context and mediated by two functionally distinct intramolecular regions. *Mol Endocrinol* 8:21–30

Valavaara R (1990): Phase II trials with toremifene in advanced breast cancer: a review. *Breast Cancer Res Treat* 16:S31–S35

Valavaara R, Kangas L (1988): The significance of estrogen receptors in tamoxifen and toremifene therapy. *Ann Clin Res* 20:380–388

Van de Velde P, Nique F, Bouchoux F, Bremaud J, Hameau MC, Lucas D, Moratille C, Viet S, Philibert D, Teutsch G (1994): RU 58 668, a new pure antiestrogen inducing a regression of human mammary carcinoma implanted in nude mice. *J Steroid Biochem Mol Biol* 48:187–196

van Dijck P, Verhoeven G (1992): Interaction of estrogen receptor complexes with the promoter region of genes that are negatively regulated by estrogens: the α_{2u}-globulins. *Biochem Biophys Res Commun,* 182:174–181 van het Schip F, Strijker R, Samallo J, Gruber M, Geert AB (1986): Conserved sequence motifs upstream from the co-ordinately expressed vitellogenin and apoVLDLII genes of chicken. *Nucleic Acids Res* 14:8669–8680

Vignon F, Bouton MM, Rochefort H (1987): Antiestrogens inhibit the mitogenic effect of growth factors on breast cancer cells in the total absence of estrogens. *Biochem Biophys Res Commun* 146:1502–1508

Vleminck K, Vakaet L, Mareel M, Fiers W, vanRoy F (1991): Genetic manipulation of E-cadherin expression by epithelial tumor cells reveals an invasion suppressor role. *Cell* 66:107–119

Wakeling AE, Slater SR (1980): Estrogen-receptor binding and biologic activity of tamoxifen and its metabolies. *Cancer Treat Rep* 64:741–744

Wakeling AE, Bowler J (1987): Steroidal pure antioestrogens. *J Endocrinol* 112:R7–R10

Wakeling AE (1989): Comparative studies on the effects of steroidal and nonsteroidal oestrogen antagonists on the proliferation of human breast cancer cells. *J Steroid Biochem* 34:183–188

Wakeling AE (1990): Therapeutic potential of pure antioestrogens in the treatment of breast cancer. *J Steroid Biochem Mol Biol* 37:771–775

Wakeling AE, Dukes M, Bowler J (1991): A potent specific pure antiestrogen with clinical potential. *Cancer Res* 51:3867–3873

Wakeling AE (1993): Are breast tumours resistant to tamoxifen also resistant to pure antioestrogens? *J Steroid Biochem Mol Biol* 47:107–114

Walker P, Germond JE, Brown-Luedi M, Givel F, Wahli W (1984): Sequence homologies in the region preceding the transcription initiation site of the liver estrogen-responsive vitellogenin and apo-VLDLII genes. *Nucleic Acids Res,* 12:8611–8626

Ward RL, Morgan G, Dalley D, Kelly PJ (1993): Tamoxifen reduces bone turnover and prevents lumbar spine and proximal femoral bone loss in early post-menopausal women. *Bone Mineral* 22:87–94

Warri AM, Laine AM, Majasuo KE, Alitalo KK, Harkonen PL (1991): Estrogen suppression of erbB2 expression is associated with increased growth rate of ZR-75-1 human breast cancer cells in vitro and in nude mice. *Int J Cancer* 49:616–623

Weisz A, Coppola L, Bresciani F (1986): Specific binding of estrogen receptor to sites upstream and within the transcribed region of the chicken ovalbumin gene. *Biochem Biophys Res Commun,* 139:396–402

Weisz A, Rosales R (1990): Identification of an estrogen responsive element upstream of the human c-fos gene that binds the estrogen receptor and the AP-1 transcription factor. *Nucleic Acids Res,* 18:5097–5106

Welch DR, Fabra A, Nakajima M (1990): Transforming growth factor beta stimulates mammary adenocarcinoma cell invasion and metastatic potential. *Proc Natl Acad Sci USA* 87:7678–7682

Westley B, Rochefort H (1980): A secreted glycoprotein induced by estrogen in human breast cancer cell lines. *Cell* 20:353–362

White IN, de Matteis F, Davies A, Smith LL, Crofton SC, Venitt S, Hewer A, Phillips DH (1992): Genotoxic potential of tamoxifen and analogues in female Fischer F344/n rats, DBA/2 and C57BL/6 mice and in human MCL-5 cells. *Carcinogenesis* 13:2197–2203

Wiebe VJ, Osborne CK, McGuire WL, DeGregorio MW (1992): Identification of estrogenic tamoxifen metabolite(s) in tamoxifen-resistant human breast tumors. *J Clin Oncol* 10:990–994

Williams GM, Iatropoulos MJ, Djordjevic MV, Kaltenberg OP (1993): The triphenylethylene drug tamoxifen is a strong liver carcinogen in the rat. *Carcinogenesis* 14:315–317

Wilson AJ (1983): Response in breast cancer to a second hormonal therapy. *Rev Endocr Rel Cancer* 14:5–11

Winneker RC, Clark JH (1983): Estrogenic stimulation of the antiestrogen specific binding site in rat uterus and liver. *Endocrinology* 112:1910–1915

Winneker RC, Guthrie SC, Clark JH (1983): Characterization of a triphenylethylene-antiestrogen-binding site on rat serum low density lipoprotein. *Endocrinology* 112:1823–1827

Winterfeld G, Hauff P, Gorlich M, Arnold W, Fichtner I, Staab HJ (1992): Investigations of droloxifene and other hormone manipulations on N-nitrosomethylurea-induced rat mammary tumours 1 Influence on tumour growth. *J Cancer Res Clin Oncol* 119:91–96

Wiseman H, Cannon M, Arnstein HR, Halliwell B (1990a): Mechanism of inhibition of lipid peroxidation by tamoxifen and 4-hydroxytamoxifen introduced into liposomes Similarity to cholesterol and ergosterol. *Febs Lett* 274:107–110

Wiseman H, Laughton MJ, Arnstein HR, Cannon M, Halliwell B (1990b): The antioxidant action of tamoxifen and its metabolites Inhibition of lipid peroxidation. *Febs Lett* 263:192–194

Wiseman H, Cannon M, Arnstein HR, Halliwell B (1993a): Tamoxifen inhibits lipid peroxidation in cardiac microsomes Comparison with liver microsomes and potential relevance to the cardiovascular benefits associated with cancer prevention and treatment by tamoxifen. *Biochem Pharmacol* 45:1851–1855

Wiseman H, Paganga G, Rice EC, Halliwell B (1993b): Protective actions of tamoxifen and 4-hydroxytamoxifen against oxidative damage to human low-density lipoproteins: a mechanism accounting for the cardioprotective action of tamoxifen? *Biochem J* 292:635–638

Wiseman LR, Johnson MD, Wakeling AE, Lykkesfeldt AE, May FE, Westley BR (1993c): Type I IGF receptor and acquired tamoxifen resistance in oestrogen-responsive human breast cancer cells. *Eur J Cancer* 29A:2256–2264

Wolf DM, Langan FS, Parker CJ, McCague R, Jordan VC (1993): Investigation of the mechanism of tamoxifen-stimulated breast tumor growth with non-isomerizable analogues of tamoxifen and metabolites. *J Natl Cancer Inst* 85: 806–12

Wolf DM, Jordan VC (1994): The estrogen receptor from a tamoxifen stimulated MCF-7 tumor variant contains a point mutation in the ligand binding domain. *Breast Cancer Res Treat* 31:129–138

Wong ST, Winchell LF, McCune BK, Eurp HS, Teidixo J, Massague J, Herman B Lee DC (1989): The TGF-a precursor expressed on the cell surface binds to the EGF receptor on adjacent cells, leading to signal transduction. *Cell* 56:495–506

Wosikowski K, Kung W, Hasmann M, Löser R, Eppenberger U (1993): Inhibition of growth-factor-activated proliferation by anti-estrogens and effects on early gene expression of MCF-7 cells. *Int J Cancer* 53:290–297

Wright C, Nicholson S, Angus B, Sainsbury JRC, Farndon J, Cairns J, Harris AL, Horne CHW (1992): Relationship between c-erbB-2 protein product expression and response to endocrine therapy in advanced breast cancer. *Br J Cancer* 65:118–121

Yamamoto T, Ikawa S, Akiyama T, Semba K, Nomura N, Miyajima N, Saito N, Toyoshima K (1986): Similarity of protein encoded by the human c-erb-B-2 gene to epidermal growth factor receptor. *Nature* 319:230–234

Yee D, Cullen KJ, Paik S, Perdue JF, Hampton B, Schwartz A, Lippman ME, Rosen N (1988): Insulin-like growth factor II mRNA expression in human breast cancer. *Cancer Res* 48:6691–6696

Yee D, Paik S, Lebovic GS, Marcus RR, Favoni RE, Cullen KJ, Lippman ME, Rosen N (1989): Analysis of IGF-I gene expression in malignancy-evidence for paracrine role in human breast cancer. *Mol Endocrinol* 3:509–517

12

Aromatase Inhibitors and Breast Cancer

ANGELA M.H. BRODIE

HORMONE-DEPENDENT BREAST CANCER

The influence of estrogens on the growth of breast cancer has been recognized since Beatson first performed ovariectomies on breast cancer patients a century ago (Beatson, 1896). Pharmacological approaches have now largely replaced surgical procedures as means of reducing estrogen concentrations and limiting its actions. Two treatment strategies for reducing the effects of estrogens are proving to be therapeutically effective. These are inhibition of estrogen action by antiestrogens that interact with estrogen receptors in the tumor, and inhibition of estrogen production by inhibitors of aromatase (estrogen synthetase). Until recently, all antiestrogens were known to be weak or partial estrogen agonists in addition to being antagonists. Inhibitors of aromatase, acting by a different mechanism, are not associated with estrogenic activity. We therefore postulated that more complete estrogen blockade might result in greater tumor response.

Breast cancer is more prevalent among older women. Moreover, a higher proportion of postmenopausal than premenopausal patients have hormone-sensitive tumors (McGuire, 1980). Following menopause, when ovarian steroid production declines, estrogens produced at peripheral sites, such as adipose tissue, are increased and provide the major contribution to circulating estrogen concentrations. Therefore, total blockade of estrogen is more likely to be accomplished with systemic methods than by surgical removal of endocrine glands. Renewed interest in regulating estrogens as a means of treating breast cancer has been stimulated by reports that better response rates are achieved with the antiestrogen tamoxifen than with cytotoxic agents in postmenopausal patients with estrogen receptor-positive breast cancer. Tamoxifen has been found to extend the disease free interval and significantly

Hormones and Cancer
Wayne V. Vedeckis, Editor

increase patient survival (Early Breast Cancer Trialists' Collaborative Group, 1992). In addition, the low toxicity associated with tamoxifen treatment is an important advantage over chemotherapy.

Although many patients initially respond to tamoxifen, their breast tumors often develop resistance to the drug, which results in disease progression. New strategies for treatment, therefore, have an important place in extending disease-free survival for such patients. A number of years ago, we reported the antitumor effects of several selective aromatase inhibitors we had developed (Brodie et al., 1977, 1982a; Marsh et al., 1985). Following extensive endocrine and antitumor studies of these inhibitors, one compound, 4-hydroxyandrostenedione (4-OHA), has now been evaluated in tamoxifen-resistant breast cancer patients and found to be effective (Coombes et al., 1984; Goss et al., 1986). Other new inhibitors are under development. By reducing estrogen production, aromatase inhibitors can elicit further responses in some patients whose disease recurs because tamoxifen no longer effectively blocks the growth stimulating-effects of estrogen in their tumors. Because of its comparatively mild side effects, tamoxifen can be administered at early stages of the disease, as adjuvant therapy and for long duration. Aromatase inhibitors may become first-line therapy when disease recurs in such patients. At the present time, it is not known which of these agents, the antiestrogens or the aromatase inhibitors, alone or in combination, is superior in first-line treatment in terms of rates of response or duration of effectiveness. However, it is clear from recent studies with the aromatase inhibitor 4-OHA that it is effective as second-line therapy in patients with advanced disease. Thus, this new class of compounds is beginning to provide additional benefits by extending the duration of response and quality of life for breast cancer patients.

Aromatase distribution

Aromatase mediates the conversion of ovarian and adrenal androgen substrates to estrogens. The enzyme is expressed in high levels by the placenta during pregnancy and by the ovaries of premenopausal women (McNatty et al., 1976). The granulosa cells are the major source of ovarian estrogen synthesis, although low levels of aromatase are expressed in the thecal compartment of the developing follicle in the human ovary (Inkster and Brodie, 1991). As indicated above, adipose tissue is considered to be the main site of extragonadal estrogen synthesis. Estrogen synthesis is increased in adipose tissue in postmenopausal women (Hemsell et al., 1974). However, there appears to be no abnormal production in breast cancer patients. In normal breast tissue, aromatase is localized around the lobular units (Weisz et al., 1995). A number of reports indicate that aromatase activity is also present in breast tumors (Miller et al., 1982; Perel et al., 1982; James et al., 1987; Killinger et al., 1987). We and others have detected aromatase mRNA in human breast tissue (Peice et al., 1992; Koos et al., 1993) and in some breast cancers (Koos et al., 1993). Increase in aromatase in breast adipose tissue was noted in 10 of 15 patients studied by Simpson et al. (1994), with the highest level of transcripts being found in the breast quadrant where the tumor was located. However, the importance of local tumor aromatization remains uncertain. Recently, Reed et al. (1990, 1994) found that aromatase activity and DNA polymerase α were inhibited in tumor samples removed from patients treated with an aromatase inhibitor. In a few of these tumors where aromatase activity or estrone levels were not decreased despite almost complete inhibition of peripheral aromatase inhibition, there was no decrease in DNA polymerase α. This suggests that local aromatase may have a role in stimulating tumor growth. Thus, inhibition of estrogen synthesis in all tissues is a logical approach for the effective treatment of breast cancer.

Mechanism of Aromatization

Aromatase is an enzyme complex consisting of a cytochrome P-450 hemoprotein (P-450$_{arom}$) and a flavoprotein, NADPH cytochrome P-450 reductase. The latter is ubiquitous to most cell types and acts to provide electrons to the cytochrome P-450 enzymes. P-450$_{arom}$ catalyzes a series of three hydroxylations of the androgen substrates, androstenedione and testosterone (Thompson et al., 1974). These hydroxylations appear to be characteristic of steroidogenic P-450 enzymatic reactions. However, the aromatization of ring A of the steroid nucleus is unique in steroid biosynthesis. Studies of the human aromatase sequence using the VGAP alignment program have shown that there are only 17.9–23.5% identical amino acids in common with the human adrenal side chain cleavage enzyme, the 11β-hydroxylase, 17α-hydroxylase, or other steroidogenic P-450 enzymes (Corbin et al., 1988). Because P-450$_{arom}$ has little sequence identity to other P-450 enzymes, it has been assigned to a separate gene family, designated CYP19 (Nebert et al., 1991). This low homology with other enzymes suggests that P-450$_{arom}$ is a good candidate for selective inhibition. In addition, P-450$_{arom}$ is a particularly suitable enzyme target for selective inhibition, as aromatization is the last step in the biosynthetic sequence of steroid production. Therefore, selective blockade of P-450$_{arom}$ will not interfere with the production of other steroids, such as adrenal corticoids.

The first hydroxylation of the androgen substrate occurs at C-19 (Meyer, 1955; Morato et al., 1961). The 19-hydroxylated intermediate then hydrogen bonds to an acidic side chain residue Glu-302 (Graham-Lorence et al., 1991; Zhou et al., 1992) within the enzyme's active site. This is thought to be of critical importance in the process of aromatization (Oh et al., 1993). Hydrogen bonding of the 3-ketone may also occur at a polar active site (His-128 residue) to anchor the intermediate. This assures stereospecific removal of the C-19 pro-R hydrogen by a heme iron-oxo species during the second hydroxylation step. The ferric peroxide breakdown that then usually occurs may be circumvented because of the high electrophilicity of the aldehyde. This would alter the normal hydroxylation cycle (Cole and Robinson, 1990).

The mechanisms involved in the last step are not completely elucidated. Although a number of hypotheses have been proposed, the evidence available to date (Akhtar et al., 1982; Cole and Robinson, 1988; Oh and Robinson, 1993) favors the idea that the unstable intermediate produced by the series of hydroxylations collapses, yielding estrogens and formic acid. This last step is postulated to involve hydride shift, proton transfer, or free-radical pathways (Stevenson et al., 1988), which results in the removal of the angular methyl group at C-19, *cis* elimination of the 1β and 2β hydrogens (Morato et al., 1962), and aromatization of ring A of the androgens to form the estrogens.

Regulation of Aromatase

The complex nature of aromatization, the use of the different substrates in the reaction, and distribution of the enzyme in a variety of tissues suggested the possibility that there may be multiple forms of the enzyme.

Studies of the molecular mechanisms of aromatase regulation are now possible with the isolation and characterization of genomic clones that contain the entire P-450_{arom} structural gene (CYP19) as well as the flanking genomic DNA (Corbin et al., 1988; Harada, 1988). The CYP19 gene has been mapped to chromosome 15 (Chen et al., 1988) and is more than 70 kbp, although the region encoding the aromatase protein spans only 35 kbp of DNA and contains 9 exons (II–X). The translation initiation site is contained in exon II. Regulation of aromatase is tissue specific, due in part to the use of alternative transcriptional start sites that arise as a consequence of the use of tissue-specific promoters. In the ovary, CYP19 gene transcription initiation occurs 120 bp upstream of the start site and is called promoter II. This promoter appears to be cyclic adenosine monophosphate (cAMP)-mediated and is used in the ovary in the regulation of aromatase by follicle stimulating hormone (FSH) (Hseuh et al., 1984). Placental transcripts have $5'$-termini containing sequences encoded in exon I.1, which is 40 kbp upstream of the translation initiation site. In adipose tissue, exon 1.3 is expressed and is 306 bp upstream of the exon II splice junction. Another untranslated region 20 kbp downstream of exon I.1 is exon 1.4 which is expressed in human breast adipose tissue and skin fibroblasts (Mahendroo et al., 1993). These untranslated exons are all spliced into the same $3'$-slice junction, which is located upstream of the initiation site in exon II. Therefore, the sequence-encoding region is identical and the protein expressed in each tissue is the same, irrespective of the splicing pattern.

AROMATASE INHIBITORS

Steroidal Inhibitors

Mechanism of Inhibition

Structure activity relationship. It was initially postulated that, because of the unique features of the aromatization reaction catalyzed by P-450_{arom}, selective inhibition might be achieved with substrate analogues (Schwarzel et al., 1973). A number of androstenedione derivatives with chemical substituents at various positions on the steroid nucleus have been identified as selective aromatase inhibitors (Schwarzel et al., 1973; Brodie et al., 1977, 1978, 1982a,b; Marsh et al., 1985). Although the spacial requirements of the A-ring for binding steroidal inhibitors to aromatase appear to be rather restrictive and permit only small structural changes, modifications at C-4 have produced several potent inhibitors. These include 4-hydroxyandrostenedione (Fig. 1) and 4-acetoxyandrostenedione. Bulky groups at the 1α position are poor inhibitors (Brueggemeier et al., 1978), whereas small changes, for example a methylene replacing the ketone at C-3, provide good inhibition (Miyairi and Fishman, 1986). Thus, 1-methylandrosta-1,4-diene-3,17-dione (SH 489) (Henderson et al., 1986) is also a potent inhibitor. The B-ring of

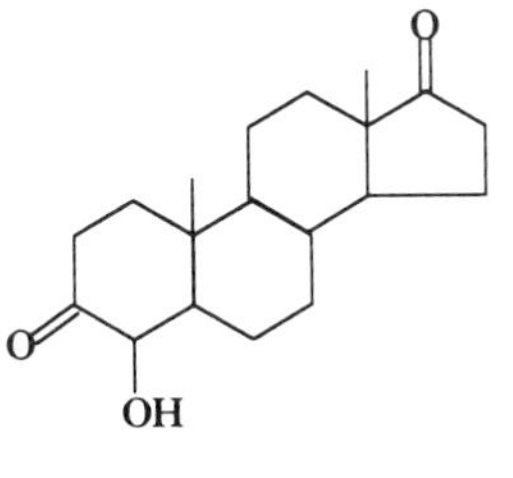

O
OH
4-Hydroxyandrostenedione

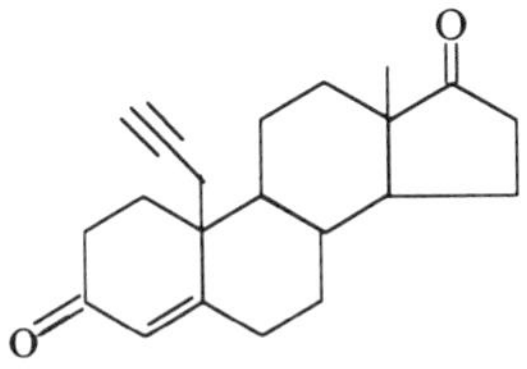

O
O
MDL 18962

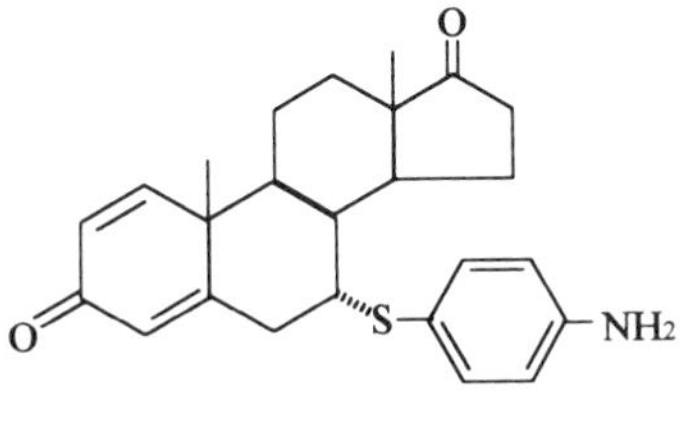

O
O
S
NH2
7α-APTADD

Figure 1. Steroidal aromatase inhibitors. Three examples of inhibitors that have been demonstrated to cause enzyme inactivation are shown.

the steroid permits greater structural modifications. Substitutions at the C-6 position produced some good inhibitors, such as androstene-3,6,17-trione (Schwarzel et al., 1973) and 6-methyleneandrosta-1,4-diene-3,17-dione (FCE 24304) (DiSalle et al., 1994). A number of potent aromatase inhibitors with bulky substitutions at the C-7 position, such as 7α-(4′-amino)phenylthio-1,4-androstadienedione (7α-APTADD; Fig. 1), have been synthesized by Brueggemeier et al. (1987, 1990). These investigators have suggested (Brueggemeier, 1994) that interactions that result in enhanced affinity take place between the phenyl ring at the 7α-position and amino acids at or near the active site (Li et al., 1990a,b). The C-14 is also a position that can be altered with small substituents and has resulted in 14-hydroxyandrostene-3,6,17-trione (NKS01) (Yoshihama et al., 1990).

As the C-19 methyl is the site of enzymatic oxidation during aromatization, considerable attention has been focused on modifying this position. A number of 19-substituted inhibitors including thiiranes, oxiranes (Childers et al., 1987; Childers and Robinson, 1987; Kellis et al., 1987), epoxysteroids (Shih et al., 1987), and alkylthiol and azido analogues have been reported (Wright et al., 1985; Deckers et al., 1989). Modifications at the C-19 position that produce potent inhibitors have geometrically small substituents, suggesting that the active site of the enzyme can accommodate only small changes in structure. Kellis et al. (1987) and Wright et al. (1991) have described some interesting steroid analogues with C-19 heteroatoms, which are strong competitive inhibitors. Based on their binding spectra, thiiranes and oxiranes appear to bind with the heme iron of P-450_{arom}, forming hexa-coordinated species. The 19R isomers of these inhibitors bind to the heme iron and show 36- to 80-fold greater affinity than the corresponding 19S isomers. Very tight competitive binding which also involves the heme iron, has been reported for 2,19 bridged androstenedione derivatives (Burkhart et al., 1991, 1992; Peet et al., 1992, 1993). Some inhibitors with substituents at C-19 appear to interact with the enzyme in a manner analogous to the nonsteroidal inhibitors, which are discussed below.

Enzyme inactivation. There are many examples of mechanism-based inhibitors that are highly effective therapeutics. Inhibitors of this type are not intrinsically reactive; rather, they compete with the substrate and are converted by the enzyme during the normal catalytic process to a form that binds at the active site either very tightly or irreversibly and inactivate the enzyme (Sjoerdsma, 1981). Thus, the continued presence of the compound is not required to maintain inhibition and the effect should be sustained until new enzyme is produced. These so-called suicide inhibitors would be expected to be highly specific because they bind to the same site as the substrate. Both these properties should reduce the risk of side effects to the patient. Many of the steroidal inhibitors mentioned above have been found to cause inactivation of aromatase (Brodie et al., 1981; Covey and Hood,

1981; Metcalf et al., 1981). Some C-19 compounds, such as 10-(2-propynyl) estr-4-ene-3,17-dione (MDL 18962; Fig. 1), were designed to act as mechanism-based inhibitors. This inhibitor was subsequently shown to have significant and lasting biochemical and pharmacological activity (Longcope et al., 1988; Johnston et al., 1989). Studies in our laboratory have demonstrated that these mechanisms are involved in the inhibition of aromatase by 4-hydroxyandrostenedione (4-OHA) (Brodie et al., 1981) as well as 1,4,6-androstatriene-3,17-dione (Schwarzel et al., 1973; Brodie et al., 1976, 1982a) and 4-acetoxyandrostenedione (Brodie et al., 1978). When human placental or rat ovarian microsomes, or human choriocarcinoma cells (JEG-3) that express aromatase are preincubated with the inhibitor in the absence of substrate, 4-OHA causes time-dependent loss of enzyme activity, which follows pseudo-first-order kinetics (Brodie et al., 1981). This inactivation of aromatase requires the presence of NADPH in microsomal preparations, consistent with an oxidative process being involved. If high concentrations of substrate are present during the reaction, inactivation of the enzyme can be prevented. Thus, binding of 6,7[^{3}H]4-OHA to aromatase purified from human placenta can be prevented by preincubating the enzyme with androstenedione. This finding suggests that the inhibitor interacts with the enzyme at the same site as the substrate, that is, the active site of the enzyme. In addition, the interaction appears to be irreversible as the radiolabel could not be displaced by excess concentrations of androstenedione (Yue, 1994). Formation of a covalent bond between the reactive group of an inhibitor and a nucleophilic group of the enzyme, such as the amine of a lysine or the sulfhydryl of a cysteine, could lead to time-dependent inactivation. However, mechanisms involved in the inactivation of aromatase by steroid inhibitors have not been adequately investigated to date.

Early studies *in vivo* with 4-OHA demonstrated significant reduction in ovarian aromatase activity and estradiol levels, and marked tumor regression of 7,12-dimethylbenz(a)anthracene (DMBA) induced mammary tumors in the rat (Brodie et al., 1977, 1982b). Peripheral aromatization was also found to be significantly inhibited in male rhesus monkeys treated with 4-OHA (Brodie and Longcope, 1980). This effect is of importance because peripheral tissues are the major source of estrogens in postmenopausal patients with breast cancer. Peripheral aromatization is measured using a constant infusion of 7[^{3}H]androstenedione and 4[^{14}C]estrone. After reaching steady state conditions, blood samples are collected and analyzed for plasma radioactivity of the infused and product steroids (Longcope et al., 1978). Early studies with 4-OHA first used this procedure and demonstrated marked inhibition of peripheral aromatization in treated monkeys (Brodie and Longcope, 1980). The metabolic clearance rates (MCR) of androstenedione and of estrone determined in the same experiment were not altered by treatment with 4-OHA, nor were the conversion rates between androstenedione and testosterone, or between estrone and estradiol. This lack of effects on MCR or interconversion of androgens and estrogens is consistent with selective inhibition of aromatization. Similar studies have been performed with MDL 18962 in monkeys and in patients (Longcope et al., 1988).

There are a number of other examples of inhibitors that cause enzyme inactivation and have good activity *in vivo*. In the rat mammary tumor model, 7α-(4′-amino)phenylthio-1,4-androstadienedione (7α-APTADD) (Brueggemeier et al., 1987, 1990) was shown to cause tumor regression. The 1-methylandrosta-1,4-diene-3,17-dione (SH 489) (Henderson et al., 1986) has been studied extensively in several animal models (Habenicht, 1994) and was demonstrated to cause significant reductions in serum estradiol levels in men. Tumor response as well as decreased serum estrogen concentrations have been shown in a small number of breast cancer patients treated with 6-methyleneantrosta-1,4-diene-3,17-dione (FCE 24304) (DiSalle et al., 1994). Further clinical studies with this compound and 14-hydroxyandrostene-3,6,17-trione (NKS01) are in progress (Yoshihama et al., 1990).

Clinical Efficacy of 4-OHA in Patients with Advanced Breast Cancer

4-OHA was the first selective aromatase inhibitor to be studied in breast cancer patients. Although its antitumor effects were reported in 1976 (Brodie et al., 1976), clinical studies did not begin until approximately 10 years later. A number of trials have now been carried out that have led to its availability in many countries as a second line agent for breast cancer treatment.

Two doses of 4-OHA have been evaluated in breast cancer patients for their effects on peripheral estradiol levels and tumor response. Comparison between 250 mg and 500 mg injected intramuscularly every 2 weeks showed similar degrees of estrogen suppression. Although there was a minor recovery of the estradiol levels just before the next injection of the lower dose, this was apparent mainly in those patients with higher estradiol levels before treatment. Also, measurements of peripheral aromatization *in vivo* were slightly lower (15.2% of baseline in comparison with 8.1%) in patients receiving the 500-mg dose 4-OHA. Nevertheless, there were no significant differences in clinical efficacy between the two doses. Oral administration of 125 and 250 mg 4-OHA daily suppressed estradiol levels to about the same extent as the injected regimens (Cunningham et al., 1987; Dowsett et al., 1989). However, recent studies found that peripheral aromatization was inhibited less effectively by oral administration of 4-OHA than by injections. Therefore, 250 mg injected intramuscularly every 2 weeks, which was better tolerated locally than 500 mg intramuscularly, is the recommended dose for breast cancer treatment (Dowsett and Coombes, 1994).

Treatment of 465 postmenopausal breast cancer patients with advanced disease by intramuscular injections of 4-OHA (lentaron, formestane) (CGS 32349) every 2 weeks caused complete or partial tumor regression in 28% of patients, and disease stabilization in an additional 22% of patients

(Hoffken et al., 1990; Pickles et al., 1990; Dowsett and Coombes, 1994). The disease progressed in the remaining women. Side effects were mild and occurred in 17% of patients. Local reactions at the site of injection occurred in less than 10% of the patients and were mainly a feature of the higher dose (500 mg). Treatment was discontinued in only 3–5% of patients.

In those patients who responded to 4-OHA treatment after they had relapsed from tamoxifen, it seems likely that their tumors remained sensitive to estrogens. ER status of the *primary* tumor was the major determinant of response of patients to 4-OHA treatment. Serum estradiol levels in patients were suppressed by 4-OHA without apparent differences between regimens and were maintained for at least several months (Dowsett et al., 1987). Thus, in a trial of 186 patients, 93% of objective responders of known ER status had ER+ tumors (Coombes et al., 1992). Favorable responses occurred in 11 of 33 patients who had previously responded to endocrine therapy, compared to only 2 of 35 patients who had failed to benefit from prior endocrine therapy. Thus, some patients are apparently no longer hormone responsive, which may explain the lack of response of some patients to 4-OHA treatment, despite significant reduction in plasma estradiol levels. This suggests they were receiving optimal doses of 4-OHA.

Recently, in an international, multicenter, trial (Perez et al., 1994), 4-OHA was compared with tamoxifen as first-line therapy in postmenopausal breast cancer patients previously untreated with endocrine therapy. Women of comparable age and disease characteristics with measurable lesions, according to modified UICC criteria, were randomized to a group of 173 patients who received 4-OHA (250 mg injected intramuscularly every 2 weeks), or to a group of 175 patients who were administered tamoxifen 30 mg orally. No significant difference was found between the two treatments. Similar response rates resulted from both 4-OHA and tamoxifen treatment. Thus, complete or partial tumor regression occurred in 28% of patients treated with 4-OHA and in 31% of patients treated with tamoxifen. With 4-OHA, the median duration of response lasted for 458 days, whereas the response lasted 604 days with tamoxifen. The mean survival time was 997 days for the 4-OHA-treated patients and 1020 days for the tamoxifen treated group. However, the time to disease progression and time to treatment failure was significantly longer with tamoxifen. Both treatments caused few side effects. Although some patients (7%) experienced local reactions at the site of the intramuscular injection of 4-OHA, this route of administration was favored by the patients, as they routinely received more frequent medical attention and it insured compliance (Dowsett and Coombes, 1994).

Beneficial results have been obtained in a small number of premenopausal breast cancer patients treated with a combination of 4-OHA and the gonadotropin-releasing hormone (GnRH) agonist, goserelin. Plasma estradiol concentrations were not consistently suppressed by 4-OHA alone in younger patients. Ovarian estrogen synthesis is under the regulation of gonadotropins. GnRH agonists are effective in reducing plasma gonadotropin levels and consequently ovarian estrogen production. Since regulation of aromatase is tissue specific, as indicated above, GnRH agonists are without effect on peripheral aromatase. A further reduction in estradiol levels was gained by adding 4-OHA to the treatment of premenopausal breast cancer patients receiving goserelin. Additional responses were also produced when 4-OHA was added to the treatment of patients who initially responded but later became resistant to goserelin (Dowsett et al., 1992).

NON-STEROIDAL AND REVERSIBLE INHIBITORS

Inhibitors of steroidogenic P-450 enzymes, such as aminoglutethimide [3-(4-aminophenyl)-3-ethylpiperidine-2,6-dione] (AG), ketoconazole, and cyanoketone also inhibit aromatase. These nonsteroidal compounds appear to inhibit cytochrome P-450 enzymes by interacting with the heme iron atom.

Several steroidogenic steps are inhibited by AG, for example, adrenal production of aldosterone (18-hydroxylase) and cortisol (11β-hydroxylase), as well as androgens. For this reason, it was used to produce medical "adrenalectomies" in breast cancer patients. Later, it was found that normal levels of androstenedione were maintained even though estrone levels were reduced in patients treated with AG (Samojlik et al., 1977). This observation suggested that AG *in vivo* has a greater effect on aromatase than on other P-450 enzymes. AG has also been reported to enhance conversion of Δ^5 to Δ^4 steroids (Badder et al., 1983) and reduce plasma levels of estrone sulfate by increasing steroid metabolism in the liver (Lonning et al 1987b). Induction of hepatic cytochrome P-450 mixed function oxidases by AG may account for this effect (Lonning et al., 1987a). Estrone sulfate, which is derived from circulating estrone and estradiol (Ruder et al., 1972), may be an important source of estrogen within breast tumors (Santen et al., 1984). However, inhibition of aromatase would be expected to reduce the production of estrone before its sulfation. Measurement of peripheral aromatization by the isotopic method indicates that almost complete inhibition of the conversion of androstenedione to estrone occurs with AG (Dowsett et al., 1985; Santen et al., 1982). It is therefore unclear why plasma estrone and estradiol concentrations are reduced only to 50% of pretreatment values. AG used with hydrocortisone replacement was an effective treatment for breast cancer. However, the lack of potency and specificity and the significant side effects limited its usefulness. Therefore, attempts have been made to develop more effective analogs. Although pyridoglutethimide has similar potency to AG, it appears to be rather more specific and less toxic (Dowsett et al., 1990). The 3-(cyclohexylmethyl)-1-(4-aminophenyl)-3-azabicyclo[3.1.0]hexane-2,4-dione is a much more potent inhibitor and has been reported to have 140-fold greater activity than AG (Stanek et al., 1991).

Analogs of the imidazole drugs that are used to inhibit fungal P-450 enzymes have been more successful. Much greater selectivity as well as

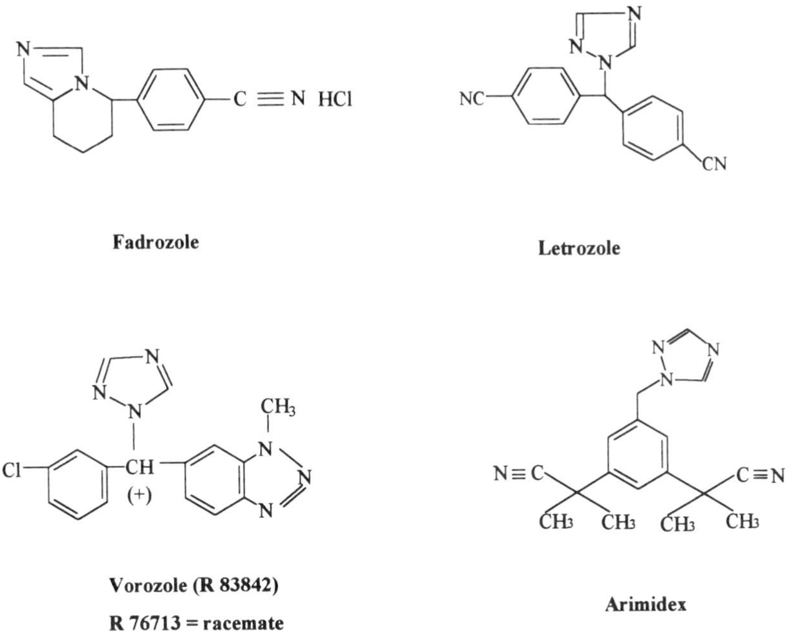

Fadrozole

Letrozole

Vorozole (R 83842)

R 76713 = racemate

Arimidex

Figure 2. Nonsteroidal aromatase inhibitors. Some of the more potent inhibitors currently in clinical trails are depicted.

potency has been accomplished with fadrozole [4-(4,5,6,7,8-tetrahydroimidazo[1,5α]pyridine-5-yl)benzonitrile monohydrochloride (CGS 16949A) (Schieweck et al., 1993) and (3αR)-trans-1-[(3α-ethyl-9-(ethylthio)-2,3,3α,4,5,6-hexahydro-1H-phenalen-2-yl)methyl]-1H-imidazole HCl) (ORG 33201) (Geelen et al., 1993; Fig. 2) These compounds were shown to coordinate with the iron of the porphyrin nucleus, a feature in common with aminoglutethimide. Nevertheless, their higher potency and selectivity cannot be explained solely by this interaction. Molecular modeling studies have revealed the similarity of the structure of fadrozole to androstenedione. Thus, the cyano function of fadrozole could be superimposed on the D-ring of an androstenedione analogue. Furthermore, the phenyl ring of fadrozole matches the steroid C-ring and the molecular volume of the saturated piperidine ring is similar to that of the A-ring of the steroid (Furet et al., 1993).

Triazole analogs (Fig. 2) are also based on antifungal agents that inhibit P-450 enzymes. These are potent and selective aromatase inhibitors and include letrozole (4-[1-(cyanophenyl)-1-(1,2,4-triazolyl)methyl]benzonitril) (CGS 20267) (Bhatnagar et al., 1994), arimidex 2,2′[5-(1H-1,2,4-triazol-1-yl methyl)-1,3-phenylene]bis(2-methylpropiononitrile) (ZD1033) (Plourde et al., 1994), and vorozole (6-[(4-chlorophenyl)(1H-1,2,4-triazol-1-yl)methyl]1-methyl-1H-benzotriazole (R 76 713) (Wouters et al., 1990). The greater selectivity of these compounds for P-450_{arom} has been investigated in studies of vorozole. Based on the kinetic data and a reverse Type I spectral change, it appears that competitive inhibition involving the substrate binding region occurs when vorozole is added to placental microsomes. Difference spectral measurements indicate that the N-4 nitrogen of the triazole ring coordinates with the heme iron atom and that its N-1 substituent occupies a lipophilic region of the apoprotein moiety of the P-450_{arom}. The S-enantiomer of vorozole is responsible for most of the aromatase inhibitory activity. This enantiomer corresponds better to androstenedione and the active site than the R-enantiomer. In the best fit model, the chlorine substituted phenyl ring of vorozole appears to interact with the gamma-carboxyl group of Glu-302. As indicated above, this is part of the substrate binding site and of critical importance in aromatization. Furthermore, the triazole ring of S-vorozole interacts with Thr-310, whereas the 1-methyl-benzotriozole moiety binds near Asp-309 (Vanden Bossche et al., 1994). This triazole derivative is selective and was found to be devoid of effects on cholesterol, progesterone, androgen, and mineralocorticoid biosynthesis *in vivo* and is without estrogenic or androgenic activity (De Coster et al., 1990).

Although AG initially inhibits ovarian estradiol levels in the rat, this results in increasing LH levels, which in turn stimulate estrogen production (Wing et al., 1985). A similar situation appears to occur in premenopausal patients treated with AG, as reduction in plasma estrogen levels are not maintained (Santen et al., 1980). In the rat, it was found that treatment with the very potent triazole inhibitor CGS 20267 suppressed serum estrogen concentrations to 12% of pretreatment values even though luteinizing hormone (LH) levels were increased three- to fourfold (Schieweck et al., 1993). CGS 20,276, CGP 45,688, and CGP 47,645 are also potent and selective nonsteroidal inhibitors that cause marked inhibition of tumors in the DMBA rat model. Complete or partial regression occurred in 95% of tumors in animals treated with 100 μg/kg CGS 20,267 orally. Suppression of estradiol levels is also maintained with racemic vorozole (R76713) in the intact rat. However, in studies with cycling female cynomolgus monkeys treated with vorozole, plasma androgens and LH gradually increased and estrogen levels were not consistently suppressed. At necropsy, enlarged ovaries and cystic follicles were observed (Wouters et al., 1994). These results suggest that, because of the normal hypothalamic-pituitary-ovarian feedback regulation, inhibition of ovarian estrogen synthesis may not be adequately maintained by nonsteroidal aromatase inhibitors.

Effects of Nonsteroidal Aromatase Inhibitors in Patients

The imidazole and triazole aromatase inhibitors are orally active and highly effective in reducing estrogen concentrations in breast cancer patients. Fadrozole (CGS 16949) suppressed plasma and urinary estrogen levels effectively in doses ranging between 2 and 4 mg/day (Santen et al., 1989; Lipton et al., 1990). However, some inhibition of adrenal steroids was indicated by a blunted response in aldosterone and cortisol secretion during adrenocorticotropin hormone (ACTH) stimulation tests (Santen et al., 1989). Greater selectivity and potency are observed with the three triazole derivatives vorozole, arimidex, letrozole (Vanden Bossche et al., 1994). No inhibition of adrenal steroid synthesis or of other endocrine parameters is observed with these inhibitors. Plasma estradiol levels were reduced to the limit of detection in postmenopausal breast cancer patients treated with 2.5 or 5 mg/day vorozole racemate for 1 month. Maximum estradiol suppression occurred in 24 hours with these doses in male volunteers, and peripheral aromatization was inhibited 93–94% with 1 to 5 mg vorozole (Wouters et al., 1994). Similar degrees of estrogen suppression were observed with doses of 1 to 10 mg arimidex. Plasma estradiol concentrations were reduced to the lowest detectable level within 2 hours of oral administration (Plourde et al., 1994). Letrozole (CGS 20267) is about 100 times more potent than fadrozole (CGS 16949A) in reducing serum estradiol levels. This may be because of the significantly longer half-life of the triazole in patients than of the imidazole (Bhatnagar et al., 1994). Within a few hours of the first administration of CGS 20267, there was marked suppression of plasma estradiol, estrone, and estrone sulfate levels (Demers, 1994). Estradiol and estrone levels declined to within 5% of the baseline level within 2 weeks of therapy with the lowest dose tested (0.1 mg/day letrozole) and were suppressed to below detection limits with other doses, despite the use of a highly sensitive assay. It had previously been thought that estrone sulfate could not be suppressed until much later in the course of aromatase inhibitor therapy (Demers, 1994). However, estrone sulfate levels also reached maximum suppression within 2 weeks of CGS 20267 therapy. All three of the triazole compounds appear to be highly selective and potent aromatase inhibitors and are well tolerated by patients. Further studies are currently in progress to determine their antitumor efficacy in breast cancer patients who have been previously treated with tamoxifen or other forms of endocrine therapy.

CONCLUSIONS

Studies indicate that 4-OHA is effective in postmenopausal breast cancer patients who have relapsed from other hormonal treatments. This aromatase inhibitor is now available in the United Kingdom, Canada, and many other countries. Clinical trials to define the antitumor efficacy of letrozole, vorozole, and arimidex are in progress. It will now be important to compare the effects of these aromatase inhibitors with one another and with antiestrogens. The diversity of structures are likely to produce unrelated effects that could influence their efficacy and tolerability. In the last few years, the beneficial effects of tamoxifen have become apparent and have brought about changes in the clinical practice of treating postmenopausal breast cancer patients. It is becoming increasingly difficult to find patients who have not already been treated or who have ended adjuvant therapy at least 12 months before metastatic relapse. We have therefore developed a mouse model that simulates the situation in the postmenopausal breast cancer patient. Tumors of human breast carcinoma cells (MCF-7) transfected with the aromatase gene grow in response to locally produced estrogens in ovariectomized nude mice. Both antiestrogens and aromatase inhibitors can be compared in this model (Yue et al., 1993). Studies indicate that CGS 20267 is highly effective in preventing human breast cancer cell growth in these animals (Yue et al., 1995).

Using the highly potent and selective triazole inhibitors, it may now be possible to determine the effects of complete estrogen suppression in postmenopausal women with breast cancer and whether additional clinical responses can be achieved. Aromatase inhibitors of different structures, such as steroids, imidazoles, and triazoles, may have further advantages in providing a series of alternatives when drug resistance develops. It is anticipated that aromatase inhibitors will contribute to providing significant benefits to breast cancer patients in the next few years.

ACKNOWLEDGMENTS

The author's research was supported by NIH grants CA-27440 and HD-13909.

REFERENCES

Akhtar M, Calder MR, Corina DL, Wright JN (1982): Mechanistic studies on C-19 demethylation in oestrogen biosynthesis. *Biochem J* 201:569–580

Badder EM, Lerman S, Santen RJ (1983): Aminoglutethimide stimulates extra-adrenal delta-4 androstenedione production. *J Surg Res* 34:380–387

Beatson GT (1896): On the treatment of inoperable cases in carcinoma of the mamma: suggestion for new method of treatment with illustrative cases. *Lancet* 2:104–107

Bhatnagar AS, Batzl CH, Hausler A, Schieweck K, Lang M (1994): Endocrine and antitumor effects of nonsteroidal aromatase inhibitors. In: Sex Hormones and Antihormones in Endocrine Dependent Pathology: Basic and Clinical Aspects. Amsterdam: Elsevier.

Brodie AMH, Schwarzel WC, Brodie HJ (1976): Studies on the mechanism of estrogen biosynthesis in the rat ovary—1. *J Steroid Biochem* 7:787–793

Brodie AMH, Schwarzel WC, Shaikh AA, Brodie HJ (1977): The effect of an aromatase inhibitor, 4-hydroxy-4-androstene-3,17-dione, on estrogen-dependent processes in reproduction and breast cancer. *Endocrinology* 100:1684–1695

Brodie AMJ, Wu JT, Marsh DA, Brodie HJ (1978): Aromatase inhibitors III. Studies on the antifertility effects of 4-acetoxy-4-androstene-3,17-dione. *Biol Reprod* 18:365–370

Brodie AMH, Longcope C (1980): Inhibition of peripheral aromatization by aromatase inhibitors, 4-hydroxy- and 4-acetoxyandrostenedione. *Endocrinology* 106:19–21

Brodie AMH, Hendrickson JR, Tsai-Morris CH, Garrett WM, Marcotte PA, Robinson CH (1981): Inactivation of aromatase *in vitro* by 4-OHA and 4-acetoxyandrostenedione and sustained effects *in vivo*. *Steroids* 38:693–702

Brodie AMH, Garrett W, Hendrickson JM, Marsh DA, Brodie HJ (1982a): The effect of 1,4,6-androstatriene-3,17-dione (ATD) on DMBA-induced mammary tumors in the rat and its mechanism of action in vivo. *Biochem Pharmacol* 31: 2017–2023

Brodie AMH, Garrett WM, Hendrickson JR, Tsai-Morris CH (1982b): Effects of 4-hydroxyandrostenedione and other compounds in the DMBA breast carcinoma model. *Cancer Res* 42:3360s–3364s

Brueggemeier RW, Floyd EE, Counsell RE. (1978): Synthesis and biochemical evaluation of inhibitors of estrogen biosynthesis. *J Med Chem* 21:1007–1011

Brueggemeier RW, Li P-K, Snider CE, Darby MV, Katlic NE (1987): 7 α-substituted androstenediones as effective *in vitro* and *in vivo* inhibitors of aromatase. *Steroids* 50:163–178

Bruggemeier RW, Li PK, Chen HH, Moh PP, Katlic NE (1990): Biochemical pharmacology of new 7-substituted androstenediones as inhibitor of aromatase. *J Steroid Biochem* 37:379–385

Brueggemeier RW, Robert W (1994): Aromatase inhibitors-mechanisms of steroidal inhibitors. *Breast Cancer Res Treat* 30:31–42

Burkhart JP, Peet NP, Wright CL, Johnston JO (1991): Novel time-dependent inhibitors of human placental aromatase. *J Med Chem* 34:1748–1750

Burkhart JP, Huber EW, Laskovic FM, Peet NP (1992): Synthesis of 2,19-bridged; androstenediones. *J Org Chem* 57:5150–5154

Chen S, Besman MJ, Sparkes RS, Zollman S, Klisak I, Mohandas T, Hall PF, Shiveley JE (1988): Human aromatase: cDNA cloning, Southern blot analysis, and assignment of the gene to chromosome 15. *DNA Mol Biol* 7:27–38

Childers WE, Robinson CH (1987): Novel 10β-thiiranyl steroids as aromatase inhibitors. *J Chem Soc Chem Comm* 320–321

Childers WE, Shih MJ, Furth PS, Robinson CH (1987): Stereoselective inhibition of human placental aromatase. *Steroids* 50:121–134

Cole PA, Robinson CH (1988): A peroxide model reaction for placental aromatase. *J Am Chem Soc* 110:1284–1285

Cole PA, Robinson CH (1990): Mechanism and inhibition of cytochrome P-450 aromatase. *J Med Chem* 33:2933–2942

Coombes RC, Goss P, Dowsett M, Gazet JC, Brodie AMH (1984): 4-Hydroxyandrostenedione treatment of postmenopausal patients with advanced breast cancer. *Lancet* 2:1237–1239

Coombes RC, Hughes SWM, Dowsett M (1992): 4-OHA: a new treatment for breast cancer. *Eur J Cancer* 28A:1941–1945

Corbin CJ, Graham-Lorence S, McPhaul M, Mason JI, Mendelson CR, Simpson ER. (1988): Isolation of a full-length cDNA insert encoding human aromatase system cytochrome P-450 and its expression in non-steroidogenic cells. *Proc Natl Acad Sci USA* 85:8948–8952

Covey DF, Hood WF (1981): Enzyme generated intermediates derived from 4-androstene-3,6,17-dione and 1,4,6-androstatriene3,17-dione cause time-dependent decrease in human placental aromatase activity. *Endocrinology* 108:1597–1599

Cunningham D, Powles TJ, Dowsett M, Hutchinson G, Brodie AMH, Ford HT, Gazet JC Coombes RC (1987): Oral 4-hydroxyandrostenedione, a new endocrine treatment for disseminated breast cancer. *Cancer Chemother Pharmacol* 20:253–255

De Coster R, Wouters W, Bowden CR, Van den Bossche H, Bruynseels J, Tuman RW, Van Ginckel R, Snoeck E, Van Peer A, Janssen PAJ (1990): New non-steroidal aromatase inhibitors: focus on R76713. *J Steroid Biochem* 37:335–341

Deckers GH, Schuurs AHWM (1989): Aromatase inhibition in hypophysectomised female rats: a novel animal model for *in vivo* screening. *J Steroid Biochem* 32:625–631

Demers LM (1994): Effects of fadrozole (CGS 16949A) and letrozole (CGS 20267) on the inhibition of aromatase activity in breast cancer patients. *Breast Cancer Res Treat* 30:95–102

DiSalle E, Ornati G, Paridaens R, Coombes C, Lobelle JP, ZurloMG (1994): Preclinical and clinical pharmacology of the aromatase inhibitor exemestane (FLE 24304). In: *Sex Hormones and Antihormones in Endocrine Dependent Pathology: Basic and Clinical Aspects*. Amsterdam: Elsevier

Dowsett M, Harris AL, Stuart-Harris R, Hill M, Cantwell BM, Smith IE, Jeffcoate SL (1985): A comparison of the endocrine effects of low dose aminoglutethimide with and without hydrocortisone in postmenopausal breast cancer patients. *Br J Cancer* 52:525–529

Dowsett M, Goss PE, Powles TJ, Brodie AMH, Jeffcoate SL, Coombes RC (1987): Use of aromatase inhibitor 4-hydroxyandrostenedione in post-menopausal breast cancer: optimization of therapeutic dose and route. *Cancer Res* 47:1957–1961

Dowsett M, Cunningham DC, Stein RC, Evans S, Dehennin L, Hedley A, Coombes RC (1989): Dose-related endocrine effects and pharmacokinetics of oral and intramuscular 4-hydroxyandrostenedione in postmenopausal breast cancer patients. *Cancer Res* 49:1306–1312

Dowsett M, Jarman M, Mehta A, Haynes B, Lonning PE, Jones A, McNeil F, Powles TJ, Coombes RC (1990): Endocrine pharmacology of a new aromatase inhibitor 3-ethyl-3-(4-pyridyl)piperidine-2,6-dione (PG). *J Steroid Biochem* 36 (Suppl.) (Abstr. 336)

Dowsett M, Stein RC, Coombes RC (1992): Aromatization inhibition alone or in combination with GnRH agonists for the treatment of premenopausal breast cancer patients. *J Steroid Biochem* 43:155–159

Dowsett M, Coombes RC (1994): Second generation aromatase inhibitor—4-OHA *Breast Cancer Res Treat* 30:81–87

Early Breast Cancer Trialists' Collaborative Group (1992): Systemic treatment of early breast cancer by hormonal, cytotoxic, or immune therapy. *Lancet* 339: 1–15

Furet P, Batzl C, Bhatnagar A, Francotte E, Lang M (1993): Aromatase inhibitors: synthesis, biological activity and binding mode of triazole type compounds. *J Med Chem* 36:1393–1400

Geelen JAA, Loozen HJJ, Deckers GH, Leeue RD, Kloosterboer HJ, Lamberts SWJ (1993): ORG 33201: A new highly selective orally active aromatase inhibitor. *J Steroid Biochem Mol Biol* 44:681–682

Goss PE, Coombes RL, Powles TJ, Dowsett M, Brodie AMH (1986): Treatment of advanced postmenopausal breast cancer with aromatase inhibitor, 4-hydroxyandrostenedione—Phase 2 report. *Cancer Res* 46:4823–4826

Graham-Lorence S, Khalil, MW, Lorence MC, Mendelson CR, Simpson ER (1991): Structure function relationships of human aromatase cytochrome P-450 using molecular modeling and site-directed mutagenesis. *J Biol Chem* 266: 11030–11946

Habenicht UF (1994): Aromatase inhibitors and benign prostatic hyperplasia: chance and limitation. In: *Sex Hormones and Antihormones in Endocrine Dependent Pathology: Basic and Clinical Aspects*. Amsterdam: Elsevier

Harada N. (1988): Cloning of a complete cDNA encoding human aromatase: immunochemical identification and sequence analysis. *Biochem Biophys Res Commun* 156:725–732

Hemsell DL, Grodin J, Breuner PF, Siiteri PK, MacDonald PC (1974): Plasma precursors of estrogen. II Correlation of the extent of conversion of plasma androstenedione to estrone with age. *J Clin Endocrinol Metab* 38:476–479

Henderson D, Norbirath G, Kerb U (1986): 1-Methyl-1,4-androstadiene-3,17-dione (SH 489): Characterization of an irreversible inhibitor of estrogen biosynthesis. *J Steroid Biochem* 24:303–306

Höffken K, Jonat W, Possinger K, Kolbel M, Kunz Th, Wagner H, Becher R, Callies R, Freiderich P, Willmanns W, Maass H, Schmidt CG (1990): Aromatase inhibition with 4-hydroxyandrostenedione in the treatment of postmenopausal patients with advanced breast cancer: a phase II study. *J Clin Oncol* 8:875–880

Hseuh AJW, Adashi EY, Jones PBC, Welsh Jr TH. (1984): Hormonal regulation of the differentiation of cultured ovarian granulosa cells. *Endocr Rev* 5:76–127

Inkster SE, Brodie AMH (1991): Expression of aromatase cytochrome P450 in premenopausal and postmenopausal human ovaries: an immunocytochemical study. *J Clin Endocrinol Metab* 73:717–726

James VHT, McNeill JM, Lai LC, Newton CJM, Ghilchik MW, Reed MJ (1987): Aromatase activity in normal breast and breast tumor tissue: *in vivo* and *in vitro* studies. *Steroids* 50:269–279.

Johnston JO, Wright CL, Schumaker RC (1989): Human trophoblast xenografts in athymic mice: a model for peripheral aromatization. *J Steroid Biochem* 33: 521–529

Kellis JT, Childers WE, Robinson CH, Vickery LE (1987): Inhibition of aromatase cytochrome P-450 by 10-oxirane and 10-triiane substituted androgens. *J Biol Chem* 262:4421–4426

Killinger DW, Perel E, Daniilescu D, Kharlip L, Blackstein ME (1987): Aromatase activity in the breast and other peripheral tissues and its therapeutic regulation. *Steroids* 50:523–535

Koos RD, Banks PK, Inkster SE, Yue W, Brodie AMH (1993): Detection of aromatase and keratinocyte growth factor expression using reverse transcription-polymerase chain reaction. *J Steroid Biochem Mol Biol* 45:217–225

Li P-K, Brueggemeier RW (1990a): Synthesis and biochemical studies of 7-substituted 4,6-androstadiene-3,17-diones as aromatase inhibitors. *J Med Chem* 33: 101–105

Li P-K, Brueggemeier RW (1990b): 7-subsituted steroidal aromatase inhibitors: structure-activity relationships and molecular modeling. *J Enzyme Inhib* 4:113–120

Longcope C, Pratt JH, Schneider SH, Fineberg SE (1978): Aromatization of androgens by muscle and adipose tissue *in vivo*. *J Clin Endocrinol Metab* 46: 146–152

Longcope C, Femino A, Johnston JO (1988): Inhibition of peripheral aromatization in baboons by an enzyme-activated aromatase inhibitor (MDL 18,962). *Endocrinology* 122:2007–2011

Lonning PE, Kvinnsland S, Bakke OM (1987a): Effect of aminoglutethimide on antipyrine, theophylline and digitoxin disposition in breast cancer. *Clin Pharmacol Ther* 36:796–802

Lonning PE, Kvinnsland S, Thoren T, Ueland PM (1987b): Alterations in the metabolism of oestrogens during treatment with aminoglutethimide in breast cancer patients: preliminary findings. *Clin Pharmacokinet* 13:393–406

Mahendroo MS, Means GD, Mendelson CR, Simpson ER (1991): Tissue-specific expression of human P-459$_{arom}$: the promoter responsible for expression in adipose is different from that utilized in placenta. *J Biol Chem* 266:11276–11281

Marsh DA, Brodie HJ, Garrett WM, Tsai-Morris CH, Brodie AMH (1985): Aromatase inhibitors-synthesis and biological activity of androstenedione derivatives. *J Med Chem* 28:788–795

McNatty KP, Baird DT, Bolton A, Chambers P, Coker CS, MacLean H (1976): Concentrations of oestrogen and androgens in human ovarian venous plasma and follicular fluid. *J Endocrinol* 71:77–85

McGuire WL (1980): An update on estrogen and progesterone receptors in prognosis for primary and advanced breast cancer. In: *Hormones and Cancer*, Iacobelli S et al., eds. New York: Raven Press

Metcalf BW, Wright CL, Burkhart JP, Johnston JO (1981): Substrate-induced inactivation of aromatase by allenic and acetylenic steroids. *J Am Chem Soc* 103:3221–3222

Meyer AS (1955): Conversion of 19-hydroxy-4-androstene-3,17-dione to estrone by endocrine tissue. *Biochim Biophys Acta* 17:441–442

Miller WR, Hawkins RA and Forrest APM (1982): Significance of aromatase activity in human breast cancer. *Cancer Res* 42 (suppl):3365s–3368s

Miyairi S, Fishman J (1986): 2-methylene-substituted androgens as novel aromatization inhibitors. *J Biol Chem* 261:6772–6777

Morato T, Hayano M, Dorfman RI, Axelrod LR (1961): The intermediate steps in the biosynthesis of estrogens from androgens. *Biochem Biophys Res Commun* 6:334–338

Morato T, Raab K, Brodie HJ, Hayano M, Dorfman RI (1962): The mechanism of estrogen biosynthesis. *J Am Chem Soc* 84:3764–3765

Nebert DW, Nelson DR, Coon MJ, Estabrook RW, Feyereisen R, Fujii-Kuriyama Y, Gonzalez FJ, Guengerich FP, Gunsalas IC, Johnson EF, Loper JC, Sato R, Waterman MR, Waxman DJ (1991): The P-450 superfamily: update on new sequences, gene mapping and recommended nomenclature. *DNA Mol Biol* 10:1–14

Oh SS, Robinson CH (1993): Mechanism of placental aromatase: A new active site model. *J Steroid Biochem Mol Biol* 44:389–397

Peet NP, Burkhart JP, Wright CL, Johnston JO (1992): Time-dependent inhibition of human placental aromatase with a 2,19-methyleneoxy bridged androstenedione. *J Med Chem* 35:3303–3306

Peet NP, Johnston JO, Burkhart JP, Wright CL (1993): A-ring bridged steroids as potent inhibitors of aromatase. *J Steroid Biochem Mol Biol* 44:409–420

Peice T, Aitken J, Head J, Mehendroo M, Means G, Simpson E (1992): Determination of aromatase cytochrome P-450 messenger ribonucleic acid in human breast tissue by competitive polymerase chain reaction amplification. *J Clin Endocrinol Metab* 174:1247–1252

Perel E, Blackstein ME, Killinger DW (1982): Aromatase in human breast carcinoma. *Cancer Res* 42(suppl):3369s–3372s

Perez Carrion R, Alberola Candel V, Calabresi F, Th. Michel R, Santos R, Delozier T, Goss P, Mauriac L, Feuihade F, Freue M, Pannuti F, Van Belle S, Martinez J, Wehrle E, Royce CM (1994): Comparison of the selective aromatase inhibitor formestane with tamoxifen as first-line hormonal therapy in postmenopausal women with advanced breast cancer. *Ann Oncol* 5(suppl 7): 519–524

Pickles T, Perry L, Murray P, Plowman P (1990): 4-Hydroxyandrostenedione—further clinical and extended endocrine observations. *Br J Cancer* 62:309–313

Plourde PV, Dyroff M, Dukes M (1994): Arimidex: a potent and selective fourth-generation aromatase inhibitor. *Breast Cancer Res Treat* 30:103–111

Reed MJ, Lai LC, Owen AM, Singh A, Coldham NG, Purohit A, Ghilchik MW, Shaikh NA, James VHT (1990): The effect of treatment with 4-hydroxyandrostenedione on the peripheral conversion of androstenedione to oestrone and *in vitro* tumor aromatase activity in postmenopausal women with breast cancer. *Cancer Res* 50:193–196

Reed MJ (1994): The role of aromatase in breast tumors. *Breast Cancer Res Treat* 30:7–17

Ruder H, Loriaux L, Lambert MB (1972): Estrone sulfate production rates and metabolism in man. *J Clin Invest* 51:1021–1033

Samojlik E, Santen RJ, Wells SA (1977): Adrenal suppression with aminoglutethimide, II. Differential effects of aminoglutethimide on plasma androstenedione and estrogen levels. *J Clin Endocrinol Metab* 45:480–487

Santen RJ, Samojlik E, Wells SA (1977): Adrenal suppression with aminoglutethimide II. Differential effects of aminoglutethimide on plasma androstenedione and estrogen levels. *J Clin Endocrinol Metab* 45:480–487

Santen RJ, Samojlik E, Well SA (1980): Resistance of the overy to blockade of aromatization with aminoglutethimide. *J Clin Endocrinol Metab* 51:473–477

Santen RJ, Worgul TJ, Lipton A, Harvey HA, Boucher A, Samojlik E, Wells SA (1982): Aminoglutethimide as treatment of postmenopausal women with advanced breast carcinoma: correlation of clinical and hormonal responses. *Ann Intern Med* 96:94–101

Santen RJ, Leszcynski D, Tilson-Mallet N, Feil PD, Wright C, Manni A, Santner SJ (1984): Enzymatic control of estrogen production in human breast cancer: relative significance of the aromatase versus sulfatase pathway. In: *Endocrinology of the Breast: Basic and Clinical Aspects,* Angeli A, Bradlow HL, Dogliotti L, eds. New York: New York Academy of Science

Santen RJ, Demers L, Lipton A, Harvey HA, Hanagan J, Mulagha M, Navari RM, Henderson IC, Garber JE, Miller AA (1989): Phase II study of the potency and specificity of a new aromatase inhibitor—CGS 16949A. *Clin Res* 37:535A

Schieweck K, Bhatnagar AS, Batzl Ch, Lang M (1993): Antitumor and endocrine effects of nonsteroidal aromatase inhibitors on estrogen-dependent rat mammary tumors. *J Steroid Biochem Mol Biol* 44:633–636

Schwarzel WC, Kruggel W, Brodie HJ (1973): Studies on the mechanism of estrogen biosynthesis. VII. The development of inhibitors of the enzyme system in human placenta. *Endocrinology* 92:866–880

Shih MJ, Carrell MH, Carrell HL, Wright CL, Johnston JO, Robinson CH. (1987): Stereoselective inhibition of aromatase by novel epoxysteroids. *J Chem Soc Chem Comm* 213–214

Simpson ER, Mehendroo MS, Means GD, Kilgore MN, Corbin CJ, Mendelson CR (1993): Tissue-specific promoters regulate aromatase cytochrome P-450 expression. *J Steroid Biochem Mol Biol* 44:321–330

Sjoerdsma A (1981): Suicide inhibitors as potential drugs. *Clin Pharm Therapeut* 30:3–22

Stanek J, Alder A, Bellus D, Bhatnagar AS, Hausler A, Schieweck K (1991): Synthesis and aromatase inhibitory activity of novel 1-(4-aminophenyl)-3-azabicyclo [3.1.0]hexane- and -[3.1.1]heptane-2,4-diones. *J Med Chem* 34:1329–1334

Stevenson DE, Wright JN, Akhtar M (1988): Mechanistic consideration of P-450 dependent enzymatic reactions: Studies on oestriol biosynthesis. *J Chem Soc Perkin Trans* I:2043–2052

Thompson EA Jr, Siiteri PK (1974): Utilization of oxygen and reduced nicotinamide adenine dinucleotide phosphate by human placental microsomes during aromatization of androstenedione. *J Biol Chem* 249:5364–5372

Van Wauwe JP, Janssen PAJ (1989): Is there a case for P-450 inhibitors in cancer treatment? *J Med Chem* 32:2231–2239

Vanden Bossche H, Moereels H, Koymans LMH (1994): Aromatase inhibitors: mechanisms for non-steroidal inhibitors *Breast Cancer Res Treat* 30:43–55

Weisz J, Frits-Wolz G, Dabbs D, Shelley K, Brodie A, Brown T (1995): Expression of aromatase in epithelial and stomal cells in terminal lobular ductal units and in epithelial cells of cysts in normal human mammary gland. *Proc Am Assoc Cancer Res* 36:1631

Wing LY, Garrett, WM, Brodie AMH (1985): The effect of aromatase inhibitors, aminoglutethimide and 4-hydroxyandrostenedione in cyclic rats with DMBA-induced mammary tumors. *Cancer Res* 45:2425–2428

Wouters W, DeCoster R, Beerens D, Doolaege R, Gruwez JA, Van Camp K, Van Der Pas H Van Herendael B (1990): Potency and selectivity of the aromatase inhibitor R 76 713. A study in human ovarian, adipose stromal, testicular and adrenal cells. *J Steroid Biochem* 36:57–65

Wouters W, Snoek E, DeCoster R (1994): Vorozole, a specific non-steroidal aromatase inhibitor. *Breast Cancer Res Treat* 30:89–94

Wright JN, Calder MR, Akhtar M (1985): Steroidal C-19 sulfur and nitrogen derivatives designed as aromatase inhibitors. *J Chem Soc Chem Comm* 1733–1735

Wright JN, Slachter G, Akhtar M (1991): "Slow-binding" sixth-ligand inhibitors of cytochrome P-450 aromatase. *Biochem J* 273:533–5391

Yoshihama M, Tamura K, Nakakoshi M, Nakamura J, Fujise N, Kawanishi G (1990): 14α-hydroxyandrost-4-ene-3,6,17-trione as a mechanism-based irreversible inhibitor of estrogen biosynthesis. *Chem Pharmaceut Bull* 38:2834–2837

Yue W, Zhou D, Chen S, Brodie A (1993): A new nude mouse model for postmenopausal breast cancer using MCF-7 cells transfected with the human aromatase gene. *Cancer Res* 54:5092–5095

Yue W, Wang J, Savinov A, Brodie A (1995): The effect of aromatase inhibitors on growth of mammary tumors in a nude mouse model. *Cancer Res* 55:3073–3077

Zhou D, Korzekwa KR, Poulos T, Chen S (1992): A site-directed mutagenesis study of human placental aromatase. *J Biol Chem* 262:4421–4426

Part III

Prostate Cancer

Nearly a quarter of a million American men will be diagnosed with prostate cancer this year, and 40,000 will die from this disease. Prostate cancer is second only to lung cancer in mortality rates by site in males. Indeed, the mortality rate from prostate cancer is comparable to that from breast cancer in females, and it is continuing to increase. Despite this, progress in the treatment of prostate cancer, and the amount of research activity in this field, have lagged far behind that of breast cancer. There may be many reasons for this. First, life-threatening prostate cancer often occurs at an advanced age. Second, although it has been known for many decades that prostate cancer is often androgen-dependent, treatment modalities based on this principle have been far less successful than, for example, tamoxifen treatment of breast cancer. Finally, the interaction of stroma and epithelium, and the probable involvement of numerous growth factors in prostate cancer growth and progression, has made a straightforward hormone-therapy approach difficult, if not impossible. These facts are elaborated upon in the chapters found in this section.

Chapter 13 describes prostate cancer and its diagnosis, staging, and treatment. It emphasizes the primary role of surgical intervention in the treatment of this disease. Because many prostate cancers are indolent, "watchful waiting" is an approach to the treatment of prostate cancer that is being discussed avidly in the field, and this chapter raises the salient features of this argument. Because of the advanced age of patients and complications associated with medical intervention, the physician must weigh carefully the benefits and risks of treatment. That is, the clinician is faced with the dilemma of guessing if the prostate cancer is life-threatening, or if the patient will most likely die of some other physical condition before he succumbs to the effects of the cancer.

Chapter 14 focuses on the myriad chromosomal loci that have been implicated in the development of prostate cancer. As with all neoplasia, the inactivation of tumor suppressor genes and the activation of proto-oncogenes take center stage. Although some of these genes have been identified, the involvement of unidentified genes at new loci provides an opportunity to dissect, at the molecular level, the biochemical pathways involved in prostate cancer. One would hope that the identification of these genes will allow molecular diagnostics to emerge as a central focus for determining if a particular prostate cancer is likely to be indolent or life-threatening.

In Chapter 15, the complex interactions between growth factors and their receptors, and prostatic stromal and epithelial tissue, are discussed. The impinging of numerous intracellular signaling pathways on the process of prostate cancer cell growth perhaps yields a clue as to why medical intervention is so difficult. The synthesis of growth factors by the stroma and epithelium, and the response of each tissue to the growth factors synthesized by the other, and perhaps themselves, will need to be understood better so as to provide reasonable targets for future treatment modalities. The combined actions of growth factors and androgens will also be an area that can be exploited, once our knowledge of both areas matures more.

Chapter 16 reviews the role of the androgen receptor in prostate cancer. Much progress has been made recently in this area; our understanding of the nuclear receptor superfamily, including the androgen receptor, has increased remarkably within the last decade. Recent results on the possible role of variable amino acid repeat lengths in the androgen receptor and their implication for prognosis are discussed. Of particular importance is the possible role of the amino acid repeats in the known association of race with prognosis—African-American men have a higher risk of contracting prostate cancer and dying from the disease than do white men.

This section, then, provides a comprehensive review of the clinical, genetic, and biochemical features of prostate cancer. With the great strides being made in the molecular genetic arena and in the field of signal transduction, it is likely that rational treatment designs will emerge, with a resultant improvement in the diagnosis, prognosis, and management of prostate cancer.

13

Current Concepts in the Treatment of Cancer of the Prostate

JOSEPH A. SMITH, JR.

EPIDEMIOLOGY

Carcinoma of the prostate is the most common cancer in men in the United States and the second most common cause of cancer death. In addition, both these figures are increasing at an alarming rate. The incidence of prostate cancer goes up with advancing age but the disease is being recognized increasingly in younger men.

Despite these concerning statistics, there are few identifiable and controllable factors that contribute to the development of prostate cancer. Undoubtedly, there is a familial if not hereditary association (Carter et al., 1993; see Chapter 14, this volume). Men with a first-order relative who has been diagnosed with prostate cancer have a nearly fourfold increase in their own risk of developing prostate cancer in their lifetime (McWhorter et al., 1992). High-fat diet has been implicated in some studies but the data are still preliminary. Within the last couple of years, there have been some epidemiologic studies that have linked vasectomy to prostate cancer. There is, however, no identifiable causal relationship between the two, and other studies have not shown this link.

NATURAL HISTORY

An extremely important concept in the evaluation and treatment of men with prostate cancer is understanding of the natural history of the disease. Almost one-fourth of the men diagnosed with prostate cancer in this country eventually succumb to the disease. This is approximately the same ratio as

Hormones and Cancer
Wayne V. Vedeckis, Editor

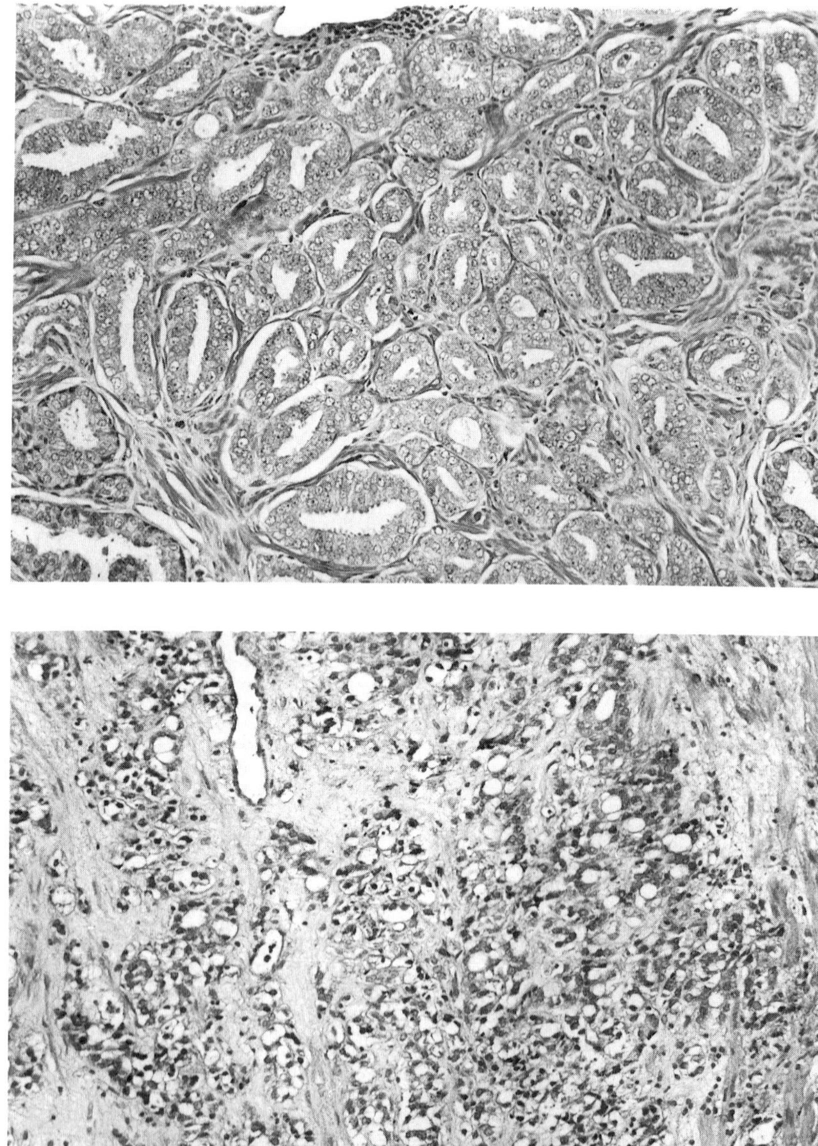

Figure 1. Tumor grade is an important prognostic factor for carcinoma of the prostate. Well-differentiated cancers (top) generally progress more slowly and present at a less advanced stage than poorly differentiated lesions (bottom). Changes in grade may be observed with this.

is seen in women with breast cancer. It is a mistake, then, to dismiss prostate cancer as an insignificant disease. On the other hand, prostate cancer may be slowly progressive and clinically insignificant in some men (Fig. 1).

There is a wide disparity between the autopsy incidence of prostate cancer and the number of men who eventually have a clinical diagnosis of the disease. Furthermore, there is an almost equal rate of cancer found at autopsy in various populations throughout the world, although there is a large difference in the clinical incidence of the disease between these same populations. A key factor, then, is making a distinction between insignificant cancers and those that may progress to cause clinical morbidity or, perhaps, patient death.

METHODS FOR EARLY DETECTION

The purpose of screening for any cancer is not simply to detect more cancers but to decrease the death rate from the disease. The two do not necessarily go hand-in-hand. A decrease in the death rate depends on the ability to detect and treat adequately the cancers that ultimately cause mortality. Debates about the value of prostate cancer screening have raged in both the medical literature and the lay press. Increasingly, though, it appears that the prostate cancers being detected through screening methods generally are those likely to become clinically significant. None of the current methods for screening seem capable of finding the autopsy types of cancers that are probably best left undiagnosed. Currently, the two most commonly used and recommended methods for early detection of prostate cancer are digital rectal examination and serum prostate-specific antigen.

Digital Rectal Examination

Most prostate cancers arise in the peripheral portion of the gland and, therefore, are accessible to palpation by digital rectal examination. Cancers generally are hard and nodular and are distinguishable from the more spongy feel of the normal prostate gland. Prostate size alone is not a criterion for cancer. Early prostate cancers often are detectable only as a relatively subtle induration in the prostate. Once the tumor is obvious by palpation, it is usually extracapsular and may be too far advanced for curative therapy (Chodak et al., 1989).

Prostate-Specific Antigen

Prostate specific antigen (PSA) is an enzyme produced by normal prostate epithelial cells. Its function is to lyse postejaculatory semen. In pathological conditions, PSA "leaks" into the serum.

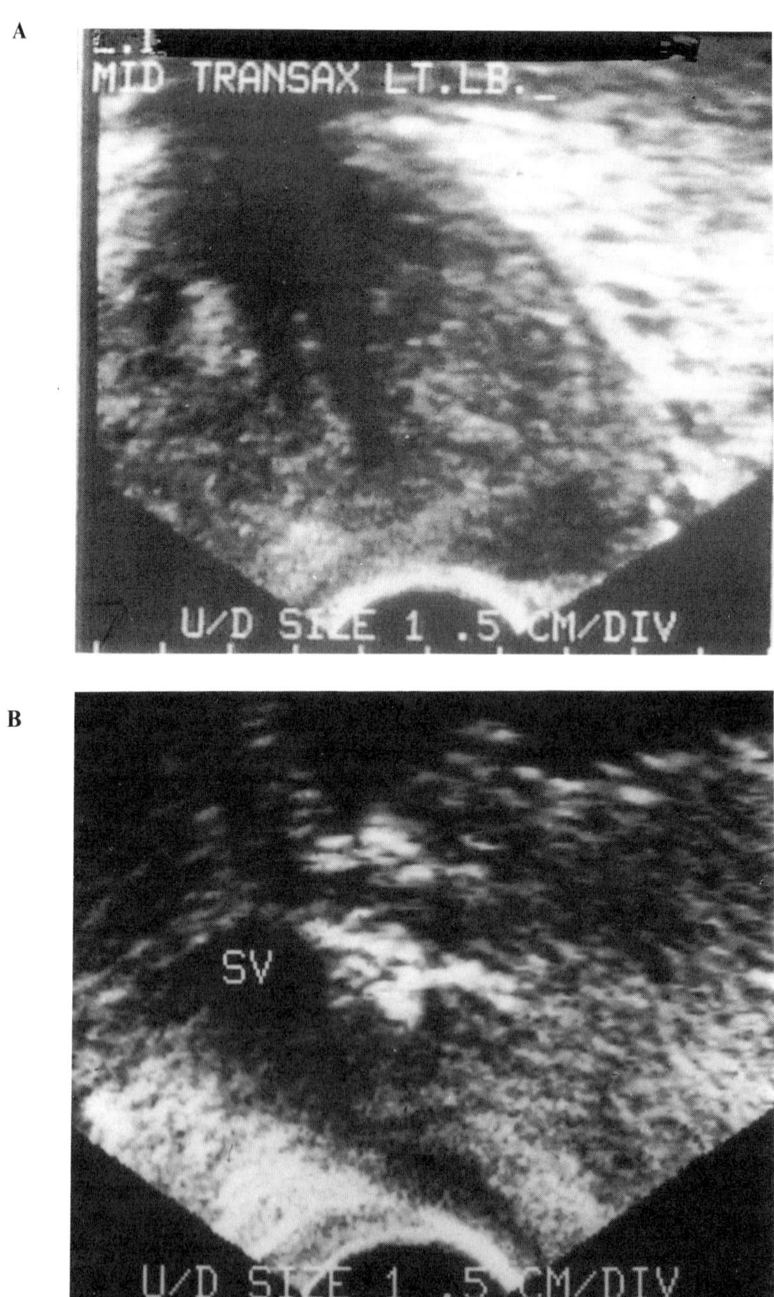

Figure 2. **A**: Transverse ultrasound image of the prostate showing a hypoechoic lesion at the left base. **B**: Sagittal view shows the same lesion extending near the base of the left seminal vesicle (*S.V.*).

PSA is prostate specific but not cancer specific (Oesterling, 1991). Any pathological condition of the prostate, such as prostatitis or benign prostatic hypertrophy, can cause an elevation in serum PSA. On the other hand, localized prostate cancer can occur in the face of a normal serum PSA. PSA levels of less than 4.0 ng/ml (Hybritech method) generally are considered normal (Catalona et al., 1991). Levels between 4 and 10 are suspicious for cancer and nearly 20% of men in this group will be found to have prostate cancer on biopsy. Most men with a PSA of greater than 10 ng/ml are proven to have prostate cancer (Catalona et al., 1994). PSA may be "age indexed" (Oesterling et al., 1993). A cut-off level of 4.0 ng/ml may be too high for a man 50 years of age and too low for someone aged 70 years. Nomograms are being established that may increase the accuracy of PSA in predicting the presence of prostate cancer.

Ultrasound

Transrectal ultrasonography is the preferred imaging modality for prostate anatomy (Fig. 2) (Shinohara et al., 1989). Originally, ultrasound had been promoted as a method for early detection of prostate cancer (Lee et al., 1988). Ultrasound is rarely used now for detection of prostate cancer, but it is used by urologists to direct biopsies in men with palpable abnormalities of the prostate or abnormal serum PSA levels (Cooner et al., 1990; Hernandez and Smith, 1990).

TUMOR STAGING

Accurate tumor staging is essential in determining prognosis and treatment of prostate cancer. Table 1 shows the TNM classification system for prostate cancer. Determining the extent of local disease is of extreme importance but notoriously inaccurate. Digital rectal palpation remains the best single test for determining local extent, but almost half of large, palpable tumors thought to be intracapsular are proven to have microscopic extracapsular extension (Rosen et al., 1992). Transrectal ultrasound, although the most commonly used imaging modality for the prostate, has not provided improved staging accuracy over digital rectal examination. Magnetic resonance imaging of the prostate using endorectal coils is being evaluated, but suffers from some of the same limitations as digital rectal exam and ultrasonography in detecting microscopic amounts of extracapsular extension.

Usually, the first detectable site of metastatic disease from prostate cancer is the pelvic lymph nodes. The lymph nodes are best imaged by pelvic computerized tomography (CT) scanning. However, nodes are rarely enlarged to the point that they are detectable on CT scan unless there is a large volume or high-grade tumor. CT scanning is not a necessary nor a

Table 1. TNM staging Ssystem for Pprostate Ccancer

Stage	Description
T1a	Incidentally detected tumor, ≤5% of resected tissue
T1b	Incidentally detected tumor, >5% of resected tissue
T1c	Impalpable tumor, detected by needle biopsy (usually because of elevated prostate-specific antigen level)
T2a	Tumor confined to prostate, <50% of lobe involved
T2b	Tumor confined to prostate, >50% of lobe involved but not both lobes
T2c	Tumor confined to prostate, involves both lobes
T3a	Unilateral extension through prostate capsule
T3b	Bilateral extension through prostate capsule
T3c	Tumor invades seminal vesicle(s)
T4a	Tumor invades bladder neck, external sphincter, or rectum
T4b	Tumor invades levator muscles and/or fixed to pelvic wall
N1	Metastasis to lymph node, ≤2 cm in greatest diameter
N2	Metastasis to lymph node, >2 cm but ≤5 cm in greatest diameter
N3	Metastasis to lymph node(s), >5 cm in greatest diameter

useful routine staging procedure for apparently localized prostate cancer (Fig. 3). Most often, staging of the pelvic lymph nodes is performed by either open or laparoscopic lymph node dissection. The probability of lymph node metastasis can be predicted with some accuracy from the size and grade of the primary tumor as well as the serum PSA level.

Increasingly, prostate cancer is being diagnosed because of a biopsy prompted by an abnormal PSA level in the face of a normal digital rectal examination. This has been categorized as stage T_1C disease. Questions have arisen as to the clinical significance of stage T_1C prostate cancer and the volume of disease within the prostate (Terris et al., 1992). Studies have shown, however, that more than 85% of men with T_1C prostate cancer have tumor volumes that, at least pathologically, seem to be of clinical relevance (Epstein et al., 1994; Scaletscky et al., 1994).

Distant metastasis from prostate cancer occurs most often in the axial skeleton (Fig. 4). The most sensitive test for detecting bone metastasis from prostate cancer is a radionuclide bone scan (Fig. 5). Bone scans rarely are positive in men with a PSA of less than 10 ng/ml.

TREATMENT OPTIONS FOR LOCALIZED DISEASE

Once clinically localized prostate cancer is diagnosed, there are a number of potential treatment options available. In many situations, there is genuine controversy about the best treatment, and a patient may receive several different opinions. Much of the controversy, though, can be diminished by attempting to tailor treatment toward characteristics of the tumor itself, the patient's chronologic and physiological age and medical history, and patient wishes.

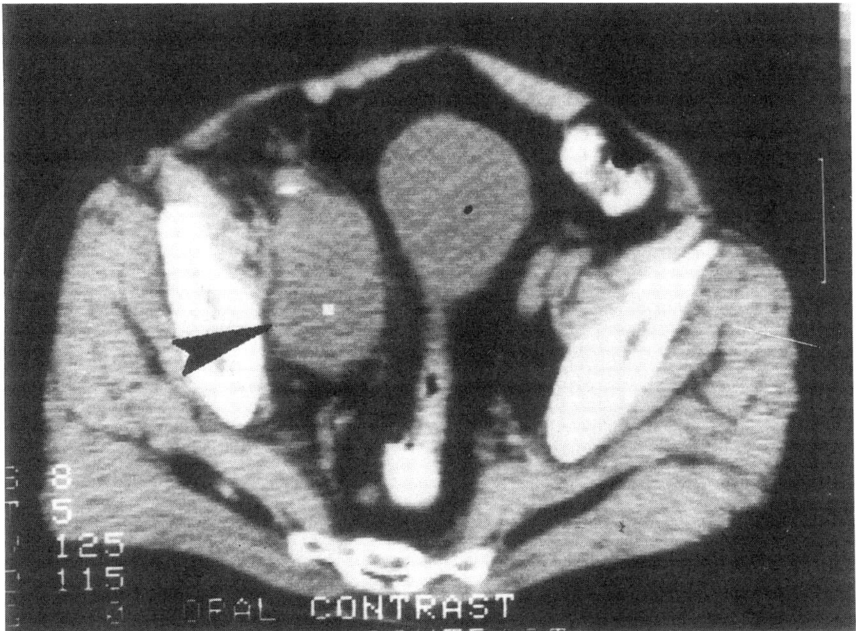

Figure 3. Pelvic CT scan in a patient with carcinoma of the prostate showing a large mass along the right iliac node chain (*arrow*). However, CT scans rarely identify nodal metastasis in most patients with apparently localized prostate cancer.

For many men with localized prostate cancer, watchful waiting is the most appropriate posture. Few men with a tumor that would otherwise be curable by treatment will die from prostate cancer within 5 years of diagnosis. The median time to death or serious morbidity is usually 8 years or more if the tumor is localized at the time of diagnosis. Thus, for most men with a life expectancy otherwise of less than 10 years, watchful waiting is the preferred approach.

It is impossible to predict how long any man will live, although certain information (age, health status, family history) can provide general information. Similarly, it is not possible to predict prognosis with certainty in an individual tumor. However, useful information can be used to categorize tumors into those most likely to have an indolent clinical course and those that may grow more rapidly. Tumor volume at the time of diagnosis, grade, and DNA ploidy status are all generally predictive of future tumor behavior.

Recently, there have been a series of studies suggesting that watchful waiting should be used for all men with localized prostate cancer (Whitmore et al., 1991; Johansson et al., 1992; Chodak et al., 1994). It is interesting to note that Scandinavia has the highest death rate from prostate cancer in the world, somewhat bringing into question the wisdom of a nontherapeutic

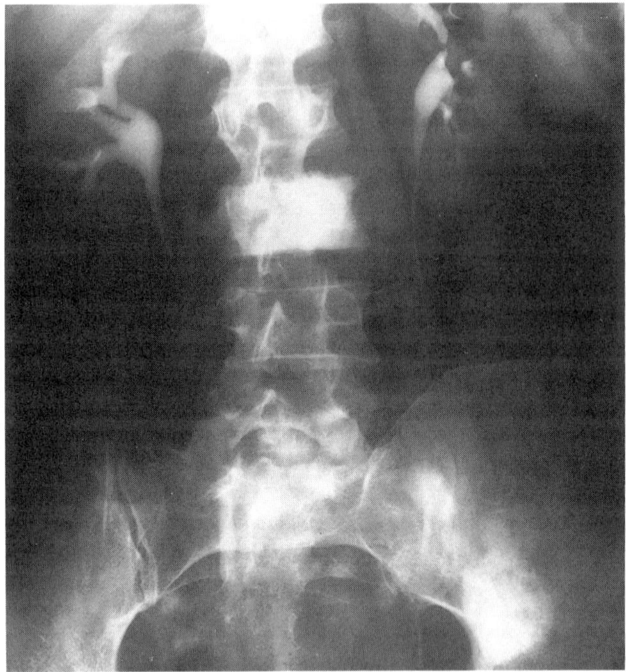

Figure 4. Typical blastic metastasis from prostate cancer seen on radiograph of the lumbar spine.

approach. Moreover, the patients in the presented series generally were over the age of 70 years and had low volume and low-grade tumors. There is little question that watchful waiting is appropriate for most of these men. However, these series do not address the dilemma of how to handle men younger than 70 to 75 years with more aggressive tumors.

Radical Prostatectomy

Radical prostatectomy involves surgical removal of the entire prostate and seminal vesicles and is associated with a high cure rate in men with clinically and pathologically localized tumors (Middleton and Larsen, 1990). As with any treatment, proper patient selection is paramount. Radical prostatectomy does not cure men with extracapsular tumor extension.

The morbidity of radical prostatectomy has declined significantly in recent years. In most series, the mortality rate from the procedure is less than 1%. We plan hospital discharge for the second or third day after surgery and the operative procedure is generally $1\frac{1}{2}$ to 2 hours long (Koch et al., 1994). In our own patients, blood transfusion is required in fewer than 1% of them.

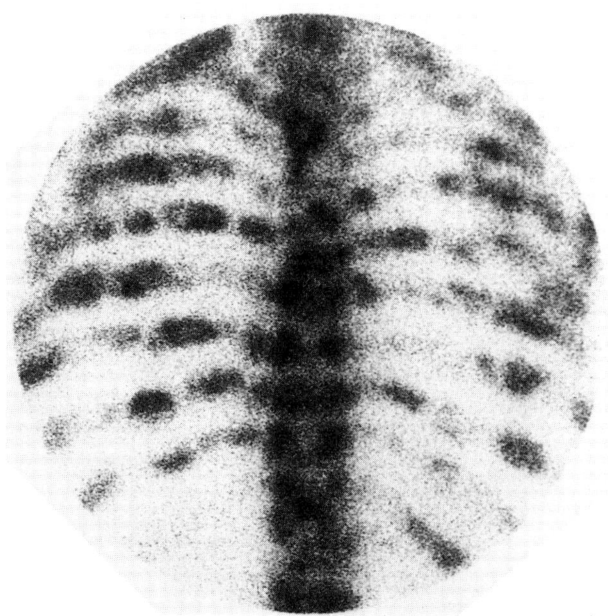

Figure 5. Radionuclide bone scan showing multiple areas of abnormal uptake in the ribs and thoracic spine in a patient with metastatic carcinoma of the prostate.

The primary drawbacks to radical prostatectomy are the risks of incontinence and impotence. Severe incontinence requiring treatment with artificial sphincter implantation is unusual (less than 2–3%), although 10–15% of men have some degree of leakage that may require them to wear a protective pad. A "nerve-sparing" surgical procedure can be used in men who are potent preoperatively. This can preserve sexual function in up to two-thirds of younger men, but the success of nerve sparing radical prostatectomy is related directly to patient age and potency status preoperatively (Walsh et al., 1987).

After radical prostatectomy, PSA levels should be 0 because there is no longer any benign or malignant prostate tissue present. Detectable levels of PSA generally imply residual cancer. Thus, after radical prostatectomy, PSA becomes one of the most sensitive and accurate tumor markers available for any cancer.

External Irradiation

External irradiation has been the primary therapeutic alternative to radical prostatectomy for the curative treatment of localized carcinoma of the prostate (Bagshaw et al., 1990). Five- and 10-year survival rates seem similar, but comparative data beyond that point are lacking (Bagshaw et al., 1990; Hanks et al., 1991). Chronic radiation cystitis and proctitis are reported in up to 10% of patients undergoing external irradiation. Potency is preserved in the early period after radiation but impotence is common 6 months to 1 year after treatment. A major concern after radiation has been the high rate of positive biopsies after treatment (Schellhammer et al., 1989; Wheeler et al., 1993).

Cryosurgery

Cryosurgical freezing of the prostate has been used in one form or another for decades. Previous experience has been relatively unfavorable and associated with a high risk of complications and high recurrence rates. More recently, cryosurgical probes have been developed for use with transrectal ultrasound imaging. The ice ball that forms can be visualized with ultrasonography and some control of the treatment extent maintained. Whether or not cryosurgery will prove to be a competitive treatment for clinically localized prostate cancer remains to be seen. Early results have shown some efficacy, but persistently positive biopsies are noted in a quarter to one-third of patients. Larger patient series and longer term follow-up will be necessary before cryosurgery can be considered anything other than an investigational technique.

Hormonal Therapy

The substantial majority of prostate cancers are sensitive to androgen deprivation. Both objective and subjective tumor response can be expected. The primary methodology revolves around suppression of serum testosterone levels. This can be achieved either by surgical bilateral orchiectomy or monthly administration of a luteinizing hormone-releasing hormone (LHRH) analog. Eulexin (flutamide), an antiandrogen, may provide a somewhat prolonged duration of response in some patients when used in conjunction with orchiectomy or an LHRH analog (Wheeler et al., 1993).

Hormonal therapy is not considered curative therapy for prostate cancer. Eventually, tumor progression occurs despite hormonal therapy. Whether the duration of response is longer when hormonal therapy is instituted early in the course of disease compared to when metastatic disease is evident is uncertain.

TREATMENT OPTIONS FOR METASTATIC DISEASE

Historically, more than 25% of men with newly diagnosed prostate cancer had metastatic disease at the time of presentation. In most contemporary, screened populations, metastatic disease is evident in only a few percent of men (Mettlin et al., 1991). Hopefully, these figures eventually will translate into a significant decrease in the death rate from prostate cancer. However, more than 40,000 men will die from carcinoma of the prostate in the United States this year. Effective treatment for metastatic prostate cancer remains one of the challenges in the treatment of this disease.

Endocrine Manipulation

Huggins and Hodges first demonstrated the partial androgen dependence of prostate cancer in 1941 (Huggins and Hodges, 1941). They showed a favorable tumor response when testicular androgens were withdrawn by either orchiectomy or the oral administration of estrogens. In turn, restoration of testosterone caused tumor growth. More than 50 years later, endocrine manipulation with the aim of androgen ablation remains the initial treatment and the cornerstone in the management of metastatic carcinoma of the prostate.

With androgen deprivation, both objective and subjective response occurs in men with prostate cancer. A decrease in tumor size is observed both in the primary lesion in the prostate as well as at metastatic sites (Smith, 1987). Serum PSA levels decrease by more than 90% in most men and reach normal levels in almost half. Patients with advanced disease and bone pain may have rapid and impressive relief of pain.

Endocrine treatment of prostate cancer generally has been accomplished by ablating testosterone output from the testis. Bilateral orchiectomy is a simple and safe surgical procedure and castrate levels of testosterone are reached within 24 hours. Diethylstilbestrol administered orally provides a comparable decrease in serum testosterone levels by feedback inhibition of pituitary gonadotropins. Although estrogens are effective and inexpensive, their use has been limited by associated cardiovascular and thromboembolic complications, which may occur in up to 20% of treated men (Garmick et al., 1985).

LHRH analogs, when administered in high doses, provide the paradoxical effect of a decrease in pituitary gonadotropins and, consequently, testicular androgens (Smith, 1984). After the initial administration of a monthly depot injection of a synthetic LHRH analog, serum testosterone levels rise to supernormal levels. However, by 1 to 2 weeks, testosterone is below baseline and castrate levels are achieved reliably by 3 to 4 weeks.

The side effects of endocrine treatment are a consequence of lowered serum levels of testosterone. Decreased libido and impotence develop, although this may not be observed for many months after treatment. Vasomotor hot flashes occur in almost two-thirds of patients but can be suppressed by orally administered progesterones (Smith, 1994).

Antiandrogens

A number of antiandrogens are undergoing clinical testing of their role in the treatment of men with prostate cancer. Flutamide, a pure, nonsteroidal anti-androgen, is approved by the Food and Drug Administration for treatment of prostate cancer in combination with an LHRH analog or orchiectomy (Labrie, 1991). Used alone, flutamide provides a suboptimal hormonal response.

Flutamide does cause some degree of gynecomastia. Also, diarrhea can be a treatment-limiting side effect. The drug is administered orally in three divided daily doses of 250 mg.

Combination Hormonal Therapy

Although most circulating androgens are of testicular origin, the androgen production from the adrenal gland may be of some clinical significance. For this reason, combination hormonal therapy consisting of a treatment to lower testosterone (orchiectomy or LHRH analogs) with an antiandrogen to block adrenal androgen effects has been explored. A number of randomized studies from around the world have been published with somewhat conflicting data (Crawford et al., 1989; Labrie, 1991). For the most part, though, it appears that combination hormonal therapy provides a modest but measurable prolongation in duration of response and survival compared to treatment methods aimed at lowering testicular androgens alone. The observed effect seems greatest in men with minimal amounts of metastatic disease at the time of presentation.

Progressive Disease After Hormonal Therapy

Despite the excellent initial response rates observed after initiation of hormonal therapy, progressive disease is almost inevitable after a certain time period. The duration of response is variable but can be predicted partly by the degree of suppression of PSA levels. At any rate, most men develop progressive disease 18 to 24 months after the initiation of hormonal therapy.

There is no treatment that has convincingly been shown to prolong survival in patients who have failed hormonal therapy. Response rates with cytotoxic chemotherapy have been disappointing (Eisenberger et al., 1987). There are studies showing a prolonged survival for responders to chemotherapy compared to nonresponders. However, this does not necessarily mean that treated patients live longer than untreated ones.

Although there are a number of studies addressing therapeutic strategies in patients with hormonal refractory prostate cancer (LaRocca et al., 1991; Yagoda et al., 1991; Seidman et al., 1992), most clinical management centers around palliation of symptomatic disease (Tannock et al., 1989). Bone pain is one of the prominent features in men with progressive disease. External irradiation is useful for localized areas of painful bone metastasis. For patients with diffuse disease, intravenous strontium therapy can be helpful (Laing, 1991; Lewington, 1991). Soft tissue metastasis is usually a late manifestation of disease and is relatively unusual in the absence of identifiable bone metastasis.

FUTURE DIRECTIONS

Enormous changes in prostate cancer management have occurred over the last decade and even more can be anticipated. Although PSA is one of the best tumor markers in human oncology, further refinements may increase the sensitivity and specificity of PSA. The most important of these seems to be determination of the ratio of free versus bowel PSA. A higher proportion of serum PSA is bound to alpha antichymotrypsin in men with benign prostatic hyperplasia compared to those with prostate cancer. In addition, new methods are being developed that allow even greater sensitivity in detecting minute amounts of PSA in the serum.

Increasing evidence suggests that mutations in the androgen receptor contribute to the development of clinical androgen-independent prostate cancer growth (see Chapter 16, this volume). Intermittent hormonal therapy is being explored as a means to decrease the selection of tumors with genetic mutations. The hope is that intermittent therapy will prolong the overall hormonal response and patient survival.

The response of prostate cancer to cytotoxic chemotherapy remains disappointing. New combinations of drugs are being explored. Genetic manipulation aimed at preventing hormone independence provides promise for future treatment schemes.

REFERENCES

Bagshaw MA, Cox RS, Rambach JE (1990): Radiation therapy for localized prostate cancer. *Urol Clin N Am* 17:787–802

Carter BS, Bova GS, Beaty TH, Steinberg GD, Childs B, Isaacs WB, Walsh PC (1993): Hereditary prostate cancer: epidemiologic and clinical features. *J Urol* 150(3):797–802

Catalona WJ, Smith DS, Ratliff TL, Dodds KM, Coplen DE, Yuan JJ, Petros JA, Andriole GL (1991): Measurement of prostate specific antigen in serum as a screening test for prostate cancer. *N Engl J Med* 324:1156–1161

Catalona WJ, Richie JP, Ahmann FR, Hudson MA, Scardino PT, Flanigan RC, deKernion JB, Ratliff TL, Kavoussi LR, Dalkin BL, Waters WB, MacFarlane MT, Southwick PC (1994): Comparison of digital rectal examination and serum prostate specific antigen in the early detection of prostate cancer: results of a multicenter clinical rial of 6630 men. *J Urol* 151(5):1283–1290

Chodak GW, Keller P, Schoenberg HW (1989): Assessment of screening for prostate cancer using the digital rectal examination. *J Urol* 141(5):1136–1138

Chodak GW, Thisted RA, Gerber GS, Johansson JE, Adolfsson J, Jones GW, Chisholm DD, Moskovitz B, Livne PM, Warner J (1994): Results of conservative management of clinically localized prostate cancer. *N Engl J Med* 330: 242–248

Cooner WH, Mosley BR, Rutherford CL, Beard JH, Pond HS, Terry WJ, Igel TC, Kidd DD (1990): Prostate cancer detection in a clinical urological practice by ultrasonography, digital rectal examination and prostate specific antigen. *J Urol* 143(6):1146–1152; discussion 1152–1154

Crawford ED, Eisenbergen MA, McLeod DG, Spaulding JT, Benson R, Dorr FA (1989): A controlled trial of leuprolide with and without flutamide in prostatic carcinoma. *N Engl J Med* 321:419–424.

Eisenberger MA, Bezerdigan L, Kalash S (1987): A critical assessment of the role of chemotherapy for endocrine resistant prostatic carcinoma. *Urol Clin N Am* 14: 695–706

Epstein JI, Walsh PC, Carmichael M, Brendler CB (1994): Pathologic and clinical findings to predict tumor extent of nonpalpable (stage T1c) prostate cancer. *JAMA* 271(5):368–374

Garmick MB, Glode LM, Smith JA Jr (1985): Leuprolide versus diethylstilbestrol for metastatic prostate cancer. *N Engl J Med* 311:1281–1286

Hanks GE, Asbell S, Krall JM, Perez CA, Doggett S, Robin P, Sause W, Pilepich MV (1991): Outcome for lymph node dissection negative T1b, T2 prostate cancer treated with external beam radiation therapy in RTOG 77–06. *Int J Rad Oncol Biol Physiol* 21:1009–1103

Hernandez AD, Smith JA Jr (1990): Transrectal ultrasonography for the early detection and staging of prostate cancer. *Urol Clin N Am* 17:745–757

Huggins C, Hodges CV (1941): Studies on prostatic cancer. I. The effect of castration, estrogen and androgen injections or serum phosphatases in metastatic carcinoma of the prostate. *Cancer Res* 1:293–295

Johansson JE, Adami HO, Andersson SO, Bergstrom R, Holmberg L, Krusemo UB (1992): High 10 year survival rate in patients with early untreated prostate cancer. *JAMA* 267:2191–2196

Koch MO, Smith JA, Hodge EM, Brandell RA (1994): Prospective development of a cost efficient program for radical retropubic prostatectomy. *Urology* 44(3):311–318

Labrie F (1991): Endocrine therapy of prostate cancer. *Endocrinol Metab Clin N Am* 20:845–872

Laing AH, Ackery DM, Bayly RJ, Buchanan RB, Lewington VJ, McEwan AJ, Macleod PM, Zivanovic MA (1991): Strontium-89 chloride for pain palliation in prostatic skeletal malignancy. *Br J Radiol* 64:816–822

LaRocca RV, Cooper MR, Uhrich M, Danesi R, Walther MM, Linehan WM, Myers CE (1991): Use of suramin in treatment of prostatic carcinoma refractory to conventional hormonal manipulation. *Urol Clin N Am* 18:123–129

Lee F, Littrup PJ, Torp-Pedersen ST, Mettlin C, McHugh TA, Gray JM, Kumasaka GH, McLeary RD (1988): Prostate cancer: comparison of transrectal US and digital rectal examination for screening. *Radiology* 168(2):389–394

Lewington VJ, McEwan AJ, Ackery DM, Bayly RJ, Keeling DH, Macleod PM, Porter AT, Zivanovic MA (1991): A prospective, randomized double-blind cross-over study to examine the efficacy of strontium-89 in pain palliation in patients with advanced prostate cancer metastatic to bone. *Eur J Cancer* 27:954–958

McWhorter WP, Hernandez AD, Meikle AW, Terreros DA, Smith JA Jr, Skolnick MH, Cannon-Albright LA, Eyre HJ (1992): A screening study of prostate cancer in high risk families. *J Urol* 148:826–828

Mettlin C, Lee F, Drago J, Murphy GP (1991): The American Cancer Society National Prostate Cancer Detection Project. Findings on the detection of early prostate cancer in 2425 men. *Cancer* 67(12):2949–2958

Middleton RG, Larsen RH (1990): Selection of patients with stage B prostate cancer for radical prostatectomy. *Urol Clin N Am* 17:779–786

Oesterling JE (1991): Prostate specific antigen: A critical assessment of the most useful tumor marker for prostate cancer. *J Urol* 145:907–923

Oesterling JE, Cooner WH, Jacobsen SJ, Guess HA, Lieber MM (1993): Influence of patient age on the serum PSA concentration. An important clinical observation. *Urol Clin N Am* 20(4):671–680

Rosen MA, Goldstone L, Lapin S, Wheeler T, Scardino PT (1992): Frequency and location of extracapsular extension and positive surgical margins in radical prostatectomy specimens. *J Urol* 148:331–337

Scaletscky R, Koch MO, Eckstein CW, Bicknell SL, Gray GF Jr, Smith JA Jr (1994): Tumor volume and stage in carcinoma of the prostate detected by elevation in prostate specific antigen. *J Urol* 152:129–131

Schellhammer PF, Whitmore RB, Kuban DA, El-Mahdi AM, Ladaga LA (1989): Morbidity and mortality of local failure after definitive therapy for prostate cancer. *J Urol* 141:567–571

Seidman AD, Scher HI, Petrylak D, Dershaw DD, Curley T (1992): Estramustine and vinblastine: use of prostate specific antigen as a clinical trial endpoint in hormone-refractory prostatic cancer. *J Urol* 147:931–934

Shinohara K, Wheeler TM, Scardino PT (1989): The appearance of prostate cancer on transrectal ultrasonography: correlation of imaging and pathological examinations. *J Urol* 142(1):76–82

Smith JA Jr (1984): Androgen suppression by a gonadotropin releasing hormone analog in patients with metastatic carcinoma of the prostate. *J Urol* 131:1110–1112

Smith JA Jr (1987): Endocrine treatment. In: *Clinical Management of Prostate Cancer*. Smith JA Jr, Middleton RG (eds). Chicago: Year Book Medical Publishers, pp 134–153

Smith JA Jr (1994): A prospective comparison of treatments for symptomatic hot flushes following endocrine therapy for carcinoma of the prostate. *J Urol* 152: 132–134.

Tannock I, Gospodarowicz M, Meakin W, Panzarella T, Stewart L, Rider W (1989): Treatment of metastatic prostatic cancer with low-dose prednisone: evaluation of pain and quality of life as pragmatic indices of response. *J Clin Oncol* 7: 590–597

Terris MK, McNeal JE, Stamey TA (1992): Detection of clinically significant prostate cancer by transrectal ultrasound-guided systematic biopsies. *J Urol* 148(3):829–832

Walsh PE, Epstein JI, Lowe FC (1987): Potency following radical prostatectomy with wide unilateral excision of the neurovascular bundle. *J Urol* 138:823–827

Wheeler JA, Zagars GK, Ayala AG (1993): Dedifferentiation of locally recurrent prostate cancer after radiation therapy. Evidence for tumorprogression. *Cancer* 71:3783–3787

Whitmore WF Jr, Warner JA, Thompson IM Jr (1991): Expectant management of localized prostate cancer. *Cancer* 67:1091–1096

Yagoda A, Smith JA Jr, Soloway MS, Tomera K, Seidmon EJ, Olsson C, Wajsman Z (1991): Phase II study of estramustine phosphate in advanced hormone refractory prostatic cancer with increasing prostate specific antigen levels. *J Urol* 145:384A

14

Tumor Suppressor Genes and Oncogenes in Human Prostate Cancer

JILL A. MACOSKA

TUMOR AND METASTASIS SUPPRESSOR GENES

A tumor suppressor gene may be broadly defined as a gene whose inactivation is permissive to tumorigenesis. Inactivation may occur through deletion or mutation of the DNA base sequence. Two hypotheses have been proposed to explain how mutation or deletion of tumor suppressor gene DNA sequences inactivates normal genic functions. In the Knudson hypothesis, deletion or mutation must affect both alleles of the gene in order to disable tumor suppression (Knudson, 1971). As might be expected, the effect of "two hits" on tumor suppressor gene integrity, for example, a combination of mutation or deletion affecting both alleles, would be to disable the gene from encoding gene product. The retinoblastoma gene is an example of a tumor suppressor gene that fulfills the Knudson hypothesis, for example, one mutant allele is inherited in the germline, and the other is mutated or deleted in affected tissues (reviewed by Weinberg, 1992). In the second, or "dominant-negative" hypothesis, mutation of one allele with retention of a normal allele is sufficient to disable tumor suppressor gene function. This rationale is appropriate for tumor suppressor genes that encode proteins that normally form homomeric complexes (e.g., dimers or other oligomers). In this hypothesis, mutation in one allele results in a heteromeric, dysfunctional gene product (e.g., a wild-type and mutant protein complex). The p53 gene is an example of a tumor suppressor gene that fulfills the "dominant-negative" hypothesis (reviewed by Vogelstein and Kinzler, 1992).

The studies summarized below describe genes or genetic loci that appear to suppress the promotion of various aspects of prostate tumor progression.

Hormones and Cancer
Wayne V. Vedeckis, Editor

For example, the inactivation of tumor suppressor genes may create a cellular environment permissive to tumorigenesis, whereas the inactivation of other genes may enhance tumor invasiveness or metastasis (see the E-cadherin gene and metastasis suppressor genes below).

Recent studies of tumor suppressor gene inactivation in prostate cancer have focused on 1) the inactivation of known tumor suppressor genes and 2) the identification of novel genes localized to frequently deleted chromosomal regions.

Tumor Suppressor Genes in Sporadic Prostate Cancer

Inactivation of Known Tumor Suppressor Genes

RB1. The 200-kb RB1 gene maps to chromosomal region 13q14 and encodes a protein active in cell cycle regulation. The "prototype" tumor suppressor gene, RB1, was originally identified through homozygous inactivation in retinoblastomas, for example, deletion of a normal allele with retention of an inherited "defective" allele (commonly characterized by base pair deletions) (Friend et al., 1986; Mihara et al., 1989). The role of RB1 gene inactivation in prostate tumorigenesis is somewhat controversial. Retroviral-mediated transfer of a normal RB1 gene into the pRb-deficient DU145 prostate cancer cell line (originally cultured from a prostate tumor brain metastasis; Stone et al., 1978) did not inhibit cell growth, but did inhibit the ability of the cells to form tumors in nude mice (Bookstein et al., 1990a). Similar results were obtained by Banerjee et al. (1992) using microcell-mediated transfer of a normal chromosome 13 into DU145 cells.

In prostate tumor specimens, Bookstein et al. (1990a) used immunohistochemical techniques to evaluate the pattern of pRb protein expression in prostate tumors (one primary and six metastatic). In the metastatic tumors, three expressed pRb strongly, one moderately, and two weakly or negatively. Interestingly, one of the tumors that did not express pRb demonstrated allelic loss of one RB1 allele and a promoter deletion in the remaining allele. However, Sarkar et al. (1992) failed to find promoter deletions in 23 primary adenocarcinomas and one small cell carcinoma of the prostate, suggesting that inactivation of the RB1 gene by promoter deletion may occur only rarely in prostate cancer. Finally, Phillips et al. (1994a) demonstrated loss of four intragenic polymorphic markers in 24 of 40 (60%) prostatic tumors and 1 of 10 benign prostatic hyperplasias.

Other studies describe deletion of 13q loci in 3 of 13 informative tumors, for example, tumors heterozygous at those loci (Carter et al., 1990b). Comparative genomic hybridization (CGH) demonstrated deletion of 13q21-q31 sequences in 32% of 31 primary tumors, and of 13cen-q21 sequences (inclusive of the RB1 locus at 13q14) in 56% of nine recurrent tumors (Visakorpi et al., 1995).

Although these studies disagree on the type and extent of deletion of RB1 sequences in human prostate tumors, it is clear that inactivation of RB1 is important for tumorigenesis in at least a subset of prostate tumors. Some of these studies also raise the possibility that other tumor suppressor genes important for prostate tumorigenesis reside on 13q.

p53. Recent studies have shown that three putative tumor suppressor genes localize to 17p13.1 (Coles et al., 1990; Biegel et al., 1992). The best characterized is the p53 gene, which encodes a protein thought to function as a negative regulator of the cell cycle (e.g., it arrests the cell cycle in G1 to permit DNA repair; see Chapter 1, this volume). Inactivation of the p53 gene through any combination of allelic deletion or mutation creates a cellular environment permissive to proliferation and malignant progression (reviewed by Vogelstein and Kinzler, 1992)

In prostate tumors, Carter et al. (1990b) described loss of 17p13 sequences in 3 of 18 informative tumors. Visakorpi et al. (1992) detected intense p53 protein staining in 8 of 132 (6%) primary tumors examined. All eight tumors were poorly differentiated, and all eight patients died of disease (six within 2 years of diagnosis). Bookstein et al. (1993) reported immunohistochemical analysis demonstrating abnormal nuclear p53 accumulation in 19 of 150 (13%) tumors examined, mostly from advanced disease. Single-strand conformational polymorphism (SSCP) analysis of exons 4–9 of 14 aberrant tumors revealed nine with abnormal band patterns; when directly sequenced through exons 4–9, all nine tumors demonstrated point mutations. Effert et al. (1993), using SSCP and direct sequencing, reported mutations in 1 of 10 primary and 1 of 2 metastatic tumors (both involving codon 172, a mutational "hot spot"). Interestingly, although allelic loss of p53 was not evident in this metastasis, a more distal locus at 17p13.3 was deleted. Navone et al. (1994) detected mutant p53 protein in 20/87 (22%) cases, most frequently in bone marrow metastases (18 of 20); all p53+ tumors were poorly differentiated (combined Gleason scores of 9 or 10) (Gleason, 1992). In contrast to these studies, Dinjens et al. (1994) found no significant difference between the frequencies of aberrant p53 protein accumulation, SSCP pattern, or base sequence mutations in exons 5–8 between primary tumors (2 of 20) and lymph node metastases (2 of 15). Finally, several *in situ* hybridization (ISH) and fluorescence *in situ* hybridization (FISH) studies have described loss of chromosome 17 in prostate tumors: Baretton et al. (1992) found loss in 16 of 35 cases examined using archival sections, Brothman et al. (1992) described loss in 7 of 20 short-term cultures of primary prostate cancers and in 16 of 20 (80%) sectioned archival specimens, Brothman et al., (1994), and Brown et al. (1994) demonstrated loss in 3 of 40 touch preparations of primary prostate tumors.

Most of these studies indicate that one or more tumor suppressor genes, including the p53 gene, localize to chromosome 17, and that deletion or mutation of these genes may play a role in the progression of a subset of prostate carcinomas.

DCC. The DCC (deleted in colon carcinoma) gene was cloned from a chromosomal region, 18q21.3, deleted with high frequency in colorectal cancers, and encodes a protein that demonstrates homology with cell surface glycoproteins (Fearon et al., 1990). Using reverse transcriptase-polymerase chain reaction (RT-PCR) analysis, Gao et al. (1993) demonstrated variably reduced levels of DCC expression in 11 of 14, and absent expression in 3 of 14, tumors examined. They also observed allelic loss of DCC in 5 of 11 (45%) informative cases. Others have reported loss of 18q21 sequences in 2 of 12 informative cases (Carter et al. 1990b) and of 18q22-qter in 19% of 31 primary tumors (Visakorpi et al., 1995). In addition, ISH and FISH studies have described loss of chromosome 18 in 19 of 32 cases using archival sections (Baretton et al., 1992), 4 of 20 short-term cultures of primary prostate cancers (Brothman et al., 1992), and 1 of 40 touch preparations of primary prostate tumors (Brown et al., 1994).

Taken together, these studies suggest that allelic deletion and/or loss of expression of DCC characterizes progression in some prostate tumors.

E-cadherin. E-cadherin is a Ca^{2+}-dependent cell adhesion molecule that mediates cell–cell interaction, and its gene maps to chromosomal region 16q22.1 (Umbas et al., 1992). Utilizing the Dunning R-3327 rat prostate cancer model system, Bussemakers et al. (1992) found differential mRNA expression of the E-cadherin gene: lines with low invasive potential expressed high levels of E-cadherin transcript, whereas lines with high invasive potential did not. Immunohistochemical studies by Umbas et al. (1992) described heterogeneous or negative E-cadherin protein expression in 40 of 84 (48%) primary prostate tumors and 6 of 8 (75%) metastatic lesions. The observation of a decreasing level of expression with increasing Gleason score was highly statistically significant.

The chromosomal domain of E-cadherin, 16q22, is frequently deleted in prostate tumors. Carter et al. (1990b) first described loss of 16q22–24 sequences in 5 of 16 informative tumors, and Bergerheim et al. (1991) observed loss of loci on distal 16q in 10 of 18 (56%) informative tumors examined. Sakr et al. (1994) described allelic loss of 16q22.1-23.1 in 5 of 12 (42%) primary and 1 of 12 (8%) metastatic tumors. This combination of functional and genetic studies suggest that E-cadherin may be an "invasion" suppressor gene important for prostate tumor progression.

Interestingly, other tumor suppressor loci may map distal to E-cadherin on 16q. For example, Phillips et al. (1994b) demonstrated loss of 16q24 sequences in 3 of 17 cases, and Cher et al. (1994) described loss of 16q24 in 15 of 30 (50%) touch preparations examined with cosmid probes. Visakorpi et al. (1995) found loss of 16cen-q23 in 19% of primary tumors and of 16q22-qter (perhaps inclusive of the E-cadherin locus) in 56% of recurrent tumors.

Therefore, inactivation of two or more tumor suppressor loci at 16q may be involved in prostate tumor progression.

Deletion of Chromosomal Regions and Chromosomes: Potential Tumor Suppressor Genes

Evidence that other tumor suppressor genes are inactivated in prostate cancer may be inferred from cytogenetic, molecular, and molecular cytogenetic findings of specific chromosomal regions and chromosomes frequently deleted in prostate tumors.

Numerical and structural alterations of chromosome 1 have been noted in several karyotypes of short-term prostate cultures. Atkin and Baker (1985) demonstrated deletion or rearrangement of chromosome 1 sequences in four of four prostate adenocarcinomas; Lundgren et al. (1988, 1992) described translocations involving 1p22, 1p32, and 1q24, and multiple breakpoints along 1p and 1q, in several tumors; and Arps et al. (1993) demonstrated deletion or rearrangement of chromosome 1 sequences in 3 of 32 (9%) cases. Molecular cytogenetic studies revealed deletion of chromosome 1 in 25%, and of 1p in 23%, of 32 tumors examined using archival sections (Baretton et al., 1992).

Molecular biological techniques have demonstrated loss of 3p sequences in 1 of 10 informative tumors (Carter et al., 1990b), and cytogenetic analysis has revealed der(4)t(4;?)(q27;?) in a short-term culture of primary prostate carcinoma (where "der" describes a derivative chromosome) (Brothman et al., 1990). Cytogenetic analysis demonstrated a translocation, t(5;7)(q14;q31), in a short-term culture of primary prostate carcinoma (Brothman et al., 1990), and molecular techniques have revealed loss of the 5q21 locus [within 1 centimorgan of the adenomatous polyposis coli (APC) gene locus] in three of seven cases (Phillips et al., 1994b) and loss of the 5q14-q23 chromosomal domain in 44% of nine recurrent tumors (Visakorpi et al., 1995). This last study also described loss of 6cen-q21 in 22% of 31 primary tumors, and loss of 6q13-q21 in 44% of nine recurrent tumors (Visakorpi et al., 1995).

Deletion of portions of the long arm of chromosome 7 have been frequently observed in prostate tumors. Atkin and Baker (1985) first described del(7)(q22) in three of four adenocarcinomas. Lundgren et al. (1988) observed the same structural alteration in a prostate tumor karyotype, as well as duplication or deletion of this region in 4 of 30 karyotypes of short-term prostate tumor cultures (Lundgren et al., 1988, 1992). Brothman et al. (1990) reported a t(5;7)(q14;q31) in a short-term primary prostate carcinoma culture and loss of chromosome 7 in 2 of 20 short-term cultures of primary prostate cancers (Brothman et al., 1992). Loss of chromosome 7 has been observed cytogenetically (5 of 62 cases; Micale et al., 1992). Molecular cytogenetic analyses have revealed loss of chromosome 7 in 8 of 35 archival sectioned specimens (Baretton et al., 1992), 15 of 25 (60%) patients with poor prognosis in disaggregated nuclei from archival specimens (Alcarez et al., 1994), and 6 of 41 sectioned archival or touch preparations of primary prostate tumors (Zitzelsberger et al., 1994). Using molecular biology techniques, Carter et al. (1990b) described loss of 7q sequences in 2 of 19 informative tumors. More recently, Zenklusen et al. (1994) demonstrated frequent (five of six cases) loss of heterozygosity at 7q31.1-q31.2 in prostate tumors. These studies indicate that one or more tumor suppressor genes important for prostate tumorigenesis map to 7q.

Great effort has been expended in recent years to physically map and isolate one or more potential tumor suppressor genes that localize to chromosomal region 8p in prostate tumors. Interestingly, cytogenetic studies provided very little evidence that tumor suppressor genes important for prostate tumorigenesis resided on 8p. Only one study, by Lundgren et al. (1988), karyotypically demonstrated deletion or rearrangment of 8p sequences, at 8p11, 8p21, and 8p22, in 5 of 57 cases. Other evidence was provided by molecular biology studies, first by Bergerheim et al. (1991), who observed loss of 8p sequences between 8pter and the neurofilament (NEFL) locus in 11 of 17 (65%) cases. Subsequently, Bova et al. (1993) described loss of sequences that map to 8p22 in 20 of 29 [69%; macrophage scavenger receptor (MSR) locus] and 15 of 32 [47%; lipoprotein lipase (LPL) locus], and of 8p21.2-21.3 in 16 of 27 (59%; D8S220 locus) informative tumors. Other studies have confirmed high frequencies of loss of 8p22 sequences (LPL locus) in prostate tumors: 3 of 11 (27%) primary tumors (Macoska et al., 1993), 4 of 14 (29%) primary and 3 of 12 (25%) metastatic tumors (Sakr et al., 1994), and 8 of 19 (42%) informative tumors (Chang et al., 1994). Utilizing FISH techniques, Macoska et al. (1994) described loss of 8p22 sequences in seven of nine (78%) frozen sections of prostate tumors hybridized with cosmid probe sequences; losses were widespread in five, and focal in two, cases. A candidate gene, PRLTS, was recently isolated from the 8p21.3-p22 region that may comprise the chromosome 8-specific tumor-suppressor gene most frequently deleted in prostate and other cancers (Fujiwara et al., 1995). Recent studies have described allelic loss of 8p22 and more proximal sequences. Trapman et al. (1994) described loss of 8p21-pter sequences in 42–69% of tumors and of 8p11-12 sequences in 36–39% of tumors, and Visakorpi et al. (1995) found 32% of 31 primary tumors demonstrating loss of 8p12-pter, and 78% of nine recurrent tumors demonstrating loss of 8p21-pter. A recent study described differential deletion of three distinct regions on 8p in prostate tumors (Macoska et al., 1995). Therefore, it appears that three tumor suppressor genes mapping to 8p22, 8p21 and 8p11-12 are inactivated in prostate tumor progression. An example of an experiment using PCR techniques to identify allelic loss on 8p is shown in Figure 1.

Data relating to the inactivation of potential tumor suppressor genes on chromosome 9 in prostate tumors are somewhat ambiguous. Using FISH techniques, Brown et al. (1994) detected loss of chromosome 9 in 1 of 40 touch preparations of primary prostate tumors. More specifically, Carter et al. (1990b) observed loss of 9q34 sequences in 2 of 13 informative

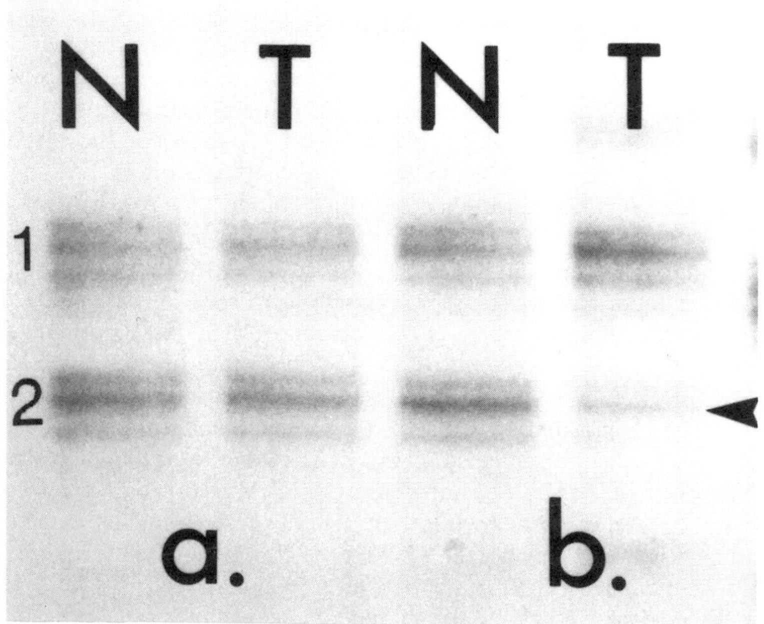

Figure 1. Polymerase chain reaction analysis of allelic loss at the LPL locus in human prostate cancer. DNA purified from microdissected areas of normal and tumor specimens was amplified with primers specific to dinucleotide repeat sequences at the LPL locus on 8p22, gel electrophoresed and autoradiographed as previously described (Macoska et al., 1994). **A:** Amplification products of normal (*N*) and tumor (*T*) DNA from the prostate transitional zone. Note that alleles 1 and 2 appear with the same intensity in both normal and tumor specimens, indicating retention of both alleles at the LPL locus. **B:** Amplification products of normal and tumor DNA from the prostate peripheral zone from the same patient. Note that alleles 1 and 2 appear with the same intensity in the normal specimen, but that allele 2 is greatly reduced in intensity compared to allele 1 in the tumor specimen (*arrow*). This reduction in signal intensity indicates loss of heterozygosity, or allelic loss, at the LPL locus in this tumor.

tumors, and Visakorpi et al. (1995) found 16% of 31 primary tumors with loss of 9p23-pter sequences. Taken together, these data suggest that sequences on both the long and short arms of chromosome 9 may harbor tumor-suppressor genes important for prostate tumorigenesis.

Another area of interest in the search for tumor-suppressor genes inactivated in prostate tumors concerns potential sites on chromosome 10. Early cytogenetic findings from Atkin and Baker (1985) described del(10)(q24) in four of four adenocarcinomas. These observations were substantiated by Lundgren et al. (1988, 1992), who demonstrated del(10)(q23), t(1;10)(p22;q24) and t(1;10)(q24;q22) in a short-term culture of a primary prostate carcinoma, and duplication or deletion of 10q23 or 10q24 in 5 of 30 karyotypes of short-term prostate tumor cultures. Also, Arps et al. (1993)

recurrently observed the del(10)(q24) chromosomal structural alteration in 3 of 32 cases examined. Molecular biological data have confirmed and expanded these cytogenetic findings. Carter et al. (1990b) first noted loss of 10q sequences, localized primarily to the 10q24-qter region, in 7 of 24 informative tumors. Bergerheim et al. (1991) observed loss of loci on chromosome 10 in 9 of 18 (50%) cases, including independent deletion of 10p or 10q sequences, as well as evidence for reduction to monosomy for chromosome 10. Macoska et al. (1993) noted loss of 10q11.2-qter sequences in 4 of 11 (35%) primary prostate tumors, and Sakr et al. (1994) described allelic loss of this same region in 3 of 16 (18%) primary and 2 of 17 (12%) metastatic tumors. Finally, Phillips et al. (1994b) demonstrated loss of 10pter-p13 sequences in 4 of 19 tumors, and 10q26 loss in 8 of 19 tumors; three cases appeared to represent reduction to monosomy for chromosome 10. Other evidence that chromosome 10 is sometimes reduced to monosomy in prostate tumors was provided by molecular cytogenetic studies using pericentromeric chromosome 10 probes. For example, using archival sections, Baretton et al. (1992) described chromosome 10 loss in 15 of 32 cases, and Macoska et al. (1993) demonstrated loss in 1 of 11 tumors. Using short-term cultures of primary prostate cancers, Brothman et al. (1992) described loss in 4 of 20 tumors. From these studies, it appears that a tumor-suppressor gene localized to 10q24 and perhaps another gene localized to 10p are frequently inactivated in prostate tumors.

Studies examining deletion of regions of chromosome 12 have found whole chromosome loss using FISH techniques in 11 of 20 (55%) sectioned archival specimens (Brothman et al., 1994) and in 2 of 40 touch preparations of primary prostate tumors (Brown et al., 1994). More specifically, Berube et al. (1994) have shown that tumor formation is suppressed in nude mice injected with DU145 cells containing a portion of chromosome 12, 12pter-12q13. Subsequent loss of the 12q13 fragment led to re-expression of the malignant phenotype; therefore, this portion of DNA may encode one or more tumor-suppressor genes. Further studies should reveal whether loss of heterozygosity for this region of chromosome 12 is evident in human prostate tumors.

Many observers have noted loss of the Y chromosome in prostate tumors, although the biological significance of this finding (e.g., whether a tumor suppressor maps to this chromosome), is debatable. Only one study has noted a structural alteration of the Y chromosome, a t(Y;22)(q11.2;p12) (Micale et al., 1992). Molecular cytogenetic studies have shown loss of the Y chromosome in 2 of 11 (18%) archival sectioned prostate tumor specimens (Macoska et al., 1993), and have specifically described loss in neoplastic but not stromal cells (Van Dekken and Alers, 1993). Cytogenetic studies have reported loss of the Y chromosome in 4 of 4 (Atkin and Baker, 1985); 6 of 30 (Lundgren et al., 1992); 5 of 62 (Micale et al., 1992), and 4 of 32 (Arps et al., 1993) short-term prostate tumor cultures. However, the cytogenetic finding of Y chromosome loss in 5 of 10 cases of benign prostatic hyperplasia (BPH) (Casalone et al., 1993) raises questions of whether this finding is an artifact of tissue culture, and whether it represents a significant event in prostate tumor progression.

Metastasis Suppressor Genes in Sporadic Prostate Cancer

Unlike tumor suppressor genes, metastasis suppressor genes are thought to normally repress tumor progression to the metastatic state. When these genes are inactivated (through mutation or deletion), the result is a cellular environment permissive to metastasis.

nm23 Family. The nm23 genes, nm23-H1 and nm23-H2, encoding subunits of the nucleoside diphosphate kinase enzyme, are down-regulated in metastatic cells, and localize to chromosomal region 17q21.3 (Steeg et al., 1988; Gilles et al., 1991; Leone et al., 1991). Analysis of 13 prostate tumors by RT-PCR failed to correlate expression of the nm23-H1 gene with tumor stage. However, expression of the nm23-H2 gene decreased significantly with increasing tumor stage and was lowest in the most "advanced" (e.g., metastatic) tumors (Fishman et al., 1994). This study indicates that nm23-H2 expression may be down-regulated in prostate metastases, and may therefore contribute to the metastatic phenotype.

Other Loci. Using microcell-mediated transfer of human chromosome 11 into AT6.2 cells (a highly metastatic rat prostatic cancer cell line) followed by injection of these cells into SCID mice, Ichikawa et al. (1992) demonstrated that the 11p11.2-p13 chromosomal region suppresses the metastatic ability of these cells (although no effect was noted on cell proliferation or tumorigenicity). Rinker-Schaeffer et al. (1994) then showed that this suppression is prostate cell–specific. A recent study reported the isolation of a gene, KAI1, that is capable of suppressing metastasis of rat AT6.1 prostate cancer cells, and is likely the metastasis-suppressor gene from the 11p11.2 chromosomal region (Dong et al., 1995). Using similar techniques, Ichikawa et al. (1994) demonstrated that human chromosome 8 sequences suppress the metastatic ability of AT6.1 cells, suggesting that chromosome 8 may harbor a metastasis-suppressor gene, and Rinker-Schaeffer et al. (1994) demonstrated that the 17pter-q23 chromosomal region suppresses the metastatic ability of AT6.1 cells. Furthermore, this effect is p53-independent and is not due to enhanced expression of nm23 protein. Taken together, these studies indicate that novel metastasis suppressor genes important in prostate tumor progression may map to chromosome 8 and to chromosomal regions 11p11.2 and 17pter-q23.

Tumor Suppressor Genes in Familial Prostate Cancer

There is evidence that some individuals—perhaps as many as 9% of all prostate cancer patients—inherit a predisposition to develop the disease.

These individuals are diagnosed at a relatively early age, for example, the mean age at onset of familial forms of prostate cancer is 59 years versus 74 years for those diagnosed with sporadic forms, and appear to inherit affected alleles in an autosomal dominant manner (Carter et al., 1992). If so, familial risk genes may adhere to the Knudson, or "two-hit," hypothesis of tumor suppressor gene inactivation. Recent work has suggested that a familial breast cancer tumor suppressor gene, BRCA1 (localized to chromosomal region 17q; see Chapter 7, this volume), may be affected in familial prostate cancers. One study demonstrated an increased risk for prostate cancer (9 of 13 cases) in men belonging to families with evidence of familial breast cancer (Arason et al., 1993), and another study estimates a relative risk for prostate cancer in men carrying defective BRCA1 alleles as three times the risk for noncarriers (Ford et al., 1994). Interestingly, an unrelated study reports loss of the 15q chromosomal region in a patient with eosinophilia and cancers of the lung, cerebral meningioma, prostate, and thyroid, belonging to a family with excessive lymphoproliferative neoplasms and prostate cancers (Goffman et al., 1983). Thus, it appears that multiple tumor suppressor loci may be affected in familial forms of prostate cancer.

ONCOGENES

The first oncogenes were identified as part of the genetic array encoded by DNA and RNA tumor viruses. Further studies showed that these oncogenes were, in fact, normal cellular genes inappropriately "activated" through their association with tumor virus activities (reviewed in Pitot, 1993). The definition of oncogene has since evolved to include normal cellular genes that are usually expressed at certain levels, during specific stages of development, or in specific tissues, but become activated, or inappropriately expressed, at higher levels, during other stages of development, or in other tissues. Oncogenes may become activated through various mechanisms, including base sequence mutations, dysregulation of transcriptional control, gene rearrangment, or gene amplification (reviewed in Suarez, 1989).

Known Oncogenes

ras Genes. The ras family of oncogenes (H-, K- and N-*ras*) encode G protein–like peptides active in signal transduction pathways. They are activated through point mutations at codons 12, 13, or 61 that appear to maintain the proteins in a perpetually dynamic, or activated, state (reviewed by Bos, 1989). Several studies have used model systems to investigate activated *ras* in prostate cancer. For example, infection of mouse urogenital sinus epithelial cells with retrovirus carrying activated viral H-*ras* resulted in acquisition of the dysplastic phenotype (Thompson et al., 1989). Infection of cultured nontumorigenic human fetal prostate cells with a murine sarcoma virus carrying activated K-*ras* resulted in acquisition of the malignant phenotype (Parda et al., 1993). Finally, transfection of a nonmetastatic rat prostate cancer cell line with activated viral H-*ras* resulted in acquisition of the metastatic phenotype (Ichikawa et al., 1991). These studies suggest that activated *ras* genes can induce phenotypic transformations in prostate cells. However, several studies agree that *ras* gene point mutations are rare in prostate cancer. These studies have described a single codon 61 mutation of the H-*ras* gene in 1 of 24 (Carter et al., 1990a) and 1 of 19 primary prostate tumors (Gumerlock et al., 1990), or no mutations in 24 tumors (Moul et al., 1992). Interestingly, there is some evidence that the frequency of *ras* mutations in prostate cancers may depend on ethnic heritage. *Ras* mutation rates of Japanese men living in Japan have been reported at 24% (16 of 68 cases), mostly involving mutation of H-*ras* codon 61 (Anwar et al., 1992). A comparison of *ras* mutation rates in latent prostate carcinomas of Japanese men living in Japan with caucasian, African-American, or Hawaiian Japanese men living in the United States, found the highest rates in native Japanese men; none of the mutations involved H-*ras* codon 61 (Watanabe et al., 1994). Taken together, these studies suggest that *ras* gene mutations are rare in clinical prostate carcinomas, although their occurrence may be prognostic of worse disease.

myc. The product of the *myc* oncogene is a DNA binding protein that appears to regulate both cell growth and apoptosis (recently reviewed by Schimke and Mihich, 1994). The *myc* oncogene can be amplified at the DNA level in human tumors, sometimes in the form of chromosomal homogeneously staining regions (HSRs) or double minute chromosomes (dmin) (Alitalo et al., 1983). Dmin identified and isolated from a short-term prostate primary tumor culture were found to contain sequences mapping to 8q24 that were homologous to *myc*; thus, this tumor contained amplified *myc* (Van Den Berg et al., 1995). An example of amplified DNA in the form of dmin chromosomes is shown in Figure 2. *myc* DNA is amplified and overexpressed in cultured human prostate cells (Nag and Smith, 1989). Overexpression of *myc* is associated with prostatic dysplasia in neonatally estrogenized mice (Pylkkanen et al., 1993), and infection of mouse urogenital sinus epithelial cells with retrovirus carrying activated *myc* resulted in acquisition of the hyperplastic phenotype (Thompson et al., 1989). These studies suggest that overexpression of *myc* may induce phenotypic responses in prostate cells, but that the role of *myc* overexpression in human prostate tumors remains largely undefined.

HER2/neu. The *HER2/neu* oncogene was originally isolated on the basis of its homology to the v-*erb*B retroviral oncogene. It was found to share sequence identity with the epidermal growth factor receptor (EGF-R) gene and the rat *neu* oncogene, and to possess a tyrosine kinase domain that could, theoretically, activate other proteins through phosphorylation (Semba et al., 1985).

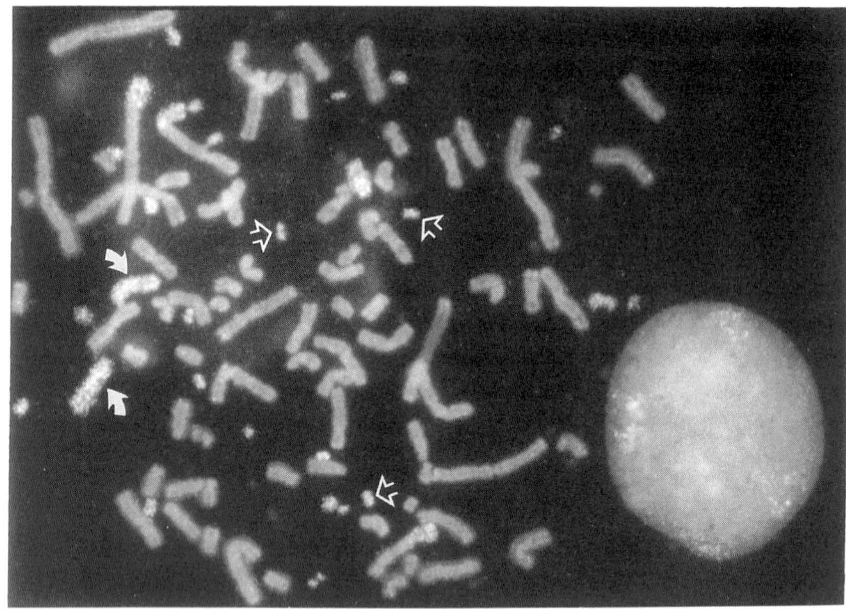

Figure 2. Fluorescence *in situ* hybridization analysis of cultured prostate tumor cells showing double minute chromosomes. PC3 human prostatic carcinoma cells (Kaighn et al., 1979) were propagated in RPMI 1640/10% fetal calf serum/1% penicillin-streptomycin, and prepared for metaphase analysis as previously described (Micale et al., 1992). Metaphase chromosomes were then hybridized with chromosome 8-specific paint probes (Oncor). *Solid white arrows* indicate two completely "painted" chromosomes (presumably chromosome 8); other partially painted chromosomes and marker chromosomes are also evident. *Hollow white arrows* indicate three of many "painted" double minute chromosomes (*Dmin*) visible in the photograph, apparent amplifications of chromosome 8 sequences.

In prostate cancer, transfection and subsequent overexpression of the rat *HER2/neu* gene in a nontumorigenic rat ventral prostate epithelial cultured cell line has been shown to convert the cells to a tumorigenic phenotype (Sikes and Chung, 1992). Overexpression of *HER2/neu* has been identified in a subset of higher grade human prostate tumors (Ross et al., 1993; Sadasivan et al., 1993). These few studies indicate that *HER2/neu* overexpression may be associated with phenotypically more aggressive prostate tumors.

Potential Oncogenic Loci: Duplicated/Amplified Chromosomal Regions or Chromosomes

If amplification at the DNA level is a means of oncogene activation, then the duplication of chromosomal domains or entire chromosomes may affect the activation of known or novel oncogenes. For example, gain of chromosome 1 has been observed in 30% of 23 primary tumors utilizing FISH techniques with disaggregated nuclei from archival tissue (Visakorpi et al., 1994). Conventional karyotypic analysis has demonstrated trisomy 3 in 3 of 32 (9%) and trisomy 7 in 4 of 32 (13%) short term prostate cultures (Arps et al., 1993). FISH analysis has also revealed gain of chromosome 7 in prostate tumors (Brothman et al., 1992; Macoska et al., 1993; Brown et al., 1994; Zitzelsberger et al., 1994), and this finding has been associated with poor prognosis (Alcarez et al., 1994; Bandyk et al., 1994). Finally, Visakorpi et al. (1995) has shown gain of the entire chromosome 7 or 7p13 domain in 56% of nine recurrent tumors by CGH, potentially defining 7p as the critical chromosomal domain.

Several studies have demonstrated gain of chromosome 8 or 8q in prostate tumors. FISH techniques have identified chromosome 8 gain in 4 of 20 short-term cultures of primary prostate cancers (Brothman et al., 1992); 7 of 13 sectioned archival tumor specimens (Macoska et al., 1993); 5 of 40 touch preparations of primary prostate tumors (Brown et al. 1994); and, in archival disaggregated nuclei from 48% of 23 primary tumors (Visakorpi et al., 1994). Molecular biological techniques have shown gain of 8q sequences in 5 of 32 (16%) informative tumors (Bova et al., 1993). More specific techniques, for example, cosmid FISH, have demonstrated gain of chromosome 8q concurrent with loss of 8p sequences in prostate tumors (Macoska et al., 1994). Finally, CGH analysis has revealed gain of the entire 8q arm in 89% of recurrent tumors, inclusive of myc sequences at 8q24; seven of these cases also demonstrated loss of 8p sequences (Visakorpi et al., 1995). Thus, it appears that gain of chromosome 8, especially concurrent with loss of 8p sequences, may identify tumors of worse prognosis, potentially because of the combined effect of tumor-suppressor deletion and oncogene activation.

Gain of chromosome 10 has been observed in several FISH studies. Brothman et al. (1992) demonstrated gain in 5 of 20 short-term cultures of primary prostate cancers. Macoska et al. (1993) observed gain in 1 of 11 sectioned archival prostate tumors; Brown et al. (1994) in 2 of 40 touch preparations of primary prostate tumors, and Visakorpi et al. (1994) in 30% of 23 primary tumors using disaggregated nuclei from archival tissue. Brown et al. (1994) reported gain of chromosome 11 in 1 of 40 touch preparations of primary prostate tumors, and gain of chromosome 16 has been demonstrated using FISH techniques by Brothman et al. (1992). Gain of chromosome 17 in 3 of 20 short-term cultures of primary prostate cancers was reported by Brothman et al. (1992), and was also observed by Brown et al. (1994) and Visakorpi et al. (1994). Brothman et al. (1992) and Brown et al. (1994) demonstrated gain of chromosome 18 (in 1 of 20 short-term cultures and 1 of 40 touch preparations of primary prostate tumors, respectively). Using molecular cytogenetic analysis, gain of the X chromosome has been demonstrated by Baretton et al. (1994) in 15 of 35 (44%) using archival sections; by Brothman et al. in 8 of 20 (40%) sectioned archival specimens, and by Visakorpi et al. (1994) in 43% of 23 primary tumors using disaggregated nuclei from archival tissue. CGH analysis has revealed gain of Xp11-q13 and Xq23-qter in 56% of nine recurrent tumors (Visakorpi et al., 1995). Finally, gain of the Y chromosome has been observed utilizing ISH and FISH techniques in 2 of 11 (18%) archival sectioned prostate tumor specimens (Macoska et al., 1993); in 10 of 35 (44%) tumor archival sections (Baretton et al., 1994); in 1 of 40 touch preparations of primary prostate tumors (Brown et al., 1994), and in 30% of 23 primary tumors using disaggregated nuclei from archival tissue (Visakorpi et al., 1994).

Although it is unclear at best if gain of entire chromosomes or chromosomal regions results in the activation of otherwise quiescent genes into oncogenes, further studies should reveal the biological consequences of increased chromosomal dosage in prostate tumorigenesis.

SUMMARY

Elucidation of the molecular genetic changes associated with prostate tumorigenesis has proven to be a formidable task. A primary obstacle has been the histologic heterogeneity of prostate cancer, where malignant glands are frequently intermixed with atrophic, normal, hyperplastic, or dysplastic glands, and intervening stromal tissue. Such heterogeneous tissue does not easily lend itself to molecular genetic analysis and has prompted the use of tissue microdissection and various *in situ* techniques to overcome issues of contamination of malignant epithelium with nonmalignant cells. Other obstacles include the paucity of spontaneously immortalized prostate cancer cell lines, lack of good animal models (although a newly developed transgenic mouse may prove useful; Greenberg et al., 1995), and difficulty identifying kindreds for use in linkage studies to identify genes involved in familial prostate cancer. Despite these problems, progress has been made toward identifying some molecular genetic events in prostate tumorigenesis. Deletion or mutation of the RB1, p53, and E-cadherin genes, and of currently unidentified genes from chromosomal regions 7q, 8p, and 10q (which likely contain tumor-suppressor genes), is probably permissive for tumorigenesis in the prostate. The development of prostate tumor metastases may depend on the inactivation of other genes localized to regions on chromosomes 8, 11 (KAI1), and 17. There is some evidence that activation of the *ras*-family, *myc*, and *HER2/neu* oncogenes contributes to prostate tumorigenesis, and that gain of chromosomes 7, 8, and the X are relatively common events associated with tumor recurrence. Taken together, these events comprise spontaneous genetic defects that contribute to malignant progression in the prostate. Further work is required to better define these, and to potentially identify other, genetic events relevant to prostate tumorigenesis.

REFERENCES

Alcarez A, Takahashi S, Brown JA, Herath JF, Bergstralh J, Larson-Keller JJ, Lieber MM, Jenkins RB (1994): Aneuploidy and aneusomy of chromosome 7 detected by fluorescence *in situ* hybridization are makers of poor prognosis in prostate cancer. *Cancer Res* 54:3998–4002

Alitalo K, Schwab M, Lin CC, Varmus HE, Bishop JM (1983): Homogeneously staining chromosomal regions contain amplified copies of an abundantly expressed cellular oncogene (c-*myc*) in malignant neuroendocrine cells from a human colon carcinoma. *Proc Natl Acad Sci USA* 80:1707–1711

Anwar K, Nakakuki K, Shiraishi T, Naiki H, Yatani R, Inuzuka M (1992): Presence of *ras* oncogene mutations and human papillomavirus DNA in human prostate carcinomas. *Cancer Res* 52:5991–5996

Arason A, Barkardottir R, Egilsson V (1993): Linkage analysis of chromosome 17q markers and breast-ovarian cancer in Icelandic families, and possible relationship to prostatic cancer. *Am J Hum Genet* 52:711–717

Arps S, Rodewald A, Schmalenberger B, Carl P, Bressel M, Kastendieck H (1993): Cytogenetic survey of 32 cancers of the prostate. *Cancer Genet Cytogenet* 66: 93–99

Atkin NB, Baker MC (1985): Chromosome study of five cancers of the prostate. *Hum Genet* 70:359–364

Bandyk JG, Zhoa L, Troncoso P, Pisters LL, Palmer JL, von Eschenbach AC, Chung LWK, Liang JC (1994): Trisomy 7: A potential cytogenetic marker of human prostate cancer progression. *Genes Chromosomes Cancer* 9:19–27

Banerjee A, Xu H-J, Hu S-H, Araujo D, Takahashi R, Stanbridge EJ, Benedict WF (1992): Changes in growth and tumorigenicity following reconstitution of retinoblastoma gene function in various human cancer cell types by microcell transfer of chromosome 13. *Cancer Res* 52:6297–6304

Baretton GB, Valina C, Vogt T, Schneiderbanger K, Diebold J, Lohrs U (1994): Interphase cytogenetic analysis of prostatic carcinomas by use of nonisotopic *in situ* hybridization. *Cancer Res* 54:4472–4480

Bergerheim USR, Kunimi K, Collins VP, Ekman P (1991): Deletion mapping of chromosomes 8, 10 and 16 in human prostatic carcinoma. *Genes Chromosomes Cancer* 3:215–220

Berube NG, Speevak MD, Chevrette M (1994): Suppression of tumorigenicity of human prostate cancer cells by introduction of human chromosome del(12)(q13). *Cancer Res* 54:3077–3081

Biegel JA, Burk CD, Barr FG, Emmanuel BS (1992): Evidence for a 17p tumor related locus distinct from p53 in neuroectodermal tumors. *Cancer Res* 52:3391–3395

Bookstein R, Rio P, Madreperla SA, Hong F, Allred C (1990a): Promoter deletion and loss of retinoblastoma gene expression in human prostate carcinoma. *Proc Natl Acad Sci USA* 87:7762–7766

Bookstein R, Shew J-Y, Chen P-L, Scully P, Lee W-H (1990b): Suppression of tumorigenicity of human prostate carcinoma cells by replacing a mutated RB gene. *Science* 247:712–715

Bookstein R, MacGrogan D, Hilsenbeck SG, Sharkey F, Allred DC (1993): p53 is mutated in a subset of advanced-stage prostate cancers. *Cancer Res* 53: 3369–3373

Bos JL (1989): *ras* oncogenes in human cancer: a review. *Cancer Res* 49:4682–4689

Bova SG, Carter BS, Bussemakers JG, Emi M, Fujiwara Y, Kyprianou N, Jacobs SC, Robinson JC, Epstein JI, Walsh PC, Isaacs WB (1993): Homozygous deletion and frequent allelic loss of chromosome 8p22 loci in human prostate cancer. *Cancer Res* 53:3869–3873

Brewster SF, Browne S, Brown KW (1994): Somatic allelic loss at the DCC, APC, nm23-H1 and p53 tumor suppressor gene loci in human prostatic carcinoma. *J Urol* 151:1073–1077

Brothman AR, Peehl DM, Patel AM, McNeal JE (1990): Frequency and pattern of karyotypic abnormalities in human prostate cancer. *Cancer Res* 50:3795–3803

Brothman AR, Patel AM, Peehl DM and Schellhammer PF (1992): Analysis of prostatic tumor cultures using fluorescence in situ hybridization (FISH). *Cancer Genet Cytogenet* 62:180–185

Brothman AR, Watson MJ, Zhu XL, Williams BJ, Rohr LR (1994): Evaluation of 20 archival prostate tumor specimens by fluorescence *in situ* hybridization (FISH). *Cancer Genet Cytogenet* 75:40–44

Brown JA, Alcaraz A, Takahashi S, Persons DL, Lieber MM, Jenkins RB (1994): Chromosomal aneusomies detected by fluorescent *in situ* hybridization analysis in clinically localized prostate carcinoma. *J Urol* 152:1157–1162

Bussemakers MJG, Van Moorselaar RJA, Giroldi LA, Ischikawa T, Isaacs JT, Takeichi M, Debruyne FMJ, Schalken JA (1992): Decreased expression of E-cadherin in the progression of rat prostatic cancer. *Cancer Res* 52:2916–2922

Carter BS, Epstein JI, Isaacs WB (1990a): *ras* gene mutations in human prostate cancer. *Cancer Res* 50:6830–6832

Carter BS, Ewing CM, Ward WS, Trieger BF, Aalders TW, Schalken JA, Epstein JI, Isaacs WB (1990b): Allelic loss of chromosomes 16q and 10q in human prostate cancer. *Proc Natl Acad Sci USA* 87:8751–8755

Carter BS, Beaty TH, Steinberg GD, Childs B, Walsh PC (1992): Mendelian inheritance of familial prostate cancer. *Proc Natl Acad Sci USA* 89:3367–3371

Casalone R, Portentoso P, Granata P, Minelli E., Righi R, Meroni E, Pozzi E, Chiaravalli AM (1992): Chromosome changes in benign prostatic hyperplasia and their significance in the origin of prostatic carcinoma. *Cancer Genet Cytogenet* 68: 126–130

Chang M, Tsuchiya K, Batchelor RH, Rabinovitch PS, Kulander BG, Haggitt RC Burmer GC (1994): Deletion mapping of chromosome 8p in colorectal carcinoma and dysplasia arising in ulcerative colitis, prostatic carcinoma and malignant fibrous histiocytomas. *Am J Pathol* 144: 1–6

Cher ML, Ito T, Weidner N, Carroll PR, Jensen RH (1995): Mapping of regions of physical deletion on chromosome 16q in prostate cancer cells by fluorescence *in situ* hybridization (FISH). *J Urol* 153:249–254

Coles C, Thompson AM, Elder PA, Cohen BB, MacKenzie IM, Crnaston G, Chetty U, MacKay J, MacDonald M, Nakamura Y, Hoyheim B, Steel CM (1990): Evidence implicating at least two genes on 17p in breast carcinogenesis. *Lancet* 336:761–763

Collins VP, Kunimi K, Bergerheim U, Ekman P (1990): Molecular genetics and human prostate carcinoma. *Acta Oncol* 30:181–185

Dinjens WNM, van der Weiden MM, Schroeder FH, Bosman FT, Trapman J (1994): Frequency and characterization of p53 mutations in primary and metastatic human prostate cancer. *Int J Cancer* 56:630–633

Dong J-T, Lamb PW, Rinker-Schaeffer CW, Vukanovic J, Ichikawa T, Isaacs JT, Barrett JC (1995): KAI1, a metastasis suppressor gene for prostate cancer on human chromosome 11p11.2. *Science* 268:884–886

Effert PJ, McCoy RH, Walther PJ, Liu ET (1993): p53 gene alterations in human prostate carcioma. *J Urol* 150:257–261

Fearon ER, Cho KR, Nigro JM, Kern SE, Simons JW, Ruppert JM, Hamilton SR, Preisinger AC, Thomas G, Kinzler KW, Vogelstein B (1990): Identification of a chromosome 18q gene that is altered in colorectal cancers. *Science* 247:49–56

Fishman JR, Gumerlock PH, Meyers FJ, deVere White RW (1994): Quantitation of nm23 expression in human prostate tissues. *J Urol* 152:202–207

Ford D, Eason DF, Bishop DT, Narod SA, Goldgar DE, Breast Cancer Linkage Consortium (1994): Risks of cancer in BRCA1-mutation carriers. *Lancet* 343: 692–695

Friend SH, Bernards R, Rogell S, Weinberg RA, Rapaport JM, Albert DM, Dryja TP (1986): A human DNA segment with properties of the gene that predisposes to retinoblastoma and osteosarcoma. *Nature* 323:643–646

Fujiwara Y, Ohata H, Kuroki T, Koyama K, Tsuchiya E, Monden M, Nakamura Y (1995) Isolation of a candidate tumor suppressor gene on chromosome 8p21.3-p22 that is homologous to an extracellular domain of the PDGF receptor beta gene. *Oncogene* 10:891–895

Gao X, Honn KV, Grignon D, Sakr W, Chen YQ (1993): Frequent loss of expression and loss of heterozygosity of the putative tumor suppressor gene DCC in prostatic carcinomas. *Cancer Res* 53:2723–2727

Gilles AM, Presecan E, Vonica A, Lascu I (1991): Nucleoside disphosphate kinase from human erythrocytes. *J Biol Chem* 266:8784–8789

Giroldi LA, Schalken JA (1993): Decreased expression of the intercellular adhesion molecule E-cadherin in prostate cancer: Biological significance and clinical implications. *Cancer Met Rev* 12:29–37

Gleason DF (1992) Histologic grading of prostate cancer: a perspective. *Hum Pathol* 23:273–279

Goffman TE, Mulvihill JJ, Carnery DN, Triche TJ, Whang-Peng J (1983): Fatal hypereosinophilia with chromosome 15q- in a patient with multiple primary and familial neoplasms. *Cancer Genet Cytogenet* 8:197–202

Greenberg NM, DeMayo F, Finegold MJ, Medina D, Tilley WD, Aspinall JO, Cunha GR, Donjacour AA, Matusik RJ, Rosen JM (1995): Prostate cancer in a transgenic mouse. *Proc Natl Acad Sci USA* 92:3439–3443

Greenblatt MS, Bennett WP, Hollstein M, Harris CC (1994): Mutations in the p53 tumor suppressor gene: Clues to cancer etiology and molecular pathogenesis. *Cancer Res* 54:4855–4878

Gumerlock PH, Poonamallee UR, Meyers FJ, deVere White RW (1991): Activated *ras* alleles in human carcinoma of the prostate are rare. *Cancer Res* 51:1632–1637

Ichikawa T, Schalken JA, Ichikawa Y, Steinberg GD, Isaacs JT (1991): H-*ras* expression, genetic instability, and acquisition of metastatic ability by rat prostatic cancer cells following v-Ha-*ras* oncogene transfection. *Prostate* 18:163–172

Ichikawa T, Ichikawa Y , Dong J, Hawkins AL, Griffin CA, Isaacs WB, Oshimura M, Barrett JC, Isaacs JT (1992): Localization of metastasis suppressor gene(s) for prostatic cancer to the short arm of human chromosome 11. *Cancer Res* 52: 3486–3490

Ichikawa T, Nihei N, Suzuki H, Oshimura M, Emi M, Nakamura Y, Hayata I, Isaacs JT, Shimazaki J (1994): Suppression of metastasis of rat prostatic cancer by introducing human chromosome 8. *Cancer Res* 54:2299–2302

Kaighn ME, Narayan KS, Ohnuki Y, Lechner JF, Jones LW (1979): Establishment and chracterization of a human prostatic carcinoma line (PC-3). *Invest Urol* 17:16–23

Knudson AG Jr. (1971): Mutation and cancer: statistical study of retinoblastoma. *Proc Natl Acad Sci USA* 68:820–823

Lalle P, De Latour M, Rio P, Bignon Y-J (1994): Detection of allelic losses on 17q12-q21 chromosomal region in benign lesions and malignant tumors occurring in a familial context. *Oncogene* 9:437–442

Leone A, McBride OW, Weston A, Wang MG, Anglard P, Cropp CS, Goepel JR, Lidereau R, Calahan R, Linehan WM, Rees RC, Harris CC, Liotta LA, Steeg PS (1991): Somatic allelic deletion of nm23 in human cancer. *Cancer Res* 51: 2490–2493

Lundgren R, Kristoffersson U, Heim S, Mandahl N, Mitelman F (1988): Multiple structural rearrangements, including del(7q) and del(10q), in an adenocarcinoma of the prostate. *Cancer Genet Cytogenet* 35:103–108

Lundgren R, Mandahl N, Heim S, Limon J, Henrikson H, Mitelman F (1992): Cytogenetic analysis of 57 primary prostatic adenocarcinomas. *Genes Chromosomes Cancer* 4:16–24

Macoska JA, Micale MA, Sakr WA, Benson PD, Wolman S R (1993): Extensive genetic alterations in prostate cancer revealed by dual PCR and FISH analysis. *Genes Chromosomes Cancer* 8:88–97

Macoska JA, Trybus TM, Sakr WA, Wolf MC, Benson PD, Powell IJ, Pontes JE (1994): Fluorescence *in situ* hybridization (FISH) analysis of 8p allelic loss and chromosome 8 instability in human prostate cancer. *Cancer Res* 54:3824–3830

Macoska JA, Trybus TM, Benson PD, Sakr WA, Grignon DJ, Wojno KD, Pietruk T, Powell IJ (1995): Evidence for three tumor suppressor gene loci on chromosome 8p in human prostate cancer. *Cancer Research* 55:5390–5395

Micale MA, Mohammed A, Sakr W, Powell IJ, Wolman Sr (1992): Cytogenetics of primary prostatic adenocarcinoma. *Cancer Genet Cytogenet* 61:165–173

Mihara K, Cao R, Yen A, Chandler S, Driscoll B, Murphree AL, T'Ang A, Fung Y-KT (1989): Cell cycle-dependant regulation of phosphorylation of the human retinoblastoma gene product. *Science* 246:1300–1303

Moul JW, Friedrichs PA, Lance RS, Theune SM, Chang EH (1992): Infrequent *ras* oncogene mutations in human prostate cancer. *Prostate* 20:327–338

Nag A, Smith RG (1989): Amplification, rearrangement and elevated expression of c-*myc* in the human prostatic carcinoma cell line LNCaP. *Prostate* 15:115–122

Navone NM, Troncoso P, Pisters LL, Goodrow TL, Palmer JL, Nichols WW, von Eschenbach AC, Conti CJ (1994): p53 protein accumulation and gene mutation in the progression of human prostate carcinoma. *J Natl Cancer Inst* 85:1657–1669

Parda DS, Thraves PJ, Kuettel MR, Lee M-S, Arnstein P, Kaighn ME, Rhim JS, Dritschilo A (1993): Neoplastic transformation of a human prostate epithelial cell line by the v-Ki-*ras* oncogene. *Prostate* 23:91–98

Persons DL, Gibney DJ, Katzmann JA, Lieber MM, Farrow GM, Jenkins RB (1993): Use of fluorescence *in situ* hybridization for deoxyribonucleic acid ploidy analysis of prostatic adenocarcinoma. *J Urol* 150:120–125

Persons DL, Takai K, Gibney DJ, Katzmann JA, Lieber MM, and Jenkins RB (1994): Comparison of fluorescence *in situ* hybridization with flow cytometry and static image analysis in ploidy analysis of paraffin-embedded prostate adenocarcinoma. *Hum Pathol* 25:678–683

Phillips SMA, Barton CM, Lee SJ, Morton DG, Wallace DMA, Lemoine NR, Neoptolemos JP (1994a): Loss of the retinoblastoma susceptibility gene (RB1) is a frequent and early event in prostatic tumorigenesis. *Br J Cancer* 70:1252–1257

Phillips SMA, Morton DG, Lee SJ, Wallace DMA, Neoptolemos JP (1994b): Loss of heteozygosity of the retinoblastoma and adenomatous polyposis susceptibility gene loci and in chromosomes 10p, 10q and 16q in human prostate cancer. *Br J Urol* 73:390–395

Pitot HC (1993): The molecular biology of carcinogenesis. *Cancer Suppl* 72:962–970

Pylkkanen L, Makela S, Valve E, Harkonen P, Toikkanen S, Santti R (1993): Prostatic dysplasia associated with increased expression of c-*myc* in neonatally estrogenized mice. *J Urol* 149:1593–1601

Rinker-Schaeffer CW, Hawkins AL, Ru N, Dong J, Stoica G, Griffin CA, Ichikawa T, Barrett JC, Isaacs JT (1994): Differential suppression of mammary and prostate cancer metastasis by human chromosomes 17, 11. *Cancer Res* 54:6249–6256

Ross JS, Nazeer T, Church K, Amato C, Figg H, Rifkin MD, Fisher HAG (1993): Contribution of *HER-2/neu* oncogene expression to tumor grade and DNA content analysis in the prediction of prostate carcinoma metastasis. *Cancer* 72: 3020–3028

Sadasivan R, Morgan R, Jennings S, Austenfeld M, Van Veldhuizen P, Stephens R, Noble M (1993): Overexpression of *HER-2/neu* may be an indicator of poor prognosis in prostate cancer. *J Urol* 150:126.131

Sakr WA, Macoska JA, Benson P, Grignon DJ, Wolman SR, Pontes JE, Crissman JD (1994): Allelic loss in locally metastatic, multisampled prostate cancer. *Cancer Res* 54:3273–3277

Sarkar FH, Sakr W, Li Y-W, Macoska J, Ball DE, Crissman JD (1992): Analysis of retinoblastoma (Rb) gene deletion in human prostatic carcinomas. *Prostate* 21: 145–152

Schimke RT, Mihich E (1994): Fifth annual pezcoller symposium: Apoptosis. *Cancer Res* 54:302–305

Semba K, Kamata N, Toyoshima K, Yamamoto T (1985): A v-erbB-related proto-oncogene, c-erbB-2, is distinct from the cerbB-1/epidermal growth factor receptor gene andis amplified in a human salivary gland adenocarcinoma. *Proc Natl Acad Sci USA* 82:6497–6501

Sikes RA, Chung LWK (1992): Acquisition of a tumorigenic phenotype by a rat ventral prostate epithelial cell line expressing a transfected activated *neu* oncogene. *Cancer Res* 52:3174–3181

Steeg PS, Bevilacqua G, Kopper L, Thorgeirsson UP, Talmadge J, Liotta LA, Sobel ME (1988): Evidence for a novel gene associated with low tumor metastatic potential. *J Natl Cancer Inst* 80:200–204

Stone KR, Mickey DD, Wunderli H, Mickey GH, Paulson DF (1978): Isolation of a human prostate carcinoma cell line (DU 145). *Int J Cancer* 21:274–281

Suarez HG (1989): Activated oncogene in human tumors. *Anticancer Res* 9:1331–1344

Thompson TC, Southgate J, Kitchener G, Land H (1989): Multistage carcinogenesis induced by *ras* and *myc* oncogenes in a reconstituted organ. *Cell* 56:917–930

Trapman J, Sleddens HFBM, van der Weiden MM, Dinjens WNM, Konig JJ, Schroder FH, Faber PW, Bosman FT (1994): Loss of heterozygosity of chromosome 8 microsatellite loci implicates a candidate tumor suppressor gene between the loci d8S87 and d8S133 in human prostate cancer. *Cancer Res* 54:6061–6064

Umbas R, Schalken JA, Aalders TW, Carter BS, Karthaus HFM, Schaafsma EW, Debruyne FMJ, Isaacs WB (1992): Expression of the cellular adhesion molecule E-cadherin in reduced or absent in high-grade prostate cancer. *Cancer Res* 52: 5104–5109

Van Dekken H, Alers J (1993): Loss of chromosome Y in prostatic cancer cells but not in stromal tissue. *Cancer Genet Cytogenet* 66:131–132

Van Den Berg C, Guan XY, Von Hoff D, Jenkins R, Bittner M, Griffin C, Kallioniemi O, Visakorpi T, McGill J, Herath J, Epstein J, Sarosdy M, Meltzer P, Trent J (1995): DNA sequence amplification in human prostate cancer identified by chromosome microdissection: Potential prognostic implications. *Clin Cancer Res* 1:11–18

Visakorpi T, Kallioniemi O-P, Heikkinen A, Koivula T, Isola J (1992): Small subgroup of aggressive, highly proliferative prostatic carcinomas defined by p53 accumulation. *J Natl Cancer Inst* 84:883–887

Visakorpi T, Hyytinen E, Kallioniemi A, Isola J, Kallioniemi O-P (1994): Sensitive detection of chromosome copy number aberratins in prostate cancer by fluorescence *in situ* hybridization. *Am J Pathol* 145:624–630

Visakorpi T, Kallioniemi AH, Syvanen A-C, Hyytinen ER, Karhu R, Tammela T, Isola JJ, Kallioniemi O-P (1995): Genetic changes in primary and recurrent prostate cancer by comparative genomic hybridization. *Cancer Res* 55:342–347

Vogelstein B, Kinzler KW (1992): p53 function and dysfunction. *Cell* 70:523–526

Watanabe M, Shiraishi T, Yatani R, Nomura AMY and Stemmermann GN (1994): International comparison on ras gene mutations in latent prostate carcinoma. *Int J Cancer* 58:174–178

Weinberg RA (1992): The retinoblastoma gene and gene product. *Cancer Surv* 12:43–57

Zenklusen JC, Thompson JC, Troncoso P, Kagan J, Conti CJ (1994): Loss of heterozygosity in human primary prostate carcinomas: a possible tumor suppressor gene at 7q31.1. *Cancer Res* 54:6370–6373

Ziztselsberger H, Szucs S, Weier H-U, Lehmann L, Braselmann H, Enders S, Schilling A, Breul J, Hofler H, Bauchinger M (1994): Numerical abnorrmalities of chromosome 7 in human prostate cancer detected by fluorescence *in situ* hybridization (FISH) on paraffin-embedded tissue sections with centromere-specific DNA probes. *J Path* 172:325–335

15

Prostatic Growth Factors, Cancer, and Steroid Hormone Activity

DONNA M. PEEHL

INTRODUCTION

Several reviews on the role of peptide growth factors in the prostate were recently published (Davies and Eaton, 1991; Habib and Chisholm, 1991; Story, 1991; Steiner, 1993). Those reviews primarily focused on several of the major and most extensively studied growth factor families, including the epidermal growth factor (EGF) family, the fibroblast growth factor (FGF) family, and the transforming growth factor-β (TGF-β) family. This chapter presents new information regarding EGF, FGF, and TGF-β, and other growth factors that have received less attention, with particular attention given to the interactions of steroid hormones and peptide growth factors.

CYTOKINES

Cytokines are sometimes defined as "any nonenzymatic mediator of immuno-inflammatory responses" (Harrison and Campbell, 1988). Epithelial cells are now recognized as producers of a number of cytokines, and the functions of these cytokines may extend beyond modulation of immune cell function to regulation of growth and differentiation of epithelial cells themselves (Stadnyk, 1994). The involvement of several cytokines in the prostate is described below.

Granulocyte Macrophage-Colony Stimulating Factor (GM-CSF)

This has shown some effectiveness as the cytokine of choice in experiments designed to demonstrate the feasibility of immunotherapy for prostate

Hormones and Cancer
Wayne V. Vedeckis, Editor

cancer (Sanda et al., 1994; Vieweg et al., 1994). This approach involves the introduction of a cytokine, usually via a viral vector, into cancer cells that have been removed from a patient. After irradiation, the cytokine-secreting cancer cells are then inoculated back into the patient and an immune response directed at the cytokine-producing cells, and other cancer cells with the same antigens, is triggered.

The effectiveness of GM-CSF in a tumor vaccine is not based on any role of GM-CSF in prostate cells themselves. Yet, GM-CSF may function in several ways in prostate cancer. Several prostate cancer cell lines secrete GM-CSF and are stimulated to grow by GM-CSF (Lang et al., 1994). Furthermore, GM-CSF is known to stimulate osteoblasts, so production of GM-CSF by prostate cancer cells could be involved in the production of the osteoblastic lesions frequently associated with prostatic metastases to bone. The role of GM-CSF and other cytokines in prostate biology thus deserves further study.

Interferon (IFN)

IFN therapy has been applied to prostate cancer with little success so far. This has been disappointing because *in vitro* studies and animal models indicated that interferon could exert antiproliferative effects on prostate cancer (Deshpande et al., 1989; Liu et al., 1991; Okutani et al., 1991). The observed lack of therapeutic efficacy may have multiple explanations, but relevant factors may be the apparent reversibility of IFN's antiproliferative effects and relative lack of potency.

Interleukin-6 (IL-6)

IL-6 is associated with processes involved in infection and inflammation. It has been reported that prostate cancer cell lines have IL-6 receptors and secrete IL-6 (Siegall et al., 1990; Siegsmund et al., 1994). Transcripts for IL-6 receptors have been found in benign prostatic hyperplasia (BPH) and prostate cancer tissues (Siegsmund et al., 1994), but expression was not localized to specific cells so the cell type expressing the receptor is unknown.

It has been suggested that the function of cytokines such as IL-6 that are secreted by epithelial cells may be to support growth and development of lymphocytes in the microenvironment (Stadnyk, 1994). This may occur in the prostate, but it is also possible that IL-6 may have autocrine growth stimulatory activity on prostate cells, as has been demonstrated for keratinocytes (Grossman et al., 1989). IL-6 present in bone could also activate IL-6 receptors on prostate cancer cells and promote the development of metastases. Regardless of the biological activity of IL-6 on prostate cells, Siegall et al. (1990) showed how the presence of IL-6 receptors on prostate cancer cells might be targeted for therapy. A recombinant chimeric protein of

IL-6–*Pseudomonas* exotoxin that bound to IL-6 receptors successfully killed several human prostate cancer cell lines. Further characterization of IL-6 and other interleukins in the prostate may provide unexpected avenues for gene or other therapies.

Tumor Necrosis Factor (TNF)

TNF has pleiotropic actions (Vilcek and Lee, 1991). The potential of TNF to exert cytotoxic activity on tumor cells has been of interest to prostate cancer researchers. TNF appears to inhibit the growth of most prostatic cancer cell lines and primary cultures of epithelial cells derived from normal or malignant tissues (Sherwood et al., 1988; Peehl et al., 1989; Liu et al., 1991). Cell lines derived from other types of tumors may be less sensitive to TNF than prostatic cancer cell lines, as indicated in a study by Spriggs et al. (1988) in which 11 of 14 human epithelial tumor cell lines were resistant to TNF. The sensitivity of normal prostatic epithelial cells to TNF is also in contrast to normal breast epithelial cells, which are not inhibited by TNF (Dollbaum et al., 1988). SV40-transformed prostatic epithelial cells, however, are not growth-inhibited by TNF (Lee et al., 1994), demonstrating that oncogenic events can indeed alter responsiveness to TNF.

EPIDERMAL GROWTH FACTOR

The status of our current knowledge of the EGF family and its receptor has recently been reviewed (Steiner, 1993). Therefore, this chapter briefly summarizes past work and focuses on updating that information; for more detailed references, please refer to Steiner's publication (1993).

Numerous studies indicate that EGF and its relative, transforming growth factor-α (TGF-α), are autocrine growth stimulatory factors in the prostatic epithelium. This conclusion has been drawn mainly from analyses of prostatic cancer cell lines (LNCaP, DU 145, or PC-3), which synthesize and secrete EGF and/or TGF-α, express the EGF receptor, and respond to EGF/TGF-α as mitogens (Wilding et al., 1989a; Connolly and Rose, 1991; Hofer et al., 1991; Ching et al., 1993). Of these lines, DU 145 cells have an exceptionally high level of expression of both TGF-α and the EGF receptor (MacDonald and Habib, 1992; Ching et al., 1993). The association of an autocrine EGF/EGF receptor growth stimulatory loop with prostate cancer cell lines has led some to propose that autocrine expression of EGF/TGF-α is intrinsic to the malignant phenotype. However, cultures of prostatic epithelial cells derived from normal tissues also express TGF-α and the EGF receptor, so this phenomenon is not specific to cancer cells (Nickas and Peehl, unpublished observations).

Are EGF/TGF-α and the EGF receptor expressed in the prostate? Several reports indicate that these proteins are both synthesized and present in the prostate. Jacobs and Story (1988) measured EGF in secretions of the rat prostate. The prostatic synthesis of EGF was indicated by the finding of RNA transcripts for EGF, TGF-α, and EGF receptor in a large number of prostatic tissue specimens (Ching et al., 1993). EGF, TGF-α, and EGF receptor have been immunologically detected in normal, BPH, and malignant tissues (Fowler et al., 1988; Harper et al., 1993; Yang et al., 1993).

Differential expression of EGF/TGF-α or the EGF receptor between benign and malignant prostatic tissues has been investigated. Some studies, such as that of Fowler et al. (1988), found more cancer than BPH specimens that were labeled by antibodies against EGF. Harper et al. (1993) also observed more prominent labeling of cancer than BPH tissues with antibody against TGF-α. In the latter study, the most intense labeling for TGF-α occurred in lesions associated with infection. That the increased expression of TGF-α might indeed promote a malignant phenotype was demonstrated by Bae et al. (1994), who found increased TGF-α expression in SV40-transformed prostatic epithelial cells selected for increasing tumorigenicity.

In normal or benign prostatic epithelia, EGF receptor expression is apparently localized to the basal epithelial cells (Maddy et al., 1987; Mayarden et al., 1992; Ibrahim et al., 1993). Because the receptor is expressed only by a subset of cells in normal or benign tissues, direct comparison of levels of EGF receptor between benign and malignant tissues is not straightforward. Perhaps for this reason, different investigations have concluded that the EGF receptor level is either higher, lower, or not changed between benign and cancer specimens (Davies and Eaton, 1989; Maddy et al., 1989; Morris and Dodd, 1990; Mayarden et al., 1992; Mellon et al., 1992; Ibrahim et al., 1993; Turkeri et al., 1994). As a whole, these studies do not suggest any prognostic value for EGF receptor levels.

Researchers have also examined c-*erb*B-2 (*neu*) expression in the prostate for prognostic value. *Neu* is a truncated form of the EGF receptor whose ligand is not EGF (Maihle and Kung, 1988). The number of studies is still quite limited, but all have found some degree of *neu* expression in a subset of prostate cancers (Mellon et al., 1992; Zhau et al., 1992; Kuhn et al., 1993; Sadasivan et al., 1993; Fox et al., 1994; Ross et al., 1994). However, some of those studies also found *neu* expression in normal or BPH tissues, and although the results of several studies found a trend toward *neu* expression in tumors of higher grade or stage, only one study found that *neu* expression was associated with a significantly worse prognosis (Fox et al., 1994). Perhaps further study of the mechanism of action of *neu* using *in vitro* models such as that of Sikes and Chung (1992), who created a tumorigenic cell line by transfecting *neu* into rat prostatic epithelial cells, will help clarify the role of *neu* in prostate carcinogenesis.

Neu expression may be constitutive in cells that activate this gene, but the expression of the EGF receptor and EGF/TGF-α is regulated by many factors, perhaps including androgen. Some investigators have found that androgen increases the expression of EGF/TGF-α in prostate cancer cell lines (Liu et al., 1993), although others did not (Connolly and Rose, 1990). Most studies have found that the EGF receptor, though, is increased by androgen in responsive cells (Schuurmans et al., 1988; Mulder et al., 1989; Liu et al., 1993). Conversely, EGF may decrease androgen receptor (AR) levels (Henttu and Vikho, 1993). Whether any of these observations reflect direct regulation of EGF/TGF-α or EGF receptor genes by androgen, or interaction of EGF with the AR, is not yet clear.

An interesting development concerning the latter possibility, direct interaction of EGF with the AR, has recently occurred. DU 145 cells were transfected with a vector containing the AR and with a construct containing androgen response elements linked to the CAT reporter gene (Culig et al., 1994). Peptide growth factors, among them EGF, induced CAT expression. The AR had to be present for this to occur, and induction by EGF was prevented by the AR antagonist casodex, proving direct interaction of the peptide growth factor with the AR. This finding was reminiscent of that of Power et al. (1991), who noted that estrogen and progesterone receptors were activated not only by their ligands, estrogen and progesterone, but by peptide growth factors and other factors as well. Thus, a new element of "cross-talk" between growth factors and steroid hormone receptors has been discovered that will dramatically alter our concepts of steroid hormone and peptide growth factor regulation of cell growth.

A final point to consider in this discussion of EGF and the prostate is that EGF/TGF-α are potent mitogens for prostate cells (Peehl et al., 1989; Nickas and Peehl, unpublished observations). If EGF/TGF-α are critical factors for proliferation of prostatic cancer cells, then inhibition of these factors might be an effective anticancer therapy. *In vitro* studies have demonstrated that the growth of prostatic cancer cell lines can be blocked by antibodies against the EGF receptor or EGF/TGF-α (Fong et al., 1992; Limonta et al., 1994). Therapeutic approaches to take advantage of the presence of EGF receptor on prostate cancer cells, or their dependence on EGF/TGF-α for proliferation, might include application of recombinant toxins using EGF to target exotoxins to EGF receptor-expressing cells (Pastan et al., 1992), or the use of antisense therapy. Rubenstein et al. (1994) demonstrated the potential of this latter approach by causing the complete remission of PC-3 tumors in host animals by treatment with TGF-α or EGF receptor antisense oligonucleotides. The efficacy of suramin, a drug with promising clinical activity against prostate cancer, may in part be due to blocking of growth factors, including EGF/TGF-α (Kim et al., 1991). The potential clinical value of growth factors remains to be fully exploited (Canalis, 1992).

FIBROBLAST GROWTH FACTOR

The FGF family has nine known members (Baird, 1993). Of these, basic FGF (FGF-2) and keratinocyte growth factor (KGF, or FGF-7) may be the most important in the prostate. Hypotheses regarding the functions of these factors and their involvement in BPH and cancer are becoming more developed as new information arises. The FGF receptor family is also complex and the pattern of receptor expression may be a critical determinant of malignant behavior (McKeehan, 1991).

Basic FGF

Basic FGF and the prostate have recently been reviewed (Story, 1991). In the normal prostate, basic FGF is apparently a stromal factor with potential growth stimulatory activity. Transcripts for basic FGF have been detected in normal, BPH, and malignant tissues (Jacobs et al., 1988; Mydlo et al., 1988). Cultured prostatic stromal cells derived from normal tissues synthesize basic FGF (Story et al., 1989). However, basic FGF lacks a signal peptide and is not actively secreted. Therefore, although prostatic stromal cells have receptors for basic FGF and respond to basic FGF as a mitogen (Story et al., 1989; Levine et al., 1992; Sherwood et al., 1992), an autocrine growth stimulatory role is not demonstrable. Recently, though, it has been proposed that basic FGF and other factors have "intercrine" activity, in that they may stimulate the cells of origin by an as-yet-unspecified mechanism not involving active secretion. Normal prostatic epithelial cells lack the basic FGF receptor and the minimal growth stimulation of these cells elicited by basic FGF is probably due to slight reactivity with other FGF receptors (Sherwood et al., 1992; Peehl et al., 1995b).

In various ways, basic FGF has been associated with both BPH and cancer. Some studies have suggested that expression of basic FGF is higher in BPH than in normal tissues (Mori et al., 1990). However, if basic FGF expression is confined to the stroma, then a higher stromal/epithelial ratio in BPH compared to normal tissues could explain this finding, and basic FGF expression on a per stromal cell basis may not be increased. In cancer, however, basic FGF expression may not be confined to the stroma. This possibility is raised because of the observation that some prostate cancer cell lines synthesize basic FGF (Mansson et al., 1989; Nakamoto et al., 1992; Yan et al., 1993). These cells also have basic FGF receptors and respond to basic FGF. Since expression of basic FGF and its receptor is not typical of normal prostatic epithelial cells, activation of an alternate FGF receptor gene and its ligand may be an event involved in progression towards malignancy (Yan et al., 1993).

KGF

KGF in general (Rubin et al., 1995) and in the prostate (Peehl and Rubin, 1995) has recently been reviewed. KGF has a unique pattern of stromal production and epithelial response and therefore has attracted much attention as a mediator of stromal-epithelial interactions.

Synthesis and secretion of KGF are restricted to mesenchymal cells, and KGF transcripts and peptide have been documented in a number of fibroblastic cells from embryonic, neonatal, and adult sources, including rat and human prostate. In addition to KGF expression in cultured prostatic stromal cells (Yan et al., 1992; Rubin et al., unpublished observations), KGF transcripts have been detected in normal, BPH, and malignant prostatic tissues (Lin et al., 1994; Rubin et al., unpublished observations).

KGF is not made by epithelial cells. However, epithelial cells from diverse tissues, including prostate, have receptors and respond to KGF in a variety of ways. Proliferation of prostatic epithelial cells is stimulated by KGF, and therefore KGF is a paracrine mediator of growth in the prostate (Yan et al., 1992; Peehl et al., 1995b). KGF also modifies the phenotype of prostatic epithelial cells in other ways (Peehl et al., 1995b). If normal prostate cells are grown in medium with KGF instead of an alternate mitogen, EGF, the colonies that form are composed of cells that are closely adherent to each other. In medium with EGF, the cells are migratory and intercellular adhesions are greatly reduced. In conjunction with these differing morphologies, cells grown in KGF are more sensitive to the growth-inhibitory and differentiation-inducing properties of retinoic acid (see the Retinoic Acid section in this chapter) than cells grown in EGF. Since KGF is postulated to be a paracrine factor in the prostate, whereas EGF may be an epithelial autocrine factor (see EGF section of this chapter), the state of the epithelium might depend on the source of mitogen at a given time. In situations where proliferation was driven by the epithelial factor, EGF, overall proliferation could be greater than if the proliferative stimulus were the stromal factor, KGF, because growth would be less limited by differentiation in the former situation.

An important aspect of KGF is that its synthesis by prostatic stromal cells is increased by androgen (Yan et al., 1992; Rubin et al., unpublished observations). There is much evidence that androgenic stimulation of prostatic epithelial growth is not mediated by direct androgen action on epithelial cells, but rather by indirect action on stromal cells. Androgen would mediate its activity by inducing expression of a growth factor by stromal cells that had specificity for epithelial cells. KGF's properties fit this description well and KGF is believed to be at least one of the factors mediating androgen action in the prostate. Strengthening this hypothesis are results from Cunha et al. (1994), who found that neutralizing antibody against KGF inhibited the growth and morphogenesis of newborn rat prostate in organ culture. Androgen, required for development of these cultures, was partially replaced by the addition of KGF.

As described for EGF (see the EGF section of this chapter), KGF itself may activate an androgen signaling mechanism in epithelial cells by stimulating AR activity in the absence of androgen ligand (Culig et al., 1994). Such interactions may be relevant to the processes occurring during the development of hormone-independent metastatic disease.

INSULIN-LIKE GROWTH FACTOR

Only recently has the insulin-like growth factor (IGF) system begun to be characterized in the prostate (for a recent review see Peehl et al., 1995a). Nevertheless, a substantial amount of information has already accumulated regarding the pattern of expression of the IGFs and their binding proteins and receptors in both prostatic stromal and epithelial cells.

Prostatic cancer cell lines synthesize and secrete IGF-I and possibly IGF-II (Matuo et al., 1988; Kaicer et al., 1991; Pietrzkowski et al., 1993); in contrast, synthesis of either IGF by normal prostatic epithelial cells has not been detected (Cohen et al., 1991). Prostatic stromal cells, on the other hand, make readily measurable quantities of IGF-II (Cohen et al., 1994a).

Primary cultures of prostatic epithelial cells (from normal, BPH, and malignant tissues), as well as prostatic cancer cell lines, have the type I-IGF receptor, and IGFs are potent mitogens for these cells (Cohen et al., 1991; Kaicer et al., 1991; Iwamura et al., 1993b; Pietrzkowski et al., 1993). Depending on the source of IGF in the prostate, IGF could act in either an autocrine, paracrine, or endocrine manner (IGF-I is the major form of circulating IGF). Prostatic stromal cells also express the type I-IGF receptor (Cohen et al., 1994a).

The mitogenic potential of IGF in the prostate is probably limited by IGF binding proteins (IGFBP). These proteins interact with the IGFs and usually inhibit the stimulatory activity of the growth factors, although enhancing effects have also been noted (Cohick and Clemmons, 1993). Prostatic epithelial and stromal cells make a number of IGFBPs, including IGFBP-2, -3, and -4 (Perkel et al., 1990; Cohen et al., 1991; Kaicer et al., 1993; Birnbaum et al., 1994; Cohen et al., 1994a). The biological effect of one of these IGFBPs, IGFBP-3, was tested in an *in vitro* system with human prostatic epithelial cells (Cohen et al., 1994b). In this experiment, no growth occurred in the absence of IGF. The addition of IGFBP-3 in the absence of IGF also had no effect. Proliferation was induced by the addition of IGF-I or -II, and this activity was blocked by IGFBP-3.

Since IGFs and IGFBP-3 are probably present in the prostate, the *in vitro* results indicate that IGF activity would be inhibited by IGFBP-3. But another element must be taken into consideration, and that is prostate-specific antigen (PSA). Proteases with activity against IGFBPs have recently been identified, and PSA is one of them (Cohen et al., 1992a). PSA cleaves IGFBP-3 into small fragments that have greatly reduced affinity for IGFs, effectively freeing the IGFs for mitogenic activity. In the cell culture experiment described in the preceding paragraph, inhibition of IGF by IGFBP-3 was relieved by addition of PSA, which cleaved IGFBP-3 and released

IGF. The significance of PSA's interaction with IGFBP-3 in the prostate can only be conjectured at this time. Since PSA is apparently confined to the lumens of epithelial acini in the normal prostate, and since those lumens are surrounded by differentiated cells with little if any proliferative potential, the release of IGF from IGFBP-3 by PSA may have little impact. In contrast, PSA is apparently leaked into the tissue surrounding tumors because the acinar architecture and polarized secretion of PSA is lost. If PSA is enzymatically active at such sites of primary or metastatic cancers, then it could cleave IGFBP-3 and release IGF for direct mitogenic activity on prostate cancer cells. In this manner, inappropriate release of a protease might indirectly stimulate growth.

In other ways, aberrations involving the IGF system may also be involved in BPH. Stromal cells cultured from BPH tissues have higher levels of IGF-II transcription and type I-IGF receptor expression than cells cultured from normal tissues (Cohen et al., 1994a). Furthermore, BPH-derived stromal cells have an altered pattern of IGFBP expression compared to normal stromal cells. The etiology of BPH is not known, but it has been postulated that the origin of BPH is a reversion of the stroma to its embryonic phenotype, with a resurgence of proliferative potential and ability to induce *de novo* epithelial glandular structures (McNeal, 1990). The findings of Cohen et al. (1994a) suggest that elements of the IGF system may be involved in the initiation and/or progression of BPH.

LUTEINIZING HORMONE-RELEASING HORMONE

Luteinizing hormone-releasing hormone (LHRH) analogs are frequently used for androgen ablation therapy of prostate cancer. Although the major mechanism of action of LHRH in eliciting androgen ablation is systemic, some suggest that LHRH may also be a direct inhibitor of prostate cancer growth. LHRH receptors have been reported in rat and human tumor cell lines and in prostate cancer tissues (Kadar et al., 1988; Fekete et al., 1989; Milovanovic et al., 1992). Furthermore, LHRH mRNA was detected in the LNCaP cell line and an LHRH antagonist stimulated growth of LNCaP cells, implying that the autocrine production of LHRH was inhibitory (Limonta et al., 1993). Whether LHRH production is a general feature of prostate cells and has an impact on prostate cancer growth needs further investigation.

NERVE GROWTH FACTOR

Nerve growth factor (NGF) has only recently begun to be investigated in the prostate. However, evidence for the presence of NGF and receptors and biological function is mounting. *In vitro* studies suggest that both prostatic stromal and epithelial cells may make NGF (Djakiew et al., 1991). However, examination of prostate tissue sections by immunohistochemistry indicates that the stroma is the major source of NGF in the prostate, whereas the epithelial cells have NGF receptors (Graham et al., 1992; Pflug et al., 1992). These findings indicate a paracrine role for NGF in the prostate, which might involve growth stimulation of epithelial cells by the stromal NGF as well as induction of chemotaxis (Djakiew et al., 1991; Djakiew et al., 1993).

NEUROENDOCRINE-ASSOCIATED FACTORS

Several years ago, cells with neuroendocrine-like features were recognized in the prostate. In normal and BPH tissues, these neuroendocrine cells are rare but, nevertheless, easily demonstrated by immunohistochemical labeling for products such as neuron-specific enolase, serotonin, and chromogranin (Abrahamsson et al., 1987; Aprikian et al., 1993; Cohen et al., 1993; Schmid et al., 1994). Neuroendocrine differentiation also occurs in prostatic adenocarcinomas, and some investigators have suggested that the extent of neuroendocrine differentiation has prognostic value (Di Sant'Agnese, 1992; Schmid et al., 1994).

The relationship of neuroendocrine and secretory epithelial cells in the prostate is not clear; it is possible that they have a common stem cell and represent end products of alternate differentiation pathways. Adlakha and Bostwick (1994) found neuroendocrine cells that expressed chromogranin and PSA, but Cohen et al. (1992b) reported that chromogranin-positive cells were negative for PSA. Similarly, the AR status of neuroendocrine cells is disputed. Nakada et al. (1993) found that most neuroendocrine cells, whether in benign or malignant tissues, expressed AR. Several other studies (Bonkhoff et al., 1993; Krijnen et al., 1993), in contrast, concluded that neuroendocrine cells were androgen receptor-negative, and that the increasing neuroendocrine component of progressive prostate cancer might provide a mechanism for the development of androgen independence.

Autocrine or paracrine effects of neuroendocrine products in the prostate are potentially significant. Calcitonin, parathyroid hormone-related protein, and vasoactive intestinal peptide are three factors associated with neuroendocrine activities that have received some attention with regard to their biological roles in the prostate.

Calcitonin

Calcitonin is a peptide hormone secreted by the thyroid. Related calcitonin-like peptides are found in other organs, including the prostate. Immunohistochemical studies indicated that the cells producing calcitonin-like peptides in the prostate are neuroendocrine cells (Shah et al., 1992). In the same study, calcitonin-like activity was also measured by radioimmunoassay in the conditioned media of primary cell cultures derived from BPH or cancer. The addition of calcitonin to cultures of prostatic cancer cell lines stimulated growth and increased cyclic adenosine monophosphate (cAMP) (Shah et al., 1994). Calcitonin and related peptides are present in large concentrations in the semen and are believed to regulate sperm activity, but it is possible that these peptides also have autocrine or paracrine functions in the prostate.

Parathyroid Hormone-Related Protein (PTHrP)

PTHrP expression by tumor cells is considered to be the primary factor responsible for humoral hypercalcemia of malignancy. However, recent identification of PTHrP production in a wide variety of normal cell types, expression of PTHrP in diverse tissues during development, and the presence of PTHrP receptors in many tissues suggest that PTHrP has a role in normal growth and differentiation (Gillespie and Martin, 1994; Muff et al., 1994; Orloff et al., 1994).

PTHrP has been detected in prostatic tissues, prostatic cancer cell lines, and primary cultures of prostatic epithelial cells (Kramer et al., 1991; Iwamura et al., 1993a,b; Cramer et al., 1995). Immunohistochemical detection of PTHrP by Iwamura et al. (1993a,b) localized expression to cancer cells and to neuroendocrine cells of normal and BPH tissues. However, similar studies by Kramer et al. (1991) and Cramer et al. (1995) found PTHrP immunoreactivity throughout the glandular epithelia of normal and BPH tissues, not solely in neuroendocrine cells. Expression of PTHrP by primary cultures of prostatic epithelial cells, which generally have features of epithelial rather than neuroendocrine cells, also indicates that PTHrP expression is not confined to neuroendocrine cells (Cramer et al., 1995).

Regulation of PTHrP expression by prostatic epithelial cells has only begun to be investigated, but one study (Cramer et al., 1995) has identified EGF as a major inducer of prostatic synthesis and secretion of PTHrP. The significance of PTHrP expression by prostatic cells also remains to be determined. Iwamura et al. (1994a) indicated that PTHrP might have autocrine growth activity in prostatic cancer cell lines. Another possibility to consider would be paracrine activities of PTHrP secreted by prostatic cancer cells on metastatic growth in bone. Other studies link PTHrP expression with the ability of cancer cells to metastasize to bone and with enhancement of osteolytic processes (Bouizar et al., 1993).

Vasoactive Intestinal Peptide (VIP)

VIP, also referred to interchangeably as bombesin, its homolog, is a product of neuroendocrine cells. Effects of VIP on rat and human prostatic epithelial cells include stimulation of cAMP and growth (Carmena et al., 1987; Bologna et al., 1989). That VIP might function as an autocrine growth factor is indicated by reports that the growth of prostatic cancer cell lines was inhibited by antibodies against VIP (Bologna et al., 1989) or by bombesin antagonists (Milovanovic et al., 1992). Efforts are in progress to characterize bombesin receptors on prostate cells and to determine whether bombesin antagonists could have therapeutic value against prostatic cancer (Reile et al., 1994).

PLATELET-DERIVED GROWTH FACTOR

Platelet-derived growth factor (PDGF) is a 30-kDa, dimeric protein composed of two polypeptide subunits (A and B) held together by disulfide bonds (Heldin, 1992). The A subunit is the product of the PDGF-1 gene, whereas the B subunit is transcribed from the PDGF-2 gene. The former gene is commonly expressed in normal mesenchymal tissues, but the latter is an oncogene and is also known as c-*sis*. PDGF acts through a membrane receptor that has tyrosine kinase activity.

Activation of the c-*sis* gene is associated with certain malignancies, particularly osteosarcomas. A role for c-*sis* in prostate cancer has not been obvious, although a few publications have suggested some involvement. Using antipeptide antibodies against the c-*sis* gene product, Niman et al. (1985) observed that PDGF was elevated in the urine of prostate cancer patients. Although this elevation was not specific for prostate cancer and occurred in patients with certain other types of malignancies as well, other oncogene products (*ras* or *fes*) were not increased in prostate cancer.

Several human and rat prostatic cancer cell lines express PDGF-1 and/or PDGF-2 (Smith and Nag, 1987; Sitaras et al., 1988). However, at least DU 145 and PC-3 apparently do not express the PDGF receptor, so a role for PDGF as an autocrine growth factor in prostatic cancer cells cannot be substantiated (Sitaras et al., 1988; Okutani et al., 1991). A more likely possibility would be activity of PDGF as a paracrine factor. Many mesenchymal cells have PDGF receptors, and cultured prostatic stromal cells reportedly have PDGF receptors and respond to PDGF's mitogenic activity (Gleason et al., 1993; Vlahos et al., 1993). Apart from a report of Smith and Nag (1987) describing c-*sis* transcripts in BPH tissue, PDGF expression in prostatic tissues has not been carefully examined and a role for PDGF as a growth factor in the prostate is still undefined.

PROLACTIN

Several studies have suggested that prolactin, a peptide hormone, augments androgen-mediated growth of the prostate (Arunakaran et al., 1987; Prins, 1987; Kadar et al., 1988; Schacht et al., 1992). The question is whether this effect of prolactin is a direct one, mediated by prolactin receptors on prostate cells, or indirect, perhaps via action on the adrenal glands.

There is some evidence suggesting direct interaction of prolactin in the prostate. Fekete et al. (1989) reported that 7 of 13 BPH tissue specimens specifically bound labeled prolactin. *In vitro* studies using BPH explants indicated that prolactin regulated growth, although both stimulatory and inhibitory effects were seen depending on the concentration of prolactin and whether testosterone or dihydrotestosterone was provided (Syms et al., 1985; de Launoit et al., 1988). Growth regulation of rat prostatic tissues in organ culture by prolactin has also been described (Nevalainen et al., 1991).

On the other hand, growth regulation by prolactin of isolated prostatic epithelial cells grown in defined, serum-free medium has been difficult to document. Although Webber and Perez-Ripoll (1986) reported that prolactin stimulated the growth of DU 145 cells, McKeehan et al. (1984) found only a modest, if any, stimulatory effect of prolactin on rat prostatic cells, and we have not observed any effect of prolactin on primary cultures of human prostatic cells (unpublished observations). Thus, the effect of prolactin on prostatic cells remains uncertain.

RETINOIC ACID

Vitamin A is an important morphogen and regulator of growth and differentiation of epithelia (Cunliffe and Miller, 1984). Receptors for the active metabolite of vitamin A, retinoic acid, were first cloned in 1987 by Giguere et al., and since then the field of study of vitamin A has exploded (de Luca, 1991). Retinoic acid receptors are discussed in Chapters 4 and 5 of this volume. Accordingly, investigations of the role of vitamin A in the prostate have increased, and fundamental information is being gathered regarding the pattern of retinoic acid receptor (RAR) expression in prostatic cells and the effects of retinoic acid on growth and differentiation. Studies completed so far point toward the possible use of vitamin A analogs as therapeutic agents for prostate cancer.

The importance of physiological levels of vitamin A in the maintenance of normal prostatic structure and function was demonstrated many years ago. In 1952, Bern showed that vitamin A deficiency in rats led to squamous metaplasia of the prostate. In 1955, Lasnitzki placed explants of mouse prostatic tissues in culture and treated them with the carcinogen methylcholanthrene. Hyperplasia and squamous metaplasia induced by the carcinogen were prevented by the addition of vitamin A to the culture medium. Recently, in 1993, Lohnes et al. used the latest technology to create mice that were mutant for all RARγ isoforms. RARγ is the main type of RAR in the mouse prostate, and in the absence of this receptor, prostates exhibited squamous metaplasia. All these studies clearly demonstrate that vitamin A is essential for the maintenance of the normal, differentiated, secretory phenotype of the prostate.

In vitro studies with cultured prostatic epithelial cells have validated the whole animal studies of vitamin A's activities. At physiological levels, retinoic acid added to serum-free medium inhibited the growth of human prostatic epithelial cells cultured from normal, BPH, or malignant tissues (Peehl et al., 1989, 1993). Growth inhibition was irreversible and, in conjunction with growth inhibition, the expression of differentiation-associated keratins was increased (Peehl et al., 1993, 1994a). If retinoic acid was added to cultures of prostatic epithelial cells maintained at confluency for long periods, the abnormal squamous phenotype that develops under these conditions was reversed (Peehl et al., 1993). Other investigators also found that the proliferation of prostatic cancer cell lines was inhibited by retinoic acid or other forms of vitamin A, and that various aspects of differentiation were induced (Reese et al., 1983; Fong et al., 1993; Igawa et al., 1994; Lee et al., 1994). In short, the effect of physiological concentrations of vitamin A on prostatic epithelial cells *in vitro* is just as *in vivo*—proliferation and the abnormal development of squamous metaplasia are prevented while the normal differentiated phenotype is promoted.

At nonphysiological levels, vitamin A has been observed to stimulate growth of prostatic epithelial cells (Chaproniere and Webber, 1985; Peehl et al., 1993). The significance of this is not clear, but similar observations have been made with other types of cells (Tong et al., 1986; Varani et al., 1990). The explanation may be that different concentrations of retinoic acid differentially regulate transcription of different genes (Simeone et al., 1990).

Current research focuses on characterizing the types of RAR expressed by prostatic cells and determining the mechanism of action of retinoic acid. RNA transcripts for the receptor RARα cloned by Giguere et al. in 1987 were found in human prostatic tissue, but at very low levels. RARβ transcripts, on the other hand, are present at high levels in human and rat prostate (Benbrook et al., 1988; de The et al., 1989). The human prostatic cancer cell line, LNCaP, expresses RARα, -β, and -γ (Dahiya et al., 1994). Binding proteins for retinoic acid, distinct from the RARs, also exist, and earlier reports of retinoic acid binding proteins in prostatic tissues (Boyd et al., 1985; Jutley et al., 1987) must be re-evaluated in light of our current concepts. The relationship of RARs and binding proteins has recently been reviewed by Giguere (1994).

Because retinoic acid is a potent regulator of proliferation and differentiation in prostatic epithelial cells, a role for vitamin A as a preventative agent against the development of prostate cancer has been sought. Epidemiologic evidence for or against vitamin A as a protective agent in the prostate has been reviewed by Willett and Hunter (1994), who concluded that, all in all, vitamin A does not seem to protect against prostate cancer.

Nevertheless, vitamin A analogs may be useful chemotherapeutic agents. Although analogs with less toxicity than those currently available need to be developed (see Chapter 5, this volume), preliminary studies using various experimental models of prostate cancer look promising. 13-*cis* retinoic acid inhibited tumor growth of PC-3 (Stearns et al., 1993) and LNCaP (Dahiya et al., 1994) human prostatic cell lines in host animals. Slawin et al. (1993) showed that fenretinide, a synthetic retinoid, decreased the tumor incidence and mass of oncogene-induced carcinomas in a mouse prostate model. This latter result is especially relevant because fenretinide is currently in phase I/II trials for prostate cancer.

Since RARs belong to the same extended family of steroid hormone receptors as androgen receptors, interactions of retinoic acid and androgen in prostate cells may be a possibility. Antagonism of androgen action in LNCaP cells by retinoic acid has been reported (Young et al., 1994), and this may be due to inhibition of 5α-reductase activity by retinoic acid (Halgunset et al., 1987; Jutley et al., 1990). Further interactions of retinoic acid and androgen remain to be explored.

TRANSFORMING GROWTH FACTOR-β

TGF-β is a member of a large superfamily of structurally related factors with multifunctional activities (Kingsley, 1994). Because of its diverse properties and several isoforms, defining the role of TGF-β in prostate biology has not been straightforward. Several recent reviews have described the expression of TGF-β in the prostate and its effects on the phenotypes of epithelial and stromal cells (McKeehan, 1991; Story, 1991; Wilding, 1991; Steiner, 1993).

TGF-β is a potent growth inhibitor of many types of epithelial cells, including normal, BPH, and malignant epithelial cells cultured from the prostate (Peehl et al., 1989). However, certain cancer cell lines are unresponsive to TGF-β, or are only transiently inhibited (Wilding et al., 1989b). Ironically, in such cell lines TGF-β is often expressed at higher than normal levels (Ware and Watts, 1992; Steiner et al., 1994). This high level of TGF-β secretion may actually promote tumor formation, as demonstrated by increased tumorigenicity of a cell line induced to produce high levels of TGF-β by DNA transfection (Steiner and Barrack, 1992). In these situations, it would certainly appear that cancer cells escape inhibition by TGF-β and that TGF-β perhaps promotes tumorigenicity by modulating angiogenesis, the immune system, or the extracellular matrix.

The response of prostatic stromal cells to TGF-β is not typical of most other types of mesenchymal cells. Whereas mesenchymal cells are generally stimulated by TGF-β, prostatic stromal cells, like normal prostatic epithelial cells, are inhibited by TGF-β (Story et al., 1993). The mechanism or biological significance of this inhibitory effect has not been clarified.

Immunohistochemical studies of TGF-β expression in the prostate have presented a puzzling picture. In the developing prostate of the fetal mouse, the highest level of TGF-β was observed in the mesenchyme in areas of active epithelial duct formation (Timme et al., 1994). In a study of BPH and cancer tissues, Glynne-Jones et al. (1994) observed TGF-β only in BPH, and only in areas of infection. In contrast, Eklov et al. (1993) saw TGF-β in both benign and malignant prostate tissues, and in both stromal and epithelial cells. In another study, extracellular TGF-β was more extensive in focal areas of cancer compared to BPH (Truong et al., 1993). In BPH, intracellular TGF-β was more prominent in the stroma than in the epithelium, whereas in malignant tissue, TGF-β was most often present in the cancer cells (Truong et al., 1993). Undoubtedly, the conflicting results seen in these various studies may be attributed to the use of different antibodies and whether the TGF-β that was detected was intracellular or extracellular.

Some studies have pointed to a role for TGF-β in programmed cell death (apoptosis). Upon castration of rats, TGF-β and its receptor were increased in prostatic tissues undergoing apoptosis, and androgen replacement in turn decreased TGF-β binding to prostatic tissues (Kyprianou and Isaacs, 1988). If TGF-β was administered to noncastrated animals, $\sim$25% of the glandular prostatic epithelial cells were killed (Martikainen et al., 1990). Yet, the administration of TGF-β along with androgen replacement did not prevent the prostatic proliferation induced by androgen (Martikainen et al., 1990), and Steiner et al. (1994) did not find TGF-β levels changed in androgen-responsive rat prostatic tumor lines after castration. TGF-β also did not block androgen stimulation of prostatic cells *in vitro* (Martikainen et al., 1990). Therefore, the participation of TGF-β in apoptosis is not clear, and certainly complex interactions are involved in cell death. Other factors can modulate TGF-β action, as demonstrated by McKeehan and Adams (1988), who found that basic FGF could attenuate TGF-β's growth-inhibitory functions. In addition, Sutkowski et al. (1992) observed that TGF-β induced death rather than growth inhibition of cultured prostatic cells, but only in growth factor-deficient medium.

VITAMIN D

Epidemiologic studies suggest that vitamin D deficiency may be a risk factor for clinical prostate cancer (Schwartz and Hulka, 1990; Hanchette and Schwartz, 1992; Corder et al., 1993). The active metabolite of vitamin D, 1,25-dihydroxyvitamin D_3 $[1,25(OH)_3D_3]$, exerts its actions via a specific cellular vitamin D receptor (VDR) belonging to the steroid receptor gene family (Norman et al., 1991; see Chapter 4, this volume). VDR has been detected in many tissues and cells, suggesting a broad physiological role for vitamin D. Recent reports demonstrate that prostatic cells have VDR and that proliferation and differentiation of prostatic cells is regulated by $1,25(OH)_2D_3$. These findings provide a biological basis for vitamin D's role as a preventative agent against prostate cancer.

An immunohistochemical study by Berger et al. in 1988 suggested the presence of VDR in normal human prostatic tissue. Subsequent investigations focused on established prostate cancer cell lines, or on primary cell cultures derived from either the epithelium or stroma of the prostate. Miller et al. (1992) first reported VDR in the LNCaP cell line. This finding was confirmed by Skowronski et al. (1993), who also found VDR in PC-3 and DU 145 cell lines. Peehl et al. (1994c) used radioligand binding methods to demonstrate the presence of VDR in fresh prostatic tissues removed at surgery. Primary cultures of stromal or epithelial cells derived from normal, BPH, or malignant tissues also specifically bound $[^3H]1,25(OH)_2D_3$. Binding among different strains of epithelial cells varied from 13–79 fmol/mg of protein, but there was no apparent association between extent of binding and histology of origin. Stromal cell strains generally exhibited lower binding (6–19 fmol/mg of protein) than epithelial cells. The presence of VDR in cultured epithelial and stromal cells was also confirmed by analysis of RNA transcripts and immunocytochemistry (Peehl et al., 1994b,c).

When prostatic cells were exposed to physiological levels of $1,25(OH)_2D_3$, growth of either established cell lines or primary cultures was generally inhibited. Of the cell lines, LNCaP was most sensitive to inhibition by $1,25(OH)_2D_3$, with half-maximal growth inhibition occurring with $\sim$1 nM of $1,25(OH)_2D_3$ in medium containing 10% serum (Skowronski et al., 1993). Growth repression of PC-3 cells in serum-supplemented medium was somewhat less than that of LNCaP cells (Skowronski et al., 1993), but in serum-free medium, half-maximal growth inhibition of PC-3 cells occurred at $<$ 1 nM (Peehl et al., 1994b). In contrast, DU 145 cells were not significantly inhibited by $1,25(OH)_2D_3$ even at concentrations as high as 100 nM, despite the confirmed presence of VDR in these cells (Skowronski et al., 1993). SV40-transformed prostatic epithelial cells were also resistant to vitamin D (Lee et al., 1994), whereas cells transformed by human papillomavirus retained the ability of the parental cells to be inhibited by $1,25(OH)_2D_3$ (Peehl et al., 1995b).

Primary cultures of epithelial and stromal cells were also inhibited by $1,25(OH)_2D_3$ (Peehl et al., 1994c). Stromal cells required higher levels of $1,25(OH)_2D_3$ for maximal growth inhibition than epithelial cells, perhaps because of the lower level of VDR on stromal cells. An important aspect of the inhibitory effect of $1,25(OH)_2D_3$ on epithelial cells is that it is irreversible (Peehl et al., 1994c). Even limited exposure of only a few hours to $1,25(OH)_2D_3$ prevented the ability of the cells to resume growth after removal of the hormone. This finding may be relevant to endeavors to use vitamin D analogs for prostate cancer therapy (discussed below).

Only one situation has been reported in which vitamin D seems to exert a stimulatory, rather than inhibitory, effect on the growth of prostatic cells. When LNCaP cells were grown in charcoal-stripped rather than whole serum, $1,25(OH)_2D_3$ was slightly stimulatory (Miller et al., 1992; Skowronski et al., 1993). No explanation for this phenomenon has been found.

In conjunction with growth inhibition, vitamin D may induce differentiation of prostatic epithelial cells. PSA is a differentiation product of prostate cells, and PSA secretion by LNCaP cells was increased by exposure to growth-inhibitory concentrations of $1{,}25(OH)_2D_3$ (Miller et al., 1992; Skowronski et al., 1993).

The antiproliferative and differentiating actions of vitamin D on prostatic cells suggest that vitamin D treatment could be a therapeutic option for prostate cancer. Recently, synthetic analogs of vitamin D that retain antiproliferative activity but exhibit less calcemic activity *in vivo* have been developed. Skowronski et al. (1995) demonstrated that some of these analogs exhibit strong antiproliferative effects against human prostatic cancer cell lines, raising the possibility of achieving therapeutically useful activity *in vivo* with less risk of hypercalcemia.

PERSPECTIVE

There is still much to be learned about growth factors in the prostate. Undoubtedly new factors remain to be identified, and the temporal and spatial localization of growth factors in the normal and diseased prostate still requires a great deal of effort. *In vitro* studies of cultured cells, in which cellular components and interactions can be simplified and defined, will continue to provide the basis for our understanding of growth factor functions. New tools, including antibodies and *in situ* hybridization, will improve our abilities to correlate growth factor expression with malignant processes. Novel therapeutic strategies may be designed from our increased knowledge of growth factors in the prostate and from new ideas on how to block stimulatory factors or take advantage of inhibitory ones. The mechanism of steroid hormone action in the prostate remains murky, but newly discovered interactions between hormones, peptide growth factors, and their receptors provide a conceptual framework for delineating the myriad autocrine, paracrine, and endocrine pathways that govern growth and differentiation.

REFERENCES

Abrahamsson P-A, Alumets J, Wadström LB, Falkmer S, Grimelius L (1987): Peptide hormones, serotonin, and other cell differentiation markers in benign hyperplasia and in carcinoma of the prostate. *Prog Clin Biol Res* 243A:489–502

Adlakha H, Bostwick DG (1994): Paneth cell-like change in prostatic adenocarcinoma represents neuroendocrine differentiation: report of 30 cases. *Hum Pathol* 25:135–139

Aprikian AG, Cordon-Cardo C, Fair WR, Reuter VE (1993): Characterization of neuroendocrine differentiation in human benign prostate and prostatic adenocarcinoma. *Cancer* 71:3952–3965

Arunakaran J, Aruldhas MM, Govindarajulu P (1987): Effect of prolactin and androgens on the prostate of Bonnet monkeys, *Macaca radiata*: 1. Nucleic acids, phosphatases, and citric acid. *Prostate* 10:265–273

Bae VL, Jackson-Cook CK, Brothman AR, Maygarden SJ, Ware JL (1994): Tumorigenicity of SV40 T antigen immortalized human prostate epithelial cells: association with decreased epidermal growth factor (EGFR) expression. *Int J Cancer* 58:721–729

Baird A (1993): Fibroblast growth factors: what's in a name? *Endocrinology* 132:487–488

Benbrook D, Lernhardt E, Pfahl M (1988): A new retinoic acid receptor identified from a hepatocellular carcinoma. *Nature* 333:669–672

Berger U, Wilson P, McClelland RA, Colston K, Haussler MR, Pike JW, Coombes RC (1988) Immunocytochemical detection of 1,25-dihydroxyvitamin D receptors in normal human tissues. *J Clin Endocrin Metab* 67:607–613

Bern HA (1952): Alkaline phosphatase activity in epithelial metaplasia. *Cancer Res* 12:85–91

Birnbaum RS, Ware JL, Plymate SR (1994): Insulin-like growth factor-binding protein-3 expression and secretion by cultures of human prostate epithelial cells and stromal fibroblasts. *J Endocrin* 141:535–540

Bologna M, Festuccia C, Muzi P, Biordi L, Ciomei M (1989): Bombesin stimulates growth of human prostatic cancer cells in vitro. *Cancer* 63:1714–1720

Bonkhoff H, Stein U, Remberger K (1993): Androgen receptor status in Endocrinole-paracrine cell types of the normal, hyperplastic, and neoplastic human prostate. *Virch Arch A Pathol Anat* 423:291–294

Bouizar Z, Spyratos F, Deytieux S, de Vernejoul MC, Jullienne A (1993): Polymerase chain reaction analysis of parathyroid hormone-related protein gene expression in breast cancer patients and occurrence of bone metastases. *Cancer Res* 53: 5076–5078

Boyd D, Copestake P, Chisholm GD, Habib FK (1985): A comparison of retinol binding in human hyperplastic and malignant prostate. *Br J Cancer* 51:903–905

Canalis E (1992): Growth factors and their potential clinical value. *J Clin Endocrin Metab* 75:1–4

Carmena MJ, Sancho JI, Fernández-Gonzalez MA, Escudero F, Prieto JC (1987) Somatostatin inhibits VIP- and isoproterenol-stimulated cyclic AMP accumulation in rat prostatic epithelial cells. *FEBS Lett* 218:73–76

Chaproniere DM, Webber MM (1985): Dexamethasone and retinyl acetate similarly inhibit and stimulate EGF- or insulin-induced proliferation of prostatic epithelium. *J Cell Physiol* 122:249–253

Ching KZ, Ramsey E, Pettigrew N, D'Cunha R, Jason M, Dodd JG (1993): Expression of mRNA for epidermal growth factor, transforming growth factor-alpha and their receptor in human prostate tissue and cell lines. *Mol Cell Biochem* 126:151–158

Cohen P, Peehl DM, Lamson G, Rosenfeld RG (1991): Insulin-like growth factors (IGFs), IGF receptors and IGF binding proteins in primary cultures of prostate epithelial cells. *J Clin Endocrin Metab* 73:401–407

Cohen P, Graves HCB, Peehl DM, Kamarei M, Giudice LC, Rosenfeld RG (1992a): Prostate-specific antigen (PSA) is an insulin-like growth factor binding protein-3 protease found in seminal plasma. *J Clin Endocrin Metab* 75:1046–1053

Cohen RJ, Glezerson G, Haffejee Z (1992b): Prostate-specific antigen and prostate-specific acid phosphatase in neuroendocrine cells of prostate cancer. *Arch Pathol Lab Med* 116:65–66

Cohen RJ, Glezerson G, Taylor LF, Grundle HAJ, Naude JH (1993): The neuroendocrine cell population of the human prostate gland. *J Urol* 150:365–368

Cohen P, Peehl DM, Baker B, Liu F, Hintz RL, Rosenfeld RG (1994a): Insulin-like growth factor axis abnormalities in prostatic stromal cells from patients with benign prostatic hyperplasia. *J Clin Endocrin Metab* 79:1410–1415

Cohen P, Peehl DM, Graves HCB, Rosenfeld RG (1994b): Biological effects of prostate specific antigen as an insulin-like growth factor binding protein-3 protease. *J Endocrin* 142:407–415

Cohick WS, Clemmons DR (1993): The insulin-like growth factors. *Annu Rev Physiol* 55:131–153

Connolly JM, Rose DP (1990): Production of epidermal growth factor and transforming growth factor -α by the androgen-responsive LNCaP human prostate cancer cell line. *Prostate* 16:209–218

Connolly JM, Rose DP (1991): Autocrine regulation of DU 145 human prostate cancer cell growth by epidermal growth factor-related polypeptides. *Prostate* 19:173–180

Corder EH, Guess HA, Hulka BS, Friedman GD, Sadler M, Vollmer RT, Lobaugh B, Drezner MK, Vogelman JH, Orentreich N (1993) Vitamin D and prostate cancer: a prediagnostic study with stored sera. *Cancer Epidemiol Biomarkers Prev* 2:467–472

Cramer SD, Peehl DM, Edgar MG, Wong ST, Deftos LJ, Feldman D (1995): Parathyroid hormone-related protein (PTHrP) is an epidermal growth factor-regulated secretory product of human prostatic epithelial cells. *Prostate* (in press)

Culig Z, Hobisch A, Cronauer MV, Radmayr C, Trapman J, Hittmair A, Bartsch G, Klocker H (1994): Androgen receptor activation in prostatic tumor cell lines by insulin-like growth factor-I, keratinocyte growth factor, and epidermal growth factor. *Cancer Res* 54:5474–5478

Cunha GR, Foster B, Donjacour A, Rubin S, Sugimura Y, Finch PW, Brody J, Aaronson SA (1994): Keratinocyte growth factor as mediator of mesenchymal-epithelial interactions in the development of androgen target organs. In: *Sex Hormones and Antihormones in Endrocrine Dependent Pathology: Basic and Clinical Aspects,* Motta M, Serio M, eds. Amsterdam: Elsevier Science BV

Cunliffe WJ, Miller AJ (1984): *Retinoid Therapy.* Boston: MT Press, Ltd

Dahiya R, Park HD, Cusick J, Vessella RL, Fournier G, Narayan P (1994): Inhibition of tumorigenic potential and prostate-specific antigen expression in LNCaP human prostate cancer cell line by 13-cis-retinoic acid. *Int J Cancer* 59:126–132

Davies P, Eaton CL (1989): Binding of epidermal growth factor by human normal, hypertrophic, and carcinomatous prostate. *Prostate* 14:123–132

Davies P, Eaton CL (1991): Regulation of prostate growth. *J Endocrin* 131:5–17

de Launoit Y, Kiss R, Jossa V, Coibion M, Paridaens RJ, De Backer E, Danguy AJ, Pasteels J-L (1988) Influences of dihydrotestosterone, testosterone, estradiol, progesterone, or prolactin on the cell kinetics of human hyperplastic prostatic tissue in organ culture. *Prostate* 13:143–153

de Luca LM (1991): Retinoids and their receptors in differentiation, embryogenesis, and neoplasia. *FASEB J* 5:2924–2933 de The H, Marchio A, Tiollais P, Dejean A (1989): Differential expression and ligand regulation of the retinoic acid receptor α and β genes. *Embo J* 8:429–433

Deshpande N, Hallowes RC, Cox S, Mitchell I, Hayward S, Towler JM (1989): Divergent effects of interferons on the growth of human benign prostatic hyperplasia cells in primary culture. *J Urol* 141:157–160

Di Sant'Agnese PA (1992): Neuroendocrine differentiation in human prostatic carcinoma. *Hum Pathol* 23:287–296

Djakiew D, Delsite R, Pflug B, Wrathall J, Lynch JH, Onoda M (1991) Regulation of growth by a nerve growth factor-like protein which modulates paracrine interactions between a neoplastic epithelial cell line and stromal cells of the human prostate. *Cancer Res* 51:3304–3310

Djakiew D, Pflug BR, Delsite R, Onoda M, Lynch JH, Arand G, Thompson EW (1993): Chemotaxis and chemokinesis of human prostate tumor cell lines in response to human prostate stromal cell secretory proteins containing a nerve growth factor-like protein. *Cancer Res* 53:1416–1420

Dollbaum C, Creasey AA, Dairkee SH, Hiller AJ, Rudolph AR, Lin L, Vitt C, Smith HS (1988) Specificity of tumor necrosis factor toxicity for human mammary carcinomas relative to normal mammary epithelium and correlation with response to doxorubicin. *Proc Natl Acad Sci USA* 85:4740–4744

Eklov S, Funa K, Nordgren H, Olofsson A, Kanzaki T, Miyazono K, Nilsson S (1993): Lack of the latent transforming growth factor β binding protein in malignant but not benign prostatic tissue. *Cancer Res* 53:3193–3197

Fekete M, Redding TW, Comaru-Schally AM, Pontes JE, Connelly RW, Srkalovic G, Schally AV (1989): Receptors for luteinizing hormone-releasing hormone, somatostatin, prolactin, and epidermal growth factor in rat and human prostate cancers and in benign prostate hyperplasia. *Prostate* 14:191–208

Fong C-J, Sherwood ER, Mendelsohn J, Lee C, Kozlowski JM (1992): Epidermal growth factor receptor monoclonal antibody inhibits constitutive receptor phosphorylation, reduces autonomous growth, and sensitizes androgen-independent prostatic carcinoma cells to tumor necrosis factor α. *Cancer Res* 52:5887–5892

Fong C-J, Sutkowski DM, Braun EJ, Bauer KD, Sherwood ER, Lee C, Kozlowski JM (1993): Effect of retinoic acid on the proliferation and secretory activity of androgen-responsive prostatic carcinoma cell. *J Urol* 149:1190–1194

Fowler JE Jr, Lau JLT, Ghosh L, Mills SE, Mounzer A (1988): Epidermal growth factor and prostatic carcinoma: an immunohistochemical study. *J Urol* 139: 857–861

Fox SB, Persad RA, Coleman N, Day CA, Silcocks PB, Collins CC (1994): Prognostic value of c-erbB-2 and epidermal growth factor receptor in stage A1 (T1a) prostatic adenocarcinoma. *Br J Urol* 74:214–220

Giguere V (1994): Retinoic acid receptors and cellular retinoid binding proteins: complex interplay in retinoid signaling. *Endocr Rev* 15:61–79

Giguere V, Ong ES, Segui P, Evans RM (1987): Identification of a receptor for the morphogen retinoic acid. *Nature* 330:624–629

Gillespie MT, Martin TJ (1994): The parathyroid hormone-related protein gene and its expression. *Mol Cell Endocrinol* 100:143–147

Gleason PE, Jones JA, Regan JS, Salvas DB, Eble JN, Lamph WW, Vlahos CJ, Huang W-L, Falcone JF, Hirsch KS (1993): Platelet derived growth factor (PDGF), androgens and inflammation: possible etiologic factors in the development of prostatic hyperplasia. *J Urol* 149:1586–1592

Glynne-Jones E, Harper ME, Goddard L, Eaton CL, Matthews PN, Griffiths K (1994): Transforming growth factor beta 1 expression in benign and malignant prostatic tumors. *Prostate* 25:210–218

Graham CW, Lynch JH, Djakiew D (1992): Distribution of nerve growth factor-like protein and nerve growth factor receptor in human benign prostatic hyperplasia and prostatic adenocarcinoma. *J Urol* 147:1444–1447

Grossman RM, Krueger J, Yourish D, Granelli-Piperno A, Murphy DP, May LT, Kupper TS, Sehgal PB, Gottlieb AB (1989): Interleukin 6 is expressed in high levels in psoriatic skin and stimulates proliferation of cultured human keratinocytes. *Proc Natl Acad Sci USA* 86:6367–6371

Habib FK, Chisholm GD (1991): The role of growth factors in the human prostate. *Scand J Urol Nephrol Suppl* 126:53–58

Halgunset J, Sunde A, Lundmo PI (1987): Retinoic acid (RA): an inhibitor of 5α-reductase in human prostatic cancer cells. *J Steroid Biochem* 28:731–736

Hanchette CL, Schwartz GG (1992): Geographic patterns of prostate cancer mortality. *Cancer* 70:2861–2869

Harper ME, Goddard L, Glynne-Jones E, Wilson DW, Price-Thomas M, Peeling WB, Griffiths K (1993): An immunocytochemical analysis of TGFα expression in benign and malignant prostatic tumors. *Prostate* 23:9–23

Harrison LC, Campbell IL (1988): Cytokines: an expanding network of immuno-inflammatory hormones. *Mol Endocrinol* 2:1151–1156

Heldin C-H (1992): Structural and functional studies on platelet-derived growth factor. *EMBO J* 11:4251–4259

Henttu P, Vihko P (1993): Growth factor regulation of gene expression in the human prostatic carcinoma cell line LNCaP. *Cancer Res* 53:1051–1058

Hofer DR, Sherwood ER, Bromberg WD, Mendelsohn J, Lee C, Kozlowski JM (1991): Autonomous growth of androgen-independent human prostatic carcinoma cells: role of transforming growth factor α. *Cancer Res* 51:2780–2785

Ibrahim GK, Kerns BM, MacDonald JA, Ibrahim SN, Kinney RB, Humphrey PA, Robertson CN (1993): Differential immunoreactivity of epidermal growth factor receptor in benign, dysplastic and malignant prostatic tissues. *J Urol* 149:170–173

Igawa M, Tanabe T, Chodak GW, Rukstalis DB (1994): N-(4-hydroxyphenyl) retinamide induces cell cycle specific growth inhibition in PC3 cells. *Prostate* 24:299–305

Iwamura M, di Sant'Agnese PA, Wu G, Benning CM, Cockett ATK, Deftos LJ, Abrahamsson P-A (1993a): Immunohistochemical localization of parathyroid hormone-related protein in human prostate cancer. *Cancer Res* 53:1724–1726

Iwamura M, Sluss PM, Casamento JB, Cockett ATK (1993b): Insulin-like growth factor I: action and receptor characterization in human prostate cancer cell lines. *Prostate* 22:243–252

Iwamura M, Abrahamsson P-A, Foss KA, Wu G, Cockett ATK, Deftos LJ (1994a): Parathyroid hormone-related protein: a potential autocrine growth regulator in human prostate cancer cell lines. *Urology* 43:675–679

Iwamura M, Wu G, Abrahamsson P-A, di Sant'Agnese PA, Cockett ATK, Deftos LJ (1994b): Parathyroid hormone-related protein is expressed by prostatic neuroendocrine cells. *Urology* 43:667–674

Jacobs SC, Story MT (1988): Exocrine secretion of epidermal growth factor by the rat prostate: effect of adrenergic agents, cholinergic agents, and vasoactive intestinal peptide. *Prostate* 13:79–87

Jacobs SC, Story MT, Sasse J, Lawson RK (1988): Characterization of growth factors derived from the rat ventral prostate. *J Urol* 139:1106–1110

Jutley JW, Kelleher J, Whelan P, Mikel J (1987): Cytosolic retinoic acid-binding protein in human prostatic dysplasia and neoplasia. *Prostate* 11:127–132

Jutley JK, Reaney S, Kelleher J, Whelan P (1990): Interactions of retinoic acid and androgens in human prostatic tissue. *Prostate* 16:299–304

Kadar T, Redding TW, Ben-David M, Schally AV (1988): Receptors for prolactin, somatostatin, and luteinizing hormone-releasing hormone in experimental prostate cancer after treatment with analogs of luteinizing hormone-releasing hormone and somatostatin. *Proc Natl Acad Sci USA* 85:890–894

Kaicer EK, Blat C, Harel L (1991): IGF-I and IGF-binding proteins: stimulatory and inhibitory factors secreted by human prostatic adenocarcinoma cells. *Growth Factors* 4:231–237

Kaicer EK, Blat C, Imbenotte J, Troalen F, Cussenot O, Calvo F, Harel L (1993): IGF binding protein-3 secreted by the prostate adenocarcinoma cells (PC-3): differential effect on PC-3 and normal prostate cell growth. *Growth Reg* 3:180–189

Kim JH, Sherwood ER, Sutkowski DM, Lee C, Kozlowski JM (1991): Inhibition of prostate tumor cell proliferation by suramin: alterations in TGF alpha-mediated autocrine growth regulation and cell cycle distribution. *J Urol* 146:171–176

Kingsley DM (1994): The TGF-β superfamily: new members, new receptors, and new genetic tests of function in different organisms. *Genes Dev* 8:133–146

Kramer S, Reynolds FH Jr, Castillo M, Valenzuela DM, Thorikay M, Sorvillo JM (1991): Immunological identification and distribution of parathyroid hormone-like protein polypeptides in normal and malignant tissues. *Endocrinology* 128: 1927–1937

Krijnen JLM, Janssen PJA, Riuzveld de Winter JA, van Krimpen H, Schroder FH, van der Kwast TH (1993): Do neuroendocrine cells in human prostate cancer express androgen receptor? *Histochemistry* 100:393–398

Kuhn EJ, Kurnot RA, Sesterhenn IA, Chang EH, Moul JW (1993): Expression of the c-erbB-2 (HER-2/neu) oncoprotein in human prostatic carcinoma. *J Urol* 150: 1427–1433

Kyprianou N, Isaacs JT (1988): Identification of a cellular receptor for transforming growth factor-β in rat ventral prostate and its negative regulation by androgens. *Endocrinology* 123:2124–2131

Lang SH, Miller WR, Duncan W, Habib FK (1994): Production and response of human prostate cancer cell lines to granulocyte macrophage-colony stimulating factor. *Int J Cancer* 59:235–241

Lasnitzki I (1955): The influence of a hypervitaminosis on the effect of 20-methylcholanthrene on mouse prostate glands grown *in vitro*. *Br J Cancer* 9:434–441

Lee M-S, Garkovenko E, Yun JS, Weijerman PC, Peehl DM, Chen L-S, Rhim JS (1994): Characterization of adult human prostatic epithelial cells immortalized by polybrene-induced DNA transfection with a plasmid containing an origin-defective SV40 genome. *Int J Oncol* 4:821–830

Levine AC, Ren M, Huber GK, Kirschenbaum A (1992): The effect of androgen, estrogen, and growth factors on the proliferation of cultured fibroblasts derived from human fetal and adult prostates. *Endocrinology* 130:2413–2419

Limonta P, Dondi D, Moretti RM, Fermo D, Garattini E, Motta M (1993): Expression of luteinizing hormone-releasing hormone mRNA in the human prostatic cancer cell line LNCaP. *J Clin Endocrin Metab* 76:797–800

Limonta P, Moretti RM, Dondi D, Marelli MM, Motta M (1994): Androgen-dependent prostatic tumors: biosynthesis and possible actions of LHRH. *J Steroid Biochem Mol Biol* 49:347–350

Lin YC, Canatan H, Chang CJG, Hu YF, Chen R, Yu CY, Brueggemeier RW, Somers WJ (1994): Detection of keratinocyte growth factor (KGF) transcripts from normal human and archival canine benign prostatic hyperplastic tissues. *J Med* 25:41–64

Liu S, Ewing MW, Anglard P, Trahan E, LaRocca RV, Myers CE, Linehan WM (1991): Effect of suramin, tumor necrosis factor and interferon gamma on human prostate carcinoma. *J Urol* 145:389–392

Liu X-H, Wiley HS, Meikle AW (1993): Androgens regulate proliferation of human prostate cancer cells in culture by increasing transforming growth factor-α (TGF-α) and epidermal growth factor (EGF)/TGF-α receptor. *J Clin Endocrin Metab* 77:1472–1478

Lohnes D, Kastner P, Dierich A, Mark M, LeMeur M, Chambon P (1993): Function of retinoic acid receptor c in the mouse. *Cell* 73:643–658

MacDonald A, Habib FK (1992): Divergent responses to epidermal growth factor in hormone sensitive and insensitive human prostate cancer cell lines. *Br J Cancer* 65:177–182

Maddy SQ, Chisholm GD, Hawkins RA, Habib FK (1987): Localization of epidermal growth factor receptors in the human prostate by biochemical and immunocytochemical methods. *J Endocrin* 113:147–153

Maddy SQ, Chisholm GD, Busuttil A, Habib FK (1989): Epidermal growth factor receptors in human prostate cancer: correlation with histological differentiation of the tumor. *Br J Cancer* 60:41–44

Maihle NJ, Kung H-J (1988): c-erbB and the epidermal growth factor receptor: a molecule with dual identity. *Biochim Biophys Acta* 948:287–304

Mansson P-E, Adams P, Kan M, McKeehan WL (1989): Heparin-binding growth factor gene expression and receptor characteristics in normal rat prostate and two transplantable rat prostate tumors. *Cancer Res* 49:2485–2494

Martikainen P, Kyprianou N, Isaacs JT (1990): Effect of transforming growth factor-β_1 on proliferation and death of rat prostatic cells. *Endocrinology* 127: 2963–2968

Matuo Y, Nishi N, Tanaka H, Sasaki I, Isaacs JT, Wada F (1988): Production of IGF-II-related peptide by an anaplastic cell line (AT-3) established from the Dunning prostatic carcinoma of rats. *In Vitro* 24:1053–1060

Mayarden SJ, Strom S, Ware JL (1992): Localization of epidermal growth factor receptor by immunohistochemical methods in human prostatic carcinoma, prostatic intraepithelial neoplasia, and benign hyperplasia. *Arch Pathol Lab Med* 116:269–273

McKeehan WL (1991): Growth factor receptors and prostate cell growth. *Cancer Surv* 11:165–175

McKeehan WL, Adams PS (1988): Heparin-binding growth factor/prostatropin attenuates inhibition of rat prostate tumor epithelial cell growth by transforming growth factor type beta. *In Vitro* 24:243–246

McKeehan WL, Adams PS, Rosser MP (1984): Direct mitogenic effects of insulin, epidermal growth factor, glucocorticoid, cholera toxin, unknown pituitary factors and possibly prolactin, but not androgen, on normal rat prostate epithelial cells in serum-free, primary cell culture. *Cancer Res* 44:1998–2010

McNeal JE (1990): Pathology of benign prostatic hyperplasia: insight into etiology. *J Urol Clin N Am* 17:477–486

Mellon K, Thompson S, Charlton RG, Marsh C, Robinson M, Lane DP, Harris AL, Horne CHW, Neal DE (1992): p53, c-erbB-2 and the epidermal growth factor receptor in the benign and malignant prostate. *J Urol* 147:496–499

Miller GJ, Stapleton GE, Ferrara JA, Lucia MS, Pfister S, Hedlund TE, Upadhya P (1992): The human prostatic carcinoma cell line LNCaP expresses biologically active, specific receptors for 1 α, 25-dihydroxyvitamin D_3. *Cancer Res* 52:515–520

Milovanovic SR, Radulovic S, Groot K, Schally AV (1992): Inhibition of growth of PC-82 human prostate cancer line xenografts in nude mice by bombesin antagonist RC-3095 or combination of agonist [D-trp^6]-luteinizing hormone-releasing hormone and somatostatin analog RC-160. *Prostate* 20:269–280

Mori H, Maki M, Oishi K, Jaye M, Igarashi K, Yoshida O, Hatanaka M (1990): Increased expression of genes for basic fibroblast growth factor and transforming growth factor type ß2 in human benign prostatic hyperplasia. *Prostate* 16:71–80

Morris GL, Dodd JG (1990): Epidermal growth factor receptor mRNA levels in human prostatic tumors and cell lines. *J Urol* 143:1272–1274

Muff R, Born W, Kaufmann M, Fischer JA (1994): Parathyroid hormone and parathyroid hormone-related protein receptor update. *Mol Cell Endocrinol* 100:35–38

Mulder E, van Loon D, de Boer W, Schuurmans ALG, Bolt J, Voorhorst MM, Kuiper GGJM, Brinkmann AO (1989): Mechanism of androgen action: recent observations on the domain structure of androgen receptors and the induction of EGF-receptors by androgens in prostate tumor cells. *J Steroid Biochem* 32: 151–156

Mydlo JH, Michaeli J, Heston WDW, Fair WR (1988): Expression of basic fibroblast growth factor mRNA in benign prostatic hyperplasia and prostatic carcinoma. *Prostate* 13:241–247

Nakada SY, di Sant'Agnese PA, Moynes RA, Hiipakka RA, Liao S, Cockett ATK, Abrahamsson P-A (1993): The androgen receptor status of neuroendocrine cells in human benign and malignant prostatic tissue. *Cancer Res* 53:1967–1970

Nakamoto T, Chang C, Li A, Chodak GW (1992): Basic fibroblast growth factor in human prostate cancer cells. *Cancer Res* 52:571–577

Nevalainen MT, Valve EM, Mäkelä SI, Bläuer M, Tuohimaa PJ, Härkönen PL (1991): Estrogen and prolactin regulation of rat dorsal and lateral prostate in organ culture. *Endocrinology* 129:612–622

Niman HL, Thompson AMH, Yu A, Markman M, Willems JJ, Herwig KR, Habib NA, Wood CB, Houghten RIA, Lerner RA (1985): Anti-peptide antibodies detect oncogene-related proteins in urine. *Proc Natl Acad Sci USA* 82:7924–7928

Norman AW, Bouillon R, Thomasset M (1991): *Vitamin D—Gene Regulation, Structure—Function Analysis and Clinical Application.* In: *Proceedings of the Eighth Workshop on Vitamin D.* New York: Walter de Gruyter

Okutani T, Nishi N, Kagawa Y, Takasuga H, Takenaka I, Usui T, Wada F (1991): Role of cyclic AMP and polypeptide growth regulators in growth inhibition by interferon in PC-3 cells. *Prostate* 18:73–80

Orloff JJ, Reddy D, de Papp AE, Yang KH, Soifer NE, Stewart AF (1994): Parathyroid hormone-related protein as a prohormone: posttranslational processing and receptor interactions. *Endocr Rev* 15:40–60

Pastan I, Chaudhary V, Fitzgerald D (1992): Recombinant toxins as novel therapeutic agents. *Annu Rev Biochem* 61:331–354

Peehl DM, Rubin JS (1995): Keratinocyte growth factor: an androgen-regulated mediator of stromal-epithelial interactions in the prostate. *World J Urol* 13:312–317

Peehl DM, Wong ST, Bazinet M, Stamey TA (1989): In vitro studies of human prostatic epithelial cells: attempts to identify distinguishing features of malignant cells. *Growth Factors* 1:237–250

Peehl DM, Wong ST, Stamey TA (1993): Vitamin A regulates proliferation and differentiation of human prostatic epithelial cells. *Prostate* 23:69–78.

Peehl DM, Leung GK, Wong ST (1994a): Keratin expression: a measure of phenotypic modulation of human prostatic epithelial cells by growth inhibitory factors. *Cell Tissue Res* 277:11–18

Peehl DM, Skowronski RJ, Feldman D (1994b): Role of vitamin D receptors in prostate cancer. In: *Sex Hormones and Antihormones in Endocrine Dependent Pathology: Basic and Clinical Aspects,* Motta M, Serio M, eds. Amsterdam: Elsevier Science BV

Peehl DM, Skowronski RJ, Leung GK, Wong ST, Stamey TA, Feldman D (1994c): Antiproliferative effects of 1,25-dihydroxyvitamin D_3 on primary cultures of human prostatic cells. *Cancer Res* 54:1–6

Peehl DM, Cohen P, Rosenfeld RG (1995a): The insulin-like growth factor system in the prostate. *World J Urol* 13:306–311

Peehl DM, Wong ST, Rhim JS (1995b): Altered growth regulation of prostatic epithelial cells by human papillomavirus-induced transformation. *Int J Oncol* 6:1177–1184

Peehl DM, Wong ST, Rubin JS (1995c): Autocrine versus paracrine factors differentially regulate the phenotype of prostatic epithelial cells. *Growth Factors* 13:306–311

Perkel VS, Linkhart TA, Mohan S, Baylink DJ (1990): An inhibitory insulin-like growth factor binding protein (IN-IGFBP) from human prostatic cell conditioned medium reveals N-terminal sequence identity with bone derived IN-IGFBP. *J Clin Endocrin Metab* 71:533–535

Pflug BR, Onoda M, Lynch JH, Djakiew D (1992): Reduced expression of the low affinity nerve growth factor receptor in benign and malignant human prostate tissue and loss of expression in four human metastatic prostate tumor cell lines. *Cancer Res* 52:5403–5406

Pietrzkowski Z, Mulholland G, Gomella L, Jameson BA, Wernicke D, Baserga R (1993): Inhibition of growth of prostatic cancer cell lines by peptide analogues of insulin-like growth factor I. *Cancer Res* 53:1102–1106

Power RF, Mani SK, Codina J, Conneely OM, O'Malley BW (1991): Dopaminergic and ligand-independent activation of steroid hormone receptors. *Science* 254: 1636–1639

Prins GS (1987): Prolactin influence on cytosol and nuclear androgen receptors in the ventral, dorsal, and lateral lobes of the rat prostate. *Endocrinology* 120:1457–1464

Reese DH, Gordon B, Gratzner HG, Claflin AJ, Malinin TI, Block NL, Politano, VA (1983): Effect of retinoic acid on the growth and morphology of a prostatic adenocarcinoma cell line cloned for the retinoid inducibility of alkaline phosphatase. *Cancer Res* 43:5443–5450

Reile H, Armatis PE, Schally AV (1994): Characterization of high-affinity receptors for bombesin/gastrin releasing peptide on the human prostate cancer cell lines PC-3 and DU 145: internalization of receptor bound ^{125}I-(Tyr^4) bombesin by tumor cells. *Prostate* 25:29–38

Ross JS, Nazeer T, Church K, Amato C, Figge H, Rifkin MD, Fisher HAG (1994): Contribution of HER-2/neu oncogene expression to tumor grade and DNA content analysis in the prediction of prostatic carcinoma metastasis. *Cancer* 72: 3020–3028

Rubenstein M, Muchnik S, Dunea G, Chou P, Guinan P (1994): Inoculation of prostatic tumors with antisense oligonucleotides against mRNA encoding growth factors and receptors. In: *Recent Advances in Chemotherapy*, Einhorn J, Nord CE, Norrby SR, eds. Washington, DC: American Society for Microbiology

Rubin JS, Bottaro DP, Chedid M, Miki T, Ron D, Cunha GR, Finch PW (1995): Keratinocyte growth factor as a cytokine that mediates mesenchymal-epithelial interaction. In: *Epithelial-Mesenchymal Interactions in Cancer*. Goldberg I, Rosen E (eds) (in press)

Sadasivan R, Morgan R, Jennings S, Austenfeld M, van Veldhuizen P, Stephens R, Noble M (1993): Overexpression of HER-2/neu may be an indicator of poor prognosis in prostate cancer. *J Urol* 150:126–131

Sanda MG, Ayyagari SR, Jaffee EM, Epstein JI, Clift SL, Cohen LK, Dranoff G, Pardoll DM, Mulligan RC, Simons JW (1994): Demonstration of a rational strategy for human prostate cancer gene therapy. *J Urol* 151:622–628

Schacht MJ, Niederberger CS, Garnett JE, Sensibar JA, Lee C, Grayhack JT (1992): A local direct effect of pituitary graft on growth of the lateral prostate in rats. *Prostate* 20:51–58

Schmid KW, Helpap B, Tötsch M, Kirchmair R, Dockhorn-Dworniczak B, Böcker W, Fischer-Colbrie R (1994): Immunohistochemical localization of chromogranins A and B and secretogranin II in normal, hyperplastic and neoplastic prostate. *Histopathology* 24:233–239

Schuurmans ALG, Bolt J, Mulder E (1988): Androgens stimulate both growth rate and epidermal growth factor receptor activity of the human prostate tumor cell LNCaP. *Prostate* 12:55–63

Schwartz GG, Hulka BS (1990): Is vitamin D deficiency a risk factor for prostate cancer? (Hypothesis). *Anticancer Res* 10:1307–1312

Shah GV, Noble MJ, Austenfeld M, Weigel J, Deftos LJ, Mebust WK (1992): Presence of calcitonin-like immunoreactivity (iCT) in human prostate gland: evidence for iCT secretion by cultured prostate cells. *Prostate* 21:87–97

Shah GV, Rayford W, Noble MJ, Austenfeld M, Weigel J, Vamos S, Mebust WK (1994): Calcitonin stimulates growth of human prostate cancer cells through receptor-mediated increase in cyclic adenosine 3, 5-monophosphates and cytoplasmic Ca^{2+} transients. *Endocrinology* 134:596–602

Sherwood ER, Fike W, Kozlowski JM, Lee C (1988): The cytotoxic/cytostatic effect of recombinant human tumor necrosis factor alpha on experimental human prostate cancer. *Proc Am Urol Assoc* 139:175A, abstract 50

Sherwood ER, Fong C-J, Lee C, Kozlowski JM (1992): Basic fibroblast growth factor: a potential mediator of stromal growth in the human prostate. *Endocrinology* 130:2955–2963

Siegall CB, Schwab G, Nordan RP, Fitzgerald DJ, Pastan I (1990): Expression of the interleukin 6 receptor and interleukin 6 in prostate carcinoma cells. *Cancer Res* 50:7786–7788

Siegsmund MJ, Yamazaki H, Pastan I (1994): Interleukin 6 receptor mRNA in prostate carcinomas and benign prostate hyperplasia. *J Urol* 151:1396–1399

Sikes RA, Chung LWK (1992): Acquisition of a tumorigenic phenotype by a rat ventral prostate epithelial cell line expressing a transfected activated neu oncogene. *Cancer Res* 52:3174–3181

Simeone A, Acampora D, Arcioni L, Andrews PW, Boncinelli E, Mavilio F (1990): Sequential activation of HOX2 homeobox genes by retinoic acid in human embryonal carcinoma cells. *Nature* 346:763–766

Sitaras NM, Sariban E, Bravo M, Pantazis P, Antoniades HN (1988): Constitutive production of platelet-derived growth factor-like proteins by human prostate carcinoma cell lines. *Cancer Res* 48:1930–1935

Skowronski RJ, Peehl DM, Feldman D (1993): Vitamin D and prostate cancer: 1,25 dihydroxyvitamin D_3 receptors and actions in human prostate cancer cell lines. *Endocrinology* 132:1952–1960

Skowronski RJ, Peehl DM, Feldman D (1995): Actions of vitamin D_3 analogs on human prostate cancer cell lines: comparison with 1,25-dihydroxyvitamin D_3. *Endocrinology* 136:1–7

Slawin K, Kadmon D, Park SH, Scardino PT, Anzano M, Sporn MB, Thompson TC (1993): Dietary fenretinide, a synthetic retinoid, decreases the tumor incidence and the tumor mass of ras + myc-induced carcinomas in the mouse prostate reconstitution model system. *Cancer Res* 53:4461–4465

Smith RG, Nag A (1987): Regulation of c-sis expression in tumors of the male reproductive tract. *Prog Clin Biol Res* 239:113–122

Spriggs DR, Imamura K, Rodriguez C, Sariban E, Kufe DW (1988): Tumor necrosis factor expression in human epithelial tumor cell lines. *J Clin Invest* 81:455–460

Stadnyk AW (1994): Cytokine production by epithelial cells. *FASEB J* 8:1041–1047

Stearns ME, Wang M, Fudge K (1993): Liarozole and 13-cis-retinoic acid antiprostatic tumor activity. *Cancer Res* 53:3073–3077

Steiner MS (1993): Role of peptide growth factors in the prostate: a review *Urology* 42:99–110

Steiner MS, Barrack ER (1992): Expression of transforming growth factor-β1 overproduction in prostate cancer: effects on growth in vivo and in vitro. *Mol Endocrinol* 6:15–25

Steiner MS, Zhou Z-Z, Tonb DC, Barrack ER (1994): Expression of transforming growth factor-β1 in prostate cancer. *Endocrinology* 135:2240–2247

Story MT (1991): Polypeptide modulators of prostatic growth and development. *Cancer Surv* 11:123–146

Story MT, Livingston B, Baeten L, Swartz SJ, Jacobs SC, Begun FP, Lawson RK (1989): Cultured human prostate-derived fibroblasts produce a factor that stimulates their growth with properties indistinguishable from basic fibroblast growth factor. *Prostate* 15:355–365

Story MT, Hopp KA, Meier DA, Begun FP, Lawson RK (1993): Influence of transforming growth factor β1 and other growth factors on basic fibroblast growth factor level and proliferation of cultured human prostate-derived fibroblasts. *Prostate* 22:183–197

Sutkowski DM, Fong C-J, Sensibar JA, Rademaker AW, Sherwood ER, Kozlowski JM, Lee C (1992): Interaction of epidermal growth factor and transforming growth factor beta in human prostatic epithelial cells in culture. *Prostate* 21: 133–143

Syms AJ, Harper ME, Griffiths K (1985): The effect of prolactin on human BPH epithelial cell proliferation. *Prostate* 6:145–153

Timme TL, Truong LD, Merz VW, Krebs T, Kadmon D, Flanders KC, Park SH, Thompson TC (1994): Mesenchymal-epithelial interactions and transforming growth factor-β expression during mouse prostate morphogenesis. *Endocrinology* 134:1039–1045

Tong P, Mayes D, Wheeler L (1986): Extracellular calcium alters the effects of retinoic acid on DNA synthesis in cultured murine keratinocytes. *Biochem Biophys Res Commun* 138:483–488

Truong LD, Kadmon D, McCune BK, Flanders KC, Scardino PT, Thompson TC (1993): Association of transforming growth factor-β1 with prostate cancer: an immunohistochemical study. *Hum Pathol* 24:4–9

Turkeri LN, Sakr WA, Wykes SM, Grignon DJ, Pontes JE, Macoska JA (1994): Comparative analysis of epidermal growth factor receptor gene expression and protein product in benign, premalignant, and malignant prostate tissue. *Prostate* 25:199–205

Varani J, Shayevitz J, Perry D, Mitra RS, Nickoloff BJ, Voorhees JJ (1990): Retinoic acid stimulation of human dermal fibroblast proliferation is dependent on sub-optimal extracellular Ca^{2+} concentration. *Am J Pathol* 136:1275–1281

Vieweg J, Rosenthal FM, Bannerji R, Heston WDW, Fair WR, Gansbacher B, Gilboa E (1994): Immunotherapy of prostate cancer in the Dunning rat model: use of cytokine gene modified tumor vaccines. *Cancer Res* 54:1760–1765

Vilcek J, Lee TH (1991): Tumor necrosis factor. *J Biol Chem* 266:7313–7316

Vlahos CJ, Kriauciunas TD, Gleason PE, Jones JA, Eble JN, Salvas D, Falcone JF, Hirsch KS (1993): Platelet-derived growth factor induces proliferation of hyperplastic human prostatic stromal cells. *J Cell Biochem* 52:404–413

Ware JL, Watts RG (1992): Isolation and characterization of transforming growth factor beta response variants from human prostatic tumor cell lines. *Prostate* 21:223–237

Webber MM, Perez-Ripoll EA (1986): Prolactin in the etiology and progression of human prostate carcinoma. *J Urol Suppl* 135:348A

Wilding G (1991): Response of prostate cancer cells to peptide growth factors: transforming growth factor-β. *Cancer Surv* 11:147–163

Wilding G, Valverius E, Knabbe C, Gelman EP (1989a): Role of transforming growth factor-α in human prostate cancer cell growth. *Prostate* 15:1–12

Wilding G, Zugmeier G, Knabbe C, Flanders K, Gelmann E (1989b): Differential effects of transforming growth factor β on human prostate cancer cells in vitro. *Mol Cell Endocrinol* 62:79–87

Willett WC, Hunter DJ (1994): Vitamin A and cancers of the breast, large bowel, and prostate: epidemiologic evidence. *Nutr Rev* 52:(II)S53–S59

Yan G, Fukabori Y, Nikolaropoulos S, Wang F, McKeehan WL (1992): Heparin-binding keratinocyte growth factor is a candidate stromal to epithelial cell andromedin. *Mol Endocrinol* 6:2123–2128

Yan G, Fukabori Y, McBride G, Nikolaropolous S, McKeehan WL (1993): Exon switching and activation of stromal and embryonic fibroblast growth factor (FGF)—FGF receptor genes in prostate epithelial cells accompany stromal independence and malignancy. *Mol Cell Biol* 13:4513–4522

Yang Y, Chisholm GD, Habib FK (1993): Epidermal growth factor and transforming growth factor α concentrations in BPH and cancer of the prostate: their relationships with tissue androgen levels. *Br J Cancer* 67:152–155

Young CY-F, Murtha PE, Andrews PE, Lindzey JK, Tindall DJ (1994): Antagonism of androgen action in prostate tumor cells by retinoic acid. *Prostate* 25:39–45

Zhau HE, Wan DS, Zhou J, Miller GJ, von Eschenbach AC (1992): Expression of c-erbB-2/neu proto-oncogene in human prostatic cancer tissues and cell lines. *Mol Carcinogen* 5:320–327

16

Androgen Receptors in Human Prostate Cancer: Heterogeneous Expression, Gene Mutations, and Polymorphic Variants

JANETTE M. HAKIMI, RACHEL H. RONDINELLI, MARK P. SCHOENBERG AND EVELYN R. BARRACK

RATIONALE FOR ANDROGEN-RECEPTOR STUDIES IN PROSTATE CANCER

Prostate cancer is the most commonly diagnosed cancer in men in the United States; it is estimated that there will be 244,000 new cases of prostate cancer in 1995, accounting for 36% of new cancer cases in men (Wingo et al., 1995). Prostate cancer is the second leading cause of cancer deaths in American men, with 40,000 deaths from prostate cancer in the United States estimated for 1995 (Wingo et al., 1995). Most prostate cancer patients have no known risk factors, neither for disease development, nor to predict whether the tumor will be slow growing or aggressive. Based on the belief that many prostate cancers are slow growing and not aggressive, one approach to the management of patients diagnosed to have prostate cancer is watchful waiting (Catalona, 1994; Chodak, 1994). With the advent of screening, increased numbers of prostate cancer will be diagnosed (Carter and Coffey, 1990; Oesterling, 1991), thereby increasing the need for indicators of risk assessment. In general, high-grade (based on histologic appearance) is associated with increased aggressiveness, however, most tumors have moderate grade, where predictive power is lacking (Whitmore, 1973; Partin et al., 1989). There is an urgent need for indicators to distinguish between indolent and aggressive forms of prostate cancer so that the latter group can be aggressively treated when tumor volume is low and the opportunity for cure is best.

Hormones and Cancer
Wayne V. Vedeckis, Editor

The natural history of prostate cancer is not well defined. One model assumes that progression from low-grade, well-differentiated, organ-confined (stage A or B) prostate cancer to high-grade, poorly differentiated, metastatic cancer (stage D) is coincident with an increase in tumor volume (Whitmore, 1973). However, an alternative model is that there are different types of prostate cancer, some that progress as predicted by the first model, versus others that instead progress early in their evolution, becoming high-grade or metastatic while still small (Whitmore, 1973). In radical prostatectomy specimens removed for nonpalpable early prostate cancer that is detected by screening techniques, 45% of high-grade tumors are low volume (<1 cc), and 8.3% of low-volume tumors are high-grade (Epstein et al., 1994). The data support the view that prostate cancer indeed has the potential to be high-grade early in its course and may not necessarily arise from low-grade cancer (Epstein et al., 1994).

Clinical stage A prostate cancer is disease that is clinically undetected but is discovered incidentally at the time of surgery for urinary obstruction due to benign prostatic hyperplasia (BPH). Stage A1 prostate cancer is low volume disease that is well differentiated or moderately differentiated, and is thought to be nonaggressive because only 2% progress to clinical disease within 4 years after diagnosis (Christensen et al., 1990); yet those small tumors that do progress are clearly aggressive. Stage A2 is high-volume or high-grade disease. Stage A2 is thought to be an aggressive form of the disease, but it is actually a heterogeneous group of tumors, because 30% progress within 4 years, yet 30% of men who are left untreated will be free of progression 10 years after diagnosis (Christensen et al., 1990).

Clinical stage B prostate cancer is detected clinically, and it appears to be organ-confined because there is no evidence of metastatic spread. Some patients who undergo radical prostatectomy are cured of their disease; however, the possibility cannot be ruled out that some of these tumors may be indolent and would not be lethal even if left untreated. On the other hand, others who undergo radical prostatectomy are not cured because their tumor had indeed already metastasized, even though metastases were not detected (Diamond et al., 1982).

About 50% of prostate cancers are diagnosed as stage D (metastatic), and not curable. It has been suggested that a proportion of these may have been more virulent since their inception, accounting for a lack of symptoms at an earlier time when the tumor load was smaller and perhaps more amenable to cure (Whitmore, 1973). If some tumors that arise are inherently more aggressive than others, it would be important to identify factors that might affect the susceptibility to develop aggressive disease.

The etiology of prostate cancer is unknown, but there is compelling evidence to support the hypothesis of a hormonal etiology involving androgen action (Ross et al., 1983; Catalona and Scott, 1986; Henderson et al., 1988; Wilding, 1992; Whittemore, 1994). Androgen, acting via the androgen receptor (AR), plays a crucial role in the differentiation and growth of the prostate *in utero* and at puberty (Cunha et al., 1987; Luke and Coffey, 1994). It is also presumed to play a role in prostate carcinogenesis because men castrated before puberty do not develop prostate cancer, and because the experimental induction of prostate carcinogenesis in animals requires androgen (Ross et al., 1983; Catalona and Scott, 1986; Henderson et al., 1988; Wilding, 1992; Whittemore, 1994). However, it is not known whether the putative role of androgen in prostate cancer etiology is in the process of carcinogenesis, progression to clinically manifested disease, progression from nonaggressive to aggressive disease, or simply the growth of prostate cells that have become tumorigenic as a result of other causes.

The AR also plays a role in prostate cancer growth because most metastatic prostate tumors respond to androgen ablation. Recognition that androgen is required for prostate growth and function is the basis for androgen ablative therapy for prostate cancer (Huggins and Hodges, 1941; Scott et al., 1980). Standard treatment of metastatic prostate cancer is suppression of androgenic stimuli, such as by castration or the administration of estrogen, antiandrogen, or a luteinizing hormone–releasing hormone (LHRH) analog (Scott et al., 1980; Catalona and Scott, 1986). About 80% of these patients experience a beneficial response to hormonal therapy; however, the extent and duration of response are variable and unpredictable, and their prostate cancer eventually becomes androgen-independent (Scott et al., 1980).

Recent data suggest that the AR may even play a role in progression to androgen-independence, since most hormone refractory prostate cancers express AR (van der Kwast et al., 1991; Ruizeveld de Winter et al., 1994), and the transcriptional activity of the normal AR can be activated under certain conditions in the absence of androgen (Culig et al., 1994). In addition, some tumors contain mutated AR (Newmark et al., 1992; Culig et al., 1993a; Suzuki et al., 1993; Gaddipati et al., 1994; Schoenberg et al., 1994), and some of these mutations allow an antiandrogen, estrogen, or a weakly androgenic adrenal steroid to act paradoxically as an agonist (Culig et al., 1993a; Suzuki et al., 1993; Gaddipati et al., 1994).

If prostate tumor epithelial cells have access to the circulation (e.g., because the basement membrane is disrupted or because the cells have metastasized), then prostate-specific antigen (PSA) produced by these cells appears in the circulation; this makes serum PSA a useful marker for prostate cancer (Oesterling, 1991). PSA is an androgen-regulated gene product (Riegman et al., 1991), so the responsiveness of metastatic prostate cancer to androgen ablation is associated with a decrease in the serum PSA level, and one of the first signs of progression to androgen independence is elevation of the serum PSA level (Oesterling, 1991). However, at least some tumors that progress during hormonal therapy are actually still androgen-dependent, but are growing in the presence of antiandrogen plus castrate levels of testosterone because the antiandrogen (e.g., flutamide) is acting paradoxically as an AR agonist instead of as an AR antagonist (Scher and Kelly, 1993; Sartor et al., 1994). This paradoxical response to antiandrogen therapy is manifested by a rise in serum PSA that abates when the antiandrogen is discontinued (Scher and Kelly, 1993; Sartor et al., 1994). This clinical phenomenon has been termed flutamide withdrawal syndrome; it is the unexpected fall in an androgen-regulated prostate marker, PSA, that may occur when flutamide administration is discontinued, and it is due to the paradoxical agonistic effect of the antiandrogen. Paradoxical responses to the antiandrogens casodex and megestrol acetate have also been reported recently (Small and Carroll, 1994; Dawson and McLeod, 1995; Nieh, 1995). The underlying molecular basis has not been defined, but the response of an androgen-regulated gene product implies an AR-mediated mechanism.

Our goal is to understand the role of the AR in prostate cancer, and whether this role is mediated by wild-type AR or by mutant AR. This chapter reviews the current state of our knowledge about AR in human prostate cancer.

ANDROGEN-RECEPTOR QUANTITATION

An early approach to studying AR in prostate cancer was to determine whether knowledge of the AR content could be used to predict the androgen responsiveness of prostate cancer (Diamond and Barrack, 1984; Barrack et al., 1987; Barrack and Tindall, 1987; Sadi et al., 1991; Sadi and Barrack, 1991, 1993). Because androgens act via AR, it seemed reasonable to hypothesize that AR might be a marker of androgen-dependent cells, and that the AR content in prostate cancer might predict the extent of its responsiveness to androgen withdrawal. A prediction of the hypothesis was that the higher the percentage of AR-positive prostate cancer cells, the more androgen-dependent and the more responsive the tumor to androgen withdrawal. Thus, it would take longer for androgen-independent tumor cells to repopulate the tumor to a size equal to or greater than that which existed before therapy (i.e., the longer the time to progression). We studied patients who had metastatic prostate cancer and underwent needle biopsy of the primary prostate tumor just before androgen ablation; we assessed the AR status of this tumor tissue. Response to hormonal therapy was monitored by patient symptomatology (i.e., pain, appetite, weight loss, activity), bone scan and bone X-rays to monitor bone metastases, serum prostatic acid phosphatase (an indicator of metastatic prostate tumor cells with access to the circulation; this study was done before the widespread use of PSA as a marker), intravenous urogram or ultrasonography of the urinary tract, chest X-rays, complete blood counts, blood urea nitrogen, serum creatinine, electrolytes, and liver function tests. Relapse was defined by evidence of any of the following: deterioration of symptoms, worsening of the bone lesions or appearance of new ones, appearance of soft tissue masses, or onset of upper urinary tract obstruction due to growth of the primary tumor. Time to progression was defined as the time between androgen ablative therapy and tumor relapse.

Tumors that grow in an androgen-depleted environment are referred to as hormone refractory or androgen independent.

Before the advent of AR gene cloning or the creation of AR-specific antibodies, AR measurements were based on [^{3}H]labeled-ligand binding. These studies were seriously hampered by the need to use homogenized tissue, which provides an average AR content of all cells in the homogenized specimen. The presence of nonmalignant cells that are AR-positive or AR-negative yields a misleadingly high or low level of "tumor" AR content, respectively. Therefore, differences in AR content between patients due to the presence of different proportions of nonmalignant prostate tissue cannot be distinguished from differences in AR content actually due to different percentages of AR-positive tumor cells and AR-negative tumor cells. Indeed, the ability of radioligand binding AR assays to predict the androgen responsiveness of advanced prostate cancer has been controversial (see Barrack and Tindall, 1987, for a detailed review). Some studies found lower levels of nuclear AR associated with a shorter duration of response (Trachtenberg and Walsh, 1982; Brendler et al., 1984; Fentie et al., 1986; Barrack et al., 1987; Benson et al., 1987), but other studies found no correlation (van Aubel et al., 1988). In no study was the relationship accurate enough to predict the response of individual tumors.

We therefore recognized the need to devise methods to localize AR at the light microscope level. Autoradiography and immunohistochemistry circumvent the problem of admixture of malignant and nonmalignant cells by allowing selective evaluation of malignant cell AR status. By AR autoradiography of high affinity [^{3}H]androgen binding to frozen sections (Peters and Barrack, 1987a,b,c; Sadi and Barrack, 1991) and by AR immunohistochemistry (Sadi et al., 1991; Sadi and Barrack, 1993), we were able for the first time to directly test the hypothesis that AR content in the malignant cells themselves correlates with the time to progression after hormonal therapy.

Both AR-positive and AR-negative malignant epithelial nuclei were present in metastatic prostate cancer before androgen ablation (Sadi et al., 1991). However, there was no relationship between the percentage of AR-positive malignant prostate epithelial cells and the time to progression of metastatic prostate cancer following hormonal therapy (time between treatment and relapse) (Fig. 1). Contrary to expectations, some tumors that were very AR-rich were poorly responsive to androgen ablation; conversely, some tumors that were AR-poor were nevertheless very responsive. These data suggested that AR content before therapy was not an accurate predictor of androgen responsiveness. Animal models also provide evidence that AR-negative cells can actually be androgen-dependent, and that AR-positive cells can nevertheless be androgen-independent. For example, AR-negative epithelial cells in the fetal mouse prostate undergo androgen-dependent proliferation; it is thought that an androgen-regulated growth factor produced by AR-positive stroma acts on the AR-negative epithelial cells (Shannon and Cunha, 1983, 1984). Proteins secreted by rat prostate stromal cells can

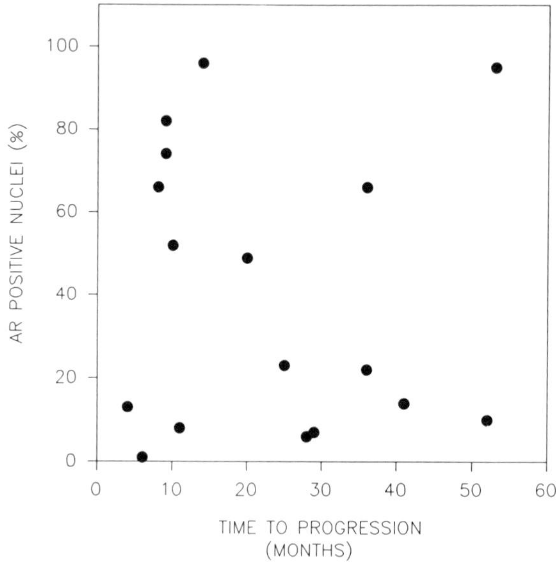

Figure 1. Relationship between the percentage of malignant epithelial nuclei that are AR-positive in stage D2 prostate cancer biopsies taken prior to androgen ablation therapy and the time to tumor progression following therapy. Time to progression is defined as the time between therapy and relapse (deterioration of symptoms). (Reprinted with permission of the J.B. Lippincott Company, from Sadi et al., 1991).

indeed affect epithelial cell proliferation and function (Swinnen et al., 1990; Djakiew et al., 1990; Yan et al., 1992; see Chapter 15, this volume), and the production of some of these stroma-derived proteins is regulated by androgen (Swinnen et al., 1990; Yan et al., 1992). Conversely, AR-positive cells do not necessarily die following androgen withdrawal because some normal and malignant rat prostate cells are still AR-positive following castration (Lippert and Keefer, 1987; Prins, 1989), and progression of the androgen-dependent Shionogi S115 mouse mammary tumor to an androgen-independent state occurs despite the continued presence of AR (Darbre and King, 1987). Androgen-independent human prostate cancers also are predominantly and extensively AR-positive following castration (van der Kwast et al., 1991; Ruizeveld de Winter et al., 1994). It is not known whether AR expression in these androgen-independent cells is upregulated following androgen ablation or whether these cells are AR-positive even before therapy. If before therapy there was a significant proportion of AR-positive malignant prostate epithelial cells that was not dependent on androgen for survival, then the AR content of a tumor would not be a reliable predictor of response to androgen withdrawal, explaining why AR measurements in metastatic prostate cancer before androgen withdrawal therapy have been unsuccessful as predictors of the androgen dependence or responsiveness of prostate cancer.

Image Analysis of AR Immunostaining: Heterogeneity as a Predictor of Response

Evaluating the AR status of malignant epithelial nuclei as AR-positive or AR-negative does not take into account variations in staining intensity within a specimen or between specimens. Therefore, we quantitated the intensity of nuclear AR immunostaining by computer-assisted image analysis (Sadi and Barrack, 1993). Optical density measurements of nuclear AR immunostaining intensity were expressed as the average optical density of each nucleus evaluated, and therefore represent AR concentration. For each patient, we plotted the average optical densities of each of 200 nuclei/specimen as a frequency distribution (Figs. 2 and 3). We also calculated the mean, variance, skewness, and kurtosis of the optical density of 200 nuclei for each patient.

Based on staining intensity, AR content varied among nuclei (Figs. 2 and 3). Interestingly, the variance of nuclear AR immunostaining in poor responders (1334 ± 318; $n = 8$ patients) was significantly greater than that in good responders (568 ± 126; $n = 9$ patients) ($p = 0.03$). This is reflected in the shapes of the frequency distribution curves. Poor responders had flattened, highly variable, multimodal distributions, indicating a wide range of nuclear AR concentrations within each tumor (Fig. 2). In striking contrast, good responders had frequency distributions of nuclear AR content that were quite uniform in shape (Fig. 3). Thus, the AR concentration per cell is significantly more heterogeneous in poor responders than in good responders. The kurtosis was also significantly different in good responders versus poor responders ($p = 0.04$), reflecting the more flattened (platykurtic) frequency distribution curves in poor responders.

When patients were divided into two groups based on the average variance of nuclear AR immunostaining, the low variance group (below average variance) had a mean time to tumor progression of 28 ± 5 months, whereas the high variance group had a mean time to progression of 13 ± 3 months. There was a significant association between the variance of AR staining and the time to tumor progression (Fisher's exact test, $p = 0.027$). Five of six patients with a high variance progressed within <20 months after therapy, whereas 8 of 11 patients with a low variance experienced a prolonged response to therapy (time to progression ≥ 20 months). Kaplan-Meier curves revealed a significantly different progression-free probability for patients with a high versus low variance of AR staining intensity ($p = 0.037$) (Fig. 4). Therefore, the variance (i.e., heterogeneity) of AR concentrations within a given specimen appears to be a valuable predictor of tumor behavior following therapy (Sadi and Barrack, 1993). This exciting finding needs to be confirmed in a larger population of patients. This study was the first to describe statistical variance as a significant discriminator of response to hormonal therapy.

By contrast, the mean nuclear AR staining intensity was not significantly different between poor responders (mean optical density, 120 ± 14; $n = 8$)

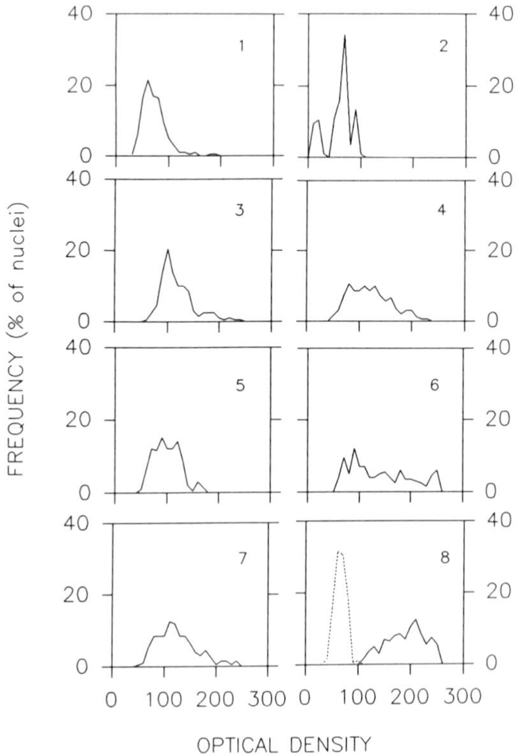

Figure 2. Frequency distributions of androgen receptor staining intensity in poor responders. Prostate biopsies were obtained prior to hormonal therapy; all patients illustrated in this figure showed evidence of tumor progression within 20 months after therapy. AR immunostaining intensity, expressed as average optical density/nucleus, was quantitated in 200 malignant nuclei by computer-assisted image analysis. Optical densities were grouped in increments of 10, and the number of nuclei (percentage of total evaluated) with optical densities in these ranges was plotted as a frequency distribution. Each distribution curve represents AR staining in a different patient. For patient #8, both AR antibody staining (solid line) and control IgG staining (dotted line) are shown. (Reprinted with permission of the J.B. Lippincott Company, from Sadi and Barrack, 1993).

and good responders (mean optical density, 102 ± 11; $n = 9$) (Sadi and Barrack, 1993). Thus, tumor behavior following androgen ablation seems not to depend simply on the presence of AR in a cell or on the average amount of AR in a tumor.

Since it is the variance of AR content that correlates with tumor behavior, we speculate that the greater intratumor AR heterogeneity in poor responders may be a manifestation of greater genetic instability. Genetic instability can give rise to tumor heterogeneity. Genetic instability indeed occurs in prostate cancer, as reflected by the development of aneuploidy

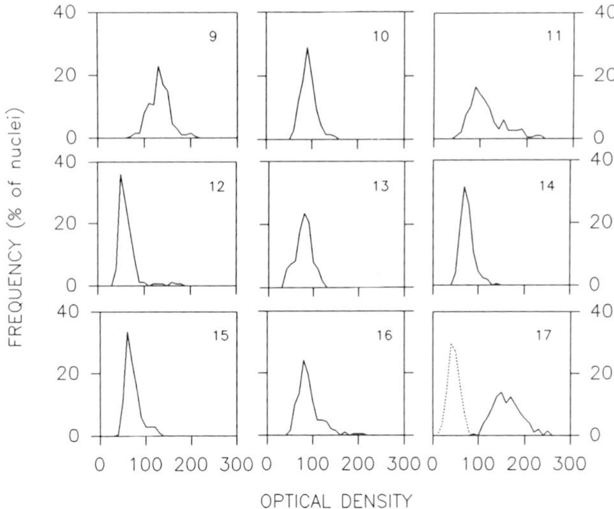

Figure 3. Frequency distributions of androgen receptor staining intensity in good responders. All patients illustrated in this figure had a time to progression $\geq$20 months after hormonal therapy. All other conditions were as described in the legend to Figure 2. For patient #17, both AR antibody staining (solid line) and control IgG staining (dotted line) are shown. (Reprinted with permission of the J.B. Lippincott Company, from Sadi and Barrack, 1993).

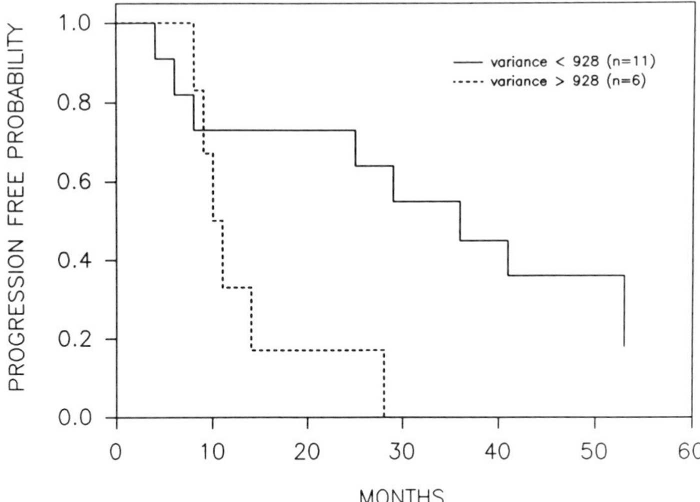

Figure 4. Kaplan-Meier estimates of progression-free probability for patients with a high variance versus a low variance of AR concentration per cell were significantly different ($p = 0.037$). (Reprinted with permission of the J.B. Lippincott Company, from Sadi and Barrack, 1993).

(Greene et al., 1991; Alcaraz et al., 1994), chromosomal abnormalities (Sandberg, 1992), allelic loss (Sakr et al., 1994; Isaacs et al., 1995; Visakorpi et al., 1995b), gene inactivation by hypermethylation (Lee et al., 1994), gene amplification (Visakorpi et al., 1995a,b), and gene mutations (Bookstein et al., 1993; Dobashi et al., 1994; Isaacs et al., 1995). Because genetic instability appears to increase as tumors progress (Heppner, 1984; Nowell, 1986; Fearon and Vogelstein, 1990), a tumor with greater genetic instability might indicate that it has progressed further toward androgen independence, and thus will result in a shorter time to progression after hormonal therapy. Thus, AR heterogeneity (Sadi and Barrack, 1993) may be a marker of genetic instability rather than strictly of androgen dependence. Using *in situ* hybridization of chromosome-specific probes, Sesterhenn et al. (1992) found frequent aneuploidy of the X chromosome (the chromosomal location of the AR gene) in interphase nuclei of prostate cancers (less than one or more than one X chromosome per cell). In addition, a recent study reported AR gene amplification in 7 of 23 (30%) hormone refractory prostate cancers, and AR gene amplification showed a great deal of intratumor variability (Visakorpi et al., 1995a). Therefore, it is conceivable that the heterogeneous AR expression in tumors that are presumably already more aggressive [i.e., because they show a poor response to androgen ablation and progress within a short time interval (Sadi and Barrack, 1993)], may be due to X chromosome aneuploidy or AR gene amplification. Visakorpi et al. (1995a) found no AR gene amplification in untreated primary tumors available from 16 of the 23 patients, an observation that was interpreted to suggest that AR amplification may be selected for during androgen ablation; however, only 6 of the 23 patients had metastatic disease at the time of diagnosis. Therefore, the possibility has not been ruled out that AR gene amplification may occur during progression to metastatic disease, before hormonal therapy. All the tumors we assessed that had heterogeneous AR content were from patients with distant metastases before hormonal therapy (Sadi and Barrack, 1993).

Genetic instability in tumors is also frequently associated with somatic mutations (Bookstein et al., 1993; Dobashi et al., 1994), raising the possibility of AR gene mutations in prostate cancer.

MOLECULAR ANALYSIS OF AR GENE STRUCTURE AND FUNCTION IN PROSTATE CANCER

There is compelling evidence that carcinogenesis results from somatic alterations in genes that are involved in the control of cell proliferation and cell function (Greenblatt et al., 1994). On the other hand, a susceptibility to develop cancer can be inherited (Hall et al., 1990; Malkin et al., 1990; Carter et al., 1992, 1993; Gronberg et al., 1994). Both mechanisms are probably involved in prostate carcinogenesis. Somatic molecular changes have been found in many prostate cancers (Bookstein et al., 1993; Lee et al., 1994; Isaacs et al., 1995; Visakorpi et al., 1995a,b), and documentation of hereditary prostate cancer and familial prostate cancer (Carter et al., 1992, 1993) provides evidence for inherited factors that may predispose to prostate carcinogenesis (see Chapter 14, this volume).

Mutations in several important signal transducing molecules confer oncogenic potential; for example, (1) mutated plasma membrane-associated guanosine triphosphate (GTP)-binding proteins $G_{s\alpha}$ and $G_{i\alpha}$ are oncogenes (Landis et al., 1989; Lyons et al., 1990); (2) mutated α_{1B}-adrenergic receptors enhance mitogenesis and tumorigenicity (Allen et al., 1991); (3) the v-*erb*A oncogene is a truncated and mutated form of the thyroid hormone receptor c-*erb*A gene (Evans 1988; Sharif and Privalsky, 1991); and (4) *ras* gene mutations are oncogenic (Bos, 1989). In light of the important signal transducing role of androgen as a regulator of prostate growth and function (Huggins and Hodges, 1941; Cunha et al., 1987; Luke and Coffey, 1994), the requirement of androgen for prostate carcinogenesis (Ross et al., 1983; Catalona and Scott, 1986; Wilding, 1992; Whittemore, 1994), and the persistent expression of AR in androgen-independent prostate cancer (van der Kwast et al., 1991; Ruizeveld de Winter et al., 1994), we reasoned that knowledge of the integrity of the AR gene or gene product may be a more accurate index of potential AR function in prostate cancer than immunoreactivity or ligand binding, which do not distinguish between functional and nonfunctional AR, and we hypothesized that mutations in the AR gene might be involved in the development or progression of prostate cancer.

Normal (Wild-Type) Androgen Receptor Structure and Function

The AR is a member of the steroid receptor superfamily of ligand-activated nuclear transcription factors that includes receptors for progesterone, glucocorticoid, mineralocorticoid, estrogen, vitamin D, thyroid hormone, and retinoic acid (Chang et al., 1988; Evans, 1988; Lubahn et al., 1988; Jenster et al., 1991; Simental et al., 1991; Tsai and O'Malley, 1994; see Chapter 4, this volume). The AR consists of a C-terminal hormone binding domain that confers ligand specificity, a central DNA binding domain that binds to androgen-responsive target genes, a "hinge" region between the DNA and hormone binding domains, and an N-terminal domain that affects transcription efficiency (Fig. 5). Some functions require more than one domain, and each domain can affect the function of other domains. For example, nuclear localization of the AR is regulated by sequences in the hormone binding domain and in the N-terminal domain (Zhou et al., 1994). The N-terminal domain inhibits receptor dimerization and DNA binding, and androgen binding relieves this inhibition (Wong et al., 1993). The N-terminal domain also plays a role in stabilizing the AR by slowing the rate of ligand dissociation and AR degradation, indicating that the

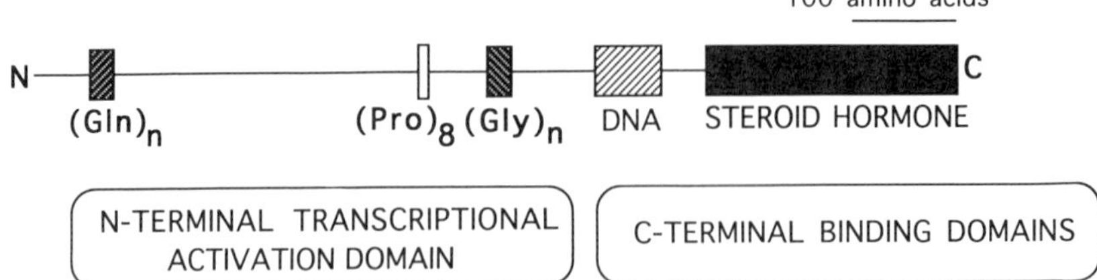

Figure 5. Schematic of human AR domain structure. The polymorphic glutamine $(Gln)_n$ and glycine $(Gly)_n$ repeats are located within the N-terminal transcriptional activation domain.

conformation of the full length wild-type AR allows the N-terminal domain to interact with the hormone binding domain (Zhou et al., 1995). Sequences in the N-terminal domain can even affect the specificity of DNA binding, as demonstrated recently for the thyroid hormone receptor and retinoic acid receptor (Wong and Privalsky, 1995).

In vitro mutagenesis experiments have generated mutant AR proteins that can still bind steroid but are nonfunctional, or AR proteins that do not bind steroid but are now constitutively active (Rundlett et al., 1990; Jenster et al., 1991; Simental et al., 1991). Deletion of all or part of the N-terminal domain or of the DNA binding domain renders the AR transcriptionally inactive despite its ability still to bind androgen with high affinity (Jenster et al., 1991; Simental et al., 1991). Most remarkably, deletion of the hormone binding domain renders the AR unable to bind steroid but constitutively transcriptionally active even in the absence of hormone (Rundlett et al., 1990; Jenster et al., 1991; Simental et al., 1991). Prostate cancer cells with such a mutation would be AR-positive (e.g., based on immunohistochemistry) but would be androgen-independent.

A distinguishing feature of the AR protein is that the N-terminal domain contains homopolymeric repeats of glutamines, prolines, and glycines (Chang et al., 1988; Lubahn et al., 1988; Faber et al., 1989; Tilley et al., 1989) (see Fig. 5). A fascinating feature of the AR glutamine repeat, which starts at amino acid 58 and is encoded by a CAG repeat in the gene, is that its length varies (11–31 CAGs) in the population (i.e., is polymorphic) (La Spada et al., 1991; Edwards et al., 1992). The polymorphic CAG repeat is followed by an invariant CAA (glutamine) codon (Chang et al., 1988; Lubahn et al., 1988; Faber et al., 1989; Tilley et al., 1989); thus, an AR with 18 CAGs actually codes for an AR with a 19-glutamine repeat. The glycine repeat (GGN; N = T, G, or C) also is polymorphic (Macke et al., 1993; Sleddens et al., 1993; Irvine et al., 1995); it starts at amino acid 440–460 (depending on the length of the glutamine CAG repeat) and is encoded by an invariant six-glycine repeat (GGT/GGG), followed by a GGC (glycine) repeat that is polymorphic in length. AR with different CAG and GGN repeat lengths represent different AR alleles or AR variants. The proline repeat appears to be invariant in length.

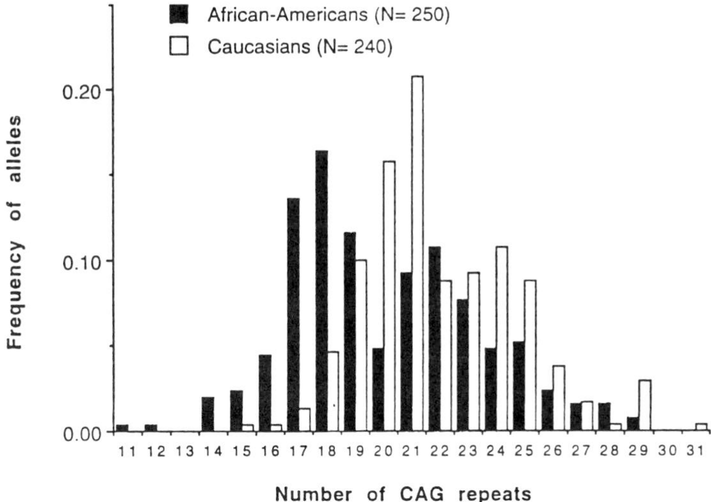

Figure 6. Frequency distributions of AR CAG repeat length in the general population of African-Americans and Caucasians (based on data from Edwards et al., 1992).

It is interesting that the frequency distribution of the AR CAG repeat length in the general population varies in different racial groups (Edwards et al., 1992) (Fig. 6). In Caucasians, the most common AR allele has 21 CAGs, whereas in African-Americans, who have a significantly higher prostate cancer incidence and mortality rate (Ross et al., 1983), the most common allele has 18 CAGs (Edwards et al., 1992) (Fig. 6). The frequency distributions actually appear to be bimodal (Fig. 6; Edwards et al., 1992). It appears that the AR glycine repeat length frequency distribution also differs among racial groups, with short glycine repeats being more common in African-Americans than in Caucasians (Irvine et al., 1995).

An interaction between two amino acid locations in a protein can have dramatic consequences, as illustrated by (1) the prion protein, in which a mutation at amino acid 178 causes two diseases with distinct phenotypes (fatal familial insomnia versus Creutzfeldt-Jakob disease subtype 178) depending on which amino acid (methionine versus valine) is present at the normal polymorphic codon 129 (Monari et al., 1994), and (2) the AR, such that the transcriptional activity of an AR protein with a mutation in the hormone binding domain is markedly affected by the length of the polymorphic glutamine repeat (12 gln versus 20 gln) in the N-terminal domain (McPhaul et al., 1991). These precedents raise the intriguing possibility that the transcriptional activity of an AR protein with 18 glutamines and 17 glycines may be different from that of an AR with 24 glutamines and 17 glycines.

Polymorphisms are not uncommon; there are numerous proteins for which genetic polymorphisms affect protein levels or protein function without causing disease (Thompson et al., 1991). Thus, there may be many forms of "normal." For example, polymorphisms of the vitamin D receptor, a member of the steroid receptor superfamily, account for varying levels of serum osteocalcin (a vitamin D target gene) (Morrison et al., 1992) and bone density (Morrison et al., 1994) among individuals. Nonetheless, "normal" variants of one protein may exhibit different interactions with other factors and thereby affect the biological phenotype, and conceivably also affect disease susceptibility (Thompson et al., 1991). This phenomenon is elegantly illustrated by the example of α_1-antitrypsin polymorphism, which leads to a several-fold variation in enzyme activity among apparently normal persons who have 1 or 2 of the more than 20 different α_1-antitrypsin alleles (Thompson et al., 1991). However, people with two *Z* alleles of α_1-antitrypsin have $<15\%$ of the normal plasma concentration of α_1-antitrypsin, and have a higher risk of obstructive lung disease, the progression of which is greatly augmented by smoking (Thompson et al., 1991). Thus, α_1-antitrypsin allelotyping of individuals may be viewed as a potential marker for lung disease risk assessment. An intriguing question is whether AR glutamine and/or glycine repeat length polymorphisms may influence androgen action, and thereby influence prostate cancer development or progression. This could represent a potential marker for prostate cancer risk assessment.

Glutamine repeats and proline repeats confer transcriptional activity on the DNA binding domain of the yeast transcription factor GAL4, and a certain number of repeats is required for optimal transcriptional activity of this engineered chimeric protein (Gerber et al., 1994). Gerber et al. (1994) hypothesize that repeats may be the main cause for modulation of transcription factor activity. It is interesting that polymeric stretches of ≥ 20 glutamines occur predominantly in transcriptional regulatory proteins (Gerber et al., 1994).

Glutamine repeats, glutamine-rich domains, and proline repeats are present in several transcription factors (Emili et al., 1994; Gerber et al., 1994; Gill et al., 1994; Madden et al., 1991). In the transcription factor Sp1, the glutamine-rich domain is critical for transcriptional activity, and it mediates interaction with another component of the transcription apparatus, namely the TATA box-binding protein (TBP, a component of the TFIID complex; Greenblatt, 1992; Courey et al., 1989; Emili et al., 1994; Gill et al., 1994); the strength of this interaction correlates with the potency of Sp1 as a transcriptional activator (Emili et al., 1994; Gill et al., 1994). In the Wilms tumor suppressor gene product (WT1), the glutamine- and proline-rich N-terminal domain is required for repression of transcription (Madden et al., 1991). Glutamine repeats form β-sheets held together strongly by hydrogen bonds (Perutz et al., 1994); it is proposed that they may thus function as polar zippers by bringing together transcription factors bound to separate DNA segments (Perutz et al., 1994).

The glutamine repeat also affects the transcriptional activity of the AR. Deletion of the glutamine repeat of the rat and human AR N-terminal domain yields an AR with increased transcriptional activity, and an increase in glutamine repeat length decreases transcriptional activity (Mhatre et al., 1993; Chamberlain et al., 1994; Kazemi-Esfarjani et al., 1995). This indicates that the repeat can repress transcription of an AR responsive reporter gene. The ability of glutamine repeats to function as polar zippers (Perutz et al., 1994) suggests a potential role of glutamine repeats in AR dimerization or in the assembly and stabilization of the AR with the multiple transcription factors that constitute the transcription machinery. The role of the glycine repeat and the proline repeat in AR function is unknown.

Steroid receptors function as transcription factors. This function requires more than ligand binding, receptor dimerization, and receptor binding to specific DNA sequences. Steroid receptor interactions with accessory proteins are required for high affinity binding to DNA (Murray and Towle, 1989; Kupfer et al., 1993, 1994; Tsai and O'Malley, 1994), and for activation of transcription of steroid responsive genes (Tsai and O'Malley, 1994; see Chapter 4, this volume). The parts of the receptor protein that are involved in these interactions are just now being identified. A receptor accessory factor (RAF), subsequently identified as insulin-degrading enzyme (IDE), enhances specific binding of a truncated fragment of the rat AR (comprising the DNA binding domain and a portion of the N-terminal domain) to an androgen response element (Kupfer et al., 1993, 1994). Proteins have been identified recently that interact with the estrogen receptor (ER) and progesterone receptor (PR) and are required for receptor-mediated transcriptional activation (Ing et al., 1992; Cavailles et al., 1994; Halachmi et al., 1994). The transcription factor TFIIB binds to an ER fragment consisting of the DNA binding domain and the hormone binding domain, but not to an ER fragment that contains the DNA binding domain and the N-terminal domain (Ing et al., 1992). TFIIB binds to a TFIID-DNA complex and facilitates subsequent binding of the transcription apparatus; in this way, TFIIB binding to the ER may facilitate ER-stimulated transcription (Ing et al., 1992). TFIIB also binds to PR, but the preferred region of interaction differs from that of the ER (Ing et al., 1992); this is not surprising because the activation domains of different members of the steroid receptor superfamily are located in different parts of the receptor molecule (Jenster et al., 1991; Simental et al., 1991; Ing et al., 1992; Tsai and O'Malley, 1994). The hormone binding domain of the ER, which contains a transcription activation domain (referred to as TAF-2; Ing et al., 1992), also interacts with a 160 kDa protein (ER-associated protein p160, ERAP160) (Cavailles et al., 1994; Halachmi et al., 1994). The interaction between ER and ERAP160 is estrogen-dependent; it is also required for transcriptional activity because ER hormone binding domain mutants that lack transcriptional activity do not bind to ERAP160 (Cavailles et al., 1994; Halachmi et al., 1994). The transactivation domains of the AR are located in the N-terminal domain (Jenster et al., 1991, 1995; Simental et al., 1991); based on the studies described above, we predict that proteins that interact with this domain are likely to modulate AR function.

The length of the glutamine repeat affects the transcriptional activity of the AR: AR with longer glutamine repeats have lower activity (Mhatre et al., 1993; Chamberlain et al., 1994; Sobue et al., 1994; Kazemi-Esfarjani et al., 1995). Thus, AR glutamine repeat length polymorphism in the general population might cause population differences in AR activity and, given the important role of androgen in development, growth, and behavior, may help to account for variations in the degree of androgenization and manifestations of androgenic activity in the population. Potential differences in the AR activity of "normal" AR alleles could even affect the role of androgen in prostate carcinogenesis, progression, or growth. We speculate that the existence of multiple allelic variants, with a range of activity that may depend on the glutamine and/or glycine repeat length, may help to account for the fact that racial groups with different AR allele frequency distributions (Edwards et al., 1992) have different incidence rates of prostate cancer (Ross et al., 1983). AR with shorter glutamine repeats may have more activity, and the effect of androgen in prostate carcinogenesis may be greater in individuals with such AR alleles; if short repeats enhance AR activity and if AR activity is required for prostate carcinogenesis, then the higher frequency of shorter AR CAG repeats (e.g., CAG_{18}) in African-Americans (see Fig. 6) may account for the higher frequency of prostate cancer in African-American men. Thus, the effect of glutamine and glycine repeat length on AR function may determine the sensitivity of tumor cells to existing tissue levels of dihydrotestosterone (DHT). Tissue DHT levels depend on circulating androgen levels and the amount of 5α-reductase activity in the tissue. Although potential AR activity may be affected by the length of the glutamine and/or glycine repeat, actual AR activity will depend also on these other factors.

The AR gene is on the X chromosome (Brown et al., 1989), so males have a single copy of the AR gene. The human AR cDNA that has been cloned, from different sources by different groups, codes for an AR protein with 910 to 919 amino acids, this difference in length due to different glutamine and/or glycine repeat lengths (Chang et al., 1988; Lubahn et al., 1988; Faber et al., 1989; Tilley et al., 1989). We refer to a 919 amino acid AR as full-length (Lubahn et al., 1988, 1989). There is only one AR gene, but two forms of the AR protein have been documented recently: AR_A and AR_B (Wilson and McPhaul, 1994). AR_B (110 kDa) is full-length, whereas AR_A (87 kDa) lacks amino acids 1 to 188 (which include the glutamine repeat), the result of alternative translation initiation at met 189 (numbering according to Lubahn et al., 1989; Zoppi et al., 1993; Wilson and McPhaul, 1994).

It seems unlikely that the A and B forms of the human AR are translated from the same mRNA transcript, because cells transfected with an AR cDNA expression vector that encodes the full-length AR express only form B (Zoppi et al., 1993). If the first translation initiation site is mutated, then the full-length cDNA does direct translation of the shorter A form (Zoppi et al., 1993). This indicates that the B and A forms can arise from translation of the same transcript, but that utilization of the two translation start sites is controlled by some mechanism. On the other hand, however, when the AR cDNA was transcribed and translated in a rabbit reticulocyte lysate *in vitro*, multiple protein products were produced, with sizes consistent with translation initiation at internal ATG/met codons (Chang et al., 1988). The AR promoter has two transcription start sites 13 bp apart (Tilley et al., 1990a; Faber et al., 1991, 1993); therefore, the A and B protein forms might be translated from different AR mRNA transcripts, the production of which may be under the control of different promoters. If one of these transcripts lacked the AUG_B but contained the AUG_A translation start site, then this transcript would encode the AR_A form. By analogy, the two forms of the human progesterone receptor (PR), PR_B (full-length; 933 amino acids) and PR_A (N-terminally truncated, lacking amino acids 1–164), arise by translation from two different transcripts, and the PR gene has two transcription start sites, 15 bp apart, that are under the control of different promoters (Kastner et al., 1990a,b). In contrast, two functional forms of the Xenopus ER are produced from a single mRNA species by alternative translation initiation at two methionines 42 codons apart (Claret et al., 1994).

AR_A has a higher ligand dissociation rate than AR_B (Zoppi et al., 1993), perhaps because AR_A lacks the part of the N-terminal domain that slows the rate of ligand dissociation from the hormone binding domain and the rate of AR degradation (Zhou et al., 1995). AR_A is also less effective than AR_B in stimulating transcription from an androgen-responsive MMTV promoter in CV-1 monkey kidney cells (Zoppi et al., 1993); however, the activity with other promoters and in other cell types has not been tested. By analogy, the A and B forms of the PR exhibit different activities with different promoters and in different cell types (Vegeto et al., 1993). In some cell types, PR_A is less active than PR_B, but in other cell types, PR_A is more active than PR_B (Vegeto et al., 1993). Because steroid receptor transcriptional activity is highly promoter-specific and cell-specific (Vegeto et al., 1993; Fujimoto and Katzenellenbogen, 1994; Ince et al., 1994; Tzukerman et al., 1994), and given the role of the N-terminal domain (especially the glutamine repeat) of the AR in transactivation (Mhatre et al., 1993; Chamberlain et al., 1994; Kazemi-Esfarjani et al., 1995), it is tempting to speculate that the two forms of the AR might have different effects in prostate cells, different effects in prostate cancer cells than in normal prostate cells, or different interactions with agonists versus antagonists.

In normal genital skin fibroblasts, AR_A represents about 10% of total AR (Wilson and McPhaul, 1994); in other normal tissues, it represents 5–20% (Wilson, 1994). Expression patterns of the two AR forms in prostate cancer have not been studied. The A and B forms differ in activity (Zoppi et al., 1993); if the ratio of the two forms in cancer differs from that in normal prostate, then AR action could affect the phenotype of the cancer cell, without a change in the structure of the AR gene. If AR_A and AR_B form heterodimers, then AR_A might modify AR_B activity, analogous to the way PR_A can modify PR_B activity by forming a heterodimer (Tung et al., 1993; Vegeto et al., 1993; see Chapter 10, this volume). Thus, the ability to produce two forms of the AR may provide the cell with an opportunity to control signaling in ways that we have not previously appreciated.

Human benign prostatic hyperplasia (BPH) contains two prominent AR mRNA transcripts, 7 kb and 10 kb, whereas genital skin fibroblasts express predominantly a 10-kb species (Lubahn et al., 1988). Similarly in the rat, some tissues produce a single 11-kb AR mRNA, whereas others produce two transcripts, 11 kb and 9.3 kb (McLachlan et al., 1991). In normal rat tissues, these two AR transcripts differ in their 5′-untranslated region (UTR), and appear to arise from alternative splicing (McLachlan et al., 1991). In the LNCaP human prostate cancer cell line, the two AR mRNA transcripts differ in their 3′-UTR, the shorter transcript lacking 3 kb within the 3′-UTR as a result of alternative splicing (Faber et al., 1991). The two AR mRNA species in human BPH have not been characterized, and AR mRNA transcripts have not been studied in prostate cancer tissue specimens. The 5′- and 3′-UTRs of many genes affect mRNA stability, translational efficiency, and translation start site selection (Jackson, 1993; Sachs and Wahle, 1993; Scotto and Assoian, 1993; Claret et al., 1994). Thus, alternative splicing is a normal process that could affect the level or activity of the AR protein product. It is not known whether the 10-kb AR mRNA species in the human prostate encodes both the A and B forms of AR. Available evidence indicates that forms A and B of the AR are not both translated from the same cloned AR cDNA (Zoppi et al., 1993), which is 2943 bp in length and contains 44 bp of the 5′-UTR and 148 bp of the 3′-UTR (Tilley et al., 1989). However, the possibility has not been ruled out that the full-length 10-kb AR mRNA may control alternate translation start site utilization, or that the 10-kb band detected in a Northern blot might contain more than one transcript. By analogy, the 11.4-kb human PR mRNA species in breast cancer cells actually consists of four transcripts of similar size, two of which have a different 5′ transcription start site and code for an N-terminally truncated PR protein (Wei et al., 1990).

The AR is encoded by 8 exons (Fig. 7), as are the other members of the steroid receptor family, and the splice site junctions are highly conserved in the different steroid receptor genes (Lubahn et al., 1989). Thus, the potential exists for alternative RNA splicing to produce transcripts that lack one or more exons and thereby encode variant AR proteins (e.g., Ris-Stalpers et al., 1990; Nigro et al., 1991). By analogy, ER mRNA splice variants that encode ER proteins with altered activity are produced both in normal tissue and in breast cancer (Fuqua et al., 1991; McGuire et al., 1991; Wang and Miksicek, 1991; Castles et al., 1993). Breast cancer cells that express an exon 5 deletion variant of the ER, a variant that encodes a constitutively active ER protein, also express a wild-type, full-length ER mRNA, but the ratio of these transcripts differs in different cell lines. It has been suggested that an increased variant/wild-type ratio may be involved in breast cancer progression to hormone independence (Castles et al., 1993). The production

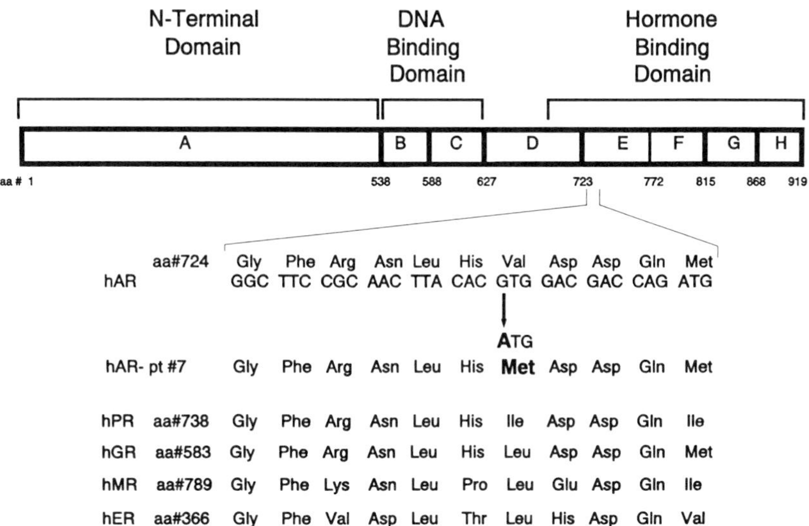

Figure 7. (*Upper*): Schematic of human AR structure. Amino acid positions of each exon are noted below the bar (numbering according to Lubahn et al., 1989). (*Lower*): Comparison of mutant AR sequence in stage B prostate cancer (patient #7) with that of wild type AR and the equivalent region of other steroid hormone receptors: human progesterone receptor (hPR), glucocorticoid receptor (hGR), mineralocorticoid receptor (hMR), and estrogen receptor (hER). Wild type sequences are from Lubahn et al. (1989). Note that the AR gene mutation at codon 730 in patient #7 is in a region of the hormone binding domain highly conserved among steroid receptors. [data from Newmark et al., 1992].

of AR mRNA splice variants in prostate cancer has not been studied extensively. In an analysis of AR in 10 specimens of prostate cancer by reverse transcriptase-PCR, the presence of splice variants was not ruled out, because some samples contained unexpected PCR products that were not identified, because the resolution of the gels may not have been adequate to distinguish products that differed in size by only 117 bp (the size of exon C), and because primers were not used that could have detected exon A splice variants (Culig et al., 1993b).

A central tenet of hormone action is that agonists are required for receptor activity, and that antagonists block this effect but have no effect on their own. This is the basis for the rational therapy of hormone-dependent cancers with steroid receptor antagonists. Remarkably, however, several recent studies challenge this dogma, with evidence that under certain conditions: (1) receptor-dependent transcriptional activity can be achieved in the absence of ligand if certain intracellular protein phosphorylation signaling pathways are activated [e.g., by protein kinase A (PKA) activators such as 8-Br-cAMP or peptide growth factors (Denner et al., 1990; Power et al., 1991; Aronica and Katzenellenbogen, 1993; Ignar-Trowbridge et al., 1993; Culig et al., 1994)];

(2) receptor antagonists can act as agonists in cells treated with PKA activators (Beck et al., 1993; Nordeen et al., 1993; Sartorius et al., 1993; Tung et al., 1993; Fujimoto and Katzenellenbogen, 1994; Ince et al., 1994); and (3) receptor mutants that are transcriptionally inactive in the presence of ligand (agonist or antagonist) can be activated in cells treated with ligand plus PKA activator, and the magnitude of the activity is dependent on promoter- and cell-contexts (Ince et al., 1994). Many different receptor types exhibit such unconventional activities, including the PR (Denner et al., 1990; Power et al., 1991; Beck et al., 1993; Sartorius et al., 1993; Tung et al., 1994), GR (Nordeen et al., 1993), and ER (Aronica and Katzenellenbogen, 1993; Ignar-Trowbridge et al., 1993; Fujimoto and Katzenellenbogen, 1994; Ince et al., 1994). Recent studies indicate that the AR may also be included in this list (Culig et al., 1994). Transcriptional activation of the rat AR by androgen is enhanced significantly by modulators of protein phosphorylation (Ikonen et al., 1994), and androgen-independent activation of the human AR can be achieved by treating cells with IGF-I, KGF, or EGF (Culig et al., 1994; see Chapter 15, this volume, for a discussion of these growth factors). It will be important to determine whether AR_A and/or AR_B can be regulated anomalously by antagonists and by activators of phosphorylation pathways, whether the A form can modulate the activity of the B form [as can occur with the two PR forms (Tung et al., 1993; Vegeto et al., 1993)], how AR mutations affect these activities, and whether the glutamine and glycine repeat length in AR_B may affect anomalous receptor activity and thereby account for variable responses to therapy within a population.

The phenomena of ligand-independent receptor activity (Denner et al., 1990; Power et al., 1991; Aronica and Katzenellenbogen, 1993; Ignar-Trowbridge et al., 1993; Culig et al., 1994), modulation of ligand-dependent activity by activators of protein phosphorylation (Fujimoto and Katzenellenbogen, 1994; Ikonen et al., 1994; Ince et al., 1994), antagonists acting as agonists (Beck et al., 1993; Nordeen et al., 1993; Sartorius et al., 1993; Tung et al., 1993; Fujimoto and Katzenellenbogen, 1994; Ince et al., 1994), and recovery of mutant receptor activity by activation of protein phosphorylation pathways (Ince et al., 1994) are intriguing. If, as described for other members of the steroid receptor family, they are applicable as well to the AR, this could help explain the clinical behavior of prostate cancers that are androgen-independent, as well as prostate cancers that exhibit a paradoxical agonistic response to flutamide or other antiandrogens such as casodex or megestrol acetate (Scher and Kelly, 1993; Sartor et al., 1994; Small and Carroll, 1994; Dawson and McLeod, 1995; Nieh, 1995), even if these tumors express wild-type AR.

Role of the AR in Proliferation

The dogma of androgen action in prostate epithelial cells is that androgen, acting via the AR, stimulates cell proliferation, perhaps by stimulating the expression or activity of growth-stimulating peptide hormones. This is certainly true for the normal prostate *in vivo*, because castration causes prostate epithelial cell death, and androgen alone can restore the prostate to its normal size (Isaacs, 1984). It may even be true for hormone-refractory prostate cancers *in vivo*, which are AR-rich (van der Kwast et al., 1991; Ruizeveld de Winter et al., 1994) and continue to proliferate in an androgen-poor environment; AR may function under these conditions as a result of ligand-independent AR activation (Culig et al., 1994) or as a result of activation by residual levels of androgen, the effect of which may be amplified by AR gene amplification (Visakorpi et al., 1995a).

On the other hand, however, the proliferative behavior of prostate cancer cells *in vitro* challenges the role of androgen as a positive regulator of proliferation, and it raises the question of whether the unexpected effect of androgen on proliferation *in vitro* is an artifact of culture conditions or whether it is an indicator of an unsuspected mechanism of growth control. Thus, AR-positive LNCaP prostate cancer cells are growth-stimulated *in vitro* by androgen but only by low concentrations (about 0.1–0.3 nM DHT); higher concentrations have no effect (Olea et al., 1990; Lee et al., 1995). LNCaP cells propagated long-term in the absence of androgen express higher levels of AR protein and are growth inhibited by concentrations of androgen that previously stimulated proliferation (Kokontis et al., 1994). When normal rat prostate epithelial cells are cultured, the cells that proliferate are AR-negative (Rundlett et al., 1992), suggesting that AR expression may select against growth *in vitro*. All but one of the human prostate cancer cell lines are AR-negative (Iizumi et al., 1987; Tilley et al., 1990b), whereas most prostate cancers *in vivo* are AR-positive (Sadi et al., 1991; van der Kwast et al., 1991; Ruizeveld de Winter et al., 1994), and the AR positive PC-82 and PC-EW human prostate cancer xenografts cannot be propagated *in vitro* (Ruizeveld de Winter et al., 1992; Hoehn et al., 1984). Moreover, stable transfection of the AR into AR-negative prostate cancer cells results in androgen-mediated inhibition of cell proliferation (Yuan et al., 1993; Suzuki et al., 1994). Such data suggest that, if a similar phenomenon occurred *in vivo*, then hormone-refractory prostate cancers might be growth-inhibited by androgen therapy. Since this is not the experience of clinical practice (Fowler and Whitmore, 1981), the relevance of these *in vitro* studies is not apparent at this time. On the other hand, however, the paradoxical growth-inhibitory effect of androgen on prostate cancer cells *in vitro* (Yuan et al., 1993; Kokontis et al., 1994; Suzuki et al., 1994) is not unique because introduction of the ER gene into ER-negative breast cancer cells renders the cells growth-inhibitable by estradiol (Garcia et al., 1992; Zajchowski et al., 1993), and breast cancer cells that overexpress EGF receptors (Armstrong et al., 1994) or FGF receptors (McLeskey et al., 1994) are growth-inhibited by EGF or FGF, respectively. A resolution of the paradox that a hormone that stimulates growth *in vivo* can inhibit growth *in vitro* may provide important clues to our understanding of the mechanism of growth control.

Loss-of-Function AR Gene Mutations

Androgen, acting via the AR, is required *in utero* and at puberty for the differentiation and growth of male external genitalia and male sex accessory tissues (Cunha et al., 1987; Kalloo et al., 1993; Luke and Coffey, 1994). Germline AR gene mutations interfere with these processes to varying degrees (Kazemi-Esfarjani et al., 1993; Sultan et al., 1993). This is best illustrated by complete androgen insensitivity syndrome (complete AIS), an X chromosome–linked inherited disorder that causes XY genotypic males to develop as phenotypic females because of defective AR (Kazemi-Esfarjani et al., 1993; Sultan et al., 1993). These patients have female external genitalia and lack a prostate and seminal vesicles, despite the presence of testes and male levels of circulating androgen (Sultan et al., 1993). Germline AR gene mutations in patients with complete AIS have been found in the N-terminal domain, the DNA binding domain, or the hormone binding domain (Sultan et al., 1993). These mutations inactivate AR function, even though some mutant AR proteins still bind androgen (Kazemi-Esfarjani et al., 1993; Sultan et al., 1993). AR mutations have also been found in males with partial AIS (Klocker et al., 1992; Kazemi-Esfarjani et al., 1993; Sultan et al., 1993; Tsukada et al., 1994), including those with Reifenstein syndrome (Klocker et al., 1992) or undervirilized male syndrome (Tsukada et al., 1994); these mutations do not completely abrogate AR function (Klocker et al., 1992; Kazemi-Esfarjani et al., 1993; Sultan et al., 1993; Tsukada et al., 1994). Some AR gene mutations that compromise AR activity are located in the introns instead of in the coding sequence; these mutations result in aberrant splicing (Ris-Stalpers et al., 1990, 1994).

The AR glutamine repeat length is expanded in the germline DNA to twice normal in men with X-linked spinal and bulbar muscular atrophy (SBMA, also known as Kennedy's disease) (La Spada et al., 1991). This AR gene mutation does not interfere with the role of AR in masculinization *in utero* or at puberty because these patients undergo normal male external genital development, but it does interfere with normal AR function in specific tissues (e.g., motor neurons) at adulthood. Some patients also develop signs of mild androgen insensitivity at adulthood, such as gynecomastia and infertility (La Spada et al., 1991). Androgen action is blunted in men with this AR defect because the ability of a synthetic androgen to lower plasma testosterone, luteinizing hormone (LH), and follicle-stimulating hormone (FSH) levels is significantly lower in SBMA patients than in controls (Sobue et al., 1994). In addition, an AR with a glutamine repeat that is twice the normal length (as found in patients with SBMA) has half as much activity in stimulating androgen-responsive gene transcription (Mhatre et al., 1993). In the genital skin fibroblasts of patients with SBMA, the affinity of AR for androgen is significantly lower than normal (MacLean et al., 1995); this indicates that the N-terminal domain can influence the three-dimensional structure of the hormone binding domain. It also predicts that in the presence of a subsaturating level of androgen, as may occur as men age and serum testosterone levels decrease, receptor occupancy and hence transcriptional activity may be less than normal in men with SBMA. There is, indeed, independent evidence that the N-terminal domain of the AR affects the rate of ligand dissociation (Zhou et al., 1995). A different study showed no effect of glutamine repeat length on the affinity of the AR for androgen; however, it was done using transiently transfected monkey kidney cells (Chamberlain et al., 1994).

Patients with AIS or SBMA have germline AR gene mutations, which thus exist in every cell in the body. Therefore, in addition to effects on the development of male external genitalia and reproductive tract organs, AR gene mutations in men with partial AIS may also affect other organ systems. For example, AR gene mutations may also affect mammary gland growth and development, as illustrated by the development of gynecomastia in men with Reifenstein syndrome (Klocker et al., 1992) and in men with undervirilized male syndrome (Tsukada et al., 1994). In addition, breast cancer has been documented in three men with partial AIS that is due to a germline mutation in the DNA binding domain of the AR coding sequence (Wooster et al., 1992; Lobaccaro et al., 1993). Two of these men were brothers who had Reifenstein syndrome with severe hypospadias (a defect of male external genital development in which the urethral opening is not at the tip of the glans penis) and other features of inadequate androgenization (Wooster et al., 1992). The third patient had partial AIS (Lobaccaro et al., 1993). Although it is not a certainty that these AR mutations were a risk factor for breast cancer development, Wooster et al. (1992) argue cogently that because breast cancer in brothers is extremely rare, it would be highly improbable to find a pair of male breast cancer patients both clinically and genetically androgen insufficient if the mutation were unrelated to breast cancer susceptibility. They also argue that two cases of male breast cancer in brothers are likely to be the result of inheriting some predisposing mutation, and that the AR gene mutation is the likely candidate gene (Wooster et al., 1992). The subsequent report by Lobaccaro et al. (1993) of another man with breast cancer, partial AIS, and a germline AR gene mutation strengthens this association. These data support the hypothesis that it may be androgen in men that protects them from developing breast cancer, rather than the relative absence of estrogen (Thomas et al., 1992). Thus, androgen action via an intact AR may protect against breast carcinogenesis or breast cancer progression, and AR mutations that decrease AR function may compromise the protective effect of androgen in the mammary gland and thereby contribute to carcinogenesis or disease progression.

Although androgen action is required for prostate development, growth, and function, it is not known whether men who have partial AIS, and a germline AR gene mutation that decreases AR activity, also have a lower than normal incidence of BPH or prostate cancer. On the other hand, however, 16 of 74 Japanese men with latent prostate cancer were found to have somatic AR gene mutations that were predicted to ablate AR activity (Takahashi et al., 1995). Latent prostate cancer refers to cancer that is detected solely by histologic evaluation (Carter et al., 1990). Based on autopsy studies, about 15–30% of men over the age of 45 to 50 years have latent prostate cancer, that is, histologic evidence of prostate cancer. This leads to an estimate that about 10 million men in the United States have latent prostate cancer (Carter and Coffey, 1990). Only a small percentage of these tumors ever become clinically manifested because only 244,000 new cases of clinical prostate cancer are diagnosed per year (Wingo et al., 1995). Therefore, it is assumed that most histologic prostate cancers are truly latent and will never progress (see Chapter 13, this volume). Of course, it is impossible to know whether a given specimen of histologically diagnosed prostate cancer might have eventually become clinically manifested if it had remained in the patient for a longer period of time. It is fascinating that the prevalence of latent prostate cancer appears to be the same in Japanese men living in Japan and American men living in the United States (Carter et al., 1990); however, the incidence of clinical prostate cancer is many times higher in American men than in Japanese men (Ross et al., 1983). It is tempting to speculate that the occurrence of somatic AR gene mutations in latent cancers of Japanese men may preclude progression to clinical disease (Takahashi et al., 1995).

AR Gene Mutations in Clinical Prostate Cancer

The first AR gene mutation found in prostate cancer was in the LNCaP human prostate cancer cell line; amino acid 877 in the hormone binding domain is alanine instead of threonine (Harris et al., 1990; Veldscholte et al., 1990, 1992; Kokontis et al., 1991). [Codon 877 of a 919-codon AR cDNA (Lubahn et al., 1989) is equivalent to codon 868 of a 910-codon AR cDNA (Faber et al., 1989; Veldscholte et al., 1990).] This mutant AR binds androgen with an affinity like that of wild-type AR (Harris et al., 1990; Veldscholte et al., 1990; Kokontis et al., 1991) and, like wild-type AR, it is not transcriptionally activated by the antiandrogen casodex (Veldscholte et al., 1992). Remarkably, however, this mutant AR binds and is transcriptionally activated by androgen, estrogen, progesterone, or the antiandrogens cyproterone acetate or hydroxyflutamide (the active metabolite of flutamide) (Harris et al., 1990; Veldscholte et al., 1990, 1992; Kokontis et al., 1991). The effect of this AR gene mutation on AR action explains the unusual phenotype of the LNCaP human prostate cancer cell line, the growth of which is androgen-independent (Horoszewicz et al., 1980) but stimulated by androgen, estrogen, progesterone, or antiandrogen (Harris et al., 1990; Veldscholte et al., 1990, 1992). These cells contain mutant AR, but no ER or PR. Although the possibility has not been ruled out that this AR gene mutation occurred during establishment of the LNCaP cell line *in vitro*, the mutation could well have contributed to continued growth of the androgen-independent lymph node metastasis from which the LNCaP cells were aspirated while the patient was receiving estrogen therapy, which did not interfere with tumor growth (Horoszewicz et al., 1980). Prostate cancers with an AR gene mutation like that in LNCaP cells would be AR-positive but would progress on androgen ablative therapies that include estrogen or antiandrogen, and would manifest a paradoxical response to flutamide withdrawal. Thus, not only does the LNCaP AR gene mutation not cause a loss of function, but it also results in a gain of function. This is strikingly different from the loss-of-function AR gene mutations found in AIS and SBMA.

We therefore initiated studies to determine whether mutations in the AR gene might occur in prostate cancer surgical specimens from patients with clinically manifest prostate cancer. By comparing prostate cancer DNA to nontumor DNA, we could determine whether mutations were somatic (present only in the tumor DNA) or germline (present in both). We use PCR to amplify genomic DNA and analyze PCR products for the presence of mutations by using methods that can distinguish mutant AR DNA from wild-type AR DNA [e.g., DNA sequencing, denaturing gradient gel electrophoresis (DGGE), or single stranded DNA conformation polymorphism analysis (SSCP)]. AR mutations found in AIS and SBMA were discovered in specimens that had already been characterized biochemically as having defective or abnormal AR (Lubahn et al., 1989; Brown et al., 1990; Harris et al., 1990; Veldscholte et al., 1990; Kokontis et al., 1991; La Spada et al., 1991; Wilson, 1992; Sultan et al., 1993). In contrast, we are able to screen for the presence of mutations in large numbers of prostate cancer specimens having unknown AR properties by analyzing the mobility of PCR-amplified DNA fragments by DGGE or SSCP. We determined that a somatic mutation can be detected by DGGE if it represents at least 10% of the sample (Newmark et al., 1992).

We have screened for the presence of AR gene mutations in clinically localized prostate cancers that were excised by radical prostatectomy (Bova et al., 1993). Based on hematoxylin and eosin staining of a frozen section of the tumor nodule, blocks are trimmed to remove nonmalignant tissue and to enrich the proportion of tumor in the specimen to >75%. We use PCR to amplify genomic DNA using primers that flank the intron/exon boundaries; this allows us to detect mutations in the coding sequence and splice sites (Lubahn et al., 1989; Batch et al., 1992). We have used both DGGE and SSCP analysis to screen for the presence of DNA mutations. DGGE can separate DNA fragments that differ in sequence by only a single base (Sheffield et al., 1989; Abrams and Stanton, 1992; Newmark et al., 1992). When a DNA fragment enters a concentration of denaturant that causes melting, its mobility is slowed. Because the T_m (melting temperature) of DNA is sequence dependent, mutant fragments have a different T_m and therefore melt at a different temperature or, in DGGE, at a different concentration of denaturant than does the wild type DNA (Sheffield et al., 1989; Abrams and Stanton, 1992). Thus, the presence of a mutation in DNA alters its mobility under these conditions, making DGGE a very sensitive method of detecting mutations (Brown et al., 1990; De Bellis et al., 1992; Newmark et al., 1992).

In SSCP analysis, double-stranded PCR products are denatured in formamide at 95°C, then rapidly cooled and loaded onto nondenaturing gels and electrophoresed (Orita et al., 1989; Batch et al., 1992; Soto and Sukumar, 1992; Hayashi and Yandell, 1993; Glavac and Dean, 1993). Single-stranded DNA molecules assume conformations that depend on their sequence, and any given single-stranded molecule may assume one or more different conformations that may or may not have different mobilities, depending on the DNA sequence, conditions of electrophoresis, and sensitivity of the method to discriminate between these forms. By including [^{32}P]labeled dATP in the PCR reaction mixture to produce [^{32}P]labeled PCR products, the ability to detect mutated fragments that may represent minor products is enhanced, increasing the sensitivity of the method. SSCP analysis of some exons of the AR gene yields two major bands, each presumably representing one of the strands of the double helix, whereas other exons yield more than two major bands, presumably because some sequences can assume multiple conformations. Each exon has a characteristic SSCP pattern. Because DNA sequence affects single-stranded DNA conformation, SSCP is capable of detecting the presence of point mutations that result in a change in conformation and a change in electrophoretic mobility. Unfortunately, not all sequence changes result in a detectable change in mobility. Several parameters influence the sensitivity of SSCP (Glavac and Dean, 1993; Hayashi and Yandell, 1993).

We reported the first example of a somatic mutation in the AR gene in clinical prostate cancer (Newmark et al., 1992); the mutation was a single base change in codon 730 [val (GTG) → met (ATG)] (see Fig. 7). The mutation was detected both by DGGE and SSCP, and its location was defined by sequencing the PCR product (Fig. 8). The first nucleotide position of codon 730 is a G, as seen in the patient's peripheral lymphocytes (Fig. 8), but the tumor had two bases, G and A, at this position, indicating the presence of a mixture of DNA fragments (wild-type and mutant), diagnostic of a somatic mutation. All other nucleotides were identical to wild-type, indicating that tumor exon E differed from normal by only a single base substitution. The mutation disrupted a *Pml*I restriction site, and the presence of the mutation was confirmed by *Pml*I digestion of tumor and nontumor DNA, which yielded DNA fragments of the expected length (Newmark et al., 1992).

Based on the relative intensity of bands in DGGE gels, we estimated that the mutation was in about 50% of the cells in the tumor specimen (Newmark et al., 1992). Thus, a substantial proportion of cells contained wild-type AR; some are likely to be nonmalignant cells (e.g., supporting stroma and blood vessels) and others may be malignant cells that lack the mutation. We inferred that the mutation was present in cells that had a growth advantage; otherwise it would not have been detectable. If the original cell with the

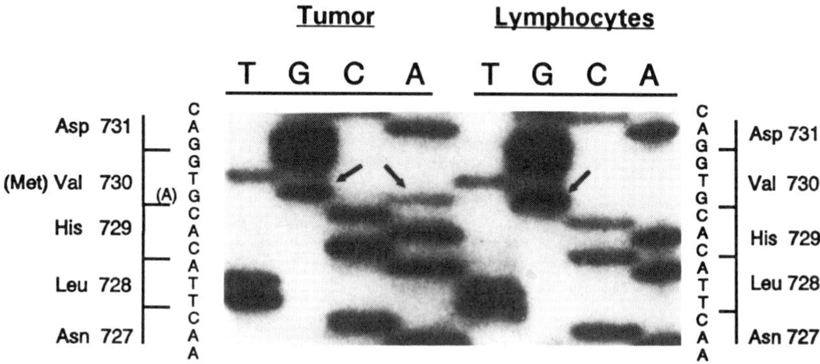

Figure 8. DNA sequencing gel of exon E in patient #7 tumor and normal lymphocytes. The tumor shows a G and A in codon 730 where lymphocytes contain only G at the position noted by the arrow. (Reprinted from Newmark et al., 1992, with permission).

somatic mutation in the AR gene had not had a growth advantage, it would have remained an insignificant percentage of the tumor, becoming diluted by other cells with a growth advantage, and it would not have been detected by screening. Whether the AR gene mutation itself conferred a growth advantage on the cell in which it occurred is not known.

Amino acid 730 is located in a region of the hormone binding domain that is highly conserved among different members of the steroid receptor family (see Fig. 7), although codon 730 itself is not conserved in different receptor classes. Only the AR contains valine at this codon, and valine is found at this site in human, rat, mouse, and canary AR (Lubahn et al., 1988; Charest et al., 1991; Nastiuk and Clayton, 1994). Thus, valine at this site may have a critical role in AR function and/or androgen binding, and the mutation of this amino acid in prostate cancer might affect some aspect of AR function. This conserved region of the receptor superfamily is involved in binding to ligand and to hsp90 (Lubahn et al., 1988; Housley et al., 1990). It has been suggested that unliganded steroid receptors (full-length wild-type) are transcriptionally inactive because hsp90 binding prevents receptor binding to target genes, and that hormone binding domain deletion mutants are constitutively active because hsp90 cannot bind (Picard et al., 1988). Deletion of this conserved region in the glucocorticoid receptor abrogates binding to hsp90 and creates a constitutively active receptor (Housley et al., 1990).

The amino acid sequence of the hormone binding domain and DNA binding domain of the AR are 100% conserved in human, mouse, and rat AR (Charest et al., 1991). This suggests that the conservation of wild type sequence is required for the conservation of wild-type AR function. Indeed, all single amino acid mutations in the AR hormone binding domain reported to date alter AR function, but in different ways. In complete AIS, AR mutants are unable to activate transcription at physiological androgen concentrations, although not all mutations abrogate androgen binding (Brown et al., 1990;

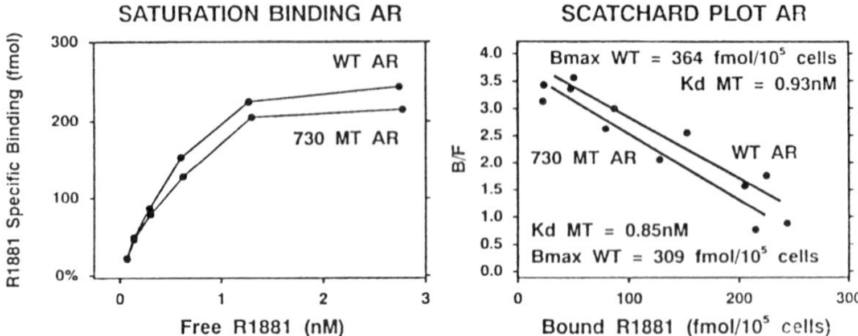

Figure 9. Analysis of androgen binding to wild-type AR (codon 730 is valine) and mutant AR (codon 730 is methionine) expressed in COS-7 cells.

Charest et al., 1991; Wilson, 1992; Sultan et al., 1993; Zhou et al., 1995). The AR mutation in the LNCaP human prostate cancer cell line alters the steroid specificity of the AR (i.e., increases its affinity for nonandrogenic steroids), but it is still an active transcription factor and, like wild-type AR, its transcriptional activity is still ligand-dependent (Harris et al., 1990; Veldscholte et al., 1990, 1992; Kokontis et al., 1991).

Interestingly, the AR_{730} mutant also exhibits an androgen binding affinity (K_d for R1881, a synthetic AR agonist) (Fig. 9), and androgen-dependent transcriptional activity (not shown) that are not different from wild-type AR (Schoenberg et al., 1993). In those experiments, site-directed mutagenesis was used to recreate the mutation in a human AR cDNA expression vector that codes for an AR with a 21-glutamine repeat and a 24-glycine repeat. The AR was then transiently expressed in COS cells (which lack endogenous AR) to assess ligand binding properties, and in CV-1 cells, along with an androgen-responsive MMTV-CAT reporter gene, to assess transcriptional activity of the mutant AR. There is now extensive evidence that the transcriptional activity of steroid receptors is highly promoter-dependent and cell type–specific (Vegeto et al., 1993; Fujimoto and Katzenellenbogen, 1994; Ince et al., 1994; Tzukerman et al., 1994), and that glutamine repeat length affects AR function (Mhatre et al., 1993; Chamberlain et al., 1994; Sobue et al., 1994; Kazemi-Esfarjani et al., 1995). Therefore, the activity of the AR_{730} mutant on an MMTV viral promoter in monkey kidney cells may not be representative of potential effects of this mutant AR in the prostate cancer cells of the patient (patient #7) who harbors the mutation at codon 730 and has a 25-glutamine repeat and a 23-glycine repeat. A mutation at codon 730 has not been found in any patient with complete or partial AIS; this is consistent with our functional characterization, and it suggests that AR gene mutations in prostate cancer [codon 730 (Newmark et al., 1992) and codon 877 (Veldscholte et al., 1990)] may represent a new class of AR mutations that do not cause a loss of function. Our data do not exclude the possibility that the codon 730 mutation may affect the steroid specificity of transactivation, as described for the LNCaP cell line AR mutation at

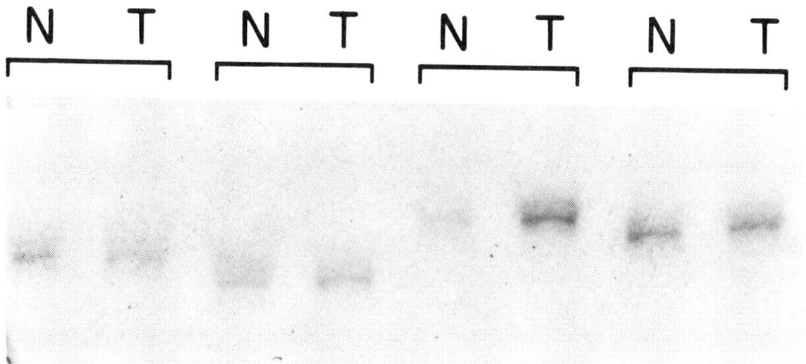

Figure 10. Comparison of the AR gene polymorphic CAG repeat length in prostate cancer. In the 4 patients illustrated the PCR products of each tumor (T) and nontumor (N) pair did not differ from each other. (Reprinted with permission of Academic Press, Inc., from Schoenberg et al., 1994).

codon 877. A preliminary report suggests that this codon 730 mutation may indeed allow transactivation by a high concentration of hydroxyflutamide (Stober et al., 1994).

In addition to its role in ligand binding, the hormone binding domain also contains a nuclear localization signal, a dimerization domain, hsp90 interaction domain(s), and transcription activation domains (Lubahn et al., 1988; Rundlett et al., 1990; Jenster et al., 1991; Simental et al., 1991; Wong et al., 1993; Zhou et al., 1994). Therefore, mutations in the hormone binding domain may affect more than just ligand binding.

We found a different clinical prostate cancer that had a somatic mutation in the length of the AR glutamine (CAG) repeat (Schoenberg et al., 1994). We compared the CAG repeat length of the AR gene in prostate cancer and in nonmalignant control tissue by PCR amplification of the region containing the repeat and subsequent electrophoretic analysis of PCR fragment length (Fig. 10). Because the AR gene is on the X chromosome (Brown et al., 1989), males have only a single allele per cell; this accounts for the presence of a single major PCR product from each specimen (Fig. 10). In one patient, nontumor DNA yielded a single PCR product as expected, but the tumor DNA yielded two discrete products, one identical to normal, and a second smaller one (Fig. 11), indicating the presence of cells that contain an AR gene with a deletion. Direct sequencing of each PCR fragment revealed that the nontumor tissue contained 24 CAGs, whereas the tumor contained one fragment with 24 CAGs (wild-type) and a second fragment with 18 CAGs (mutant), representing a somatic contraction of the AR CAG repeat ($CAG_{24} \rightarrow CAG_{18}$) in the tumor (Fig. 12). The remainder of the DNA sequence of the upper and lower bands and of tumor and nontumor DNA was identical to published wild-type sequence.

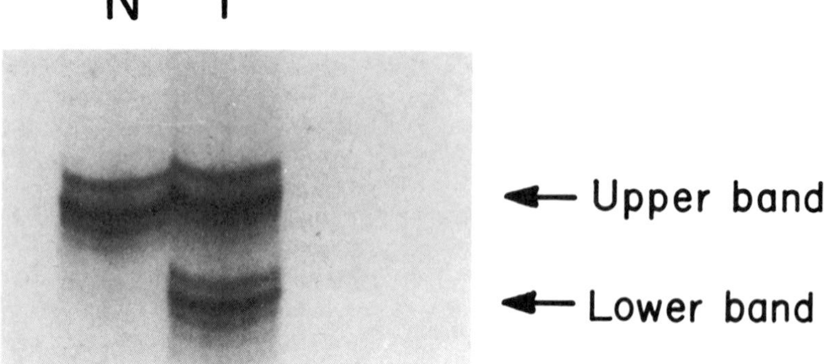

Figure 11. Prostate cancer (patient #P-6) with a somatic mutation in the CAG repeat of the AR gene. Nontumor DNA yielded a single major PCR product (*upper band*), but tumor DNA yielded 2 major products (*upper band and lower band*), indicating the presence of cells with a wild-type allele (*upper band*) and cells with a mutant allele (*lower band*) that contains a deletion. Accompanying minor bands are a common feature in denaturing gels of PCR products of repeat sequences. (Reprinted with permission of Academic Press, Inc., from Schoenberg et al., 1994).

The AR CAG microsatellite repeat length mutation in prostate cancer is especially interesting because microsatellite DNA sequences are potential sites of genetic instability (Loeb, 1994). Expansion to 40–52 CAGs occurs in the AR gene in patients with SBMA (La Spada et al., 1991). Several other neurologic diseases (fragile X syndrome, myotonic dystrophy, Huntington's disease, spinocerebellar ataxia type 1, and dentatorubral-pallidoluysian atrophy) have also been linked to microsatellite expansions in the genes that are thought to be involved in disease development (Martin, 1993). In addition, widespread microsatellite instability due to defects in mismatch repair enzymes occurs in many colon cancers and other types of cancers (Loeb, 1994).

We therefore considered the possibility that the microsatellite AR mutation identified in prostate cancer might be a marker of widespread genetic instability. However, we think this is unlikely because microsatellite instability appears to be uncommon in sporadic prostate cancer (Bussemakers et al., 1994). Interestingly, the prostate cancer with the AR codon 730 mutation has microsatellite instability, exhibiting somatic microsatellite mutations at 9 of 12 loci (GS Bova and WB Isaacs, unpublished), but it does not have a mutation at the AR CAG microsatellite (Schoenberg et al., unpublished).

It is interesting that the tumor with the AR CAG repeat length mutation manifested a paradoxical agonistic response to hormonal therapy with the antiandrogen flutamide (Schoenberg et al., 1994). The clinical features of a paradoxical response to flutamide withdrawal are initial response to androgen blockade, followed by evidence of disease progression that subsequently

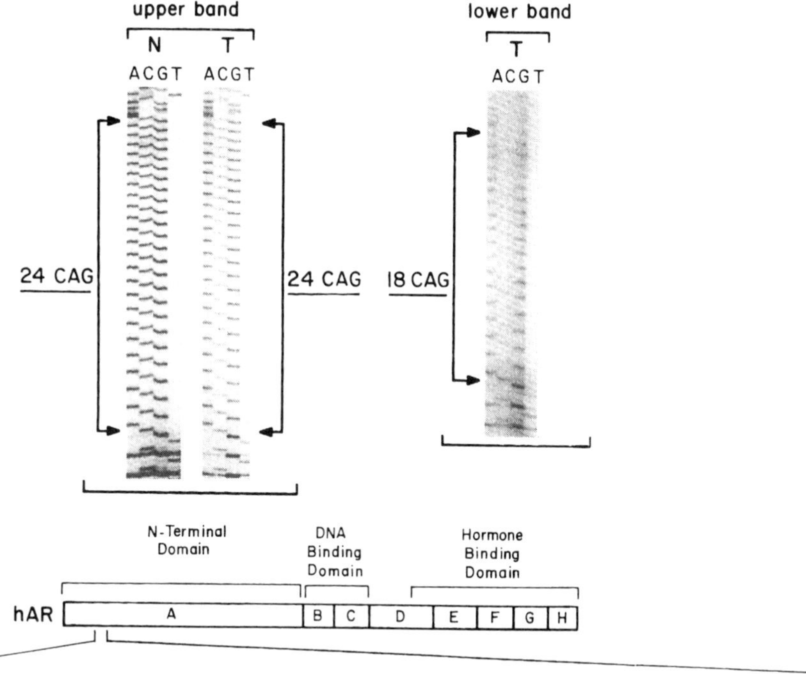

Figure 12. DNA sequence analysis of PCR products from tumor (T) and nontumor (N) DNA of patient #P-6. The lanes marked *upper band* or *lower band* refer to the PCR products in Figure 11. The upper band products from nontumor and tumor-derived DNA contain 24 CAGs; the lower band product, present only in tumor-derived DNA, contains 18 CAG repeats. A schematic representation of the AR gene, the location of the CAG repeat and the wild-type and mutant sequences appear below the sequencing gel autoradiograms. (Reprinted with permission of Academic Press, Inc., from Schoenberg et al., 1994).

abates after cessation of flutamide therapy (Scher and Kelly, 1993; Sartor et al., 1994). The patient was a 53-year-old Caucasian man with high-grade (Gleason 8) prostate adenocarcinoma who underwent radical prostatectomy followed by radiation therapy because the surgical margins contained tumor. Three years later his serum PSA level rose to 190 ng/ml (normal range, 1–4 ng/ml), and a bone scan revealed evidence of metastatic disease. After treatment with an LHRH agonist and the antiandrogen flutamide, his PSA fell to castrate levels (<0.2 ng/ml). Eighteen months later his PSA rose to 35 ng/ml, despite a serum testosterone level still in the castrate range, apparently signaling tumor progression to androgen independence. Unexpectedly, however, cessation of flutamide therapy led to a fall in serum PSA to 7.5 ng/ml within 1 month, suggesting that flutamide was acting paradoxically as an agonist. His PSA remains low (13–16 ng/ml) 24 months after flutamide withdrawal. The existence of an AR gene mutation in a patient with prostate cancer who had a paradoxical response to flutamide withdrawal implies that an AR-mediated mechanism for this syndrome may exist.

Glutamine repeat length affects transcriptional activity, AR with shorter repeats being more active (Mhatre et al., 1993; Chamberlain et al., 1994; Kazemi-Esfarjani et al., 1995). Therefore, it is conceivable that the CAG_{18} mutation enhanced AR function in his prostate cancer cells. Whether his tumor contains additional mutations in the rest of the AR gene, that alone or in concert with the CAG mutation affect AR function or account for the paradoxical response to flutamide, remains to be determined. Since this tumor was evaluated before therapy (including radiation), it is possible that additional mutations in the AR gene may have occurred subsequently. We are unable to obtain tumor tissue from this patient because he is in remission, so we are unable to analyze the tumor AR in its current state. The basis for paradoxical responses to flutamide remains to be determined, as well as whether glutamine repeat length could affect binding and transcriptional activity by hydroxyflutamide.

Of 54 clinical stage B prostate tumors we screened, we have found several additional AR gene mutations. All mutations were reproducibly seen in at least three independent PCR reaction products. Experiments are still in progress to identify the nature of these mutations. Failure to detect mutations in most of the prostate cancers may be due to the limit of sensitivity of the method used to screen for the presence of mutations. Alternatively, if mutations occur during tumor progression, the mutation frequency may be low in early-stage disease, and higher in late-stage disease. The presence of AR gene mutations in only a subset of early-stage prostate cancers suggests that the putative role of androgen in prostate carcinogenesis is mediated in most cases by a wild-type AR gene. On the other hand, however, we cannot rule out the possibility that a wild-type AR gene may give rise to AR variants, such as by alternative splicing of AR transcripts or by alternative translation initiation site utilization, that might play a role in prostate carcinogenesis. By analogy, ER variants in breast cancer result primarily from alternative splicing (Fuqua et al., 1991; Wang and Miksicek, 1991; Castles et al., 1993), rather than from DNA mutations (Karnik et al., 1994).

The presence of AR gene mutations in a subset of early-stage prostate cancers raises the question of whether there is something unique about these patients. One possibility is that these tumors are more aggressive, either because the mutation helped to confer this behavior, or because the mutation occurred in cells with a growth advantage. Of the tumors we have screened to date, the follow-up interval after prostatectomy is too short at this time to be able to say whether the patients with mutations represent a group with more aggressive disease. A proportion of early-stage prostate cancers progress following radical prostatectomy, even in the absence of evidence of metastatic spread (Diamond et al., 1982; Epstein et al., 1993; Veltri et al., 1994; Partin et al., 1995). It will be of interest to determine whether AR gene mutations are more frequent in such tumors.

AR gene mutations have also been found in hormone refractory prostate cancers (Culig et al., 1993; Suzuki et al., 1993; Gaddipati et al., 1994; Taplin et al., 1995). These affect codons 701 (Suzuki et al., 1993), 715 (Culig et al., 1993), and 877 (Suzuki et al, 1993; Gaddipati et al., 1994; Taplin et al., 1995) in the hormone binding domain. The codon 715 mutation (val → met) is a gain-of-function mutation, having no effect on the steroid specificity of ligand binding and not disrupting the androgen-dependent transcriptional activity, but increasing the efficacy with which progesterone and adrenal steroids like dehydroepiandrosterone (DHEA) and androstenedione can activate transcriptional activity of the AR (Culig et al., 1993). The codon 701 mutation (leu → his), properties unknown, was found in a primary tumor, whereas a codon 877 mutation, identical to that found in the LNCaP cell line, was found in the liver, kidney, and lymph node metastases of the same patient (Suzuki et al., 1993). Codon 877 may be a hot-spot for mutations, having been found in 6 of 24 hormone-refractory tumors (Gaddipati et al., 1994); the possibility that PCR products amplified from prostate specimens might become contaminated in the laboratory with LNCaP mutant AR PCR product is recognized and must be rigorously avoided. Reports of additional AR gene mutations in hormone refractory prostate cancer have been presented at recent meetings (Sharief et al., 1995; Tilley et al., 1995). Preliminary characterization of some of these indicates that they too are gain-of-function mutations (Sharief et al., 1995; Taplin et al., 1995). It is not known whether AR gene mutations in hormone-refractory disease arose during therapy or were already present before therapy. It will be interesting to determine whether AR mutations found in early-stage and late-stage disease affect AR function in similar ways.

Although studies are still in progress to determine the frequency of AR gene mutations in prostate cancer, the presence of AR gene mutations in only a subset of early-stage prostate cancers suggests that the putative role of androgen in prostate carcinogenesis is mediated in most cases by a wild-type AR gene. However, the wild-type AR gene exists as multiple alleles that differ in glutamine and glycine repeat length, and these potentially differ in signal transducing activity. In light of the role of androgen in prostate cancer, we speculate that the polymorphic variation of the AR may bear some relationship to prostate carcinogenesis.

SUMMARY

About 1 in 10 American men develop clinical prostate cancer, making it the most commonly diagnosed cancer in this group, and about 50% of men over the age of 70 years have histologic evidence of prostate cancer (Carter and Coffey, 1990). This high incidence is indicative of a common risk factor. The highest risk factor for prostate cancer is androgen action. Therefore, we propose that the AR may be a susceptibility gene. The existence of multiple allelic variants, with a range of activity that may depend on the glutamine and/or glycine repeat length, may account for the fact that clinical prostate cancer affects a subset of men, rather than all men. It may also account for the fact that racial groups that have different frequency distributions of AR alleles (Edwards et al., 1992) have different incidence rates of prostate cancer (Ross et al., 1983). We propose that the effect of glutamine and glycine repeat length on AR function determines the sensitivity of tumor cells to existing tissue levels of dihydrotestosterone (DHT). Tissue DHT levels depend on circulating androgen levels and the amount of 5α-reductase activity in the tissue. Thus, actual AR activity will depend also on these other factors.

Our goal is to understand the role(s) of the AR in prostate cancer, and to determine whether this role is mediated by a normal AR or by abnormal AR. AR gene mutations indeed occur both in early-stage prostate cancers (Newmark et al., 1992; Schoenberg et al., 1994) and in late-stage disease (Culig et al., 1993; Suzuki et al., 1993; Gaddipati et al., 1994; Taplin et al., 1995). This suggests that AR mutations may characterize a more aggressive disease, and it raises the question of whether these mutations affect AR function in the same way. A large percentage of tumors may have no AR gene mutations, but the possibility has not been ruled out that tumors without AR gene mutations may nonetheless produce variant AR, for example, by aberrant splicing. Alternatively, the presence of AR mutations in only a subset of tumors may indicate that the role of AR in prostate cancer is mediated predominantly by normal AR.

About 50% of prostate cancers have already metastasized at the time of initial diagnosis. At this time the only effective treatment available is androgen ablation, but it is not curative. As new therapeutic approaches are developed, reliable predictors of response to hormonal therapy will be needed, so that patients who would experience a poor response (i.e., short time to progression) could receive alternative therapies at a time when they may be better able to tolerate them. The heterogeneity (variance) of AR concentrations within a specimen appears to be significantly greater in poor responders than in good responders (Sadi and Barrack, 1993). This heterogeneity may be a manifestation of greater genetic instability in poor responders, and an indicator that the tumor has progressed further toward androgen independence, thereby resulting in a shorter time to progression.

ACKNOWLEDGMENTS

Supported by the NCI, Grant CA58236. We gratefully acknowledge the contributions of our colleagues: Dr. M.V. Sadi who did the immunohistochemistry studies, Drs. J.I. Epstein, W.B. Isaacs, and G.S. Bova who provided prostate cancer DNA samples, Dr. P.C. Walsh who brought to our attention patients with an interesting clinical history, Dr. T.B. Brown who taught us about DGGE and helped us characterize the transcriptional activity of mutant AR, and Dalal C. Tonb, who provided superb technical assistance.

REFERENCES

Abrams ES, Stanton VP Jr (1992): Use of denaturing gradient gel electrophoresis to study conformational transitions in nucleic acids. *Meth Enzymol* 212:71–104

Alcaraz A, Takahashi S, Brown JA, Herath JF, Bergstralh EJ, Larson-Keller JJ, Lieber MM, Jenkins RB (1994): Aneuploidy and aneusomy of chromosome 7 detected by fluorescence in situ hybridization are markers of poor prognosis in prostate cancer. *Cancer Res* 54:3998–4002

Allen LF, Lefkowitz RJ, Caron MG, Cotecchia S (1991): G-protein-coupled receptor genes as protooncogenes: constitutively activating mutation of the α_{1B}-adrenergic receptor enhances mitogenesis and tumorigenicity. *Proc Natl Acad Sci USA* 88: 11354–11358

Armstrong DK, Kaufmann SH, Ottaviano YL, Furuya Y, Buckley JA, Isaacs JT, Davidson NE (1994): Epidermal growth factor-mediated apoptosis of MDA-MB-468 human breast cancer cells. *Cancer Res* 54:5280–5283

Aronica SM, Katzenellenbogen BS (1993): Stimulation of estrogen receptor-mediated transcription and alteration in the phosphorylation state of the rat uterine estrogen receptor by estrogen, cyclic adenosine monophosphate, and insulin-like growth factor-I. *Mol Endocrinol* 7:743–752

Barrack ER, Tindall DJ (1987): A critical evaluation of the use of androgen receptor assays to predict the androgen responsiveness of prostatic cancer. *Prog Clin Biol Res* 239:155–187

Barrack ER, Brendler CB, Walsh PC (1987): Steroid receptor and biochemical profiles in prostatic cancer: correlation with response to hormonal treatment. *Prog Clin Biol Res* 243A:79–97

Batch JA, Williams DM, Davies HR, Brown BD, Evans BAJ, Hughes IA, Patterson MN (1992): Androgen receptor gene mutations identified by SSCP in fourteen subjects with androgen insensitivity syndrome. *Hum Mol Genet* 1:497–503

Beck CA, Weigel NL, Moyer ML, Nordeen SK, Edwards DP (1993): The progesterone antagonist RU486 acquires agonist activity upon stimulation of cAMP signaling pathways. *Proc Natl Acad Sci* 90:4441–4445

Benson Jr RC, Gorman PA, O'Brien PC, Holicky EL, Veneziale CM (1987): Relationship between androgen receptor binding activity in human prostate cancer and clinical response to endocrine therapy. *Cancer* 59:1599–1606

Bookstein R, MacGrogan D, Hilsenbeck SG, Sharkey F, Allred DC (1993): p53 is mutated in a subset of advanced-stage prostate cancers. *Cancer Res* 53:3369–3373

Bos JL (1989): *ras* oncogenes in human cancer: a review. *Cancer Res* 49:4682–4689

Bova GS, Fox WM, Epstein JI (1993): Methods of radical prostatectomy specimen processing: a novel technique for harvesting fresh prostate cancer tissue and review of processing techniques. *Mod Pathol* 6:201–207

Brendler CB, Isaacs JT, Follansbee AL, Walsh PC (1984): The use of multiple variables to predict response to endocrine therapy in carcinoma of the prostate: a preliminary report. *J Urol* 131:694–700

Brown CJ, Goss SJ, Lubahn DB, Joseph DR, Wilson EM, French FS, Willard HF (1989): Androgen receptor locus on the human X chromosome: regional localization to Xq11–12 and description of a DNA polymorphism. *Am J Hum Genet* 44:264–269

Brown TR, Lubahn DB, Wilson EM, French FS, Migeon CJ, Corden JL (1990): Functional characterization of naturally occurring mutant androgen receptors from subjects with complete androgen insensitivity. *Mol Endocrinol* 4:1759–1772

Bussemakers MJG, Bova GS, Schoenberg MP, Hakimi JM, Barrack ER, Isaacs WB (1994): Microsatellite instability in human prostate cancer. *J Urol* 151:469A (abstract 967)

Carter BS, Beaty TH, Steinberg GD, Childs B, Walsh PC (1992): Mendelian inheritance of familial prostate cancer. *Proc Natl Acad Sci USA* 89:3367–3371

Carter BS, Bova GS, Beaty TH, Steinberg GD, Childs B, Isaacs WB, Walsh PC (1993): Hereditary prostate cancer: epidemiologic and clinical features. *J Urol* 150:797–802

Carter HB, Coffey DS (1990): The prostate: an increasing medical problem. *Prostate* 16:39–48

Carter HB, Piantadosi S, Isaacs JT (1990): Clinical evidence for and implications of the multistep development of prostate cancer. *J Urol* 143:742–746

Castles CG, Fuqua SAW, Klotz DM, Hill SM (1993): Expression of a constitutively active estrogen receptor variant in the estrogen receptor-negative BT-20 human breast cancer cell line. *Cancer Res* 53: 5934–5939

Catalona WJ (1994): Editorial: expectant management and the natural history of localized prostate cancer. *J Urol* 152:1751–1752

Catalona WJ, Scott WW (1986): Carcinoma of the prostate. In: *Campbell's Urology*, 5th ed, Walsh PC, Gittes RF, Perlmutter AD, Stamey TA, eds. Philadelphia: WB Saunders, pp 1463–1534

Cavailles V, Dauvois S, Danielian PS, Parker MG (1994): Interaction of proteins with transcriptionally active estrogen receptors. *Proc Natl Acad Sci USA* 91:10009–10013

Chamberlain NL, Driver ED, Miesfeld RL (1994): The length and location of CAG trinucleotide repeats in the androgen receptor N-terminal domain affect transactivation function. *Nucleic Acids Res* 22:3181–3186

Chang C, Kokontis J, Liao S (1988): Structural analysis of cDNA and amino acid sequences of human and rat androgen receptors. *Proc Natl Acad Sci USA* 85: 7211–7215

Charest NJ, Zhou Z-x, Lubahn DB, Olsen KL, Wilson WM, French FS (1991): A frameshift mutation destabilizes androgen receptor messenger RNA in the Tfm mouse. *Mol Endocrinol* 5:573–581

Chodak GW (1994): The role of watchful waiting in the management of localized prostate cancer. *J Urol* 152:1766–1768

Christensen WN, Partin AW, Walsh PC, Epstein JI (1990): Pathologic findings in clinical stage A2 prostate cancer. Relation of tumor volume, grade, and location to pathologic stage. *Cancer* 65:1021–1027

Claret F-X, Chapel S, Garces J, Tsai-Pflugfelder M, Bertholet C, Shapiro DJ, Wittek R, Wahli W (1994): Two functional forms of the Xenopus laevis estrogen receptor translated from a single mRNA species. *J Biol Chem* 269:14147–14055

Courey AJ, Holtzman DA, Jackson SP, Tjian R (1989): Synergistic activation by the glutamine-rich domains of human transcription factor Sp1. *Cell* 59:827–836

Culig Z, Hobisch A, Cronauer MV, Cato ACB, Hittmair A, Radmayr C, Eberle J, Bartsch G, Klocker H (1993a): Mutant androgen receptor detected in an advanced-stage prostatic carcinoma is activated by adrenal androgens and progesterone. *Mol Endocrinol* 7:1541–1550

Culig Z, Klocker H, Eberle J, Kaspar F, Hobisch A, Cronauer MV, Bartsch G (1993b): DNA sequence of the androgen receptor in prostatic tumor cell lines and tissue specimens assessed by means of the polymerase chain reaction. *Prostate* 22:11–22

Culig Z, Hobisch A, Cronauer MV, Radmayr C, Trapman J, Hittmair A, Bartsch G, Klocker H (1994): Androgen receptor activation in prostatic tumor cell lines by insulin-like growth factor-I, keratinocyte growth factor, and epidermal growth factor. *Cancer Res* 54:5474–5478

Cunha GR, Donjacour AA, Cooke PS, Mee S, Bigsby RM, Higgins SJ, Sugimura Y (1987): The endocrinology and developmental biology of the prostate. *Endocr Rev* 8:338–362

Darbre PD, King RJB (1987): Progression to steroid insensitivity can occur irrespective of the presence of functional steroid receptors. *Cell* 51:521–528

Dawson NA, McLeod DG (1995): Dramatic prostate specific antigen decrease in response to discontinuation of megestrol acetate in advanced prostate cancer: expansion of the antiandrogen withdrawal syndrome. *J Urol* 153:1946–1947

De Bellis A, Quigley CA, Cariello NF, El-Awady MK, Sar M, Lane MV, Wilson EM, French FS (1992): Single base mutations in the human androgen receptor gene causing complete androgen insensitivity: rapid detection by a modified denaturing gradient gel electrophoresis technique. *Mol Endocrinol* 6:1909–1920

Denner LA, Weigel NL, Maxwell BL, Schrader WT, O'Malley BW (1990): Regulation of progesterone receptor-mediated transcription by phosphorylation. *Science* 250:1740–1743

Diamond DA, Barrack ER (1984): The relationship of androgen receptor levels to androgen responsiveness in the Dunning R3327 rat prostate tumor sublines. *J Urol* 132:821–827

Diamond DA, Berry SJ, Jewett HJ, Eggleston JC, Coffey DS (1982): A new method to assess metastatic potential of human prostate cancer: relative nuclear roundness. *J Urol* 128:729–734

Djakiew D, Tarkington MA, Lynch JH (1990): Paracrine stimulation of polarized secretion from monolayers of a neoplastic prostatic epithelial cell line by prostatic stromal cell proteins. *Cancer Res* 50:1966–1974

Dobashi Y, Shuin T, Tsuruga H, Uemura H, Torigoe S, Kubota Y (1994): DNA polymerase beta gene mutation in human prostate cancer. *Cancer Res* 54:2827–2829

Edwards A, Hammond HA, Jin L, Caskey CT, Chakraborty R (1992): Genetic variation at five trimeric and tetrameric tandem repeat loci in four human population groups. *Genomics* 12:241–253

Emili A, Greenblatt J, Ingles CJ (1994): Species-specific interaction of the glutamine-rich activation domains of Sp1 with the TATA box-binding protein. *Mol Cell Biol* 14:1582–1593

Epstein JI, Carmichael MJ, Pizov G, Walsh PC (1993): Influence of capsular penetration on progression following radical prostatectomy: a study of 196 cases with long-term followup. *J Urol* 150:135–141

Epstein JI, Carmichael MJ, Partin AW, Walsh PC (1994): Small high grade adenocarcinoma of the prostate in radical prostatectomy specimens performed for nonpalpable disease: pathogenetic and clinical implications. *J Urol* 151:1587–1592

Evans RM (1988): The steroid and thyroid hormone receptor superfamily. *Science* 240:889–895

Faber PW, Kuiper GGJM, van Rooij HCJ, van der Korput JAGM, Brinkmann AO, Trapman J (1989): The N-terminal domain of the human androgen receptor is encoded by one, large exon. *Mol Cell Endocrinol* 61:257–262

Faber PW, van Rooij HCJ, van der Korput HAGM, Baarends WM, Brinkmann AO, Grootegoed JA, Trapman J (1991): Characterization of the human androgen receptor transcription unit. *J Biol Chem* 266:10743–10749

Faber PW, van Rooij HCJ, Schipper HJ, Brinkmann AO, Trapman J (1993): Two different, overlapping pathways of transcription initiation are active on the TATA-less human androgen receptor promoter. The role of Sp1. *J Biol Chem* 268:9296–9301

Fearon ER, Vogelstein B (1990): A genetic model for colorectal tumorigenesis. *Cell* 61:759–767

Fentie DD, Lakey WH, McBlain WA (1986): Applicability of nuclear androgen receptor quantification to human prostatic adenocarcinoma. *J Urol* 135:167–173

Fowler JE Jr, Whitmore WF Jr (1981): The response of metastatic adenocarcinoma of the prostate to exogenous testosterone. *J Urol* 126:372–375

Fujimoto N, Katzenellenbogen BS (1994): Alteration in the agonist/antagonist balance of antiestrogens by activation of protein kinase A signaling pathways in breast cancer cells: antiestrogen selectivity and promoter dependence. *Mol Endocrinol* 8:296–304

Fuqua SAW, Fitzgerald SD, Chamness GC, Tandon AK, McDonnell DP, Nawaz, Z, O'Malley BW, McGuire WL (1991): Variant human breast tumor estrogen receptor with constitutive transcriptional activity. *Cancer Res* 51:105–109

Gaddipati JP, McLeod DG, Heidenberg HB, Sesterhenn IA, Finger MJ, Moul JW, Srivastava S (1994): Frequent detection of codon 877 mutation in the androgen receptor gene in advanced prostate cancer. *Cancer Res* 54:2861–2864

Garcia M, Derocq D, Freiss G, Rochefort H (1992): Activation of estrogen receptor transfected into a receptor-negative breast cancer cell line decreases the metastatic and invasive potential of the cells. *Proc Natl Acad Sci USA* 89: 11538–11542

Gerber H-P, Seipel K, Georgiev O, Hofferer M, Hug M, Rusconi S, Schaffner W (1994): Transcriptional activation modulated by homopolymeric glutamine and proline stretches. *Science* 263:808–811

Gill G, Pascal E, Tseng ZH, Tjian R (1994): A glutamine-rich hydrophobic patch in transcription factor Spl contacts the $dTAF_{II}110$ component of the *Drosophila* TFIID complex and mediates transcriptional activation. *Proc Natl Acad Sci USA* 91:192–196

Glavac D, Dean M (1993): Optimization of the single-strand conformation polymorphism (SSCP) technique for detection of point mutations. *Hum Mutation* 2:404–414

Greenblatt J (1992): Riding high on the TATA box. *Nature* 360:16–17

Greenblatt MS, Bennett WP, Hollstein M, Harris CC (1994): Mutations in the p53 tumor suppressor gene: Clues to cancer etiology and molecular pathogenesis. *Cancer Res* 54:4855–4878

Greene DR, Taylor SR, Wheeler TM, Scardino PT (1991): DNA ploidy by image analysis of individual foci of prostate cancer: a preliminary report. *Cancer Res* 51:4084–4089

Gronberg H, Damber L, Damber J-E (1994): Studies of genetic factors in prostate cancer in a twin population. *J Urol* 152:1484–1489

Halachmi S, Marden E, Martin G, MacKay H, Abbondanza C, Brown M (1994): Estrogen receptor-associated proteins: possible mediators of hormone-induced transcription. *Science* 264:1455–1458

Hall JM, Lee MK, Newman B, Morrow JE, Anderson LA, Huey B, King M-C (1990): Linkage of early-onset familial breast cancer to chromosome 17q21. *Science* 250:1684–1689

Harris SE, Rong Z, Harris MA, Lubahn DB (1990): Androgen receptor in human prostate carcinoma LNCAP/ADEP cells contains a mutation which alters the specificity of the steroid-dependent transcriptional activation region. The Endocrine Society 72nd Annual Meeting Program and Abstracts, Abstract 275, p. 93.

Hayashi K, Yandell DW (1993): How sensitive is PCR-SSCP? *Hum Mutation* 2:338–346

Henderson BE, Bernstein L, Ross RK, Depue RH, Judd HL (1988): The early *in utero* oestrogen and testosterone environment of blacks and whites: Potential effects on male offspring. *Br J Cancer* 57:216–218

Heppner GH (1984): Tumor heterogeneity. *Cancer Res* 44:2259–2265

Hoehn W, Wagner M, Riemann JF, Hermanek P, Williams E, Walther R, Schrueffer R. (1984): Prostatic adenocarcinoma PC EW, a new human tumor line transplantable in nude mice. *Prostate* 5:445–452

Horoszewicz JS, Leong SS, Chu TM, Wajsman ZL, Friedman M, Papsidero L, Kim U, Chai LS, Kakati S, Arya SK, Sandberg AA (1980): The LNCaP cell line—a new model for studies on human prostatic carcinoma. *Prog Clin Biol Res* 37:115–132

Housley PR, Sanchez ER, Danielsen M, Ringold GM, Pratt WB (1990): Evidence that the conserved region in the steroid binding domain of the glucocorticoid receptor is required for both optimal binding of hsp90 and protection from proteolytic cleavage. *J Biol Chem* 265:12778–12781

Huggins C, Hodges CV (1941): The effect of castration, of estrogens and of androgen injection on serum phosphatase in metastatic carcinoma of the prostate. *Cancer Res* 1:293–297

Ignar-Trowbridge DM, Teng CT, Ross KA, Parker MG, Korach KS, McLachlan JA (1993): Peptide growth factors elicit estrogen receptor-dependent transcriptional activation of an estrogen-responsive element. *Mol Endocrinol* 7:992–998

Iizumi T, Yazaki T, Kanoh S, Kondo I, Koiso K (1987): Establishment of a new prostatic carcinoma cell line (TSU-Pr1). *J Urol* 137:1304–1306

Ikonen T, Palvimo JJ, Kallio PJ, Reinikainen P, Janne OA (1994): Stimulation of androgen-regulated transactivation by modulators of protein phosphorylation. *Endocrinology* 135:1359–1366

Ince BA, Montano MM, Katzenellenbogen BS (1994): Activation of transcriptionally inactive human estrogen receptors by cyclic adenosine 3′,5′-monophosphate and ligands including antiestrogens. *Mol Endocrinol* 8:1397–1406

Ing NH, Beekman JM, Tsai SY, Tsai M-J, O'Malley BW (1992): Members of the steroid hormone receptor superfamily interact with TFIIB (S300-II). *J Biol Chem* 267:17617–17623

Irvine RA, Yu MC, Ross RK, Coetzee GA (1995): The CAG and GGC microsatellites of the androgen receptor gene are in linkage disequilibrium in men with prostate cancer. *Cancer Res* 55:1937–1940

Isaacs JT (1984): Antagonistic effect of androgen on prostatic cell death. *Prostate* 5:545–557

Isaacs WB, Bova GS, Morton RA, Bussemakers MJG, Brooks JD, Ewing CM (1995): Molecular genetics and chromosomal alterations in prostate cancer. *Cancer* 75:2004–2012

Jackson RJ (1993): Cytoplasmic regulation of mRNA function: the importance of the 3′ untranslated region. *Cell* 74:9–14

Jenster G, van der Korput HAGM, van Vroonhoven C, van der Kwast TH, Trapman J, Brinkmann AO (1991): Domains of the human androgen receptor involved in steroid binding, transcriptional activation, and subcellular localization. *Mol Endocrinol* 5:1396–1404

Jenster G, van der Korput HAGM, Trapman J, Brinkmann AO (1995): Identification of two transcription activation units in the N-terminal domain of the human androgen receptor. *J Biol Chem* 270:7341–7346

Kalloo NB, Gearhart JP, Barrack ER (1993): Sexually dimorphic expression of estrogen receptors, but not of androgen receptors in human fetal external genitalia. *J Clin Endocrinol Metab* 77:692–698

Karnik PS, Kulkarni S, Liu X-P, Budd GT, Bukowski RM (1994): Estrogen receptor mutations in tamoxifen-resistant breast cancer. *Cancer Res* 54:349–353

Kastner P, Boquel M-T, Turcotte B, Garnier J-M, Horwitz KB, Chambon P, Gronemeyer H (1990a): Transient expression of human and chicken progesterone receptors does not support alternative translational initiation from a single mRNA as the mechanism generating two receptor isoforms. *J Biol Chem* 265: 12163–12167

Kastner P, Krust A, Turcotte B, Stropp U, Tora L, Gronemeyer H, Chambon P (1990b): Two distinct estrogen-regulated promoters generate transcripts encoding the two functionally different human progesterone receptor forms A and B. *EMBO J* 9:1603–1614

Kazemi-Esfarjani P, Beitel LK, Trifiro M, Kaufman M, Rennie P, Sheppard P, Matusik R, Pinsky L (1993): Substitution of valine-865 by methionine or leucine in the human androgen receptor causes complete or partial androgen insensitivity, respectively with distinct androgen receptor phenotypes. *Mol Endocrinol* 7:37–46

Kazemi-Esfarjani P, Trifiro MA, Pinsky L (1995): Evidence for a repressive function of the long polyglutamine tract in the human androgen receptor: possible pathogenetic relevance for the $(CAG)_n$-expanded neuronopathies. *Hum Mol Genet* 4:523–527

Klocker H, Kaspar F, Eberle J, Uberreiter S, Radmayr C, Bartsch G (1992): Point mutation in the DNA binding domain of the androgen receptor in two families with Reifenstein syndrome. *Am J Hum Genet* 50:1318–1327

Kokontis J, Ito K, Hiipakka RA, Liao S (1991): Expression and function of normal and LNCaP androgen receptors in androgen-insensitive human prostatic cancer cells. *Receptor* 1:271–279

Kokontis J, Takakura K, Hay N, Liao S (1994): Increased androgen receptor activity and altered c-myc expression in prostate cancer cells after long-term androgen deprivation. *Cancer Res* 54:1566–1573

Kupfer SR, Marschke KB, Wilson EM, French FS (1993): Receptor accessory factor enhances specific DNA binding of androgen and glucocorticoid receptors. *J Biol Chem* 268:17519–17527

Kupfer SR, Wilson EM, French FS (1994): Androgen and glucocorticoid receptors interact with insulin degrading enzyme. *J Biol Chem* 269:20622–20628

Landis CA, Masters SB, Spada A, Pace AM, Bourne HR, Vallar L (1989): GTPase inhibiting mutations activate the α chain of G_s and stimulate adenylyl cyclase in human pituitary tumours. *Nature* 340:692–696

La Spada AR, Wilson EM, Lubahn DB, Harding AE, Fischbeck KH (1991): Androgen receptor gene mutations in X-linked spinal and bulbar muscular atrophy. *Nature* 352:77–79

Lee W-H, Morton RA, Epstein JI, Brooks JD, Campbell PA, Bova GS, Hsieh W-S, Isaacs WB, Nelson WG (1994): Cytidine methylation of regulatory sequences near the p-class glutathione S-transferase gene accompanies human prostate carcinogenesis. *Proc Natl Acad Sci* 91:11733–11737

Lee C, Sutkowski DM, Sensibar JA, Zelner D, Kim I, Amsel I, Shaw N, Prins GS, Kozlowski JM (1995): Regulation of proliferation and production of prostate-specific antigen in androgen-sensitive prostatic cancer cells, LNCaP, by dihydrotestosterone. *Endocrinology* 136:796–803

Lippert MC, Keefer DA (1987): Prostate adenocarcinoma: effects of castration on in situ androgen uptake by individual cell types. *J Urol* 137:140–145

Lobaccaro J-M, Lumbroso S, Belon C, Galtier-Dereure F, Bringer J, Lesimple T, Heron J-F, Pujol H, Sultan C (1993): Male breast cancer and the androgen receptor gene. Nature Genet 5:109–110

Loeb LA (1994): Microsatellite instability: marker of a mutator phenotype in cancer. *Cancer Res* 54:5059–5063

Lubahn DB, Joseph DR, Sar M, Tan J-a, Higgs HN, Larson RE, French FS, Wilson EM (1988): The human androgen receptor: complementary deoxyribonucleic acid cloning, sequence analysis and gene expression in prostate. *Mol Endocrinol* 2:1265–1275

Lubahn DB, Brown TR, Simental JA, Higgs HN, Migeon CJ, Wilson EM, French FS (1989): Sequence of the intron/exon junctions of the coding region of the human androgen receptor gene and identification of a point mutation in a family with complete androgen insensitivity. *Proc Natl Acad Sci USA* 86:9534–9538

Luke MC, Coffey DS (1994): The male sex accessory tissues. Structure, androgen action, and physiology. pp 1435–1487 In: *The Physiology of Reproduction*, 2nd edn, Knobil E, Neill JD, eds. New York: Raven Press

Lyons J, Landis CA, Harsh G, Vallar L, Grunewald K, Feichtinger H, Duh Q-Y, Clark OH, Kawasaki E, Bourne HR, McCormick F (1990): Two G protein oncogenes in human endocrine tumors. *Science* 249:655–659

Macke JP, Hu N, Hu S, Bailey M, King VL, Brown T, Hamer D, Nathans J (1993): Sequence variation in the androgen receptor gene is not a common determinant of male sexual orientation. *Am J Hum Genet* 53:844–852

MacLean HE, Choi W-T, Rekaris G, Warne GL, Zajac JD (1995): Abnormal androgen receptor binding affinity in subjects with Kennedy's disease, (spinal and bulbar muscular atrophy). *J Clin Endocrinol Metab* 80:508–516

Madden SL, Cook DM, Morris JF, Gashler A, Sukhatme VP, Rauscher FJ III (1991): Transcriptional repression mediated by the WT1 Wilms tumor gene product. *Science* 253:1550–1553

Malkin D, Li FP, Strong LC, Fraumeni JF Jr, Nelson CE, Kim DH, Kassel J, Gryka MA, Bischoff FZ, Tainsky MA, Friend SH (1990): Germ line p53 mutations in a familial syndrome of breast cancer, sarcomas, and other neoplasms. *Science* 250:1233–1238

Martin JB (1993): Molecular genetics of neurological diseases. *Science* 262:674–676

McGuire WL, Chamness GC, Fuqua SAW (1991): Estrogen receptor variants in clinical breast cancer. *Mol Endocrinol* 5:1571–1577

McLachlan RI, Tempel BL, Miller MA, Bicknell JN, Bremner WJ, Dorsa DM (1991): Androgen receptor gene expression in the rat central nervous system: Evidence for two mRNA transcripts. *Mol Cell Neurosci* 2:117–122

McLeskey SW, Ding IYF, Lippman ME, Kern FG (1994): MDA-MB-134 breast carcinoma cells overexpress fibroblast growth factor (FGF) receptors and are growth-inhibited by FGF ligands. *Cancer Res* 54:523–530

McPhaul MJ, Marcelli M, Tilley WD, Griffin JE, Isidro-Gutierrez RF, Wilson JD (1991): Molecular basis of androgen resistance in a family with a qualitative abnormality of the androgen receptor and responsive to high-dose androgen therapy. *J Clin Invest* 87:1413–1421

Mhatre AN, Trifiro MA, Kaufman M, Kazemi-Esfarjani P, Figlewicz D, Rouleau G, Pinsky L (1993): Reduced transcriptional regulatory competence of the androgen receptor in X-linked spinal and bulbar muscular atrophy. *Nature Genet* 5:184–188

Monari L, Chen SG, Brown P, Parchi P, Petersen RB, Mikol J, Gray F, Cortelli P, Montagna P, Ghetti B, Goldfarb LG, Gajdusek DC, Lugaresi E, Gambetti P, Autilio-Gambetti L (1994): Fatal familial insomnia and familial Creutzfeldt-Jakob disease: Different prion proteins determined by a DNA polymorphism. *Proc Natl Acad Sci USA* 91:2839–2842

Morrison NA, Yeoman R, Kelly PJ, Eisman JA (1992): Contribution of trans-acting factor alleles to normal physiological variability: vitamin D receptor gene polymorphisms and circulating osteocalcin. *Proc Natl Acad Sci USA* 89:6665–6669

Morrison NA, Qi JC, Tokita A, Kelly PJ, Crofts L, Nguyen TV, Sambrook PN, Eisman JA (1994): Prediction of bone density from vitamin D receptor alleles. *Nature* 367:284–287

Murray MB, Towle HC (1989): Identification of nuclear factors that enhance binding of the thyroid hormone receptor to a thyroid hormone response element. *Mol Endocrinol* 3:1434–1442

Nastiuk KL, Clayton DF (1994): Seasonal and tissue-specific regulation of canary androgen receptor messenger ribonucleic acid. *Endocrinology* 134:640–649

Newmark JR, Hardy DO, Tonb DC, Carter BS, Epstein JI, Isaacs WB, Brown TR, Barrack ER (1992): Androgen receptor gene mutations in prostate cancer. *Proc Natl Acad Sci USA* 89:6319–6323

Nieh PT (1995): Withdrawal phenomenon with the antiandrogen casodex. *J Urol* 153:1070–1073

Nigro JM, Cho KR, Fearon ER, Kern SE, Ruppert JM, Oliner JD, Kinzler KW, Vogelstein B (1991): Scrambled exons. *Cell* 64:607–613

Nordeen SK, Bona BJ, Moyer ML (1993): Latent agonist activity of the steroid antagonist, RU486, is unmasked in cells treated with activators of protein kinase A. *Mol Endocrinol* 7:731–742

Nowell PC (1986): Mechanisms of tumor progression. *Cancer Res* 46:2203–2207

Oesterling JE (1991): Prostate specific antigen: a critical assessment of the most useful marker for adenocarcinoma of the prostate. *J Urol* 145:907–923

Olea N, Sakabe K, Soto AM, Sonnenschein C (1990): The proliferative effect of "anti-androgens" on the androgen-sensitive human prostate tumor cell line LNCaP. *Endocrinology* 126:1457–1463

Orita M, Suzuki Y, Sekiya T, Hayashi K (1989): Rapid and sensitive detection of point mutations and DNA polymorphisms using the polymerase chain reaction. *Genomics* 5:874–879

Partin AW, Walsh PC, Pitcock RV, Mohler JL, Epstein JI, Coffey DS (1989): A comparison of nuclear morphometry and Gleason grade as a predictor of prognosis in stage A2 prostate cancer: a critical analysis. *J Urol* 142:1254–1258

Partin AW, Piantadosi S, Sanda MG, Epstein JI, Marshall FF, Mohler JL, Brendler CB, Walsh PC, Simons JW (1995): Selection of men at high risk for recurrence for experimental adjuvant therapy following radical prostatectomy. *J Urol* 153:449A (abstract #884)

Perutz MF, Johnson T, Suzuki M, Finch JT (1994): Glutamine repeats as polar zippers: their possible role in inherited neurodegenerative diseases. *Proc Natl Acad Sci USA* 91:5355–5358

Peters CA, Barrack ER (1987a): A new method for labeling and autoradiographic localization of androgen receptors. *J Histochem Cytochem* 35:755–762

Peters CA, Barrack ER (1987b): Androgen receptor localization in the human prostate: Demonstration of heterogeneity using a new method of steroid receptor autoradiography. *J Steroid Biochem* 27:533–541

Peters CA, Barrack ER (1987c): Androgen receptor localization in the prostate using a new method of steroid receptor autoradiography. pp 175–187 In: *Benign Prostatic Hyperplasia*, vol II, Rodgers CH, Coffey DS, Cunha G, Grayhack JT, Hinman Jr F, Horton R, eds. United States Department of Health and Human Services, NIH Publication #87–2881

Picard D, Salser SJ, Yamamoto KR (1988): A movable and regulable inactivation function within the steroid binding domain of the glucocorticoid receptor. *Cell* 54:1073–1080

Power RF, Mani SK, Codina J, Conneely OM, O'Malley BW (1991): Dopaminergic and ligand-independent activation of steroid hormone receptors. *Science* 254: 1636–1639

Prins GS (1989): Differential regulation of androgen receptors in the separate rat prostate lobes: androgen independent expression in the lateral lobe. *J Steroid Biochem* 33:319–326

Riegman PHJ, Vlietstra RJ, van der Korput JAGM, Brinkmann AO, Trapman J (1991): The promoter of the prostate-specific antigen gene contains a functional androgen responsive element. *Mol Endocrinol* 5:1921–1930

Ris-Stalpers C, Kuiper GGJM, Faber PW, Schweikert HU, van Rooij HCJ, Zegers ND, Hodgins MB, Degenhart HJ, Trapman J, Brinkmann AO (1990): Aberrant splicing of androgen receptor mRNA results in synthesis of a nonfunctional receptor protein in a patient with androgen insensitivity. *Proc Natl Acad Sci USA* 87:7866–7870

Ris-Stalpers C, Verleun-Mooijman MCT, de Blaeij TJP, Degenhart HJ, Trapman J, Brinkmann AO (1994): Differential splicing of human androgen receptor pre-mRNA in X-linked Reifenstein syndrome, because of a deletion involving a putative branch site. *Am J Hum Genet* 54:609–617

Ross RK, Paganini-Hill A, Henderson BE (1983): The etiology of prostate cancer: What does the epidemiology suggest? *Prostate* 4:333–344

Ruizeveld de Winter JA, van Weerden WM, Faber PW, van Steenbrugge GJ, Trapman J, Brinkmann AO, van der Kwast TH (1992): Regulation of androgen receptor expression in the human heterotransplantable prostate carcinoma PC-82. *Endocrinology* 131:3045–3050

Ruizeveld de Winter JA, Janssen JA, Sleddens HMEB, Verleun-Mooijman MCT, Trapman J, Brinkmann AO, Santerse AB, Schroder FH, van der Kwast TH (1994): Androgen receptor status in localized and locally progressive hormone refractory human prostate cancer. *Am J Pathol* 144:735–746

Rundlett SE, Wu X-P, Miesfeld R L (1990): Functional characterizations of the androgen receptor confirm that the molecular basis of androgen action is transcriptional regulation. *Mol Endocrinol* 4:708–714

Rundlett SE, Gordon DA, Miesfeld RL (1992): Characterization of a panel of rat ventral prostate epithelial cell lines immortalized in the presence or absence of androgens. *Exp Cell Res* 203:214–221

Sachs A, Wahle E (1993): Poly(A) tail metabolism and function in eucaryotes. *J Biol Chem* 268:22955–22958

Sadi MV, Barrack ER (1991): Androgen receptors and growth fraction in metastatic prostate cancer as predictors of the time to tumour progression following hormonal therapy. *Cancer Surv* 11:195–215

Sadi MV, Barrack ER (1993): Image analysis of androgen receptor immunostaining in metastatic prostate cancer: Heterogeneity as a predictor of response to hormonal therapy. *Cancer* 71:2574–2580

Sadi MV, Walsh PC, Barrack ER (1991): Immunohistochemical study of androgen receptors in metastatic prostate cancer: Comparison of receptor content and response to hormonal therapy. *Cancer* 67:3057–3064

Sakr WA, Macoska JA, Benson P, Grignon DJ, Wolman SR, Pontes JE, Crissman JD (1994): Allelic loss in locally metastatic, multisampled prostate cancer. *Cancer Res* 54:3273–3277

Sandberg AA (1992): Chromosomal abnormalities and related events in prostate cancer. *Hum Pathol* 23:368–380

Sartor O, Cooper M, Weinberger M, Headlee D, Thibault A, Tompkins A, Steinberg S, Figg WD, Linehan WM, Myers CE (1994): Surprising activity of flutamide withdrawal, when combined with aminoglutethimide, in treatment of "hormone refractory" prostate cancer. *J Natl Cancer Inst* 86:222–227

Sartorius CA, Tung L, Takimoto GS, Horwitz KB (1993): Antagonist-occupied human progesterone receptors bound to DNA are functionally switched to transcriptional agonists by cAMP. *J Biol Chem* 268:9262–9266

Scher HI, Kelly WK (1993): Flutamide withdrawal syndrome: its impact on clinical trials in hormone-refractory prostate cancer. *J Clin Oncol* 11:1566–1572

Schoenberg MP, Tonb DC, Brown TR, Barrack ER (1993): Functional analysis of an androgen receptor gene mutation in localized human prostate cancer. Abstract presented at the 7th Annual Meeting of the Society for Basic Urologic Research, San Antonio, Texas, May 14–15

Schoenberg MP, Hakimi JM, Wang S, Bova GS, Epstein JI, Fischbeck KH, Isaacs WB, Walsh PC, Barrack ER (1994): Microsatellite mutation ($CAG_{24 \rightarrow 18}$) in the androgen receptor gene in human prostate cancer. *Biochem Biophys Res Commun* 198:74–80

Scott WW, Menon M, Walsh PC (1980): Hormonal therapy of prostatic cancer. *Cancer* 45:1929–1936

Scotto L, Assoian RK (1993): A GC-rich domain with bifunctional effects on mRNA and protein levels: Implications for control of transforming growth factor $\beta 1$ expression. *Mol Cell Biol* 13:3588–3597

Sesterhenn I, Mostofi F, Davis C, McCarthy J, George J (1992): Numerical chromosomal aberrations in paraffin embedded sections of prostatic carcinoma. *J Urol* 147:215A

Shannon JM, Cunha GR (1983): Autoradiographic localization of androgen binding in the developing mouse prostate. *Prostate* 4:367–373

Shannon JM, Cunha GR (1984): Characterization of androgen binding and deoxyribonucleic acid synthesis in prostate-like structures induced in the urothelium of testicular feminized (Tfm/Y) mice. *Biol Reprod* 31:175–183

Sharief Y, Wilson EM, Hall SH, Hamil KG, Tan J-A, French FS, Mohler JL, Pretlow TG, Gumerlock PH, White Rd (1995): Androgen receptor gene mutations associated with prostatic carcinoma. *Proc Am Assoc Cancer Res* 36:269 (abstract 1605)

Sharif M, Privalsky ML (1991): v-erbA oncogene function in neoplasia correlates with its ability to repress retinoic acid receptor action. *Cell* 66:885–893

Sheffield VC, Cox DR, Lerman LS, Myers RM (1989): Attachment of a 40-base-pair G + C-rich sequence (GC-clamp) to genomic DNA fragments by the polymerase chain reaction results in improved detection of single-base changes. *Proc Natl Acad Sci USA* 86:232–236

Simental JA, Sar M, Lane MV, French FS, Wilson EM (1991): Transcriptional activation and nuclear targeting signals of the human androgen receptor. *J Biol Chem* 266:510–518

Sleddens HFBM, Oostra BA, Brinkmann AO, Trapman J (1993): Trinucleotide (GGN) repeat polymorphism in the human androgen receptor (AR) gene. *Hum Mol Genet* 2:493

Small EJ, Carroll PR (1994): Prostate-specific antigen decline after casodex withdrawal: evidence for an antiandrogen withdrawal syndrome. *Urology* 43:408–410

Sobue G, Doyu M, Morishima T, Mukai E, Yasuda T, Kachi T, Mitsuma T (1994): Aberrant androgen action and increased size of tandem CAG repeat in androgen receptor gene in X-linked recessive bulbospinal neuronopathy. *J Neurol Sci* 121: 167–171

Soto D, Sukumar S (1992): Improved detection of mutations in the p53 gene in human tumors as single-stranded conformation polymorphs and double-stranded heteroduplex DNA. *PCR Meth Applic* 2:96–98

Stober J, Culig Z, Gast A, Hobisch A, Radmayr C, Bartsch G, Klocker H, Cato ACB (1994): Ligand specific modulation of activity of mutant androgen receptors from human prostate cancer: a model for invesigating hormone-receptor interactions. Abstract presented at AACR Special Conference on Basic and Clinical Aspects of Prostate Cancer, Palm Springs, CA, Dec 8–12, 1994 [abstract]

Sultan C, Lumbroso S, Poujol N, Belon C, Boudon C, Lobaccaro J-M (1993): Mutations of androgen receptor gene in androgen insensitivity syndromes. *J Steroid Biochem Mol Biol* 46:519–530

Suzuki H, Sato N, Watabe Y, Masai M, Seino S, Shimazaki J (1993): Androgen receptor gene mutations in human prostate cancer. *J Steroid Biochem Mol Biol* 46:759–765

Suzuki H, Nihei N, Sato N, Ichikawa T, Mizokami A, Shimazaki J (1994): Inhibition of growth and increase of acid phosphatase by testosterone on androgen-independent murine prostatic cancer cells transfected with androgen receptor cDNA. *Prostate* 25:310–319

Swinnen K, Cailleau J, Heyns W, Verhoeven G (1990): Prostatic stromal cells and testicular peritubular cells produce similar paracrine mediators of androgen action. *Endocrinology* 126:142–150

Takahashi H, Furusato M, Allsbrook WC Jr, Nishii H, Wakui S, Barrett JC, Boyd J (1995): Prevalence of androgen receptor gene mutations in latent prostatic carcinomas from Japanese men. *Cancer Res* 55:1621–1624

Taplin M-E, Bubley GJ, Shuster TD, Frantz ME, Spooner AE, Ogata GK, Keer HN, Balk SP (1995): Mutation of the androgen-receptor gene in metastatic androgen-independent prostate cancer. *N Engl J Med* 332:1393–1398

Thomas DB, Jiminez LM, McTiernan A, Rosenblatt K, Stalsberg H, Stemhagen A, Thompson WD, Curnen MGM, Satariano W, Austin DF, Greenberg RS, Key C, Kolonel LN, West DW (1992): Breast cancer in men: risk factors with hormonal implications. *Am J Epidemiol* 135:734–748

Thompson MW, McInnes RR, Willard HF (1991): *Genetics in Medicine*, 5th ed. Philadelphia: WB Saunders

Tilley WD, Marcelli M, Wilson JD, McPhaul MJ (1989): Characterization and expression of a cDNA encoding the human androgen receptor. *Proc Natl Acad Sci USA* 86:327–331

Tilley WD, Marcelli M, McPhaul MJ (1990a): Expression of the human androgen receptor gene utilizes a common promoter in diverse human tissues and cell lines. *J Biol Chem* 265:13776–13781

Tilley WD, Wilson CM, Marcelli M, McPhaul MJ (1990b): Androgen receptor gene expression in human prostate carcinoma cell lines. *Cancer Res* 50:5382–5386

Tilley WD, Buchanan G, Hickey TE, Horsfall DJ (1995): Detection of androgen receptor mutations in human prostate cancer by immunohistochemistry and SSCP. *Proc Am Assoc Cancer Res* 36:266 (abstract 1586)

Trachtenberg J, Walsh PC (1982): Correlation of prostatic nuclear androgen receptor content with duration of response and survival following hormonal therapy in advanced prostatic cancer. *J Urol* 127:466–471

Tsai M-J, O'Malley BW (1994): Molecular mechanisms of action of steroid/thyroid receptor superfamily members. *Annu Rev Biochem* 63:451–486

Tsukada T, Inoue M, Tachibana S, Nakai Y, Takebe H (1994): An androgen receptor mutation causing androgen resistance in undervirilized male syndrome. *J Clin Endocrinol Metab* 79:1202–1207

Tung L, Mohamed MK, Hoeffler JP, Takimoto GS, Horwitz KB (1993): Antagonist-occupied human progesterone receptor B-receptors activate transcription without binding to progesterone response elements and are dominantly inhibited by A-receptors. *Mol Endocrinol* 7:1256–1265

Tzukerman MT, Esty A, Santiso-Mere D, Danielian P, Parker MG, Stein RB, Pike JW, McDonnell DP (1994): Human estrogen receptor transactivational capacity is determined by both cellular and promoter context and mediated by two functionally distinct intramolecular regions. *Mol Endocrinol* 8:21–30

van Aubel O, Bolt-deVries J, Blankenstein MA, Schroder FH (1988): Prediction of time to progression after orchiectomy by the nuclear androgen receptor content from multiple biopsy specimens in patients with advanced prostate cancer. *Prostate* 12:191–198

van der Kwast TH, Schalken J, Ruizeveld de Winter JA, van Vroonhoven CCJ, Mulder E, Boersma W, Trapman J (1991): Androgen receptors in endocrine-therapy-resistant human prostate cancer. *Int J Cancer* 48:189–193

Vegeto E, Shahbaz MM, Wen DX, Goldman ME, O'Malley BW, McDonnell DP (1993): Human progesterone receptor A form is a cell- and promoter-specific repressor of human progesterone receptor B function. *Mol Endocrinol* 7:1244–1255

Veldscholte J, Berrevoets CA, Brinkmann AO, Grootegoed JA, Mulder E (1992): Anti-androgens and the mutated androgen receptor of LNCaP cells: differential effects on binding affinity, heat shock protein interactions, and transcriptional activation. *Biochemistry* 31:2393–2399

Veldscholte J, Ris-Stalpers C, Kuiper GGJM, Jenster G, Berrevoets C, Claassen E, van Rooij HCJ, Trapman J, Brinkmann AO, Mulder E (1990): A mutation in the ligand binding domain of the androgen receptor of human LNCaP cells affects steroid binding characteristics and response to anti-androgens. *Biochem Biophys Res Commun* 173:534–540

Veltri RW, Partin AW, Epstein J, Marley GM, Miller CM, Singer DS, Patton KP, Criley SR, Coffey DS (1994): Quantitative nuclear morphometry, Markovian texture descriptors, and DNA content captured on a CAS-200 image analysis system, combined with PCNA and Her-2/neu immunohistochemistry for prediction of prostate cancer progression. *J Cell Biochem Suppl* 19:249–258

Visakorpi T, Hyytinen E, Koivisto P, Tanner M, Keinanen R, Palmberg C, Palotie A, Tammela T, Isola J, Kallioniemi O-P (1995a): In vivo amplification of the androgen receptor gene and progression of human prostate cancer. *Nature Genet* 9: 401–406

Visakorpi T, Kallioniemi AH, Syvanen A-C, Hyytinen ER,Karhu R, Tammela T, Isola JJ, Kallioniemi O-P (1995b): Genetic changes in primary and recurrent prostate cancer by comparative genomic hybridization. *Cancer Res* 55:342–347

Wang Y, Miksicek RJ (1991): Identification of a dominant negative form of the human estrogen receptor. *Mol Endocrinol* 5:1707–1715

Wei LL, Gonzalez-Aller C, Wood WM, Miller LA, Horwitz KB (1990): 5′-Heterogeneity in human progesterone receptor transcripts predicts a new amino-terminal truncated "C"-receptor and unique A-receptor messages. *Mol Endocrinol* 4:1833–1840

Whitmore WF Jr (1973): The natural history of prostatic cancer. *Cancer* 32:1104–1112

Whittemore AS (1994): Prostate cancer. *Cancer Surv* 19/20:309–322

Wilding G (1992): The importance of steroid hormones in prostate cancer. *Cancer Surv* 14:113–130

Wilson CM (1994): Two forms of the androgen receptor protein are expressed in a variety of human tissues. 76th Annual Endocrine Society Meeting, Anaheim, CA, June 15–18, p 629, abstract 1713.

Wilson JD (1992): Syndromes of androgen resistance. *Biol Reprod* 46:168–173

Wilson CM, McPhaul MJ (1994): A and B forms of the androgen receptor are present in human genital skin fibroblasts. *Proc Natl Acad Sci USA* 91:1234–1238

Wingo PA, Tong T, Bolden S (1995): Cancer statistics, 1995. *CA Cancer J Clin* 45:8–30

Wong C-i, Zhou Z-x, Sar M, Wilson EM (1993): Steroid requirement for androgen receptor dimerization and DNA binding. Modulation by intramolecular interactions between the NH_2-terminal and steroid-binding domains. *J Biol Chem* 268:19004–19012

Wong C-W, Privalsky ML (1995): Role of the N terminus in DNA recognition by the v-erb A protein, an oncogenic derivative of a thyroid hormone receptor. *Mol Endocrinol* 9:551–562

Wooster R, Mangion J, Eeles R, Smith S, Dowsett M, Averill D, Barrett-Lee P, Easton DF, Ponder BAJ, Stratton MR (1992): A germline mutation in the androgen receptor gene in two brothers with breast cancer and Reifenstein syndrome. Nature Genet 2:132–134

Yan G, Fukabori Y, Nikolaropoulos S, Wang F, McKeehan WL (1992): Heparin-binding keratinocyte growth factor is a candidate stromal to epithelial cell andromedin. *Mol Endocrinol* 6:2123–2128

Yuan S, Trachtenberg J, Mills GB, Brown TJ, Xu F, Keating A (1993): Androgen-induced inhibition of cell proliferation in an androgen-insensitive prostate cancer cell line (PC-3) transfected with a human androgen receptor complementary DNA. *Cancer Res* 53:1304–1311

Zajchowski DA, Sager R, Webster L (1993): Estrogen inhibits the growth of estrogen receptor-negative, but not of estrogen receptor-positive, human mammary epithelial cells expressing a recombinant estrogen receptor. *Cancer Res* 53: 5004–5011

Zhou Z-x, Sar M, Simental JA, Lane MV, Wilson EM (1994): A ligand-dependent bipartite nuclear targeting signal in the human androgen receptor. Requirement for the DNA-binding domain and modulation by NH_2-terminal and carboxyl-terminal sequences. *J Biol Chem* 269:13115–13123

Zhou Z-x, Lane MV, Kemppainen JA, French FS, Wilson EM (1995): Specificity of ligand-dependent androgen receptor stabilization: receptor domain interactions influence ligand dissociation and receptor stability. *Mol Endocrinol* 9:208–218

Zoppi S, Wilson CM, Harbison MD, Griffin JE, Wilson JD, McPhaul MJ (1993): Complete testicular feminization caused by an amino-terminal truncation of the androgen receptor with downstream initiation. *J Clin Invest* 91:1105–1112

Part IV

HEMATOLOGICAL MALIGNANCIES

Hematopoiesis begins with the pluripotent, or primitive hematopoietic, stem cell that gives rise to the cellular components of blood and marrow. Two major lineages emerge from the stem cell: the myeloid and the lymphoid. The myeloid lineage gives rise to the erythroid lineage, which ends with the erythrocyte. It also produces megakaryotes that give rise to platelets. The other major branch of the myeloid pathways bifurcates into the monocyte/macrophage and granulocyte pathways. The lymphoid lineage splits into the B-cell and T-cell lineages. Although somewhat oversimplified, this scheme outlines hematopoiesis in humans. Numerous genetic defects along any of these pathways can give rise to a multitude of hematologic malignancies. Many of the genes involved have been identified by specific translocations that occur in each leukemia or lymphoma, and many of these genes are proto-oncogenes.

Leukemias and lymphomas comprise a relatively small proportion of all human cancers (~7% incidence and 8% mortality from cancer in the United States per year), however, they nonetheless are of great importance. First, they are important to the patients who are afflicted with these conditions. Second, because most of them are not solid tumor masses, they respond very well to chemotherapy. Finally, and of importance to the current treatise, a number of them provide paradigms for hormonal control of human malignancies, particularly at the molecular level. We will focus on two well understood situations.

Lymphoid malignancies are sensitive to glucocorticoid therapy. Prednisone is a common glucocorticoid analogue that is used in treating acute T-cell lymphoblastic leukemia, and it is active against some B-cell cancers as well. Chapter 17 reviews the clinical literature on the efficacy of glucocorticoid therapy in humans and correlates these results to mechanisms that have been elucidated in human and rodent cell culture systems. The role of the glucocorticoid receptor protein in this process is key.

Glucocorticoids kill T cells via the mechanism of apoptosis (programmed cell death). Apoptosis is an extremely active area of current research, and hematopoiesis provides an ideal model system for its study. Because hematopoietic cells in the various lineages self-renew as well as differentiate, it is essential that mature cells die via apoptosis. Otherwise, the rate of self-renewal will exceed turnover, leading to leukemia. Chapter 18 discusses the current biochemical and molecular processes that may contribute to apoptosis. Although the precise mechanisms are far from clear, current research is rapidly progressing in solving these problems. At present, calcium appears to be a central point of focus, but its precise role in causing apoptosis is unknown.

Acute promyelocytic leukemia (APL) is rare. However, the basic science knowledge that has been obtained from its recent study has been truly phenomenal. APL derives from a t(15; 17) reciprocal translocation that fuses one of the nuclear receptor family members, retinoic acid receptor α (RARα), to another putative transcription factor, PML. Most remarkably, APL patients treated with all-*trans* retinoic acid (ATRA) routinely achieve clinical complete remission. Chapter 19 presents the clinical manifestations of APL and the hemostatic disorders that are characteristic of this disease. The roles of the t(15;17) and RARα in this syndrome are introduced here. The utility of ATRA therapy in the clinic is extensively reviewed and then coalesced into an overview of patient management and expected outcomes. Chapter 20 summarizes the immense amount of information that we have obtained about retinoic acid receptor action and leukemogenesis from the study of APL. It also highlights how clinical disease can be translated into basic science knowledge—studies on APL have uncovered a new structure in the nucleus of cells, and they may lead to a better understanding of apoptosis.

The chapters presented in this section feature the immense patient benefit that has been obtained from hormonal approaches to the treatment of leukemia. The inclusion of glucocorticoids in standard chemotherapy regimens in lymphoid leukemias and the nearly 100% remission rate achieved in APL patients treated with ATRA are truly uplifting success stories. One hopes that such successes can someday be realized with the more prevalent solid tumor cancers as well.

17

Basic and Clinical Studies of Glucocorticosteroid Receptors in Lymphoid Malignancies

CLARK W. DISTELHORST

INTRODUCTION

Glucocorticosteroids, including prednisone and dexamethasone, are among the most effective agents employed in the treatment of lymphoid malignancies. The efficacy of glucocorticoids in these malignancies is based on their ability to induce programmed cell death, or apoptosis, of immature lymphocytes (see Chapter 18, this volume). Biochemical and molecular studies indicate that glucocorticoid-induced lymphocytolysis is mediated through the glucocorticoid receptor and involves regulation of gene expression, although genes that mediate the lethal effect of glucocorticoids are unidentified.

Efforts have been made to translate basic understanding of mechanisms of glucocorticoid action into clinically useful information that will guide the use of glucocorticoids as therapeutic agents. Although studies have demonstrated that glucocorticoid receptor levels in malignant cells are useful prognostic indicators in patients with lymphoid malignancies, information as to the mechanisms of glucocorticoid resistance in patients with lymphoid malignancies is quite limited. Application of molecular techniques to identify and characterize glucocorticoid receptor mutants in glucocorticoid resistant malignancies, and a more complete understanding of the mechanism of glucocorticoid-induced lymphocytolysis, should contribute to improved use of glucocorticoids as therapeutic agents and to the development of novel forms of therapy based on manipulation of the cell death signal transduction pathway at both receptor and postreceptor levels.

Hormones and Cancer
Wayne V. Vedeckis, Editor

LYMPHOID MALIGNANCIES: AN OVERVIEW

The term "lymphoid malignancy" encompasses a heterogeneous group of disorders, all of which represent a monoclonal proliferation of abnormal lymphocytes. Monoclonality, defined by either genetic or immunologic markers, is the essential feature that differentiates lymphoid malignancy from benign lymphoproliferation such as might occur in response to a viral infection. Lymphoid malignancies are generally grouped into one of two categories: lymphomas and leukemias. In lymphomas, the primary site of lymphoproliferation is the lymph node, whereas in leukemias, the primary site of lymphoproliferation is the bone marrow and blood. The distinction between these two categories may at times be obscure because of overlapping features shared by both. For example, lymphomas often involve the bone marrow, whereas leukemias are often accompanied by some degree of involvement of lymph nodes. Moreover, it is important to recognize that lymphomas can arise in virtually any organ, including bone, brain, and liver; such lymphomas are referred to as extranodal.

In human lymphoid lineages, cells progress from immature precursors to fully functional mature forms through sequential levels of differentiation, defined by monoclonal antibodies and molecular probes that characterize cells according to membrane antigenic phenotype, and by the rearrangement and expression of immunoglobulin and T-cell receptor genes. The particular clone of malignant lymphocytes that constitutes a leukemia or a lymphoma in a given patient often appears to be frozen at a particular level of differentiation (Greaves, 1986). Hence, on the basis of immunologic and molecular phenotypic parameters, it is possible to identify subsets of lymphoid malignancies and align them with their approximate normal counterparts in lymphopoiesis. For example, the scheme of B lymphoid differentiation shown in Fig. 1 identifies a variety of lymphoid malignancies of B lymphocyte derivation with their corresponding level of normal cellular differentiation. This ranges from the earliest recognizable progenitor B cell, having undergone immunoglobulin gene rearrangement but without immunoglobulin production, to the mature plasma cell that synthesizes and secretes immunoglobulin molecules.

In acute lymphoblastic leukemia, which is one of the most common malignancies of childhood, cell populations correspond to immature B lymphocyte precursors and appear to be frozen in the act of immunoglobulin gene rearrangement, without the surface membrane antigenic phenotype or immunoglobulin expression that characterizes mature, immunocompetent cells. Similar immunologic and molecular analyses of chronic lymphocytic leukemia and multiple myeloma, two of the most common lymphoid malignancies in adults, have shown that these neoplasms usually represent clonal expansions of relatively mature lymphocyte populations corresponding to immunocompetent B lymphocyte subsets detectable in normal

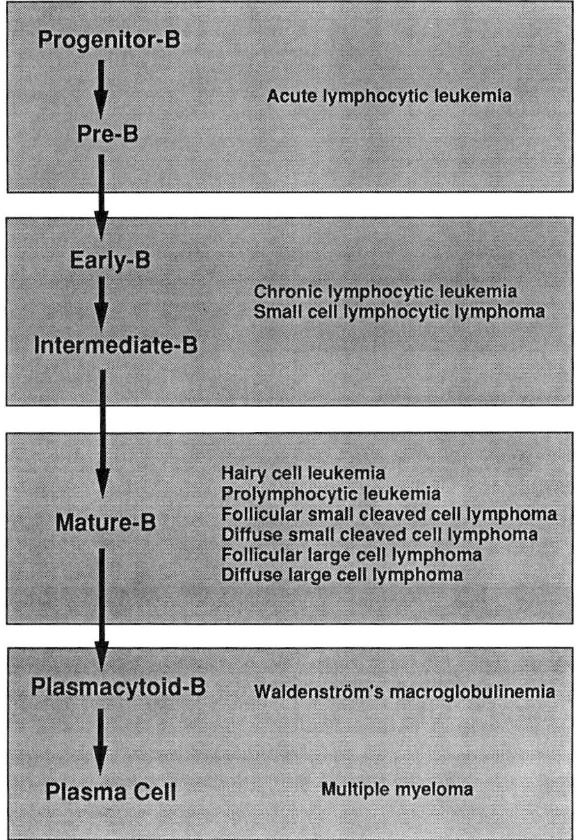

Figure 1. Differentiation scheme of common lymphoid malignancies of B-cell derivation.

lymphoid tissue (Greaves, 1986). The cells of B lymphocyte lineage produce immunoglobulin, which is either expressed on the cell surface, typically the case in chronic lymphocytic leukemia, or secreted into the serum, typically the case in multiple myeloma. However, the immunoglobulin production in these neoplasms is abnormal. For example, in chronic lymphocytic leukemia, the amount and type of immunoglobulin expressed on the cell surface is abnormal, and patients with this disease suffer from a state of immunodeficiency leading to increased susceptibility to infection and autoimmune destruction of normal cellular elements. In multiple myeloma, the cells are often capable of secreting large amounts of abnormal, monoclonal immunoglobulin, which can contribute directly to the morbidity associated with this neoplasm.

Figure 1 is typical of broad classification schemes currently in use for lymphoid malignancies of B lymphoid derivation, but further subdivision of diagnostic categories has been useful. For example, acute lymphoblastic leukemias are often subdivided into various phenotypic categories predictive of clinical response and therapeutic outcome. This represents a major advance that has enabled clinicians to recognize patients who are at particularly high risk of relapse and, therefore, need more intensive therapy, compared to those who are at lower risk of relapse and, therefore, require less intensive therapy (Pui et al., 1993).

Based on immunologic and molecular phenotype, as well as pathoogical features detected by light and electron microscopy, lymphomas are generally classified into one of a large number of different diagnostic categories (Harris et al., 1994). Table 1 summarizes the more common subtypes of lymphomas according to the Working Formulation. To the uninitiated, the classification of lymphomas may appear to be complex. Nevertheless, this classification is of considerable prognostic and therapeutic importance, as it characterizes individual lymphomas as either high-grade, intermediate-grade or low-grade. These subcategories differ considerably in their natural history and response to therapy. The high-grade lymphomas, which correspond to a relatively immature level of T or B lymphocyte differentiation, tend to be clinically more aggressive (Armitage, 1993), whereas the low-grade lymphomas, which correspond

Table 1. Characteristics of non-Hodgkin's lymphoma subtypes

Subtype	Frequency %	Immunophenotype *B-cell* %	*T-cell* %	Median (yrs)	Curability
Low-grade					
Small lymphocytic	4	99	1	61	Unproved
Follicular small cleaved cell	23	100	0	54	Unproved
Follicular mixed	8	100	0	56	Controversial
Intermediate-grade					
Follicular large cell	4	100	0	55	Controversial
Diffuse small cleaved cell	7	90	10	58	Controversial
Diffuse mixed cell	7	60	40	58	Yes
Diffuse large cell	20	90	10	57	Yes
High-grade					
Immunoblastic	8	80	20	51	Yes
Lymphoblastic	4	5	95	17	Yes
Small noncleaved cell	5	100	0	30	Yes
Miscellaneous	10				

Adapted with permission of the New England Journal of Medicine, from Armitage, 1993.

to a more mature level of B or T lymphocyte differentiation, tend to behave in a more indolent fashion (Horning and Rosenberg, 1984). Somewhat paradoxically, the high-grade lymphomas are the most responsive to chemotherapy, and may in many cases be cured, whereas the low-grade lymphomas are generally less responsive to chemotherapy and are rarely, if ever, cured (Armitage, 1993).

The relationship between level of differentiation and clinical behavior is also characteristic of different types of leukemia. For example, children who have acute lymphoblastic leukemia are often cured by chemotherapy, reflecting the propensity of immature lymphocytes to be killed by chemotherapeutic agents (Rivera et al., 1993). In contrast, patients with chronic lymphocytic leukemia or multiple myeloma are only rarely cured, reflecting the relative resistance of more mature lymphoid cells to the lethal effects of chemotherapy (Alexanian and Dimopoulos, 1994).

The last decade has produced remarkable advances in understanding the pathogenesis of lymphoid malignancies at a molecular level. Two classes of oncogenes, those that promote cell proliferation (e.g., c-*myc*) and those that suppress programmed cell death (e.g., *bcl*-2) have been implicated in the pathogenesis of lymphoid malignancies. For example, in Burkitt's lymphoma, a type of aggressive or high-grade lymphoma, the t(8;14) chromosomal translocation leads to abnormal activation of c-*myc* (Spencer and Groudine, 1991), while in the most common type of low-grade lymphoma, the t(14;18) translocation produces increased expression of *bcl*-2 (Ngan, et al, 1988).

Although lymphoid malignancies are relatively common in both children and adults, insight into the etiology of these malignancies is generally lacking. Immunosuppression, as in organ transplant recipients (Penn, 1990) and patients with acquired immunodeficiency syndrome (Harnly et al., 1988), predisposes to an increased incidence of lymphoma. However, a predisposing immunodeficiency is not evident in the vast majority of patients with lymphoid malignancy. Moreover, while certain viruses, including the Epstein-Barr virus and the human lymphotropic virus type I (HTLV-1), are involved in the etiology of certain types of lymphoma (Hanto et al., 1982; Hollsberg and Hafler, 1993), a viral etiology has not been established for the vast majority of lymphoid malignancies. One of the most curious etiologies of lymphoma is the relationship between *Heliobacter pylori* infection and lymphoid malignancy, in which infection with this bacterial organism has been implicated in the causation of gastric lymphoma (Parsonnet et al., 1994).

Overall, lymphoid malignancies represent important clinical entities that have contributed, perhaps more than most other diseases, to both basic understanding and clinical therapeutic advancement. Nevertheless, their incidence has increased much more rapidly than most other malignancies (Rabkin et al., 1993), emphasizing the need for continued investigation into etiology, pathogenesis, and treatment.

ROLE OF GLUCOCORTICOSTEROIDS IN TREATMENT OF LYMPHOID MALIGNANCIES

The use of glucocorticosteroids in the treatment of lymphoid malignancies dates to the late 1940s, when Olaf Pearson and co-workers discovered that treating lymphoma patients with either adrenocorticotropic hormone (ACTH) or cortisone induced tumor regression, an observation that they quickly extended to acute and chronic leukemias (Pearson et al., 1949; Pearson and Eliel, 1950). This was not a chance discovery, but was based on earlier evidence in laboratory animals in which increased adrenal cortical function resulted in involution of normal lymphoid tissues (Dougherty and White, 1943), and on evidence that glucocorticoid administration resulted in regression of lymphoid tumors in mice (Heilman and Kendall, 1944). This was indeed a seminal contribution, as glucocorticoids continue today to be one of the most effective agents employed in the treatment of virtually all varieties of lymphoid malignancy.

Perhaps the most striking example is the treatment of childhood acute lymphoblastic leukemia, where administration of a glucocorticosteroid, usually prednisone, in combination with the vinca alkaloid vincristine, induces rapid and complete remission of disease in more than 90% of children (Rivera et al., 1993). In this disease, glucocorticoid administration can lead to such rapid destruction of malignant lymphoblasts that patients may develop tumor lysis syndrome, a consequence of the rapid release of purine metabolites, potassium and phosphate into the circulation (Silverman and Distelhorst, 1989). The rapid destruction of malignant lymphoblasts by glucocorticoids is reminiscent of the similar rapid destruction of normal thymocytes by glucocorticoids, a prototype of what is now known as programmed cell death or apoptosis (Wyllie, 1990). Indeed, *in vitro* exposure of malignant lymphoblasts, freshly isolated from the blood or bone marrow of children with acute lymphoblastic leukemia, induces an apoptotic form of cell death (Distelhorst, 1988).

Glucocorticosteroids also command a major role in the treatment of lymphoid malignancies that correspond to more mature levels of lymphocyte differentiation, including chronic lymphocytic leukemia and multiple myeloma. In the case of multiple myeloma, for example, recent studies have established that intermittent treatment of multiple myeloma patients with the glucocorticoid compound dexamethasone produces a reduction in tumor mass that equals or even exceeds that achieved with combinations of chemotherapeutic agents traditionally employed in multiple myeloma, including melphan and prednisone, and vincristine and doxorubicin (Alexanian et al., 1992). Moreover, glucocorticoids are currently employed in all effective regimens for the treatment of lymphomas, including both the more indolent and aggressive varieties (Armitage, 1993).

Glucocorticoids are almost never employed as single agents in treating lymphomas, but rather are integral components of multidrug regimens. For example, the present standard form of treatment for aggressive lymphomas employs monthly cycles, usually a total of at least six cycles, of the CHOP regimen (cytoxan, hydroxydaunomycin, oncovin, prednisone). This regimen has stood the test of time and compares favorably with a number of newer, more toxic regimens (Fisher et al., 1993). In this regimen, the glucocorticoid prednisone is administered orally for 5 days at monthly intervals. This is a form of treatment that has been arrived at through a sequential series of carefully controlled clinical trials. Disappointingly, however, not much emphasis has been placed on comparing different glucocorticoid compounds or different modes of administration (e.g., oral vs. intravenous) in patients with lymphoid malignancies. Hence, optimal application of glucocorticoid pharmacology may not yet have been achieved in clinical trials.

Glucocorticoids are at center stage in the treatment of lymphoid malignancies; however the major caveat in their use is the frequent development of glucocorticoid resistance by malignant lymphocytes. This factor is currently often overlooked because glucocorticoids are invariably used in combination with other types of cytolytic agents. However, the fact that development of glucocorticoid resistance is a major limiting factor in the treatment of lymphoid malignancies was well described in early clinical trials which employed the glucocorticoid prednisone as the sole therapeutic modality for remission induction in childhoood acute leukemia (Vietti et al., 1965; Wolff et al., 1967). Although these studies included patients with nonlymphoid acute leukemia, which is now known to be essentially unresponsive to glucocorticoid treatment, the important observation is that even though most children with acute leukemia respond initially to prednisone therapy, most eventually relapse and the response rate to a second course of prednisone therapy is reduced. Hence, the initial course of prednisone therapy, while successful at reducing the bulk of malignant tumor, selects for or induces development of malignant cells that are glucocorticoid resistant. Certainly, the same problem arises in all types of lymphoid malignancies where glucocorticoids are employed as therapeutic agents.

As discussed subsequently, considerable progress has been made toward a fundamental understanding of mechanisms of glucocorticoid resistance in laboratory models of lymphoid malignancy, but the mechanisms of glucocorticoid resistance in humans with lymphoid malignancy have not been as well defined. It would seem that a better understanding of the fundamental mechanism by which glucocorticoids induce programmed cell death, or apoptosis, in lymphocytes could lead not only to a clearer understanding of glucocorticoid resistance, but also to the development of new, more effective agents for treating lymphoid malignancies.

MECHANISM OF GLUCOCORTICOSTEROID ACTION IN LYMPHOID MALIGNANCIES

The natural lympholytic action of glucocorticosteroids has been recognized for many years and provides the basis for using glucocorticoids in the treatment of lymphoid malignancies. The lympholytic action of glucocorticoids is well described as a property of normal thymocytes. Extensive electron microscopic studies have characterized in detail the morphologic evolution of thymocytes following exposure to glucocorticoids (Cowan and Sorenson, 1964). The major findings include condensation of chromatin around the periphery of the nucleus, fragmentation of nucleoli into coarse osmiophilic particles, dilation of the endoplasmic reticulum under the cell membrane, and contraction of the cytoplasm; mitochondria and other organelles appear unaffected. These morphologic features are distinct from those of necrosis and are now recognized as characteristic of a form of cell death referred to as apoptosis (Wyllie, 1990). This form of cell death is not unique to glucocorticoid-induced lymphocyte cell death, but it is commonly induced in a variety of cells by a variety of treatments, including growth factor withdrawal or treatment with toxic agents that interfere with normal cell growth and cell division. As in other forms of apoptosis, glucocorticoid-induced lymphocytolysis is accompanied by extensive DNA fragmentation produced by activation of constitutively expressed endonucleases (Wyllie, 1990; see Chapter 18, this volume).

However, both the morphologic changes of apoptosis and DNA fragmentation lie downstream from the initial stimulating event in lymphocyte cell death. The nature of the initiating event remains unclear, but appears to involve glucocorticoid receptor-mediated transcriptional regulation. This concept is based on a number of experimental findings, the most significant of which is that glucocorticoid-induced lymphocytolysis is mediated by glucocorticoid receptor, a ligand-activated transcription factor (Beato, 1989).

Glucocorticoid receptors were initially detected in thymocytes (Munck, 1968), and subsequently demonstrated to be present in both normal lymphocytes and malignant lymphoblasts (Baxter et al., 1971; Lippman et al., 1973). In a series of elegant studies, Tomkins and co-workers analyzed wild-type and mutant glucocorticoid receptors in mouse lymphoma cells (Baxter et al., 1971; Rosenau et al., 1972; Sibley and Tomkins, 1974a,b; Gehring and Tomkins, 1974; Yamamoto et al., 1974). Their findings demonstrated that glucocorticoid-induced lymphocytolysis indeed is mediated through the glucocorticoid receptor, and that a deficiency of glucocorticoid receptors leads to glucocorticoid resistance. A variety of glucocorticoid receptor mutations were identified that accounted for development of glucocorticoid resistance in mouse lymphoma cells maintained in tissue culture. Some of the defective receptors failed to undergo nuclear translocation, whereas other mutants displayed an increased level of nuclear translocation, demonstrating that nuclear uptake and DNA binding were necessary, but not necessarily sufficient, to confer glucocorticoid-induced cell death.

Analysis of glucocorticoid receptors in a human T lymphoblastic leukemia cell line, CEM, has confirmed the essential role of the glucocorticoid receptor in mediating glucocorticoid-induced lymphocytolysis (Harmon and Thompson, 1981). These studies have also detected receptor mutations, all of which appear to be recessive, that account for development of glucocorticoid resistance. One mutation leads to a deficiency of functional receptor, whereas other mutations lead to receptors that are labile upon activation to a DNA-binding state (Harmon et al., 1985). One derivative of the CEM line, CEM-C1, is particularly valuable, as it expresses glucocorticoid receptors that appear to be normal in functional assays, yet the CEM-C1 cells are glucocorticoid-resistant (Zawydiwski et al., 1983). Hence, the block in glucocorticoid-induced cell death in the CEM-C1 line appears to lie downstream of the glucocorticoid receptor. Further elucidation of the defect in glucocorticoid-induced lymphocytolysis in this particular line could generate valuable insight into the postreceptor signal transduction pathway that mediates glucocorticoid-induced lymphocytolysis. Importantly, the phenotype of the CEM-C1 line illustrates that glucocorticoid resistance in human lymphoid malignancies need not necessarily be due to abnormalities of glucocorticoid receptor number or function.

Cloning of the glucocorticoid receptor gene from both murine and human cells produced considerable insight into the structural and functional properties of the glucocorticoid receptor (Beato, 1989). This knowledge has contributed considerably to a clearer understanding of both the role of glucocorticoid receptors in mediating glucocorticoid-induced lymphocytolysis and the molecular basis for receptor defects that underlie glucocorticoid-resistant phenotypes. The results of cDNA cloning and the comparison of amino acid sequences of different hormone receptors confirmed predictions that all steroid receptors are structurally organized in a series of three major domains: a variable N-terminal region; a short and well conserved cysteine-rich central domain; and a relatively well conserved C-terminal half (Fig. 2). The central domain exhibits an array of cysteine residues compatible with the formation of two zinc fingers, each of which is composed of a zinc atom tetrahedrally coordinated to four cysteines. This type of structure accounts for the DNA-binding activity of the glucocorticoid receptor. Immediately adjacent to the C-terminal zinc finger is a nuclear localization signal that appears to be in part responsible for intranuclear localization of the receptor following hormone binding. A second nuclear localization

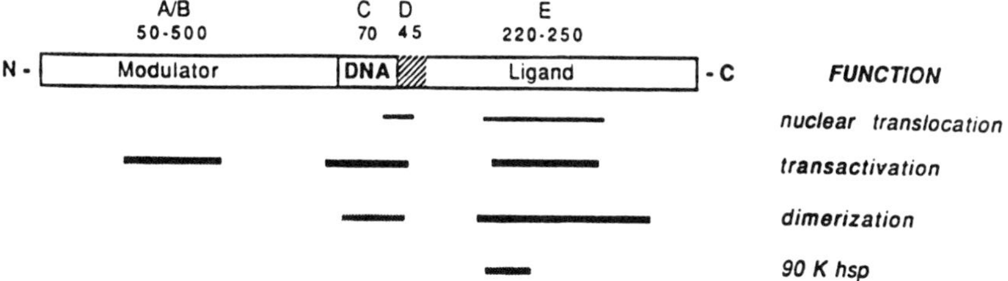

Figure 2. General structure and functional organization of the glucocorticoid receptor. (Reprinted with permission of Cell Press, from Beato, 1989).

signal is located within the carboxy terminal half of the receptor, which incorporates the hormone binding domain. The hormone binding domain is mainly responsible for interaction of the glucocorticoid receptor molecule with heat shock proteins (hsp90, hsp70, hsp56), which play a crucial role in maintaining the receptor in a conformation capable of interaction with hormonal ligand and in mediating nuclear translocation (Pratt, 1993). The central action of the glucocorticoid receptor is the regulation of gene transcription. The DNA-binding zinc finger region of the receptor is principally involved in interaction with the hormone response element, DNA sequences responsible for induction of specific genes by steroid hormones. Both the N-terminal and the C-terminal domains of the receptor molecule appear to have a modulatory effect on transactivation by the glucocorticoid receptor, apparently through interaction with complex arrays of transcription factors (see Chapter 4, this volume).

Utilization of molecular genetic techniques to analyze glucocorticoid receptor mutants in murine and human lymphoid cell lines has provided considerable insight into the nature of glucocorticoid-induced lymphocytolysis, confirming the essential role of the glucocorticoid receptor in mediating lymphocytolysis and the probability that lymphocytolysis is triggered by receptor-mediated transcriptional activation (Miesfeld et al., 1984). One of the glucocorticoid receptor mutations biochemically defined in mouse lymphoma cells by studies of Tomkins and co-workers, discussed earlier, produces a glucocorticoid-resistant phenotype in which an increased proportion of receptors are translocated into the nucleus. Miesfeld and co-workers have demonstrated that these mutant receptors lack sequences encoding the amino-terminal transcriptional modulatory domain (Dieken et al., 1990; Dieken and Miesfeld, 1992). In glucocorticoid resistant CEM cells, Harmon and co-workers have identified an amino acid substitution in the carboxy terminal half of the receptor that appears to alter both ligand binding and receptor transactivation activity (Powers et al., 1993). These findings provide strong support for the hypothesis that glucocorticoid-induced lymphocytolysis is controlled at the level of gene expression, and probably involves activation of a gene or group of genes. A different view has been presented by Thompson and co-workers (Nazareth et al., 1991), based on an analysis of glucocorticoid receptor deletion mutants in relation to cell viability. Their findings indicated that the DNA-binding domain of the glucocorticoid receptor is, by itself, sufficient to trigger death of human leukemic lymphoblasts. Deletion of all other domains of the receptor, including transactivation and ligand binding domains, leaves a receptor fragment highly potent for cell killing. Based on these findings, it has been suggested that occupancy of the glucocorticoid response element by the receptor DNA-binding domain may inhibit transcription.

The preceding findings provide strong evidence that glucocorticoid-induced lymphocytolysis is initiated by a transcriptional regulatory event.

It has been proposed that a key primary event in glucocorticoid-induced lymphocytolysis involves receptor-mediated induction of a hypothetical "suicide gene" (Rosenau et al., 1972; Sibley and Tomkins, 1974). A number of glucocorticoid-regulated genes have been identified in lymphoid cells by subtractive hybridization techniques (Owens et al., 1991; Harrigan et al., 1991; Dieken and Miesfeld, 1992), but the relationship of these genes to cell death induction is unknown. Two genes are of particular interest, however, in view of considerable evidence implicating calcium in cell death signaling. One is the calmodulin gene, whose expression has been found to increase early in glucocorticoid-mediated lysis of mouse lymphoma cells (Dowd et al., 1991). The other is the RP-2 gene (Owens et al., 1991). Increased expression of the RP-2 gene is observed early in the course of glucocorticoid-induced thymocyte death. The functional significance of this gene was not recognized until recently, when it was found to encode an adenosine triphosphate (ATP)-gated cation channel (Brake et al., 1994; Valera et al., 1994). Hence, it is possible that glucocorticoid-mediated lymphocytolysis may involve increased expression of a plasma membrane channel that facilitates entry of extracellular calcium ions.

Most observations to date favor the concept that glucocorticoid-mediated lymphocytolysis involves transcriptional induction, but an alternative view favors transcriptional repression (Helmberg et al., 1995). Indeed, there is considerable evidence that glucocorticoid treatment represses expression of the c-*myc* proto-oncogene, and that enforced expression of c-*myc* inhibits glucocorticoid-induced lymphocytolysis (Eastman-Reks and Vedeckis, 1986; Ramachandran et al., 1993; Helmberg et al., 1995).

In summary, functional glucocorticoid receptors are required for induction of lymphocyte death by glucocorticoids, and there is considerable evidence to support a model of glucocorticoid-induced lymphocytolysis in which glucocorticoid receptor-mediated regulation of gene transcription is a central step. However, the actual genes involved and the mechanism(s) by which their products signal lymphocyte cell death requires further elucidation.

Glucocorticoid receptor concentration is one well established determinant of glucocorticoid-dependent enhancer activity (Vanderbilt et al., 1987). Thus, it follows that glucocorticoid receptor concentration should be one factor in determining the sensitivity of lymphoid cells to lysis by glucocorticoids. Normal lymphocytes appear to be heterogeneous with respect to glucocorticoid receptor concentration (Distelhorst and Benutto, 1981). One factor contributing to this heterogeneity is cell cycle phase distribution. In mouse lymphoma cells, glucocorticoid binding capacity is several-fold higher during S phase of the cell cycle than during G1 phase (Distelhorst et al., 1984). Interestingly, cellular susceptibility to glucocorticoid-induced lysis appears to correlate with the level of receptor elevation during S phase. Also, cAMP increases glucocorticoid binding capacity in murine lymphoma cells and the cytolytic action of glucocorticoids appears to be modulated by cAMP-dependent protein kinase activity (Gruol et al., 1989). Glucocorticoid receptors are under autoregulatory control. In certain cell types, including human lymphoid cells, glucocorticoids appear to down-regulate their own receptor through repression of glucocorticoid receptor gene transcription (Okret et al., 1986; Rosewicz et al., 1988). Post-transcriptional mechanisms have been proposed to regulate glucocorticoid receptor levels in certain types of cells, but receptor half-life appears unaffected by hormone binding in murine lymphoma cells (Distelhorst and Howard, 1989). In human T lymphocytes, however, glucocorticoids up-regulate receptor levels, and this may be important in insuring that the appropriate apoptotic response ensues in this cell type (Denton et al., 1993).

Although considerable attention has been placed on glucocorticoid receptor expression and function as a major determinant of the sensitivity of the lymphoid cells to glucocorticoid-induced lysis, mechanisms unrelated to receptors *per se* have been proposed. Membrane permeability can be a rate-limiting step in glucocorticoid action, and may be the basis for glucocorticoid resistance in certain mouse lymphoma cell lines (Johnson et al., 1984). Moreover, there is evidence that glucocorticoid insensitivity in murine lymphoma lines can be due to elevated expression of the multidrug resistance gene (Bourgeois et al., 1993).

GLUCOCORTICOID RECEPTOR STUDIES IN LYMPHOID MALIGNANCIES

The preceding discussion has emphasized progress in understanding the mechanism of glucocorticoid action in lymphoid malignancies through studies in murine and human lymphoid cell lines. The following discussion deals with efforts over the past two decades to translate this information into a clinically relevant understanding of the mechanism of glucocorticoid resistance in human lymphoid malignancies, and to develop laboratory tests that would be clinically useful in predicting whether patients will or will not respond to glucocorticoid therapy.

Shortly after the essential role of the glucocorticoid receptor in mediating glucocorticoid-induced lymphocytolysis became recognized, Lippman and co-workers demonstrated glucocorticoid receptors in lymphoblasts isolated directly from patients with acute lymphoblastic leukemia (Lippman et al., 1973). Intriguingly, they documented six patients who initially had glucocorticoid receptors in their lymphoblasts and were clinically responsive to glucocorticoid therapy, but who no longer had glucocorticoid receptors detectable after they had responded to combinations of drugs, including glucocorticosteroids. Hence, in at least some patients previously exposed to glucocorticoids during the course of therapy, glucocorticoid resistance developed because of emergence of lymphoblasts that have reduced glucocorticoid binding capacity. Although the series of patients was relatively small, it is interesting that none of the lymphoblasts that contained glucocorticoid receptors failed to respond to glucocorticoids in *in vitro* assays. This suggested that receptor mutations, such as those detected frequently in lymphoid cell lines, or postreceptor defects in the lympholytic pathway, might be uncommon in acute lymphocytic leukemia. However, subsequent results from several laboratories have made it clear that patients with acute lymphoblastic leukemia who have receptor positive tumor cells may nonetheless be resistant to glucocorticoids, and that some of the classic techniques used to assess steroid sensitivity *in vitro* may fail to correlate with actual cell killing in patients (Thompson, 1979). This point is illustrated by the work of Mastrangelo and co-workers (1980), who performed a short-term clinical trial to analyze the relationship between glucocorticoid receptor levels in childhood acute lymphoblastic leukemia cells and clinical response to glucocorticoid therapy. In a relatively small study population, patients were initially treated with glucocorticoids alone to assess clinical responsiveness. Three important conclusions were reached: (1) a substantial number of glucocorticoid receptors did not necessarily confer a glucocorticoid-sensitive phenotype; (2) a low number of glucocorticoid receptors correlated with a poor clinical response to glucocorticoids; and (3) *in vitro* assays of cell killing, based on measuring [^{3}H]thymidine incorporation into DNA, did not correlate well with clinical responsiveness.

These findings underscore the fact that glucocorticoid-induced lymphocytolysis is mediated through the glucocorticoid receptor; however, the presence of glucocorticoid receptors does not necessarily confer sensitivity to the lethal effects of glucocorticoids. This conclusion is based on work by Crabtree and co-workers (1978), who analyzed the relationship between glucocorticoid binding capacity and sensitivity of isolated human leukemia and lymphoma cells to glucocorticoids. Their findings demonstrated similar numbers of glucocorticoid receptors in glucocorticoid-sensitive malignancies, such as acute lymphoblastic leukemia, chronic lymphocytic leukemia, and lymphoma, and in glucocorticoid-insensitive malignancies, such as acute nonlymphocytic leukemias. These observations reflect a striking difference in the basic biological behavior of nonlymphoid malignancies compared to lymphoid malignancies, and this is borne out from therapeutic trials demonstrating that glucocorticoids are of limited value as therapeutic agents in nonlymphoid leukemias. Moreover, these observations emphasize that lymphoid cells are genetically programmed to die in response to glucocorticoids, whereas nonlymphoid cells, including myeloid cells, are not.

It is important to realize that, over the past two decades, standard therapy for the most common lymphoid malignancies, including acute lymphoblastic leukemia and lymphoma, has included multiple chemotherapeutic agents given more or less simultaneously in combination with glucocorticosteroids. Hence, it is uncommon for patients to be treated with glucocorticoids alone, limiting direct correlations between glucocorticoid

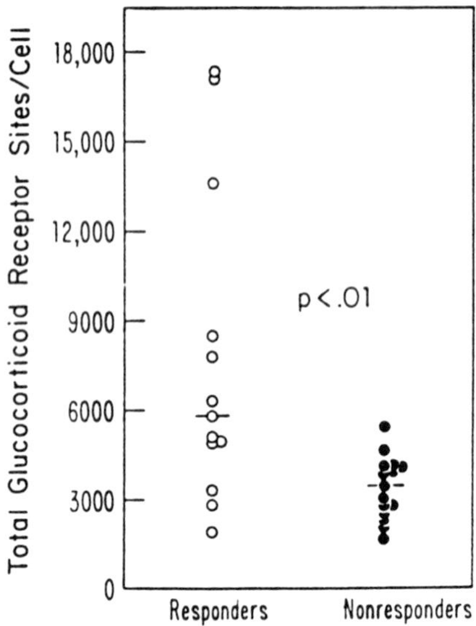

Figure 3. Relationship between receptor level and clinical response. Total glucocorticoid receptors in lymphomas from 13 patients whose tumors clinically decreased by 50% on glucocorticoid therapy and 15 nonresponders. The medians (*horizontal bars*) are significantly different ($p < 0.01$). (Reprinted with permission of the Lancet Ltd, from Bloomfield et al., 1980).

receptor levels and glucocorticoid responsiveness *per se*. However, this goal was achieved in a study by Bloomfield and co-workers (1980), in which 28 adults with lymphoma were treated initially with the glucocorticoid dexamethasone as a single agent for up to 14 days, allowing assessment of clinical glucocorticoid responsiveness. Lymphoma cells from patients who responded had a higher mean number of glucocorticoid receptors per cell than did tumor cells from nonresponders (Fig. 3). Although the authors of this study concluded that measurement of glucocorticoid receptors in lymphoma tissue may enable selection of patients most likely to respond to glucocorticoid therapy, the overlap of receptor values in responders and nonresponders was considerable. Such measurements are not employed in the current management of patients with lymphoma, all of whom routinely are treated with glucocorticoid-containing regimens. However, the findings of this study clearly indicate that the presence of glucocorticoid receptors in lymphoma tissue is not necessarily a reflection of a glucocorticoid-sensitive phenotype, as tumor cells from all the nonresponders contained glucocorticoid receptors. This conclusion is similar to the situation described earlier in acute lymphoblastic leukemia cells, and suggests that glucocorticoid resistance in lymphomas may be due to mutations that inhibit nuclear uptake of receptors, perhaps similar to that described in lymphoma cell lines, or at a postreceptor step in the cell death pathway.

Most clinical studies have analyzed the relationship between glucocorticoid receptor levels and clinical responsiveness in patients with acute lymphoblastic leukemia. Most commonly, acute lymphoblastic leukemia cells phenotypically resemble early B lymphocyte precursors. T-cell acute lymphoblastic leukemia is a less common subset, in which lymphoblasts correspond phenotypically to early T-cell precursors. Patients with non-T-cell leukemias generally have a better response to chemotherapy and a longer duration of survival than patients with T-cell leukemia. Lippman and coworkers (1978) have shown that T-cell lymphoblasts have a lower glucocorticoid receptor content compared to their non-T counterparts. Moreover, within each subset, patients with a high lymphoblast receptor content had a better prognosis than patients with a low lymphoblast receptor content. This concept has been confirmed in several studies analyzing data from large numbers of patients with acute lymphoblastic leukemia, demonstrating that a low glucocorticoid receptor level in lymphoblasts isolated at the time of diagnosis is significantly correlated with a poor response to therapy and a poor overall prognosis (Costlow et al., 1982; Quddus et al., 1985; Kato et al., 1993). On the other hand, high receptor levels do not necessarily confer clinical responsiveness, consistent with results of earlier, smaller studies. Whether these relationships are mediated entirely by glucocorticoid responsiveness is not known. Another possibility is that glucocorticoid receptor levels could be an important biological marker associated with other factors that are related to chemotherapy responsiveness. Finally, it should be noted that glucocorticoid receptor levels are influenced significantly by recent glucocorticoid therapy; hence, whether or not patients have received recent glucocorticoid therapy needs to be taken into account in clinical studies where glucocorticoid receptor measurements are used as a prognostic index (Shipman et al., 1981).

The preceding studies confirm that glucocorticoid-induced lymphocytolysis is mediated through the glucocorticoid receptor, and that glucocorticoid receptor deficiency is one mechanism of glucocorticoid resistance in human lymphoid malignancies, but they have not elucidated the mechanism of glucocorticoid resistance in the vast majority of patients who fail glucocorticoid therapy. In these patients, glucocorticoid resistance lies either at the level of the glucocorticoid receptor, presumably due to receptor mutations that abrogate the transcriptional regulatory properties of the receptor without significantly affecting hormone binding, or at the level of postreceptor signal transduction. Biochemical characterization of glucocorticoid receptors isolated from malignant lymphoblasts has revealed abnormalities in biophysical properties of receptors, although a cause-effect relationship between these abnormalities and glucocorticoid resistance has not been established (McCaffrey et al., 1982). Moreover, characterization of glucocorticoid receptors isolated from malignant lymphocytes may be complicated by proteolytic cleavage of the receptor in cell extracts (Stevens et al., 1979; Sherman et al., 1984; Distelhorst and Miesfeld,

1987). In many leukemic samples, receptor cleavage is due to an elastase-like serine protease present in either lymphoblasts or neutrophils that invariably contaminate isolated lymphoblast samples to a small degree (Distelhorst et al., 1987). Hence, it is important when characterizing glucocorticoid receptors from tumor tissues to include potent protease inhibitors prior to preparing cell extracts. When this was done in one study of 52 bone marrow samples from children with acute lymphoblastic leukemia, normal size glucocorticoid receptors were invariably present (Distelhorst et al., 1987). Moreover, glucocorticoid receptors in isolated acute lymphoblastic leukemia cells are functionally normal in terms of their capacity to undergo nuclear translocation, even in cases where glucocorticoid resistance is evident (Costlow and Pui, 1987). Thus, if glucocorticoid resistance is due to a glucocorticoid receptor defect, the defect is likely to arise because of point mutations in receptor structure, rather than gross deletions. Whether or not such mutations occur is unknown.

An intriguing glucocorticoid receptor abnormality is under investigation by Rosen and co-workers in glucocorticoid resistant multiple myeloma cells (Moalli et al., 1992, 1993). These investigators have found evidence of truncated glucocorticoid receptor mRNAs that arise as splicing abnormalities, resulting in deletion of specific exons of the hormone binding domain. The abnormal receptors that are produced would not be expected to bind hormone but might cause glucocorticoid resistance by heterodimerization with normal glucocorticoid receptors, or by sequestering accessory transcription factors needed for glucocorticoid receptor-mediated transcriptional regulation. Intriguingly, these abnormally spliced receptor messages appear to be detected in some normal tissues, as well as in myeloma cells. Hence, the possibility arises that these abnormal receptor forms might have a normal physiological function, perhaps in regulating the transcriptional regulatory properties, of normal, intact receptors.

CONCLUSIONS

Often in clinical medicine, practical application precedes basic understanding. Nowhere is this more true than in the use of glucocorticosteroids in treating lymphoid malignancies. Glucocorticoids were used as therapeutic agents for these malignancies for almost three decades before it became understood that their action was mediated through receptors that function by regulating gene transcription. To this day, the postreceptor signaling events that mediate glucocorticoid-induced lymphocytolysis, including the suicide genes that encode this process, have not been identified. Whereas clinical investigators have identified relationships between levels of glucocorticoid receptors in malignant lymphocytes and prognosis, modern molecular techniques have not yet revealed the mechanism(s) of glucocorticoid resistance that so frequently limit the clinical efficacy of glucocorticoid therapy.

Hence, the study of glucocorticoid action as it relates to lymphoid malignancy remains an important and fertile area for both basic and clinical investigation.

REFERENCES

Alexanian R, Dimopoulos M (1994): The treatment of multiple myeloma. *N Engl J Med* 330:484–490

Alexanian R, Dimopoulos MA, Delasalle K, Barlogie B (1992): Primary dexamethasone treatment of multiple myeloma. *Blood* 80:887–890

Armitage JO (1993): Treatment of non-Hodgkin's lymphoma. *N Engl J Med* 328:1023–1030

Baxter JD, Harris AW, Tomkins GM, Cohn M (1971): Glucocorticoid receptors in lymphoma cells in culture: relationship to glucocorticoid killing activity. *Science* 171:189–191

Beato M (1989): Gene regulation by steroid hormones. *Cell* 56:335–344

Bloomfield CD, Peterson BA, Zaleskas J, Frizzera G, Smith KA, Hildebrandt L, Gail-Peczalska KJ, Munck A (1980): In vitro glucocorticoid studies for predicting response to glucocorticoid therapy in adults with malignant lymphoma. *Lancet* 1:1952–1956

Bourgeois S, Groul DJ, Newby RF, Rajah FM (1993): Expression of an mdr gene is associated with a new form of resistance to dexamethasone-induced apoptosis. *Mol Endocrinol* 7:840–851

Brake AJ, Wagenbach MJ, Julius D (1994): New structural motif for ligand-gated ion channels defined by an ionotropic ATP receptor. *Nature* 371:519–523

Costlow ME, Pui C-H (1987): Nuclear translocation of lymphoblast glucocorticoid receptors in childhood leukemia does not predict steroid responsiveness *J Steroid Biochem* 26:15–18

Costlow ME, Pui C-H, Dahl GV (1982): Glucocorticoid receptors in childhood acute lymphocytic leukemia. *Cancer Res* 42:4801–4806

Cowan WK, Sorenson GD (1964): Electron microscopic observations of acute thymic involution produced by hydrocortisone. *Lab Invest* 13:353–370

Crabtree GR, Smith KA, Munck A (1978): Glucocorticoid receptors and sensitivity of isolated human leukemia and lymphoma cells. *Cancer Res* 38:4268–4272

Denton RR, Eisen LP, Elsasser MS, Harmon JM (1993): Differential autoregulation of glucocorticoid receptor expression in human T- and B-cell lines. *Endocrinology* 133:248–256

Dieken ES, Meese EU, Miesfeld RL (1990): nt^i glucocorticoid receptor transcripts lack sequences encoding the amino-terminal transcriptional modulatory domain. *Mol Cell Biol* 10:4574–4581

Dieken ES, Miesfeld RL (1992): Transcriptional transactivation functions localized to the glucocorticoid receptor N terminus are necessary for steroid induction of lymphocyte apoptosis. *Mol Cell Biol* 12:589–597

Distelhorst CW (1988): Glucocorticosteroids induce DNA fragmentation in human lymphoid leukemia cells. *Blood* 72:1305–1309

Distelhorst CW, Benutto BM (1981): Glucocorticoid receptor content of T lymphocytes: Evidence for heterogeneity. *J Immunol* 126:1630–1634

Distelhorst CW, Benutto BM, Bergamini RA (1984): Effect of cell cycle position on dexamethasone binding by mouse and human lymphoid cell lines: correlation between an increase in dexamethasone binding during S phase and dexamethasone sensitivity. *Blood* 63:105–113

Distelhorst CW, Janiga KE, Howard KJ, Strandjord SE, Campbell EJ (1987): Neutrophil elastase produces 52-kD and 30-kD glucocorticoid receptor fragments in the cytosol of human leukemia cells. *Blood* 70:860–868

Distelhorst CW, Miesfeld R (1987): Characterization of glucocorticoid receptors and glucocorticoid receptor mRNA in human leukemia cells: stabilization of the receptor by diisopropylfluorophosphate. *Blood* 69:750–756

Distelhorst CW, Howard KJ (1989): Kinetic pulse-chase labeling study of the glucocorticoid receptor in mouse lymphoma cells. Effect of glucocorticoid and antiglucocorticoid hormones on intracellular receptor half-life. *J Biol Chem* 264:13080–13085

Dougherty TF, White A (1943): Effect of pituitary adrenotropic hormone on lymphoid tissue. *Proc Soc Exp Biol Med* 53:132–133

Dowd DR, MacDonald PN, Komm BS, Haussler MR, Miesfeld R (1991): Evidence for early induction of calmodulin gene expression in lymphocytes undergoing glucocorticoid-mediated apoptosis. *J Biol Chem* 266:18423–18426

Eastman-Reks SB, Vedeckis WV (1986): Glucocorticoid inhibition of c-*myc*, c-*myb*, and c-*Ki-ras* expression in a mouse lymphoma cell line. *Cancer Res* 46:2457–2462

Fisher RI, Gaynor ER, Dahlberg S, Oken MM, Grogan TM, Mize EM, Glick JH, Coltman CA, Miller TP (1993): Comparison of a standard regimen (CHOP) with three intensive chemotherapy regimens for advanced non-Hodgkin's lymphoma. *N Engl J Med* 328:1002–1006

Gehring U, Tomkins GM (1974): A new mechanism for steroid unresponsiveness: loss of nuclear binding activity of a steroid hormone receptor. *Cell* 3:301–306

Greaves MF (1986): Differentiation-linked leukemogenesis in lymphocytes. *Science* 234:697–704

Gruol DJ, Rajah FM, Bourgeois S (1989): Cyclic AMP-dependent protein kinase modulation of the glucocorticoid-induced cytolytic response in murine T-lymphoma cells. *Mol Endocrinol* 3:2119–2127

Hanto DW, Frizzera G, Gajl-Peczalska KJ, Sakamoto K, Purtilo DT, Balfour HH, Simmons RL, Najarian JS (1982): Epstein-Barr virus-induced B-cell lymphoma after renal transplantation. *N Engl J Med* 306:913–918

Harmon JM, Thompson EB (1981): Isolation and characterization of dexamethasone-resistant mutants from human lymphoid cell line CEM-C7. *Mol Cell Biol* 1:512–521

Harmon JM, Thompson EB, Baione KA (1985): Analysis of glucocorticoid-resistant human leukemic cells by somatic cell hybridization. *Cancer Res* 45:1587–1593

Harnly ME, Swan SH, Holly EA (1988): Temporal trends in the incidence of non-Hodgkin's lymphoma and selected malignancies in a population with high incidence of acquired immunodeficiency syndrome. *Am J Epidemiol* 128:261–267

Harrigan MT, Campbell NF, Bourgeois S (1991): Identification of a gene induced by glucocorticoids in murine T-cells: A potential G protein-coupled receptor. *Mol Endocrinol* 5:1331–1338

Harris NL, Jaffe ES, Stein H, Banks PM, Chan JKC, Cleary ML, Delsol G, Wolf-Peeters CD, Falini B, Gatter KC, Grogan TM, Isaacson PG, Knowles DM, Mason DY, Muller-Hermelink H-K, Pileri SA, Piris MA, Ralfkiaer E,

Warnke RA (1994): A revised European-American classification of lymphoid neoplasms: A proposal from the international lymphoma study group. *Blood* 84:1361–1392

Heilman FR, Kendall EC (1944): The influence of 11-dehydro-17-hydroxycorticosterone (compound E) on the growth of a malignant tumor in the mouse. *Endocrinology* 34:416–420

Helmberg A, Auphan N, Caelles C, Karin M (1995): Glucocorticoid-induced apoptosis of human leukemic cells is caused by the repressive function of the glucocorticoid receptor. *EMBO J* 14:452–460

Hollsberg P, Hafler DA (1993): Pathogenesis of diseases induced by human lymphotropic virus type I infection. *N Engl J Med* 328:1173–1182

Horning SJ, Rosenberg SA (1984): The natural history of initially untreated low-grade non-Hodgkin's lymphomas. *N Engl J Med* 311:1471–1476

Johnson DM, Newby RF, Bourgeois S (1984): Membrane permeability as a determinant of dexamethasone resistance in murine thymoma cells. *Cancer Res* 44:2435–2440

Kato GJ, Quddus FF, Shuster JJ, Boyett J, Pullen JD, Borowitz MJ, Whitehead WM, Crist WM, Leventhal BG (1993): High glucocorticoid receptor content of leukemic blasts is a favorable prognostic factor in childhood acute lymphoblastic leukemia. 82:2304–2309

Lippman ME, Halterman RH, Leventhal BG, Perry S, Thompson EB (1973): Glucocorticoid-binding proteins in human acute lymphoblastic leukemic blast cells. *J Clin Invest* 52:1715–1725

Lippman ME, Yarbro GK, Leventhal BG (1978): Clinical implications of glucocorticoid receptors in human leukemia. *Cancer Res* 38:4251–4256

Mastrangelo R, Malandrino R, Riccardi R, Longo F, Ranelletti FO, Lacobelli S (1980): Clinical implications of glucocorticoid receptor studies in childhood acute lymphoblastic leukemia. *Blood* 56:1036–1040

McCaffrey R, Lillquist A, Bell R (1982): Abnormal glucocorticoid receptors in acute leukemia cells. *Blood* 59:393–400

Miesfeld R, Okret S, Wikstrom AC, Wrange O, Gustafsson JÅ ,Yamamoto KR (1984): Characterization of a steroid hormone receptor gene and mRNA in wild-type and mutant cells. *Nature* 312:779–781

Moalli PA, Pillay S, Weiner D, Leikin R, Rosen ST (1992): A mechanism of resistance to glucocorticoids in multiple myeloma: transient expression of a truncated glucocorticoid receptor mRNA. *Blood* 79:213–222

Moalli PA, Pillay S, Krett NL, Leikin R, Rosen ST (1993): Alternatively spliced glucocorticoid receptor messenger RNAs in glucocorticoid-resistant human multiple myeloma cells. *Cancer Res* 53:3877–3879

Munck A (1968): Metabolic site and time course of cortisol action on glucose uptake, lactic acid output, and glucose 6-phosphate levels of rat thymus cells in vitro. *J Biol Chem* 243:1039–1042

Nazareth LV, Harbour DV, Thompson EB (1991): Mapping the human glucocorticoid receptor for leukemic cell death. *J Biol Chem* 266:12976–12980

Ngan B-Y, Chen-Levy Z, Weiss LM, Warnke RA, Cleary ML (1988): Expression in non-Hodgkin's lymphoma of the *bcl*-2 protein associated with the t(14;18) chromosomal translocation. *N Engl J Med* 318:1638–1644

Okret S, Poellinger L, Dong Y, Gustafsson JÅ (1986): Down-regulation of glucocorticoid receptor mRNA by glucocorticoid hormones and recognition by the

receptor of a specific binding sequence within a receptor cDNA clone. *Proc Natl Acad Sci USA* 83:5899–5903
Owens GP, Hahn WE, Cohen JJ (1991): Identification of mRNAs associated with programmed cell death in immature thymocytes. *Mol Cell Biol* 11:4177–4188
Parsonnet J, Hansen S, Rodriquez L, Gelb AB, Warnke RA, Jellum E, Orentreich N, Vogelman JH, Friedman GD (1994): Heliobacter pylori infection and gastric lymphoma. *N Engl J Med* 330:1267–1271
Pearson OH, Eliel LP (1950): Use of pituitary adrenocorticotropic hormone (ACTH) and cortisone in lymphomas and leukemias. *J Am Med Assoc* 144:1349–1353
Pearson OH, Eliel LP, Rawson RW, Dobriner K, Rhoads CP (1949): ACTH- and cortisone-induced regression of lymphoid tumors in man: a preliminary report. *Cancer* 2:943–945
Penn I (1990): Cancers complicating organ transplantation. *N Engl J Med* 323:1767–1768
Powers JH, Hillmann AG, Tang DC, Harmon JM (1993): Cloning and expression of mutant glucocorticoid receptors from glucocorticoid-sensitive and -resistant human leukemic cells. *Cancer Res* 53:4059–4056
Pratt WB (1993): The role of heat shock proteins in regulating the function, folding and trafficking of the glucocorticoid receptor. *J Biol Chem* 268:21455–21458
Pui C-H, Behm FG, Crist WM (1993): Clinical and biologic relevance of immunologic marker studies in childhood acute lymphoblastic leukemia. *Blood* 82:343–362
Quddus FF, Leventhal BG, Boyett JM, Pullen DJ, Crist WM, Borowitz MJ (1985): Glucocorticoid receptors in immunological subtypes of childhood acute lymphocytic leukemia cells: a pediatric oncology group study. *Cancer Res* 45:6482–6486
Rabkin CS, Decesa SS, Zahm SH, Gail MH (1993): Increasing incidence of non-Hodgkin's lymphoma. *Semin Hematol* 30:286–296
Ramachandran T, Harbour DV, Thompson EB (1993): Suppression of c-*myc* is a critical step in glucocorticoid-induced human leukemic cell lysis. *J Biol Chem* 268:18306–18312
Rivera GK, Pinkel D, Simone JV, Hancock ML, Crist WM (1993): Treatment of acute lymphoblastic leukemia. *N Engl J Med* 329:1289–1296
Rosenau W, Baxter JD, Rousseau GG, Tomkins GM (1972): Mechanism of resistance to steroids: r^- receptor defect in lymphoma cells. *Nature (New Biol)* 237:20–23
Rosewicz S, McDonald AR, Maddux BA, Goldfine ID, Miesfeld RL, Logsdon CD (1988): Mechanism of glucocorticoid receptor down-regulation by glucocorticoids. *J Biol Chem* 263:2581–2584
Sherman MR, Stevens Y, Tuazon FB (1984): Multiple forms and fragments of cytosolic glucocorticoid receptors from human leukemic cells and normal lymphocytes. *Cancer Res* 44:3783–3795
Shipman GF, Bloomfield CD, Smith KA, Peterson BA, Munck A (1981): The effects of glucocorticoid therapy on glucocorticoid receptors in leukemia and lymphoma cells. *Blood* 58:1198–1202
Sibley CH, Tomkins GM (1974a): Isolation of lymphoma cell variants resistant to killing by glucocorticoids. *Cell* 2:213–220
Sibley CH, Tomkins GM (1974b): Mechanisms of steroid resistance. *Cell* 2:221–227
Silverman P, Distelhorst CW (1989): Metabolic emergencies in clinical oncology. *Semin Oncol* 16:504–15

Spencer CA, Groudine M (1991): Control of c-*myc* regulation in normal and neoplastic cells. *Adv Cancer Res* 56:1–48

Stevens J, Stevens Y, Rosenthal RL (1979): Characterization of cytosolic and nuclear glucocorticoid-binding components in human leukemic lymphocytes. *Cancer Res* 39:4939–4948

Thompson EB (1979): Report on the International Union Against Cancer workshop on steroid receptors in leukemia. *Cancer Treat Rep* 63:189–194

Valera S, Hussy N, Evans RJ, Adami N, North RA, Surprenant A, Buell G (1994): A new class of ligand-gated ion channel defined by P2x receptor for extracellular ATP. *Nature* 371:516–519

Vanderbilt JN, Miesfeld R, Maler BA, Yamamoto KR (1987): Intracellular receptor concentration limits glucocorticoid-dependent enhancer activity. *Mol Endocrinol* 1:68–74

Vietti TJ, Sullivan MP, Berry DH, Haddy TB, Haggard ME, Blattner RJ (1965): The response of acute childhood leukemia to an initial and a second course of prednisone. *Pediatrics* 66:18–26

Wolff JA, Brubaker CA, Murphy ML, Pierce MI, Severo N (1967): Prednisone therapy of acute childhood leukemia: prognosis and duration of response in 330 treated patients. *J Pediatr* 70:629–631

Wyllie AH (1990): Glucocorticoid-induced thymocyte apoptosis is associated with endonuclease activation. *Nature* 284:555–556

Yamamoto KR, Stampfer MR, Tomkins GM (1974): Receptors from glucocorticoid-sensitive lymphoma cells and two classes of insensitive clones: physical and DNA-binding properties. *Proc Nat Acad Sci USA* 71:3901–3905

Zawydiwski R, Harmon JM, Thompson EB (1983): Glucocorticoid-resistant human acute lymphoblastic leukemic cell line with functional receptor. *Cancer Res* 43:3865–3873

18

Glucocorticoid Actions on Normal and Neoplastic Lymphocytes: Activation of Apoptosis

JENNIFER W. MONTAGUE AND JOHN A. CIDLOWSKI

INTRODUCTION

The glucocorticoid steroid hormones are used as part of treatment plans for some forms of cancer because of their lytic and growth suppressive effects (see Chapter 17, this volume). It was believed for a long time that the lytic effect of glucocorticoids occurred only in thymocytes; however, it has now been shown that B cells are also sensitive to glucocorticoid treatment (Distelhorst, 1988; McConkey et al., 1991). Glucocorticoids are prescribed in the cases of chronic myeloid leukemia, Hodgkin's lymphoma, non-Hodgkin's lymphoma, and multiple myeloma (Polliack, 1991; Cidlowski and Schwartzman, 1993), although they also demonstrate growth suppressive effects on solid tumors (Alexander et al., 1993; Evers et al., 1993). Additionally, they are recommended as adjunct treatment for patients with cancerous complications such as metastatic spinal cord compression and brain metastases (Weissman, 1988; Sorensen et al., 1994). The first report concerning glucocorticoids and cancer was published in 1944 (Heilman and Kendall, 1944). Heilman and Kendall (1944) discovered that corticosterone treatment resulted in regression of a lymphatic tumor that had been transplanted into mice. Shortly after that, Dougherty and White (1945) observed that rat thymi underwent involution in response to injections of adrenal cortical hormones and adrenocorticotropin. Such information was suggestive that hormone treatment might also be useful in treating certain tumors in humans; however, further studies showed that humans did not respond in precisely the same manner to glucocorticoid treatment as mice and rats. For example, early work by Schrek (1961, 1964) demonstrated that although

Hormones and Cancer
Wayne V. Vedeckis, Editor

lymphocytes from patients with chronic lymphocytic leukemia or lymphosarcomas (CLL) tended to be more sensitive (i.e., underwent lysis) to glucocorticoid treatment compared to normal human lymphocytes, rodent lymphocytes, in general, displayed a much more rapid and complete response to glucocorticoids. This discrepancy in response to glucocorticoids prompted the labeling of different species as either glucocorticoid-sensitive or glucocorticoid-insensitive. Based on the methods of that time for comparing reactions to hormones, humans were labeled as a glucocorticoid-insensitive species. Data gathered since then, however, have shown that subpopulations of lymphocytes do, indeed, display inhibition of proliferation or a lytic response to glucocorticoid treatment. Nieto et al. (1992) observed that human cortical and medullary thymocytes are equally lysed by dexamethasone, whereas Tuosto et al. (1994) revealed that mature T lymphocytes were susceptible to dexamethasone if cultured for a long time. Other subpopulations of nonmalignant lymphocytes that will lyse after exposure to glucocorticoids include unactivated B-cells, as well as certain activated T lymphocytes (Galili, 1983). The lytic response to glucocorticoids was later identified as apoptosis, or programmed cell death (Kerr et al., 1972). The seminal paper by Kerr et al. describes apoptosis as a distinct form of cell death that occurs in a wide variety of cells and can be stimulated by an array of different signals, including glucocorticoid addition to lymphocytes. The mechanisms involved in glucocorticoid-induced death are still unknown, however.

APOPTOSIS

There are at least two methods by which a cell can die: necrosis or apoptosis. Necrosis is primarily a response to injury by nonphysiological effectors or conditions, such as ischemia or toxin administration. During necrosis, the cell undergoes irreversible swelling followed by lysis. As the cell swells, caused by loss of ion pump activity with the resulting influx of ions and fluid, membrane disruption occurs. Hydrolases are released from lysosomes and hasten the destruction of the cell. Histones are degraded and the exposed portions of DNA are digested, resulting in DNA fragments of varying lengths. This disruption of the plasma membrane allows the intracellular milieu to be exposed to the environment, generating an immune response and disrupting more cells in the general vicinity. Apoptosis, on the other hand, allows one cell within a tissue to die without affecting any of the surrounding cells. Apoptosis is an integral part of cellular processes and occurs during many different physiological activities, such as embryogenesis, differentiation, metamorphosis, and normal tissue turnover. Apoptosis provides an important balance to mitosis. In the immune system itself, apoptosis is observed during positive and negative selection of maturing thymocytes, cell death caused by cytotoxic T cells, activation-induced death, and antibodies directed against the T cell receptor (reviewed in Schwartzman and Cidlowski, 1993; Gruber et al., 1994). A popular system used to study apoptosis is glucocorticoid-treated thymocytes. Glucocorticoid-induced thymocyte apoptosis has been termed "inducible" apoptosis by one group because of its apparent requirement for RNA and protein production, as opposed to other models that do not seem to require new protein products ("release" and "transduction" models) (Owens and Cohen, 1992).

Cells undergoing the apoptotic process display characteristic morphologic and biological changes (Fig. 1). Once the apoptotic signal is received, the cell begins to pull away from neighboring cells as its cytoplasm condenses. The organelles maintain their integrity and remain free in solution throughout the process, but the chromatin condenses and migrates to the margins of the nuclear envelope. This apoptotic chromatin is known to undergo some form of cleavage in addition to condensation. DNA fragmentation alone is enough to cause cell death (McConkey et al., 1989). Two different patterns of cleavage are recognized. One is internucleosomal cleavage, such that DNA fragments that are integer multiples of 180–200 bp in size are formed (Wyllie et al., 1984); the other is the "large break" cleavage, in which

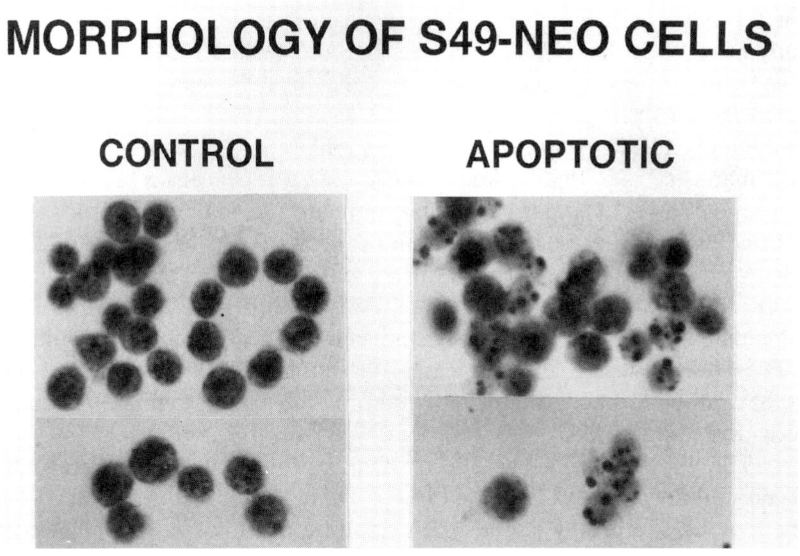

MORPHOLOGY OF S49-NEO CELLS

CONTROL

APOPTOTIC

Figure 1. Glucocorticoid-induced apoptosis of S49 cells. Hematoxylin and eosin staining of mouse thymoma cells (S49). The left panel (control) shows the normal S49 cells while the right panel (apoptotic) shows cells treated with 1×10^{-6} M dexamethasone for 24 hours. The dexamethasone-treated cells demonstrate shrinkage and chromatin condensation typical for apoptosis (Photo courtesy of Dr. Rosemary B. Evans-Storms, NIEHS, RTP, NC).

fragments ranging from 50–300 kbp are formed (Oberhammer et al., 1993). This second pattern of DNA degradation is a recent discovery, and it is unclear if large break cleavage occurs in other apoptotic systems. The large break fragments are postulated to result from cleavage of the loops of DNA at the level of the rosette structure (50 kbp), or release of the entire rosette structure (300 kbp), based on a chromatin organization described by Filipski et al. (1990). The plasma membrane becomes convoluted, and portions of the cell bleb off to form apoptotic bodies, which are membrane-bound vesicles containing various organelles and chromatin fragments. These apoptotic bodies are then neatly and efficiently removed via phagocytosis by either neighboring cells or resident macrophages. In stark contrast to necrosis, no immune response is elicited (Kerr et al., 1972; Schwartzman and Cidlowski, 1993).

Although at first cancer and apoptosis may seem to be completely opposite mechanisms, they are intimately associated with one another in several ways. First, lack of apoptosis can result in cancer and, second, many anticancer drugs, such as glucocorticoids, exert their effect by stimulating the cells to undergo apoptosis (Williams, 1991; Oren, 1992; Marx, 1993; Kerr et al., 1994). Chronic myeloid leukemia (CML) is an example of how lack of apoptosis contributes to malignancy. This disease is characterized by the overaccumulation of mature cells throughout the bloodstream and in the bone marrow. The myeloid precursors to CML do not proliferate any faster than the normal cells (Koeffler and Golde, 1981), suggesting that the cause of malignancy is a decrease in cell death rather than an increase of cell growth (McGahon et al., 1994). Rat hepatocellular carcinoma also appears to result from inhibition of cell death (Schulte-Hermann et al., 1990). Additionally, certain cancerous cells, such as leukemias or colon cancers, can be stimulated to undergo apoptosis when normal copies of the p53 gene are introduced (Yonish-Rouach et al., 1991; Shaw et al., 1992), suggesting that the only reason for this tumorigenicity was a disruption in the apoptotic pathway caused by a defective p53 gene. Premalignant breast hyperplasia and colorectal adenomas, in which p21 *ras* is overexpressed in both cases (Going et al., 1992; Williams et al., 1985), provide other examples of tumors that result from the "rescue" of cells from apoptosis.

The apoptotic-inducing effects of many forms of cancer therapy have been well documented (reviewed in Hickman, 1992). In this chapter we will be looking specifically at the death response of tumors and malignant cell lines to treatment with glucocorticoid hormones.

Cells respond in a variety of ways when exposed to glucocorticoids, but it is not known how, and if, all these responses are related to the induction and progression of apoptosis. Therefore, it is of interest first to describe the general metabolic and gene regulatory responses to glucocorticoids, and then discuss in greater detail how some of these responses may be involved in various aspects of apoptosis.

CELLULAR RESPONSES TO GLUCOCORTICOIDS—AN OVERVIEW

The changes in metabolism as a result of glucocorticoid treatment have been studied for many years. One of the first responses noted was a decrease in glucose uptake by cells (Munck and Leung, 1977). Other cases of altered metabolism include decreased amino acid transport (Morita and Munck, 1964) and the accumulation of nucleosides (Makman et al., 1968). Adenosine triphosphate (ATP) production (Makman et al., 1971) and RNA polymerase activity (Bell and Borthwick, 1975) are also decreased in response to glucocorticoids. Although both protein and nucleic acid biosynthesis are diminished overall following hormone treatment (Nordeen and Young, 1976), there are examples of increases in certain protein or nucleic acid levels. Glucocorticoids have also been shown to stimulate catabolic processes by enhancing protein and RNA degradation (MacDonald et al., 1980; Cidlowski, 1982). The activity of several hydrolytic enzymes, including acid phosphatase (Clarke and Wills, 1978), serine hydrolases (MacDonald and Cidlowski, 1981), ribonuclease (Wiernik and MacLeod, 1965; Ambellan and Hollander, 1966; Mashburn et al., 1969), and deoxyribonucleases (Wiernik and MacLeod, 1965), are stimulated by glucocorticoids. In addition to altering a variety of general metabolic processes, glucocorticoids can have an effect on specific genes, in some cases stimulating gene expression and in other cases blocking gene expression. The interplay between gene induction, gene inhibition, and metabolic alterations necessary for the process of glucocorticoid-induced apoptosis is not yet fully defined; however, exciting progress is being made in several different arenas. We will discuss a variety of cellular events that are generated in response to glucocorticoid treatment, focusing on how they can result in the lytic/apoptotic response.

GLUCOCORTICOIDS AND CALCIUM LEVELS

Many cell types respond to glucocorticoid treatment with an increase in intracellular calcium levels. This response has been documented in immune cells, for example, certain cell lines (Lynn et al., 1989), mouse lymphoma cells (Lam et al., 1993) and rat thymocytes (McConkey et al., 1989), as well as vascular smooth muscle cells (Kato et al., 1992; Kornel et al., 1993), hippocampal neurons (Elliot et al., 1993), and pituitary growth hormone 3 cells (Fomima et al., 1993). Calcium is important in normal cellular functioning, such as gene expression, ion transport, and energy metabolism, and disturbances in calcium homeostasis can cause the cell to become dysfunctional. The importance of calcium increases in hormone-induced apoptosis is still controversial and is actively being investigated.

A calcium requirement in glucocorticoid-induced death of rat thymocytes was suggested in the work by Kaiser and Edelman (1977). These researchers were able to show that thymocytes treated with the ionophore A23187 displayed a similar response, that is, inhibition of uridine metabolism and cytolysis, as the thymocytes exposed to the steroid hormone triamcinolone acetonide. The thymocytes in their experiments displayed a decreased sensitivity to hormone-induced death when in a calcium-free medium. Indeed, calcium increases created by stimuli other than hormone treatment have also resulted in cell death, such as that seen with calcium ionophores A23187 (McConkey et al., 1989; Caron-Leslie and Cidlowski, 1991), thapsigargin (Lam et al., 1993), and 2,3,7,8-tetrachlorodibenzo-*p*-dioxin (McConkey et al., 1988). Additionally, overexpression of the high-affinity calcium-binding protein calbindin-D_{28K} in WEHI 7.2 cells (a thymoma cell line) provided those cells protection from dexamethasone- and A23187-induced apoptosis (Dowd et al., 1992). The level of survival correlated to the level of calbindin-D_{28K} expression.

The significance of increased intracellular calcium concentration following addition of hormone is based on the second messenger nature of calcium ions. In this capacity, calcium may activate one or several key components of the apoptotic pathway. However, it is still unclear whether the observed increase in intracellular calcium following hormone treatment is the direct cause of apoptosis or a secondary effect, resulting from some other apoptotic initiator (our unpublished observations). Whether calcium is required early or late in apoptosis (Bansal et al., 1990; Iseki et al., 1993), if a calcium influx is necessary (Nicholson and Young, 1979; Alnemri and Litwack, 1990), or if release of intracellular stores are sufficient to activate apoptotic machinery (Bansal et al., 1990) are additional questions that need to be addressed. Calcium binding and calcium-dependent regulatory proteins include proteases, nucleases, transglutaminases, calmodulin and calmodulin-dependent kinases, phosphatases, phospholipases, ion channels, and ATPases, any of which may play a role in apoptosis. The next three sections discuss several specific components of the apoptotic process and the possible role calcium may serve.

APOPTOTIC NUCLEASES

One of the more controversial components of apoptosis is the activation of the nuclease or nucleases responsible for the characteristic cleavage of DNA. The two apoptotic patterns involve either the internucleosomal fragmentation of DNA into multiples of approximately 200 bp, or the large break fragmentation of DNA into strands of 50–300 kbp, as discussed previously. The "ladder pattern," observed following electrophoresis of internucleosomally cleaved DNA, was first described in connection with apoptosis by Wyllie (1980) and was thereafter considered the hallmark for cells undergoing apoptosis. This internucleosomal cleavage was soon associated with a calcium-dependent endogenous endonuclease (Cohen and Duke, 1984). Several possible candidates for a calcium-dependent nuclease have been reported, such as NUC18 (Gaido and Cidlowski, 1991) and DNAse I (Peitsch et al., 1993), but no one definitive nuclease has yet been determined. The more recent discovery of a second pattern of DNA fragmentation representative of apoptosis has prompted further questions and investigations.

The observation of Kaiser and Edelman that thymocytes display a decreased sensitivity to hormone when they are incubated in the absence of calcium has been confirmed by McConkey et al. (1989), who showed that incubation of rat thymocytes in a calcium-free medium or in the presence of the calcium buffer quin-2 prevents DNA fragmentation and cell death. These authors attribute the lack of DNA fragmentation to the prevention of nuclease activation by the absence of calcium. The extent of chromatin cleavage is dependent on an increase in cytoplasmic calcium levels, again supporting the idea of a calcium-requiring nuclease in the apoptotic process. Nuclei isolated from thymocytes or liver cells can be stimulated to break down their own DNA in a manner similar to apoptotic cells (autodigestion) when incubated with calcium ions (Vanderbilt et al., 1982; Cohen and Duke, 1984; Jones et al., 1989). In fact, this technique is used to define the ionic requirements that distinguish large break from internucleosomal cleavage activity. Recently, three reports that deal with this topic were published, two of which employed liver nuclei and another that employed thymocyte nuclei (Cain et al., 1994; Sun and Cohen, 1994; Zhivotovsky et al., 1994). All three came to different conclusions as to the ion requirements for large break activity, but one finding common to these investigations was the necessity for calcium for internucleosomal cleavage to occur. Thus, it is apparent that calcium is a requirement for at least one part of apoptotic DNA degradation. One possible exception includes the human T-cell leukemia line, CEM-C7 (Alnemri and Litwack, 1990). Although these cells will die in response to glucocorticoids, they require a much longer incubation time with hormone than thymocytes and do not require extracellular calcium for DNA degradation. This difference may be cell-specific, as the transformed aspect of CEM-C7 cells may involve loss of a critical component of the apoptotic pathway. Alternatively, release of calcium from internal stores may provide enough of a change in intracellular calcium levels so that uptake of external calcium is not necessary. This possibility was suggested by Lam et al. (1993), who observed the release of internal stores of calcium in W7MG1 mouse lymphoma cells following treatment with hormone. Thus, the internal calcium release may provide enough of a signal so that nuclease activation—and apoptosis—can occur even in the absence of extracellular calcium.

Despite these conflicting data, there is significant evidence that increased calcium levels alone are enough to stimulate many of the apoptotic-specific changes in chromatin, including condensation of chromatin, internucleosomal and large break DNA fragmentation, and even membrane blebbing (Zhivotovsky et al., 1994).

APOPTOTIC PROTEASES

Increased proteolysis as a result of glucocorticoid treatment has been studied for some time (MacDonald et al., 1980; MacDonald and Cidlowski, 1982). These workers demonstrated that the previously observed inhibition of labeled amino acid incorporation into proteins following hormone treatment was a result of active protein degradation and not simply an inhibition of protein synthesis. They postulated that this protein degradation might play a role in glucocorticoid-induced death. Interestingly, more recent research is also implicating proteolysis in the process of apoptosis. For example, the interleukin-1 (IL-1) protein is known to undergo proteolysis during apoptosis. Murine peritoneal macrophages induced to undergo apoptosis by incubation in the presence of ATP have been observed to proteolyze and eventually release IL-1α and IL-1β (Hogquist et al., 1991). Additionally, apoptotic macrophages (targeted by cytotoxic T cells) also process and release IL-1. Conversely, when the proteolytic action of IL-1β-converting enzyme (ICE) is blocked by the product of the crmA gene, which is a specific inihibitor of ICE, apoptosis is prevented (Gagliardini et al., 1994). In this experiment, an expression vector containing the crmA cDNA was injected into dorsal root ganglion neurons. After nerve growth factor was removed from the culture, 80% of the control neurons died, whereas 60% of the injected neurons survived up to 6 days. The prevention of apoptosis by inhibiting ICE activity provides just one specific example of proteolysis during apoptosis, although the role of this activity during apoptosis has not yet been explored. Whether IL-1β itself is important for apoptosis or if ICE (or an ICE-related protein; Darmon et al., 1994) can act on another, perhaps apoptotic-specific protein, is not known. There is a possibility for such an apoptotic-specific protein, because ICE mRNA has been identified in tissues that do not express IL-1β (Cerretti et al., 1992; Nett et al., 1992). There are several possible roles for protease activity during apoptosis, the mechanics of which are currently being investigated. Such possibilities include the protease acting in conjunction with a nuclease to cause DNA degradation, or the protease acting on its own to disrupt the plasma membrane and cause the blebbing and eventual formation of apoptotic bodies. Thus, a scenario in which glucocorticoids stimulate an increase in calcium levels, which then activate a protease to participate in the progression of apoptosis, can easily be imagined.

DNA Degradation and Proteases

How might proteases be involved in DNA fragmentation? One possibility involves proteases attached to the nuclear scaffold. Recent work by Clawson et al. (1992) reveals that there is a calcium-regulated serine protease that is associated with the nuclear scaffold. This calcium-regulated protease, which displays a chymotryptic-like activity, is activated with an increase in calcium levels, and it cleaves lamins A and C. The lamins are a component of the nuclear scaffold that are attached to other proteins that, in turn, are bound to chromatin loops. The cleavage of lamins A and C results in the release of large pieces of DNA, which could increase the susceptibility to nuclease digestion. Another possible role for proteases in DNA degradation is the activation of the nuclease itself, either by cleaving a pro-form of the nuclease or by freeing the inherently active nuclease from an inhibitory molecule (Smyth et al., 1994; Montague and Cidlowski, 1995). Although such specific roles are being investigated, many researchers are trying first to obtain general information on what types of proteases are involved in apoptosis and at what stage in the process they exert their effect.

Protease activity in apoptosis can be studied by adding a protease inhibitor to the autodigestion technique mentioned in the section above. This experimental technique has been employed to determine possible roles for calcium-dependent cysteine proteases, such as calpains, and serine proteases. The use of specific calpain inhibitors with rat liver nuclei did result in the attenuation of the production of 50 kbp DNA fragments, but this effect was not as complete as that seen with inhibitors of serine proteases (N-tosyl-phenylalanine chloromethyl ketone and N^{α}-*p*-tosyl-L-lysine chloromethyl ketone, TPCK and TLCK, respectively) (Zhivotovsky et al., 1994). When serine protease inhibitors were included in the autodigestion reaction, formation of fragments of DNA from 50 kbp to smaller sizes was completely abolished. A similar inhibition of DNA degradation occurred when rat thymocytes were treated with prednisolone plus TPCK and TLCK. Importantly, it was also noted that DNA replication, although retarded, was not concomitantly inhibited by these agents, demonstrating that DNA degradation is a process separate from DNA replication (Bruno et al., 1992). This conclusion is supported by cell death studies of lymphocytes (Caron-Leslie et al., 1994). In another set of experiments, where protease activity was studied by incubating 2B4 murine T cell hybridoma cells with corticosteroids and protease inhibitors, the addition of both serine- and cysteine-protease inhibitors resulted in a slight enhancement of the dexamethasone-induced apoptosis in all cases, as measured by ^{51}Cr release, nuclear morphology, and DNA fragmentation (Sarin et al., 1993). Interestingly, when these same cells were induced to undergo T-cell receptor-mediated apoptosis, the presence of protease blockers proved inhibitory for the process of apoptosis. Again, the differences observed as to the effect of protease inhibition on hormone-induced death may result from the use of freshly isolated cells versus an immortalized line. Another piece of evidence that implicates serine proteases in DNA degradation is seen with three serine proteases isolated from natural killer cells. These lymphocyte granule proteases are capable of inducing target cells to degrade DNA (Shi et al., 1992, 1994). It is this degradation of DNA that is apparently the cause of death in the target cells (Helgason et al., 1993). Further research and comparisons are necessary before any conclusions can be made as to the role of cysteine- or serine-proteases in apoptosis.

Membrane Blebbing and Proteases

In addition to being found on the nuclear scaffold, proteases are also associated with the cytoskeleton and plasma membrane. The activation of these membrane-associated proteases by calcium may serve a purpose in apoptosis. In this case, proteolysis would result in the membrane blebbing characteristic of apoptosis. There are numerous reports on the effects of calcium on the cytoskeleton. In fact, high calcium levels alone can cause disruption of cytoskeletal organization (Orrenius et al., 1992). Most reports concerning the membrane-disrupting effects of calcium involve toxic chemicals or ischemic agents, and the results are not necessarily transferrable to the more physiological situation of glucocorticoid-induced apoptosis. Some of the results suggest an interesting correlation, however. One study designed to examine the effect of calcium on membrane structure in hepatocytes used several agents known to increase levels of intracellular calcium: ATP, cystamine, and A23187 (Nicotera et al., 1986). In each case, the amount of membrane blebbing and proteolytic activity, in general, were markedly reduced when calpain-specific inhibitors, leupeptin and antipain, were employed. Thus, calpains were again investigated, this time with an emphasis on membrane effects. In some cases, calpains have been implicated in the shedding of microvesicles from aggregated platelets by hydrolyzing the skeleton of the platelet membrane (Fox et al., 1991). More specifically, these enzymes are able to hydrolyze actin-binding proteins, thereby disrupting associations between actin and membrane glycoproteins. This effect, determined by shedding of microvesicles, can be blocked by calpain inhibitors (Fox et al., 1990). Although this process is not technically apoptosis, these results suggested that calpains play a role in altering membrane structures. A recent study directly addressed the issue of calpain activation in glucocorticoid-induced apoptosis in murine thymocytes, showing that an increase in calpain activity preceded the onset of apoptotic morphology (changes in cell shape and DNA fragmentation) by approximately 4 hours (Squier et al., 1994). Additionally, thymocytes preincubated with calpain inhibitors, Calpain Inhibitor I and MDL 28,170, did not undergo apoptosis when dexamethasone was added to the cells and, in fact, the cells retained a viability consistent with that of control cells. The apoptotic-inhibitory effect by calpain blockers was also observed when thymocytes were irradiated, thereby revealing one event common to two induction pathways. This work did not look at effects on membrane blebbing specifically, and it suggested a variety of possible substrates, including protein kinase C and transcription factors, as well as cytoskeletal proteins and enzymes involved in signal transduction.

TRANSGLUTAMINASES

A third family of potentially apoptotic enzymes that are calcium-dependent are the transglutaminases. These enzymes form $\epsilon(\gamma$-glutamyl) lysine cross-links between strands of polypeptides, which provide great resistance against a variety of damaging agents, whether chemical, mechanical, or enzymatic (Fesus et al., 1989). Such cross-linkages are found in apoptotic bodies, which are highly resistant to disruption by a variety of chaotropic agents. The cross-linking in apoptotic bodies is created specifically by tissue transglutaminase, which has been shown to be induced and activated during apoptosis (Fesus et al., 1987). An involucrin-like protein was identified as a substrate for tissue transglutaminase in apoptotic hepatocytes (Tarcsa et al., 1992); involucrin is the substrate for keratinocyte transglutaminase activity during the process of cornification of terminally differentiating keratinocytes and squamous epithelial cells. Thus, it appears that similar mechanisms are involved, at least in part, in the death processes of apoptosis and terminal differentiation. The formation of cross-links is thought to serve as a means for preserving the integrity of the apoptotic body, preventing unwanted reactions caused by leakage of organelles and chromatin into the extracellular environment.

As the roles for these three categories of calcium-dependent enzymes in apoptosis are being more clearly defined, there are still many other calcium-dependent enzymes that may prove vital for apoptosis. We can look forward to more research in these areas over the coming years.

GLUCOCORTICOIDS AND GENE EXPRESSION

In addition to regulating protein activity through second messengers such as calcium, glucocorticoids regulate protein synthesis through induction of RNA synthesis. Glucocorticoid-stimulated induction of mRNAs has been well documented (Yamamoto and Alberts, 1976; Ringold, 1985). Exactly what these induced genes are, and what purpose they may serve during apoptosis, is currently being investigated. Because apoptosis stimulated by agents other than glucocorticoids has also been shown to require RNA and protein synthesis, it is likely that some of the genes turned on by hormone treatment are responsible for initiating apoptosis in these other systems as well. For this reason, glucocorticoid-treated lymphocytes are a common model system used to investigate apoptotic genes. Giant 2-D gel electrophoresis of proteins from glucocorticoid-treated thymus cells was used to identify seven proteins that are rapidly synthesized following glucocorticoid treatment (Colbert and Young, 1986). It has yet to be determined if the observed induction of these proteins is a result of increased mRNA synthesis or decreased RNA degradation; however, based on the thymocyte requirement for RNA induction during apoptosis, it is likely that increased production of mRNA is the reason. Thus, these genes and gene products may be responsible for the progression of apoptosis after glucocorticoid treatment. These interesting observations have apparently not been investigated further, as the nature of these rapidly synthesized proteins has not been reported.

Another laboratory has used the subtractive hybridization technique to identify two mRNAs produced in apoptotic rat thymocytes. A portion of the sequence of one mRNA, RP-8, codes for a zinc finger domain, suggesting that the resulting protein binds to DNA. The other mRNA, RP-2, appears to encode an integral membrane protein (Owens et al., 1991). The expression of RP-2 is coincident with the appearance of fragmented DNA (Owens and Cohen, 1992). The recent finding that RP-2 codes for an ATP-gated cation channel (Brake et al., 1994; Valera et al., 1994) again suggests the intriguing possibility that calcium influx is an important glucocorticoid-mediated event involved in apoptosis.

Thirteen genes induced in glucocorticoid-treated WEHI-7TG cells (a murine thymoma line) were also detected by the method of subtractive hybridization (Harrigan et al., 1989; Baughman et al., 1991). One of these messages, designated Tcl-30, has been characterized; the gene is expressed exclusively in the thymus, but shows similarity with a placental protein called PP11, whose function is unknown (Baughman et al., 1992). The delineation of apoptotic genes is still in its early stages. It will also be interesting to find genes that are stimulated by a wide range of apoptosis inducers, and therefore are not cell-type specific.

Because of the relationship between apoptosis and cancer, researchers are investigating the effects of glucocorticoid-induced apoptosis on oncogene and growth gene expression. For instance, glucocorticoid-sensitive CEM-C7 cells display a suppression of c-*myc* expression that is correlated with growth arrest (Yuh and Thompson, 1989). Additionally, glucocorticoid-induced down-regulation of growth related genes, such as c-*myc*, c-*myb* and c-K-*ras*, has been reported in S49 cells (Eastman-Reks and Vedeckis, 1986), as well as the down-regulation of c-*myc* in P1798 cells (Forsthoefel and Thompson, 1987).

A gene that is not actually induced by glucocorticoid treatment, but whose translation product has a major influence on glucocorticoid-induced apoptosis, is *bcl*-2. The discovery that overproduction of bcl-2 prevented cell death was the first proven connection between a specific gene and

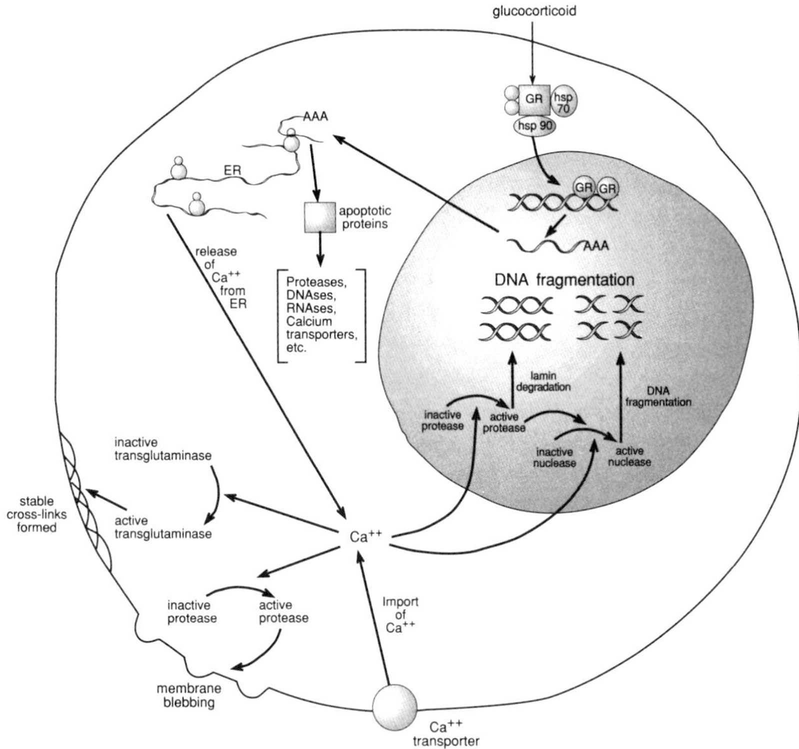

Figure 2. Model for glucorticoid-induced apoptosis. The glucocorticoid hormone binds to its cytoplasmic receptor, causing the receptor to become activated and enter the nucleus. The activated receptor then binds as a dimer to specific sites on the DNA, resulting in either inhibition or induction of gene transcription. The changes in transcriptional activity are dependent on the specific gene. Another result of glucocorticoids entering the cell, though less well understood, is an increase in intracellular calcium levels, either through an extracellular influx or by release from endoplasmic reticulum stores. The increase in calcium is thought to be responsible for activation of many of the enzymes involved in the apoptotic process. For example, cytoplasmic proteases become activated and cleave the cytoskeleton, resulting in membrane blebbing. Concomitantly, calcium-dependent transglutaminases would begin forming the cross-links found in apoptotic bodies. Several nuclear enzymes, such as proteases or nucleases, also become activated during glucocorticoid-induced apoptosis, perhaps through an increase in nuclear calcium levels. The proteases could act to fragment DNA through two mechanisms: either by a direct approach on scaffold protein disruption, releasing large DNA fragments, or by an indirect approach, in which the protease activates a nuclease which can then act directly on the DNA. Additionally, an increase in calcium alone may be enough to activate the nuclease and result in DNA degradation typical of apoptosis.

apoptosis. Several laboratories report that bcl-2 expression can prevent glucocorticoid-induced apoptosis in several systems, including WEHI 7.1 (Miyashita and Reed, 1992), S49 (Caron-Leslie et al., 1994; Miyashita and Reed, 1992), human pre-B-leukemias (Alnemri et al., 1992) and cortical thymocytes (Sentman et al., 1991). There are some proposals that bcl-2 acts as an antioxidant (Hockenberry et al., 1993; Kane et al., 1993), but the exact mechanism of the protection provided by bcl-2 is still unclear. Caron-Leslie et al. (1994) have shown bcl-2 protection to be specific for certain inducers, but not as effective for others, that is, bcl-2 prevents glucocorticoid-induced apoptosis but not cycloheximide- or A23187-induced apoptosis in S49 cells, thus suggesting a more specific role for bcl-2 in blocking certain apoptotic signals rather than blocking the apoptotic pathway itself.

Another model system that has provided important information concerning the genes involved in apoptosis is the developing roundworm *Caenorhabditis elegans*. Horvitz and co-workers (Ellis et al., 1991) have discovered several cell death (ced) genes in *C. elegans* that have homologs in the mammalian systems. For instance, the ICE gene discussed earlier was identified as being involved in apoptosis because of its similarity to the ced-3 gene. The two protein products display a 28% match in amino acids, five of which are believed to comprise the active site (Barinaga, 1994). Additionally, the ced-9 gene, whose corresponding protein protects the cells from apoptosis (Hengartner et al., 1992), was shown to be an analog of *bcl*-2, a mammalian gene whose expression can prevent apoptosis (Hengartner and Horvitz, 1994). In addition to providing insight into key apoptotic genes, exciting homologies such as these suggest a conserved mechanism for programmed cell death despite the controversies that are now reported.

CONCLUDING REMARKS

Although glucocorticoids were originally thought to have little or no effect in humans, current research now reveals a multitude of cellular responses upon glucocorticoid treatment in both normal and neoplastic cells. One well-studied response is the induction of apoptosis, especially noted in immune cells. The mechanism through which glucocorticoids exert their effect is the subject of intense investigation. Several specific characteristics of glucocorticoid-induced apoptosis have been discussed in this chapter, including the increase in intracellular calcium levels, with its subsequent effects on calcium-activated enzymes, and the effect on gene transcription (Fig. 2). Although much research has been done already, the diversity of systems studied, and the variety of results from each system, clearly demonstrate the complexity of the apoptotic process and emphasize the need for further investigations.

REFERENCES

Alexander D, Goya L, Webster M, Haraguchi T, Firestone G (1993): Glucocorticoids coordinately disrupt a transforming growth factor alpha autocrine loop and suppress the growth of 13762NF-derived Con8 rat mammary adenocarcinoma cells. *Cancer Res* 53:1808–1815

Alnemri E, Litwack G (1990): Activation of internucleosomal DNA cleavage in human CEM lymphocytes by glucocorticoid and novobiocin. *J Biol Chem* 265:17323–17333

Alnemri E, Fernandes T, Haldar S, Croce C, Litwack G (1992): Involvement of bcl-2 in glucocorticoid-induced apoptosis of human pre-B-leukemias. *Cancer Res* 52:491–495

Ambellan E, Hollander V (1966): The role of ribonuclease in regression of lymphosarcoma P1798. *Cancer Res* 26:903–908

Bansal N, Houle A, Melnykovych G (1990): Dexamethasone-induced killing of neoplastic cells of lymphoid derivation: lack of early calcium involvement. *J Cell Physiol* 143:105–109

Barinaga M (1994): Cell suicide: by ICE, not fire. *Science* 263:754–756

Baughman G, Harrigan M, Campbell N, Nurrish S, Bourgeois S (1991): Genes newly identified as regulated by glucocorticoids in murine thymocytes. *Mol Endocrinol* 5:637–644

Baughman G, Lesley J, Trotter J, Hyman R, Bourgeois S (1992): Tcl-30, A new T cell-specific gene expressed in immature glucocorticoid-sensitive thymocytes. *J Immunol* 149:1488–1496

Bell P, Borthwick N (1975): Glucocorticoid effects on DNA-dependent RNA polymerase activity of rat thymocytes. *J Steroid Biochem* 7:1147–1150

Brake A, Wagenbach M, Julius D (1994): New structural motif for ligand-gated ion channels defined by an ionotropic ATP receptor. *Nature* 371:519–523

Bruno S, Del Bino G, Lassota P, Giaretti W, Darzynkiewicz Z (1992): Inhibitors of proteases prevent endonucleolysis accompanying apoptotic death of HL-60 leukemic cells and normal thymocytes. *Leukemia* 6:1113–1120

Cain K, Inayat-Hussein S, Wolfe J, Cohen G (1994): DNA fragmentation into 200–250 and/or 30–50 kilobase pair fragments in rat liver nuclei is stimulated by Mg^{2+} alone and Ca^{2+}/Mg^{2+} but not by Ca^{2+} alone. *FEBS Lett* 349:385–391

Caron-Leslie L, Cidlowski J (1991): Similar actions of glucocorticoids and calcium on the regulation of apoptosis in S49 cells. *Mol Endocrinol* 5:1169–1179

Caron-Leslie L-A, Evans R, Cidlowski J (1994): Bcl-2 inhibits glucocorticoid-induced apoptosis but only partially blocks calcium ionophore or cycloheximide-regulated apoptosis in S49 cells. *FASEB J* 8:639–645

Cerretti D, Kozlosky C, Mosley B, Nelson N, Van Ness K (1992): Molecular cloning of the interleukin-1 beta converting enzyme. *Science* 256:97–100

Cidlowski J, Schwartzman R (1993): Corticosteroids. In: *Cancer Medicine,* Holland J, Frei E, Bast R, Kufe D, Morton D, Weichselbaum R, eds. Philadelphia: Lea & Febiger

Cidlowski J (1982): Glucocorticoids stimulate ribonucleic acid degradation in isolated rat thymic lymphocytes *in vitro*. *Endocrinology* 111:184–190

Clarke C, Wills E (1978): The activation of lymphoid tissue lysosomal enzymes by steroid hormone. *J Steroid Biochem* 9:135–139

Clawson G, Norbeck L, Hatem C, Rhodes C, Amiri P, McKerrow J, Patierno S, Fiskum G (1992): Ca^{2+}-regulated serine protease associated with the nuclear scaffold. *Cell Growth Diff* 3:827–838

Cohen J, Duke R (1984): Glucocorticoid activation of a calcium-dependent endonuclease in thymocyte nuclei leads to cell death. *J Immunol* 132:38–42

Colbert R, Young D (1986): Glucocorticoid-induced messenger ribonucleic acids in rat thymic lymphocytes: rapid primary effects specific for glucocorticoids. *Endocrinology* 119:2598–2605

Darmon A, Ehrman N, Caputo A, Fujinaga J, Bleackley R (1994): The cytotoxic T cell proteinase granzyme B does not activate interleukin-1β-converting enzyme. *J Biol Chem* 269:32043–32046

Distelhorst C (1988): Glucocorticoids induce DNA fragmentation in human lymphoid leukemia cells. *Blood* 72:1305–1309

Dougherty T, White A (1945): Functional alterations in lymphoid tissue induced by adrenal cortical secretion. *Am J Anat* 77:81–116

Dowd D, MacDonald P, Komm B, Haussler M, Miesfeld R (1992): Stable expression of the calbindin-D28K complementary DNA interferes with the apoptotic pathway in lymphocytes. *Mol Endocrinol* 6:1843–1848

Eastman-Reks S, Vedeckis W (1986): Glucocorticoid inhibition of c-*myc*, c-*myb*, and c-Ki-*ras* expression in a mouse lymphoma cell line. *Cancer Res* 46:2457–2462

Elliot E, Mattson M, Vanderklish P, Lynch G, Chang I, Sapolsky R (1993): Corticosterone exacerbates kainate-induced alterations in hippocampal tau immunoreactivity and spectrin proteolysis in vivo. *J Neurochem* 61:57–67

Ellis R, Yuan J, Horvitz H (1991) Mechanisms and functions of cell death. *Annu Rev Cell Biol* 7:663–698

Evers B, Thompson E, Townsend C, Lawrence J, Johnson B, Srinivasan G, Thompson J (1993): Cortivazol increases glucocorticoid receptor expression and inhibits growth of hamster pancreatic cancer (H2T) in vivo. *Pancreas* 8:7–14

Fesus L, Thomaszy V, Autuori F, Ceru M, Tarcsa E, Piacenti M (1989): Apoptotic hepatocytes become insoluble in detergents and chaotropic agents as a result of transglutaminase action. *FEBS Lett* 245:150–154

Fesus L, Thomaszy V, Falus A (1987): Induction and activation of tissue transglutaminase during programmed cell death. *FEBS Lett* 224:104–108

Filipski J, Leblanc J, Youdale T, Sikorska M, Walker P (1990): Periodicity of DNA folding in higher order chromatin structures. *EMBO J* 9:1319–1327

Fomima A, Kostyuk P, Sedova M (1993): Glucocorticoid modulation of calcium currents in growth hormone 3 cells. *Neuroscience* 55:721–725

Forsthoefel A, Thompson E (1987): Glucocorticoid regulation of transcription of the c-myc cellular protooncogene in P1798 cells. *Mol Endocrinol* 1:899–907

Fox J, Austin C, Reynolds C, Steffen P (1991): Evidence that agonist-induced activation of calpain causes the shedding of procoagulant-containing microvesicle from the membrane of aggregating platelets. *J Biol Chem* 266:13289–13295

Fox J, Austin C, Boyles J, Steffen P (1990): Role of the membrane skeleton in preventing the shedding of procoagulant-rich microvesicles from the platelet plasma membrane. *J Cell Biol* 111:483–493

Gagliardini V, Fernandez P-A, Lee R, Drexler H, Rotello R, Fishman M, Yuan J (1994): Prevention of vertebrate neuronal death by the *crmA* gene. *Science* 263:826–828

Gaido M, Cidlowski J (1991): Identification, purification, and characterization of a calcium-dependent endonuclease (NUC18) from apoptotic rat thymocytes. *J Biol Chem* 266:18580–18585

Galili U (1983): Glucocorticoid induced cytolysis of human normal and malignant lymphocytes. *J Steroid Biochem* 19:483–490

Going J, Anderson T, Wyllie A (1992): *Ras* p21 in breast tissue: association with pathology and cellular localization. *Br J Canc* 65:45–50

Gruber J, Sgonc R, Hu Y-H, Beug J, Wick G (1994): Thymoctye apoptosis induced by elevated endogenous corticosterone levels. *Eur J Immunol* 24:1115–1121

Harrigan M, Baughman G, Campbell N, Bourgeois S (1989): Isolation and characterization of glucocorticoid- and cyclic AMP-induced genes in T lymphocytes. *Mol Cell Biol* 9:3438–3446

Heilman R, Kendall E (1944): The influence of 11-dehydro-17-hydroxycorticosterone (compound E) on the growth of malignant tumor in the mouse. *Endocrinology* 34:416–420

Helgason C, Shi L, Greenberg A, Shi Y, Bromley P, Cotter T, Green D, Bleackely R (1993): DNA fragmentation induced by cytotoxic T lymphocytes can result in target cell death. *Exp Cell Res* 206:302–310

Hengartner M, Ellis R, Horvitz R (1992): *Caenorhabiditis elegans* gene ced-9 protects cells from programmed cell death. *Nature* 356:494–499

Hengartner M, Horvitz R (1994): C. elegans cell survival gene ced-9 encodes a functional homolog of the mammalian proto-oncogene bcl-2. *Cell* 76:665–676

Hickman J (1992): Apoptosis induced by anticancer drugs. *Cancer Metast Rev* 11:121–139

Hockenberry D, Oltvai Z, Yin X-M, Milliman C, Korsmeyer S (1993): Bcl-2 functions in an antioxidant pathway to prevent apoptosis. *Cell* 75:241–251

Hogquist K, Nett M, Unanue E, Chaplin D (1991): Interleukin 1 is processed and released during apoptosis. *Proc Natl Acad Sci USA* 88:8485–8489

Iseki R, Kudo Y, Iwata M (1993): Early mobilization of Ca^{2+} is not required for glucocorticoid-induced apoptosis in thymocytes. *J Immunol* 151:5198–5207

Jones D, McConkey D, Nicotera P, Orrenius S (1989): Calcium-activated DNA fragmentation in rat liver nuclei. *J Biol Chem* 264:6398–6403

Kaiser N, Edelman I (1977): Calcium dependence of glucocorticoid-induced lymphocytolysis. *Proc Natl Acad Sci USA* 74:638–642

Kane D, Sarafian T, Anton R, Hahn H, Gralla E, Valentine J, Ord T, Bredesen D (1993): Bcl-2 inhibition of neural death: decreased generation of reactive oxygen species. *Science* 262:1274–1277

Kato H, Hayashi T, Koshino Y, Kutsumi Y, Nakai T, Miyabo S (1992): Glucocorticoids increase Ca^{2+} influx through dihydropyridine-sensitive channels linked to activation of protein kinase C in vascular smooth muscle cells. *Biochem Biophys Res Comm* 188:934–941

Kerr J, Wyllie A, Currie A (1972): Apoptosis: a basic biological phenomenon with wide-ranging implications in tissue kinetics. *Br J Cancer* 26:239–257

Kerr J, Winterford C, Harmon B (1994): Apoptosis: its significance in cancer and cancer therapy. *Cancer* 73:2013–2026

Koeffler P, Golde D (1981): Chronic myelogenous leukemia—new concepts. *N Engl J Med* 304:1201–12??

Kornel L, Nelson W, Manisundaram B, Chigurupati R, Hayashi T (1993): Mechanism of the effects of glucocorticoids and mineralocorticoids on vascular smooth muscle contractility. *Steroids* 58:580–587

Lam M, Dubyak G, Distelhorst C (1993): Effect of glucocorticosteroid treatment on intracellular calcium homeostasis in mouse lymphoma cells. *Mol Endocrinol* 7:686–693

Lynn W, Mathews D, Cloyd M, Wallwork J, Thompson A, Sachs CJ (1989): Intracellular Ca^{2+} and cytotoxicity. *Arch Envir Health* 44:323–330

MacDonald R, Martin T, Cidlowski J (1980): Glucocorticoids stimulate protein degradation in lymphocytes: A possible mechanism of steroid-induced cell death. *Endocrinology* 107:1512–1524

MacDonald R, Cidlowski J (1981): Glucocorticoid regulation of two serine hydrolases in rat splenic lymphocytes *in vitro*. *Bioc Biop Acta* 678:18–26

MacDonald R, Cidlowski J (1982): Glucocorticoids inhibit precursor incorporation into protein in splenic lymphocytes by stimulating protein degradation and expanding intracellular amino acid pools. *Bioc Biop Acta* 717:236–247

Makman M, Dvorkin B, White A (1968): Influences of cortisol on the utilization of precursors of nucleic acids and protein by lymphoid cells *in vitro*. *J Biol Chem* 243:1485–1497

Makman M, Dvorkin B, White A (1971): Evidence for induction by cortisol *in vitro* of a protein inhibitor of transport and phosphorylation processes in rat thymocytes. *Proc Natl Acad Sci USA* 68:1269–1273

Marx J (1993): Cell death studies yield cancer clues. *Science* 259:760–761

Mashburn B, Freeman C, Hollander V (1969): Effect of *in vitro* glucocorticoid treatment on acid ribonuclease activity in P1798 lymphosarcoma cells. *Proc Soc Exp Biol Bed* 131:108–111

McConkey D, Aguilar-Santelises M, Hartzell P, Eriksson I, Mellstedt H, Orrenius S, Jondal M (1991): Induction of DNA fragmentation in chronic B-lymphocytic leukemia cells. *J Immunol* 146:1072–1076

McConkey D, Hatzell P, Nicotera P, Orrenius S (1989a): Calcium-activated DNA fragmentation kills immature thymocytes. *FASEB J* 3:1843–1849

McConkey D, Nicotera P, Hartzell P, Bellomo G, Wyllie A, Orrenius S (1989b): Glucocorticoids activate a suicide process in thymocytes through an elevation of cytosolic Ca^{2+} concentration. *Arch Biochem Biophys* 269:365–370

McConkey D, Hartzell P, Duddy S, Hakansson H, Orrenius S (1988): 2,3,7,8-Tetrachlorodibenzo-p-dioxin kills immature thymocytes by Ca^{2+}-mediated endonuclease activation. *Science* 242:256–259

McGahon A, Cotter T, Green D (1994): The abl oncogene family and apoptosis. *Cell Death Diff* 1:77–83

Miyashita T, Reed J (1992): bcl-2 gene transfer increases relative resistance of S49.1 and WEHI7.2 lymphoid cells to cell death and DNA fragmentation induced by glucocorticoids and multiple chemotherapeutic drugs. *Cancer Res* 52:5407–5411

Montague J, Cidlowski J (1995): Glucocorticoid-induced death of immune cells: mechanisms of action. *Curr Top Microbiol Immunol* 200:51–65

Morita Y, Munck A (1964): Effects of glucocorticoids *in vivo* and *in vitro* on net glucose uptake and amino acid incorporation in rat thymus cells. *Biochem Biophys Acta* 93:150–157

Munck A, Leung K (1977) Glucocorticoid receptors and mechanisms of action. In: *Receptors and mechanisms of action of steroid hormones*, J Pasqualini, eds. New York: Marcel Decker, Inc

Nett M, Cerretti D, Berson D, Seavitt F, Gilbert D, Jenkins N, Copeland N, Black R, Chaplin D (1992): Molecular cloning of the murine IL-1 beta converting enzyme cDNA. *J Immunol* 149:3254–3259

Nicholson M, Young D (1979): Independence of the lethal actions of glucocorticoids on lymphoid cells from possible hormone effects of calcium uptake. *J Supramol Struct* 10:165–174

Nicotera P, Hartzell P, Davis G, Orrenius S (1986): The formation of plasma membrane blebs in hepatocytes exposed to agents that increase cytosolic Ca^{++} is mediated by the activation of a non-lysosomal proteolytic system. *FEBS Lett* 209:139–144

Nieto M, Gonzalez A, Gambon F, Diaz-Espada F, Lopez-Rivas A (1992): Apoptosis in human thymocytes after treatment with glucocorticoids. *Clin Exp Immunol* 88:341–344

Nordeen S, Young D (1976): Glucocorticoid action on rat thymic lymphocytes: experiments utilizing adenosine to support cellular metabolism lead to a reassessment of catabolic hormone action. *J Biol Chem* 251:7295–7303

Oberhammer F, Wilson J, Dive C, Morris I, Hickman J, Wakeling A, Walker P, Sikorska M (1993): Apoptotic death in epithelial cells: cleavage of DNA to 300 and/or 500 kb fragments prior to or in the absence of internucleosomal fragmentation. *EMBO J* 12:3679–3684

Oren M (1992): The involvement of oncogenes and tumor suppressor genes in the control of apoptosis. *Cancer Metast Rev* 11:141–148

Orrenius S, McCabe MJ, Nicotera P (1992): Ca^{2+}-dependent mechanisms of cytotoxicity and programmed cell death. *Toxic Lett* 64/65:357–364

Owens G, Cohen J (1992): Identification of genes involved in programmed cell death. *Cancer Metast Rev* 11:149–156

Owens G, Hahn W, Cohen J (1991): Identification of mRNAs associated with programmed cell death in immature thymocytes. *Mol Cell Biol* 11:4177–4188

Peitsch M, Polzar B, Stephan H, Crompton L, MacDonald H (1993): Characterization of the endogenous deoxyribonuclease involved in nuclear DNA degradation during apoptosis (programmed cell death). *EMBO J* 12:371–377

Polliack A (1991): *A Handbook of Essential Drugs and Regimens in Hematological Oncology*. Switzerland: Harvad Academic Publishers

Ringold G (1985): Steroid hormone regulation of gene expression. *Annu Rev Pharmacol Tox* 25:529–566

Sarin A, Adams D, Henkart P (1993): Protease inhibitors selectively block T cell receptor-triggered programmed cell death in a murine T cell hybridoma and activated peripheral T cells. *J Exp Med* 178:1693–1700

Schrek R (1961): Cytotoxicity of adrenal cortex hormones on normal and malignant lymphocytes of man and rat. *Proc Soc Exp Biol Med* 108:328–332

Schrek R (1964): Prednisolone sensitivity and cytology of viable lymphocytes as tests for chronic lymphocytic leukemia. *J Natl Cancer Inst* 33:837–847

Schulte-Hermann R, Timmermen-Trosiener I, Barthel G, Bursch W (1990): DNA synthesis, apoptosis and phenotypic expression as determinants of growth of altered foci in rat liver during phenobarbital production. *Cancer Rsch* 50:5127–5135

Schwartzman R, Cidlowski J (1993): Apoptosis: The biochemistry and molecular biology of programmed cell death. *Endocr Rev* 14:133–151

Sentman C, Shutter J, Hockenberry D, Kanagawa O, Korsmeyer S (1991): bcl-2 inhibits multiple forms of apoptosis but not negative selection in thymocytes. *Cell* 67:879–888

Shaw P, Bovey R, Tardy S, Sahli R, Sordat B, Costa J (1992): Induction of apoptosis by wild-type p53 in a human colon tumor-derived cell line. *Proc Natl Acad Sci USA* 89:4495–4499

Shi L, Kam C-M, Powers J, Aebersold R, Greenberg A (1992): Purification of three cytotoxic lymphocyte granule serine proteases that induce apoptosis through distinct substrate and target cell interaction. *J Exp Med* 176:1521–1529

Shi L, Nishioka W, Th'ng J, Bradbury E, Litchfield D, Greenberg A (1994): Premature $p34^{cdc2}$ activation required for apoptosis. *Science* 263: 1143–1145

Smyth M, Browne K, Thia K, Apostolidis V, Kershaw M, Trapani J (1994): Hypothesis: Cytotoxic lymphocyte granule serine proteases activate target cell endonucleases to trigger apoptosis. *Clin Exp Pharm Phys* 21:67–70

Sorensen P, Helweg-Larsen S, Mouridsen H, Hansen H (1994): Effect of high-dose dexamethasone in carcinamatous metastatic spinal cord compression treated with radiotherapy: a randomized trial. *Eur J Cancer* 30A:22–27

Squier M, Miller A, Malkinson A, Cohen J (1994): Calpain activation in apoptosis. *J Cell Physiol* 159:229–237

Sun X-M, Cohen G (1994): Mg^{2+}-dependent cleavage of DNA into kilobase pair fragments is responsible for the initial degradation of DNA in apoptosis. *J Biol Chem* 269:14857–14860

Tarcsa E, Kedei N, Thomazsy V, Fesus L (1992): An involucrin-like protein in hepatocytes serves as a substrate for tissue transglutaminase during apoptosis. *J Biol Chem* 267:25648–25651

Tuosto L, Cundari E, Montani M, Piccolella E (1994): Analysis of susceptibility of mature human T lymphocytes to dexamethasone-induced apoptosis. *Eur J Immunol* 24:1061–1065

Valera S, Hussy N, Evans R, Adami N, North R, Surprenant A, Buell G (1994): A new class of ligand-gated ion channel defined by P_{2x} receptor for extracellular ATP. *Nature* 371:516–519

Vanderbilt J, Bloom K, Anderson J (1982): Endogenous nuclease: properties and effects on transcribed genes in chromatin. *J Biol Chem* 257:13009–13017

Weissman D (1988): Glucocorticoid treatment for brain metastases and epidural spinal cord compression: a review. *J Clin Oncol* 6:543–555

Wiernik P, MacLeod R (1965): The effect of a single large dose of 9α-fluoroprednisolone on nucleodepolymerase activity and nucleic acid content of the rat thymus. *Acta Endocrinol* 49:138–144

Williams G (1991): Programmed cell death: apoptosis and oncogenesis. *Cell* 65:1097–1098

Williams A, Piris J, Spandidos A, Wyllie A (1985): Immunohistochemical detection of the *ras* oncogene p21 product in an experimental tumour and in human colorectal neoplasms. *Br J Canc* 52:687–693

Wyllie A, Morris R, Smith A, Dunlop D (1984): Chromatin cleavage in apoptosis: association with condensed chromatin morphology and dependence on macromolecular synthesis. *J Pathol* 142:67–77

Wyllie A (1980): Glucocorticoid-induced thymocyte apoptosis is associated with endogenous nuclease activation. *Nature* 284:555–556

Yamamoto K, Alberts B (1976): Steroid receptors: elements for modulation of eukaryotic transcription. *Annu Rev Biochem* 45:721–746

Yonish-Rouach E, Resnitzky D, Lotem J, Sachs L, Kimchi A, Oren M (1991): Wild-type p53 induces apoptosis of myeloid leukemic cells that is inhibited by interleukin-6. *Nature* 352:345–347

Yuh Y-S, Thompson E (1989): Glucocorticoid effect on oncogene/growth gene expression in human T lymphoblastic leukemic cell line CCRF-CEM. *J Biol Chem* 264:10904–10910

Zhivotovsky B, Wade D, Gahm A, Orrenius S, Nicotera P (1994): Formation of 50 kbp chromatin fragments in isolated liver nuclei is mediated by protease and endonuclease activation. *FEBS Lett* 351:150–154

Zhivotovsky B, Cedervall B, Jiang S, Nicotera P, Orrenius S (1994): Involvement of Ca^{2+} in the formation of high molecular weight DNA fragments in thymocyte apoptosis. *Biochem Biophys Res Comm*un 202:120–127

19

Clinical Aspects of Acute Promyelocytic Leukemia and Response to Retinoid Therapy

MARTIN S. TALLMAN

INTRODUCTION

Acute promyelocytic leukemia (APL) [M3 in the French-American-British (FAB) classification] is an uncommon, but distinctive, subtype of acute myeloid leukemia (AML) that has recently received increased attention. It warrants separate discussion because of its remarkable responsiveness to the vitamin A derivative, all-*trans* retinoic acid (ATRA). Evaluation of the role of ATRA in the treatment of patients with APL has fostered significant progress in our understanding of the molecular biology of leukemia. Rearrangement of the retinoic acid receptor α (RARα) gene locus (located on chromosome 17), as part of the balanced reciprocal translocation between chromosomes 15 and 17 characteristic of APL, represents the first direct link between a mutation in a nuclear receptor and human cancer. This suggests that abnormalities in other nuclear receptors may play a role in the pathogenesis of other human malignancies. Finally, the success of ATRA in APL represents the first example of effective differentiation therapy in human cancer and suggests that similar kinds of selective targeted treatment may be possible in other malignancies.

DISTINCTIVE CHARACTERISTICS OF APL

APL represents 5–15% of cases of AML (Groopman and Ellman, 1979) and is recognized morphologically by an apparent maturation arrest at the promyelocytic stage of myeloid differentiation with abnormal, often bizarre, primary azurophilic granules, a kidney-shaped (reniform) nucleus, and

Hormones and Cancer
Wayne V. Vedeckis, Editor

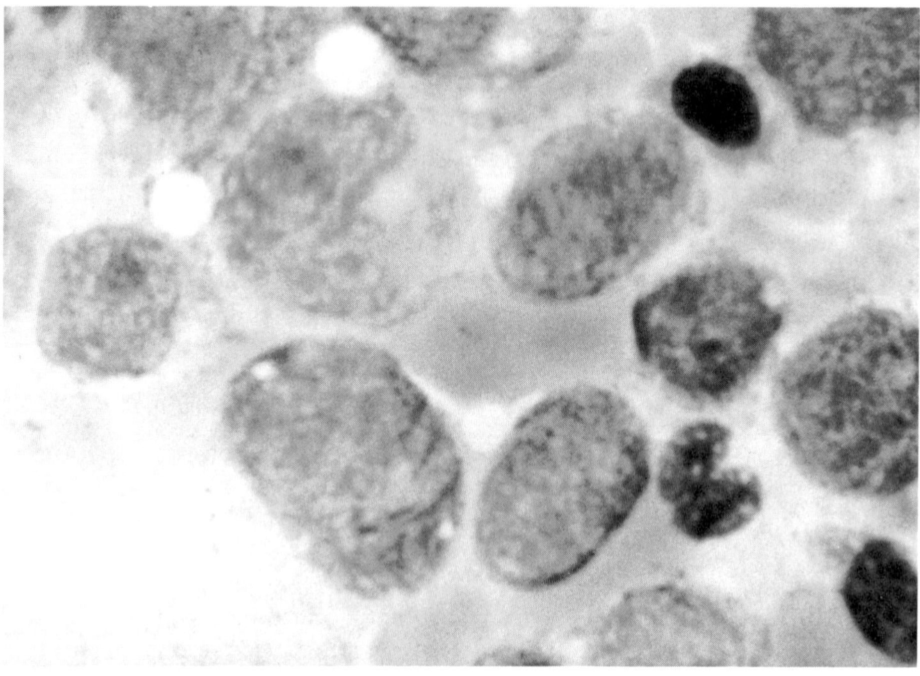

Figure 1. Bone marrow aspirate showing leukemic promyelocytes with abundant + granules, reniform (bilobed) nuclei. (May-Gruenwald Giemsa stain; magnification ×1000).

abundant Auer rods (Fig. 1). Promyelocytic leukemic cells with numerous Auer rods in bundles are called faggot cells (Fig. 2). In addition to this hypergranular form, three variant subtypes have been observed. A microgranular variant (M3v) with granules often below the limit of resolution of the light microscope has been well described and accounts for approximately 25% of cases (Groopman and Ellman, 1979; Bennett et al., 1980). Patients with this morphologic variant have a higher absolute blast cell count and similar bleeding tendency or perhaps a higher early hemorrhagic death rate (Bassan et al., 1995) (discussed below) than patients with the hypergranular type. A basophilic variant has been reported in isolated case reports, in which the leukemic promyelocytes show basophilic differentiation (Koike et al., 1992; Tallman et al., 1993). Toluidine blue staining identifies intracytoplasmic metachromatic granules. The number of basophils may be increased; however, it is not clear whether the basophils are derived from the malignant clone. This variant may be associated with manifestations of severe hyperhistaminemia. Leukemic cells from a patient with basophilic APL have been found to have a 12p13 chromosomal abnormality in addition to the t(15;17) (Tallman et al., 1993), and this abnormality is also present in cells from patients with FAB-AML-M2 and basophilia (Daniel et al., 1985).

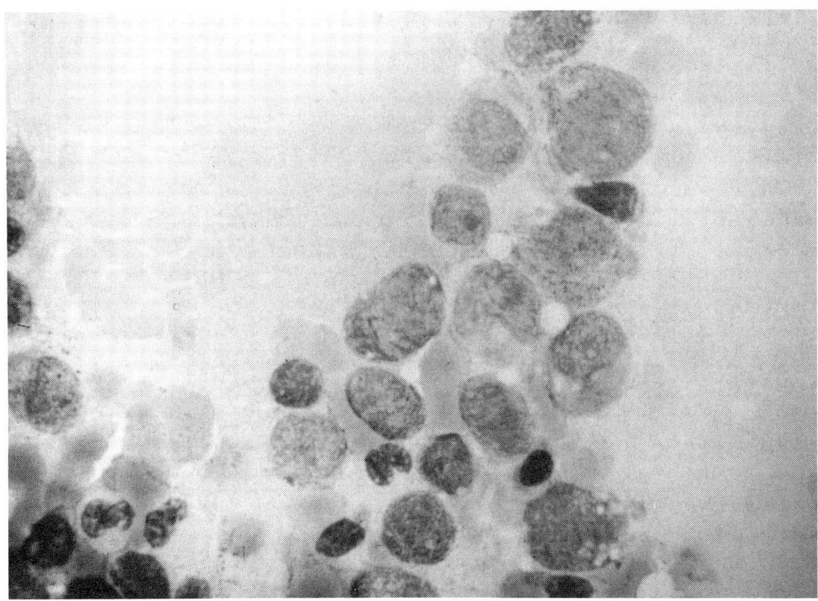

Figure 2. Leukemic promyelocyte showing multiple bundles of Auer rods (faggot cell).

Finally, a variant with hand-mirror morphology, but typical clinical features, has also been reported (Sun and Weiss, 1991).

A characteristic immunophenotype of promyelocytic leukemia cells has been observed and includes $CD34^-$, $CD11b^-$, $CD15^-$, $CD9^+$, $CD33^+$, and $CD13^+$ (San Miguel et al., 1986; Davey et al., 1989; Lo Coco et al., 1991). Patients with APL are virtually always $HLA\text{-}DR^-$, whereas patients with other subtypes of AML are uniformly positive (DeRossi et al., 1990). This distinguishing feature may be helpful in establishing an early diagnosis before the results of the cytogenetic analysis are available. Cells from patients with the microgranular variant, particularly children, express the CD2 antigen (Krause et al., 1989; Biondi et al., 1993; Rovelli et al., 1992). Cytochemistry studies show some heterogeneity in staining. Cells from some patients stain strongly with myeloperoxidase and naphthol-ASD chloroacetate esterase but lack the monocyte enzyme, NaF-sensitive alpha-naphthyl-butyrate esterase. Cells from other patients stain with all three stains. This heterogeneity is in addition to the cases of basophilic APL (Liso et al., 1975; Tomonago et al., 1985; Biondi et al., 1993). Other characteristic features of APL include a younger age at onset (Mertelsman et al., 1980), frequent presentation with leukopenia (Collins et al., 1978), and a relatively favorable prognosis compared to other subtypes of AML once clinical complete remission (CR) is attained (Bernard et al., 1973;

Cunningham et al., 1989; Fenaux et al., 1991). An apparent increased incidence among Latinos with AML has recently been recognized, suggesting a potential genetic predisposition (Malta Corea et al., 1993; Douer et al., 1994; Keung et al., 1994).

APL is distinguished from all other AMLs by the presence of a balanced reciprocal translocation between chromosomes 15 and 17 in virtually every patient (Larson et al., 1984), the persistence of a hypercellular bone marrow during induction of CR with either chemotherapy or ATRA rather than obligatory bone marrow aplasia (Kantarjian et al., 1985; Stone et al., 1988), the association with a life-threatening hemorrhagic syndrome (Baker et al., 1964; Cooperberg, 1967; Rand et al., 1969; Gralnick and Sultan, 1975), the presence of a unique genetic marker, the PML/RARα fusion mRNA and protein produced as a result of the specific t(15;17) chromosomal translocation (Chomienne et al., 1990; de The et al., 1990), and the remarkable ability of the leukemic promyelocytes to differentiate with exposure to ATRA (Huang et al., 1988; Castaigne et al., 1990; Warrell et al., 1991).

t(15;17) TRANSLOCATION AND MOLECULAR CONSEQUENCES

In 1976, Rowley and colleagues described a balanced reciprocal translocation between chromosomes 15 and 17 that was subsequently found to be present in virtually every patient with APL (Golomb et al., 1976; Rowley et al., 1977; Larson et al., 1984). Variant translocations may also occur. Recently, APL has been described associated with a t(11;17) translocation that results in the fusion of RARα to a novel gene, promyelocytic leukemia zinc finger (PLZF), a transcription factor (Chen SJ et al., 1993; Chen Z et al., 1993). This recognition is important because, although rearrangement of the RARα gene in these cases results in impaired myeloid differentiation, such patients do not respond to ATRA (Guidez et al., 1994; Licht et al., 1995).

Several important genes have been mapped to one of these two chromosomes. The t(15;17) involves the rearrangement of the RARα gene located at chromosome band 17q21 (Mattei et al., 1988). As a result of this translocation, the RARα is translocated to and fuses with a chromosome 15-derived transcription unit initially called myl (for *myel*oid) but later renamed PML (for *pro*myelocyte) (de The et al., 1990). Consequently, an aberrant PML/RARα fusion mRNA is produced. The presence of the fusion PML/RARα mRNA predicts responsiveness to ATRA. Several different species of PML/RARα mRNA are expressed because of variations of exon splicing (Chen et al., 1992; Pandolfi et al., 1992; Gallagher et al., 1995). This fact is important because unique mRNA species due to out-of-frame transcripts in some cases of APL are associated with diminished sensitivity to differentiation therapy with ATRA (Gallagher et al., in press). Gene rearrangements involving PML and RARα may occur in the absence of detectable t(15;17) (Lo Coco et al., 1992). Cases with submicroscopic 15;17 recombinations detected by fluorescence *in situ* hybridization that generate the PML/RARα or RARα/PML fusion genes have been described (Lafage-Pochitaloff et al., 1995). A more detailed analysis of the t(15;17) and resultant PML/RARα protein is found in Chapter 20.

The gene coding for the production of granulocyte colony-stimulating factor (G-CSF), a hematopoietic growth factor critical for differentiation and proliferation of myeloid cells, is located near the chromosome 17 breakpoint, but it does not appear to be part of the translocation (Simmers et al., 1987). Finally, the gene coding for the production of myeloperoxidase, a bactericidal enzyme synthesized in the azurophilic granules of promyelocytes and present in abundance in mature granulocytes and promyelocytes, is translocated from chromosome 17 to chromosome 15 as part of the characteristic translocation (Weil et al., 1988).

RETINOID BIOLOGY

It has been known for many years that vitamin A is important in maintaining the integrity of epithelial tissues. Vitamin A together with its natural and synthetic derivatives constitutes the group of compounds known as retinoids. Vitamin A is not naturally synthesized and must be either ingested as a dietary provitamin such as beta-carotene or as preformed retinol or its ester. The retinoid structure is divided into three moieties: a polar terminal end, a conjugated side chain, and a trimethylcyclohexenyl terminal group (Fig. 3). The naturally occurring retinoids include retinol, the alcohol of vitamin A; retinoic acid (RA), the carboxylic acid; retinal, the aldehyde; 13-*cis*-RA, an isomer of RA; and 3-dehydroretinal or vitamin A_2 (Fig. 4).

Retinoid activity is mediated in a manner similar to that of steroid hormones (Evans, 1988). The retinoid molecule binds to a specific receptor, and the complex acts in the cell nucleus where modification of gene transcription occurs. Cellular retinol- and retinoic acid-binding proteins (CRBP and

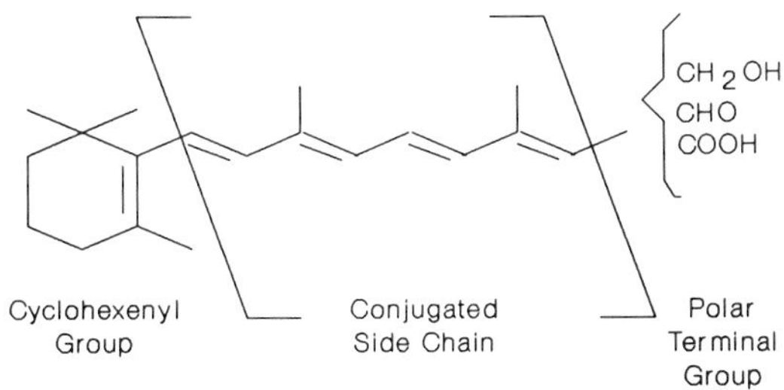

Figure 3. The retinoid structure with a polar terminal end, a conjugated side chain, and trimethylcyclohexenyl group.

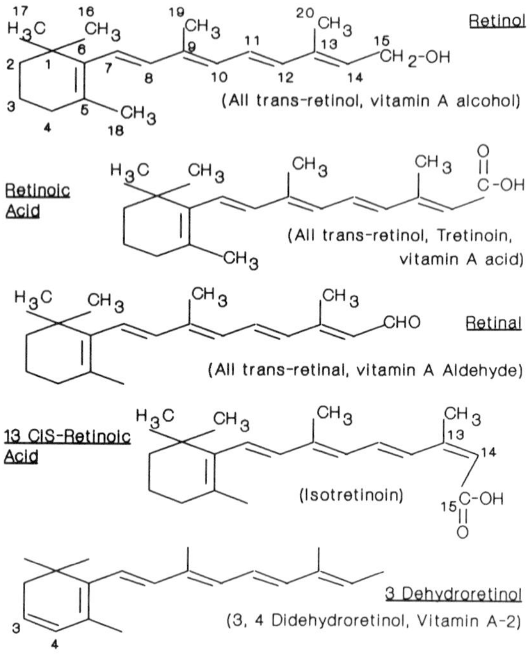

Figure 4. The naturally occurring retinoids.

CRABP) bind retinol or *trans*-retinoic acid and transport the retinoid to the nucleus (Takase et al., 1986; Cornic et al., 1992; Giguere, 1994). Such binding proteins are actually induced following prolonged *in vivo* exposure to ATRA (Evans, 1988). However, the role of CRABP is not completely clear because myeloid leukemia cell lines usually lack CRABP but may differentiate in the presence of ATRA (Breitman et al., 1983).

A nuclear receptor for retinoic acid (RARα) was first identified by screening cDNA libraries, using an oligonucleotide probe corresponding to the highly conserved DNA-binding region of the steroid hormone superfamily (including steroid hormone receptors, receptors for vitamin D_3, thyroid hormone, RARs and retinoid X receptors) that have distinct DNA-binding and ligand-binding domains (Petkovich et al., 1987; Evans, 1988). A second nuclear receptor (RARβ) was identified in a hepatocellular carcinoma, using the conserved DNA-binding domain of the steroid and thyroid hormone receptors (de The et al., 1987). A third nuclear receptor (RARγ) was initially identified in mice and subsequently cloned from a human breast cancer cDNA library (Zelent et al., 1989). The RARs are most homologous to the thyroid hormone receptors. Three retinoid X receptors (RXRα, β, γ) have been identified and respond to ATRA only in high concentrations (Zhang et al., 1992; Allenby et al., 1993). A stereoisomer of ATRA, 9-*cis* retinoic acid, activates both RARs and RXRs and is the naturally occurring ligand of the RXRs (Levin et al., 1992; Zhang et al., 1992). Ligands activate RAR/RXR heterodimers or RXR homodimers which bind to specific DNA sequences called retinoic acid response elements, where they act as transcription factors to control the expression of target genes (de The et al., 1990; Leroy et al., 1991; Smith et al., 1991; Vivanco-Ruiz et al., 1991; Leid et al., 1993). Retinoid receptors are discussed extensively in Chapter 5.

The final mechanism by which leukemic promyelocytes are eliminated appears to be apoptosis, or programmed cell death (Martin et al., 1990). Following exposure to N-(4-hydroxyphenyl)-all-*trans* retinamide, the level of expression of *bcl*-2, which is one of the genes controlling apoptosis, markedly diminishes in NB4 cells (Delia et al., 1995). The mechanism by which this leads to apoptosis may be related to oxidative injury (Hockenberry et al., 1993; Kane et al., 1993; Delia et al., 1995).

DECIPHERING THE PATHOGENESIS OF THE HEMORRHAGIC DIATHESIS IN APL

Approximately 10–20% of patients with APL die of early fatal hemorrhage (Gralnick et al., 1972; Jones and Saleem, 1978; Cordonnier et al., 1985; Kantarjian et al., 1986; Cunningham et al., 1989) because of a hemorrhagic diathesis, which has been characterized as disseminated intravascular coagulation (DIC) (Gralnick et al., 1972; Sultan et al., 1973; Gouault-Heilmann et al., 1975; Bauer and Rosenberg, 1984). Bleeding is often present at the time of presentation because of normal cell turnover, and it is often exacerbated by cytoxic chemotherapy. The bleeding disorder has traditionally been attributed to DIC induced by the release of one or more procoagulants from the leukemic promyelocytes, but recent data suggest that the pathogenesis may involve activation of a general proteolytic cascade involving proteolysis and fibrinolysis (Tallman and Kwaan, 1992; Tallman et al., 1993). Clinically, manifestations of the coagulopathy may include petechiae, oral mucosal bleeding, widespread ecchymoses, retinal hemorrhages, or life-threatening intracerebral hemorrhage (Fig. 5).

The characteristic pattern of abnormal basic coagulation studies includes thrombocytopenia, prolongations of the prothrombin time, partial thromboplastin time and thrombin times, increased levels of fibrin degradation products, and hypofibrinogenemia (Rand et al., 1969; Gralnick and Sultan, 1975; Groopman and Ellman, 1979). These findings are consistent with both DIC and fibrinolysis, with the exception of thrombocytopenia, which may be caused by leukemic infiltration of the marrow itself. It is not clear whether the activation of clotting or fibrinolysis is generated by the leukemic promyelocyte itself, a reactive inflammatory cell, or the endothelial cell, which may express mediators with procoagulant and/or profibrinolytic properties.

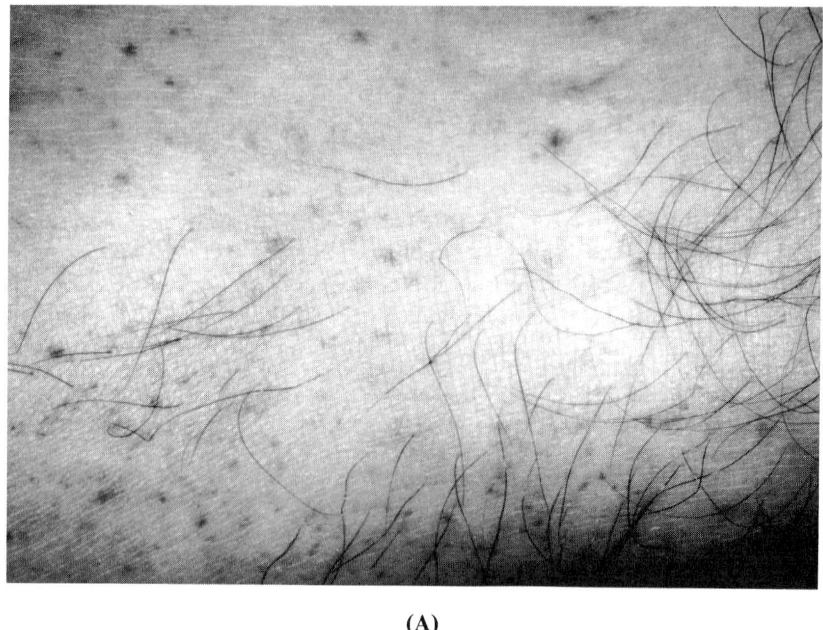

(A)

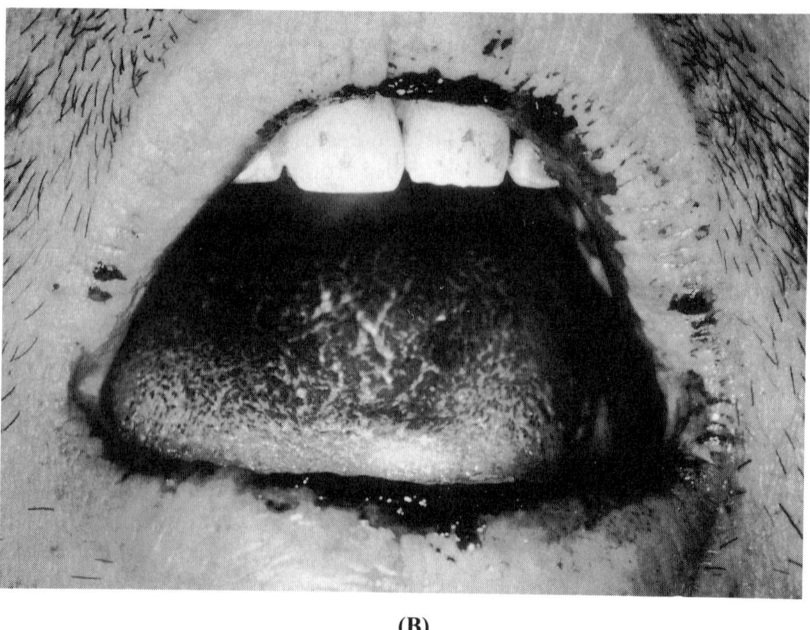

(B)

Figure 5. Manifestations of the bleeding disorder showing lower extremity petechiae (**A**), oral mucosal bleeding (**B**), widespread ecchymoses (**C**), retinal hemorrhage on funduscopic examination (**D**), and intracerebral hemorrhage on computed tomography scan (**E**).

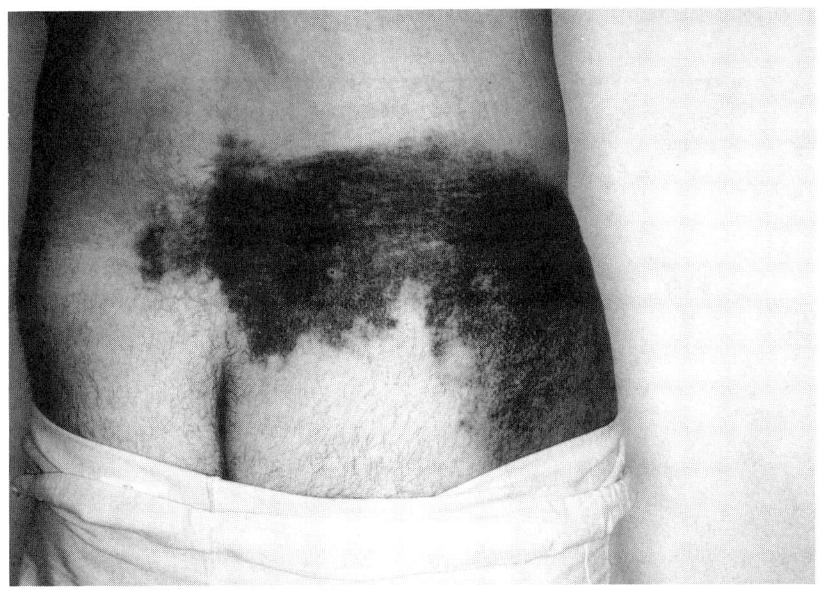

(C)

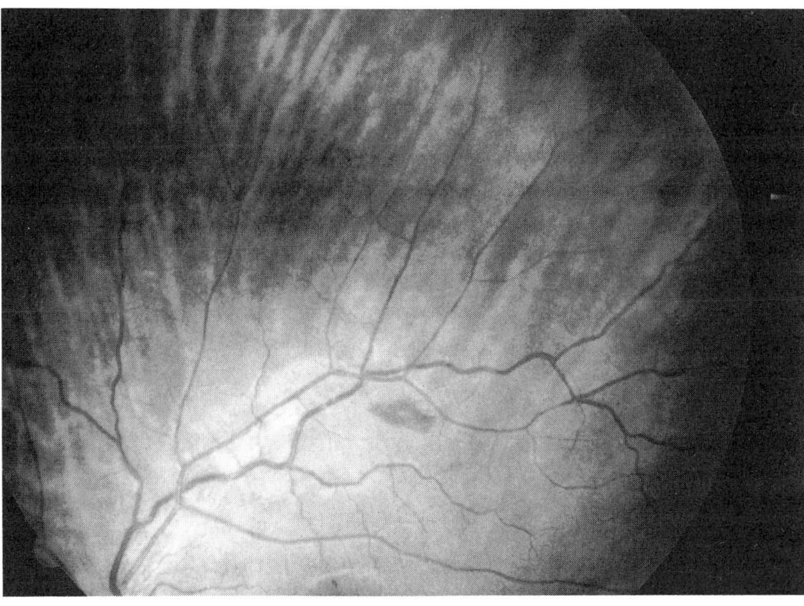

(D)

Figure 5. Continued.

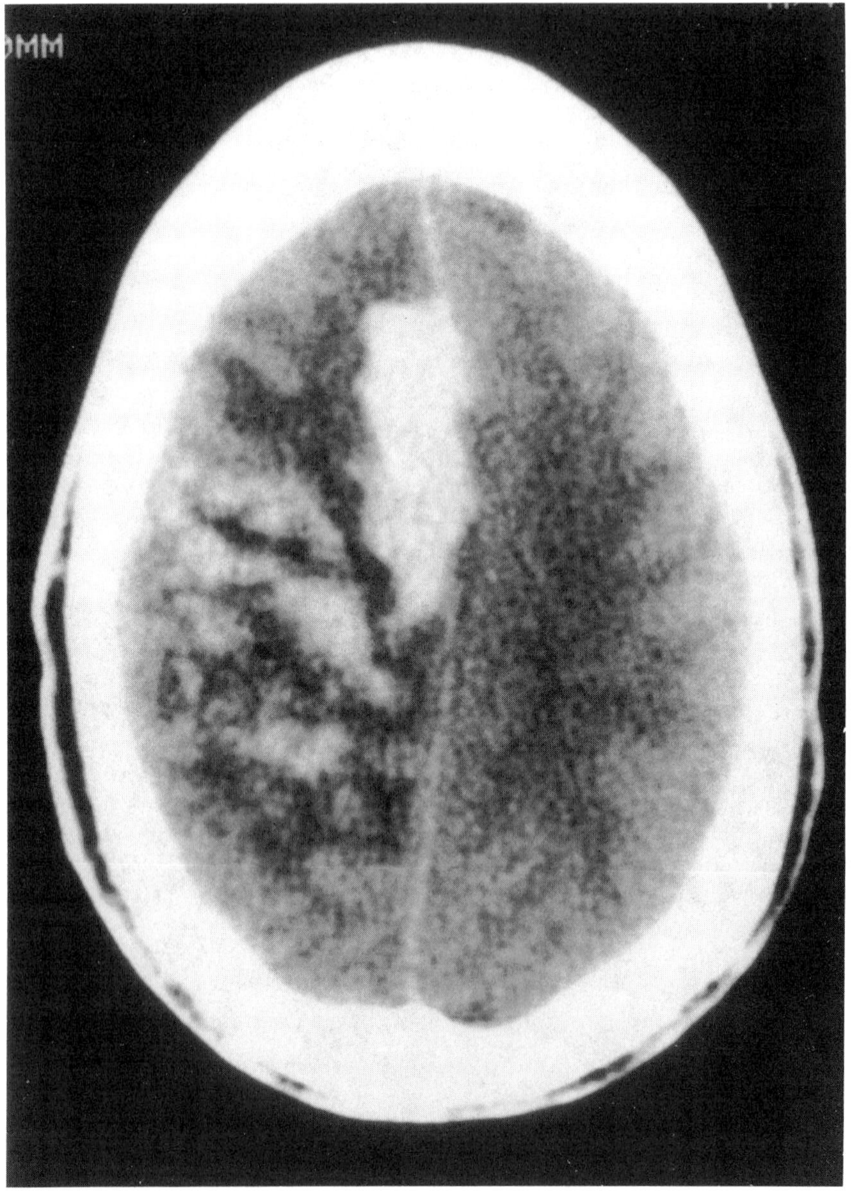

(E)

Figure 5. Continued.

PUTATIVE PROCOAGULANT MEDIATORS IN APL THAT GENERATE THROMBIN AND DIC

Several potential procoagulants have been described in malignant cells, and two have been described in APL cells (Table 1). Tissue factor (TF), the major initiator of blood coagulation *in vivo*, is the membrane protein receptor for factor VII (Bach, 1988; Nemerson, 1988; Edgington et al., 1991). The resulting factor VIIa that is generated activates factors IX and X, leading to thrombin generation and fibrin formation. TF may be released from leukemic cells (Kubota et al., 1991), and the severity of DIC may correlate with the concentration of TF (Andoh et al., 1987). The TF gene is expressed in cells from patients with APL (Bauer et al., 1989). However, there is no definite evidence that mature granulocytes or purified populations of APL cells can generate TF activity. Studies of normal human peripheral blood cell subpopulations have identified TF expression in monocyte fractions (Bauer et al., 1989).

Cancer procoagulant (CP) is a protein with properties of a cysteine proteinase enzyme. It was described initially by Gordon and colleagues (1975) and isolated by Falanga and Gordon (1985). CP initiates coagulation by directly activating factor X in the absence of factor VII and is uniquely expressed in tumors and trophoblastic tissues (Gordon et al., 1985; Donati et al., 1986). CP is expressed in AML cells and in particularly high levels in APL cells (Falanga et al., 1988). Furthermore, CP levels are essentially undetectable in cells from patients in CR (Donati et al., 1990). However, the relative contribution of CP to procoagulant activity *in vivo* is not clear, because much of the factor X cleaving activity in tumor extracts may be attributable to a complex of TF and factor VII (Francis et al., 1988).

Cytokines such as interleukin-1, tumor necrosis factor (TNF), and vascular permeability factor (VPF) are indirect procoagulants because they

Table 1. Procoagulant and profibrinolytic mediators of coagulopathy in APL

Putative procoagulant mediators
Tissue factor
Cancer procoagulant
Interleukin-1
Other cytokines
Putative profibrinolytic mediators
Tissue-type plasminogen activator
Urokinase-type plasminogen activator
Plasminogen activated inhibitors I and II
Putative anticoagulants
Annexin VIII

initiate coagulation by inducing TF in endothelial cells and monocytes (Bevilacqua et al., 1984, 1986a; Nawroth et al., 1986a,b; Clauss et al., 1990). Leukemic cells release IL-1, which induces endothelial cell TF and suppresses thrombomodulin, a cell surface receptor for the activation of protein C (Bevilacqua et al., 1986b). Cozzolino and colleagues (1988) have shown that IL-1 secreted by leukemic cells leads to DIC. Furthermore, cytokines can generate plasminogen activator inhibitors, which inhibit vessel wall fibrinolytic activity (Emeis and Kooistra, 1986; Nachman et al., 1986). Interferon-γ and VPF-like mediators can also induce endothelial cell procoagulant activity (Miyauchi et al., 1988; Moon and Greczy, 1988; Noguchi et al., 1989). Complex interactions undoubtedly exist between inflammatory cytokines and procoagulant-producing cells capable of initiating thrombotic events in patients with APL. Although thrombotic events such as arterial occlusions, hepatic and portal vein occlusions, and pulmonary emboli are only infrequently recognized clinically (Parker and Lowney, 1965; Pittman et al., 1966; Cooperberg, 1967; Albarracin and Haust, 1971; Polliack, 1971), postmortem examinations show widespread thromboses in 15–25% of patients (Bervengo et al., 1962; Albarracin and Haust, 1971; Polliack, 1971; Tan et al., 1972).

PUTATIVE PROFIBRINOLYTIC MEDIATORS IN APL

There is evidence that proteolysis, particularly excessive fibrinolysis, is important in the pathogenesis of the coagulopathy in APL (Chan et al., 1984; Sterrenberg et al., 1985). Levels of both plasminogen and alpha-2-plasmin inhibitor, which binds free plasmin, are reduced in patients with APL (Velasco et al., 1984; Kahle et al., 1985; Schwartz et al., 1986). Furthermore, leukemic promyelocytes release plasminogen activators which cleave plasminogen and initiate fibrinolysis (Gralnick and Abrell, 1973; Velasco et al., 1984; Kahle et al., 1985). Circulating tissue-type plasminogen activator can be found in the plasma of some patients with APL (Wilson et al., 1983). In other patients decreased levels of circulating plasminogen-activator inhibitor type 1 (PAI-1) have been observed (Sakata et al., 1991). In addition, APL cells contain elastases that inactivate alpha-2-plasmin inhibitor (Kahle et al., 1985).

Annexin VIII is yet another potential mediator of the coagulopathy in APL (Hirata et al., 1980). Annexin VIII is one of a group of naturally occurring proteins that bind phospholipids and have both anticoagulant and phospholipase-A_2 inhibitory properties. The annexin VIII gene is highly expressed in cells from patients with APL compared to cells from patients with other AMLs (Chang et al., 1992). Furthermore, annexin VIII is highly expressed in the APL cell line NB4, and its expression is significantly reduced after exposure to ATRA. This observation suggests that over-expression of the annexin VIII gene may contribute to the bleeding disorder characteristic of APL.

EVOLVING THERAPEUTIC APPROACHES TO APL

Induction Chemotherapy

The CR rate with anthracycline-based induction chemotherapy is approximately 70% (Arlin et al., 1984; Kantarjian et al., 1985; Kingsley et al., 1987; Bobbio Pallavicini et al., 1988; Stone et al., 1988; Cunningham et al., 1989; Feldman et al., 1989; Vivanco-Ruiz et al., 1991). The 5-year disease-free survival is 35–45% (Kantarjian et al., 1986; Stone et al., 1988; Cunningham et al., 1989). Although the standard induction chemotherapy program has traditionally included an anthracycline antibiotic combined with cytosine arabinoside, the same standard regimen used for all AMLs, APL cells appear to be particularly sensitive to anthracyclines alone. With single-drug anthracycline therapy, a high percentage of CRs with a relatively long duration compared to other subtypes of AML can be achieved (Bernard et al., 1973; Marty et al., 1984; Petti et al., 1987; Bobbio Pallavicini et al., 1988; Sanz et al., 1988). Furthermore, larger doses of daunorubicin in combination with other agents than are usually administered in the conventional induction program may be beneficial (Fenaux et al., 1991; Hewlett et al., 1993). In addition, there is evidence that the newer anthracycline, idarubicin, either alone or in combination with standard cytosine arabinoside, may result in improved disease-free survival and overall survival (Avvisati et al., 1990; Berman, 1993). Early death due to hemorrhage remains the most common cause of induction failure among patients receiving chemotherapy alone. The role of heparin in this setting has not been established (Tallman and Kwaan, 1992; Tallman et al., 1993). Aggressive platelet support and transfusion of fresh frozen plasma together with heparin has been advocated by some to reduce the risk of hemorrhage during induction (Feldman et al., 1989).

Remission Induction with ATRA

The *in vitro* effect of retinoic acid on HL-60 cells provided the rationale to explore the therapeutic role of retinoids in patients with APL (Breitman et al., 1980). Initially, several anecdotal reports appeared describing the achievement of remission with *cis* retinoic acid (Flynn et al., 1983; Nilsson, 1984; Fontana et al., 1986). The remarkable activity of ATRA in APL was first reported by Huang and colleagues (1988), who observed CR in 23 of 24 patients with previously treated (8 patients) and untreated (16 patients) APL. The single patient not responding to ATRA alone achieved CR with the addition of low-dose cytosine arabinoside, a putative differentiating agent. CRs were achieved in approximately 30 to 50 days (range, 22–119) and with rapid resolution of both clinical and biochemical manifestations of the coagulopathy. Six of the patients were maintained with ATRA alone following CR, and four remained in remission for 5 to 10 months at the time of publication. Two patients relapsed at 2 and 4 months. The bone marrow aspirate remained hypercellular throughout the treatment. In France, Castaigne and colleagues (1990) subsequently confirmed the Chinese experience in that among 22 patients, 14 achieved CR, 4 achieved partial remission (PR), 1 failed and 3 died early of acute respiratory failure, intracranial hypertension, and both cardiac and respiratory failure. Frankel and colleagues (1994) expanded the initial report from Memorial Sloan-Kettering Cancer Center (Warrell et al., 1991) and also observed that approximately 89% of patients achieve CR with ATRA alone at a dose of 45 mg/m^2 given daily. A distinctive respiratory syndrome, the retinoic acid syndrome, manifested by fever, respiratory distress, pulmonary infiltrates, pleural effusions, and weight gain, developed in 13 patients (23%) during treatment and was fatal in 5 (9%). Several additional phase II studies have reported similar results (Table 2) (Chen et al., 1991; Fenaux et al., 1992; Ohno et al., 1993; Kanamaru et al., 1995). Most series report an early mortality rate of approximately 5–15%.

Evidence strongly supports the concept that remission is achieved by terminal differentiation of the leukemic clone rather than direct cytotoxicity (Fig. 6) (Elliott et al., 1992). When the differentiative effects of ATRA and *cis* RA are compared to the clonogenic and self-renewal capacity of freshly isolated AML cells, cell lines with the t(15;17) and/or NB4 cells are found to be much more sensitive to ATRA than *cis* RA, yet equivalent to other AML blasts (Tohda et al., 1992). The successful induction of remission with ATRA appears related to the expression of hematopoietic growth factors, including TNF-α, G-CSF, granulocyte-macrophage colony-stimulating factor (GM-CSF), and interleukin-3 (Dubois et al., 1994). Leukemic cells from APL patients not expressing these growth factors do not differentiate with ATRA exposure. In all studies reported, the remissions induced with ATRA alone were of short duration, and relapses frequently occurred despite continued administration of ATRA. Resistance to retinoids has been implicated to explain the lack of durable remissions (Warrell, 1993).

Induction therapy with ATRA has several potential advantages compared to conventional cytotoxic chemotherapy, including the potential for outpatient treatment, fewer blood product transfusions, less antibiotic requirements, and rapid resolution of the coagulopathy (discussed below). The expression of hematopoietic growth factors by leukemia promyelocytes and the modulation by ATRA may explain two important limitations to the success of this approach: the retinoic acid syndrome and hyperleukocytosis (Dubois et al., 1994).

TOXICITIES OF ATRA

The most common side effects of ATRA are cutaneous and include dryness of the skin and mucous membranes, cheilitis, and skin rash. Unusual cutaneous reactions include erythema nodosum and florid dermatosis (Toh and Winfield, 1992; Hakimian et al., 1993). Other toxicities include headache, hypertriglyceridemia, bone pain, and pseudotumor cerebri, particularly in pediatric patients.

Two life-threatening side effects have been consistently reported with variable incidence. The French group initially reported life-threatening pulmonary and central nervous system complications attributable to hyperleukocytosis (Castaigne et al., 1990; Dombret et al., 1992). Investigators at Memorial Sloan-Kettering Cancer Center have reported that the white blood cell (WBC) count rises above 20,000/μL between days 6 and 20 in 40% of patients (Warrell et al., 1991). Frequently, following a peak, the WBC declines without incident. A similar syndrome has been described in patients with AML receiving GM-CSF priming (Wiley et al., 1993).

Hyperleukocytosis is often, but not exclusively, associated with the second major toxicity, the retinoic acid syndrome (Frankel et al., 1992). This is clearly the most dangerous complication of ATRA therapy and occurs in approximately 25% of patients. There appears to be geographic variability in the incidence of this complication in that Ohno and collaborators (1993) did not observe any cases. This syndrome includes a constellation of findings similar to a pulmonary capillary leak syndrome, including fever, respiratory distress, weight gain, edema, pleural and pericardial effusions, and hypotension. The syndrome usually resolves rapidly if dexamethasone is given at the earliest signs of its development. Autopsies of patients succumbing to this complication show interstitial pulmonary infiltrate, which includes both mature and immature myeloid cells but no true leukostasis. The course of the syndrome is not completely understood. Preliminary evidence suggests that expression of CD13 antigen (aminopeptidase N) has been identified as a poor prognostic factor in APL (Vahdat et al., 1994) and, interestingly, it has been associated with an unfavorable outcome in AML (Griffin et al., 1986) and invasive potential in some human tumor cell lines (Menrad et al., 1993). Increased expression of myeloid cell surface adhesion molecules such as ICAM-1 may play a role and can be blocked by dexamethasone (Dedhar et al., 1991; Zhang et al., 1993). The monoclonal antibody CD18 identifies the beta chain of leukocyte-specific adhesion molecules, and gene expression of this subunit is up-regulated by retinoic acid (Hickstein et al., 1988; Agura et al., 1992).

To prevent the hyperleukocytosis and retinoic acid syndrome, the French investigators have adopted a strategy of introducing early conventional antileukemic chemotherapy based on the timing and degree of WBC elevation (Fenaux et al., 1992). Full-dose cytotoxic chemotherapy has been recommended if the peripheral WBC reaches 5.0×10^9/L by day 6 of ATRA, 10.0×10^9/L by day 10, and 15.0×10^9/L by day 15. Others have reported the benefits of intravenous corticosteroids (dexamethasone, 10 mg twice daily, for at least 3 days) at the earliest signs or symptoms (Menrad et al., 1993; Frankel et al., 1994).

Table 2. Studies of all-*trans* tetinoic acid in acute promyelocytic leukemia

Authors	# Pts	Dose	CR Rate (%)	Median time to CR (days)	Early deaths	Toxicities
Huang et al. (1988)	24 16 untreated 8 unresponsive or resistant	45–100 mg/m²/day	24/24 (100)	53 (20–119)	0	Dryness of lips and skin, occasional headache, digestive symptoms
Castaigne et al. (1990a)	22 4 untreated 2 resistant 11 in 1st relapse 4 in 2nd relapse 1 in 3rd relapse	45 mg/m²/day	14/22 (64)	34 (30–90)	3/22 (14%)	Skin and mucosal dryness, hypertriglyceridemia, increase in hepatic transaminases, hyperleukocytosis, bone pain
Fenaux et al. (1992)	26[a]	45 mg/m²/day	25/26 (96)		1/26 (4%)	Hyperleukocytosis, dryness of skin and mucosae, fever
Warrell et al. (1991)	11 6 untreated 5 in 1st or 2nd relapse	45 mg/m²/day	9/11 (92)	41 (24–53)		Leukocytosis, headache nasal congestion, hypertriglyceridemia, pseudotumor cerebri, skin rash
Chen et al. (1991)	50 47 untreated 3 relapsed	60–80 mg/day	47/50 (94)	39 (25–70)	3/50 (6%)	Dryness of mouth and lips, hyperkeratosis, cutaneous infection

Frankel et al. (1992)	56[b] 34 untreated 22 relapsed	45 mg/m^2/day	44/51 (89)	41 (18–78) untreated 35 (24–77) relapsed	5/51 (9%)	Retinoic acid syndrome, headache, skin reactions, nausea, emesis, bone pain, ear/nasal congestion, hypertriglyceridemia
Ohno et al. (1993)	64 21 refractory to induction 10 refractory to salvage 26 relapsed 7 untreated	45 mg/m^2/day	54/64 (84)	29 (8–53)	7/64 (11%)	Cheilitis, xerosis, dermatitis, bone pain, elevation in hepatic transaminases, hypertriglyceridemia
Kanamaru et al. (1995)	110 untreated[c]	45 mg/m^2/day	97/109 (89) 25/28 (89) ATRA alone	35 (12–69)	17/109 (15%)	Retinoic acid syndrome, hyperleukocytosis, cheilitis, muscle pain, hypertriglyceridemia

[a]Eleven patients required daunorubicin and cytosine arabinoside during induction with ATRA starting between 2 and 30 days to control hyperleukocytosis (9 patients), resistance to ATRA (1 patient) or organomegaly (1 patient).
[b]Four patients did not have the PML/RARα gene rearrangement and were not included in the response analysis, and one patient with the PML/RARα gene rearrangement was removed from the study prematurely and given cytotoxic chemotherapy. Five patients received cytosine arabinoside to control hyperleukocytosis.
[c]Twenty-eight (26%) of 109 patients (1 death before therapy) received ATRA alone, 51 received ATRA plus initial chemotherapy, 30 received ATRA plus later chemotherapy.

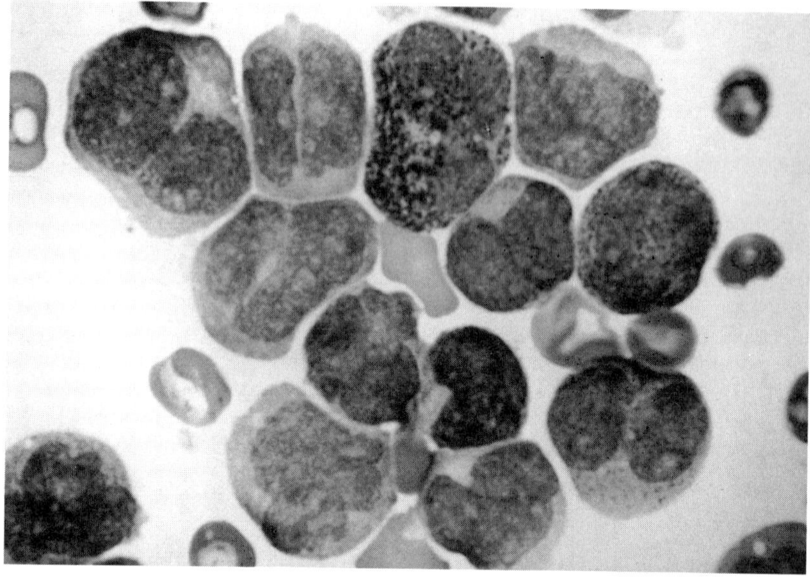

(A)

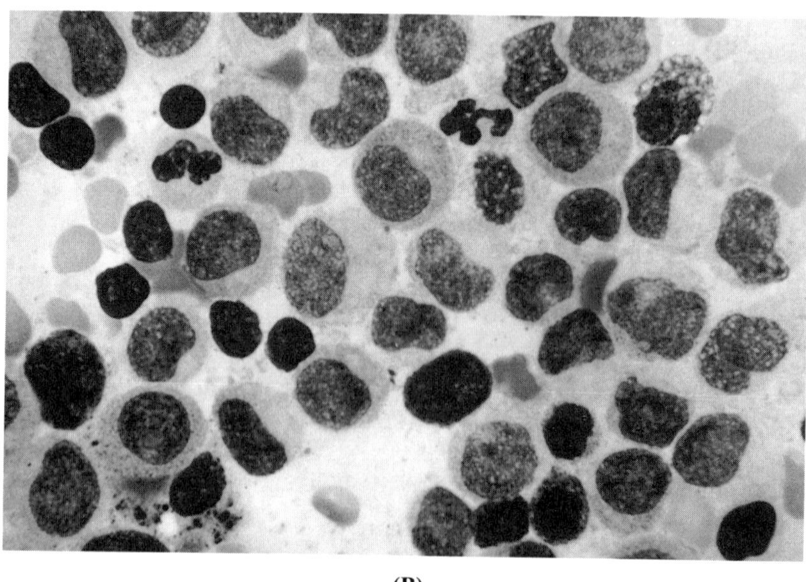

(B)

Figure 6. Acute promyelocytic leukemia microgranular type (AML-M3v) in bone marrow aspirate before therapy (×1000) (**A**), bone marrow aspirate 12 days post therapy with ATRA showing early differentiation of promyelocytes (×625) (**B**), and bone marrow aspirate 28 days post therapy with ATRA, showing maturing granulocytes, including segmented neutrophils (×625) (**C**). (Photographs supplied courtesy of Susan D. Wheaton, M.D.)

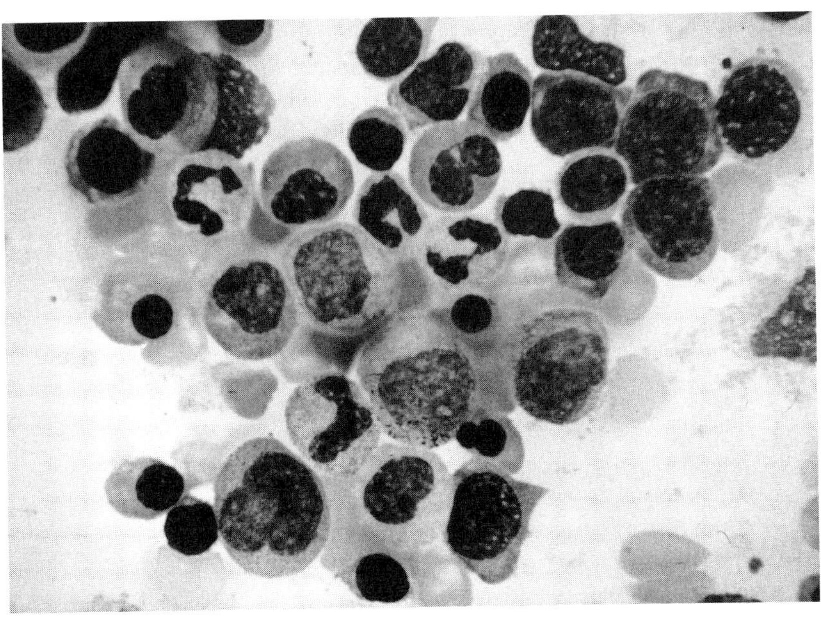

(C)

Figure 6. Continued.

Retinoid Pharmacology

ATRA is the active metabolite of vitamin A. Following ingestion in healthy human volunteers, ATRA is absorbed and conjugated in the liver and then enters the enterohepatic circulation. Isomerization of ATRA to the stereoisomer follows, and both isomers become metabolized to 4-oxo-all-*trans* and 14-oxo-13-*cis* retinoic acid (Blomhoff et al., 1990). The median time to attain maximal plasma concentration in patients with APL following a single oral dose of 45 mg/m^2 is 90 minutes (range, 60–210), and the median plasma elimination half-life is 30 minutes (range, 16.8–77.4) (Lefebre et al., 1991). However, this half-life is likely an underestimate because the limit of detection of ATRA is 0.003 mg/ml. Second peak concentrations may occur with food intake.

There is significant interpatient variability in bioavailability, which may explain the somewhat different results reported by other pharmacokinetic studies. The New York group reported that, in patients with APL, following a single oral dose of 45 mg/m^2, the maximal plasma concentration is reached after 1 or 2 hours (Muindi et al., 1992). The concentration declines thereafter and is $<$10 mg/ml by 8 hours. The peak plasma concentration of ATRA is 346 ± 266 mg/ml (mean, $\pm$SD). The parent drug is rapidly eliminated from the plasma in a monoexponential way, with a half-life of 0.8 ± 0.1 hours.

The bioavailability of retinoids is increased with food ingestion (Colburn et al., 1985), leading to the recommendation to administer ATRA with meals. Despite this recommendation, the plasma concentration declines significantly when the drug is administered on a chronic daily schedule (Muindi et al., 1992).

Therefore, Adamson and colleagues (1995) administered ATRA at an oral dose of $40\,mg/m^2$ on a repetitive cycle of 7 consecutive days of the drug, followed by 7 days without the drug. Plasma exposure to ATRA, as determined by the area under the plasma concentration time curve (AUC), decreased significantly during the first week of drug administration, from a mean of $145 \pm 26\,\mu mol/L$-min on day 1 to $18 \pm 4\,\mu mol/L$-min by day 7. Plasma ATRA concentrations at the start of weeks 3 and 11 with this every-other-week schedule were equivalent to those achieved on day 1 of administration, with mean AUCs of 177 ± 39 and $128 \pm 30\,\mu mol/L$-min, respectively. These data suggest that an intermittent administration schedule of ATRA may circumvent the low plasma ATRA exposure commonly observed with a chronic daily administration schedule.

Retinoid Resistance

Resistance to ATRA in previously untreated patients in whom either the t(15;17) chromosomal translocation or the presence of PML/RARα can be detected is extremely uncommon. However, resistance is always observed in patients relapsing early after discontinuation of ATRA (Delva et al., 1993). Although the precise mechanism(s) of apparent retinoid resistance is not fully understood, Delva and colleagues and others found high levels of intracellular CRABP in some patients with relapsed APL that was not detectable before ATRA exposure (Adamson et al., 1993; Delva et al., 1993). Increases in CRABP and hepatic cytochrome P-450 oxidase enzymes induce catabolism of ATRA, accounting for the observed progressive decline in plasma levels. Since pharmacokinetic studies indicate that continuous daily ATRA exposure is associated with a marked decrease in plasma drug concentration at the time of relapse compared to the initial day of treatment (Muindi et al., 1992), attempts at preventing this decline in plasma concentrations by modulating retinoid catabolism have been pursued, including attempts to inhibit the hepatic cytochrome P-450 oxidase enzyme system with drugs such as ketoconazole and liarozole, and such maneuvers are partially effective in increasing plasma levels (Francis et al., 1993; Rigas et al., 1993; Miller et al., 1994). Another proposed mechanism of resistance involves mutations in nuclear retinoid receptors (Pratt et al., 1990; Collins et al., 1990).

RESOLUTION OF THE COAGULOPATHY WITH ATRA

Resolution of both the clinical and biochemical signs of coagulopathy is the earliest indication of response to ATRA (Huang et al., 1988; Castaigne et al.,

1990; Chen et al., 1991; Warrell et al., 1991; Fenaux et al., 1992; Ohno et al., 1993; Frankel et al., 1994). In one of the largest studies published, the mean time to resolution of the bleeding disorder was 4 days, and no patient died by hemorrhage (Chen et al., 1991).

Several preliminary studies have been carried out to determine the specific changes in coagulation parameters occurring with ATRA exposure (Table 3). Dombret and colleagues (1993) reported evidence of both DIC and fibrinolysis in patients with APL, with complete correction of the fibrinolysis but persistent procoagulant and proteolytic activity, during differentiation therapy (Dombret et al., 1993). Such findings may account for reports of thrombolic events during treatment with ATRA (Runde et al., 1992; de Lacerda et al., 1993; Escudier et al., 1993). Other investigators have recently demonstrated that the down-regulation of the thrombomodulin gene and up-regulation of the TF gene (induced by TNF-α) in human endothelial cells can be blocked by ATRA (Ishii et al., 1992). This is consistent with the observations by Koyama and colleagues, who have reported that ATRA up-regulates thrombomodulin and down-regulates TF expression in NB4 cells (Koyama et al., 1994). Expression of the TF gene in NB4 is inhibited in the presence of ATRA (Rickles et al., 1993). Others have reported that RA appears to enhance fibrinolytic activity *in vivo* by increasing tissue type plasminogen activity, accounting for a potential antithrombotic effect (Van Giezen et al., 1993).

COMPARISON OF ATRA WITH CONVENTIONAL CHEMOTHERAPY IN APL

The precise role of ATRA in the treatment of patients with APL remains to be determined but is evolving. A nonrandomized study comparing ATRA to conventional chemotherapy for remission induction (historical controls) showed a survival advantage for ATRA (Frankel et al., 1994). Patients induced into CR were given three cycles of consolidation chemotherapy with idarubicin and cytosine arabinoside (conventional induction chemotherapy). Eighty-six percent of the ATRA-treated patients achieved CR. The median survival time of the newly diagnosed patients exceeded 31 months (range, 0.4–36+) at the time of publication. This was superior to the median overall survival of 17 months in the historical population (Frankel et al., 1994). Despite initial expectations, no decrease in early mortality was observed compared to a historical control population treated with chemotherapy. Fenaux and colleagues (1993) also reported benefits of ATRA treatment. The CR rate for ATRA-treated patients was 96%, compared to 76% for a chemotherapy-treated (amsacrine or zorubicine and cytosine arabinoside) population. Patients treated in the ATRA-treated study and in the historical cohort received similar consolidation and maintenance chemotherapy, including daunorubicin and cytosine arabinoside

Table 3. Phase II studies of all-*trans* retinoic acid in acute promyelocytic leukemia, detailing response of coagulopathy

Author	# Pts	(%)	CR rate coagulopathy	Blood products	Response of Coagulopathy (days)	Early Death (hemorrhage or thrombosis)
Huang et al. (1988)	24	96	3/24	ND	7	0
Castaigne et al. (1990a)	22	64	13/22	Platelets Heparin (3%)	8	0
Frankel et al. (1992)	56	86	38/56	Platelets FFP	—	4
Chen et al. (1991)	50	94	50/50	Heparin	4.2	2
Fenaux et al. (1992)	26	96	19/26	Platelets	NA	1
Fenaux et al. (1993)	54[a]	91	41/54	Platelets Heparin (46%) Fibrinogen (17%) Transexemic acid (7%)	4	3

NA, not available. FFP, fresh frozen plasma.
[a]Represents patients receiving ATRA on French APL91 randomized trial. Seventy-one percent of patients required early chemotherapy to control hyperleukocytosis.

for four cycles with prolonged maintenance, including 6-mercaptopurine and methotrexate. The actuarial disease-free survival for the ATRA-treated patients was 87% after 18 months, compared to 59% for the historical patients (Fenaux et al.,1992). Although the incidence of early death from hemorrhage may be reduced, the potential benefits appear to be balanced by deaths from toxicities attributable to ATRA therapy, such as the retinoic acid syndrome. If the incidence of this complication can be reduced, the high CR rates may diminish early mortality.

The first prospective randomized trial (APL91) comparing chemotherapy alone to ATRA followed by chemotherapy in previously untreated APL patients was conducted in Europe, and it was terminated early because an improved event-free survival advantage was observed for the patients receiving ATRA (Fenaux et al., 1993). Events were defined as relapse or death. The event-free survival at 12 months was significantly better among 54 patients treated with ATRA than among 47 patients receiving daunorubicin and cytosine arabinoside as induction therapy (79% vs 50% of patients respectively) (Fig. 7). Mortality from induction was not different, but there was a significant difference in the relapse rate. Six patients in the ATRA group relapsed after 7 to 15.5 months, whereas in the chemotherapy group 12 patients relapsed and 2 died within 16 months. These preliminary data suggest that the combination of ATRA and chemotherapy improved the event-free survival, primarily by reducing the relapse rate.

A number of questions regarding the role of ATRA in APL remain. The optimal dose and schedule are not known. Although the most common dose used is 45 mg/m^2 daily, a smaller dose (15 mg/m^2) appears equally effective and is associated with similar toxicities (Castaigne et al., 1993). Similarly,

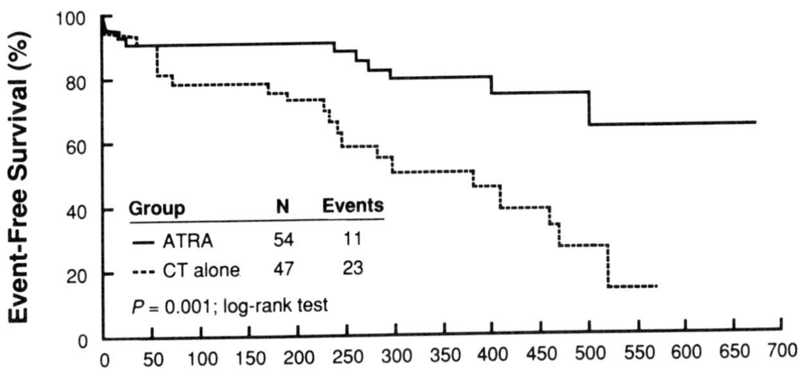

Figure 7. Kaplan-Meier estimate of event-free survival (in percentage of patients) in patients treated with ATRA compared with those treated with chemotherapy alone in the French APL91 Trial. (Reprinted with permission of W.B. Saunders Company, from Feneaux et al., 1993).

although it appears that ATRA should be given with chemotherapy, the optimal schedule and sequence are not known. The best approach to prevent hyperleukocytosis and the retinoic acid syndrome remains to be established. Further studies will be required to determine the optimal strategy.

MOLECULAR EVALUATION OF RESPONSE

Multiple studies have demonstrated that analysis of gene rearrangements can detect RARα fusion products in every patient with untreated APL, and these disappear when CR is achieved (Biondi et al., 1991; Lo Coco et al., 1991; Castaigne et al., 1992; Chang et al., 1992; Ikedi et al., 1993; Zhao et al., 1995). However, persistent occult, clonal leukemic promyelocytes exist in the bone marrow of some patients in apparent CR by morphology and karyotype analysis (Biondi et al., 1991). Both Northern blot analysis and the more sensitive reverse transcriptase-polymerase chain reaction (RT-PCR) have been used to identify RARα fusion transcripts. Following the achievement of CR and consolidation chemotherapy, these techniques may be useful to detect minimal residual disease and to monitor patients for early recurrence, which may identify a subset of patients warranting an alternative therapeutic strategy. Indeed, preliminary studies suggest that molecular detection of the persisting leukemic clone after the completion of therapy is highly predictive of relapse (Miller et al., 1994). One novel approach for patients achieving CR with ATRA involves the administration of a radiolabelled monoclonal antibody ^{131}I-M195, a mouse IgG_{2a} monoclonal antibody that binds CD33, a cell-surface glycoprotein, on most of myeloid progenitor cells (Jurcic et al., 1995). In this small pilot study, two of six patients with detectable PML/RARα mRNA after ATRA had no detectable PML/RARα mRNA after a single exposure to ^{131}I-M195.

POTENTIAL FOR SYNERGISTIC DIFFERENTIATION AND NEW RETINOIDS

9-*cis* retinoic acid is a naturally occurring stereoisomer of ATRA that is a high affinity ligand of the retinoid X receptor, transactivates target genes 40-fold more efficiently than ATRA, and is a more potent inducer of differentiation (Levin et al., 1992; Kizaki et al., 1993). Although neither ATRA nor 9-*cis* retinoic acid alone induces differentiation in RA-resistant HL-60 leukemia cells, morphologic differentiation was induced by the combination (Kizaki et al., 1994). Combinations of ATRA with other differentiating agents, including sodium butyrate, hexamethylene bisacetamide (HMBA), and dimethyl sulfoxide show synergistic differentiation in HL-60 cells (Breitman and He, 1990). In addition, hematopoietic growth factors such as G-CSF and GM-CSF potentiate differentiation induced by ATRA (Gianni et al., 1994; Imaizumi et al., 1994). Furthermore, there is evidence that although NB4 cells are resistent to non-retinoid differentiating agents such as HMBA and butyrates, prior exposure of such cells to ATRA abolishes resistance to nonretinoids and potentiates differentiation (Chen et al., 1994). Combinations of retinoids, non-retinoid differentiating agents, cytokines, and vitamin D or its analogs may have potent synergistic activity in APL or other hematologic malignancies (Kurzrock et al., 1993; Ohno et al., 1993; Bollag et al., 1994; Ganzer et al., 1994).

SUMMARY

Although APL is uncommon, important insights into leukemogenesis have emerged from intensive basic and clinical research during a short period of time. Although it has been recognized for many years that certain agents can induce differentiation of leukemic cells *in vitro*, Huang and colleagues in 1988 made the first observation that almost every patient with APL achieved CR with ATRA. Since then, a variety of studies have been conducted, ranging from clinical trials to confirm the initial Chinese observations to laboratory studies designed to elucidate the pathogenesis at the molecular level. The results of these studies are of crucial importance because the success of ATRA in APL represents the first example in a human malignancy of a targeted form of treatment directed at a specific genetic abnormality. Remarkably, all the phase II studies show rapid improvement in both the biochemical and clinical signs of life-threatening coagulopathy, so characteristic of this disease. However, none has produced an improvement in the mortality rate during induction of remission because of toxicities of ATRA.

Investigators have shown that minimal residual or recurrent disease can be detected by RT-PCR. Such patients appear to be at risk for morphologic and clinical relapse and may be candidates for further therapy, but the best strategy is not established. It is possible that differentiation therapy will have broader implications in settings other than APL. The challenge for the future will be to elucidate the pathogenesis of other subtypes of leukemia or other malignancies at the molecular level. This may permit the development of other novel differentiating agents. The activity of ATRA in APL serves as a paradigm for this approach.

REFERENCES

Adamson PC, Bailey J, Pluda J, Poplack DG, Bauza S, Murphy RF, Yarchoon R, Balis FM (1995): Pharmacokinetics of all-trans retinoic acid administered on an intermittent schedule. *J Clin Oncol* 13:1238–1241

Adamson PC, Boylan JF, Murphy RF, Godwin KA, Gudas LJ, Poplack DG (1993): Time course of induction of metabolism of all-trans-retinoic acid and the upregulation of cellular retinoic acid-binding protein. *Cancer Res* 53:472–476

Agura ED, Howard MI, Collins SJ (1992): Identification and sequence analysis of the promoter for the leukocyte integrin beta-subunit (CD18): a retinoic acid-inducible gene. *Blood* 79:603–609

Albarracin NS, Haust MD (1971): Intravascular coagulation in promyelocytic leukemia. A case study including ultrastructure. *Am J Clin Pathol* 55:677–685

Allenby G, Bocquel MT, Saunders M, Kazmer S, Speck J, Rosenberger M, Lovey A, Kastner P, Grippo JF, Chambon P, Levin A (1993): Retinoic acid receptors and retinoic X receptors: interactions with endogenous retinoic acids. *Proc Natl Acad Sci USA* 90:30–34

Andoh K, Kubota T, Takada M, Tanaka H, Kobayashi N, Maekawa T (1987): Tissue factor activity in leukemic cells. Special reference to disseminated intravascular coagulation. *Cancer* 59:748–754

Arlin Z, Kempin S, Mertelsman RT, Gee T, Higgins C, Jhanwar S, Chaganti RSK, Clarkson B (1984): Primary therapy of acute promyelocytic leukemia: results of amsacrine- and daunorubicin-based therapy. *Blood* 63:211–212

Avvisati G, Mandelli F, Petti MC, Vegna ML, Spadea A, Liso V, Specchia G, Bernasconi C, Alessandrino EP, Piatti C, Carella AM (1990): Idarubicin (4-demethoxydaunorubicin) as a single agent for remission induction of previously untreated acute promyelocytic leukemia: a pilot study of the Italian cooperative group GIMEMA. *Eur J Haematol* 44:257–260

Bach RR (1988): Initiation of coagulation by tissue factor. *Crit Rev Biochem* 23:339–368

Baker W, Bang NU, Nachman RL, Raafat RT, Hortwitz HI (1964): Hypofibrinogenemic hemorrhage in acute myelogenous leukemia treated with heparin. *Ann Intern Med* 61:116–123

Bassan R, Battista R, Viero P, d'Emilio A, Buelli M, Montaldi A, Rambaldi A, Tremul L, Dini E, Barbui T (1995): Short-term treatment of adult hypergranular and microgranular acute promyelocytic leukemia. *Leukemia* 9:238–243

Bauer KA, Rosenberg RD (1984): Thrombin generation in acute promyelocytic leukemia. *Blood* 64:791–796

Bauer KA, Conway EM, Bach R, Konigsberg WH, Griffin JD, Demetri G (1989): Tissue factor gene expression in acute myeloblastic leukemia. *Thromb Res* 56:425–430

Bennett JM, Catovsky D, Daniel MT, Flandrin G, Galton DA, Gralnick HR, Sultan C (1980): A variant form of hypergranular promyelocytic leukemia. *Ann Intern Med* 92:261–280

Berman E (1993): A review of idarubicin in acute leukemia. *Oncology* 7:91–98

Bernard J, Weil M, Boiron M, Jacquillat C, Flandrin G, Gemon MF (1973): Acute promyelocytic leukemia: results of treatment by daunorubicin. *Blood* 41:489–496

Bervengo MG, Leighets G, Zina G (1962): A case of acute promyelocytic leukemia with bullous, hemorrhagic and necrotic skin lesions. *Dermatologica* 151:1984

Bevilacqua MP, Pober JS, Majeau GR, Cotran RS, Gimbrone MA Jr (1984): Interleukin 1 (IL-1) induces biosynthesis and cell surface expression activity of procoagulant activity in human vascular endothelial cells. *J Exp Med* 160:618–623

Bevilacqua MP, Pober JS, Mageau GR, Fiers W, Cotran RS, Gimbrone MA Jr (1986a): Recombinant human tissue necrosis factor induces procoagulant activity in cultured human vascular endothelium. Characterization and comparison with interleukin-1. *Proc Natl Acad Sci USA* 83:4533–4537

Bevilacqua MP, Schleef RR, Gimbrone HA, Loskutoff DJ (1986b): Regularities of the fibrinolytic system of cultured human vascular endothelium by interleukin-1. *J Clin Invest* 78:587–591

Biondi A, Rambaldi A, Alcalay M, Pandolfi PP, Lo Coco F, Diverio D, Rossi V, Mencarelli A, Longo L, Zangrilli D, Masera G, Barbui T, Mandelli F, Grignani F, Pelicci PG (1991): RAR-α gene rearrangements as a genetic marker for diagnosis and monitoring in acute promyelocytic leukemia. *Blood* 77:1418–1422

Biondi A, Luciano A, Bassan R, Mininni D, Specchia G, Lanzi E, Castaigne S, Rainoldi AC, Liso V, Masera G, Barabui T, Rambaldi A (1993): CD2 expression correlates with microgranular acute promyelocytic leukemia (M3v) and not with PML gene breakpoint. *Blood* 82:436 (abstr)

Blomhoff R, Green MH, Frond B, Norum KR (1990): Transport and storage of vitamin A. *Science* 250:399–404

Bobbio Pallavicini E, Luliri P, Anselmetti L, Gorini M, Invernizzi R, Ascari E (1988): High-dose daunorubicin (DNR) for induction and treatment of relapse in acute promyelocytic leukemia (APL): Report of 17 cases. *Haematologica* 73:48–53

Bollag W, Majewski S, Jablonska S (1994): Cancer combination chemotherapy with retinoids: experimental rationale. *Leukemia* 8:1453–1457

Breitman T, Keene B, Hemmi H (1983): Retinoic acid-induced differentiation of fresh human leukemia cells and the human myelomonocytic cell lines HL-60, U-937 and THP-1. *Cancer Surv* 2:201–291

Breitman TR, Selonick SE, Collins SJ (1980): Induction of differentiation of the human promyelocytic leukemia cell line HL-60 by retinoic acid. *Proc Natl Acad Sci USA* 77:2936–2941

Breitman TR, He R (1990): Combinations of retinoic acid with either sodium butyrate, dimethyl sulfoxide or hexamethylene bisacetamide synergistically induce differentiation of the human myeloid leukemia cell line HL-60. *Cancer Res* 50:6268–6273

Castaigne S, Chomienne C, Daniel MT, Ballerini P, Berger R, Fenaux P, Degos L (1990a): All-trans retinoic acid as differentiation therapy for acute promyelocytic leukemia. I. Clinical results. *Blood* 76:1704–1709

Castaigne S, Chomienne C, Fenaux P, Daniel MJ, Degos L (1990b): Hyperleukocytosis during all-trans retinoic acid for acute promyelocytic leukemia. *Blood* 76:260a (supp 1, abstr)

Castaigne S, Lefebvre P, Chomienne C, Suc E, Rigal-Huguet F, Gardin C, Delmer A, Archimbaud E, Tilly H, Janvier M, Isnard F, Travade P, Montfort L, Delannoy A, Rapp MJ, Christian B, Montastruc M, Weh H, Fenaux P, Dombret H, Gourmel B (1993): Effectiveness and pharmacokinetics of low-dose all-trans retinoic acid ($25\,mg/m^2$) in acute promyelocytic leukemia. *Blood* 82:3560–3563

Castaigne S, Balitrand N, de The H, Dejean A, Degos L, Chomienne C (1992): A PML/retinoic acid receptor α fusion transcript is constantly detected by RNA-based polymerase chain reaction in acute promyelocytic leukemia. *Blood* 79:3110–3115

Castoldi GL, Liso V, Specchia G, Tomasi P (1994): Acute promyelocytic leukemia: morphological aspects. *Leukemia* 8:1441–1446

Chan TK, Chan GT, Chan V (1984): Hypofibrinogenemia due to increased fibrinolysis in two patients with acute promyelocytic leukemia. *Aust NZ J Med* 14:245–249

Chang KS, Lu J, Wang G, Trujillo JM, Estey E, Cork A, Chu DT, Freireich EJ, Stass SA (1992a): The t(15;17) breakpoint in acute promyelocytic leukemia cluster within two different sites of the myl gene: targets for the detection of minimal residual disease by the polymerase chain reaction. *Blood* 79:554–558

Chang KS, Wang G, Freireich EJ, Daly M, Naylor SL, Trujillo JM, Stass SA (1992b): Specific expression of the annexin VIII gene in acute promyelocytic leukemia. *Blood* 79:1802–1810

Chen A, Licht JD, Wu Y, Hellinger N, Scher W, Waxman S (1994): Retinoic acid is required for and potentiates differentiation of acute promyelocytic leukemia cells by nonretinoid agents. *Blood* 84:2122–2129

Chen SJ, Chen Z, Chen A, Tong JH, Dong S, Wang ZY, Waxman S, Zelent A (1992): Occurrence of distinct PML-RARα fusion gene isoforms in patients with acute promyelocytic leukemia detected by reverse transcriptase/polymerase chain reaction. *Oncogene* 7:1223–1232

Chen SJ, Zelent A, Tong JH, Yu HQ, Wang ZY, Derre J, Berger R, Waxman S, Chen Z (1993): Rearrangements of the retinoic acid receptor alpha and PLZF genes resulting from t(11;17) (q23;q21) in a patient with acute promyelocytic leukemia. *J Clin Invest* 91:2260–2266

Chen ZX, Xue YQ, Zhang R, Tao RF, Xia XM, Li C, Wang W, Zu WY, Yao XZ, Ling BJ (1991): A clinical and experimental study on all-*trans* retinoic acid-treated acute promyelocytic leukemia patients. *Blood* 78:1413–1419

Chen ZX, Brand NJ, Chen A, Chen SJ, Tong JH, Wang ZY, Waxman S, Zelent A (1993): The translocation t(11;17) fuses the retinoic acid receptor alpha to a novel zinc finger protein. *EMBO J* 12:1161–1167

Chomienne C, Ballerini P, Balitrand N, Huang ME, Krawice I, Castaigne S, Fenaux P, Tiollais P, Dejean A, Degos L, de The H (1990): The retinoic acid receptor gamma gene is rearranged in retinoic acid sensitivepromyelocytic leukemia. *Leukemia* 12:802–807

Clauss M, Gerlach M, Gerlach H, Brett J, Wang F, Familletti PC, Pan YC, Olander JV, Connolly DT, Stern D (1990): Vascular permeability factor: a tumor-derived polypeptide that induces endothelial cell and monocyte procoagulant activity and promotes monocyte migration. *J Exp Med* 172:1535–1545

Colburn WA, Gibson DM, Rodriguez LC, Bugge LJL, Blumenthal HP (1985): Food increases the bioavailability of isotretinoin. *J Clin Pharmacol* 25:583–589

Collins AJ, Bloomfield CD, Peterson BA, McKenna RW, Edson R (1978): Acute promyelocytic leukemia: management of the coagulopathy during daunorubicin-prednisone remission induction. *Arch Intern Med* 138:1677–1680

Collins SJ, Robertson KA, Mueller L (1990): Retinoic acid-induced granulocytic differentiation of the HL-60 myeloid leukemia cells is mediated directly through the retinoic acid receptor (RAR-alpha). *Mol Cell Biol* 10:2154–2163

Cooperberg AA (1967): Acute promyelocytic leukemia. *Can Med Assoc J* 97:57–63

Cordonnier C, Vernant JP, Brun B, Goualt-Heilmann M, Kuentz M, Bierling P, Farcet JP, Rodet M, Duedari N, Imbet M, Jouault H, Mannoni P, Reyes F, Dreyfus B, Rochant H (1985): Acute promyelocytic leukemia in 57 previously untreated patients. *Cancer* 55:18–25

Cornic M, Delva L, Guidez F, Balitrand N, Degos L, Chomienne C (1992): Induction of retinoic acid-binding protein in normal and malignant human myeloid cells by retinoic acid in acute promyelocytic leukemia patients. *Cancer Res* 52:3329–3334

Cozzolino F, Torcia M, Miliani A, Carossino AM, Giordani R, Cinotti S, Filimberti E, Saccardi R, Bernasei P, Guidi G, Di Guglielmo R, Pistoria V, Ferrarini M, Nawroth PP, Stern D (1988): Potential role of interleukin-1 as the trigger for diffuse intravascular coagulation in acute nonlymphoblastic leukemia. *Am J Med* 84:240–250

Cunningham I, Gee TS, Reich LM, Kempin SJ, Naval AN, Clarkson BD (1989): Acute promyelocytic leukemia: treatment results during a decade at Memorial Hospital. *Blood* 73:1116–1122

Daniel MT, Bernheim A, Flandrin G, Berger R (1985): Acute myeloblastic leukemia (M_2) with basophilic cell line involvement and abnormalities of the short arm of chromosome 12(12p). *C R Acad Sci (III)* 301:299–301

Davey FR, Davis RB, MacCallum JM, Nelson DA, Mayer RJ, Ball ED, Griffin JD, Schiffer CA, Bloomfield CD (1989): Morphologic and cytochemical characteristics of acute promyelocytic leukemia. *Am J Hematol* 30:221–227

de Lacerda JF, do Carmo JA, Guerra ML, Geraldes J, de Lacerda JM (1993): Multiple thrombosis in acute promyelocytic leukaemia after tretinoin. *Lancet* 342:114 (letter)

de The H, Marchio A, Tiollais P, Dejean A (1987): A novel steroid thyroid hormone receptor-related gene inappropriately expressed in human hepatocellular carcinoma. *Nature* 330:667–670

de The H, Chomienne C, Lanotte M, Degos L, Dejean A (1990a): The t(15;17) translocation of acute promyelocytic leukemia fuses the retinoic acid receptor alpha gene to a novel transcribed locus. *Nature* 347:558–561

de The H, Vivanco-Ruiz MM, Tiollais P, Stunnenberg H, Dejean A (1990b): Identification of a retinoic acid responsive element in the retinoic acid receptor β gene. *Nature* 343:177–180

Dedhar S, Robertson K, Ceras V (1991): Induction of expression of the a_vb_1 and a_vb_3 integrin heterodimers during retinoic acid-induced neuronal differentiation of murine embryonal carcinoma cells. *J Biol Chem* 266:21846–21857

Delia D, Aiello A, Formelli F, Fontanella E, Costa A, Miyashita T, Reed JC, Pierotti MA (1995): Regulation of apoptosis induced by the retinoid N-(4-hydroxyphenyl) retinamide and effect of deregulated bcl-2. *Blood* 85:359–367

Delva L, Cornic M, Balitrand N, Guidez F, Miclea JM, Delmer A, Teillet F, Fenaux P, Castaigne S, Degos L, Chomienne C (1993): Resistance to all-trans retinoic acid (ATRA) therapy in relapsing acute promyelocytic leukemia: study of in vitro ATRA sensitivity and cellular retinoic acid binding protein levels in leukemic cells. *Blood* 82:2175–2181

DeRossi G, Avvisati G, Coluzzi S, Fenu S, Lo Coco F, Lopez M, Lanni M, Pasqualetti D, Mandelli F (1990): Immunological definition of acute promyelocytic leukemia (FAB M3): a study of 39 cases. *Eur JHaematol* 45:168–171

Dombret H, Sutton L, Duarte M, Daniel MT, Leblond V, Castaigne S, Degos L (1992): Combined therapy with all-trans retinoic acid and high-dose chemotherapy in patients with hyperleukocytic acute promyelocytic leukemia and severe visceral hemorrhage. *Leukemia* 6:1237–1242

Dombret H, Scrobohaci ML, Ghorra P, Zini JM, Daniel MT, Castaigne S, Degos L (1993): Coagulation disorders associated with acute promyelocytic leukemia: corrective effect of all-trans retinoic acid treatment. *Leukemia* 7:2–9

Donati MB, Falanga A, Consonni R, Alessio MG, Bassan R, Buelli M, Borin L, Catani L, Pogliani E, Gugliotta L, Masera G, Barbui T (1990): Cancer procoagulant in acute nonlymphoid leukemia: relationship of enzyme detection to disease activity. *Thromb Haemost* 64:11–16

Donati MB, Gambacorti-Passerini C, Casali B, Falanga A, Vannotti P, Fossati G, Semeraro N, Gordon SG (1986): Cancer procoagulant in human tumor cells: evidence from melanoma patients. *Cancer Res* 46:6471–6474

Douer D, Preston-Martin S, Keung Y-K, Chang E, Watkins K (1994): High occurrence of FAB M3 subtype among acute myelocytic leukemia (AML) patients of Latino origin. *Blood* 84:51a (abstr)

Dubois C, Schlageter MH, de Gentile A, Guidez F, Balitrand N, Toubert ME, Krawice J, Fenaux P, Castaigne S, Najean Y, Degos L, Chomienne C (1994): Hematopoietic growth factor expression and ATRA sensitivity in acute promyelocytic leukemia blast cells. *Blood* 83:3264–3270

Edgington TS, Mackman N, Brand K, Rof W (1991): The structural biology of expression and function of tissue factor. *Thromb Haemost* 66:67–79

Edwards RL, Rickles FR (1992): The role of leukocytes in the activation of blood coagulation. *Semin Hematol* 29:202–212

Elliott S, Taylor K, White S, Rodwell R, Marlton P, Meagherl D, Wiley J, Taylor D, Wright S, Timms P (1992): Proof of differentiative made of action of all-trans retinoic acid in acute promyelocytic leukemia using X-linked clonal analysis. *Blood* 79:1916–1919

Emeis JJ, Kooistra T (1986): Interleukin-1 and lipopolysaccharides induce an inhibitor of plasminogen activator in vivo and in human cultured endothelial cells. *J Exp Med* 163:1260–1266

Escudier S, Kantarjian H, Estey E (1993): Thrombosis in acute promyelocytic leukemia in patients treated with all-trans retinoic acid. *Proc ASCO* 12:310

Evans RM (1988): The steroid and thyroid hormone receptor superfamily. *Science* 240:889–895

Falanga A, Gordon SG (1985): Isolation and characterization of cancer procoagulant A: a cysteine protease from malignant tissue. *Biochem* 24:5558–5567

Falanga A, Alessio MG, Donati MB, Barbui T (1988): A new procoagulant in acute leukemia. *Blood* 71:870–875

Feldman EJ, Arlin ZA, Ahmed T, Mittelman A, Ascensao JL, Puccio CA, Coombe N, Baskind P (1989): Acute promyelocytic leukemia: a 5-year experience with new antileukemic agents and a new approach to preventing fatal hemorrhage. *Acta Haematol* 82:117–121

Fenaux P, Pollet JP, Vandenbossche-Simon L, Morel P, Zandecki M, Jouet JP, Bauters F (1991): Treatment of acute promyelocytic leukemia: a report of 70 cases. *Leuk Lymph* 4:239–248

Fenaux P, Castaigne S, Dombret H, Archimbaud E, Duarte M, Morel P, Lamy T, Tilly H, Guerci A, Maloisel F, Bordessoule D, Sadoun A, Tiberghien P, Fegueux N, Daniel MT, Chomienne C, Degos L (1992): All-trans retinoic acid followed by intensive chemotherapy gives a high complete remission rate and may prolong remissions in a newly diagnosed acute promyelocytic leukemia: a pilot study on 26 cases. *Blood* 80:2176–2181

Fenaux P, Le Deley MC, Castaigne S, Archimbaud E, Chomienne C, Link H, Guerci A, Duarte M, Daniel MT, Bowen D, Huebner G, Bauters F, Fegueux N, Fey M, Sans M, Lowenberg B, Maloisel F, Auzanneau G, Sadoun A, Gardin C, Bastion Y, Ganzer A, Jacky E, Dombret H, Chastang C, Degos L (1993): Effect of all-trans retinoic acid in newly diagnosed acute promyelocytic leukemia. Results of a multicenter randomized trial. *Blood* 82:3241–3249

Flynn PJ, Miller WJ, Weisdorf DJ, Arthur DC, Brunning R, Branda RF (1983): Retinoic acid treatment of acute promyelocytic leukemia: in vitro and in vivo observations. *Blood* 62:1211–1217

Fontana JA, Rogers JS, Durham JP (1986): The role of 13-cis-retinoic acid in the remission induction of a patient with acute promyelocytic leukemia. *Cancer* 57:209–217

Francis JL, el-Baruni K, Roath OS, Taylor I (1988): Factor X-activating activity in normal and malignant colorectal tissue. *Thromb Res* 52:207–217

Francis PA, Rigas JR, Muindi J, Kris MG, Young CW, Warrell RP Jr (1993): Modulation of all-trans retinoic acid (RA) pharmacokinetics by a cytochrome P450 enzyme system. *J Natl Cancer Inst* 85:1921–1926

Frankel SR, Eardley A, Lauwers G, Weiss M, Warrell RP Jr (1992): The retinoic acid syndrome in acute promyelocytic leukemia. *Ann Intern Med* 17:292–296

Frankel SR, Eardley A, Heller G, Berman E, Miller WH, Dmitrovsky E, Warrell RP (1994): All-*trans* retinoic acid for acute promyelocytic leukemia. Results of the New York Study. *Ann Intern Med* 120:278–286

Gallagher R, Li Y-P, Rao S, Paietta E, Andersen J, Etkind P, Bennett J, Tallman MS, Wiernik PH (1995): Characterization of acute promyelocytic leukemia cases with PML-RARα break/fusion sites in PML exon 6: identification of a subgroup with decreased in vitro responsiveness to all-trans-retinoic acid. *Blood* 86:1540–1547

Ganzer A, Seipelt G, Verbeek W, Ottmann OG, Maurer A, Kolbe K, Hess U, Elsner S, Reutzel R, Wormann B, Hiddemann W, Hoelzer D (1994): Effect of combination therapy with all-trans-retinoic acid and recombinant human granulocyte colony-stimulating factor in patients with myelodysplastic syndromes. *Leukemia* 8:369–375

Gianni M, Terao M, Zanotta S, Barbui T, Rambaldi A, Garattini E (1994): Retinoic acid and granulocyte colony-stimulating factor synergistically induce leukocyte alkaline phosphatase in acute promyelocytic leukemia cells. *Blood* 83:1909–1921

Giguere V (1994): Retinoic acid receptors and cellular retinoid binding proteins: complex interplay in retinoid signalling. *Endocrine Rev* 15:61–79

Golomb HM, Rowley J, Vardiman J, Baron J, Locker G, Krasnow S (1976): Partial deletion of long arm of chromosome 17: a specific abnormality in acute promyelocytic leukemia? *Arch Intern Med* 136:825–828

Golomb HM, Rowley JD, Vardiman JW, Testa JR, Butler A (1980): Microgranular acute promyelocytic leukemia: a distinct clinical, ultrastructural and cytogenetic entity. *Blood* 55:253–259

Gordon SG, Franks JJ, Lewis B (1975): Cancer procoagulant A: a factor X-activating procoagulant from malignant tissue. *Thromb Res* 6:127–137

Gordon SG, Hasibu U, Cross BA, Poole MA, Falanga A (1985): Cysteine proteinase coagulant from amnion-chorion. *Blood* 66:1261–1265

Gouault-Heilmann M, Chardon E, Sultan C, Josso F (1975): The procoagulant factor of leukemic promyelocytes: demonstration of immunologic cross-reactivity with human brain tissue factors. *Br J Haematol* 30:151–158

Gralnick HR, Abrell E (1973): Studies of the procoagulant and fibrinolytic activity of promyelocytes in acute promyelocytic leukemia. *Br J Haematol* 24:89–99

Gralnick HR, Sultan C (1975): Acute promyelocytic leukemia: haemorrhagic manifestations and morphologic criteria. *Br J Haematol* 29:373–376

Gralnick HR, Bagley J, Abrell E (1972): Heparin treatment for the hemorrhagic diathesis of acute promyelocytic leukemia. *Am J Med* 52:167–174

Griffin JD, Davis R, Nelson DA, Davey FR, Mayer RJ, Schiffer C, McIntyre OR, Bloomfield CD (1986): Use of surface marker analysis to predict outcome of adult acute myeloblastic leukemia. *Blood* 68:1232–1241

Groopman J, Ellman L (1979): Acute promyelocytic leukemia. *Am J Hematol* 7:359–408

Guidez F, Huang W, Tong JH, Dubois C, Balitrand N, Waxman S, Michaux JL, Martiat P, Degos L, Chen Z, Chomienne C (1994): Poor response to all-trans retinoic acid therapy in a t(11;17) PLZF/RARα patient. *Leukemia* 8:312–317

Hakimian D, Tallman MS, Zugerman C, Caro WA (1993): Erythema nodosum associated with all-trans-retinoic acid in the treatment of acute promyelocytic leukemia. *Leukemia* 7:758–759

Hewlett J, Kopecky KJ, Head D, Eyre HJ, Elias L, Kingsbury L, Balcerzak SP, Dabich L, Hynes H, Bickers JN, Appelbaum FR (1995): A prospective evaluation of the roles of allogeneic marrow transplantation and low-dose monthly maintenance chemotherapy in the treatment of adult acute myelogenous leukemia (AML): a Southwest Oncology Group Study. *Leukemia* 9:562–569.

Hickstein DD, Hickey MJ, Collins SJ (1988): Transcriptional regulation of the leukocyte adherence protein beta subunit during human myeloid cell differentiation. *J Biol Chem* 263:13863–13867

Hirata F, Schiffmann E, Venkatasubramanian K, Salomon D, Axelrod J (1980): a phospholipase A_2 inhibitory protein in rabbit neutrophils induced by glucocorticoids. *Proc Natl Acad Sci USA* 77:2533–2536

Hockenberry DM, Olfavi ZN, Yin X-M, Milliman CL, Korsmeyer SJ (1993): Bcl-2 functions as an antioxidant pathway to prevent apoptosis. *Cell* 75:241–251

Huang ME, Ye YC, Chen SR, Chai JR, Lu JX, Zhoa L, Gu LJ, Wang ZY (1988): Use of all-trans retinoic acid in the treatment of acute promyelocytic leukemia. *Blood* 72:567–572

Ikeda K, Sasaki K, Tasaka T, Nagai M, Kawanishi K, Takahara J, Irino S (1993): Reverse transcription-polymerase chain reaction for PML/RARα fusion transcripts in acute promyelocytic leukemia and its application to minimal residual leukemia detection. *Leukemia* 7:544–548

Imaizumi M, Sato A, Koizumi Y, Inove S, Suzuki H, Suwabe N, Yoshinari M, Ichinohasama R, Endo K, Sauai T, Tadu K (1994): Potentiated maturation with a high proliferating activity of acute promyelocytic leukemia induced in vitro by granulocytic or granulocytic/macrophage colony-stimulating factors in combination with all-trans retinoic acid. *Leukemia* 8:1301–1308

Ishii H, Horie S, Kizaki K, Kazama M (1992): Retinoic acid counteracts both the downregulation of thrombomodulin and the induction of tissue factor in cultured human endothelial cells exposed to tumor necrosis factor. *Blood* 80:2556–2562

Jones ME, Saleem A (1978): Acute promyelocyte leukemia: a review of the literature. *Am J Med* 65:673–677

Jurcic JG, Caron PC, Miller WH, Yao TJ, Maslak P, Finn RD, Larson SM, Warrell RP, Scheinberg DA (1995): Sequential targeted therapy for relapsed acute promyelocytic leukemia with all-trans retinoic acid and anti-CD33 monoclonal antibody M195. *Leukemia* 9:244–248

Kahle LH, Avvisati G, Lampling RJ, Moretti T, Mandelli F, ten Cate JW (1985): Turnover of alpha-2 antiplasmin in patients with acute promyelocytic leukemia. *Scand J Clin Lab Med* 45:75–80

Kanamaru A, Takemoto Y, Tanimoto M, Murakami H, Asou N, Kobayashi T, Kuriyama K, Ohmoto E, Sakamaki H, Tsubaki K, Hiraoka A, Yamada O, Oh H, Saito K, Matsuda S, Minato K, Veda T, Ohno R (1995): All-*trans* retinoic acid for the treatment of newly diagnosed acute promyelocytic leukemia. *Blood* 85:1202–1206

Kane DJ, Sarafian TA, Anton R, Hahn H, Gralla EB, Valentine JS, Ord T, Bredesen DE (1993): Bcl-2 inhibition of neural death: decreased generation of reactive oxygen species. *Science* 262:1274–1277

Kantarjian HM, Keating MJ, McCredie KB, Beran M, Walters R, Dalton WT, Hittelman W, Freireich EJ (1985): A characteristic pattern of leukemic cell differentiation without cytoreduction during remission induction in acute promyelocytic leukemia. *J Clin Oncol* 3:793–798

Kantarjian HM, Keating MJ, Walters RS, Estey EH, McCredie KB, Smith TL, Dalton WT, Cork A, Trujillo J, Freireich EJ (1986): Acute promyelocytic leukemia. M.D. Anderson Hospital experience. *Am J Med* 80:789–797

Keung Y-K, Chen S-C, Groshen S, Douer D, Levine AM (1994): Acute myeloid leukemia subtypes and response to treatment among ethnic minorities in a large US urban hospital. *Acta Haematol* 92:18–22

Kingsley EC, Durie BGM, Garewal HS (1987): Acute promyelocytic leukemia. *West J Med* 140:322–327

Kizaki M, Ikeda Y, Tanosaki R, Nakajima H, Morikawa M, Sakashita A, Koeffler HP (1993): Effects of novel retinoic acid compound, 9-cis-retinoic acid on proliferation, differentiation and expression of retinoic acid receptor-α and retinoic X receptor-α RNA by HL-60 cells. *Blood* 82:3592–3599

Kizaki M, Nakajima H, Mori S, Koike T, Morikawa M, Ohta M, Saito M, Koeffler HP, Ikeda Y (1994): Novel retinoic acid, 9-cis retinoic acid, in combination with all-trans retinoic acid is an effective inducer of differentiation of retinoic acid-resistant HL-60 cells. *Blood* 83:3289–3297

Koike T, Tatewaki W, Aoki A, Yoshimoto H, Yagisawa K, Hashimoto S, Furukawa T, Saitoh H, Takahashi M, Li-Bo Y, Ying W, Shibata A (1992): A brief report: severe symptoms of hyperhistaminemia after the treatment of acute promyelocytic leukemia with tretinoin (all-trans retinoic acid). *N Engl J Med* 327: 385–387

Koyama T, Hirosawa S, Kawamata N, Tohda S, Aoki N (1994): All-trans retinoic acid upregulates thrombomodulin and downregulates tissue-factor expression in acute promyelocytic leukemia cells: distinct expression of thrombomodulin and tissue factor in human leukemic cells. *Blood* 84:3001–1009

Krause JR, Stolc V, Kaplan SS, Penchansky L (1989): Microgranular promyelocytic leukemia: a multiparameterexamination. *Am J Hematol* 30:158–163

Kubota T, Andoh K, Sadakata H, Tanaka H, Kobayashi N (1991): Tissue factor released from leukemic cells. *Throm Haemost* 65:59–63

Kurzrock R, Estey E, Talpaz M (1993): All-trans retinoic acid: tolerance and biologic effects in myelodysplastic syndromes. *J Clin Oncol* 11:1489–1495

Lafage-Pochitaloff M, Alcalay M, Brunel V, Longo L, Sainty D, Simonetti J, Brig F, Pelicci PG (1995): Acute promyelocytic leukemia cases with nonreciprocal PML/RARα or RARα/PML fusion genes. *Blood* 85:1169–1174

Larson R, Kondo K, Vardiman JW, Butler AE, Golomb HM, Rowley JD (1984): Evidence for a 15:17 translocation in every patient with acute promyelocytic leukemia. *Am J Med* 76:827–841

Lefebre P, Thomas G, Gourmel B, Agadir A, Castaigne S, Dreux C, Degos L, Chomienne C (1991): Pharmacokinetics of oral all-trans retinoic acid in patients with acute promyelocytic leukemia. *Leukemia* 5:1054–1058

Leid M, Kastner P, Durand B, Krust A, Leroy P, Lyons R, Mendelssohn C, Nagpal S, Nakshatti H, Reibel C (1993): Retinoic acid signal transduction pathways. *Ann NY Acad Sci* 684:19–34

Leroy P, Nakshatri H, Chambon P (1991): Mouse retinoic and receptor alpha 2 isoform is transcripted from a promotor that contains a retinoic acid response element. *Proc Natl Acad Sci USA* 88:10138–10142

Levin AA, Sturzenbecker LJ, Kazmer S, Bosakowski T, Huselton C, Allenby G, Speck J, Kratzeisen C, Rosenberger M, Lovey A, Grippo JF (1992): 9-Cis retinoic acid stereoisomer binds and activates a nuclear receptor RXR alpha. *Nature* 355:359–361

Licht JD, Chomienne C, Goy A, Chen A, Scott AA, Head DR, Michaux JL, Wu Y, DeBlasio A, Miller WH, Zelentz AD, Willman CL, Chen Z, Chen SJ, Zelent A, Macintyre E, Veil A, Cortes J, Kantarjian H, Waxman S (1995): Clinical and molecular characterization of a rare syndrome of acute promyelocytic leukemia associated with the translocation (11;17). *Blood* 85:1083–1094

Liso V, Trocolli G, Grande M (1975): Cytochemical study of acute promyelocytic leukemia. *Blut* 30:261–268

Lo Coco F, Avvisati G, Diverio D, Biondi A, Pandolfi PG, Alcalay M, De Rossi G, Petti MC, Cantu-Rajnoldi A, Pasqualetti D, Nanni M, Fenu S, Frontani M, Mandelli F (1991a): Rearrangements of the RARα gene in acute promyelocytic leukemia: correlations with morphology and immunophenotype. *Br J Haematol* 78:494–499

Lo Coco F, Avvisati G, Diverio D, Petti MC, Alcalay M, Pandolfi PP, Zangrilli D, Biondi A, Rambaldi A, Moleti ML, Mandelli F, Pelicci PG (1991b): Molecular evaluation of response to all-trans retinoic acid therapy in patients with acute promyelocytic leukemia. *Blood* 77:1657–1659

Lo Coco F, Diverio D, D'Adamo F, Avvisati G, Alimena G, Nanni M, Alcalay M, Pandolfi PP, Pelicci PG (1992): PML/RAR-α rearrangement in acute promyelocytic leukaemias apparently lacking the t(15;17) translocation. *Eur J Haematol* 48:173–176

Malta Corea A, Pacheco Espinoza C, Cant Rajnoldi A, Conter V, Lietti G, Masera G, Sessa C, Cavalli F, Biondi A, Rovelli A (1993): Childhood acute promyelocytic leukemia in Nicaragua. *Ann Oncol* 4:892–894

Martin SJ, Bradley JG, Cotter TG (1990): HL-60 cells induced to differentiate toward neutrophils subsequently die via apoptosis. *Clin Exp Immunol* 79:448–453

Marty M, Ganem G, Fischer J, Flandrin G, Berger R, Schaison G, Degos L, Boiron M (1984): Leucmie aigu promylocytaire: étude rtrospective de 119 malades traits par daunorubicine. *Nouv Rev Fr Hmatol* 26:371–378

Mattei MG, Petkovich M, Mattei JF, Brand N, Chambon P (1988): Mapping of the human retinoic acid receptor to the Q21 band of chromosome 17. *Hum Genet* 80:166–188

Menrad A, Speicher D, Wacker J, Herlyn M (1993): Biochemical and functional characterization of aminopeptidase N expressed by human melanoma cells. *Cancer Res* 53:1450–1455

Mertelsman R, Thaler HT, To L (1980): Morphological classification, response to therapy and survival in 263 adult patients with acute nonlymphoblastic leukemia. *Blood* 56:773–781

Miller VA, Rigas JR, Kris MG, Muindi JRF, Young CW, Warrell RP Jr (1994): Use of liarozole, a novel inhibitor of cytochrome P-450 oxidases, to modulate catabolism of all-trans retinoic acid. *Cancer Genet Cytogenet* 34:522–526

Miller WH Jr, Heller G, Warrell RP Jr (1994): Implications of the PML/RAR-α presentation and detection of PML/RAR-α by reverse transcription polymerase chain reaction (RT-PCR) in remission of acute promyelocytic leukemia patients. *Blood* 84:1501 (abstr)

Miyauchi S, Moroyama T, Kyoizumi S, Asakawa J, Okamoto T, Takada K (1988): Malignant tumor cell lines produce interleukin-1-like factor in vitro. *Cell Dev Biol* 24:753

Moon DK, Greczy CL (1988): Recombinant IFN-γ synergizes with lipopolysaccharide to induce macrophage membrane procoagulants. *J Immunol* 141: 1536–1542

Muindi J, Frankel R, Miller WH Jr, Jakubowski A, Scheinberg DA, Young CW, Dmitrovsky E, Warrell RP (1992a): Continuous treatment with all-trans retinoic acid causes a progressive reduction in plasma drug concentrations: implications for relapse and retinoid "resistance" in patients with acute promyelocytic leukemia. *Blood* 79:299–303

Muindi JRF, Frankel SR, Huselton C, De Grazia F, Garland WA, Young CW, Warrell RP (1992b): Clinical pharmacology of oral all-trans retinoic acid in patients with acute promyelocytic leukemia. *Cancer Res* 52:2138–2142

Nachman RL, Hajjar KA, Silverstein RL, Dinarello CA (1986): Interleukin-1 induces endothelial cell synthesis of plasminogen activator inhibitor. *J Exp Med* 163:1595–1600

Nawroth PP, Handley D, Esmon CT, Stern DM (1986a): Interleukin-1 induces cell surface anticoagulant activity. *Proc Natl Acad Sci USA* 83:3460–3464

Nawroth PP, Stern DM (1986b): Modulation of endothelial cell hemostatic properties by tumor necrosis factor. *J Exp Med* 163:740–745

Nemerson Y (1988): Tissue factor and hemostasis. *Blood* 71:1–8

Nilsson B (1984): Probable in vivo induction of differentiation by retinoic acid of promyelocytes in acute promyelocytic leukemia. *Br J Haematol* 57:365–371

Noguchi M, Sakai T, Kisiel W (1989): Identification and partial purification of a novel tumor-derived protein that induces tissue factor in cultured human endothelial cells. *Biochem Biophys Res Commun* 160:222–227

Ohno R, Naoe T, Hirano M, Kobayashi M, Hirai H, Tubaki K, Oh H (1993a): Treatment of myelodysplastic syndromes with all-trans retinoic acid. *Blood* 81:1152–1154

Ohno R, Yoshida H, Fukutani H, Naoe T, Ohshima T, Kyo T, Endoh N, Fujimoto T, Kobayashi T, Hiraoka A, Mizoguchi H, Kodera Y, Suzuki H, Hirano M, Akiyama H, Aoki N, Shindo H, Yokomaku S (1993b): Multi-institutional study of all-trans-retinoic acid as a differentiation therapy of refractory acute promyelocytic leukemia. *Leukemia* 7:1722–1727

Pandolfi P, Alcalay M, Fagioli M, Zangrilli D, Mencarelli A, Diverio D, Biondi A, Lo Coco F, Rambaldi A, Grignani F, Rochette-Egly C, Gaube M-P, Chambon P, Pelicci PG (1992): Genomic variability and alternative splicing generate multiple PML/RARα transcripts that encode aberrant PML proteins and PML/RARα isoforms in acute promyelocytic leukaemia. *EMBO J* 11:1397–1407

Parker MG, Lowney JF (1965): Acute promyelocytic leukemia. *Missouri Med* 62:374–378

Petkovich M, Brand NJ, Krust A, Chambon P (1987): A human retinoic acid receptor which belongs to the family of nuclear receptors. *Nature* 330:444–450

Petti MC, Avvisati G, Amadori S, Baccarani M, Guarini AR, Papa G, Rosti GA, Tura S, Mandelli F (1987): Acute promyelocytic leukemia: clinical aspects and results of treatment in 62 patients. *Haematologica* 72:151–155

Pittman GR, Senhauser DA, Lowney JF (1966): Acute promyelocytic leukemia. *Am J Clin Pathol* 46:214–220

Polliack A (1971): Acute promyelocytic leukemia with disseminated intravascular coagulation. *Am J Clin Pathol* 56:155–161

Pratt MAC, Kralova J, McBurney MW (1990): A dominant negative mutation of the alpha-retinoic acid receptor gene in a retinoic acid-nonresponsive embryonal carcinoma cell. *Mol Cell Biol* 10:6445–6453

Rand JJ, Moloney WC, Sise HS (1969): Coagulation defects in acute promyelocytic leukemia. *Arch Intern Med* 123:39–47

Rickles FR, Hair G, Schmeizl M, Kwaan H, Lanotte M, Tallman MS (1993): All-trans-retinoic acid (all-trans RA) inhibits the expression of tissue factor in human progranulocytic leukemia cells. *Thromb Haemost* 69:107 (abstr)

Rigas JR, Francis PA, Muindi JR, Kris MG, Huselton C, DeGrazia F, Orazem JP, Young CW, Warrell RP Jr (1993): Constitutive variability in catabolism of the natural retinoid, all-trans retinoic acid, and its modulation by ketaconazole. *J Natl Cancer Inst* 85:1921–1926

Robertson KA, Emani B, Collins SJ (1992): Retinoic acid-resistant HL-60R cells harbor a point mutation in the retinoic acid receptor ligand-binding domain that confers dominant negative activity. *Blood* 80:1885–1889

Rovelli A, Biondi A, Cantu-Rajnoldi A, Conter V, Guidici G, Jankovic M, Locasciulli A, Rizzari C, Romitti L, Rossi MR, Schiro R, Tosi S, Uderzo C, Masera G (1992): Microgranular variant of acute promyelocytic leukemia in children. *J Clin Oncol* 10:1413–1418

Rowley J, Golomb H, Dougherty C (1977): 15/17 translocation: a consistent chromosomal change in acute promyelocytic leukaemia. *Lancet* 1:549–550

Runde V, Aul C, Sudhoff T, Heyll A, Schneider W (1992): Retinoic acid in the treatment of acute promyelocytic leukemia: inefficacy of the 13-cis isomer and induction of complete remission by the all-trans isomer complicated by thrombo-embolic events. *Ann Hematol* 64:270–272

Sakata Y, Murakami T, Noro A, Mori K, Matsuda M (1991): The specific activity of plasminogen activator inhibitor-1 in disseminated intravascular coagulation with acute promyelocytic leukemia. *Blood* 77:1949–1957

San Miguel JF, Gonzales M, Canizo MC, Anta SP, Zola H, Lopez Borrasca A (1986): Surface marker analysis in acute myeloid leukaemia and correlation with FAB classification. *Br J Haematol* 64:547–560

Sanz MA, Jarque I, Martin G, Lorenzo I, Martinez J, Rafecas J, Pastor E, Sayas MJ, Sanz G, Gomis F (1988): Acute promyelocytic leukemia: therapy results and prognostic factors. *Cancer* 61:7–13

Schwartz BS, Williams EC, Conlan MG, Mosher DF (1986): Epsilonaminocaproic acid in the treatment of patients with acute promyelocytic leukemia and acquired alpha-2-plasmin inhibitor deficiency. *Ann Int Med* 105:873–877

Simmers RN, Webber LM, Shannon MF, Garson OM, Wong G, Vadas MA, Sutherland GR (1987): Localization of the G-CSF gene on chromosome 17 proximal to the breakpoint in the t(15;17) in acute promyelocytic leukemia. *Blood* 70:330–332

Smith WC, Nakshatri H, Leroy P, Rees J, Chambon P (1991): A retinoic acid response element is present in the mouse cellular retinol binding protein 1 (mCRBPI) promoter. *EMBO J* 10:2223–2230

Sterrenberg L, Haak HL, Brommer EJP, Nieuwenhuizen W (1985): Evidence of fibrinogen breakdown by leukocyte enzymes in a patient with acute promyelocytic leukemia. *Haemost* 15:126–133

Stone RM, Maguire M, Goldberg MA, Antin JH, Rosenthal DS, Mayer RJ (1988): Complete remission in acute promyelocytic leukemia despite persistence of abnormal bone marrow promyelocytes during induction therapy: experience in 34 patients. *Blood* 71:690–696

Sultan C, Heilmann-Goualt M, Tulliez M (1973): Relationship between blast-cell morphology and occurrence of a syndrome of disseminated intravascular coagulation. *Br J Haematol* 24:255–259

Sun T, Weiss R (1991): Hand-mirror variant of microgranular acute promyelocytic leukemia. *Leukemia* 5:266–269

Takase S, Ong DE, Chytil F (1986): Transfer of retinoic acid from its complex with cellular retinoic acid-binding protein to the nucleus. *Arch Biochem Biophys* 247:328–334

Tallman MS, Kwaan HC (1992): Reassessing the hemostatic disorder associated with acute promyelocytic leukemia. *Blood* 79:543–553

Tallman MS, Hakimian D, Kwaan HC, Rickles FR (1993a): New insights into the pathogenesis of coagulation dysfunction in acute promyelocytic leukemia. *Leuk Lymph* 11:27–36

Tallman MS, Hakimian D, Snower D, Rubin CM, Reisel H, Variakojis D (1993b): Basophilic differentiation in acute promyelocytic leukemia. *Leukemia* 7:521–526

Tan HR, Wages B, Gralnick HR (1972): Ultrastructural studies in acute promyelocytic leukemia. *Blood* 39:628-636

Toh CH, Winfield DA (1992): All-trans retinoic acid and side effects. *Lancet* 339:1239–1240 (letter)

Tohda S, Curtis JE, McCulloch E, Minden MD (1992): Comparison of the effects of all-trans and cis-retinoic acid on the blast stem cells of acute myeloblastic leukemia in culture. *Leukemia* 6:656–661

Tomonaga M, Yoshida Y, Tagawa M, Jinnai I, Kuriyama K, Amenomori T, Yoshioka A, Matsuo T, Nonoka H, Ichimaru M (1985): Cytochemistry of acute promyelocytic leukemia (M3): leukemic promyelocytesexhibit heterogeneous patterns in cellular differentiation. *Blood* 66:350–357

Vahdat L, Maslak P, Miller WH Jr, Eardley A, Heller G, Scheinberg DA, Warrell RP Jr (1994): Early mortality and the retinoic acid syndrome in acute promyelocytic leukemia: impact of leukocytosis, low-dose chemotherapy, PML/RAR-α

isoform, and CD13 expression in patients treated with all-*trans* retinoic acid. *Blood* 84:3843–3849

Van Giezen JJJ, Boon GdIA, Jansen JWCM, Bouma BN (1993): Retinoic acid enhances fibrinolytic activity in vivo by enhancing tissue type plasminogen activator (t-PA) activity and inhibits venous thrombosis. *Thromb Haemost* 69:381–386

Velasco F, Torres A, Andres P, Martinez F, Gomez P (1984): Changes in plasma levels of protease and fibrinolytic inhibitors induced by treatment of acute myeloid leukemia. *Thromb Haemost* 52:81–84

Vivanco Ruiz MM, Bugge TH, Hirschmann P, Bugge TH, Hirschmann P, Stunnenberg HG (1991): Functional characterization of a natural retinoic acid responsive element. *EMBO J* 10:3829–3838

Warrell RP Jr (1993): Retinoid resistance in acute promyelocytic leukemia: new mechanisms, strategies and implications. *Blood* 82:1949–1953 (editorial)

Warrell RP, Frankel SR, Miller WH, Scheinberg DA, Itri LM, Hittelman WN, Vyas R, Andreeff M, Tafuri A, Jakubowski A, Gabrilove J, Gordon MS, Dimitrovsky E (1991): Differentiation therapy of acute promyelocytic leukemia with tretinoin (all-trans-retinoic acid). *N Engl J Med* 374:1385–1393

Weil SC, Rosner BL, Reid MS, Chisholm RL, Lemons RS, Swanson MS, Carrino JJ, Diaz MO, LeBeau MM (1988): Translocation and rearrangement of myeloperoxidase gene in acute promyelocytic leukemia. *Science* 240:790–792

Wiley JS, Jamieson GP, Cebon JS, Woodruff RK, McKendrick JJ, Szer J, Gibson J, Sheridan WP, Biggs JC, Rallings MC (1993): Cytokine priming of acute myeloid leukemia may produce a pulmonary syndrome when associated with a rapid increase in peripheral blood myeloblasts. *Blood* 82:3511–3512 (letter)

Wilson EL, Jacobs P, Dowdle EB (1983): The secretion of plasminogen activators by human myeloid leukemia cells in vitro. *Blood* 61:568–574

Zelent A, Krust A, Petkovich M, Kastner P, Chambon P (1989): Cloning of murine alpha and beta retinoic acid receptors and novel receptor gamma predominantly expressed in skin. *Nature* 339:714–717

Zhang X-K, Lehmann J, Hoffmann B, Dawson MI, Cameron J, Graupner G, Hermann T, Tran P, Pfahl M (1992): Homodimer formation of retinoic X receptor induced by 9-cis retinoic acid. *Nature* 358:587–591

Zhang Z, Tarone G, Turner DC (1993): Expression of integrin a_vb_1 is regulated by nerve growth factor and dexamethasone in PCIZ cells: function consequences for adhesion and neurite outgrowth. *J Biol Chem* 268:5557–5565

Zhao L, Chang KS, Estey E, Hayes K, Deisseroth AB, Liang JC (1995): Detection of residual leukemic cells in patients with acute promyelocytic leukemia by the fluorescence in situ hybridization method: potential for predicting relapse. *Blood* 85:495–499

20

Molecular Biology of Acute Promyelocytic Leukemia

FRANCESCO GRIGNANI AND PIER GIUSEPPE PELICCI

INTRODUCTION

Acute promyelocytic leukemia (APL) is a subtype of human acute myeloid leukemia characterized by a differentiation block at the promyelocytic stage of myeloid differentiation (Bennet et al., 1976). The differentiation block is the most distinctive biological feature of the APL phenotype because the promyelocytic blasts have a modest proliferative rate (Raza et al., 1987, 1992); and the release of the block and terminal differentiation obtained by all-*trans* retinoic acid (ATRA) correlates *in vivo* with complete hematologic remission in the great majority of APL patients (Huang et al., 1988; Castaigne et al., 1990; Chomienne et al., 1990; Warrel et al., 1991; Lo Coco et al., 1992a; see Chapter 19, this volume). This effect of ATRA makes APL the best example of the effectiveness of differentiation therapy. Although attempts to treat other hematologic neoplastic disorders with differentiation therapy have led in the past to disappointing results (Degos et al., 1985; Cheson et al., 1986; Sawyers et al., 1991; Young et al., 1993), this therapeutic strategy now provides a novel approach for the control of neoplastic cell growth.

The clinical response to ATRA is in APL blasts, and it is predicted by rearrangements of the retinoic acid receptor α (RARα) gene as a result of a translocation of chromosomes 15 and 17 [t(15;17)] (Mitelman, 1988) that involves the PML (for Promyelocyte; Goddard et al., 1991; Fagioli et al., 1992) and RARα genes (de Thé et al., 1990; Borrow et al., 1990; Lemmons et al., 1990; Longo et al., 1990; Alcalay et al., 1991). The resulting PML/RARα fusion gene is transcriptionally active in all cases and encodes a PML/RARα fusion protein (Pandolfi et al., 1991; de Thé et al., 1991;

Hormones and Cancer
Wayne V. Vedeckis, Editor

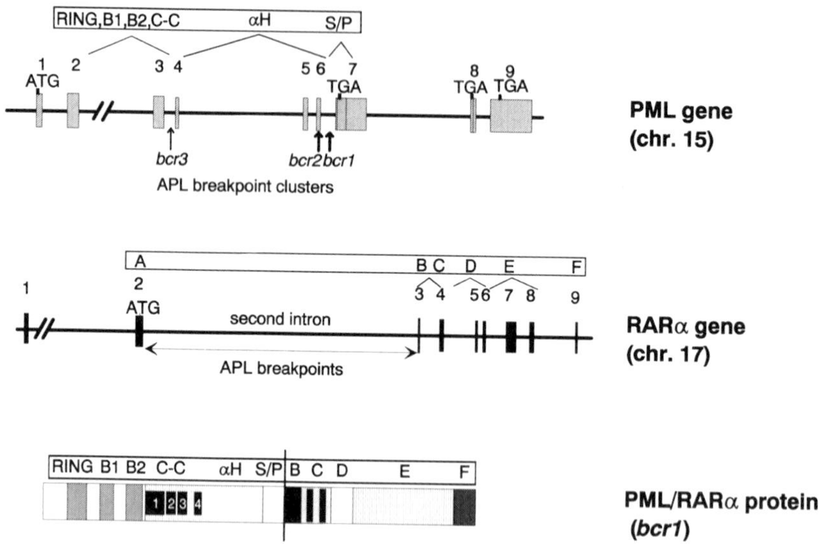

Figure 1. Schematic representation of the position of t(15;17) breakpoints on the RARα and PML genes. **A**: PML gene. The *boxes* represent PML exons. Ring, B1, B2, C–C, αH, and S/P indicate the exons of derivation of the RING finger domain, the B-boxes, the coiled-coil region, the α-helix, and the serine/proline-rich region, respectively (see text). *TGA* indicates the position of the termination codon of the different PML isoforms. The *arrows* indicate the breakpoint clusters regions (*bcr*). **B**: RARα gene. The *boxes* represent RARα exons. *A, B, C, D, E, F* indicate the exons of derivation of the functional regions of RARα. All the APL breakpoints fall into the second intron. **C**: Schematic representation of the functional regions of the PML/RARα protein. The regions derived from PML and RARα are indicated. *1, 2, 3, 4* represent the four heptad clusters that are thought to represent a dimerization interface.

Kakizuka et al., 1991; Kastner et al., 1992) (Fig. 1). Since RARα is involved in the regulation of myeloid differentiation (see below), the alteration in the (retinoic acid) signaling pathway caused by PML/RARα could contribute to the differentiation block and the outgrowth of the APL blasts. Paradoxically, the activation of the same signaling pathway by ATRA induces terminal differentiation and disappearance of the leukemic blasts, indicating APL as the first example of a leukemia that can be treated by specifically targeting therapy to the transforming protein.

15;17 TRANSLOCATION

The t(15;17) is closely associated with the pathogenesis of APL. Cytogenetically, the percentage of cases with the t(15;17) varies from 70–90% and, in 90% of cases, it is the only cytogenetic anomaly (Rowley et al., 1977;

Mitelman, 1988). Molecular analysis, however, has revealed rearrangements of PML and RARα genes in all cases with an apparently normal karyotype so far examined (Lo Coco et al., 1992). In two APL cases with apparently normal karyotypes, fluorescence *in situ* hybridization (FISH) has demonstrated a submicroscopic translocation of PML on chromosome 17 and of RARα on 15 (Lafage-Pochitaloff et al., 1995). Overall, molecular analysis detects the translocation in almost 100% of APL cases (Biondi et al., 1991). The chromosome breakpoints have been mapped by combining chromosome banding and *in situ* hybridization techniques to 15q24 and 17q21 (Mitelman, 1988; Donti et al., 1991). Two fusion genes are formed as a consequence of the translocation: the PML/RARα gene on the recombinant 15q+ chromosome and its reciprocal RARα/PML gene on the recombinant 17q− chromosome (Fig. 1) (de Thé et al., 1990; Borrow et al., 1990; Lemmons et al., 1990; Longo et al., 1990; Alcalay et al., 1991, 1992).

The t(15;17) is often the only chromosomal anomaly seen in the neoplastic metaphases. However, additional karyotypic changes may accompany the t(15;17) and, as in other types of myeloid leukemia, trisomy 8 is the most common (Berger et al., 1991 and our unpublished results).

Expression of the t(15;17) has never been reported in any other neoplasm. Cases of chronic myeloid leukemia in blast crisis with a t(15;17) displayed the promyelocytic phenotype and rearrangements in PML and RARα (Lai et al., 1987; Kadam et al., 1990; Takaku et al., 1993). Interestingly, gene transfer experiments have shown that the expression of the PML/RARα fusion protein induces cell death in most lymphoid and nonhematopoietic cell lines tested (our unpublished results). This suggests that, *in vivo,* a negative selection may take place against nonmyeloid cells developing a 15;17 translocation.

Rare variant translocations in APL involve chromosome 17 with chromosomes other than 15 (Mitelman, 1988). An 11;17 chromosome translocation has recently been cloned from a case of APL. The chromosome 17 breakpoint has been demonstrated to lie within the RARα gene, whereas in chromosome 11 it is located in a previously unknown gene named PLZF, which has structural similarities with PML (Chen SJ et al., 1993; Chen Z et al., 1993). In a t(5;17) the RARα gene is fused to a gene called NPM, which encodes a nucleolar phosphoprotein involved in RNA processing (Redner et al., 1994). The fusion gene produces a NPM/RARα fusion mRNA. In both cases, the recombination leads to the formation of fusion proteins (PLZF/RARα and NPM/RARα).

It would seem that chromosome 17 and RARα rearrangement is indispensable for expression of the promyelocytic phenotype. However, concomitant alterations in a second gene, almost always PML, and occasionally PLZF or NPM, are invariably present. This suggests that RARα is crucial for the pathogenesis of the disease, that the fusion with specific translocation partners activates its leukemogenic potential, and that the translocation partners share common properties.

In the t(15;17) the centromere-telomere orientation of both the PML and RARα loci is $5' \rightarrow 3'$ (Alcalay et al., 1991, 1992). Accordingly, in the PML/RARα chimeric gene, PML and RARα are fused in a head-to-tail configuration and are under the transcriptional control of the PML promoter (Fig. 1). The chimeric PML/RARα gene is transcriptionally active in all cases of APL (Longo et al., 1990; Miller et al., 1990). The fusion transcript can be identified by Northern blotting using either PML or RARα cDNA probes, which both hybridize to transcripts of 4.4 and 3.6 kb in approximately half of APL cases, and 4.0 and 3.2 kb in the remaining cases. The different sizes of the PML/RARα transcripts originate from the variable position of the chromosome 15 breakpoints (see below) and the alternative usage of two RARα polyadenylation sites (Pandolfi et al., 1992). The nucleotide sequence of the PML/RARα fusion transcript shows that it has the potential to encode a PML/RARα fusion protein (de Thé et al., 1991; Kakizuka et al., 1991; Pandolfi et al., 1991, 1992; Kastner et al., 1992).

THE WILD-TYPE RARα PROTEIN

RARα is a member of a retinoic acid receptor family, which also includes RARβ and RARγ (Evans, 1988). A separate retinoid receptor family, the X receptors, includes RXRα, RXRβ, and RXRγ (Mangelsdorf et al., 1990). Only the structural and functional characteristics of the RARα protein that are relevant to the pathogenesis of APL are briefly discussed here. More detailed analyses of the retinoid receptors are found in Chapters 4 and 5.

Physiological Role in Myeloid Differentiation

Retinoids include vitamin A, its natural and synthetic derivatives, and its metabolites. These molecules exert multiple fundamental effects on development, cell differentiation, and growth in vertebrates (Giguere et al., 1987; Tabin, 1991). A number of findings support the notion that RARα plays a key role in regulating cell differentiation. The differentiation response to retinoic acid (RA) is lost in two cell lines with a RARα protein truncated within the retinoid binding region, one derived from the embryo carcinoma cell line P19 (Pratt et al., 1990), the other from the HL-60 myeloid cell line (Robertson et al., 1989). Exogenous expression of the wild-type RARα restores sensitivity to RA in both cell line mutants (Collins et al., 1990; Kruyt et al., 1992), suggesting that RARα is physiologically implicated in the regulation of embryonic and myeloid differentiation.

Further evidence derives from experiments showing that a RARα dominant negative protein expressed in a multipotent hematopoietic cell line blocks myeloid differentiation at the promyelocytic stage (Tsai and Collins, 1993). However, these experiments should be interpreted with caution, since it cannot be excluded that RARα mutations interfere, directly or indirectly, with the function of other members of the nuclear receptor family (Robertson et al., 1992; see below).

Modular Organization of RARα

The biological effects of RARα derive from its ligand dependent transcription factor function. The ability of RARα to bind retinoids and DNA have been mapped to specific regions of the molecule (Fig. 2). The amino acid sequence of RARα can be divided into six regions (A–F) on the basis of their homology with the other members of the nuclear receptor superfamily (Krust et al., 1986; Petkovich et al., 1987). This modular structure coincides with the functional organization of the molecule, and it is relevant to comprehend the properties of the fusion PML/RARα protein, which retains all but the first functional domain.

The *ligand binding activity* has been mapped to region E of RARα (Fig. 2). The receptor binds a variety of different retinoids. All-*trans* RA (ATRA), a vitamin A metabolite, has very high binding affinity. The binding affinity is similarly high for some retinoids (3,4 dideydro ATRA; 9-*cis* RA) and low for others (13-*cis* RA) (Heyman et al., 1992; Levin et al., 1992; Allenby et al., 1993).

The *DNA binding* domain corresponds to region C (Fig. 2). It contains two zinc fingers and directly binds to specific *cis*-acting elements (RA responsive elements, RAREs) in the promoter region of retinoic acid target genes (de Thé et al., 1990; Leroy et al., 1991; Smith et al., 1991). RAREs are polymorphic and contain a direct repetition of two 5′-PuGGTCA-3′ consensus sequences separated by variable spacing (1–5 bp). The spacing, the integrity of the two motifs, and the surrounding bases are critical for the efficiency of RAR binding and transactivation (Leroy et al., 1991; Umesono et al., 1991; Vivanco-Ruiz et al., 1991), and may be one of the properties that are modified by the fusion with PML.

Stable binding of RARα to the RARE takes places only if RXR is present. RARα forms heterodimers with RXRs that are more stable and bind DNA more efficiently than RARα homodimers (Kliewer et al., 1991; Yu et al., 1991; Bugge et al., 1992; Leid et al., 1992; Marks et al., 1992; Zhang et al., 1992). The *dimerization properties* of RARα have been mapped to the E region, but significant contribution is provided by the second zinc finger of the C domain (Zechel et al., 1994a,b). Both these regions are retained in the PML/RARα protein, and they confer to it the ability to form complexes with RXRs (Fig. 2).

The *nuclear localization* of RARα is diffuse in the nucleoplasm and seems to be determined by region D. The fusion with PML alters significantly the intracellular localization of the PML/RARα protein with respect to both the normal RARα and PML proteins (see below).

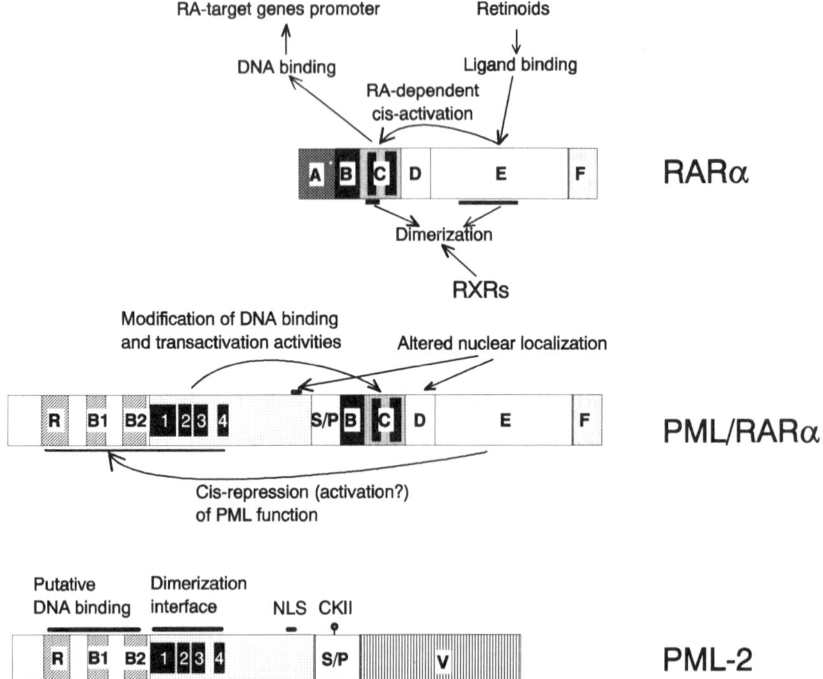

Figure 2. Modular organization and functions of the PML, RARα and PML/RARα proteins. Only the isoform 2 of PML is shown. See legend to Fig. 1 for symbols. *NLS*, nuclear localization signal; *CKII*, Casein kinase phosphorylation consensus. *V*, variable C-termini. The functions of the different regions of PML and RARα are indicated. For PML/RARα, regions are indicated for additional functions acquired by the fusion protein.

Finally, the A/B region, which is partially lost in the APL translocation, has *transcriptional activation* function (Nagpal et al., 1992).

PML PROTEIN

The biological function of the PML protein has yet to be understood. Nevertheless, a number of studies have shed some light on its pattern of expression, intracellular localization, dimerization ability, and biological activity.

The PML locus comprises 9 coding exons (see Fig. 1). The mature PML transcripts result from the alternative assemblage of exons, portions of exons or retained introns, which produce 13 separate transcripts encoding an equal number of PML isoforms (Goddard et al., 1991; Fagioli et al., 1992). Variable amounts of the PML transcripts have been identified in histologically different human cell lines (Fagioli et al., 1992).

Modular Organization of the PML Protein

The first clues to the identification of PML function have derived from its homology with other proteins. The PML protein consists of regions with distinct putative functional relevance (Fig. 2).

The N-terminal part of the PML protein includes three cysteine-histidine-rich regions. The first region consists of a motif referred to as *RING finger*, which defines a large family of proteins whose functions, where known, often involve regulation of gene expression and DNA binding (Freemont et al., 1991; Goddard et al., 1991; Reddy et al., 1992; Lovering et al., 1993). The DNA binding activity of the RING motif is suggested by the fact that a RING peptide binds zinc with a tetrahedral coordination to the cysteines and binds to DNA in a zinc-dependent fashion (Lovering et al., 1993). The other two cysteine-rich regions are referred to as *the B1 and B2 boxes* (Fig. 2). They are shared by a more restricted group of proteins whose functions include: (1) regulation of gene expression, (e.g., RPT-1, which regulates the expression of the IL-2 receptor gene); (2) regulation of development (e.g., the Xenopus XNF-7 gene); (3) repair of UV-damaged DNA (e.g. Rad18); and (4) DNA recombination (e.g., RAG-1). Interestingly, other genes of this subfamily are involved in tumorigenesis. T18 and Rfp are transforming proteins that are generated by the fusion of a RING protein and the B-Raf and Ret proteins, respectively, whereas Bmi-1 cooperates with Myc in lymphoma development (Lovering et al., 1993, and references therein).

C-terminal to the cysteine-hystidine clusters is an *α-helical region* (see Figs. 1, 2). The first portion of the α-helix can assume a coiled-coil configuration and contains four clusters of heptad repeats with hydrophobic amino acids at the first, fourth, and eighth positions (Forman et al., 1990; Perez et al., 1993). This domain is considered to be a dimerization interface (Forman et al., 1990) by analogy to similar repeats of thyroid hormone receptor (TR), RAR, and vitamin D receptor (VDR). Coimmunoprecipitation experiments have demonstrated that the heptad repeat clusters are responsible for both homodimerization and heterodimerization with PML/RARα (Perez et al., 1993, and our unpublished results, see below).

The ring finger-like region, the coiled-coil regions, and a portion of the α-helix region are retained in all the different PML isoforms and in the PML/RARα fusion protein, suggesting their critical functional role (Fagioli et al., 1992).

Carboxy-terminal to the α-helix is a *serine- and proline-rich region.* It contains several recognition sequences for serine/threonine kinase and casein kinase II (CKII) (Fagioli et al., 1992, and our unpublished results) (see Fig. 2). Phosphorylation by CKII is a post-translational modification, often associated with modifications in the biological activity of transcription factors (Hunter et al., 1992). The CKII site is retained in all PML isoforms (Fagioli et al., 1992). Four alternative C-termini have been identified.

Expression and Subcellular Localization of the PML Protein

PML expression is widely distributed in normal tissues. Immunohistological analysis (Flenghi et al., 1995) has demonstrated that the PML protein shows preferential expression in differentiated postmitotic cells. Endothelial and epithelial cells and tissue macrophages are the cell types with the highest amounts of PML protein. The highest levels of expression are found in the activated macrophages from toxoplasmosis, T-cell lymphomas, and Hodgkin's disease. PML protein expression is also increased during differentiation of the monoblastic cell line, U937, or after its activation by interferon-γ treatment (Koken et al., 1994; Flenghi et al., 1995). In the normal human bone marrow, PML appears to be preferentially expressed in the myeloid lineage (Koken et al., 1994); although low levels of expression could be detected in lymphoid cells. Overall, these data suggest that the levels of expression of PML are inversely related with the proliferative status of the cells and that PML may play a role in maturation and activation of specific cell types, for example, macrophages (Flenghi et al., 1995).

In the last few years, the intracellular distribution of the PML protein has aroused a great interest because its alteration in APL blasts has been regarded as the cellular mechanism of leukemogenesis. In addition, immunocytologic studies on the PML protein have allowed the identification of novel subnuclear structures. By immunofluorescence, immunohistochemistry, and immunoelectronmicroscopy, several groups have shown that the PML protein has both a cytoplasmic and nuclear distribution (Kastner et al., 1992; Daniel et al., 1993; Perez et al., 1993; Dyck et al., 1994; Everett and Maul, 1994; Koken et al., 1994; Maul and Everett, 1994; Weis et al., 1994; Flengi et al., 1995). The region of the PML protein that is responsible for nuclear localization has been mapped within the α-helix (Flenghi et al., 1995; Fig. 2). Some physiological isoforms of PML are exclusively or predominantly localized in the cytoplasm when overexpressed by gene transfer in human cells, and they probably make an important contribution to the cytoplasmic component of the protein (Flenghi et al., 1995).

The nuclear component of PML appears to be prevalent and has a typical distribution in discrete nuclear subdomains, referred to as POD (PML Oncogenic Domain) or nuclear bodies (Dyck et al., 1994; Koken et al., 1994; Maul and Everett, 1994; Weis et al., 1994). This distribution is responsible for the characteristic speckled pattern in immunohistochemical and immunofluorescence studies (Fig. 3). The POD/nuclear body has also been studied at the ultrastructural level, and appears to be formed by a dense fibrillar capsule, where PML is localized, and a finely tubular core (Dyck et al., 1994; Koken et al., 1994; Weis et al., 1994). The region of the PML protein responsible for the speckled distribution has been mapped within the RING finger (Kastner et al,. 1992; Perez et al., 1993). PML colocalizes in the nuclear bodies with other proteins defined by their speckled

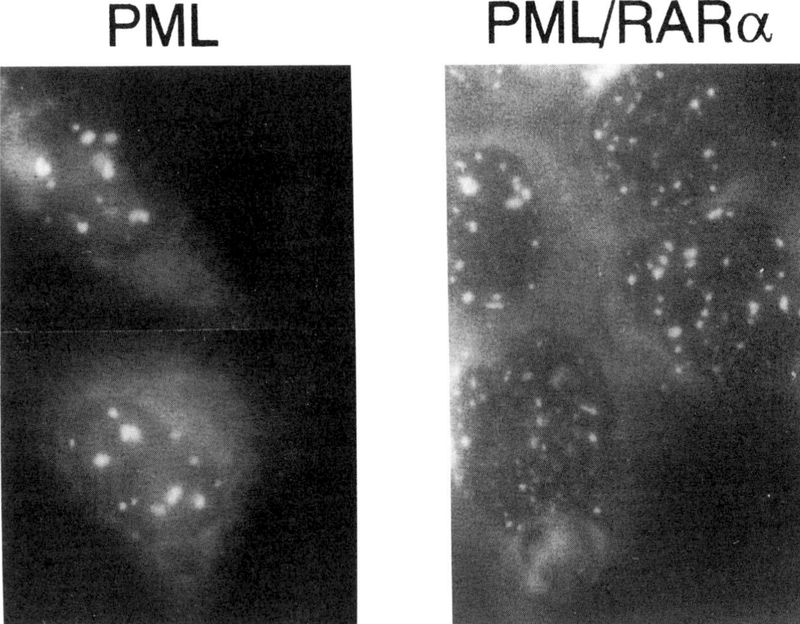

Figure 3. Intranuclear distribution of the PML and PML/RARα proteins. Indirect immunofluorescence on cells overexpressing PML or PML/RARα. A monoclonal antibody directed against the N-terminal region of PML was used in these experiments.

intranuclear distribution and with reactivity of monoclonal antibodies raised against nuclear antigens or with sera from patients affected by autoimmune diseases. These proteins include the autoantigen SP100, identified in patients with biliary cirrhosis, and two undefined proteins of 55 and 65 kDa (Dyck et al., 1994; Koken et al., 1994; Weis et al., 1994). As a whole, these proteins constitute a macromolecular complex that cofractionate with the nuclear matrix. The interaction among the components of the POD domain appears to be functionally stringent because overexpression of one of the proteins causes enlargement of the subnuclear domain and increased immunologic labeling of the other antigens. However, coimmunoprecipitation studies have failed to show a physical interaction between PML and SP100 (Koken et al., 1994).

The increasing interest in the nuclear body structure is because, in APL cells, this subnuclear domain appears to be disrupted by the expression of the mutant PML/RARα protein (see below). In addition, it has been shown that the nuclear body is also altered during infection of the cell by herpes simplex virus type 1 (HSV). The HSV immediate-early protein, Vmw110, required for efficient viral gene expression, localizes in the nuclear bodies in the early phases of infection. Subsequently, the nuclear body structure becomes undetectable by immunofluorescence staining with antibodies directed against PML or other nuclear body antigens (Maul et al., 1993, 1994; Everett and Maul, 1994). Both in APL and during viral infection, Western blot analysis has shown that the expression of nuclear body proteins is unchanged, indicating that they are simply redistributed within the cell. Overall, these observations suggest that the nuclear body and its constituent proteins, including PML, could be physiologically involved in the control of viral infection or, alternatively, in the control of the cellular functions that may influence viral infection susceptibility, such as proliferation and differentiation.

PML does not co-localize with other intranuclear macromolecular structures, that also have a speckled distribution within the nucleus (Dyck et al., 1994; Weis et al., 1994). These nuclear subdomains are sites of DNA replication and RNA processing and are defined by complexes of splicing factors, ribonucleoproteins, and RNA. The coexistence of several discrete structures of diverse function suggests a previously unrecognized elaborate functional organization within the nucleus.

Biological Activity

The biological function of PML in the normal cell remains unknown. PML has been regarded as a growth regulator because, when overexpressed in rat embryo fibroblasts (REF) transformed by H-*ras* and p53 or H-*ras* and c-*myc*, it reduces the malignant behavior of the cells. In addition, overexpression of PML in the APL cell line, NB4, suppresses its clonogenicity and tumorigenicity in nude mice, suggesting that overexpression of PML may antagonize the activity of PML/RARα (Mu et al., 1994). It has been hypothesized that the functional mechanism of this effect could be the negative regulation of the activity of the promoters of specific genes, such as the multidrug resistance gene or the EGF receptor gene. The fusion with RARα may disrupt the growth controlling activity of the normal PML protein because PML/RARα has no effect on the process of transformation of REF by activated oncogenes and on the activity of the same promoters (Mu et al., 1994).

PML/RARα PROTEIN AND ITS ROLE IN THE PATHOGENESIS OF APL

The PML/RARα protein retains specific domains of the RARα and PML proteins, whose functional properties are altered in the context of the fusion. Since the chromosome 17 break of the 15;17 translocation always occurs in RARα intron 2, the RARα component of the PML/RARα protein is the same in all cases of APL and corresponds to regions B to F (Chen SJ et al., 1991; Chen Z et al., 1991; Diverio et al., 1992; Tong et al., 1992). PML/RARα should retain the DNA binding, ligand binding, dimerization, and nuclear localization properties of RARα. On the other hand, the diverse molecular breakpoints within the PML gene can provide us with a directive of the functional regions of PML that are necessarily incorporated in the PML/RARα fusion protein to generate a leukemic phenotype (Chen SJ et al., 1992; Pandolfi et al., 1992) (see Fig. 2).

In large series of APL cases the chromosome 15 breakpoint of the t(15;17) has been mapped in three regions of the PML locus. In 90% of the cases, it is equally distributed between intron 6 (breakpoint cluster region 1; *bcr1*) and intron 3 (*bcr3*) (see Fig. 1). In the remaining 10% cases, it is located within exon 6 (*bcr2*) (Diverio et al., 1992; Pandolfi et al., 1992). In *bcrl* or *bcr3* cases, the 5′ portion of a PML intron fuses with the 3′ portion of RARα intron 2. In the fusion transcript, the chimeric intron is spliced out and the PML and RARα open reading frames become aligned. In *bcr2* a cryptic donor site of the PML exon 6 and the physiological RARα intron 2 acceptor site take part in the assemblage (Pandolfi et al., 1992). Despite the variability in the breakpoint, PML/RARα genes encoding a fusion PML/RARα protein are consistently selected by the leukemic cells. The PML region retained in all PML/RARα fusion genes encodes the putative DNA binding domain and part of the α-helix, including the four clusters of heptad repeats, that is, the putative DNA binding domain and the dimerization interface (see Fig. 2). The region of the PML gene encoding variable portions of the C-terminal region of the α-helix, the serine proline-rich region (including the CKII phosphorylation site), and the variable C-termini are contained in the other fusion product of the t(15;17), the RARα/PML gene (Alcalay et al., 1992).

Biological Effects of PML/RARα on Differentiation and Growth of Hematopoietic Precursors

The expression of the PML/RARα protein and the outburst of APL coincide, suggesting that PML/RARα is responsible for the transformed phenotype in APL. However, technical difficulties have so far prevented the expression of the fusion protein into normal hematopoietic precursors, which would provide direct proof of the leukemogenic potential of PML/RARα. Indirect indications have been obtained from the expression of PML/RARα in hematopoietic precursor cell lines. These cell lines can be used as a model system for hematopoietic differentiation, because they can be induced to mature upon exposure to various physiological and synthetic molecules. The biological activities of PML/RARα in this assay recapitulate critical features of the promyelocytic leukemia phenotype (Grignani et al., 1993).

Effects of PML/RARα on Differentiation. Vitamin D_3, or combined vitamin D_3 and TGF-β1 (TGF), treatment induces terminal monocytic differentiation of the U937 promonocytic cell line (Metcalf, 1989, Testa et al., 1993). U937 cells that express the PML/RARα protein fail to terminally differentiate, as shown by both surface marker analysis and functional tests (proliferation and phagocytosis). Differentiation is blocked when the PML/RARα expression levels are higher than those of the normal RARα protein, as occur in the APL blasts (Grignani et al., 1993). This situation suggests a dominant negative mechanism of action of PML/RARα, based on competition for the molecular pathway or by direct stoichiometrical interaction with dimerization partners. When exposed to low ATRA concentrations, comparable to physiological levels, (Muindi et al., 1992) cells containing PML/RARα do not differentiate (Grignani et al., 1993), whereas the control cells undergo limited but measurable differentiation. The block of differentiation is not restricted to the effects of Vitamin D_3 or ATRA in myelomonocytic differentiation. Exogenous expression of PML/RARα also inhibits erythroid differentiation induced by hemin in K562 cells. Since a similar effect is seen when RARα is overexpressed (Grignani et al., 1995); it appears that PML/RARα affects the activity of RARα target genes involved in the early phases or in the basic processes of differentiation.

The effect that PML/RARα exerts on differentiation is reversed when the cells expressing the fusion protein are exposed to high ATRA concentrations, comparable to the peak plasma levels achieved during ATRA therapy of APL patients (Muindi et al.,1992). The percentage of cells that enter the differentiation program is higher in the cell population where PML/RARα is expressed (Grignani et al., 1993; Rousselot et al., 1994). Thus, a high concentration of ATRA converts the activity exerted by the fusion protein from an RA-independent inhibition to a ligand-dependent stimulation of differentiation (Grignani et al., 1993). This effect is not only due to a loss of the inhibition of differentiation exerted by PML/RARα in the presence of low doses of ATRA, but also to a direct positive effect, as shown by the fact that cells that normally would not differentiate now enter the maturation cascade. The RA-dependent differentiation activity dominates over the block of vitamin D_3-induced differentiation, as shown by the fact that exposure of the APL-derived NB4 cell line or of U937 cells expressing PML/RARα to ATRA releases the block of vitamin D_3-induced differentiation. In these conditions, the maturation level achieved by the cells exposed to both agents is more advanced than that obtainable with vitamin D_3 or ATRA alone, indicating that, once the activity of PML/RARα is reversed, the two inducers have a synergistic effect (Testa et al., 1994). The PML/RARα fusion protein, in summary, could alone be responsible for two major features of the APL phenotype: the block of differentiation and the high sensitivity to RA.

Nothing is known about the differentiation stage for the target cell of PML/RARα As for other myeloid leukemias, one would expect that the target cell is a myeloid precursor. However, the APL blasts have the morphology of a relatively mature myeloid precursor, suggesting that the APL target cell is the promyelocyte itself. Either the fusion protein is generated in precursors that are already committed, or PML/RARα does not inhibit commitment, acting rather on maturation of myeloid precursors. It has recently been shown that PML/RARα expressing erythroid precursors can be identified from the bone marrow of APL patients (Takatsuki et al., 1994). Moreover, PML/RARα expressing U937 cells treated with vitamin D_3 not only lose their capacity to mature but, to a large extent, also their ability to proliferate (Grignani et al., 1994; our unpublished results). This suggests that PML/RARα allows commitment of the cells accompanied by a loss of self renewal capacity, but blocks maturation. It appears, therefore, that the translocation event occurs as early as in the erythromyeloid stem cell, but the block induced by PML/RARα occurs at a more differentiated stage.

In the presence of ATRA, on the contrary, a higher percentage of cells enter the differentiation process, but they do not complete maturation, suggesting that PML/RARα acts by increasing the frequency of commitment events (Grignani et al., 1994; our unpublished results).

Effects on Cell Growth. The effects of PML/RARα are not restricted to cell differentiation. U937 cells expressing the PML/RARα protein do not undergo programmed cell death in conditions of serum deprivation that induce apoptosis in control cells (Grignani et al., 1993). In addition, in the GM-CSF dependent cell line TF-1, where growth conditions are more strictly defined, survival in the absence of GM-CSF growth factor is prolonged by the expression of PML/RARα (Rogaia et al., 1995). Whereas control cells become irreversibly committed to apoptosis after a short GM-CSF deprivation, PML/RARα-expressing cells can regain competence to proliferate after as long as 20 days of growth factor deprivation. Thus, it seems that the PML/RARα protein promotes cell survival by inhibiting the commitment of cells to enter the genetic program of cell death.

In vivo labeling experiments have shown that APL blasts have a moderate proliferation rate (Raza et al., 1987, 1992). By prolonging cell survival, PML/RARα could support the expansion of a population of leukemic cells with a low proliferative index. This effect, combined with the differentiation block, provides a cellular mechanism to account for the oncogenic potential of PML/RARα.

Functional Properties of the PML/RARα Protein

Homo- and Heterodimerization of the PML/RARα Protein. Macromolecular Nuclear Complexes. The ability of the PML/RARα protein to dimerize with various partners is currently considered one important mechanism by which the fusion protein exerts its leukemogenic effect.

HPLC analysis of RA-binding proteins in fresh APL blasts and in the APL cell line, NB4, have demonstrated the presence of APL-specific, high molecular weight, nuclear protein complexes of 600 and 1200 kDa (the apparent molecular weight of PML/RARα is 110 kDa) (Nervi et al., 1992). These complexes contain RARα, PML, PML/RARα, and RXR (our unpublished results), and they may contain other nuclear proteins. The relationship of these complexes with nuclear bodies is unknown.

Experiments using gel retardation, co-immunoprecipitation, and protein interaction in the yeast two-hybrid system have shown that PML/RARα can form homodimers, heterodimers with PML, and heterodimers with RXR (see Fig. 2) (Kastner et al., 1992; Daniel et al., 1993; Perez et al., 1993; Dyck et al., 1994; Koken et al., 1994; Weis et al., 1994). PML/RARα homodimers bind to RAREs in the absence of RXR and with a different affinity or specificity than RAR/RXR heterodimers (see below, Perez et al., 1993); thus interfering directly with the normal RARα molecular pathway. Since PML/RARα expression is high in APL (Pandolfi et al., 1992) dimerization of PML/RARα with RXR could also influence indirectly the activity of other nuclear receptors like RARs, TR, and VDR, which physiologically dimerize with RXR, by sequestering this common dimerization partner (Kliewer et al., 1991; Yu et al., 1991; Bugge et al., 1992; Leid et al., 1992; Marks et al., 1992; Zhang et al., 1992). Accordingly, an excess of PML/RARα prevents both VDR binding to the VDRE *in vitro* and activation of a reporter gene by VDR (Perez et al., 1993).

Experiments of co-immunoprecipitation and dimerization on DNA-responsive elements have shown that PML/RARα homodimerizes and heterodimerizes with PML through the common heptad cluster dimerization domain of the coiled-coil region (see Fig. 2) (Perez et al., 1993). These interactions can conceivably alter PML function in two distinct ways. On the one hand, it is unlikely that the PML/RARα-PML heterodimers and PML/RARα homodimers perform the same biochemical functions as wild-type PML homodimers; it is therefore probable that PML/RARα interferes with PML by competing with the normal protein for the same molecular targets. On the other hand, the nuclear localization of PML/RARα homodimers and PML/RARα-PML heterodimers in APL cells is different from that of PML homodimers (see below). Through these multiple molecular interactions, PML/RARα can form multiple nuclear complexes.

TRANSACTIVATING ACTIVITY. The competence of PML/RARα to function as an RA-inducible transcription factor has been studied in co-transfection experiments on different target cells (Cos-1, HeLa, HepG2, HL-60) (de Thé et al., 1991; Kakizuka et al., 1991; Pandolfi et al., 1991; Kastner et al., 1992). The ability of the fusion protein to regulate transcription was tested on specific RA-responsive target genes (TRE-TK, RARα2, RARβ2, CRABPII, CRBPI) containing a RARE, a minimum promoter region, and a reporter gene. The results demonstrated that PML/RARα and RARα have different transactivating properties, but the effects on transcription of both PML/RARα and RARα are strongly cell type-specific. These effects can be either stimulatory or inhibitory, depending on the promoter and cell type used, either in the presence or absence of RA. However, PML/RARα is consistently more active than RARα. The high variability of the results, depending on the experimental setting, makes it difficult to correlate these data to the mechanisms of leukemogenesis because cultured promyelocytes and the RARα target genes regulating myeloid differentiation are not known.

The molecular mechanism for the different behavior of the fusion protein is unknown. The A region of RARα is lost in PML/RARα. Since the A/B region contains a transcription-activation function with promoter and cell type-dependent activity (Nagpal et al., 1992), the transcription regulation activity of the fusion protein may be altered. However, a RARα mutant that lacks the A region does not have the same activity on transcription as PML/RARα (de Thé et al., 1991; Kakizuka et al., 1991). Alternatively, it could be hypothesized that the presence of a transcriptional regulation function in PML may influence the functions of RARα (Baniahmad et al., 1992). Yet, variable portions of PML fused to the GAL4 DNA binding domain did not reveal any transactivating function of PML on GAL4 reporter gene activity (our unpublished results).

DNA Binding. The DNA binding properties of PML/RARα have been compared with those of RARα by gel retardation experiments using synthetic RAREs. Different direct repeats separated by spacers of different length were used as DNA binding sites. The data demonstrate that PML/RARα homodimers and PML/RARα-RXR heterodimers bind the responsive elements with different efficiency and specificity than the physiological heterodimers, RARα-RXR (Perez et al., 1993). On the contrary, *in vitro* experiments have demonstrated that PML/RARα-RXR complexes partially overlap with the RAR-RXR dimers in terms of binding activity to RAREs (Perez et al., 1993). In the presence of RXR, PML/RARα may bind RAREs through one RARα and one RXR DNA binding domain, as in the RAR/RXR heterodimers, and not through two RARα DNA binding domains, as in the PML/RARα homodimers (Perez et al., 1993). These observation could be the biochemical counterpart of the altered performance of the PML/RARα protein in transactivation experiments, and they suggest that (1) the fusion protein could regulate a different set of genes with respect to the wild type RARα protein; and (2) the two proteins have a different activity on the genes regulated by RARα. However, the genetic program that is regulated by RARα upon triggering of myeloid differentiation remains to be established.

Retinoid Binding. Scatchard binding experiments on Cos-1 cells transfected with PML/RARα or RARα expression vectors have shown that the binding affinity and specificity of PML/RARα for RA and various retinoids are similar to that of RARα (Nervi et al., 1992).

Intracellular Localization of the PML/RARα Protein. The intracellular localization of PML/RARα differs from that of PML and RARα. PML/RARα expression has been studied in transiently transfected cell lines, in stably transfected cell lines, in the APL cell line, NB4, and in APL blasts (Kastner et al., 1992; Daniel et al., 1993; Perez et al., 1993; Dyck et al., 1994; Everett and Maul, 1994; Koken et al., 1994; Maul and Everett, 1994; Weis et al., 1994; Flenghi et al., 1995). Immunofluorescence microscopy, immunohistochemistry, immunoelectronmicroscopy, and cell fractionation studies have revealed that PML/RARα is partially cytoplasmic, with a distribution variably described as diffuse (Koken et al., 1994) or perinuclear (Dyck et al., 1994). The cytoplasmic component would be difficult to detect because of its fine distribution as opposed to the nuclear fraction that is localized in defined domains. The nuclear component has a micropunctated nuclear pattern, with a large number of small dots, distinguished from the speckled nuclear pattern of PML and from the finely dispersed pattern of RARα (see Fig. 3) (Gaub et al., 1989). Anti-PML antibodies reveal that APL cells, which express all three proteins, display the PML/RARα-like micropunctated nuclear pattern, indicating that the fusion protein localization dominates over the two wild-type proteins. It has been demonstrated that PML co-localizes with PML/RARα in the nucleus of APL blasts; the PML protein is, accordingly, displaced from the POD-nuclear body structure. Both immunofluorescence and immunoelectronmicroscopic studies have shown that the POD-nuclear body structure is totally disrupted in APL cells, and that the PML/RARα and the PML proteins are localized in small particles with poor structural organization, tightly bound to chromatin (Daniel et al., 1993; Perez et al., 1993; Dyck et al., 1994; Koken et al., 1994; Weis et al., 1994). In addition, at least part of the SP100 protein and of the RXR cellular pools are redistributed to the small PML/RARα containing particles (Dyck et al., 1994). This observation suggests that one of the mechanisms of action of PML/RARα is the displacement and sequestration of RXR outside its normal intranuclear location. Strikingly, the treatment of APL blasts with RA converts the micropunctated nuclear pattern to the speckled, PML-like, pattern (see below).

Mechanisms of PML/RARα Activity on Differentiation. The experimental evidence accumulated on PML/RARα function and cellular distribution are still far from complete in clarifying the molecular mechanism through which the fusion protein participates in the differentiation block of APL blasts. PML/RARα may act as a dominant negative factor in the PML molecular pathway; however, the normal activity of PML on differentiation is unknown. The only leukemogenic activities of PML/RARα that can be hypothesized are those exerted on the molecular pathway of RARα

and other steroid receptors. In the presence of low, near-physiological, concentrations of RA (Muindi et al., 1992) *in vitro*, PML/RARα acts as a transcriptional repressor of certain RA-target genes (de Thé et al., 1991; Kakizuka et al., 1991; Pandolfi et al., 1991; Kastner et al., 1992). Furthermore, PML/RARα may also function as a vitamin D_3 signaling antagonist by acting directly on vitamin D_3-target genes or, indirectly, by sequestering RXR, a co-factor essential for VDR activity. As both RA and vitamin D_3 are implicated in myeloid differentiation (Reitsman et al., 1983); these effects may contribute to the inhibition of differentiation by PML/RARα. This view has received support from the ultrastructural observations on the PML/RARα-dependent RXR displacement in APL cells (Weis et al., 1994). However, PML/RARα seems to be able to inhibit myeloid differentiation induced by the phorbol ester TPA (our unpublished results), and erythroid differentiation induced by hemin in K562 cells (Grignani et al., 1995). These facts suggest that the activity of PML/RARα is not limited to RXR sequestration and may be directed toward the regulation of fundamental, differentiation-inducing master genes.

Mechanism of RA Sensitivity of APL Blasts. APL undergoes remission when patients are treated with ATRA. This result correlates with the induction of terminal differentiation of APL blasts (Nilson et al., 1984; Lo Coco et al., 1991). Indeed, APL blasts differentiate in the presence of ATRA *in vitro,* and the clinical response can be predicted by *in vitro* differentiation tests (Huang et al., 1987; Chomienne et al., 1990). The differentiation activity of RA, as opposed to a cytotoxic mechanism, is also indicated by the absence of bone marrow aplasia during treatment; in addition, in the bloodstream of treated patients it is possible to detect the presence of cells having the morphological maturation stages intermediate between promyelocytes and neutrophils, which carry PML and RARα rearrangements (Lo Coco et al., 1991b; Warrel et al., 1993).

The presence of the PML/RARα protein seems to be strictly necessary for RA sensitivity of leukemic blasts. First, there is a strict correlation between the expression of the PML/RARα fusion transcript and response to treatment (Miller et al., 1992). Second, PML/RARα expression increases the sensitivity to RA in myeloid precursors *in vitro* (Grignani et al., 1993). Third, clones of the APL NB4 cell line become ATRA-resistant when they lose PML/RARα expression (Dermime et al., 1994). Taken together, these experimental results seem to indicate an active role for PML/RARα in causing the RA sensitivity of APL.

The fusion protein retains both the RARα DNA and retinoid binding domains and, consequently, it could directly influence the RARα-dependent endogenous pathway that controls terminal myeloid differentiation. In addition, PML/RARα possesses the ability to act on RAREs as a homodimer, independently from the availability of RXR. Transactivation experiments have demonstrated that, under certain experimental conditions,

PML/RARα overstimulates the expression of RARα target genes when RA is present (see above). This PML/RARα function might explain RA sensitivity of APL blasts. Nevertheless, it should be kept in mind that the interaction between RA and the RA-binding domain of PML/RARα could *cis*-activate the PML portion of the fusion protein (Grignani et al., 1994). Alternatively, RA may exert its effects simply by the release of a dominant negative action of PML/RARα over PML, RARα, and/or other unidentified functional proteins. It has been recently shown that RA causes a redistribution of PML/RARα, PML, and RXR proteins within the nucleus (Daniel et al., 1993; Dyck et al., 1994; Koken et al., 1994; Weis et al., 1994). The dispersed and poorly structured PML/RARα-containing particles identified by immunoelectron microscopic studies in APL blasts are converted by RA to the physiological discrete nuclear body structure, defined by the presence of PML, SP100, and other unidentified proteins of 65 and 55 kDa (see above). During this redistribution, RXR also reassumes its normal diffuse location and is likely to be available again as a co-factor for the various steroid receptors (Dyck et al., 1994). In this view, the role played by PML/RARα in generating RA sensitivity of APL blasts would be less "active." The pathogenesis of APL would depend on the altered subnuclear organization and disruption of the POD-nuclear bodies, due to PML/RARα, and RA would release the differentiation block by modifying the PML/RARα molecule, reconstituting the POD-nuclear bodies and restoring a physiological distribution of important functional proteins.

It has been also suggested that, since PML/RARα is localized, at least in part, in the cytoplasm (Kastner et al., 1992; Daniel et al., 1993; Dyck et al., 1994; Koken et al., 1994; Weis et al., 1994); it may constitute a barrier for RA that can be bypassed by high concentrations of RA.

Despite the accumulating experimental evidence indicating a role for PML/RARα in the sensitivity to RA of APL blasts, it still cannot be ruled out that differentiation under the action of RA could, instead, be an intrinsic property of promyelocytes and, therefore, also of leukemic promyelocytes. The promyelocytes that accumulate in the bone marrow in congenital agranulocytosis are, in fact, sensitive to RA (Hassan et al., 1990). The translocation products would cause RA sensitivity only because they produce a differentiation block at the promyelocytic stage.

OTHER TRANSLOCATION PRODUCTS: ABERRANT PML AND RARα/PML PROTEINS

PML/RARα is not the only product of the t(15;17). Other mRNAs have been identified in APLs with the t(15;17) that have the potential to encode abnormal proteins and, in turn, could contribute to promyelocytic leukemogenesis. The chimeric PML/RARα gene encodes an aberrant PML protein by alternatively spliced PML/RARα transcripts, in which the longest open reading frames (ORFs) of PML and RARα are not aligned. In these transcripts a stop codon is found just 3′ to the PML/RARα junction. The open reading frame encodes a PML protein truncated at the C-terminal. This protein possesses the putative DNA binding and the dimerization domains of PML, and these can potentially interact with PML protein dimerization partners and DNA target sequences. This aberrant PML transcript is present in 100% APL cases (Pandolfi et al., 1992).

The RARα/PML fusion gene is expressed as a mRNA in 70–80% APL cases (Alcalay et al., 1992). This transcript encodes a RARα/PML fusion protein (Alcalay et al., 1992; Chang et al., 1992; Lafage-Pochitaloff et al., 1995). Being the reciprocal of PML/RARα, RARα/PML contains only the RARα A domain, and a variable portion of the PML α-helix tract, phosphorylation sites and variable C-termini. On the basis of these structural characteristics, it difficult to predict whether RARα/PML has a biological function because most of the identified functional regions of PML and RARα are not included in the protein. Immunostaining of cells that overexpress the RARα/PML protein with anti-RARα antiserum directed against the A domain has demonstrated that RARα/PML is a nuclear protein (Kastner et al., 1992).

In summary, the t(15;17) has the potential to generate several abnormal mRNAs that encode three abnormal proteins, PML/RARα, RARα/PML, and aberrant PML, that are apparently selected by the leukemic phenotype. In addition, the t(15;17), unlike other leukemia-associated chromosomal lesions, has never been documented in preleukemic syndromes (Mitelman, 1988); also, the genetic lesions associated with the multistep transformation processes that are common to other types of myeloblastic leukemia (additional chromosome abnormalities, *ras* and p53 mutations) have not been detected in APL (Longo et al., 1993). Taken together, these data favor the intriguing hypothesis that t(15;17), being able to encode three potentially transforming proteins, recapitulates the multistep carcinogenesis and is sufficient to fully transform the APL target cells.

VARIANT TRANSLOCATIONS IN APL

In rare cases APL blasts do not carry the t(15;17). Two other translocations, t(11;17) and t(5;17), have been identified in these cases (Chen SJ et al., 1993; Chen Z et al., 1993; Redner et al., 1994).

Cloning of the t(11;17) breakpoint has allowed the identification of a novel gene called PLZF (promyelocytic leukemia zinc finger) that is fused by the chromosomal rearrangement to RARα. As in the t(15;17), two reciprocal fusion genes are formed, both transcriptionally active, PLZF/RARα and RARα/PLZF. The ORF of the PLZF mRNA encodes a protein with nine zinc finger motifs, related to the motifs of the Drosophila transcription factor, Krüppel. Two zinc fingers are incorporated in the PLZF/RARα

protein and seven are contained in the reciprocal RARα/PLZF protein. The portion of RARα enclosed in the PLZF/RARα protein is the same as in PML/RARα, that is, it includes the DNA and ligand binding domains (Chen SJ et al., 1993; Chen Z et al., 1993). Both abnormal translocation products, therefore, contain potentially relevant functional domains.

The function of normal PLZF is unknown. The gene is expressed preferentially in myeloid cells in normal bone marrow (Chen Z et al., 1993), but its expression has been recently reported in several nonhematopoietic tissues (Li et al., 1994). The mRNA levels seem to be down-regulated in differentiated NB4 or HL-60 myeloid cell lines. From the structural point of view, PLZF has the potential to act as a transcription factor, but a direct proof of this function is still lacking. Indirect evidence is provided by the fact that the PLZF DNA-binding domain is able to bind a specific TA-rich DNA sequence when expressed in the context of a glutathione S-transferase fusion protein and immobilized on glutathione-coated agarose beads (Li et al., 1994).

The PLZF protein accommodates a region that bears homology with a recently described domain contained in several other zinc finger-based transcription factors (Bardwell and Treisman, 1994). This domain, referred to as POZ (from poxvirus zinc finger), is able to mediate protein-protein interaction and has the function of a transcriptional repressor. This activity is probably due to inhibition of DNA binding through a *cis*-acting mechanism. In the PLZF/RARα protein the POZ domain of PLZF is fused with most of the RARα protein (Chen Z et al., 1993). It can be predicted, from the biochemical activity of the POZ domain, that the fusion protein is inactive as a transcription factor, and that it can actually function as a dominant negative factor on the normal RARα protein by competing for retinoid binding and for dimerization partners like RXR (Bardwell and Treisman, 1994). This hypothesis would provide a pathogenetic model for APLs containing PLZF/RARα. Consistent with this view, APL cases carrying t(11;17) are not sensitive to the therapeutic effect of retinoic acid (Guidez et al.,1994), and PLZF/RARα is able to repress the transactivation activity of RARα in co-transfection experiments (Chen Z et al., 1994)

Finally, molecular cloning of the t(5;17) breakpoint has shown that in this translocation a portion of RARα functionally identical to that of the PML/RARα and PLZF/RARα genes is fused to a gene called NPM, or nucleophosmine, a nucleolar phosphoprotein that is involved in RNA processing (Redner et al., 1994). A NPM/RARα fusion mRNA is produced by this fusion gene. The pathogenetic activity of this latter fusion protein has still to be investigated.

CONCLUDING REMARKS

The occurrence of different translocations, all involving the same portion of the RARα gene but fused with different partners to generate fusion proteins, in cases of leukemia with the same phenotype is of extreme interest. This suggests that the RARα portion of the fusion molecules is indispensable to block myeloid differentiation at the promyelocytic stage, and that the different partners of RARα may provide a common functional activity to the fusion protein. It should be kept in mind that a mutated RARα protein with dominant negative activity on the normal protein also blocks the maturation of a multipotent hematopoietic precursor cell line at the promyelocytic stage (Tsai and Collins, 1993). In a similar fashion, the different translocations may simply create dominant negative RARα mutants.

Nevertheless, a number of questions remain unresolved. Why is PML involved in the pathogenesis of APL so much more frequently than other genes? What is the common feature of RARα partners? What roles do PML and PLZF have as putative transcription factors? Interestingly, the NPM gene can also become transforming when recombined with a gene different from RARα In the t(2;5) of anaplastic lymphomas, the NPM and the ALK genes fuse to produce a NPM/ALK fusion gene and protein. Does this fusion protein have the same mechanism of action as does NPM/RARα ? Does ALK have the same role in lymphoid differentiation as RARα in the myeloid system?

Most of these question are more easily answered assuming, as a working hypothesis, that the diverse proteins that are fused with RARα in the various APL translocations have a specific role. These questions will be the subject of intense investigation in the near future.

ACKNOWLEDGEMENTS

This research was supported by grants from the "Associazione italiana per la ricerca sul cancro" (A.I.R.C.), the Italian Council for Research (ACRO project), and the European Community (Biomed and Biotech).

REFERENCES

Alcalay M, Zangrilli D, Pandolfi PP, Longo L, Mencarelli A, Giacomucci A, Rocchi M, Biondi A, Rambaldi A, Lo Coco F, Diverio D, Donti E, Grignani F, Pelicci PG (1991): Translocation breakpoint of acute promyelocytic leukemia lies within the retinoic acid receptor α locus. *Proc Natl Acad Sci USA* 88:1977–1981

Alcalay M, Zangrilli D, Fagioli M, Pandolfi PP, Mencarelli A, Lo Coco F, Biondi A, Grignani F, Pelicci PG (1992): Expression pattern of the RARα-PML fusion gene in acute promyelocytic leukemia. *Proc Natl Acad Sci USA* 89:4840–4844

Allenby G, Bocquel MT, Saunders M, Kazmer S, Speck J, Rosenberger M, Lovey A, Kastner P, Grippo JF, Chambon P, Levin A (1993): Retinoic acid receptors (RARs) and retinoid X receptors (RXRs): interactions with endogenous retinoic acids. *Proc Natl Acad Sci USA* 90:30–34

Baniahmad A, Köhne AC, Renkawitz R (1992): A transferable silencing domain is present in the thyroid hormone receptor, in v-erbA oncogene product and in the retinoic acid receptor. *EMBO J* 11:1015–1023

Bardwell VJ, Treisman R (1994): The POZ domain: a conserved protein–protein interaction motif. *Genes Dev* 8:1664–1677

Bennet JM, Catovski D, Daniel MT, Flandrin G, Galton DAG, Gralnick HR, Sultan C (1976): Proposals for the classification of the acute leukemias. *Br J Haematol* 33:451–458

Berger R, Le Coniat M, Derré J, Vecchione D, Jonveaux P (1991): Cytogenetic studies in acute promyelocytic leukemia. *Genes Chrom Cancer* 3:332–337

Biondi A, Rambaldi A, Alcalay M, Pandolfi PP, Lo Coco F, Diverio D, Rossi V, Mencarelli A, Longo L, Zangrilli D, Masera G, Barbui T, Mandelli F, Grignani F, Pelicci PG (1991): RARα gene rearrangements as a genetic marker for diagnosis and monitoring in acute promyelocytic leukaemia. *Blood* 77:1418–1422

Borrow J, Goddard AD, Sheer D, Solomon E (1990): Molecular analysis of acute promyelocytic leukemia breakpoint cluster region on chromosome 17. *Science* 249:1577–1580

Borrow J, Goddard AD, Gibbons B, Katz F, Swirsky D, Fioretos T, Dube I, Winfield DA, Kingston J, Hagemeijer A, Solomon E (1992): Diagnosis of acute promyelocytic leukaemia by RT-PCR: detection of PML-RARα and RARα/PML fusion transcripts. *Br J Haematol* 82:529–540

Bugge TH, Pohl J, Lonnoy O, Stunnenberg HG (1992): RXRα, a promiscuous partner of retinoic acid and thyroid hormone receptors. *EMBO J* 11:1409–1418

Castaigne S, Chomienne C, Daniel MT, Berger R, Fenaux P, Degos L (1990): All-*trans* retinoic acid as a differentiating therapy for acute promyelocytic leukemia I. Clinical results. *Blood* 76:1704–1709

Chang KS, Trujillo JM, Ogura T, Castiglione CM, Kidd KK, Zhao SR, Freireich EJ, Stass SA (1991): Rearrangement of the retinoic acid receptor gene in acute promyelocytic leukemia. *Leukemia* 5:200–205

Chang KS, Stass SA, Chu DT, Deaven LL, Trujillo JM, Freireieh EJ (1992): Characterization of a fusion cDNA (RARα/myl) transcribed from the t(15;17) translocation breakpoint in acute promyelocytic leukemia. *Mol Cell Biol* 12:800–810

Chen SJ, Zhu YJ, Tong JH, Dong S, Huang W, Chen Y, Xiang WM, Zhang L, Song Li X, Qian GQ, Wang ZY, Chen Z, Larsen CJ, Berger R (1991): Rearrangements in the second intron of the RARα gene are present in a large majority of patients with acute promyelocytic leukemia and are used as molecular marker for retinoic acid induced leukemic cell differentiation. *Blood* 78:2696–2701

Chen SJ, Chen Z, Chen A, Tong JH, Dong S, Wang ZY, Waxman S, Zelent A (1992): Occurrence of distinct PML-RAR-α fusion gene isoforms in patients with acute promyelocytic leukemia detected by reverse transcriptase/polymerase chain reaction. *Oncogene* 7:1223–1232

Chen SJ, Zelent A, Tong JH, Yu HQ, Wang Z-Y, Derre J, Berger R, Waxman S, Chen Z (1993): Rearrangements of the retinoic acid receptor alpha and promyelocytic leukemia zinc finger genes resulting from t(11;17)(q23;q21) in a patient with acute promyelocytic leukemia. *J Clin Invest* 91:2260–2267

Chen Z, Chen SJ, Tong JH, Dong S, Huang W, Chen Y, Xiang WM, Zhang L, Song X, Qian XS, Wang ZY, Chen Z, Larsen CJ, Berger R (1991): The retinoic acid α receptor gene is frequently disrupted in its 5′ part in Chinese patients with acute promyelocytic leukemia. *Leukemia* 5:288–292

Chen Z, Brand NJ, Chen A, Chen SJ, Tong JH, Wang ZY, Waxman S, Zelent A (1993): Fusion between a novel *Kruppel*-like finger gene and the retinoic acid receptor-α locus due to a variant t(11;17) translocation associated with acute promyelocytic leukemia. *EMBO J* 12:1161–1167

Chen Z, Guidez F, Rousselot P, Agadir A, Chen S-J, Wang Z-Y, Degos L, Zelent A, Waxman S, Chomienne C (1994): PLZF-RARα fusion proteins generated from the variant t(11;17)(q23;q21) translocation in acute promyelocytic leukemia inhibit ligand-dependent transactivation of wild-type retinoic acid receptors. *Proc Natl Acad Sci* 91:1178–1182

Cheson BD, Jasperse DM, Simon R, Friedman MA (1986): A critical appraisal of low dose cytosine arabinoside in patients with acute non-lymphocytic leukemia and myelodysplastic syndrome. *J Clin Oncol* 4:1857–1864

Chomienne C, Ballerini P, Balitrand N, Daniel MT, Fenaux P, Castaigne S, Degos L (1990): All-trans retinoic acid in acute promyelocytic leukemias. II. In vitro studies: structure-function relationship. *Blood* 76:1710–1717

Collins SJ, Robertson K, Mueller (1990): Retinoic acid induced granulocytic differentiation of HL60 myeloid leukemia cells is mediated directly through the retinoic acid receptor (RAR-α). *Mol Cell Biol* 10:2154–2161

Daniel MT, Koken M, Romagne O, Barbey S, Bazarbachi A, Stadler M, Guillemin MC, Degos L, Chomienne C, de Thé H (1993): PML protein expression in hematopoietic and acute promyelocytic leukemia cells. *Blood,* 82: 1858–1867

de Thè H, Chomienne C, Lanotte M, Degos L, Dejean A (1990): The t(15;17) translocation of acute promyelocytic leukaemia fuses the retinoic acid receptor α gene to a novel transcribed locus. *Nature* 347:558–561,

de Thè H, Lavau C, Marehio A, Chomienne C, Degos L, Dejean A (1991): The PML-RARα fusion mRNA generated by the t(15;17) translocation in acute promyelocytic leukemia encodes a functionally altered RAR. *Cell* 66:675–684

Degos L, Castaigne S, Tilly H, Sigaux F, Daniel MT (1985): Treatment of leukemia with low dose ara-C. A study of 159 cases. *Hematol-Bluttrans* 29: 56–59

Degos L, Chomienne C, Daniel MT, Berger R, Dombret H, Fenaux P, Castaigne S (1990): Treatment of first relapse in acute promyelocytic leukemia with all-trans retinoic acid. *Lancet* 336:1440–1441

Dermime S, Grignani Fr, Clerici M, Nervi C, Sozzi G, Talamo GP, Marchesi E, Formelli F, Parmiani G, Pelicci PG, Gambacorti-Passerini C (1993): Occurrence of resistance to retinoic acid in the acute promyelocytic leukemia cell line NB306 is associated with altered expression of the PML/RARα protein. *Blood* 82: 1573–1577

Diverio D, Lo Coco F, D'Adamo F, Biondi A, Fagioli M, Grignani Fr, Rambaldi A, Rossi V, Avvisati G, Petti MC, Testi AM, Liso V, Specchia G, Fioritoni G, Recchia A, Frassoni F, Ciolli S, Pelicci PG (1992): Identification of DNA rearrangements at the RARα locus in all patients with acute promyelocytic leukemia (APL) and mapping of APL breakpoints within the RARα second intron. *Blood* 79:3331–3336

Donti E, Longo L, Pelicci PG (1991): Chromosomal localization of the APL t(15;17) breakpoints by molecular cytogenetic analysis. *Cancer Genet Cytogenet* 54: 265–266

Dyck JA, Maul GG, Miller WH Jr, Chen JD, Kakizuka A, Evans RM (1994): A novel macromolecular structure is a target of the promyelocyte-retinoic acid receptor oncoprotein. *Cell* 76:333–343

Evans RM (1988): The steroid and thyroid hormone receptor superfamily. *Science* 240:889–895

Everett RD, Maul (1994): HSV-1 IE protein Vmw110 causes redistribution of PML. *EMBO J.* 13:5062–5069

Fagioli M, Alcalay M, Pandolfi PP, Venturini L, Mencarelli A, Simeone A, Acampora D, Grignani F, Pelicci PG (1992): Alternative splicing of PML transcripts predicts coexpression of several carboxy-terminally different protein isoforms. *Oncogene* 7:1083–1091

Flenghi L, Fagioli M, Tomassoni L, Pileri S, Gambacorta M, Pacini R, Grignani Fr, Casini T, Ferrucci PF, Martelli MF, Pelicci PG, Falini B (1995): Characterization of a new monoclonal antibody (PG-M3) directed against the aminoterminal portion of the PML gene product: immunocytochemical evidence for high expression of PML proteins on activated macrophages, endothelial cells, and epithelia. *Blood 85,* 7: 1871–1880

Forman BM, Samuels HH (1990): Interaction among a subfamily of nuclear hormone receptors: The regulatory zipper model. *Mol Endocrinol* 4:1293–1301

Freemont PS, Hanson IM, Trowsdale J (1991): A novel cysteine-rich sequence motif. *Cell* 64: 483–484

Gaub MP, Lutz Y, Ruberte E, Petkovich M, Brand N, Chambon P (1989): Antibodies specific to the retinoic acid human nuclear receptors α and β. *Proc Natl Acad Sci USA* 86:3089–3094

Giguere V, Ong ES, Segui P, Evans RM (1987): Identification of a receptor for the morphogen retinoic acid. *Nature* 330:624–629

Goddard AD, Borrow J, Freemont P, Solomon E (1991): Characterization of a zinc finger gene disrupted by the t(15;17) in acute promyelocytic leukemia. *Science* 254:1371–1374

Grignani FR, Ferrucci PF, Testa U, Talamo G, Fagioli M, Alcalay M, Mencarelli A, Grignani F, Peschle C, Nicoletti I, Pelicci PG (1993): The acute promyelocytic leukaemia specific PML/RARα fusion protein inhibits differentiation and promotes survival of myeloid precursor cells. *Cell* 74: 423–431

Grignani FR, Fagioli M, Alcalay M, Longo L, Pandolfi PP, Donti E, Biondi A, Lo Coco F, Grignani F, Pelicci PG (1994): Acute promyelocytic leukemia: from genetics to treatment. *Blood* 83:10–25

Grignani FR, Testa U, Fagioli M, Barberi T, Masciulli R, Mariani G, Peschle C, Pelicci PG (1995): Promyelocytic leukemia-specific PML-retinoic acid α receptor fusion protein interferes with erythroid differentiation of human erythroleukemia K562 cells. *Cancer Res* 55:440–443

Guidez F, Huang W, Tong J-H, Bubois C, Balitrand N, Waxman S, Michaux JL, Martiat P, Degos L, Chen Z, Chomienne C (1994): Poor response to All-trans retinoic acid therapy in a t(11;17) PLZF/RARα patient. *Leukemia* 8:312–317

Hassan HT, Pearson EC, Rees JK (1990): Retinoic acid induces granulocytic differentiation of myeloid progenitors in congenital agranulocytosis bone marrow cells. *Cell Biol Intl Rep* 14:247–250

Heyman RA, Mangelsdorf DJ, Dyck JA, Stein RB, Eichele G, Evans RM, Thaller C (1992): 9-cis retinoic acid is a high affinity ligand for the retinoid X receptor. *Cell* 68:397–406

Huang M, Yu-Chen Y, Shu-Rong C, Chai J, Lin Z, Long J, Wang Z (1988): Use of all-trans retinoic acid in the treatment of acute promyelocytic leukemia. *Blood* 72:567–572

Huang ME, Ye YC, Chen SR, Zhao JC, Gu LJ, Cai JR, Zhan L, Xie JX, Shen ZX, Wang ZY (1987): All-trans retinoic acid with or without low dose cytosine arabinoside in acute promyelocytic leukemia: report of 6 cases. *Chin Med J (Engl)* 100:949–953

Hunter T, Karin M (1992): The regulation of transcription by phosphorylation. *Cell* 70:375–387

Kadam PR, Merchant AA, Advani SH (1990): Cytogenetic findings in patients with acute promyelocytic leukemia and case of CML blast crisis with promyelocytic proliferation. *Cancer Genet Cytogenet* 50:109–117

Kakizuka A, Miller WH Jr, Umesono K, Warrel RP Jr, Frankel SR, Murty VVVS, Dmitrovsky E, Evans RM (1991): Chromosomal translocation t(15;17) in human acute promyelocytic leukemia fuses RARα with a novel putative transcription factor, PML. *Cell* 66:663–674

Kastner P, Perez A, Lutz Y, Rochette-Egly C, Gaub MP, Durand B, Lanotte M, Berger R, Chambon P (1992): Structure, localization and transcriptional properties of two classes of retinoic acid receptor a fusion proteins in acute promyelocytic leukemia (APL): structural similarities with a new family of oncoproteins. *EMBO J* 11:629–642

Kliewer SA, Umesono K, Mangelsdorf DJ, Evans RM (1991): Retinoid X receptor interacts with nuclear receptors in retinoic acid, thyroid hormone and vitamin D_3 signalling. *Nature* 355:446–449

Koken MHM, Puvio-Dutilleul F, Guillemin MC, Viron A, Cruz-Linares G, Stuurman N, de Jong L, Szostecki C, Calvo F, Chomienne C, Degos L, Puvion E, de Thè H (1994): The t(15;17) translocation alters a nuclear body in a retinoic acid-reversible fashion. *EMBO J* 13:1073–1083

Krust A, Green S, Argos P, Kumar V, Walter P, Bornet JM, Chambon P (1986): The chicken oestrogen receptor sequence: homology with v-erbA and the human oestrogen and gluococorticoid receptors. *EMBO J* 5:891–897

Kruyt FAE, van der Verr LJ, Mader S, van den Brink CE, Feijen A, Jonk LJC, Kruijer W, van der Saag PT (1992): Retinoic acid resistance of the variant embryonal carcinoma cell line RAC65 is caused by expression of a truncated RARα. *Differentiation* 49:27–37

Lafage-Pochitaloff M, Alcalay M, Brunel V, Longo L, Sainty D, Simonetti J, Birg F, Pelicci PG (1995): Acute promyelocytic leukemia cases with nonreciprocal PML/RARα or RARα/PML fusion genes. *Blood* 85:1169–1174

Laï JL, Fenaux P, Zandecki M, Savary JB, Estienne MH, Jouet JP, Bauters F, Deminatti M (1987): Promyelocytic blast crisis of Philadelphia-positive thrombocythemia with translocations (9;22) and (15;17). *Cancer Genet Cytogenet* 29:311–3314

Leid M, Kastner P, Lyons R, Nakshatri H, Saunders M, Zacharewsky T, Chen JY, Staub A, Garnier JM, Mader S,Chambon P (1992): Purification, cloning and RXR identity of the HeLa cell factor with which RAR or TR heterodimerizes to bind target sequences efficiently. *Cell* 68:377–395

Lemmons RS, Eilender D, Waldmann RA, Rebentisch M, Frej AK, Ledbetter DM, Willmann C, McConnell T, O'Connell P (1990): Cloning and characterization of the t(15;17) translocation breakpoint region in acute promyelocytic leukemia. *Genes Chrom Cancer* 2:79–87

Leroy P, Nakshatri H, Chambon P (1991): Mouse retinoic acid receptor α2 isoform is transcribed from a promoter that contains a retinoic acid response element. *Proc Natl Acad Sci USA* 88:10138–10142

Levin AA, Sturzenbecker LJ, Kazmer S, Bosakowski T, Huselton C, Allenby G, Speck K, Kratzeisen C, Rosenberger, M, Lovely A, Grippo JF (1992): 9-cis retinoic acid stereoisomer binds and activates the nuclear receptor RXRα. *Nature* 355:359–361

Li J-Y, English MA, Bisht S, Waxman S, Licht JD, Brookdale (1994): DNA binding and transcription regulation by the promyelocytic leukemia zinc finger protein. *Blood* 84:152 Supplement 1, Thirty-Sixth Annual Meeting of the American Society of Hematology

Lo Coco F, Avvisati G, Diverio D, Biondi A, Pandolfi PP, Alcalay M, De Rossi G, Petti MC, Cantù-Rajnoldi A, Pasqualetti D, Nanni M, Fenu S, Frontani M, Mandelli F (1991a): Rearrangements of the RARα gene in acute promyelocytic leukemia: correlations with morphology and immunophenotype. *Br J Haematol* 78:494–499

Lo Coco F, Avvisati G, Diverio D, Petti MC, Alcalay M, Pandolfi PP, Zangrilli D, Biondi A, Rambaldi A, Moleti ML, Mandelli F, Pelicci PG (1991b): Molecular evaluation of response to all-trans-retinoic acid therapy in patients with acute promyelocytic leukemia. *Blood* 77:1657–1659

Lo Coco F, Diverio D, D'Adamo F, Avvisati G, Alimena G, Nanni M, Alcalay M, Pandolfi PP, Pelicci PG (1992a): PML/RARα rearrangement in acute promyelocytic leukaemias apparently lacking the t(l5;17) translocation. *Eur J Haematol* 48:173–176

Lo Coco F, Diverio D, Pandolfi PP, Biondi A, Rossi V, Avvisati G, Rambaldi A, Arcese W, Petti MC, Meloni G, Mandelli F, Grignani F, Masera G, Barbui T, Pelicci PG (1992b): Molecular evaluation of residual disease as a predictor of relapse in acute promyelocytic leukaemia. *Lancet* 340:1437–1438

Longo L, Pandolfi PP, Biondi A, Rambaldi A, Mencarelli A, Lo Coco F, Diverio D, Pegoraro L, Avanzi G, Tabilio A, Zangrilli D, Alcalay M, Donti E, Grignani F, Pelicci PG (1990): Rearrangements and aberrant expression of the retinoic acid receptor α gene in acute promyelocytic leukemia. *J Exp Med* 172:1571–1575

Longo L, Trecca D, Biondi A, Lo Coco F, Grignani F, Maiolo AT, Pelicci PG, Neri A (1993): Frequency of RAS and p53 mutations in acute promyelocytic leukemias: *Leukemia Lymphoma* 11:405–410

Lovering R, Hanson IM, Borden KLB, Martin S, O'Reilly NJ, Evan GI, Rahman D, Pappin DJC, Trowsdale J, Freemont P (1993): Identification and preliminary characterization of a protein motif related to the zinc finger. *Proc Natl Acad Sci USA* 90: 2112–2116

Mangelsdorf DJ, Ong ES, Dyck JA, Evans RM (1990): Nuclear receptors that identifies a novel retinoic acid response pathway. *Nature* 345:224–229

Marks MS, Hallenbeck PL, Nagata T, Segars JH, Appella E, Nikodem NM, Ozato K (1992): H-2RIIBP (RXRß) heterodimerization provides a mechanism for combinatorial diversity in the regulation of retinoic acid and thyroid hormone responsive genes. *EMBO J* 11:1419–1435

Maul GG, Everett RD (1994): The nuclear location of PML, a cellular member of the C3HC4 zinc-binding domain protein family, is rearranged during herpes simplex virus infection by the C3HC4 viral protein ICP0. *J Gen Virol* 75:1223–1233

Maul GG, Guldner HH, Spivack JG (1993): Modification of discrete nuclear domains induced by herpes simplex virus type 1 immediate early gene 1 product (ICP0). *J Gen Virol* 74:2679–2690

Metcalf D (1989): The molecular control of cell division, differentiation commitment and maturation in haemopoetic cells. *Nature* 339:27–30

Miller WH Jr, Kakizuka A, Frankel SR, Warrel RP Jr, DeBlasio A, Levine K, Evans RM, Dmitrovsky E (1992): Reverse transcription polymerase chain reaction for the rearranged retinoic acid receptor α clarifies diagnosis and detects minimal residual disease in acute promyelocytic leukemia. *Proc Natl Acad Sci USA* 89:2694–2698

Miller WH Jr, Warrell RP Jr, Frankel SR, Jakubowski A, Gabrilove Jl, Muindi J, Dmitrovsky E (1990): Novel retinoic acid receptor α transcript in acute promyelocytic leukemia responsive to all-trans-retinoic acid. *J Natl Cancer Inst* 82:1932–1933

Mitelman F (1988): *Catalog of Chromosome Aberrations in Cancer*. 3rd ed. New York: Alan R. Liss, Inc

Mu Z-M, Chin K-V, Liu J-H, Lozano G, Chang K-S (1994): PML: a growth suppressor disrupted in acute promyelocytic leukemia. *Mol Cell Biol* 14:6858–6867

Muindi J, Frankel SR, Miller WH Jr, Jakuboski A, Scheinberg DA, Young CW, Dmitrovsky E, Warrel RP Jr (1992): Continuous treatment with all-trans retinoic acid causes progressive reduction in plasma drug concentrations: implication for relapse and retinoid "resistance" in patients with acute promyelocytic leukemia. *Blood* 79:299–303

Nagpal S, Saunders M, Kastner P, Durand B, Nakshatri H, Chambon P (1992): Promoter context and response element specificity of the transcriptional activation and modulating function of retinoic acid receptors. *Cell* 70:1007–1019

Nervi C, Poindexter CE, Grignani F, Pandolfi PP, Lo Coco F, Avvisati G, Pelicci PG, Jetten AM (1992): Characterization of the PML/RARα chimeric product of the acute promyelocytic leukemia-specific t(15;17) translocation. *Cancer Res* 52:3687–3692

Nilsson B (1984): Probable in vivo induction of differentiation by retinoic acid of promyelocytes in acute promyelocytic leukaemia. *Br J Haematol* 57:365–371

Pandolfi PP, Grignani F, Alcalay M, Mencarelli A, Biondi A, Lo Coco F, Grignani F, Pelicci PG (1991): Structure and origin of the acute promyelocytic leukemia myl/RARα cDNA and characterization of its retinoid-binding and transactivation properties. *Oncogene* 6:1285–1292

Pandolfi PP, Alcalay M, Fagioli M, Zangrilli D, Mencarelli A, Diverio D, Biondi A, Lo Coco F, Rambaldi A, Grignani F, Rochette-Egly C, Gaube MP, Chambon P, Pelicci PG (1992): Genomic variability and alternative splicing generate multiple PML/RARα transcripts that encode aberrant PML proteins and PML/RARα isoforms in acute promyelocytic leukaemia. *EMBO J* 1:1397–1408

Perez A, Kastner P, Sethi S, Lutz Y, Reibel C, Chambon P (1993): PML/RAR homodimers: distinct DNA binding properties and heterodimeric interaction with RXR. *EMBO J* 12:3171–3182

Petkovich M, Brand NJ, Krust A, Chambon P (1987): A human retinoic acid receptor which belongs to the family of nuclear receptors. *Nature* 330:444–450

Pratt MAC, Kralova J, McBurney MW (1990): A dominant negative mutation of the alpha retinoic acid receptor gene in a retinoic acid-nonresponsive embryonal carcinoma cell. *Mol Cell Biol* 10:6445–6453

Raza A, Maheshwari Y, Preisler HD (1987): Differences in cell cycle characteristics amongst patients with acute nonlymphocytic leukemia. *Blood* 69:1647–1656

Raza A, Yousuf N, Abbas A, Umerani A, Mehdi A, Bokhari SKJ, Sheikh Y, Qadir K, Freeman J, Masterson M, Miller MA, Lampkin B, Browman G, Bennett J, Goldberg J, Grunwald H, Larson R, Vogler R, Preisler HD (1992): High expression of transforming growth factor-β prolongs cell cycle times and a unique clustering of S-phase cells in patients with acute promyelocytic leukemia. *Blood* 79:1037–1048

Reddy BA, Etkin LD, Freemont PS (1992): A novel zinc finger coiled-coil domain in a family of nuclear proteins. TIBS 17:344–345

Redner RL, Rush EA, Faas S, Rudert WA, Corey SJ (1994): The t(5;17) translocation in acute promyelocytic leukemia generates a nucleophosmin-RARα fusion transcript. *Blood* 84:1486 Supplement 1, Thirty-Sixth Annual Meeting of the American Society of Hematology

Reitsman PH, Rothberg PG, Astrin SM, Trial J, Bar-Shavit Z, Hall A, Teitelbaun SL (1983): Regulation of myc gene expression in HL-60 leukemia cells by a vitamin D metabolite. *Nature* 306:492–494

Robertson KA, Emami B, Collins SJ (1989): Retinoic acid-resistant HL-60R cells harbor a point mutation in the retinoic acid receptor ligand binding domain that confers dominant negative activity. *Blood* 80:1885–1889

Robertson KA, Emami B, Mueller LM, Collins SJ (1992): Multiple members of retinoic acid receptor family are capable of mediating granulocytic differentiation of HL60 cells. *Mol Cell Biol* 12:3743–3749

Rogaia D, Grignani FR, Grignani F, Pelicci PG (1995): The APL-specific PML/RARα fusion protein reduces the frequency of commitment to apoptosis upon growth factor deprivation of GM-CSF dependent myeloid cells. *Leukemia* 9:1467–1472

Rousselot P, Hardas B, Patel A, Guidez F, Gäken J, Castaigne S, Dejean A, de Thè H, Degos L,Farzaneh F, Chomienne C (1994): The PML/RARα gene product of the t(15;17) translocation inhibits retinoic acid-induced granulocytic differentiation and mediated transactivation in human myeloid cells. *Oncogene* 9:545–552

Rowley JD, Golomb HM, Dougherty C (1977): 15/17 Translocation: a consistent chromosomal change in acute promyelocytic leukaemia. *Lancet* 1:549–550

Sawyers CL, Denny CT, Witte ON (1991): Leukemia and the disruption of normal hematopoiesis. *Cell* 64:337–350

Smith WC, Nakshatri H, Leroy P, Rees J, Chambon P (1991): A retinoic acid response element is present in the mouse cellular retinol binding protein I (mCRBPI) promoter. *EMBO J* 10:2223–2230

Tabin CJ (1991): Retinoids, homeoboxes and growth factors: toward molecular models for limb development. *Cell* 66:199–217

Takatsuki H, Umemura T, Yufu Y, Nishimura J, Nawata H (1994): Transformation at the level of pluripotent stem cell in acute promyelocytic leukemia. *Blood* 84:185 Supplement 1, Thirty-Sixth Annual Meeting of the American Society of Hematology

Takaku F, Ogawa S, Hirai H, Yamada K (1993): Rearrangement of retinoic acid receptor α and PML in promyelocytic blast crisis of Ph chromosome positive chronic myelocytic leukemia with normal copies of chromosome 15. *Blood* 81:2469–2470

Testa U, Masciulli R, Tritarelli E, Pustorino R, Mariani G, Martucci R, Barberi T, Camagna A, Valtieri M, Peschle C (1993): Transforming growth factor-β potentiates vitamin D3 induced terminal monocytic differentiation of human leukemic cell lines. *J Immunol* 150:2418–2430

Testa U, Grignani Fr, Barberi T, Fagoili M, Masciulli R, Ferrucci PF, Seripa D, Camagna A, Alcalay M, Pelicci PG, Peschle C (1994): PML/RARα + U937 mutant and NB4 cell lines: retinoic acid restores the monocytic differentiation response to vitamin D_3. *Cancer Res* 54:4508–4515

Tong JH, Dong S, Geng JP, Huang W, Wang ZY, Sun GL, Chen SJ, Chen Z, Larsen CJ, Berger R (1992): Molecular rearrangements of the *MYL* gene in acute promyelocytic leukemia (APL, M3) define a breakpoint cluster region as well as some molecular variants. *Oncogene* 7:311–316

Tsai S, Collins SJ (1993): A dominant negative retinoic acid receptor blocks neutrophil differentiation at the promyelocyte stage. *Proc Natl Acad Sci USA* 90:7153–7157

Umesono K, Murakami KK, Thompson CC, Evans RM (1991): Direct repeats as selective response elements for the thyroid hormone, retinoic acid, and vitamin D_3 receptors. *Cell* 65:1255–1266

Vivanco-Ruiz M, Bugge TH, Hirschmann P, Stunnenberg HG (1991): Functional characterization of natural retinoic acid responsive elements. *EMBO J* 10:3829–3838

Warrell RP Jr, Frankel SR, Miller WH Jr, Scheinberg DA, Itri LM, Hittelman WN, Vyas R, Andreeff M, Tafuri A, Jakubowski A, Gabrilove J, Gordon MS, Dmitrovsky E (1991): Differentiation therapy for acute promyelocytic leukemia with tretinoin (all-trans-retinoic acid). *N Engl J Med* 324:1385–1392

Warrel RP Jr, de Thé H, Wang ZY Degos L(1993): Acute promyelocytic leukemia. *N Engl J Med* 329:177–189

Weis K, Rambaud S, Lavau C, Jansen J, Carcalho T, Carmo-Fonseca M, Lamond A, Dejean A (1994): Retinoic acid regulates aberrant nuclear localization of PML-RARα in acute promyelocytic leukemia cells. *Cell* 76:345–358

Young CW, Warrel RP (1993): Differentiating agents. In: *Cancer Principle & Practice of Oncology. Fourth Edition.* De Vita Vt, Hellman S, Rosemberg SA eds. JB Lippincott Co. Philadelphia 2636–2646

Yu VC, Delsert C, Andersen B, Holloway JM, Devary OV, Naar AM, Kim SY, Boutin JM, Glass CK, Rosenfeld MG (1991): RXRß: a coregulator that enhances binding of retinoic acid, thyroid hormone, and vitamin D receptors to their cognate response elements. *Cell* 67:1251–1266

Zechel C, Shen X-Q, Chambon P, Gronemeyer H (1994a): Dimerization interfaces formed between the DNA binding domains determine the cooperative binding of RXR/RAR and RXR/TR heterodimers to DR5 and DR4 elements. *EMBO J* 13:1414–1424

Zechel C, Shen X-Q, Chen J-Y, Chen Z-P, Chambon P, Gronemeyer H (1994b): The dimerization interfaces formed between the DNA binding domains of RXR, RAR and TR determine the binding specificity and polarity of the full-length receptors to direct repeats. *EMBO J* 13:1425–1433

Zhang XK, Hoffmann B, Tran PBV, Graupner G, Pfahl M (1992): Retinoid X receptor is an auxiliary protein for thyroid hormone and retinoic acid receptors. *Nature* 355:441–446

Index